ANNUAL REVIEW
OF IMMUNOLOGY

ANNUAL REVIEW OF IMMUNOLOGY

Volume 1, 1983

WILLIAM E. PAUL, *Editor*
National Institutes of Health, Bethesda, Maryland

C. GARRISON FATHMAN, *Associate Editor*
Stanford University, Stanford, California

HENRY METZGER, *Associate Editor*
National Institutes of Health, Bethesda, Maryland

ANNUAL REVIEWS INC.   4139 EL CAMINO WAY   PALO ALTO, CALIFORNIA 94306 USA

ANNUAL REVIEWS INC.
Palo Alto, California, USA

International Standard Serial Number: 0732-0582
International Standard Book Number: 0-8243-3001-3

Annual Review and publication titles are registered trademarks of Annual
Reviews Inc.

Annual Reviews Inc. and the Editors of its publications assume no responsibility
for the statements expressed by the contributors to this *Review*.

PRINTED AND BOUND IN THE UNITED STATES OF AMERICA

FOREWORD

The Annual Review of Immunology is the 25th in the series of yearly review volumes that began in 1932 with the publication of the Annual Review of Biochemistry. There followed the Annual Reviews of Physiology (1939) and Microbiology (1947), after which Annual Reviews Inc. moved beyond strictly biomedical fields with the publication of the Annual Review of Physical Chemistry (1950) and subsequent series (the full listing of Annual Reviews' titles is given on page xii). Over 200 scientists currently take part in planning the contents of the volumes through their membership on the various editorial committees, and more than 11,000 authors have contributed reviews during the past half century.

Those who deplore the increasing isolation of the subspecialties of contemporary science may question the wisdom of our embarking on the present enterprise, as the Annual Review of Immunology might seem to threaten related Annual Reviews by preempting chapters on the immunological aspects of, for example, biochemistry, microbiology and genetics. The best responses to these fears are provided in Dr. William Paul's preface (page vii), and by the expressions of relief we have received from the editors and editorial committee members of sister volumes, who can now offer to reviewers in other burgeoning subfields of their disciplines the pages previously reserved for immunological topics.

In addition, the need for Annual Reviews Inc. to undertake new review series in expanding or novel scientific fields has been aptly summed up by Dr. J. Murray Luck, founder of Annual Reviews Inc., editor of the Annual Review of Biochemistry from its inception until 1965, and now member emeritus of the Board of Directors. Dr. Luck writes[1]:

> In Annual Reviews we once hoped that a newly founded review would not overlap in subject matter with an existing review. At long last we realize that a certain amount of duplication is not only unavoidable, but is even desirable. The brain is one thing to an anatomist, something else to a physiologist, and different again to a biochemist. Genetics, biochemistry, immunology, and medicine deeply interpenetrate. The subject, "inborn errors of metabolism," as it was once called, can be reviewed most profitably by disciples of these four sciences through simultaneous publications in four different reviews ...

The present volume would not exist without the enthusiastic cooperation of those who made up the first ad hoc editorial committee, which convened

[1]Luck, J. M. 1982. A 50-Year History of Annual Reviews Inc. *BioScience* 32:868–70

v

immediately before the Midwinter Immunology Conference in Asilomar in January, 1982, to decide which topics should be covered in the first volume and who should be invited to write. The meeting was guided toward the selection of the firstrate collection of reviews that follows by the skillful chairmanship of Dr. Winslow Briggs, a member of the Board of Directors of Annual Reviews Inc. who, as editor himself of the Annual Review of Plant Physiology, is fully experienced in the art of coaxing an editorial committee to brainstorm effectively. The authors, who agreed to write for The Annual Review of Immunology on a greatly accelerated schedule (the usual time that elapses between the planning of an Annual Review volume and the appearance of the book itself is 2 years), are to be warmly thanked for helping us speed the volume through—there was not one default, something of a record even for the more senior Annual Review series—although our usual consistency in reference style lapsed because of the rush to get the manuscripts to press. In addition, much of the preliminary organization for the Annual Review of Immunology was carried out by William Kaufmann, my predecessor as Editor-in-Chief. Jean Heavener and Roberta Parmer worked unstintingly on much of the production organization and editing, under unusual pressure.

To give our second round of authors slightly more writing time than we gave to the first panel, Volume 2 of the Annual Review of Immunology will be published in June of 1984, rather than April. However, Volume 3 is slated to appear in April 1985, back on the production schedule we intend to maintain for the series from then on.

Alister Brass, Editor in Chief
Annual Reviews Inc.

PREFACE

Immunological science is in the midst of a revolution. The introduction of monoclonal antibody methodology, the application of the techniques of molecular and somatic cell genetics, and the development of methods to grow cloned lines of lymphocytes in long-term culture have made possible experiments that, even a few years ago, would have been regarded as science fiction. Together with the application of modern biochemical technology to the study of immunologic phenomena, these approaches are reshaping our understanding of the immune system. The advances achieved by immunologists are of enormous importance in understanding the biological basis of this most fascinating system. Equally important are their impact in the clinical sphere. The immune system is the most important defense against infections by pathogenic microorganisms, may be critical in prevention and therapy of neoplasms, and is intimately involved in the pathogenesis of a wide range of chronic diseases. Its primacy in these areas is tragically emphasized by the recent emergence of a new human disorder—acquired immunodeficiency syndrome.

The enormous expansion of immunology as both a basic biological discipline and as one of the most important of the clinical sciences makes it virtually impossible for even the most diligent immunologist to remain abreast of important developments outside his or her area of special interest. The problem of keeping up to date with this critical science is even greater for individuals in allied fields and for students. The Board of Directors of Annual Reviews Inc. has initiated this series, the *Annual Review of Immunology,* in an effort to deal with this problem by providing regular reviews of recent developments in the major subdisciplines that comprise the fields of immunology.

The editorial committee of the *Annual Review of Immunology* intends to invite articles in each of the principal areas of immunology by active and respected scientists so that major topics are covered on a periodic basis. In addition, the tradition of other Annual Reviews volumes of inviting personal essays by distinguished scientists will be continued in the *Annual Review of Immunology* and will, it is hoped, provide a much needed vehicle for discussions of the humanistic aspects of our field and for a presentation of the major developments of immunology as they happened. It is a personal pleasure that the first of the essays is by Elvin Kabat, who has played a critical role in the development of our understanding of the chemical basis of immune phenomena.

William E. Paul, Editor

Annual Review of Immunology
Volume 1, 1983

CONTENTS

(*continued*)

CONTENTS (*continued*)

Special Announcement: New From Annual Reviews

*Some Historical and Modern Aspects of Amino Acids, Fermentations
and Nucleic Acids,* Proceedings of a Symposium held in St. Louis,
Missouri, June 3, 1981, edited by Esmond E. Snell. Published
October, 1982. 141 pp.; softcover; $10.00 USA/$12.00 elsewhere,
postpaid per copy

UPCOMING PUBLICATION DATES OF OTHER *ANNUAL REVIEWS* SERIES

*Annual Review of Biophysics
   and Bioengineering,* Volume 12: May, 1983

Annual Review of Biochemistry, Volume 52: July, 1983

Annual Review of Microbiology, Volume 37: October, 1983

Annual Review of Genetics, Volume 17: December, 1983

Ammual Review of Physiology, Volume 45: March, 1984

Annual Review of Immunology, Volume 2: June, 1984

Elvin A. Kabat

Ann. Rev. Immunol. 1983. 1:1–32

GETTING STARTED 50 YEARS AGO—EXPERIENCES, PERSPECTIVES, AND PROBLEMS OF THE FIRST 21 YEARS

Elvin A. Kabat

Departments of Microbiology and Human Genetics and Development, and the Cancer Center/Institute for Cancer Research, Columbia University College of Physicians and Surgeons, New York, New York 10032; and the National Institute of Allergy and Infectious Diseases, National Institutes of Health, Bethesda, Maryland 20205

This first volume of the *Annual Review of Immunology* will appear just 50 years after I began working in immunochemistry in Michael Heidelberger's laboratory at Columbia University's College of Physicians and Surgeons. The differences between then and now in the path for a graduate student, postdoctoral, and beginning independent investigator are striking, and I thought for this prefatory chapter I would recount some of the experiences of my first 21 years in the field, considering not only the work but also the outside economic and political influences on my career, as well as World War II and its aftermath of loyalty and security investigations and the Senator Joseph McCarthy period.

On January 1, 1933, I began working in Michael Heidelberger's laboratory as a laboratory helper. The definitive paper on the quantitative precipitin method by Heidelberger and Forrest E. Kendall had appeared in the *Journal of Experimental Medicine* in 1929 and the laboratory was well on its way to providing analytical chemical methods for measurement of antigens and antibodies. This paper furnished the key to modern structural immunology and immunochemistry, which, not without strong resistance from the then classical immunologists, changed our way of thinking.

1

0732-0582/83/0410-0001$02.00

I had received a BS degree in September, 1932, from the City College of New York and had majored in chemistry. I was 18 years old. The country was in the depths of the depression and my family had suffered acutely. I had been walking the length and breadth of New York visiting universities and hospitals looking for a job, literally walking, since I rarely had a nickel to take the subway and if I had only a nickel I used it for lunch. My mother had started selling dresses in our apartment; by chance Mrs. Nina Heidelberger had become one of her customers and she suggested I go up to see Michael. At the end of October or early November of 1932, he offered me a job at a salary of $90 a month to begin on the first of January, and the possibility of a future for me and for my family began.

Michael had hesitated about giving me the job since I had been to college and the previous incumbent had only been a high school graduate. I assured him that I would do the routine, which involved making solutions, keeping the laboratory clean, and washing glassware when Mary O'Neill, our half-time glassware worker, was ill, etc, provided he would let me do as much technical work and research as I was capable of doing. Hans T. Clarke, the Chairman of the Biochemistry Department at P and S, had interviewed me in connection with my application to do graduate work and had accepted me as a PhD candidate. I had explained that I would be unable to begin until I found a job, but this was the norm during the depression. Many graduate students had part- or full-time jobs; some were teaching at City College or elsewhere and were doing their graduate studies part-time at Columbia. Of the $90 a month, I had to give my parents $50 toward paying the rent and I had to pay the Columbia tuition, then $10 a point per semester, for my courses, plus registration and laboratory fees.

My hours were 8:30 A.M. to 5:00 P.M., and beginning in February, 1933, I took an evening course in experimental physical chemistry with Professor Charles O. Beckman. My graduate courses were taken in the evening or late afternoon, or during the summer session, except for those given at P and S. During the summer of 1933 I took the course in medical bacteriology given by Calvin B. Coulter. We became good friends and he, Florence Stone, and I collaborated in a study of the ultraviolet absorption spectra of a number of proteins I had prepared in Michael's laboratory. Through him I met Arthur Shapiro, then a medical student, who was full of ideas. We became very close friends; he was one of the very first to see the potentialities of bacterial genetics and several years later Sol Spiegelman worked with him. One of the required courses, which was given only in the daytime at the downtown campus, was quantitative organic analysis and students had to spend lots of extra time in the lab. My job made this difficult, so Hans Clarke, who had written the classical text in the field, gave me 18 compounds to identify and I did these in Michael's lab.

The Department of Biochemistry was at its peak in reputation and productivity with Michael, Rudolf Schoenheimer, Karl Meyer, Erwin Chargaff, Erwin Brand, Sam Gurin, David Rittenberg, and Sam Graff. Graduate students worked in a large laboratory on the fifth floor and included Joe Fruton, Lew Engel, David Shemin, De Witt Stetten Jr, Sarah Ratner, Konrad Bloch, William H. Stein, Abe Mazur, and Ernest Borek, and somewhat later Seymour Cohen and David Sprinson. I got to know W. E. van Heyningen, who came to us as a postdoctoral fellow. Since I worked directly in Michael's lab my contacts with the other students were much less close. My interests also led me to be close to the Bacteriology Department, of which Frederick P. Gay was Chairman and whose members included Beatrice Seegal, Theodor Rosebury, James T. Culbertson, Claus Jungeblut, Maxim Steinbach, Calvin Coulter, Florence Stone, Sidney J. Klein, and Rose R. Feiner. I attended both the weekly biochemistry and microbiology seminars. Professor Gay, who had been closely associated with Jules Bordet and was still waging the Paul Ehrlich-Jules Bordet war of antibodies as substances rather than as vague properties of immune serum, was strongly anti-immunochemical and became very impatient with my questions at seminars. This view came across clearly to the medical students and Edward H. Reisner, P and S '39, wrote the following little ditty, which I learned about from Oscar Ratnoff.

Pasteur inspired Metchnikoff and Metchnikoff was nuts,
He had a pupil named Bordet who hated Ehrlich's guts,
Now Ehrlich had the goods you know, but this we dare not say,
For we descend from Metchnikoff through Bordet out of Gay.

The Heidelberger laboratory in 1933–37 consisted of Forrest E. Kendall, who was working on pneumococcal polysaccharides and on the mechanism of the precipitin reaction, Arthur E. O. Menzel, who was studying the proteins and polysaccharides of the tubercle bacillus, Check M. Soo Hoo, our bacteriologist, myself, and Mary O'Neill. Several visiting scientists, among whom were Henry W. Scherp, Torsten Teorell, D. L. Shrivastava, Alfred J. Weil, and Maurice Stacey, came for varying periods, the longest being 1 year. When Forrest left in 1936 to go to Goldwater Memorial Hospital with David Seegal, Henry P. Treffers, who had just completed his PhD with Louis P. Hammett at Columbia, replaced him.

Forrest Kendall worked opposite me and taught me the micro-Kjeldahl method, which in those days was the basic tool in the laboratory since all quantitative precipitin assays on the washed precipitates were finally analyzed for total nitrogen. Digestions were carried out in 100-ml flasks. The working range was between 0.1 and 1.0 mg of N. I generally was responsible for doing all of the digestions for Michael and Forrest, and after I acquired sufficient skill I also carried out the distillations and titrations for Michael. It was not unusual to run 40 distillations a day. I also ran many quantitative iodine determinations on thyroglobulin, using the method developed by Professor G. L. Foster in the Biochemistry Department, until Herbert E. Stokinger became a graduate student with Michael and was given the thyroglobulin problem.

The laboratory was a very exciting place in which to work and I generally kept Forrest busy asking him questions, no doubt interfering often with his train of thought. He was very patient and helpful, but Michael once suggested that I had better put some of my questions to him rather than to Forrest. I was always able to get answers or was told where to find them.

After a few months, as Michael has described (2), I suggested that one might use a well-washed suspension of heat-killed bacteria of known N content to remove antibody and measure antibody N as agglutinin. Michael said that I could do this so I started with type I pneumococci. The method was very successful, provided one took care in growing the organisms to avoid autolysis. What happens when this procedure is not strictly adhered to has already been published (3).

The combined application of the quantitative precipitin and quantitative agglutinin methods permitted us to establish that the antibody to the type-specific polysaccharide of pneumococci was the same whether measured as precipitin or as agglutinin. Naturally, the bacterial suspension was able to remove antiprotein, and this was measured independently with a suspension of rough unencapsulated pneumococci. In those days we were only beginning to be aware of antibody heterogeneity, that our type-specific antibody was a complex mixture of antibodies, some of which were non- or co-precipitating, and that all of these were being measured as agglutinin or as precipitin. Michael and I also studied the course of the quantitative agglutinin reaction and showed that it followed the empirical equation he and Forrest had described for the quantitative precipitin reaction, and which they later derived from the Law of Mass Action.

The quantitative agglutinin work was supposed to be my PhD dissertation; the first two papers on the method and on the identity of agglutinin and precipitin had been published in the *Journal of Experimental Medicine* in 1934 and 1936. These, as reprints, and the third part on the mechanism of agglutination were submitted early in 1937. However, the University's rules had been changed so that the student was required to be the first author on any publication and Michael had been the first author on the two papers. I could of course have used the third part as the thesis, but I had also been studying the quantitative precipitin reaction between R-salt-azobiphenylazo-serum albumin and rabbit antibody to serum albumin. This system had certain unsuspected unique aspects since the introduction of the haptenic group had not altered the reactivity with anti-serum albumin. Michael suggested that I write this up and use it as my PhD dissertation, which I did in about 3 or 4 weeks. It created a minor sensation at the Faculty of Pure Science office when I brought down the copies of the new dissertation and took back the other.

Arne Tiselius had suggested some time earlier that it would be nice to have someone from Michael's laboratory to study the physicochemical properties of purified antibodies at Uppsala and that the Rockefeller Foundation might provide a Fellowship. He had just developed the moving boundary method of electrophoresis with the rectangular cell, which first made possible precise quantitative estimates of purity and permitted resolution of mixtures of proteins. Michael had spent two summers in Uppsala working with Kai O. Pedersen and with Tiselius and had suggested me. This was an extraordinary opportunity to learn the newer methods of fractionating and characterizing macromolecules in the Svedberg laboratory. Frank Blair Hanson came to interview me from the Foundation and I was awarded the Fellowship.

The Fellowship paid $125 per month. On the original application there was a question as to whether one had any special financial obligations and I had noted that I gave my parents $50 a month. When I received the award, it specified $125 per month. I went to the Foundation's offices to discuss whether I could get along on $75 per month. They looked at my original application and assured me that $50 would be sent to my parents and that I would receive the full $125. I made sure that I did not spend much of my first month's stipend until several weeks later when a letter from my mother arrived (this was before transatlantic airmail) saying that the first check had been received promptly on September 1, 1937. After that I was able to live quite comfortably.

The Rockefeller Foundation also paid for cabin class transportation from New York to Uppsala. Since I was entitled to a month vacation for my work with Michael, I wanted to spend it in Europe and especially to visit the Soviet Union, to which at that time I was very favorably inclined, predominantly because of the great economic suffering of my family during the depression and also because of Russia's United Front policy in opposition to Hitler and their support of the Spanish loyalists. The Rockefeller Foundation agreed to give me what it would cost them to send me directly in cabin class and I could go anywhere I chose. In those days if one purchased a tourist class ticket on the Queen Mary, one could go by rail second class anywhere in Europe for $10 extra. I chose Leningrad, spending several days in Paris and Warsaw. For $50 one could spend 10 days in Leningrad and Moscow with all hotels, meals, travel between cities, and guided tours to places of interest included. In Paris I visited the Institut Pasteur, saw the

Pasteur Museum, and met Gaston Ramon, who invited me to lunch at his home where I met André Boivin and Lydia Mesrobeanu. In the Soviet Union I was unable to see any scientists and so I was essentially a tourist visiting museums, factories, etc, on standard guided tours. I then spent a day or two in Helsinki, took the boat to Stockholm and the train to Uppsala, and arrived at the Institute early in September.

Before leaving and in anticipation of my being able to go to Uppsala, Michael had suggested that we immunize several different species with suspensions of pneumococci, so Soo Hoo and I injected two pigs and a monkey. These immunizations were less traumatic than those we had done earlier, injecting two goats (4). Rabbit and horse antisera were available in the lab and a cow was immunized at Sharpe and Dohme. Torsten Teorell with Forrest and Michael had found that a given quantity of pneumococcal polysaccharide precipitated less antibody in 15% salt than in physiological saline, due to a shift in the combining proportions at equilibrium when the reaction was carried out in high salt. Michael and Forrest then used this finding to purify antibodies to polysaccharides by precipitating the antibody from a large volume of antiserum with polysaccharide under physiological conditions of 0°C, washing repeatedly with cold saline, and extracting the washed precipitate with 15% NaCl at 37°C to dissociate a portion of the antibody. With individual antisera as much as 15–30% purified antibody could be obtained and over 90% was precipitable by polysaccharide. This was the first time substantial amounts, tens of milligrams, of antibody could be prepared. Michael and I also extended the method by using a suspension of bacteria to remove the antibody followed by washing and elution with 15% salt. These purified antibodies plus various antisera and antigens were shipped to Uppsala and were available when I arrived.

The laboratory was at a peak in its productivity with Professor Svedberg popping in and out, with Arne Tiselius, then a docent, and with Kai O. Pedersen and Ole Lamm. Numerous visitors came for various periods, including Basil Record from Haworth's laboratory, Frank L. Horsfall from the Rockefeller Institute, whose interests were similar to mine, J. B. Sumner with crystalline urease and concanavalin A, G. Bressler from Leningrad, G. S. Adair from Cambridge, and Gerhard Schramm from Germany.

Professor Svedberg was very anxious to test his new separation cell, which had a membrane dividing the cell into an upper and a lower compartment. Using some of my horse antipneumococcal serum, he centrifuged it until the 18S peak had gone below the membrane. I tested both proteins and found all of the antibody in the lower compartment.

I had also brought some of the crystalline horse serum albumin used in my PhD thesis. Tiselius was very surprised when we examined it by electrophoresis at a concentration of 0.5 or 1% and saw only a single peak. He said it was the first time he had seen a protein that was a single component electrophoretically. Although new techniques provide more and often better criteria for establishing purity, they frequently only show that not enough care was taken in purifying the material. The need for such care was most forcefully demonstrated years later when Knight (8) examined by paper chromatography Emil Fischer's collection of 34 synthetic peptides made half a century earlier, which his son Hermann had brought to Berkeley. All but three, which contained a trace of one of the constituent amino acids, gave a single spot.

I got to work right away with the help of Kai Pedersen learning to run the oil turbine ultracentrifuge and the diffusion apparatus to measure the molecular weights of the antibodies I had brought. Surprisingly, the horse antibodies were quite polydisperse. We were at a loss to understand this. Fortunately, the State Serum Institute had a horse that had been under immunization for a short time, and on purifying its antibody by the salt dissociation method we found a single 18S peak. All of the other antibodies gave multiple peaks, as did a second sample from the same horse after further immunization. Therapeutic use of antipneumococcal horse sera in Sweden had not been very successful, so I was able to get serum from humans recovering from lobar pneumonia who had not received antibody, through the courtesy of Jan Waldenström. I tested the serum of quite a few convalescents, found one with about 1 mg of antibody protein per ml, purified the antibody from about 40 ml of serum, and measured its molecular weight.

Arne Tiselius and I studied the electrophoretic properties of antibodies. One of the rabbit hyperimmune anti-ovalbumin sera I had brought had 36.4% precipitable antibody. We examined its electrophoretic pattern before and after removal of the antibody and found, by the decrease, 37.2%, in area of the gamma globulin peak, that we had removed the same proportion of the total serum protein. This definitely established that these antibodies were gamma globulins [now immunoglobulin (Ig) G, one of the five major immunoglobulin classes]. The electrophoretic patterns we published appear in most textbooks of immunology.

I also spent a considerable amount of time studying specific precipitates of ovalbumin rabbit anti-ovalbumin and horse serum albumin–rabbit antiserum albumin dissolved in excess antigen both by ultracentrifugation and by electrophoresis, calculating the composition of the soluble complexes. This was all done by the Lamm scale method, which was very tedious and especially hard on the eyes. Before my return to the US I left a manuscript with Arne Tiselius, which, because of the war, never was published, although he cited our work and an electrophoretic pattern showing the soluble complexes as a schlieren band migrating behind the ovalbumin band appears in Tiselius' Harvey Lecture (10). It remained for Jonathan Singer and Dan Campbell to do the definitive study. By that time, the automated Schlieren method had replaced the laborious scale method and work was much easier.

I have already described my experiences at the Nobel Ceremony in December 1937 (6).

The terms of the Rockefeller Foundation Fellowship required that the sponsor permit the Fellow to return to a position in the laboratory from which he came and Michael had assured the Foundation he would do so. However, he told me he would do his best to find me a more independent position, and some months after I had arrived in Uppsala I received a letter from Jacob Furth offering me a position as Instructor in Pathology at Cornell University Medical College at $2400 per year beginning September 1, 1938, to work on viruses causing tumors and leukemias in chickens. Michael advised me to accept and I did.

I believed it was important for me to visit certain laboratories before returning to the US and discussed this with Harry Miller of the Rockefeller Foundation. They tended to discourage such travel but finally agreed that I could leave Uppsala about July 15 to visit Linderstrom-Lang and the State Serum Institute in Copenhagen; J. D. Bernal, I. Fankuchen, and John Marrack in London; F. G. Hopkins in Cambridge; W. N. Haworth and Maurice Stacey in Birmingham; Hans Krebs in Sheffield, where I spent 4 days learning to do some enzyme reactions in the Warburg apparatus; Gorter in Leyden; and den Dooren de Jong in Amsterdam, finally going to Zurich to attend the International Physiological Congress before returning to the US. All of these visits established lasting friendships and contacts with junior as well as senior investigators, and in my report to the Rockefeller Foundation I emphasized the desirability of providing similar opportunities to their other fellows.

In Zurich in August, 1938, everyone was very disturbed about the situation in Spain where the Franco forces were about to or had already cut Catalonia from the rest of loyalist Spain. Many of the members of the Loyalist government were physiologists, including Juan Negrin, the Prime Minister, and Cabrera, the Foreign Minister, and a good-sized delegation attended the Congress. A group of us felt we should do something to express our support and organized a dinner at the Congress, which was very well attended and raised a considerable sum. I was anxious to go to Spain to find out what people in the US could do to help, and it was arranged for me and Lew Engel to go to Spain for a few days before sailing for the US. My US passport was not valid for travel to Spain so Dr. Cabrera wrote on a sheet of paper that all border patrols should permit me to enter and leave Spain without making any marks in my passport. We took a train from Zurich to Perpignan, France, and to Cerbère, the usual crossing point, thinking it would be easy to get across. I very soon had a French soldier with a bayonet prodding me along until I got on the train back to Perpignan. Wondering what to do we decided to visit the Spanish consul in Perpignan. When he saw the Cabrera note, he told us to go out and get pictures taken, and when we returned he affixed them to a Spanish Carte d'Identité on the back of which he had written the contents of the Cabrera letter (Figure 1), summoned his car and chauffeur, and had us driven from Perpignan to the Hotel Majestic in Barcelona. The Hotel Majestic was the center for foreign correspondents and there I met Herbert L. Matthews, the *New York Times* correspondent. Food was very scarce and I was hungry all the time, most especially after the rationed meals. Indeed, when I left a French dining car, on my way back, I emptied the bread tray into my pockets. Most of my time was spent visiting hospitals where the wounded Americans of the Lincoln Brigade and other International Brigade volunteers were being treated. I vividly remember speaking to one wounded anti-fascist German soldier who felt his problems were not being understood because he spoke only German; I was able to translate his wishes into English to one of the doctors.

On returning to New York, I immediately began to work at Cornell with Jacob Furth. Eugene Opie, the Chairman of the Pathology Department, had built up an important department with Jacob, Murray Angevine, Robert A. Moore, and Richard Linton, who had been at Columbia and with whose work in India on cholera antigens I had become familiar. Vincent du Vigneaud was Chairman of Biochemistry and permitted me to attend their weekly seminars. I became close friends with W. H. Summerson and soon collaborated in studies with Dean Burk, Otto K. Behrens, Herbert Sprince, and Fritz Lipmann, who had left Germany and arrived shortly thereafter from Denmark. Cornell Medical School was not anxious to have too many Jews on its staff and du Vigneaud, who had made a definite offer to Lipmann, indicated that he would resign if the appointment did not go through.

At Cornell the main problem Jacob Furth wished me to work on was to purify the virus of one of his chicken strains that caused leukosis if injected intravenously and tumors like the Rous sarcoma if injected subcutaneously. The approach was essentially to centrifuge crude extracts in the Pickels air-driven ultracentrifuge, following the biological activity by assays in baby chicks, as well as by nitrogen to estimate the extent of purification. I wanted to buy a micro-Kjeldahl apparatus, but a previous postdoctoral fellow with Jacob had purchased a Dumas apparatus. Jacob, considering me a chemist and wishing to save money, wanted to get the Dumas working. I had to insist that this was not a suitable method if one had to run many determinations and he finally consented to buy a micro-Kjeldahl apparatus; he was

REPUBLICA ESPAÑOLA

CONSULADO DE ESPAÑA
EN
PERPIÑAN

EL CONSUL DE ESPAÑA EN PERPIÑAN
Le Consul d'Espagne à Perpignan

CONCEDE AUTORIZACION para entrar en ESPAÑA y regreso
autorise pour se rendre en Espagne

por la frontera de La Junquera á Elvin Kabat
par la frontière de   M.

nacido el 1 de septiembre de 1914
né le

en New York Provincia de Estados Unidos
à   Province de

de estado soltero y profesión Químico
état civil   et profession

Motivo del viaje particular
Motif du voyage

y por tanto ruega á las Autoridades civiles y militares,
et à cet effet prie les Autorités civiles et militaires,

nacionales y extranjeras, que no le pongan impedimento en
espagnoles et étrangères, de ne lui opposer aucune difficulté dans

su viaje, antes bien le presten todo el favor y ayuda que
son voyage, et même au besoin, lui prêter aide et protection.

necesitare.

Dado en Perpiñan á 20 de agosto de 1938
Délivré à Perpignan, le

EL CONSUL DE ESPAÑA,
Le Consul d'Espagne,
p. d. EL CANCILLER

Firma del portador,
Signature du porteur

Elvin Kabat

J. Camp

Figure 1   Spanish Carte d'Identité, with the Foreign Minister's letter.

very happy when he saw 40 to 50 analyses a day coming out immediately and told me that they had tried to use the Dumas for a whole year without a single successful result.

Albert Claude had essentially done a study with Rous sarcoma virus similar to what we were planning and had found that high-speed sedimentable materials were present in normal tissues. This was all in the days before subcellular organelles, mitochondria, microsomes, etc, and it took me a long time to convince Jacob that one could not get a pure virus just by centrifugation of tumor extracts. Among our interesting findings at that time were the demonstration that alkaline phosphatase and the Forssman and Wassermann antigens were associated with these high-molecular-weight particles. We also prepared rabbit antisera, which we used to try to characterize these materials.

Jacob also suggested that I try to work out some histochemical methods for localizing enzymes in tissues. I was just getting started when George Gomori published his elegant method for localizing alkaline phosphatase in paraffin sections of alcohol-fixed tissues. We adopted it and carried out a study of the distribution of alkaline phosphatase in normal and neoplastic tissues. The stained sections were very impressive; colored lantern slides were just coming into use and Jacob thought we should have a colored plate for our paper in the *American Journal of Pathology,* which would cost $400. This was considered too expensive. It was finally arranged that the Department of Pathology would pay $100, the research fund would pay $200, and Jacob and I would each pay $50. It was a very worthwhile investment because it stimulated much work on the subject.

In Europe I had hoped to visit Kögl's laboratory at Utrecht; he had startled the world by announcing that malignant tumors had large amounts of D-amino acids instead of L-amino acids. Much of his data could be accounted for by racemization, but from one tumor several grams of D-glutamic acid had been isolated. Naturally, everyone began to try to check this and Jacob got me some tumors, thinking that after I established that I could isolate D-amino acids I could then work on leukemic cells, which were available in much smaller amounts. However, I worked up a number of tumors but only found L-glutamic acid. Two confirmations of Kögl's work were soon reported and it seemed clear that I was not a very good chemist. The problem was solved by the genius of Fritz Lipmann, who suggested that after hydrolysis, one could add D-amino acid oxidase and estimate D-amino acid concentrations by using the Warburg apparatus. He, Dean Burk, Otto Behrens, and I soon showed that there were no significant amounts of D-amino acids. David Rittenberg at Columbia applied the isotope dilution method and also failed to find D-amino acids. This turned out to be one of the instances of falsification of data, someone having apparently added D-glutamic acid to Kögl's hydrolysates. It led to my coining an aphorism: "Every incorrect discovery has always been independently verified."

In our efforts to purify the tumor viruses, we frequently found that saline extracts of tumors were extremely viscous, and from one chicken with a tumor, we were able to get a considerable quantity of this very viscous fluid. The viscosity reminded me of the pneumococcal polysaccharides I had prepared. I added some sodium acetate to the fluid and then two volumes of ethanol and got an extraordinarily stringy precipitate, almost all of which adhered to the stirring rod and could be removed. Karl Meyer at P and S had reported the isolation of hyaluronic acid as behaving similarly and I said to Jacob, who was watching me precipitate the polysaccharide, that it looked like hyaluronic acid. He became somewhat upset that I could make such a statement from so little evidence, but when I added a little hyaluronidase provided by Karl Meyer, the viscosity disappeared. We learned it was essential to treat our tumor extracts with hyaluronidase before ultracentrifugation. David Shemin, who had been at City College with me, had received his PhD in biochemistry at Columbia, and had taken a job there in the Department of Pathology with James Jobling, the Chairman, working on Rous sarcoma, had similar findings.

Dean Burk was also very interested in tumor metabolism and he, Jacob, Herbert Sprince, Janet Spangler, Albert Claude, and I carried out studies in the Warburg on chicken tumors.

On the international scene my stay at Cornell was the period of the Munich agreement, the conquest of Czechoslovakia by Hitler, the Nazi-Soviet pact, the partition of Poland between Germany and Russia, and the Finnish-Soviet war. These events shook me and I began to worry about my political views.

About the spring of 1940, Tracy J. Putnam, Director of the Neurological Institute at Columbia, wanted an immunochemist to work on multiple sclerosis and asked Michael Heidelberger to suggest someone. Michael gave him a choice, assuring him that if he picked me he would have a strong personality on his hands. He invited me for an interview and said that he understood that I had lots of my own ideas and that I could work on whatever I wished except he hoped I would not discriminate against neurological problems. The salary was $3600 per year and Hans Clarke had agreed that I would have an appointment as Research Associate in Biochemistry (assigned to Neurology). Funds were available for a technician; two laboratories were available in the Neurological Institute, which had formerly been intern's bedrooms, plus some space shared with the clinical laboratory. It took little thought on my part to decide to accept. Robert F. Loeb, for whom I had the greatest admiration and affection, was then

Associate Director of the Neurological Institute and was a great attraction for me. I was especially thrilled when I met him at Woods Hole in the summer of 1940 and he came over to me saying, "Elvin, we need you, we can hardly wait until you get to Neuro."

I arranged to leave Cornell at the end of May, 1941, taking June as my terminal vacation. I started at P and S and the Neurological Institute on June 1, 1941, and took my vacation later in the summer.

My laboratories were temporarily occupied by Norman Weissman, a friend who had gotten his degree in biochemistry, and Murray Glusman, with whom I was to collaborate years later when he returned from a Japanese prison camp. Harold Landow, a resident in neurology, was interested in collaborating. We decided to look at the electrophoretic patterns of cerebrospinal fluid in the Tiselius electrophoresis apparatus. Dan H. Moore, who had come from Lederle laboratories where he had studied horse antisera electrophoretically, had set up a laboratory for electrophoresis and had an air-driven ultracentrifuge. For the Tiselius electrophoresis, the cerebrospinal fluids had to be concentrated by pressure dialysis to a small volume and had to be run in a 2-ml micro cell. Large amounts of fluid were available from pneumoencephalograms and we arranged to obtain these. We readily obtained good patterns on concentrated cerebrospinal fluid and showed that patients with multiple sclerosis often had substantial increases in the gamma globulin fraction, as did patients with neurosyphilis, and that positive colloidal gold tests correlated with increased gamma globulin. This led naturally to a study of serum proteins with Franklin M. Hanger, which showed high levels of gamma globulin to be responsible for positive cephalin flocculation tests in various sera. These findings provided some insight into the mechanism of these two clinical diagnostic tests.

At the same time, Harold Landow and I began a quantitative study of passive anaphylaxis in the guinea pig. Oddly enough, although the antibody N content of rabbit antisera could be determined by quantitative precipitin analysis, nobody had measured how little antibody was required to sensitize a guinea pig so that fatal anaphylaxis would result on subsequent administration of antibody. Our data showed, with anti-ovalbumin and anti-pneumococcal serum that 30 μg of antibody N sensitized 250-g guinea pigs, so that fatal anaphylaxis would result if 1 mg of ovalbumin or of pneumococcal polysaccharide were given 48 hr later. This was the beginning of a series of studies on quantitative aspects of allergic reactions. In the summer of 1940, while still at Cornell, Mary Loveless and I had worked at Woods Hole, trying to measure uptake of skin-sensitizing antibody (later recognized as IgE) in sera of ragweed-sensitive patients, using a suspension of formalinized pollen with negative results. The skin sensitizing power of the sera was removed, but no increase in N in the washed pollen was detectable.

As we now know from the work of the Ishizakas and Bennich, IgE is so much more active per unit weight than rabbit IgG that the method was not suitable. For similar reasons I also obtained negative results trying to measure uptake of bacteriophage by suspensions of bacteria.

Also while at Cornell Professor du Vigneaud had given me permission to use the Tiselius electrophoresis apparatus he had just purchased. I examined numerous leukemic sera for changes in protein pattern, but significant changes were not seen. However, a Bence Jones protein he also provided gave a beautiful homogeneous peak.

During the first weeks after my return to Columbia in June, 1941, I stopped in to see Alexander and Ethel Gutman, whose laboratories were opposite Michael's and from whom I had received advice while with Michael. They were studying myeloma sera in the Tiselius apparatus and showed me their patterns with their sharp peaks moving between β_2 and gamma globulin, the mobility of the peaks differing among individual sera. I asked them why they were not studying Bence Jones proteins since the mobility of my peak had also been between that of β_2 and gamma globulins. Al took out a pattern of a Bence Jones protein he had sent to someone several years earlier. There was a beautiful sharp peak like the one I had observed—it was labeled "boundary disturbance." We decided to add Bence Jones proteins from a myeloma patient to normal serum and compare the pattern with that of the patient's serum by itself; we found we could reproduce the patient's serum pattern in a number of instances by using Bence Jones proteins of different mobility. We then used a rabbit antiserum to urinary Bence Jones protein from a myeloma patient to measure the levels in the serum of the same patient.

I also continued the histochemical localization of enzymes, studying normal and neoplastic tissues of the nervous system with Harold Landow and William Newman, who came to us as a technician, later studied medicine, and became Professor of Pathology at George Washington University. Abner Wolf, who was Professor of Neuropathology, also was interested in this field. We collaborated closely on histochemical studies on acid and alkaline phosphatases and especially on producing and studying disseminated encephalomyelitis in monkeys by injection of brain tissue emulsified in Freund adjuvants, until the grant for this work was summarily terminated in 1953 during the hysteria generated by Senator Joseph McCarthy.

On the evening of June 22, 1941, I was at a party—everyone was discussing the rumors that Germany was going to attack Russia. I argued vigorously that Hitler would not, and when I had just about convinced everyone, we turned on the radio. This marked my retirement as a political prognosticator. It was far more tragic for mankind that Stalin was taken in than that I was. I joined with many others to set up a group for Russian War Relief at the Medical Center. The doubts generated by the Nazi-Soviet pact were stilled.

On Sunday, December 7, 1941, I was working in the laboratory and heard the news about Pearl Harbor. No one knew what was happening but we decided to move acids and hazardous chemicals to a storeroom in the basement. It was clear that medical scientists, immunologists, and immunochemists were going to make important contributions to the war effort. Michael Heidelberger, who I saw very frequently since my return to Columbia, had been working for the Pneumonia Commission of the Board for the Control of Epidemic Diseases, US Army, on immunization against lobar pneumonia, using pneumococcal polysaccharides. I proposed that the type I (now group A) meningococcal polysaccharide, which had been purified by Geoffrey Rake and Henry W. Scherp, might be used for immunization in man. Meningitis had been a serious problem during recruiting in World War I. The Meningitis Commission under the Chairmanship of J. J. Phair was centered at Johns Hopkins and gave me an initial subcontract for $1000, renewed for a year at $2500, to purify type I meningococcal polysaccharide and study its antigenicity in man. I needed to order some equipment, which I did before going on a trip. When I returned I found that the order had not been processed. I phoned Dean Willard C. Rappleye, who said they were having a problem about whether my subcontract would receive overhead. I asked how much the overhead was and when he said $40, I shrieked into the telephone, "And for forty dollars you are holding up my work on immunization against meningitis with a war going on!" There was a moment of deathly silence and I expected to hear that I was no longer employed by Columbia. Instead, he said, "I'm sorry. I guess you are right. We'll order it right away."

From today's perspective it is easy to see how the Universities' attitudes on getting the incredible amounts of money they demand in indirect costs developed, to understand what happens to the reputation of the investigator who cannot bring in what they consider his share, and to marvel at the unanimity administrators exhibit in lobbying against any proposed reductions when they can agree on almost nothing else.

I hired Hilda Kaiser, whose husband, Samuel Kaiser, had been dismissed from Brooklyn College as a consequence of the New York State Legislature's Rapp Coudert Committee and later suggested to Michael that he hire Sam for the pneumonia work, which he did. All of the individuals fired from the City Colleges during that period were reinstated with apologies 40 years later, many posthumously.

We began to immunize medical student volunteers with type I meningococcal polysaccharide following Michael's schedule for pneumococcal polysaccharides. In these and all my later studies on blood group substances, dextrans, levans, etc, I generally injected myself first with any new materials unless they were expected to cross-react with antibodies to antigens I had already received. We used Michael's precipitin assay with about 4 ml of serum per test and with Hilda Kaiser and Helen Sikorski were able clearly to demonstrate that precipitins were formed; C. Philip Miller and Alice Foster at the University of Chicago assayed the sera for protective antibody and we demonstrated that the polysaccharide was antigenic. Only 4 of 38 individuals, however, produced a significant antibody response and this seemed poor by comparison with the pneumococcal polysaccharides, so immunization was not considered promising. When immunization with meningococcal polysaccharides was taken up years later, radioimmunoassay was available and detection of an antibody response became thousands of times more sensitive. We too would have found detectable antibody by radioimmunoassay in a substantial proportion of the 38 subjects.

My laboratory was involved in two other projects during the war: one with the Office of Scientific Research and Development and the Committee on Medical Research on false-positive serological tests for syphilis; and the other for the National Defense Research Committee on the plant toxin ricin, a possible chemical-biological warfare agent, with a view toward developing methods of detecting it and protecting against its toxic effects. Our experiences in buying and immunizing two horses as part of this work have already been described (5).

Bernard D. Davis was assigned to the laboratory by the US Public Health Service to work on serological tests for syphilis. With Ad Harris and Dan Moore we studied the anticomplementary action of human gamma globulin and succeeded in purifying Wassermann antibody by absorption on and elution from lipid floccules.

For the studies on ricin, which were classified secret, Michael Heidelberger and I joined forces as co-responsible investigators. Ada E. Bezer came to work in my laboratory on this problem and we began an association that was to last over 20 years until she went to work with WHO in Ibadan, Nigeria. We were able to demonstrate by immunochemical methods that the toxic and hemagglutinating properties of ricin were due to different substances. This was later confirmed by Moses L. Kunitz and R. Keith Cannan, who independently succeeded in crystallizing ricin. One very puzzling observation was that the protective power of anti-ricin sera could be assayed readily in animals but did not correlate with the amount of precipitable antibody. We now know of course that ricin has a combining site for carbohydrates and therefore was reacting with and precipitating non-antibody gamma globulin and probably other serum glycoproteins, as well as antibody.

During the war years Manfred Mayer had replaced me in Michael's laboratory. He continued the study of the cross-reaction of types III and VIII antipneumococcal antibodies Michael, D. L. Shrivastava, and I had begun earlier. We became very close friends.

The war years had seen important changes in my personal life. Harold Landow, a brilliant neurologist and close friend, had committed suicide in 1940 because of his parent's objections to his marrying outside of his religion. In 1940 at a literary meeting I had been introduced to a young Canadian named Sally Lennick. She had come to New York from Toronto to study painting. We met again a year later. Then in the summer of 1942 we met at a party, began seeing one another frequently, and were married on November 28, 1942; our oldest son, Jonathan, was born on June 5, 1944.

It had become clear at this time that a book outlining the thinking and methodology of quantitative immunochemistry was sorely needed. Most workers using the quantitative precipitin method had learned it directly from Michael or from someone who had learned it from Michael and this was limiting growth of the field. I had been invited to write a review entitled "Immunochemistry of the Proteins" for the *Journal of Immunology* by Alfred J. Weil; it appeared in December, 1943, and I was astonished at the hundreds of requests for reprints. Manfred Mayer and I decided to write a text called *Experimental Immunochemistry,* which would not only give the methods in detail but would also outline principles and concepts so that it would be useful to students and workers who might think immunochemistry of value in attacking their problems. It would also include the preparation of materials needed for work in immunochemistry. We prepared an outline and submitted it to John Wiley, who did a survey and assured us that it would never sell even a thousand copies. Fortunately, Charles C Thomas came to visit Michael and immediately agreed to take it and invited Michael to write a preface.

We set right to work by dividing up the chapters and beginning to write. We usually met on Saturday or Sunday at one or another's apartment and read aloud what had been written, correcting, revising, and reordering sections. Our wives were left to entertain each other or to be bored by listening to what we had written. The book was sent to the publisher at the end of 1945, but because of paper shortages after the war, and various other delays, it did not appear until 1948. Fortunately, we were aware of many studies during the war years that were just being prepared for publication and insisted on doing extensive revisions in proof so that the book was up to date. It had a substantial influence on the field and the first edition went through four printings, the last being in 1958. The second edition, which appeared in 1961, also went through four printings.

After Pearl Harbor, a group of us at Columbia who were members of the

American Association of Scientific Workers devoted a considerable amount of time to considering what could be done to aid the war effort, especially with respect to defense and improving methods of immunization. The use of bacterial and viral warfare agents by the Nazis was considered a possibility and Theodor Rosebury and I undertook to prepare from the literature a report of the agents that might be used; we were assisted by Martin H. Boldt, then a medical student. We spent several months preparing the review, and Alphonse Dochez, who was an adviser to the Secretary of War, gave it to him. Naturally, it was withheld from publication by us voluntarily during the war. The Army was also very concerned about the possible military use of infectious agents and set up Camp Detrick near Frederick, Maryland, as a large-scale military research facility. Ted Rosebury started working there full time and I became a consultant, spending several days there each month. Our report was among the first items new personnel at Fort Detrick were given to read.

At the end of the war, when the Smythe Report on the Atomic Bomb was published, we felt it was in the public interest that our report on bacterial warfare should also be published and we requested clearance from the War Department. Several incidents arose. In our original request for clearance we stated in a footnote that we had withheld the report from publication during the war but that it would now be published with the approval of the War Department. In their letter stating that we could submit the review for publication, they requested that the words with the approval of the War Department be changed to "in view of the removal of war time restrictions" (Figure 2). We naturally complied. The *Journal of Immunology* indicated it would consider publishing it. In accordance with University policy we then gave the report for approval to Dr. Dochez, who was Chairman of the Bacteriology Department. Although he had originally taken the report to the Secretary of War, he said it had nonscientific implications and that he would have to take it up with Dean Rappleye. Rosebury and I were called to the Dean's office and were told the report might offend religious groups and that if we insisted on publishing it we should write our resignations on the spot. Since we were in no position to do this, and since our appeals about Freedom of Speech and the Press were unavailing, we put the report in a drawer. We also went to see Osmond K. Frankel, a leading civil liberties attorney, who told us essentially that we had unlimited freedom to publish but having done so we did not have the right to work for Columbia University. About 6 months later, Dean Rappleye came to Ted Rosebury's office and said that he had made a mistake, that the University did not intend to limit our freedom to publish, and we could submit the report for publication. We did, and it was published in the May, 1947, issue of *Journal of Immunology* (9), which also arranged to sell reprints. The report created a world-wide sensation (Figure 3).

ARMY SERVICES FORCES
Camp Detrick
Frederick, Maryland

30 March 1946

Dr. Theodor Rosebury
Department of Bacteriology
Columbia University
630 West 168th Street
New York, New York

Dear Ted:

I took up with Captain Foy the matter of publication of your review on bacterial warfare.

It is his opinion that you are free to publish this in view of the fact that all restrictions on publication have been removed, that is to say, insofar as there is no direct relation to the War Department. Although you and your associates were identified with the War Department program in this field after preparation of the manuscript, it is very obviously unrelated to those experiences and might as well have been written by any scientist of equal competency and interest.

If it is to be submitted for publication, the footnote on the first page should be revised to exclude reference to the War Department. The note should, however, indicate the date of the original preparation, the fact that it was withheld from publication at that time, and that only minor editorial changes have been made in the text. In other words, the footnote should not imply that the War Department has reviewed the article and that the article has been revised to exclude material which still remains classified.

Captain Foy believes it is likely that the journal to which you submit the article may feel it necessary to submit it to the Bureau of Public Relations for clearance, in which case it would probably be referred back to us, but we would not recommend that you suggest such action to the journal or make any reference to clearance other than such as I have indicated.

I trust that this is clear and satisfactory to you and I hope that you will be able to find a publisher for the article. Frankly, I believe that it is good propaganda for us at this time. Perhaps I should not even call it propaganda since I feel that in general it is a fairly factual and well-balanced presentation of the subject matter. We both understand, of course, that were you given complete liberty you would on the basis of your recent experiences, amend and supplement considerably, but that of course is not permissible.

Dr. Theodor Rosebury   -2-   30 March 1946

Will you please let me see a draft of the footnote as revised before the paper is submitted for publication.

Sincerely yours,

(signed) Bram

. . . . BRAM, M.D.
. . . . Director

C O P Y   (Of a photostated letter, as follows:)

ARMY SERVICES FORCES
CAMP DETRICK
Frederick, Maryland

4 April 1946

Dr. Theodor Rosebury
Department of Bacteriology
Columbia University
630 West 168th Street
New York, New York

Dear Ted:

I have looked over the revised copy in connection with your proposed article on bacterial warfare and believe that it is now in conformity with Captain Foy's suggestions. You may, therefore, proceed to submit it for publication.

While in the Washington office yesterday I saw a draft of the article which Sidney Shalett has prepared for publication in Colliers. It appears to me to be permissible under our security classification, but as is true of most of the articles that are coming out now, perhaps makes for a rather further "bulge in the line". I suppose that is to be expected. Its references therein to you, as to certain others, are perhaps slightly on the distasteful side, but I suppose that the press must be permitted some freedom in its own field, and has to work in a little "local color" in order to dilute the otherwise heavy fare of scientific data. I passed the paper along for General Loucks and General Waitt to have a look at it. You mentioned that you would like to see the copy before publication and I believe that this might be possible for you to arrange through the New York Times office there.

Sincerely yours,

(Signed) Oram

ORAM C. WOOLPERT, M.D.
Technical Director

OCW:pkm

EMPLOYEE'S EXHIBIT #16

Figure 2   Letters from the War Department.

Small World

—*Little. in The Nashville Tennessean*

Figure 3   Political cartoon (by Tom Little, in *The Tennessean*) resulting from the published bacterial warfare article. (Reprinted with permission.)

When it appeared, Time magazine implied that we were procommunist and interpreted the statement in the footnote to indicate that we had not obtained clearance but had published it on our own "in view of the removal of wartime restrictions." Michael Heidelberger wrote a letter supporting our loyalty, which they published. At the same time Howard J. Mueller, Professor of Bacteriology and Immunology at Harvard, who had also been a consultant at Camp Detrick, wrote Dean Rappleye suggesting that Rosebury and Kabat should be fired because they had timed the publication of their report to appear when Congress was considering the War Department's budget so that they would get more money for biological warfare, although the war was over. Dean Rappleye knew that if any one had determined if and when the report would appear it was he and not Rosebury and Kabat. However, we learned that our clearances for Camp Detrick were cancelled just after the report was published. The FBI was assigned to investigate us and the following episode with my landlord was told to me.

Sally and I were leaving for Europe on June 20, 1947, to attend the International Cytology Congress in Stockholm and the International Microbiology Congress in Copenhagen. I wrote my landlord, giving him the dates we would be away and enclosing two rent checks dated July 1 and August 1. We left our two children, Jonathan, age 3, and Geoffrey, born May 11, 1946, with my parents. Shortly after our departure, an FBI agent visited the landlord. The conversation as recounted to me went about as follows. Did you know the Kabats were leaving for Europe? Do you think they are ever coming back? He showed them the letter and the checks. The questions continued. If you were planning to leave the country wouldn't this be a good way to hide your intentions? My landlord naturally was very surprised when we returned on August 20.

In Stockholm at the Cytology Congress, there was substantial interest in the report on bacterial warfare and a group of Swedish microbiologists invited Sally and me to an elegant lunch at the Operkellaren; several of the guests were evidently assigned to entertain Sally while the others turned the topic of conversation to my report. Since Rosebury and I had been involved in research subsequent to our report, all of which was still secret, it would have been almost impossible for me to discuss anything without creating the impression that it represented the report plus additional classified information. Fortunately, suspecting what the lunch was about, I had put a reprint of the report in my pocket and in reply to each specific question, I merely read the pertinent sections written in 1941 and finally gave the reprint to one of the microbiologists sitting next to me.

At the end of the war, I took up two major lines of investigation. One, on the immunochemistry of blood group A and B substances, was to continue to the present day. The background and details of work in this area and the personal aspects have been described by me recently (7). The second was an attempt, together with Abner Wolf and Ada Bezer, to produce acute disseminated encephalomyelitis in monkeys because of its possible relation to multiple sclerosis. The use of the Pasteur treatment for rabies had been known to result in what were termed paralytic accidents, which involved disseminated demyelinating lesions in the brain and spinal cord. This had been shown by Tom Rivers and Francis Schwenkter to be due to the nervous tissue and they succeeded in producing the disease in monkeys by injecting suspensions of spinal cord tissue. Many injections were needed and the disease did not appear for a long period. Tracy Putnam had always been most interested in multiple sclerosis and was delighted when we got back to this problem. Jules Freund had shown that one could get a very enhanced and protracted antibody response, as well as delayed type hypersensitivity, by incorporating antigens with mineral oil and killed mycobacterium by using an emulsifying agent, and I was anxious to see if this procedure could be used to produce disseminated encephalomyelitis in monkeys more rapidly and reproducibly. We found that after only one to three injections of brain tissue emulsified in the Freund adjuvant, the monkeys developed the disease and we sent a short note to *Science* early in 1946. While having dinner with Isabel Morgan during the Federation meetings I learned that she had made similar observations in monkeys she had been injecting with spinal cord tissue containing poliomyelitis virus with Freund adjuvants and then had omitted the virus with similiar results. She was to present her work at the Society of American Bacteriologists meetings in May. We decided that we would write up our detailed papers completely independently, then send them to one another to criticize and request that they be published side by side. Since her paper arrived a few days before ours had been typed, I gave the unopened envelope to Robert Loeb for safekeeping until I had mailed our manuscript. The results were completely concordant and both papers appeared together in the *Journal of Experimental Medicine* in January 1947.

The National Multiple Sclerosis Society had just been founded and sometime after our paper appeared I received a phone call from them saying they wanted me to apply for a grant. I said that I really was not interested in applying and didn't need the money. They replied that I "just have to" since they had received an anonymous contribution on condition that they used it to support our work. When I inquired as to the amount of the contribution and was told it was $10,000, I said that this was not really sufficient to make the more intensive effort they wished. I drew up an application for $64,350 for 3 years of support, which they showed the anonymous donor who generously provided the entire sum. This was the first grant made by the National Multiple Sclerosis Society. I was able to increase the size of my monkey colony from about 10 or 12 to 40. In 1950, the Public Health Service gave me a grant to continue this work, but it was cancelled in 1953.

By the end of the war my salary had increased to $4500 per year, but with the inflation following the removal of price controls, plus my growing family, I had to take a part-time job teaching elementary chemistry at City College. The multiple sclerosis grant provided an equivalent sum and made it possible for me to give up this outside work.

We spent the summer of 1948 in Woods Hole where we first met Fred and Sally Karush and their children. This was the beginning of a life-long friendship between the two families and Fred spent 6 months in my laboratory in 1950. The laboratory was then so crowded that he had to work a second shift beginning in the late afternoon.

Toward the end of that summer I saw a note on the bulletin board at the MBL that a house was for sale by the Professor of Microbiology at Washington University, Jacques Bronfenbrenner, whom I knew quite well. It took only a few minutes to arrange to buy the house. Our children became very attached to it and we spent part of each summer there during the years when they were growing up, except when I was on sabbatical. I would either work in the library or rent a laboratory bench. In 1949, I rented a laboratory bench, which then cost $50 for the summer, so that I could prepare an enzyme from snail hepatopancreas that split blood group substances. The business office at Columbia refused to pay the bill, saying that grants stipulated that the institution was to provide laboratory facilities. I assured them that the $50 entitled me to order and receive snails collected by the MBL boats and that if they would put snails on my desk in 48 hr I would do the work in New York—they paid the $50. Woods Hole was, of course, a marvelous place for making scientific contacts. In 1949, Shlomo Hestrin of the Hebrew University had the laboratory bench opposite me and we became very close friends.

Around this time a problem developed in my relationship to the Biochemistry Department at Columbia. For several years Tracy Putnam, the Professor of Neurology, had been asking Hans Clarke to promote me to Assistant Professor, but to no avail. This was in no way directed towards me personally, nor did it reflect any doubts about my work. Hans Clarke was a very fine person, but he was completely unaware of the need to promote people. He did not consider persons in other departments with titles in biochemistry as real members of his department, although he did not promote the regular members either. A substantial number of Research Associates had been in these positions for years, requests of the chairmen of the departments in which they worked being unavailing. Indeed, Walter W. Palmer, the Professor of Medicine, told me he had asked Hans Clarke to promote Michael Heidelberger to full Professor for 17 years before he agreed.

I broke the log jam of Research Associates by transferring to the Department of Bacteriology. The Chairman, Dr. Dochez, was very anxious to introduce immunochemistry into the medical teaching. In discussing my situation he suggested that he would be willing to give me an Assistant Professorship. The appointment was delayed for a year by the objections of Hans Clarke, who evidently didn't want to promote me or lose me, but it finally took effect on July 1, 1946. Louis Levin, who for about 10 years had been Research Associate in Biochemistry assigned to Anatomy, an appointment similar to mine, had left a year earlier for the University of Chicago because the Anatomy Department could not arrange his promotion. He was brought back as Assistant Professor of Anatomy and later went to the Office of Naval Research and the National Science Foundation.

After this, Hans Clarke agreed to several promotions of other Research Associates.

As Assistant Professor of Bacteriology, I was entitled to have graduate students and my first was Sam M. Beiser, who had been in the Navy and came to Columbia under the GI Bill. He was recommended by Arthur Shapiro. His PhD dissertation was on blood group substances from bovine gastric mucosa. The Academy of Allergy was just setting up a fellowship program (1). Even though Beiser was not working in allergy, at my suggestion that the fellowship might stimulate his interest in allergy, he received it and later devoted some time to work in this area. After two postdoctoral years with Bernard Davis, he returned to Columbia, rising to Professor and was Acting Chairman at the time of his death from cancer in 1972. Among his most important contributions were studies on antibodies to nucleotides with Bernard F. Erlanger. His bovine blood group glycoproteins have remained important standards and Michael and I are still using the pneumococcal C-polysaccharide he prepared.

My position in the Department of Bacteriology, the title of which was changed to Microbiology, and in the Neurology Department was eminently satisfactory over all these years. Tracy Putnam left in 1946 and was followed by Edwin G. Zabriskie, and in 1948 by H. Houston Merritt. Professor Dochez retired, and Beatrice Seegal became Acting Chairman of the Department of Bacteriology until Harry Rose was appointed in 1951. Drs. Zabriskie and Dochez had recommended my promotion to Associate Professor in 1947 and I was promoted in 1948. Harry Rose was also very supportive. Indeed, when he was offered the position as Chairman of Microbiology in 1951 he laid down two conditions—that I be promoted to full Professor and that my salary, which until then had come from grants, be paid out of departmental funds. This took effect on July 1, 1952.

By 1947, the laboratory had three major lines of investigation—immunochemistry of blood group substances (see 7), acute disseminated encephalomyelitis, and quantitative studies on allergic reactions. Edward E. Fischel, who was in the Department of Medicine, and I measured the amounts of antibody needed to produce Arthus reactions passively in the rabbit, and Grange Coffin, a medical student, and David M. Smith, a dental student, continued studies on passive anaphylaxis. Baruj Benacerraf came to see me in 1946, having completed medical school and a 1-year internship before serving in the Army and wanting to spend a postdoctoral year in the laboratory. There were no funds, but he told me he had independent means and came, as he later put it, on the "Benacerraf Fellowship." He was a very intense and dedicated worker and we studied quantitative aspects of the latent period in passive anaphylaxis and also the passive Arthus reactions in the guinea pig. John H. Vaughan spent two postdoctoral years in the laboratory studying the skin-sensitizing properties of rabbit antibodies to ovalbumin and conalbumin in human skin.

We had found increases in the gamma globulin in the cerebrospinal fluid of patients with multiple sclerosis by electrophoresis, but this was impractical for diagnostic purposes. Murray Glusman, who had returned to the Neurological Institute after the war, Vesta Knaub, and I decided to measure the amounts of gamma globulin by using a microquantitative precipitin reaction. We also measured albumin immunochemically at the same time. It became clear that the method would give more informative results than the colloidal gold tests, and it became possible to follow gamma globulin levels in cerebrospinal fluid in patients over prolonged periods. David A. Freedman, Jean Murray, and Vesta Knaub measured the albumin and gamma globulin levels in 100 cases of multiple sclerosis and found 85 with elevated gamma globulin levels. In carrying out such a study it was essential not to report our findings until the diagnosis had been made in the absence of these data, otherwise clinicians who avidly grasp for objective data would soon gain a clinical impression of the value of the test and would not make their diagnosis until the test was in the patient's chart. Subsequent studies were carried out with Melvin D. Yahr and Sidney S. Goldensohn, and the quantitative precipitin method for determining gamma globulin in the cerebrospinal fluid of patients at the Neurological Institute became one of my routine responsibilities for 30 years, until it was superseded by more sensitive and automated immunochemical methods. I had many close friends and other colleagues at the Neurological Institute during this period, including Saul R. Korey, Harry Grundfest, and David Nachmansohn.

At the end of the war Abner Wolf had become an attending consultant in Neuropathology at the Bronx Veterans Administration Hospital. The Veterans Hospitals were encouraging research and were setting up substantial laboratory facilities. William Newman had received his MD and had gone there, as had another MD, Irwin Feigen, and Abner suggested that we continue studies on histochemical localization of enzymes there. I was appointed attending consultant and spent one afternoon a week there for several years.

While I was at the VA, President Truman issued an Executive Order initiating loyalty and security investigations of Federal employees with the following criterion: "Reasonable grounds must exist for the belief that one is disloyal to the United States." James B. Sumner, with whom I had been very friendly at Uppsala in 1937–1938 and who I had later met briefly on only one occasion in the US, went to the FBI to tell them that I had been a communist while in Uppsala. This initiated a series of FBI investigations on the basis of which I was presented with charges about the various organizations to which I had belonged or contributed during the late 1930s.

I had a hearing before the VA loyalty board, which dismissed me. I appealed finally to the Presidential Loyalty Review Board, which reversed the decision and reinstated me. I returned to the VA, but it was obvious that there were pressures to reduce further the rigidity of the criteria and the quality of the evidence upon which an individual could be dismissed, so I decided to resign. This essentially led to my giving up work on the histochemical localization of enzymes in tissues. The Presidential Loyalty Review Board was abolished by President Eisenhower, responding to pressures that it had been too lenient. Indeed, it was the only board whose members were not Federal employees and thus was not subject to pressures from Congress and from within the Executive branch. I was very pleased that I had carried out all my appeals without a lawyer.

The Bronx VA Hospital Loyalty Board that dismissed me had also written a letter to the passport office telling them that I should not be allowed to travel, and so my passport was cancelled; it was not returned when I was reinstated. I had been invited to address the First International Congress of Allergists in Zurich in the summer of 1950 but was unable to obtain a passport. I went to the passport office to discuss the matter and was told that if I gave the names of anyone I knew or thought to be a communist I could get a passport. Needless to say I was unable to travel until the decision of the US Supreme Court in 1955 that every American citizen had an unlimited right to travel, when I attended the International Congress of Allergology in Petropolis, Brazil. Baruj Benacerraf brought my situation to the attention of the Zurich Congress in his address.

I received the invitation to the Allergology Congress while in Woods Hole. It provided $1000 for travel expenses. I wrote a letter saying that I could not accept because I could not get a passport. I walked into town and dropped the letter in the box and walked across the street to the drug store to buy the New York Times. The headline announced the US Supreme Court Decision. I rushed to the Post Office and was able to retrieve my letter.

Sanford Elberg, with whom I became friendly at Camp Detrick, invited me to teach two courses at the University of California at Berkeley during the summer of 1950. Sally, Jonathan, Geoffrey, and I drove across country, stopping in Springfield, Illinois, to see the Lincolniana, in Denver, in Salt Lake City, and at Boulder (now Hoover) Dam. I had a very enjoyable time teaching—among the students or auditors were A. A. Benedict, Mel Herzberg, Fred Aladjem, and Keith Smart. I met and became friends with Ed Adelberg, Roger Stanier, Mike Doudoroff at Berkeley, and K. F. Meyer, Professor of Bacteriology at San Francisco. While in Berkeley I was invited to speak at the Naval Biological Laboratory, the Navy's equivalent of Camp Detrick. In this highly classified institution and at a time of considerable hysteria, Keith Smart, in trying to shorten an obituary-like, overly long introduction, brought down the house by saying, "Dr. Kabat is a member of a large number of organizations, which I had better not mention." Of course he then felt obligated to list them.

In 1952 Congress appropriated $100,000 for research in immunochemistry in the budget of the Office of Naval Research (ONR). To evaluate what problems were worthy of support, William V. Consolazio, whom I had never met, and Louis Levin called a conference at P and S. About 10 or 15 people were present, including myself, Michael Heidelberger, and Dan Campbell. Everyone around the table indicated what problems he thought were important. I stressed use of immunochemical criteria of purity of proteins and polysaccharides. The meeting lasted until lunch time. When it was over and everyone was leaving, Bill Consolazio said to me, "You are staying here," and then asked, "To whom would you give money on the basis of the suggestions?" I outlined the programs I thought should be supported. He then said, "You have to take some money." I responded, "I don't need any money, I'm loaded with money." He gave me a yellow pad, saying, "You are not leaving here until you write out a proposal. You can hold on to the money and you can activate the contract any time within the next five years." I wrote out a title, "Immunochemical Criteria of Purity of Proteins and Polysaccharides," my name, the University's name and address, a short abstract, and a budget that provided for a postdoctoral fellow and supplies. When I came to overhead I asked, "What are you going to do about overhead?" At that time ONR was having a battle with Columbia—they would only pay 10% overhead, believing that the investigator was interested in doing the work anyway. The Dean was very dissatisfied with 10% and had threatened to throw ONR out completely. Consolazio replied, "Oh, put down 25% overhead—you don't want the money and I had to twist your arm." Several weeks later two Navy contract negotiators came to see Dean Rappleye with the piece of paper containing my almost illegible handwriting saying, "We've come to give you this contract." The Dean said, "What contract?" They replied, "Its all written on this piece of yellow paper." I was in the process of having it typed. The Dean then said, "What about overhead?" They replied, "Twenty-five percent." My stock at Columbia rose enormously. I only activated the contract a year later when the Public Health Service cancelled my grants at the height of the McCarthy hysteria. It was like having money in the bank. ONR supported me for 17 years. Consolazio and Levin left ONR when the National Science Foundation was set up and Bill suggested that I apply for a grant. I received one of the first awards-$60,000 for 3 years, appropriated out of the first year's molecular biology budget; it represented about 8% of the total. This replaced my blood group grant, which the Public Health Service cancelled in 1953. NSF has been the main support of my laboratory since then.

In 1951 I was asked to serve on the National Research Council's Subcommittee on Shock, which was concerned about the severe allergic reactions being found on administration of dextran, which had been developed as a plasma expander in Sweden. It was generally believed that the reactions were due to contamination with bacterial protein. I naturally suggested that the dextran itself, like the pneumococcal polysaccharides, might be antigenic in man. Although clinicians were prepared to inject 30 g of dextran intravenously, no one on the committee would inject 1 mg to see if it was antigenic. I did an initial skin test, had a blood sample taken, and gave myself two injections of 0.5 mg of dextran a day apart. A second skin test 3 weeks later showed a typical wheal and erythema, and quantitative precipitin assays on the pre- and post-immunization sera showed that my antidextran level had risen from 1.5 to 25 μN/ml. At the next meeting of the Subcommittee on Shock, I demonstrated my precipitates and also did a skin test on myself and on Doug Lawrason, the Secretary of the Committee, as a control.

Deborah Berg and I studied the quantitative precipitin reaction of dextran and antidextran produced in medical student volunteers. I suggested to the National Research Council that Paul Maurer be asked to do a study to confirm our findings, which he did. To prove that the antibodies were indeed antidextran we used biosynthetically labeled [^{14}C]dextran to precipitate the antibody, and ^{14}C analyses by David Rittenberg, Laura Pontecorvo at P and S, and Leon Hellman and Maxwell Eidinoff at Sloan-Kettering established that the dextran was precipitated by the antibody.

The antigenicity of dextran made possible studies probing the size of the antibody combining site. One of the dextrans, B512, developed at the Northern Regional Research Laboratory of the Department of Agriculture at Peoria, Illinois, was built of 96% $\alpha 1 \rightarrow 6$- and 4% $\alpha 1 \rightarrow 3$-linked glucoses and so had very long stretches of $\alpha 1 \rightarrow 6$-linked glucoses. Allene Jeanes at Peoria and Turvey and Whelan in England were isolating the series of $\alpha 1 \rightarrow 6$-linked isomaltose oligosaccharides. The system, $\alpha 1 \rightarrow 6$ dextran and human antidextran and the $\alpha 1 \rightarrow 6$ oligosaccharides, essentially provided a molecular ruler, since they permitted one to compare the potency on a molar basis of the various oligosaccharides in inhibiting precipitation of antidextran by dextran. This became a fourth area of interest of the laboratory, which continues to the present day.

In the summer of 1952, Sally and I again drove across country with our three sons (David, born August 7, 1951). I wished to learn more carbohydrate chemistry and decided to work with Herman O. L. Fischer at Berkeley, having become very attracted to the Bay Area from my earlier visit. Herman asked me what I wished to do and when I said something about learning methylation of sugars, he took out a sample of galactinol, an

α-galactoside of inositol, and said that methylation and hydrolysis would show where galactose was linked to the inositol. Clinton E. Ballou and Donald L. MacDonald collaborated on the problem and taught me the technics; fortunately, the methylated compound crystallized. The sabbatical was a wonderful experience and we were able to renew old friendships and make many new ones that have lasted. One unexpected consequence of the methylation study was that galactinol turned out to be one of the best inhibitors of the blood group B-anti-B reaction until the disaccharide DGal $\alpha1\rightarrow3$DGal and larger oligosaccharides were isolated.

My sabbatical ended in February, 1953, and we drove back to New York via the southern route, visiting various friends at universities and doing some sightseeing, arriving in New York early in March. It was hard to adjust to the rest of the New York winter after Berkeley. The laboratory was thriving; the allergic encephalitis, spinal fluid gamma globulin, blood group substance, and dextran problems were all going well. I had been cleared by the top Presidential Loyalty Review Board and presumably could carry on with my activities normally. However, rumors of grants being cancelled were becoming more frequent. Linus Pauling had his Public Health Service Grants cancelled and had to appoint others as Responsible Investigator so that the work could continue. My 3-year grant for the monkey studies was running out and I had naturally applied for a renewal. I received a letter from Frederick L. Stone, Chief, Extramural Programs, National Institute of Neurological Diseases and Blindness, dated December 14, 1953, which read,

Dear Dr. Kabat:

Your application for a research grant, identified as B-9(C4), and entitled "Immuno-chemical Studies on Acute Disseminated Encephalomyelitis and on Multiple Sclerosis" has now had a final review. I regret to report that this request falls in the group of applications for which grants cannot be made.

This was followed by a visit to Houston Merritt saying that this was because of the political climate, that they didn't know exactly why, etc. He then made the suggestion that the work could go on if someone else's name was substituted as Responsible Investigator. Although many others in this situation complied, I refused, responded with the appropriate four-letter words, and began a boycott of the US Public Health Service, which had imposed the policy on NIH. Anyone wishing to work in my laboratory had to agree not to accept any funds from any unit of the USPHS for the period he was working with me, nor could any employee of the USPHS set foot in my laboratory unless he was coming for some unrelated purpose. I received much support from scientists at NIH, but the top administration did nothing.

The reactions of my colleagues were varied. Most expressed support. Columbia University through Dean Rappleye, Houston Merritt, and Harry Rose supported me unequivocally. The scientific societies fought the policy vigorously. Michael and many others refused to review grant applications. Unfortunately, not all actions were of this type: One person, whom I had considered one of my closest friends, avoided me and on one occasion even hid in the stacks of the Woods Hole library when he saw me coming toward him.

Abner Wolf, Ada Bezer, and I had to kill off the only monkey colony in the world then being devoted to the multiple sclerosis problem. Shortly thereafter I was informed that the tenth committed year of my Blood Group Grant was cancelled.

I shall have to leave the story at this point. The suspense will not be dreadful. You all know that I survived and continued to work actively, that I spent a year at NIH as a Fogarty Scholar, and that I have been spending 2 days a week at NIH for the past 7 years. The rest of the story must wait for another occasion.

ACKNOWLEDGMENT

I thank Daniel M. Singer and Harvey N. Bernstein of Fried, Frank, Harris, Shriver, and Kampelman, Washington, D. C., for their advice, assistance, and direction in obtaining various government files about me under the Freedom of Information/Privacy Act.

Literature Cited

1. Cohen, S. D. 1979. The American Academy of Allergy—An historical review. *J. Allergy Clin. Immunol.* 64:332–466
2. Heidelberger, M. 1979. A "pure" organic chemist's downward path: Chapter 2—The years at P and S. *Ann. Rev. Biochem.* 48:1–21
3. Kabat, E. A. 1978. Life in the laboratory. *Trends Biochem. Sci.* 3:N87
4. Kabat, E. A. 1978. Close encounters of the odoriferous kind. *Trends Biochem. Sci.* 3:N256
5. Kabat, E. A. 1979. Horse sense? *Trends Biochem. Sci.* 4:N204
6. Kabat, E. A. 1980. Wrong door at the Nobel banquet. *Trends Biochem. Sci.* 6:VI
7. Kabat, E. A. 1982. Contributions of quantitative immunochemistry to knowledge of blood group A, B, H, Le, I, and i antigens. *Am. J. Clin. Pathol.* 78:281–92
8. Knight, C. A. 1951. Paper chromatography of some lower peptides. *J. Biol. Chem.* 190:753–62
9. Rosebury, T., Kabat, E. A. 1947. Bacterial warfare. *J. Immunol.* 56:7–96
10. Tiselius, A. 1939–1940. Electrophoretic analysis and the constitution of native fluids. *Harvey Lectures* 35:37–70

Ann. Rev. Immunol. 1983. 1:33–62

CELLULAR MECHANISMS OF IMMUNOLOGIC TOLERANCE

G. J. V. Nossal

The Walter and Eliza Hall Institute of Medical Research, Post Office, Royal Melbourne Hospital, Victoria 3050, Australia

INTRODUCTION

Immunologic tolerance, the phenomenon whereby antigen interacts with the lymphoid system to impair its later capacity to respond to that antigen, remains one of the most fascinating problems in cellular immunology. The capacity of individuals to discriminate "self" from "not self" has assumed a new dimension since the realization that T lymphocytes only see foreign antigens in the context of "self" molecules of the major histocompatibility complex (MHC) (131), and so the newer field of MHC restriction has become intertwined with immunologic tolerance and indeed with all aspects of immunoregulation. Although the central interest of students of immunologic tolerance lies in the establishment and maintenance of self-recognition, experimental realities mandate that most work on tolerance in fact is performed with model systems where foreign antigens, rather than autologous constituents, are presented to the lymphoid system to induce a state of non-reactivity. It then becomes a matter of considerable difficulty to judge whether the effects observed mimic the physiological situation of lymphocytes coming to grips with the need to avoid *horror autotoxicus,* or whether the model illustrates some facet of immunoregulation needed to limit the cascade of immunoproliferation that follows the introduction of antigen.

Faced with this constraint, investigators of immunologic tolerance fall into three groups. The first, numerically the largest, describes particular model systems and refrains from speculation as to how major a role the cellular mechanism revealed by the study plays in physiological self-recognition. Some members of this group form a "tolerance is dead" subgroup, who believe nothing is all that special about self antigens, and that the avoidance of autoimmunity depends on myriad immunoregulatory and

33

0732-0582/83/0410-0033$02.00

feedback mechanisms, none of which is uniquely important to self-recognition and all of which may apply in particular circumstances to foreign antigens as well. The second group considers suppressor T lymphocytes as central to immunologic tolerance. On this view, the repertoire of immunocytes is perfectly competent to recognize self antigens, but the capacity to threaten health by actually switching on these immunocytes is thwarted by a powerful activation of the appropriate suppressors. The third group sees the most important bulwark against auto-reactivity as lying in some purging of the immunologic dictionary, effector precursors with receptors for self being either destroyed or inactivated in some way.

What is not revealed in most articles on tolerance is that these three major approaches are not mutually exclusive, nor are the conceptual demarcations between them as rigid as may first appear. Given the importance of avoiding uncontrolled autoimmunity, may it not make sense to buttress a repertoire-purging mechanism with a suppressor system that acts as a fail-safe device? Given the immunopathological consequences that can result from continued antibody production in the face of persisting antigen, could one not envisage the body utilizing a trick developed primarily for self-recognition to limit the immune attack against foreign antigens, such as intestinal commensal organisms? Is the thought that an anti-idiotypic suppressor T cell stops a particular response so far from the thought that the idiotype is silenced by clonal purging of the cell bearing it? The temptation to place tolerance phenomena into tight compartments must be resisted for another reason as well. A selective immune system, geared to react to the unknown, must be degenerate and redundant (47), otherwise the repertoire of unispecific cellular elements would have to be infinitely large. A necessary consequence of this fact is that the affinity of antibodies varies greatly, and that cross-reactivities among unrelated antigens and antibodies are common. This being so, and the number and diversity of self antigens being very large, clonal deletion of all cells with any reactivity to any self antigen cannot be absolutely correct, because there would be a real risk of deleting the total repertoire! Thus any thought of clonal purging as the mechanism of tolerance must immediately be qualified by exemptions for sequestered antigens, antigens present in low concentrations in extracellular fluids, and (for antigens present at intermediate concentrations) perhaps with respect to antibody affinity thresholds, below which the purging mechanism could not be expected to operate. By the same token, a limited amount of autoantibody production, e.g. to heart muscle antigens after a myocardial infarction, neither transgresses the rules of tolerance nor causes any harm.

It appears, therefore, that immunologic tolerance cannot be described or explained simplistically or in isolation from immune responses as a whole. From the viewpoint of this review, it is fortunate that two other chapters in this volume deal with idiotypic regulation (100) and suppressor pathways (53a), which therefore receive less emphasis here. The plan of this review is as follows. A brief historical introduction describes the mainstreams of thinking in the field. The major section of the review deals with the B lymphocyte, because more detailed knowledge about its receptors for antigen and more advanced technology for its study at the single cell level have given us a clearer picture than is available for T cells of how it can be signaled following encounter with antigen. Present knowledge of how activation takes place is presented to contrast with the various mechanisms of negative signaling. A key finding that emerges is that not all B cells react equally to a given signal. The more complex problem of T lymphocyte tolerance is then addressed, with special emphasis on tolerance to MHC antigens.

HISTORICAL PERSPECTIVES

The word tolerance was first applied to this field by Owen (89, 90), in his classical studies on binovular twin cattle sharing a common placenta. He noted that "erythrocyte precursors from each twin fetus had become established in the other and had conferred on their new host a tolerance towards. . . . foreign cells that lasted throughout life" (89). Even earlier, Traub (122) had noted that mice infected in utero with lymphocytic choriomeningitis virus carried the virus in their blood and tissues throughout life, without an apparent immune response, whereas mice first infected in adult life made a brisk, typical antibody response. On the basis of these two findings, Burnet & Fenner (19) predicted that an antigen introduced into the body during embryonic life, before the immune apparatus had developed, would be mistaken for self, and that it would not evoke antibody formation then or if reencountered later in life. It is of interest to recall that the somewhat awkward self-marker theory of antibody production proposed in that work in fact postulated self-recognition units in antibody-producing cells that gave them the capacity to recognize self. The postulated characteristics of these receptors are somewhat similar to those popular today for self-MHC recognition by proponents of the two-receptor theory of MHC restriction. Burnet's own attempts to provide experimental evidence for the theory were unsuccessful (20), but experimental induction of tolerance was achieved soon after by Billingham et al (9), who used living cells as the source of MHC antigens and skin graft rejection as the readout system. These authors coined the phrase "actively acquired immunological tolerance." This study ushered in an avalanche of research on the subject.

   It was perhaps fortunate that most early workers who sought to validate the concept of tolerance by using non-living antigens, and antibody production as a readout, used xenogeneic serum proteins as the test toleragen (24, 32, 56, 109, 110), as had they followed Traub's lead and used microorganisms, the results would have been less successful (20, 82). In the event, the conclusion that antigens introduced in the perinatal period preferentially caused tolerance quickly became widespread. Tolerance induced in adult animals by lethal doses of X-irradiation and allo- or xenografting (69) was explicable on the postulate that a repopulating lymphoid system recapitulates the (unknown) events of early ontogeny. The finding of Felton (44) that small but supraimmunogenic amounts of pneumococcal polysaccharide could cause nonreactivity even in normal, adult mice was generally placed into a separate conceptual basket until much later.

A major stride towards a framework for the understanding of tolerance was the development of Burnet's (18) clonal selection theory. If immune recognition were indeed encoded in a population of immunocytes with one specificity per cell, then deletion of self-reactive clones could elegantly account for self-tolerance. The notion was further elaborated by Lederberg (65), who postulated that lymphocytes passed through an obligatorily paralyzable state early in their ontogeny, during which antigenic encounters would silence or delete them; if this milepost were passed without antigenic encounter, the cell would become inducible, i.e. susceptible to activation by clonal selection processes. The discovery that one cell always made only one antibody specificity gave an early boost to clonal selection (83), and the discovery of great heterogeneity in antigen binding among lymphocytes lent strong support (80). It was not until the late 1960s that clonal selection came to be almost universally accepted, at least for the B lymphocyte. In the meantime, the notion of something unique about the newborn state that facilitated tolerance induction had received some rude shocks.

In 1962, Dresser found that if care was taken to free preparations of serum proteins of all aggregated material, intravenous injections of minute amounts into normal adult animals could induce tolerance (37). This was an important discovery, because it raised the possibility that even mature immunocytes could receive negative signals from antigen. Gradually the perception dawned that unprocessed antigen, acting alone, was insufficient to trigger an immune response, but that some accessory stimulus or characteristic was required. The term adjuvanticity was coined. The belief grew that direct encounters between lymphocytes and soluble antigens lacking this characteristic were likely to deliver a negative stimulus (46). Yet the nature of this signal remained entirely unclear.

Until the middle 1960s, writers on tolerance considered phenomena involving predominantly T cells (e.g. graft rejection) and those depending on reduced antibody responses more or less together. Then came twin revolutions in rapid succession, namely the clear separation of lymphocytes into two great families, T and B cells, and the realization that these families had to collaborate in antibody formation (reviewed in 73). The impact on research into tolerance was immediate and profound. Weigle's group (23, 127–129) soon reached the view that human gamma globulin injected into adult mice produced tolerance in the T cell compartment more rapidly and at a much lower dose of antigen than for the B cells. The greater ease of tolerance induction in T cells was confirmed in several other studies (74, 76, 99, 116). Nevertheless, one probable cause at work in these experiments still remained to be uncovered by independent work. The discovery of suppressor T cells by McCullagh (70) and Gershon & Kondo (49) gave new thrust to tolerance research. It is in some ways remarkable that suppressor T cells had been missed until then. In retrospect, the ultra-low dose tolerance caused by femtogram amounts of flagellar antigens in our laboratory (107) could hardly have had any other explanation. In any event, once discovered, suppressor T cells were found to be active in many situations. As early as 1973, a review by Droege (38) cited no fewer than 81 references on suppressor T cells, and the pace of research has certainly not slackened since.

The key role of regulatory T cells in both antibody formation and T cell-mediated immune reactions raised another issue of great importance to the tolerance field. Given that the precursors of effector cells, particularly B cells, might not be tolerant of certain self components, the possibility was raised that autoimmunization might follow if a self molecule came into association with a foreign molecule that could act as a carrier, thus being capable of inaugurating T cell help (3, 127). A similar mechanism can be invoked for breakdown of experimentally induced tolerance in some models.

Though the emphasis of research in the early 1970s swung heavily towards regulatory T cells, the concept that B cells might be negatively signaled through direct contact with antigen was kept alive. An important experimental contribution was that of Borel's group (11, 50), who showed that haptenic antigens coupled to autologous immunoglobulin acted as powerful, direct toleragens for adult B cells. A seminal theoretical work was the articulation by Bretscher & Cohn (17) of their two-signal theory of lymphocyte activation. This states that lymphocytes require two inductive signals to become activated. One is provided by antigen occupying the receptor for antigen; the other is provided by the action of a helper T cell. If the first signal acts alone, in the absence of the second, the result is a negative signal or tolerance. In many ways, this formulation still provides a useful framework.

By the mid-1970s, the last of the crucial issues mentioned in the introduction, namely that of MHC restriction, entered the arena, profoundly affecting all our concepts of the T cell and therefore also of tolerance. As Doherty

& Bennink have put it (33), the tolerance problem could no longer be seen simply as a need to distinguish between self and non-self, but rather as a requirement that the T cell be capable of distinguishing self from self plus other. The intriguing possibility emerges that within the thymus cells with a high affinity for self-MHC components are eliminated by a clonal abortion mechanism (10, 33), whereas those with low affinity for self-MHC are positively selected. The extensive literature on MHC restriction is, however, outside the scope of this review.

SIGNALS INVOLVED IN B LYMPHOCYTE TRIGGERING

Before considering the possible mechanisms of induction of tolerance among B lymphocytes, it is essential to summarize what is known about the induction of immunity in this population. Studies with purified B cells cultured at low density (40, 58) or singly (97, 114, 124, 130) have materially enlarged our understanding. It is now clear that a cross-linking of the cell's membrane immunoglobulin (mIg) receptors alone, whether by anti-Ig antibodies or by a multivalent, T-independent antigen, does not suffice to trigger the cell. Additional signals are required. For some B cells, a T-independent antigen and one or more B cell growth and differentiation factors(s) initiate proliferation and subsequent antibody production. For others, an antigen-specific, MHC-restricted, T cell-derived helper signal is also needed (4, 81). Furthermore, the stimulatory co-factors able to move the B cell from a resting G_0 state to active blastogenesis may not be identical to those required for continued cycles of multiplication, and these, in turn, may differ from factors that promote differentiation to antibody-producing status. Some evidence suggests that macrophage-derived, interleukin 1-like molecules are needed early in the cascade and at least two types of T cell-derived factors are needed somewhat later. The nature of these various co-stimulatory factors is under active investigation at the moment. From my viewpoint, the most important point is that this recent work validates the requirement for something other than Signal 1 alone, opening the door to the possibility that Signal 1 without this complex of factors that constitute Signal 2 indeed may exert a negative effect on the cell.

Some further interesting points warrant mention with respect to Signal 1. Although it appears that receptor cross-linking is an essential element in immune induction, not every cross-linker behaves equally. Dintzis' group (30, 31), by using size-fractionated linear polymers of acrylamide substituted with hapten, reached the conclusion that a minimum number of receptors (about 12 to 16) must be connected as a spatially compact cluster before an immunogenic signal is delivered. Polymers substituted with fewer hapten molecules exert an inhibitory effect. The importance of some critical degree of epitope spacing in immune signaling has also been noted in other systems (42, 43). We should therefore be aware that subtle differences may exist between the Signal 1 that combines with Signal 2 in immune induction, and the Signal 1 only, which may be operative in some forms of tolerance.

Another important question in the creation of an appropriate micropatch on the B lymphocyte surface involves the cell Fc receptor. This must concern us, particularly as so many of the toleragens used in current research feature Ig as the carrier molecule. Isolated Fc portions of human or murine gamma globulin are capable of polyclonally activating murine splenic B cells to proliferation and antibody production (8), mimicking the effects of lipopolysaccharides (LPS). Although the effect was claimed to involve only Fc-receptor-positive B cells and not macrophages or T cells, the point is difficult to prove in high-density cultures. Multivalent antigens attached to the B cell surface could easily engage the Fc receptors in a micropatch through capturing secreted antibody. Furthermore, antigen attached to dendritic follicle cells is complexed to natural or acquired antibody and so could lead to the formation of a mixed mIg-Fc receptor patch on the lymphocyte surface. We need to know much more about the characteristics of the macromolecular interactions that constitute Signal 1, which may be more complex than we currently imagine.

NEGATIVE SIGNALS TRANSDUCED VIA THE B LYMPHOCYTE MIG RECEPTOR

Many studies show that an engagement of the B cell mIg receptors can also lead to negative signaling. This section reviews evidence of this important process in immunologic tolerance. Every B cell with mIg is susceptible to this negative signal, but a dominant feature that will engage our attention is the fact that cells at different maturation stages in B cell development display markedly differing sensitivities to this influence. B lymphocytes are derived from proliferating pre-B precursor cells and exit from the mitotic cycles that generate them as small, non-dividing lymphocytes lacking mIg (88). They then develop mIg, first IgM and later IgM plus IgD, during a non-mitotic maturation phase. I argue that cells encountering antigen as the first receptors emerge are exquisitely sensitive to negative signaling. This phase is when a potentially anti-self B cell would first meet antigen. Immature B cells already possessing mIgM can be isolated from newborn spleen or adult bone marrow, and can be confronted with antigen in vitro. Note that this is subtly different from allowing the mIg receptor coat to develop on the cell surface in the constant presence of antigen. An immature B cell is less susceptible to negative signals than one caught in the pre-B to B transition, but it is much more susceptible than is a mature mIgM plus mIgD-positive B cell. Finally, even an activated B cell, engaged in large-scale antibody synthesis, can be switched off by antigen, but only at high concentration.

As I review the data on which these assertions are made, the picture that emerges is not a dramatic, all-or-none difference conferring unique properties on the immature cell, but rather a carefully programmed quantitative difference geared to help self-recognition and also to limit appropriate immune responses.

Negative Signaling During the Pre-B to B Transition

The first evidence that receptor cross-linking could have profound and long-lasting effects on B lymphocyte differentiation was produced by an elegant series of studies by Cooper's group (reviewed in 64). When antibodies to IgM are introduced into an animal before B cells have developed, B cell maturation can be prevented, and an agammaglobulinemic, B cell-deprived individual results. However, such animals still possess pre-B cells. When such lymphocytes are cultured in the absence of further anti-μ antibody, mIg-positive B cells can emerge. These findings could be explained by postulating that as soon as the first few mIgM receptors appear on the maturing B cell, contact with anti-μ immediately initiates the patching-capping-endocytosis cycle and thus the cell never really develops into a B cell. This view gained further credence by two independent studies (98, 108), which showed that maturing B cells were much more susceptible to modulation by anti-Ig than were mature B cells. Furthermore, although mIg receptor deprivation in mature B cells was readily reversible, a considerable proportion of immature B cells failed to recover their receptor coat in tissue culture following anti-Ig treatment. It seems likely that such cells would die in vivo without ever becoming mIg-positive, especially in the continuing presence of the model universal toleragen, anti-Ig.

We wished to take this line of investigation further. In some recent studies (93), Pike et al set out to determine whether or not concentrations of anti-μ chain antibodies insufficient to modulate the totality of the mIgM receptor coat could nevertheless exert functional effects on B cells. In particular, what would happen if very low concentrations of anti-μ were present as immature small lymphocytes changed from being mIgM-negative to mIgM-positive? They developed an anti-μ chain monoclonal antibody, termed E4. The fluorescence-activated cell sorter (FACS) was used to select small, mIgM-negative cells from adult murine bone marrow or spleen. These were cultured for 1–2 days, and in the absence of E4, a population consisting of 30–40% mIg-positive B cells resulted. As predicted, E4 was able to inhibit this pre-B to B transition. At a concentration of 10 μg/ml, virtually no cells with high mIg density developed. At 1 μg/ml, a slight inhibition of receptor emergence was noted. At concentrations of 0.1 μg/ml or below, the E4 failed to affect mIg appearance. With these low concentrations, FACS analysis post-culture showed the mIg status of the cells to be exactly equivalent to that of controls. Accordingly, Pike et al (93) allowed B cell maturation from the pre-B pool to proceed in the presence of 0.1 μg/ml and still lower concentrations of E4. Following 1 or 2 days of such preculture, the resulting B cells were subjected to functional tests. They were stimulated with the potent mixture of polyclonal B cell activators, LPS plus dextran sulfate. The capacity of single B cells to give rise to proliferating clusters was studied, as was their capacity to produce Ig, measured by the protein A reverse plaque technique. It was found that 0.1 μg of E4 present during maturation completely abrogated response capacity. With 10^{-2} μg/ml, 97–100% inhibition was induced, and even with 10^{-4} μg/ml, only half as many B cells responded in the clonal cultures as in controls with no E4 present during maturation. Yet, in all these groups, the number of B cells appearing during preculture, and the average number of mIg receptors per B cell, were perfectly normal. Thus, the B cells must have received and stored some negative signal, which, however, did not impair the insertion of mIg into the plasma membrane of the cell.

If this experimental design is a model for what may be occurring in vivo during the induction of self-tolerance, it ought to be possible to produce similar effects on antigen-specific B cells. We have examined this possibility in vivo by introducing toleragen into the tissues of fetal mice via the maternal placenta (94), at a time before any B cells were present in the fetus. The hapten fluorescein (FLU) linked to the carrier human gamma globulin (HGG) was chosen. Isotope tracer studies indicated free passage into the fetus and helped to establish tissue concentrations. To gauge the effects on the emerging B cell pool in a strictly quantitative manner, Pike et al (94) once again made use of B cell cloning techniques, on this occasion using the T-independent antigen FLU-polymerized flagellin to trigger the B cells. The results led to very similar conclusions to those described above. It was found that with a fetal serum concentration of around 80 pg/ml (8×10^{-5} μg/ml) or 5×10^{-13} M, half the FLU-responsive B cells were rendered incapable of responding, indicating an extraordinary susceptibility to negative signaling. One interesting feature of this study was that fully tolerant mice, given much larger doses of toleragen in utero, possessed in their spleens mIg-negative pre-B cells on the verge of receptor emergence. When these were placed in vitro in a highly stimulatory environment, they acquired their receptors and, then confronted with Signal 1 plus Signal 2, reacted normally. This is at least one reason why antigen must persist for tolerance to be maintained, and it sounds a warning for studies in which large numbers of cells are adoptively transferred to environments full of Signal 2.

The capacity of such very low concentrations of antigen to affect the cells functionally suggests that some mechanism other than modulation of the mIg receptor coat must be at work. It is therefore of considerable interest to ask whether the antigen, despite its low concentration, delivers some death blow to the maturing antigen-specific cell, or whether the functionally impaired cell can persist for some time. Nossal & Pike studied this question in vivo by inducing tolerance via the placental route and subsequently searching for antigen-binding cells (85). When small doses of toleragen were used, no significant reduction in FLU-binding cell numbers was noted. Furthermore, the antigen avidity profile of tolerant and control populations was identical.

The Concept of Clonal Anergy, and Competition between Signals

Taken together, these results suggested that clonal abortion was not really an accurate description of the events induced, at least not when critically low concentrations of antigens were used. Rather, they are more consistent with the induction of a potentially reversible anergic state in the maturing B cell. I have therefore coined the phrase clonal anergy to denote that Signal 1 only can render a cell non-responsive without actually killing it. I believe this finding settles an old controversy. Tolerant cells do seem to exist, and this allows one to pose the question of whether or not some reversal of this intracellular anergy might form part of the spectrum of autoimmunity.

Important evidence exists that even at the earliest phase of B cell neogenesis tolerance induction is not the only possible outcome of an encounter with antigen. The first hint of this was the work of Metcalf & Klinman (71, 72), who used unfractionated cell sources rich in immature B cells in a T-dependent B cell cloning system, the splenic microfocus assay. They found that when these cells were confronted with specific multivalent antigen in the absence of T cell help, the cells were rendered tolerant. However, when they saw antigen for the first time simultaneously with a strong activation of helper T cells present in the immediate vicinity, they formed antibody-forming clones. Thus even here, Signal 1 plus Signal 2 resulted in immunity. Teale et al were sufficiently interested in this result to repeat it, using FACS-fractionated mIg-negative pre-B cells from either newborn spleen or (in small numbers) adult spleen (118). In the absence of T cell help, hapten-HGG present during their maturation rendered the immature B cells tolerant. In the presence of help, a clonal response was obtained.

An analogous result was obtained in experiments where fractionated pre-B cells were placed in culture simultaneously with hapten-HGG and the polyclonal B cell activator LPS (96). This mitogen has been seen by some as representing a mixture of Signal 1 and Signal 2. In this situation, about two thirds of the cells of relevant specificity were indeed rendered tolerant, but one third went on to clonal proliferation and antibody production. In that substantial minority, the tolerance signal had been overridden. A somewhat similar intervention of a polyclonal B cell activator during tolerance induction had been reported previously (68).

Negative Signaling of Immature B Cells

If the line of reasoning postulating a gradation of progressively decreasing sensitivity to negative signaling as the B cell matures is correct, interest attaches to the behavior of B cells with mIg receptors that may still not be fully mature. A likely source for such cells is the mIg-positive cell pool from newborn spleen or adult marrow, where immature cells predominate. To study such cells, one can either resort again to the model universal toleragen, anti-μ antibody, or seek to identify, isolate, and study the immature B cells of a given specificity. Nossal et al (84, 85, 87, 92–96) have done both and have documented that such B cells do represent an intermediate stage in sensitivity to tolerance induction between the exquisite sensitivity of cells in the pre-B to B transition state and mature B cells.

Let us consider, first, the behavior of B cells from newborn spleen following a brief encounter with anti-μ chain antibody (87). These were found to be about 30-fold more susceptible to negative signaling than were adult B cells similarly treated. To achieve 60% inhibition of clonal proliferation following mitogen stimulation, 0.1 μg of anti-μ antibody per ml was required, acting for 24 h on the cells, whereas 3 μg/ml only reduced the proliferative potential of mature cells by 40%. A similar difference was noted for antigenic stimulation and antibody formation as the readout. Note, however, that the threshold concentration delivering an effective negative signal is about 1000-fold higher than for the pre-B to B transition.

As immature B cells, thus defined, possess adequate amounts of mIg, it is possible to prepare antigen-specific immature B cells by the same antigen-affinity fractionation procedures as for mature cells (55). One can then react such cells with multivalent antigen, e.g. for 24 h, and subsequently can challenge with antigen in immunogenic form. Their analysis (84) showed that with an oligovalent FLU-HGG that has 3.6 haptens per mol of carrier, hapten-specific newborn B cells could be tolerized with 1000-fold lower antigen concentrations than could adult B cells, the threshold being around 0.1 μg/ml. In other words, the sensitivity is much lower than that of cells that first see antigen before achieving full receptor status, but much higher than that of fully mature cells.

Pike et al have performed an extensive exploration of negative signaling among this intermediate population to determine the importance of molecular arrangement of haptens and of the carrier molecule (92). Spleen cells from newborn or adult mice were incubated for 24 h with FLU coupled to

HGG, the F(ab^1)$_2$ fragment of HGG, or bovine serum albumin (BSA). Conjugates were prepared that possessed a mean of 2, 5, 8, or 12 FLU residues per molecule of protein. After this putative tolerance-inducing regime, the cells were placed into a limit dilution cloning assay with FLU-POL (polymerized flagellin) as the antigenic stimulus. It was found that with all conjugates tested, the newborn cells were more sensitive to negative signaling than were adult cells. However, the gap between immature and mature cell, in terms of sensitivity, varied widely from conjugate to conjugate. For all three carriers, the higher the hapten density, the lower the molarity required for tolerance induction. The Fc piece appeared to play some role in making HGG the most effective of the three carriers. At a given hapten substitution rate, FLU conjugated to the F (ab^1)$_2$ fraction of HGG (FLU-FAB) was less toleragenic than was FLU-HGG. This difference could be overcome by crowding on more haptens. FLU-BSA was less toleragenic than either FLU-HGG or FLU-FAB. Depending on the conjugate, the sensitivity difference factor between immature and mature cells varied from 6 to 53. With highly multivalent conjugates it is easy to see how susceptibility differences could have been missed by other investigators, as the differences are less pronounced than with oligovalent antigen.

Though few authors have sought to dissect the pre-B to B transition from slightly later stages of ontogeny, the sensitivity of the immature B cell to negative signaling has been widely confirmed (21, 39, 71, 72, 105). Nevertheless, the adult B cell is not completely refractory, and we must now examine mIg as a transducer of negative signals in more mature cells of the B series.

Negative Signaling of Mature B Cells and their Progeny

I have already made reference to the work of Borel (11, 50), to studies by Nossal et al (see above), and to work by Weigle's group (35, 36, 91, 128), which amply demonstrate that a sufficient (and by no means unphysiologically high) concentration of antigen can deliver a negative signal even to mature B cells. Conjugates with Ig carriers are particularly effective, which brings us back to the question of the role of the Fc receptor. This has been addressed in an interesting series of studies by Taylor's group (117, 119–121). They prepared stable covalent antigen-antibody complexes, for example linking hapten directly to antibody, or hapten-proteins to anti-hapten antibody. The hapten-antibody conjugates were profoundly suppressive when T-independent anti-hapten responses were later measured. Evidence suggested a direct effect on B cells rather than a suppressor T cell effect, and the Fc piece of the antibody in the conjugate was required. This was closely analogous to earlier studies with non-covalent flagellin-antibody complexes (43). However, when the animal had been pre-primed against the rabbit antibody in the conjugate, the end result was an enhanced response, suggesting that the conjoint action of (normally toleragenic) conjugate and active T cell help was immunogenic. When ribonuclease was linked to antibody in this fashion, this normally poor immunogen induced priming or immunologic memory, but this immunogenicity was lost if recipient mice had been previously made tolerant to the rabbit IgG carrier. In that case the complexes were frequently toleragenic. Again, the suggestion is strong that an engagement of the Fc receptor within the micropatch on the B cell surface somehow causes the generation of a more effective Signal 1, which requires the Signal 2 of T cell help to become immunogenic and in its absence acts to produce tolerance. It would be of considerable interest to quantitate the effects of these stable antigen-antibody conjugates on immature B cells.

The regulatory role of the mIg receptors as signal transducers does not end at the moment of lymphocyte activation. In fact, activated B blasts and early antibody-forming cells including many plasma cells still bear mIg. Presumably this indicates that such cells are still under the regulatory influence of antigen. Certainly premature removal of antigen halts expansion of immunocyte clones (95). There may be circumstances, however, such as the existence of immune complexes in the circulation, where it might be wise to halt the ongoing immune proliferation, or even antibody formation at the single cell level. Some years ago, Schrader & Nossal discovered that an interaction of a single antibody-forming cell with multivalent, specific antigen can result in profound inhibition of that cell's secretory rate (104). This represents an unexpected late down-regulatory signal mediated via mIg, which may represent, though at a much less sensitive level, a mechanism analogous to the tolerance induction just described. Independent confirmation was provided by Klaus (62), whose group also soon showed that antibody production in monoclonal plasmacytoma cell cultures could be similarly inhibited (1). Antigen-antibody complexes were particularly effective in this regard (2).

My laboratory has recently sought to characterize this phenomenon, termed effector cell blockade, more fully. As a first step (12), a mouse hybridoma cell line that secretes IgM specific for FLU was developed. Treatment of these cells with highly multivalent FLU conjugates (e.g. FLU_{20}-gelatin) resulted in inhibition of hemolytic plaque formation. Biosynthetic labeling studies indicated this was a result of reduced Ig synthesis. Total protein synthesis and cell proliferation were not affected, documenting the specificity of the inhibitory effect. Fluorescence studies showed that FLU conjugates capable of causing blockade aggregated on the cell surface, and clearance of cell-associated antigen correlated with recovery from blockade. Reversibility has also been documented in other systems (78).

In later studies, Boyd & Schrader have provided further valuable insights (14, 15). Colchicine reverses effector cell blockade, provided exogenous antigen was removed. In the presence of colchicine, capping occurs, presumably aiding antigen removal. However, in the continued presence of exogenous antigen, colchicine was not able to reverse the negative signal, arguing against the notion that colchicine-sensitive structures, such as microtubules, are essential for signal transmission. It was also found that antibody-forming cells (AFC) from the spleens of 10-day-old mice are more susceptible to blockade than are adult AFC. Furthermore, blockade in neonatal AFC could not be completely reversed by enzymatic removal of surface antigen. This increased susceptibility of neonatal AFC is reminiscent of the greater sensitivity of newborn splenic B cells to negative signaling by antigen. It supports the notion of some intrinsic extra susceptibility in mIg-transduced signaling in the neonatal state.

Another model of a relatively mature B cell is a B lymphoma, such as the cell line WEHI 231, which resembles a B cell in bearing large amounts of IgM on its surface but not in secreting Ig. It was found (13) that anti-μ chain antibodies, including E4, profoundly suppressed the growth of WEHI 231 in both soft agar and liquid culture. Other antibodies binding to non-Ig cell surface antigens lacked this effect, and LPS, which induces antibody formation in normal WEHI 231 cells, could neither prevent nor reverse the growth inhibition. This negative signal differs from that involved in effector cell blockade in affecting cellular life span. It provokes the thought of whether or not receipt of the anergic signal in normal B cells may aid their early demise.

The picture that emerges, then, is a gradually diminishing susceptibility to negative signaling as the B cell matures. The monoclonal anti-μ chain antibody, E4, has allowed some quantitative comparisons in this regard (93). It takes approximately 5.5 $\log_{10}$ more E4 to cause effector cell blockade in mature antibody-forming cells than it takes to cause clonal anergy during the pre-B to B transition. Operationally, this could well mean that many exogenous antigens never reach the concentrations that might impair the B cell response to them.

CONSTRAINTS ON THE UNIVERSAL APPLICABILITY OF CLONAL ANERGY IN SELF-TOLERANCE

Although a strong case can be made for the plausibility of the clonal anergy model as being relevant to self-tolerance, particularly for antigens (e.g. cell surface molecules) present in multivalent form, it is important to mention some of the constraints on the universality of the mechanism. Perhaps the most obvious of these is that monovalent antigens do not induce clonal anergy in vitro. Most of the evidence suggests that the negative signal requires mIg receptor cross-linking. One could imagine in vivo matrix-generating mechanisms, e.g. self antigens associating with macrophage surfaces, or even one B cell presenting toleragenic antigen to another. Given the extreme sensitivity during the pre-B to B transition, even some self-association through protein-protein interactions at the B cell surface could not be entirely excluded. It is also of interest that despite the enormous amount of work on autoimmune diseases in both experimental animals and man, no one has yet reported an autoimmune disease that involves antibody production to autologous albumin, transferrin, complement components, or other soluble monomeric serum proteins present in high concentration (rheumatoid factors are an interesting exception). If tolerance to albumin did not exist within the B cell compartment, it is hard to see how antibody production could be avoided on occasions of abnormal induction (e.g. endotoxins present in septicemia). There are no albumin-binding cells in human peripheral blood (5), though B cells capable of binding autoantigens present in low concentrations in serum do exist (5, 41, 102). This suggests that for antigens present in high concentration, some clonal purging mechanism may be at work in vivo, perhaps one we have not yet been able to model in the artificial in vitro models.

Nevertheless, the possibility must be faced that some self antigens cannot evoke the clonal anergy/clonal abortion pathway, but cause tolerance through other mechanisms. From that viewpoint, recent studies from Diener's group are of interest (29, 125, 126). The sensitivity of the fetal immune system to tolerance induction in vivo was investigated by injecting various antigens into the maternal circulation. Transplacental passage was monitored, as was persistence of antigen in serum and lymphoid tissues. Despite adequate exposure, the fetuses failed to become tolerant in most instances. Of seven antigens, only two gamma globulins (HGG and bovine) induced significant tolerance. Furthermore, when haptens were introduced multivalently onto the proteins, the various non-toleragenic carriers failed to induce hapten-specific tolerance. This was despite fetal concentrations of up to 10^{-6} M. In other studies it was found that some carbohydrates, but not others, can act as a toleragenic carrier for hapten-specific B cell tolerance, and there appeared to be no clear correlation between the capacity to induce tolerance and either the molecular weight of the carrier or its epitope density. Indeed, rather subtle chemical changes were found to abolish an antigen's toleragenic potential.

These somewhat surprising findings are interpreted as suggesting that the toleragenic potential of certain extrinisic antigens reflects some obscure carrier-related property. To account for the results, Diener's group has postulated (28, 125) a dual recognition event in self-tolerance induction that bears some similarities to the dual recognition mechanism responsible for immune induction in T cells. Consider an organ-specific cell membrane self antigen, S, and a ubiquitous, polymorphic self marker or interaction structure, Z. The latter is postulated to be present on all cells including lymphocytes. When an anti-S lymphocyte arises in ontogeny its anti-S receptor meets S, and its self-marker Z meets the Z present on the S-bearing cell. This interaction is postulated to cause clonal abortion. Extrinsic, non-self antigens are seen as incapable of inserting into cell membranes in such a way as to allow correct dual recognition, with the exception of some molecules with special properties. This novel speculation, however, does not account for mechanisms by which animals become tolerant of self components not associated with cell surfaces, e.g. serum proteins.

Preliminary studies in our laboratory (B. L. Pike, G. J. V. Nossal, unpublished data) have sought to use transplacental FLU-BSA in vivo as an agent to induce clonal anergy in FLU-specific B cells emerging in fetal and early postnatal life. Even large doses achieved only moderate reduction in the numbers of clonable anti-FLU B cells. This contrasts with the relatively poor but still definite potential of FLU-BSA to give a negative signal to immature B cells in vitro (92). The result confirms that the carrier molecule is not a neutral entity in tolerance induction. Obviously this constraint will require much further investigation.

B CELL TOLERANCE AND AUTHENTIC SELF ANTIGENS

There would obviously be many advantages if students of immunologic tolerance, rather than introducing foreign model toleragens into animals or cultures, could investigate interactions between authentic autologous antigens and the lymphoid system of the body. However, surprisingly little work has been done in this way, probably because of the practical difficulties posed by omnipresent antigen, e.g. through neutralizing any antibody that may be formed or through blockading immunocyte receptors. I have already mentioned some of the studies (5, 41, 102) seeking autoantigen-binding B cells, which showed presence of B cells for self antigens occurring at low molarity but absence for those in serum at high concentration. This finding seems consistent with the rules of clonal anergy/clonal abortion. Two early studies that are frequently quoted are those of Triplett (123), who claimed that extirpating an organ early in ontogeny led to a loss of tolerance to the organ-specific antigens so that later transplantation led to rejection, and the hemoglobin studies of Reichlein (101), who concluded that rabbits could only form antibodies to such determinants of hemoglobin chains as are not represented in rabbit hemoglobin. In this section, three recent studies are reviewed, two of which argue against, while one argues for, clonal anergy/clonal abortion as an important mechanism.

The first study comes from Borel's group (57) and exploits a pair of congenic strains of mice, one of which lacks the fifth component of complement, C5, a serum protein normally present at a concentration of 50–85 μg/ml. Lymphoid cells from C5-deficient mice were adoptively transferred into X-irradiated hosts of the normal strain and vice versa. C5-deficient cells were not tolerant to C5. They formed antibody when transferred to C5-sufficient (normal) irradiated hosts. Cells from normal mice were tolerant, failing to form antibody to C5 on primary or secondary immunization of the C5-deficient hosts with C5. T and B cell mixing experiments demonstrated that the tolerance appeared to reside in the T cell compartment of the C5-sufficient (normal) lymphoid cells. T cells from C5-deficient (nontolerant) donors could collaborate either with B cells from nontolerant (C5-deficient) donors or from tolerant (normal) donors. This argues that anti-C5 B cells had not been eliminated in normal mice, nor were they functionally silenced.

This thought-provoking study exhibits two features that are somewhat worrying. First, the adoptive hosts received only 760–780 rads of irradiation in a split dose. In my experience, this would allow considerable endogenous reconstitution. Second, 70 million cells were transferred, a very large number, which raises the possibility that T cell sources might be contaminated by significant numbers of B cells and vice versa. Taken at face value, the results provide another example of a monovalent antigen not leading to B cell tolerance.

As a general point, caution must be taken in adoptive transfer studies to exclude the possibility that effects attributed to B cells in fact are not ascribable to mIg-negative pre-B cells or even earlier precursors. It is clear that such occur in spleen and bone marrow in substantial numbers, and if they mature in the concomitant presence of antigen and T cell help, they will immediately be triggered. Of course, pre-B cells that lack mIg have not yet been subjected to self-censorship.

The second studies concern an alloantigen, the cytoplasmic liver protein F. This molecule of 40,000 mol wt occurs in two forms in mice, F^1 and F^2. Serum concentrations are around 10^{-8} to 10^{-9} M. It has long been known that in certain strains of mice alloimmunization can lead to an autoantibody response directed against determinants common to F^1 and F^2, implying that self-tolerance is normally maintained by T cells (59). This lack of B cell tolerance is not too worrying, given the relatively low serum concentration. Detailed investigation of cellular interactions in this system are now possible because of the recent development of an in vitro assay (113). The system measures the proliferation of F-primed T cells exposed to F-pulsed splenic adherent cells. Syngeneic F antigen fails to activate under circumstances where allo-F works well. It has been suggested (27) that this is due to in vivo elimination of self-F reactive T cells. If so, this would be one interesting example of clonal deletion affecting T cells but not B cells, an interesting counterpoint to the usual explanations involving suppressor T cells.

The third example involves tolerance to antigenic determinants on mouse cytochrome c (60). This study from Klinman's group is of particular interest, not only because it supports clonal anergy/clonal abortion models of tolerance, but also because it depends on the careful, quantitative analytical technology so vital in tolerance research. To understand the study, it is necessary again to recall that pre-B cells can form antibody-forming cell clones if artificially transplanted into a milieu where they mature into B cells in the concomitant presence of antigen and T cell help (Signal 1 plus Signal 2 overriding the toleragenic potential of Signal 1 alone). It is possible to prepare pre-B cell-rich cell populations, e.g. by using bone-marrow depleted of mIg-positive cells. One can then contrast this pre-B cell repertoire of clonotypes with that of mature splenic B cells. If clonal anergy or analogous methods of repertoire purging are important in self-tolerance, one should find autoreactive B cells present in the pre-B cell repertoire but absent from the mature B cell pool. Jemmerson et al (60) synthesized two peptides from mouse cytochrome c and probed the T-dependent immune response against each in the clonal splenic microfocus assay. With one peptide it was found that the pre-B cell pool contained many clonal precursors against the self component, the numbers being analogous to those for a foreign peptide of equivalent size. Mature B cells showed far fewer precursors, indicating an in vivo repertoire purging during the pre-B to B transition. With the other peptide, which included an evolutionarily highly conserved region at the N-terminal end of cytochrome c, neither cell population made an anti-self response. This raises the fascinating possibility that V genes whose products might have reacted with this region may have been eliminated from the specificity repertoire by evolutionary means.

In some respects, we find this example more illuminating than the other two. We have already stated that clonal abortion/clonal anergy cannot be absolutely true for every self antigen, for fear of purging the whole repertoire through cross-reactivity. The degree of the contribution of this mechanism to self tolerance therefore becomes a quantitative question. The existence of some undetermined number of anti-self B cells, as there appear to be in the C5 and F antigen examples, cannot invalidate clonal silencing as an important quantitative contribution to the immune balance. What one needs to know is whether or not there are fewer anti-self C5 or anti-self F antigen B cells than there would be for a foreign antigen of comparable size and shape and whether or not such cells display a lower median avidity for the self antigen than might be expected. Until such information comes forward, the possibility of a partial B cell contribution to the final tolerant state in these models remains open.

TOLERANCE PHENOMENA WITHIN THE T CELL COMPARTMENT

The subject of tolerance within the T cell compartment is much more difficult to address at either the cellular or the molecular level. We still know relatively little about the T cell receptor for antigen. We have no method of enumerating effector T cells that can compare with the Jerne plaque technique. Above all, we have to face the complexity that whereas the B cell can encounter and bind antigen free in solution, and evidently can receive signals from such an event, the T cell appears to be programmed to see antigen only at cell surfaces. Furthermore, both the triggering of resting T cells into activation and the delivery by the effector progeny of some effect (cytotoxic killing, help, suppression) require both a union of T cell receptor with foreign antigen and recognition by the T cell of certain self-MHC determinants, the nature of that "self-hood" having been somatically learned by the T cell pool. The only exception to these rules appears to be the effector cell engaged in mediating suppression, which can adhere to antigen-coated surfaces that lack any MHC molecules. Cell interactions involving products of the I-J genes are probably active in getting resting suppressor precursors moving, so even for this category of cells, signal induction involves linked recognition.

Conceptually, this extra complexity poses no major dilemma for clonal anergy/clonal abortion theories. One would simply have to postulate that for those self antigens not already on a cell surface, mechanisms exist for presenting them in cell-associated form to the maturing T cell, e.g. by means of accessory cells in the thymus. On the other hand, a large body of opinion prefers to attribute self-recognition within the T cell compartment to the induction and maintenance of an anti-self-suppressor T cell population. If this special kind of immune response is placed at a post-thymic stage of T cell development, it is hard to see how the peripheral pool of (not clonally purged) T cells could operationally distinguish self from not self, activating exclusively or predominantly a suppressor response only against the former. If, on the other hand, one were to postulate an intrathymic preferential genesis of suppressors should the maturing T cell see antigen, a formulation that would give ever-present self antigens the edge they possess on a clonal abortion model, then we are left with the need to explain why foreign antigens, in the face of a partial blood:thymus barrier, can so readily generate suppressor cells under some circumstances, even in adult animals. For these reasons, I prefer to regard the generation of suppressor cells as a regulatory mechanism, an ancillary device in self-tolerance, rather than as the primary cause.

The main point, however, is that the existence of a family of immunocytes known as suppressor T cells cannot explain self-recognition. It merely shifts the problem to another level. Furthermore, the existence of an (undefined) number of suppressor cells in a T cell population that behaves as tolerant does not necessarily imply that these cells are the sole cause of the tolerance. Clonal silencing of effector precursors may be present concurrently, but they escape detection unless sought. Once again, only strictly quantitative methods can address the relative contributions of the various mechanisms.

Suppressor T Cells in Tolerance Measured by Antibody Formation

It is surprising that some antigens that are good at negatively signaling B cells in vitro or in vivo also seem to be excellent at inducing suppressor cells. For example, HGG has been used in many studies of suppression (6, 7, 35), although it has also been claimed that suppressor cells may not be the essential element in keeping the animal as a whole unresponsive (36, 91, 126). Most authors study induction of suppressor cells in adult mice, but some (66, 126) have used fetuses and the transplacental route of antigen administration and find it effective in initiating suppressor cell generation. An interesting recent series of studies (66, 67) documents that HGG can not only produce suppressor cells, but also can set up a state of suppressor T cell memory. Thus, when the first effective wave of active suppression has passed, animals may be left with memory suppressor T cells ready to spring into a secondary wave of multiplication and active suppression should antigen be encountered again. HGG in vitro can generate memory suppressors that even 6 months later can be reactivated by antigen. Such cells may have been missed in previous studies, possibly explaining some examples where tolerance appeared to persist though active suppressors could not be detected. The memory suppressor cells were antigen specific and were found to have the same phenotype as those mediating primary suppression, namely Ly-1-23$^+$, Thy-1$^+$, Ia$^+$. Moreover, though it requires more than 100 μg of HGG to generate primary suppression, as little as 1 μg induces suppressor T cell memory. Exposure to very small concentrations of self antigens throughout life might give a continuing series of boosts to the pool of memory cells without necessarily generating detectable active suppressor activity. Any unexpected pulsatile release of self antigen, e.g. from organ damage, could quickly lead to emergence of active suppressors, preventing autoimmunity. Thus, a powerful case has been made for a mechanism with physiological relevance. We can only conclude that this co-exists with B cell-silencing mechanisms as demonstrated in our studies.

Suppressor T Cells in MHC Reactions

A large literature exists on suppressor T cells in various models of T cell tolerance involving MHC antigens (e.g. 34, 45, 52, 53, 61, 75, 103). This is beyond the scope of this review. The view has been expressed (77) that in this case clone elimination is the prime process and suppressor cells operate as a fail-safe mechanism.

Functional Clonal Deletion in T Cell Tolerance

I wish now briefly to review some evidence supporting the view (16) that functional clonal deletion does occur among anti-allo-MHC T lymphocytes. Given that T cells of a given specificity cannot be enumerated by an antigen-binding technology, it is difficult to ask how many T cells active against a given self antigen exist in the body. However, the point can be approached by cloning technology. We cannot visualize the specific T cells nor isolate them by antigen-affinity binding techniques as in the case of the B cell, but we can enumerate specific cells capable of yielding progeny in vitro whose effects can be measured. The best known such technology is the enumeration of precursors of cytotoxic T lymphocytes (CTL-P), which develop from single cells by appropriate stimulation with irradiated allogeneic cells acting as an antigen source, and T cell growth factor as a co-stimulus, a methodology now in use in many laboratories. Nossal & Pike have applied this method to the classical tolerance model of newborn mice·receiving semiallogeneic spleen cells (86).

FUNCTIONAL CLONAL DELETION OF ANTI-ALLO-MHC CYTOTOXIC T LYMPHOCYTE PRECURSORS   CBA (H-2^k) mice were rendered tolerant to H-2^d antigens by injection of (CBA X BALB/c)F$_1$ spleen cells on the day of birth (86). At intervals of 2 days to 12 weeks, the frequencies of anti-H-2^d CTL-P in the thymus and spleen were determined by limiting dilution cloning technology. A profound and long-lasting functional clonal deficit was noted. This was first clear-cut by day 5 of life in the thymus and by day 8–10 in the spleen, as if to suggest that the clonal purging occurred in the thymus and moved to the spleen only as that organ came to be dominated by purged thymus migrants. The functional clonal deletion reduced the observed proportion of anti-H-2^d cells in adult spleens from a normal level of about 1 in 500 spleen cells (1 in 150 T cells) to 3% of that figure. This shows that whether or not suppression is also induced concomitantly, clonal silencing is a most important feature of the tolerant state. Its rapid occurrence in the thymus makes a clonal abortion/clonal anergy mechanism highly plausible.

Although this strict quantitative approach awaits confirmation, there is a considerable amount of indirect evidence for functional clonal deletion emerging from bulk culture studies. For example, Streilein's group (54, 112) failed to detect lymphocytes reactive with tolerated H-2 alloantigens in a wide battery of tests, including mixed lymphocyte reactions, graft versus host reactions, cell-mediated lympholysis, or concanavalin A-mediated, polyclonal activation of tolerant T cell populations. Suppressor cells were not found in these studies, and the inference was of an active process achieving specific clonal deletion of an early stage in the ontogeny of the alloreactive T cells.

Proponents of the suppressor T cell might then postulate that suppressor T cells (anti-idiotype in nature?) might actually be responsible for this functional or actual clonal deletion. A hint in this direction was provided by Gorczynski & MacRae (52, 53). Streilein (111) investigated this possibility in an original way by using MHC recombinant strains to induce tolerance. Newborn mice were rendered tolerant to F_1 semiallogeneic spleen cells in such a manner that donor and host differed over the entire H-2 region, over a region including the K or D locus alone, or over a region embracing either K or D and the mid-I (probably IJ) region. Although the first and last groups developed excellent tolerance, groups differing at the Class I antigens alone showed little or no tolerance as judged by skin graft rejection. It is as if isolated K or D antigens acted as poor tolerogens, requiring some kind of activation (presumably of suppressor cells) dependent on IJ disparity to reach the final tolerant state. If the K or D antigen is regarded as a hapten, and the mid-I (presumably IJ) antigen is regarded as a carrier, then active suppression is induced only when the immunogen differs from the host in both hapten and carrier. How this activation of suppression finally achieves clonal deletion is not clear.

These somewhat unexpected results prompted Nossal & Pike to investigate the effects of anti-IJ sera on tolerance induction in their model depending on injecting F_1 spleen cells differing at the full MHC into newborn mice (86). Repeated high doses of anti-IJ injected into newborns during the period of toleragenesis partially inhibited clonal deletion, an effect, however, that proved to be quite transient. This has been a frustrating finding to follow further. Though the phenomenon was readily reproducible, it depended on a particular batch of anti-IJ serum, which so far has not been reproduced with several other batches. The role of suppressor cells in functional clonal deletion therefore remains obscure. It must be remembered that the injection of semiallogeneic cells into newborn mice at birth represents an imperfect model for self-tolerance. The toleragen-bearing cells enter the body at a time when small but significant numbers of alloreactive T cells are already present in the thymic medulla and the periphery. It could be that suppressor cells are more involved in preventing these mature cells from getting out of hand than in clonal abortion/anergy within the thymus.

FUNCTIONAL CLONAL DELETION OF ANTI-HAPTEN CYTOTOXIC T LYMPHOCYTE PRECURSORS   Another model of T lymphocyte physiology extensively explored in recent years is the reaction of T lymphocytes to haptenic antigens presented in association with syngeneic cells (106) or isolated cell membranes (25). Tolerance can be induced in such systems and is usually ascribed to suppressor T cells (e.g. 25, 115). Alerted to the fact that suppression may co-exist with an underlying clonal deletion or anergy, Good & Nossal studied anti-hapten T cells in a model system involving adult tolerance (51). Reactive haptens were administered intravenously to adult mice and doses were carefully chosen to provide convincing non-reactivity. Presumably the operative toleragen was hapten coupled to autologous cells. The anti-hapten CTL-P numbers of control and tolerant animals were studied in a limiting dilution microculture system. After appropriate correction for anti-self activity within the population, the anti-hapten CTL-P numbers in tolerant mice were found to be reduced by 90%. In other words, functional clonal deletion was present. Kinetic studies showed that a substantial degree of tolerance existed already at 24 h. These studies document that tolerance within the T lymphocyte system, even in models where suppressor T cells have been noted, may contain an element of functional clonal deletion. Again, the relative importance of each process in physiological self-tolerance must await the development of models that more closely mimic the real-life situation.

Other Possibilities for Clonal Abortion of T Cells

So far it has appeared as though clonal purging of the T cell repertoire must take place in the thymus. However, there are some hints that bone marrow precursors of thymic cells already bear receptors for antigen (26, 63) and are susceptible to toleragenesis by self antigens. Although this would destroy the elegant notion of the thymus as the chief generator of diversity among T cells, it would not offend the central tenets of clonal abortion/-clonal anergy. An intriguing recent study (79) involves radiation chimeras in which bone marrow cells repopulated a thymus graft syngeneic to them, but within the body of a lethally irradiated semiallogeneic thymectomized host. The intrathymic lymphocytes, proven to be of bone marrow genotype, were nevertheless tolerant not only to themselves, but also to the peripheral alloantigens. No alloantigenic material was detected intrathymically, nor were suppressor cells found. Although the results were interpreted as favoring pre-thymic functional inactivation of pre-T cells, the presence of sub-detectable but still toleragenic alloantigen in the thymus cannot be excluded.

An interesting example of what appears to be clonal abortion/anergy of T cells, as judged by a limiting dilution assay, has been reported by Chieco-Bianchi et al (22). Mice inoculated at birth with Moloney murine leukemia virus were challenged as adults with Moloney murine sarcoma virus. Although other adults (regressors) rejected the resulting sarcomas, these dually infected mice (progressors) died with rapidly growing sarcoma. Progressor mice were found to lack virus-specific CTL, and this was not due to suppressor cells. A limit dilution study showed a severe defect of anti-viral CTL-P in the Moloney murine leukemia virus carrier mice, as well as in the dually infected progressors. This study has reached similar conclusions to Nossal & Pike involving MHC antigens (86).

On the basis of these results, it may be concluded that functional clonal deletion is well established as one mechanism of T lymphocyte non-responsiveness. Whether we are dealing with clonal abortion or clonal anergy will not be clear until knowledge of T cell receptors increases materially.

CONCLUSIONS

This review has concentrated on recent studies supporting the notion that functional silencing of particular elements in the immunologic repertoire is an important mechanism in immunologic tolerance for both B and T lymphocyte classes. It has stressed, however, that it is not the only mechanism, and, in particular, a substantial role is seen for suppressor T cells, not perhaps as the primary mechanism, but as an important ancillary one.

For the B cell, the view has been promoted that mIg is the vital signal transducer, and that each stage in the B cell life history that exhibits mIg displays its own particular sensitivity to such signals. Effective signaling may involve the aggregation of receptors into an array with particular characteristics, e.g. a critical minimal number before signaling can occur. Involvement of the F_c receptor, although not essential, may amplify the signal. Formation of an mIg receptor micropatch is a prelude either to activation or to a down-regulatory event. The former requires conjoint activity of T cells and perhaps macrophages, with B cell growth and differentiation-promoting factors of an antigenically nonspecific, MHC-unrestricted nature prominently involved. The latter is the end result when non-immunogenic carriers with particular characteristics multivalently present antigenic determinants to the B cell in the absence of co-stimulatory signals coming from T cells or artificial mitogens. The hierarchy of sensitivities to such negative signaling is as follows, in decreasing order of susceptibility: cells in transition from pre-B to B cell status (self antigens obviously first interact with developing B cells at this stage); immature B cells; mature B cells; activated, antibody-forming B cells. The negative signal, particularly if delivered by small concentrations of antigen, insufficient to modulate the mIg receptor coat, does not kill the cell but rather renders it anergic.

Functional clonal deletion is also a possibility within the T lymphocyte class. This has been formally demonstrated for anti-hapten, anti-viral, and anti-allo-MHC CTL-P. Whether an actual deletion or the induction of an anergic state is involved is not known. Functional deletion can co-exist with induction of suppressor cells. Although some evidence suggests that functional deletion occurs in the thymus, a pre-thymic repertoire purging remains a possibility.

The complex process of self-tolerance is probably an amalgam of repertoire purging, suppression, sequestration of self antigens, and blockade of receptors by monovalent, non-immunogenic self molecules and other regulatory loops, possibly including anti-idiotype networks. Evolutionary elimination of V genes with self-reactive potential is also conceivable, but chimeric states amply testify that much of tolerance is somatically acquired.

Carefully quantitated experiments with cloning techniques for enumeration of competent B and T cells should continue to enlarge our understanding. More experimentation on authentic self antigens is badly needed.

ACKNOWLEDGMENTS

Original work reviewed in this chapter was supported by the National Health and Medical Research Council, Canberra, Australia; by U.S. Public Health Service Grant AI-03958; and by specific donations to The Walter and Eliza Hall Institute.

Literature Cited

1. Abbas, A. K., Klaus, G. G. B. 1977. Inhibition of antibody production in plasmacytoma cells by antigen. *Eur. J. Immunol.* 7:667–74
2. Abbas, A. K., Klaus, G. G. B. 1978. Antibody-antigen complexes suppress antibody production by mouse plasmacytoma cells in vitro. *Eur. J. Immunol.* 8:217–20
3. Allison, A. C. 1971. Unresponsiveness to self antigens. *Lancet* ii:1401–3
4. Andersson, J., Melchers, F. 1981. T cell dependent activation of resting B cells: requirement for both nonspecific, unrestricted and antigen-specific, Ia-restricted soluble factors. *Proc. Natl. Acad. Sci. USA* 78:2497–501
5. Bankhurst, A. D., Torrigiani, G., Allison, A. C. 1973. Lymphocytes binding human thyroglobulin in healthy people

and its relevance to tolerance for autoantigens. *Lancet* i:226–29
6. Basten, A., Miller, J. F. A. P., Sprent, J., Cheers, C. 1974. Cell-to-cell interaction in the immune response. X. T-cell-dependent suppression in tolerant mice. *J. Exp. Med.* 140:199–217
7. Benjamin, D. C. 1975. Evidence for specific suppression in the maintenance of immunologic tolerance. *J. Exp. Med.* 141:635–46
8. Berman, M. A., Weigle, W. O. 1977. B lymphocyte activation by the F$_c$ region of IgG. *J. Exp. Med.* 146:241–56
9. Billingham, R. E., Brent, L., Medawar, P. B. 1953. Actively acquired tolerance of foreign cells. *Nature* 172:603–6
10. Blanden, R. V., Ada, G. L. 1978. A dual recognition model for cytotoxic T cells based on thymic selection of precursors

58   NOSSAL

with low affinity for self H-2 antigens. *Scand. J. Immunol.* 7:181–90

11. Borel, Y., Kilham, L. 1974. Carrier-determined tolerance in various strains of mice: the role of isogenic IgG in the induction of hapten specific tolerance. *Proc. Soc. Exp. Biol. Med.* 145:470–74

12. Boyd, A. W., Schrader, J. W. 1980. Mechanism of effector cell blockade. I. Antigen-induced suppression of Ig synthesis in a hybridoma cell line, and correlation with cell-associated antigen. *J. Exp. Med.* 151:1436–51

13. Boyd, A. W., Schrader, J. W. 1981. The regulation of growth and differentiation of a murine B cell lymphoma. II. The inhibition of WEHI 231 by anti-immunoglobulin antibodies. *J. Immunol.* 126:2466–69

14. Boyd, A. W., Schrader, J. W. 1983. Mechanism of effector cell blockade. II. Colchicine fails to block effector cell blockade but enhances its reversal. *Cell. Immunol.* In press

15. Boyd, A. W., Schrader, J. W. 1983. Mechanism of effector cell blockade. III. The increased sensitivity of neonatal PFC to effector cell blockade. *Cell. Immunol.* In press

16. Brent, L., Brooks, C., Lubling, N., Thomas, A. V. 1972. Attempts to demonstrate an in vivo role for serum blocking factors in tolerant mice. *Transplantation* 14:382–87

17. Bretscher, P., Cohn, M. 1970. A theory of self-nonself discrimination: paralysis and induction involve the recognition of one and two determinants on an antigen, respectively. *Science* 169:1042–49

18. Burnet, F. M. 1957. A modification of Jerne's theory of antibody production using the concept of clonal selection. *Aust. J. Sci.* 20:67–69

19. Burnet, F. M., Fenner, F. 1949. *The Production of Antibodies,* 2nd ed. London: Macmillan

20. Burnet, F. M., Stone, J. D., Edney, M. 1950. The failure of antibody production in the chick embryo. *Aust. J. Exp. Biol. Med. Sci.* 28:291–97

21. Cambier, J. C., Vitetta, E. S., Uhr, J. W., Kettman, J. R. 1977. B cell tolerance. II. Trinitrophenyl human gamma globulin-induced tolerance in adult and neonatal murine B cells responsive to thymus dependent and independent forms of the same hapten. *J. Exp. Med.* 145:778–83

22. Chieco-Bianchi, L., Collavo, D., Biasi, G., Zanovello, P., Ronchese, F. 1983. T lymphocyte tolerance in RNA tumor virus oncogenesis: a model for the clonal abortion hypothesis. In *Biochemical and Biological Markers of Neoplastic Transformation,* ed. P. Chandra, In press. New York: Raven

23. Chiller, J. M., Habicht, G. S., Weigle, W. O. 1970. Cellular sites of immunologic unresponsiveness. *Proc. Natl. Acad. Sci. USA* 65:551–56

24. Cinader, B., Dubert, J.-M. 1955. Acquired immune tolerance to human albumin and the response to subsequent injections of diazo human albumin. *Br. J. Exp. Pathol.* 36:515–29

25. Claman, H. N., Miller, S. D., Sy, M.-S., Moorhead, J. W. 1980. Suppressive mechanisms involving sensitization and tolerance in contact allergy. *Immunol. Rev.* 50:105–32

26. Cohn, M. L., Scott, D. W. 1979. Functional differentiation of T cell precursors. I. Parameters of carrier-specific tolerance in murine helper T cell precursors. *J. Immunol.* 123:2083–87

27. Czitrom, A., Mitchison, N. A., Sunshine, G. H. 1981. I-J, F, and hapten conjugates: canalisation by suppression. In *Immunobiology of the Major Histocompatibility Complex, 7th Internat. Convoc. Immunol.,* pp. 243–53. Basel: Karger

28. Diener, E., Waters, C. A., Singh, B. 1979. Restraints on current concepts of self-tolerance. In *T and B Lymphocytes: Recognition and Function, ICN-UCLA Symp. Molecular and Cellular Biol.,* ed. F. Bach, B. Bonavida, E. Vitetta, C. F. Fox, 16:209–27. New York: Academic

29. Diner, U. E., Kunimoto, D., Diener, E. 1979. Carboxymethyl Cellulose: a nonimmunogenic hapten carrier with tolerogenic properties. *J. Immunol.* 122:1886–91

30. Dintzis, H. M., Dintzis, R. Z., Vogelstein, B. 1976. Molecular determinants of immunogenicity: the immunon model of immune response. *Proc. Natl. Acad. Sci. USA* 73:3671–75

31. Dintzis, R. Z., Vogelstein, B., Dintzis, H. M. 1982. Specific cellular stimulation in the primary immune response: experimental test of a quantized model. *Proc. Natl. Acad. Sci. USA* 79:884–88

32. Dixon, F. J., Maurer, P. 1955. Immunologic unresponsiveness induced by protein antigens. *J. Exp. Med.* 101:245–57

33. Doherty, P. C., Bennink, J. R. 1980. An examination of MHC restriction in the context of a minimal clonal abortion model for self tolerance. *Scand. J. Immunol.* 12:271–80

34. Dorsch, S., Roser, B. 1977. Recirculating, suppressor T cells in transplantation tolerance. *J. Exp. Med.* 145:1144–57
35. Doyle, M. V., Parks, D. E., Weigle, W. O. 1976. Specific suppression of the immune response by HGG-tolerant spleen cells. I. Parameters affecting the level of suppression. *J. Immunol.* 116:1640–45
36. Doyle, M. V, Parks, D. E., Weigle, W. O. 1976. Specific, transient suppression of the immune response by HGG-tolerant spleen cells. II. Effector cells and target cells. *J. Immunol.* 117:1152–58
37. Dresser, D. W. 1962. Specific inhibition of antibody production. I. Protein over loading paralysis. *Immunology* 5:161–68
38. Droege, W. 1973. Five questions on the suppressive effect of thymus-derived cells. In *Curr. Titles Immunol., Transplant. Allergy,* 1:95–134 New York: Pergamon
39. Elson, C. J. 1977. Tolerance in differentiating B lymphocytes. *Eur. J. Immunol.* 7:6–10
40. Farrar, J. J., Benjamin, W. R., Hilfiker, M., Howard, M., Farrar, W. L., Fuller-Farrar, J. 1982. The biochemistry, biology and role of interleukin 2 in the induction of cytotoxic T cell and antibody-forming B cell responses. *Immunol. Rev.* 63:129–66
41. Feizi, T., Wernet, P., Kunkel, H. G., Douglas, S. D. 1973. Lymphocytes forming red cell rosettes in the cold in patients with chronic cold agglutinin disease. *Blood* 42:753–62
42. Feldmann, M., Howard, J. G., Desaymard, C. 1975. Role of antigen structure in the discrimination between tolerance and immunity by B cells. *Transplant. Rev.* 23:78–97
43. Feldmann, M., Nossal, G. J. V. 1972. Tolerance, enhancement and the regulation of interactions between T cells, B cells and macrophages. *Transplant. Rev.* 13:3–34
44. Felton, L. D. 1949. Significance of antigen in animal tissues. *J. Immunol.* 61:107–17
45. Fitch, F. W., Engers, H. D., Cerrottini, J.-C., Brunner, K. T. 1976. Generation of cytotoxic T lymphocytes in vitro. VII. Suppressive effect of irradiated MLC cells on CTL response. *J. Immunol.* 116:716–23
46. Frei, P. C., Benacerraf, B., Thorbecke, G. J. 1965. Phagocytosis of the antigen, a crucial step in the induction of the primary response. *Proc. Natl. Acad. Sci. USA* 53:20–23

47. Gally, J. A., Edelman, G. M. 1972. The genetic control of immunoglobulin synthesis. *Ann. Rev. Genet.* 6:1–46
48. deleted in proof
49. Gershon, R. K., Kondo, K. 1971. Infectious immunological tolerance. *Immunol.* 21:903–14
50. Golan, D. T., Borel, Y. 1971. Nonantigenicity and immunologic tolerance: the role of the carrier in the induction of tolerance to the hapten. *J. Exp. Med.* 134:1046–61
51. Good, M. F., Nossal, G. J. V. 1983. Characteristics of tolerance induction amongst adult hapten-specific T lymphocyte precursors revealed by clonal analysis. *J. Immunol.* In press
52. Gorczynski, R. M., MacRae, S. 1979. Suppression of cytotoxic response to histoincompatible cells. I. Evidence for the two types of T lymphocyte-derived suppressors acting at different stages in the induction of a cytotoxic response. *J. Immunol.* 122:737–46
53. Gorczynski, R. M., MacRae, S. 1979. Suppression of cytotoxic response to histoincompatible cells. II. Analysis of the role of two independent T suppressor pools in maintenance of neonatally induced allograft tolerance in mice. *J. Immunol.* 122:747–52
53a. Green, D. R., Flood, P. M., Gershon, R. K. 1983. Immunoregulatory T Cell Pathways. *Ann. Rev. Immunol.* 1:439–63
54. Gruchalla, R. S., Streilein, J. W. 1982. Analysis of neonatally induced tolerance of H-2 alloantigens. II. Failure to detect alloantigen specific T lymphocyte precursors and suppressors. *Immunogenetics* 15:111–27
55. Haas, W., Layton, J. E. 1975. Separation of antigen-specific lymphocytes. I. Enrichment of antigen-binding cells. *J. Exp. Med.* 141:1004–14
56. Hanan, R., Oyama, J. 1954. Inhibition of antibody formation in mature rabbits by contact with the antigen at an early age. *J. Immunol.* 73:49–53
57. Harris, D. E., Cairns, L., Rosen, F. S., Borel, Y. 1983. A natural model of immunologic tolerance: tolerance to murine C5 is mediated by T cells and antigen is required to maintain unresponsiveness. *J. Exp. Med.* In press
58. Howard, M., Farrar, J., Hilfiker, M., Johnson, B., Takatsu, K., Hamaoka, T., Paul, W. E. 1982. Identification of a T cell-derived B cell growth factor distinct from Interleukin 2. *J. Exp. Med.* 155:914–23

59. Iverson, G. M., Lindenmann, J. 1972. The role of a carrier determinant and T cells in the induction of liver-specific autoantibodies in the mouse. *Eur. J. Immunol.* 2:195–97

60. Jemmerson, R., Morrow, P., Klinman, N. 1982. Antibody responses to synthetic peptides corresponding to antigenic determinants on mouse cytochrome c. *Fed. Proc.* 41(3):420

61. Kilshaw, P. J., Brent, L., Pinto, M. 1975. Suppressor T cells in mice made unresponsive to skin allografts. *Nature* 255:489–91

62. Klaus, G. G. B. 1976. B cell tolerance induced by polymeric antigens. IV. Antigen-mediated inhibition of antibody-forming cells. *Eur. J. Immunol.* 6:200–7

63. Kubara, T., Hosono, M., Fujiwara, M. 1981. Studies on the resistance to tolerance induction against human IgG in DDD mice. IV. Transient tolerant state of T-cell precursors in bone marrow. *Cell. Immunol.* 57:377–88

64. Lawton, A. R., Cooper, M. D. 1974. Modification of B lymphocyte differentiation by anti-immunoglobulin. *Contemp. Top. Immunobiol.* 3:193–225

65. Lederberg, J. 1959. Genes and antibodies: do antigens bear instructions for antibody specificity or do they select cell lines that arise by mutation. *Science* 129:1649–53

66. Loblay, R. H., Fazekas, B., Pritchard-Briscoe, H., Basten, A. 1983. Suppressor T cell memory. II. The role of memory suppressor T cells in tolerance to human gamma globulin. *J. Exp. Med.* In press

67. Loblay, R. H., Pritchard-Briscoe, H., Basten, A. 1983. Suppressor T cell memory. I. Induction and recall of HGG-specific memory suppressor T cells and their role in regulation of antibody production. *J. Exp. Med.* In press

68. Louis, J., Chiller, J. M., Weigle, W. O. 1973. Fate of antigen-binding cells in unresponsive and immune mice. *J. Exp. Med.* 137:461–69

69. Loutit, J. F. 1959. Ionizing radiation and the whole animal. *Sci. Am.* 201:117–34

70. McCullagh, P. J. 1970. The immunological capacity of lymphocytes from normal donors after their transfer to rats tolerant of sheep erythrocytes. *Aust. J. Exp. Biol. Med. Sci.* 48:369–79

71. Metcalf, E. S., Klinman, N. R. 1976. In vitro tolerance induction of neonatal murine spleen cells. *J. Exp. Med.* 143:1327–40

72. Metcalf, E. S., Klinman, N. R. 1977. In vitro tolerance of bone marrow cells: a marker for B cell maturation. *J. Immunol.* 118:2111–16

73. Miller, J. F. A. P. 1972. Lymphocyte interactions in antibody responses. *Int. Rev. Cytol.* 33:77–130

74. Miller, J. F. A. P., Basten, A., Sprent, J., Cheers, C. 1971. Interaction between lymphocytes in immune responses. *Cell. Immunol.* 2:469–95

75. Miller, S. D., Sy, M.-S., Claman, H. N. 1977. The induction of hapten-specific T cell tolerance using hapten-modified lymphoid membranes. II. Relative roles of suppressor T cells and clone inhibition in the tolerant state. *Eur. J. Immunol.* 7:165–70

76. Mitchison, N. A. 1971. The relative ability of T and B lymphocytes to see protein antigens. In *Cell Interactions and Receptor Antibodies in Immune Responses,* ed. O. Mäkelä, A. Cross, T. U. Kosunen, pp. 249–60. New York: Academic

77. Mitchison, N. A. 1978. New ideas about self-tolerance and autoimmunity. *Clinics Rheumatic Dis.* 4:539–48

78. Moreno, C., Hale, C., Hewett, R., Esdaile, J. 1981. Induction and persistence of B cell tolerance to the thymus-dependent component of the $\alpha(1\rightarrow6)$ glucosyl determinant of dextran. Recovery induced by treatment with dextranase in vitro. *Immunology* 44:517–27

79. Morrissey, P. J., Kruisbeek, A. M., Sharrow, S. O., Singer, A. 1982. Tolerance of thymic cytotoxic T lymphocytes to allogeneic H-2 determinants encountered prethymically: Evidence for expression of anti-H2 receptors prior to entry into the thymus. *Proc. Natl. Acad. Sci. USA* 79:2003–7

80. Naor, D., Sulitzeanu, D. 1967. Binding of radioiodinated bovine serum albumin to mouse spleen cells. *Nature* 214:687–88

81. Nisbet-Brown, E., Singh, B., Diener, E. 1981. Antigen Recognition V: Requirement for histocompatibility between antigen-presenting cell and B cell in the response to a thymus-dependent antigen, and lack of allogeneic restriction between T and B cells. *J. Exp. Med.* 154:676–87

82. Nossal, G. J. V. 1957. The immunological response of foetal mice to influenza virus. *Aust. J. Exp. Biol. Med.* 35:549–58

83. Nossal, G. J. V., Lederberg, J. 1958.

Antibody production by single cells. *Nature* 181:1419–20
84. Nossal, G. J. V., Pike, B. L. 1978. Mechanisms of clonal abortion tolerogenesis. I. Response of immature, hapten-specific B lymphocytes. *J. Exp. Med.* 148:1161–70
85. Nossal, G. J. V., Pike, B. L. 1980. Clonal anergy: persistence in tolerant mice of antigen-binding B lymphocytes incapable of responding to antigen or mitogen. *Proc. Natl. Acad. Sci. USA* 77:1602–6
86. Nossal, G. J. V., Pike, B. L. 1981. Functional clonal deletion in immunological tolerance to major histocompatibility complex antigens. *Proc. Natl. Acad. Sci. USA* 78:3844–47
87. Nossal, G. J. V., Pike, B. L., Battye, F. L. 1979. Mechanisms of clonal abortion tolerogenesis. II. Clonal behaviour of immature B cells following exposure to anti-μ chain antibody. *Immunology* 37:203–15
88. Osmond, D. G., Nossal, G. J. V. 1974. Differentiation of lymphocytes in mouse bone marrow. II. Kinetics of maturation and renewal of antiglobulin-binding cells studied by double labeling. *Cell. Immunol.* 13:132–45
89. Owen, R. D. 1945. Immunogenetic consequences of vascular anastomoses between bovine twins. *Science* 102:400–1
90. Owen, R. D. 1956. Erythrocyte antigens and tolerance phenomena. *Proc. R. Soc. London Ser. B* 146:8–18
91. Parks, D. E., Doyle, M. V., Weigle, W. O. 1978. Induction and mode of action of suppressor cells generated against human gamma globulin. I. An immunologic unresponsive state devoid of demonstrable suppressor cells. *J. Exp. Med.* 148:625–38
92. Pike, B. L., Battye, F. L., Nossal, G. J. V. 1981. Effect of hapten valency and carrier composition on the tolerogenic potential of hapten-protein conjugates. *J. Immunol.* 126:89–94
93. Pike, B. L., Boyd, A. W., Nossal, G. J. V. 1982. Clonal anergy: the universally anergic B lymphocyte. *Proc. Natl. Acad. Sci. USA* 79:2013–17
94. Pike, B. L., Kay, T. W., Nossal, G. J. V. 1980. Relative sensitivity of fetal and newborn mice to induction of hapten-specific B cell tolerance. *J. Exp. Med.* 152:1407–12
95. Pike, B. L., Nossal, G. J. V. 1976. Requirement for persistent extracellular antigen in cultures of antigen-binding B lymphocytes. *J. Exp. Med.* 144:568–72

96. Pike, B. L., Nossal, G. J. V. 1979. Mechanisms of clonal abortion tolerogenesis. III. Antigen abrogates functional maturation of surface Ig-negative adult bone marrow lymphocytes. *Eur. J. Immunol.* 9:708–14
97. Pike, B. L., Vaux, D. L., Clark-Lewis, I., Schrader, J. W., Nossal, G. J. V. 1983. Proliferation and differentiation of single, hapten-specific B lymphocytes promoted by T cell factor(s) distinct from T cell growth factor. *Proc. Natl. Acad. Sci. USA.* In press
98. Raff, M. C., Owen, J. J. T., Cooper, M. D., Lawton, A. R., Megson, M., Gathings, W. E. 1975. Differences in susceptibility of mature and immature mouse B lymphocytes to anti-immunoglobulin suppression in vitro. *J. Exp. Med.* 142:1052–64
99. Rajewsky, K. 1971. The carrier effect and cellular cooperation in the induction of antibodies. *Proc. R. Soc. London Ser. B.* 176:385–92
100. Rajewsky, K., Takemori, T. 1983. Genetics, Expression, and Function of Idiotypes. *Ann. Rev. Immunol.* 1:569–606
101. Reichlein, M. 1972. Localising antigenic determinants in human haemoglobin with mutants: molecular correlations of immunologic tolerance. *J. Mol. Biol.* 64:485–96
102. Roberts, I. M., Whittingham, S., Mackay, I. R. 1973. Tolerance to an autoantigen-thyroglobulin. Antigen-binding lymphocytes in thymus and blood in health and autoimmune disease. *Lancet* ii:936–40
103. Rouse, B. T., Warner, N. L. 1974. The role of suppressor cells in avian allogeneic tolerance: implications for the pathogenesis of Marek's disease. *J. Immunol.* 113:904–9
104. Schrader, J. W., Nossal, G. J. V. 1974. Effector cell blockade. A new mechanism of immune hyporeactivity induced by multivalent antigens. *J. Exp. Med.* 139:1582–98
105. Scott, D. W., Venkataraman, M., Jandinski, J. J. 1979. Multiple pathways of B lymphocyte tolerance. *Immunol. Rev.* 43:241–80
106. Shearer, G. M. 1974. Cell-mediated cytotoxicity to trinitrophenyl-modified syngeneic lymphocytes. *Eur. J. Immunol.* 4:527–33
107. Shellam, G. R., Nossal, G. J. V. 1968. Mechanisms of induction of immunological tolerance. IV. The effects of ultra-low doses of flagellin. *Immunology* 14:273–84

108. Sidman, C. L., Unanue, E. R. 1975. Receptor-mediated inactivation of early B limphocytes. *Nature* 257:149–51
109. Smith, R. T., Bridges, R. A. 1958. Immunological unresponsiveness in rabbits produced by neonatal injection of defined antigens. *J. Exp. Med.* 108:227–50
110. Stevens, K. M., Pietryk, H. C., Ciminera, J. L. 1958. Acquired immunological tolerance to a protein antigen in chickens. *Br. J. Exp. Pathol.* 39:1–7
111. Streilein, J. W. 1979. Neonatal tolerance: towards an immunogenetic definition of self. *Immunol. Rev.* 46:125–46
112. Streilein, J. W., Gruchalla, R. S. 1981. Analysis of neonatally induced tolerance of H-2 alloantigens. I. Adoptive transfer indicates that tolerance of Class I and Class II antigens is maintained by distinct mechanisms. *Immunogenetics* 12:161–73
113. Sunshine, G. H., Cyrus, M., Winchester, G. 1982. In vitro responses to the liver antigen F. *Immunology* 4:357–63
114. Swain, S. L., Wetzel, G. D., Saubiran, P., Dutton, R. W. 1982. T cell replacing factors in the B cell response to antigen. *Immunol. Rev.* 63:111–28
115. Tagart, V. B., Thomas, W. R., Asherson, G. L. 1978. Suppressor T cells which block the induction of cytotoxic T cells in vivo. *Immunology* 34:1109–16
116. Taylor, R. B. 1969. Cellular cooperation in the antibody response of mice to two serum albumins: specific function of thymus cells. *Transplant. Rev.* 1:114–49
117. Taylor, R. B., Tite, J. P. 1979. Immunoregulatory effects of a covalent antigen-antibody complex. *Nature* 281:488–90
118. Teale, J. M., Layton, J. E., Nossal, G. J. V. 1979. In vitro model for natural tolerance to self antigens. *J. Exp. Med.* 150:205–17
119. Tite, J. P., Morrison, C. A., Taylor, R. B. 1981. Immunoregulatory effects of covalent antigen-antibody complexes. III. Enhancement or suppression depending on the time of administration of complex relative to a T-independent antigen. *Immunology* 42:355–62
120. Tite, J. P., Morrison, C. A., Taylor, R. B. 1982. Immunoregulatory effects of covalent antigen-antibody complexes. IV. Priming and tolerance in T-dependent responses. *Immunology* 46:809–17
121. Tite, J. P., Taylor, R. B. 1979. Immunoregulation by covalent antigen-antibody complexes. II. Suppression of a T cell-independent anti-hapten response. *Immunology* 38:325–31
122. Traub, E. 1938. Factors influencing the persistence of choriomeningitis virus in the blood of mice after clinical recovery. *J. Exp. Med.* 68:229–50
123. Triplett, E. L. 1962. On the mechanism of immunogenic self-recognition. *J. Immunol.* 89:505–10
124. Vaux, D. L., Pike, B. L., Nossal, G. J. V. 1981. Antibody production by single, hapten-specific B lymphocytes: an antigen-driven cloning system free of filler or accessory cells. *Proc. Natl. Acad. Sci. USA* 78:7702–6
125. Waters, C. A., Diener, E., Singh, B. 1981. Antigen-related susceptibility to tolerance induction in utero: an objection to the current dogma. In *Cellular and Molecular Mechanisms of Immunological Tolerance*, ed. T. Hraba, M. Hasek, pp. 481–91. New York: Dekker
126. Waters, C. A., Pilarski, L. M., Wegmann, T. G., Diener, E. 1979. Tolerance induction during ontogeny. I. Presence of active suppression in mice rendered tolerant to human γ-globulin in utero correlates with the breakdown of the tolerant state. *J. Exp. Med.* 149:1134–51
127. Weigle, W. O. 1971. Recent observations and concepts in immunological unresponsiveness and autoimmunity. *Clin. Exp. Immunol.* 9:437–47
128. Weigle, W. O. 1973. Immunological unresponsiveness. *Adv. Immunol.* 16:61–122
129. Weigle, W. O., Chiller, J. M., Habicht, G. S. 1972. Effect of immunological unresponsiveness on different cell populations. *Transplant. Rev.* 8:3–25
130. Wetzel, G. D., Kettman, J. R. 1981. Activation of murine B lymphocytes. III. Stimulation of B lymphocyte clonal growth with lipopolysaccharide and dextran sulfate. *J. Immunol.* 126:723–28
131. Zinkernagel, R. M., Doherty, P. C. 1974. Imunological surveillance against altered self components by sensitised T lymphocytes in lymphocytic choriomeningitis. *Nature* 251:547–48

Ann. Rev. Immunol. 1983. 1:63–86

T-CELL AND B-CELL RESPONSES TO VIRAL ANTIGENS AT THE CLONAL LEVEL

L. A. Sherman, A. Vitiello, and N. R. Klinman

Department of Immunology, Scripps Clinic and Research Foundation, La Jolla, California 92037

INTRODUCTION

Over the years it has become apparent that the immune system can invoke a multiplicity of mechanisms with which to combat infectious agents. These include the specific activation of both phagocytic and lymphoid cell sub-populations. Although the action of antibodies in neutralizing infectious agents and facilitating opsonization is well known, in recent years the central role of cellular immunity in defense mechanisms has become more apparent. With respect to immunologic defenses against viral infection per se, it has long seemed evident that even relatively low concentrations of antibodies in the serum or target organs could neutralize or immobilize virus and thus, if present, could protect against primary infection and subsequent viremic spread of the infectious agent. Under certain normal infectious circumstances, however, it appears that T-cell responses, by limiting or abolishing primary or secondary infectious foci, may serve as the main weapon in the self-defense armamentarium (9). Since, in the course of a natural infection, it is likely that the primary presentation of viral antigens is in the context of the surface of infected cells, an understanding of immune recognition of infected cells will be fundamental to our understanding of both natural and prophylactically induced anti-viral responses.

This review focuses on the recognition of viral determinants, particularly as they are expressed on infected cells, by both the B- and T-cell systems. Recent findings have indicated that B cells as well as T cells can recognize viral determinants that are uniquely expressed in the context of the infected cell. This finding provides a comparative analysis of T-cell and B-cell recognition not heretofore available. We attempt, wherever possible, to empha-

0732-0582/83/0410-0063$02.00

size studies carried out at the clonal level since it is probable that an unambiguous study of recognition will ultimately depend on the relative simplicity of monoclonal responses. No attempt is made to be comprehensive in terms of serological recognition of viruses per se, since the plethora of information concerning this subject has already been reviewed extensively (20, 46). Rather, we try to assess emerging concepts of recognition to elucidate relevant issues.

Given the enormous selective pressure that survival of virus infections must have played in the evolution of the immune system, it is likely that both mammalian viruses and the mammalian immune system, as we perceive them today, in large part, are the product of these selective forces. For example, it has been shown that encounters with the immune system are probably the selective force for variations in influenza antigens (34, 52). Similarly, since major histocompatibility (MHC) antigens are so integrally involved in immune recognition of foreign antigens expressed on cell surfaces, it is likely that the structure and function of MHC alloantigens has been influenced over evolutionary time by their essential involvement in immune recognition of infected cells. In this context, therefore, it may be inappropriate to attempt to separate immune recognition of viral determinants per se from the recognition of such determinants in the context of given MHC alloantigens. As the studies presented in this review reveal, both antigen systems increasingly appear to be integrally related in terms of the structure of determinants perceived by the immune system. Nonetheless, elements of the immune system, primarily B cells, can recognize virion determinants per se, and in the first section we review aspects of that recognition that may further our understanding of defense mechanisms in general and the specific recognition of viral determinants in the context of MHC alloantigens, which will be discussed subsequently.

THE RECOGNITION OF VIRION DETERMINANTS PER SE

The capacity of serum antibodies to recognize various virion components is well established. Over the past few years, these studies have been advanced markedly by the advent of monoclonal technology (33, 48). Studies with monoclonal antibodies, derived either by the stimulation of individual B cells in culture (33) or by hybridoma technology (48), have allowed the detailed dissection of viral determinants expressed both on the isolated virion and on infected cell surfaces. Most impressive in this regard, and a paradigm for such studies, has been the analysis of influenza hemagglutinin (HA) by large panels of monoclonal anti-influenza antibodies (18, 31–33, 47). The first studies that successfully generated monoclonal antibody responses to viral determinants and used such antibodies to delineate viral determinants were carried out with the fragment culture system in which isolated B cells were stimulated in vitro and the antibody product of their clonal progeny was collected over a period of several weeks (17, 18, 31, 33, 47). This technique can provide microgram quantities of monoclonal antibodies which have proven sufficient for detailed specificity analysis. These studies also introduced the use of a large panel of influenza variants to study both the diversity of recognition of monoclonal antibodies and to delineate the determinants recognized by such antibodies. Such an analysis is shown in Table 1.

This table shows a comparative analysis of monoclonal antibodies derived from primary versus secondary B cells. The antibodies were pre-screened on viral strains that share only HA with PR8 versus those that

Table 1 Relative frequency of reactivity patterns in primary and secondary anti-PR8 HA responses[a]

W E I S S	C A M	B E L	Heterologous virus							
			BH	+	+	+	−	+	−	−	−
			WSE	+	+	−	+	−	+	−	−
			MEL	+	−	+	+	−	−	+	−
+	+	+		7.0[b]	1.6	*[c]	1.6	*	*	*	*
				4.5	5.3	*	1.5	*	*	*	*
−	+	+		3.1	*	*	0.8	*	0.8	*	2.3
				*	3.0	*	*	*	*	*	*
+	−	+		1.6	*	0.8	*	0.8	*	*	1.6
				3.8	0.8	0.8	0.8	*	6.0	*	0.8
+	+	−		2.3	*	2.3	*	*	*	*	*
				12.0	*	5.3	3.0	0.8	3.0	0.8	*
−	−	+		*	*	0.8	*	*	2.3	0.8	2.3
				*	0.8	0.8	*	*	*	0.8	1.5
−	+	−		0.8	0.8	0.8	*	*	2.3	4.7	1.6
				0.8	0.8	0.8	*	*	0.8	0.8	1.5
+	−	−		1.6	*	*	*	0.8	5.5	1.6	1.6
				*	0.8	*	0.8	0.8	0.8	2.3	1.5
−	−	−		5.5	2.3	1.6	7.8	3.1	6.2	5.5	14.0
				1.5	*	*	*	*	*	12.8	23.3

[a] Relative frequencies are given as a percentage of total response and are based upon 129 primary and 134 secondary monoclonal anti-HA antibodies (from 18).
[b] Upper number refers to primary response; lower number refers to secondary response.
[c] *, RP not observed.

share other virion determinants, including neuraminidase (NA), and those that share only chicken host component. The table, therefore, includes only those antibodies that presumably recognize determinants on the HA of PR8. This and similar analyses established several highly important general principles. First, it is clear that a panel of viral variants is a very powerful tool in the dissection of available antibody specificities. Second, these studies give definitive evidence of the enormous degree of diversity available in the B cell repertoire. By expanding such panels (26, 84), it has now been found that more than 10^3 different recognition patterns are available within the Balb/c mouse. Since B cells responsive to the influenza HA are approximately 1 in 10^5 of all B cells (18), such studies have established that the lower limit to the size of the antibody repertoire is of the order of 10^8 specificities. Third, if each individual reactivity pattern can be equated with the recognition of a given determinant, then it may be concluded that the B cell repertoire is not redundant in its recognition of most determinants. Thus, for many determinants a fully mature mouse may have only one or no B cell clones capable of recognition. Fourth, since subsequent studies have shown that only a few limited regions of the HA molecule are responsible for this complex pattern of binding (32, 81), it is clear that small nuances in the structure of a region of a protein can be recognized and discriminated by monoclonal antibodies. Although in some instances such nuances may be created by the substituted amino acid per se it is equally likely that in some instances amino acid substitutions may be recognized as perturbations in other regions of the molecule.

Subsequent studies with monoclonal antibodies derived from hybridomas have confirmed all of the above findings (26, 32, 35, 84). In addition, it has been possible, by the use of hybridoma antibodies, to select for influenza HA variants that represent a single mutational difference (34, 52). Such newly selected variants have served as invaluable tools in further discriminating the B cell repertoire and further delineating regions of the HA molecule wherein sequence variation affects immunologic recognition (81).

Figure 1 is a schematic representation of the three-dimensional structure of influenza HA as determined by X-ray crystallography by Wyley & Wilson (81). Included in this figure are point mutations selected by monoclonal antibodies. It is apparent these mutational differences cluster in immunologically relevant regions of the HA molecule (81). The combined analysis by monoclonal antibodies, selection of mutants, and crystal structure represents an extraordinarily powerful new set of approaches for the understanding of determinant recognition and the power of selective forces on both immune expression and viral expression. Recently, several other viral systems have been subjected to analysis by monoclonal antibodies (49, 56, 58). Some of these studies have been the subject of extensive recent reviews, including comprehensive analyses of the antigenic structures of murine leukemia viruses (58), the 70S glycoproteins of murine retroviruses (56), and rabies viruses (49). In each of these instances, it has been possible by the use of monoclonal antibodies to more clearly delineate structural relationships and variations among viral strains.

VIRAL COMPONENTS RECOGNIZED ON INFECTED CELLS

It is now clear that both T cells and B cells can function, and may primarily function, by recognizing viral determinants as they are expressed on infected cells. Such recognition is particularly relevant to viruses that repli-

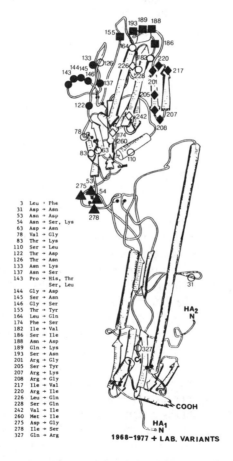

3	Leu	Phe
31	Asp	Asn
53	Asn	Asp
54	Asn	Ser, Lys
63	Asp	Asn
78	Val	Gly
83	Thr	Lys
110	Ser	Leu
122	Thr	Asp
126	Thr	Asn
133	Asn	Lys
137	Asn	Ser
143	Pro	His, Thr
		Ser, Leu
144	Gly	Asp
145	Ser	Asn
146	Gly	Ser
155	Thr	Tyr
164	Leu	Gln
174	Phe	Ser
182	Ile	Val
186	Ser	Ile
188	Asn	Asp
189	Gln	Lys
193	Ser	Asn
201	Arg	Gly
205	Ser	Tyr
207	Arg	Lys
208	Arg	Gly
217	Ile	Val
220	Arg	Ile
226	Leu	Gln
228	Ser	Gln
242	Val	Ile
260	Met	Ile
275	Asp	Gly
278	Ile	Ser
327	Gln	Arg

Figure 1 Schematic drawing of a monomer of the AICHI-1968 hemagglutinin showing locations of amino acid substitutions on HA₁. ● Site A; ■ site B; ▲ site C; ◆ site D (subunit interface); ◇ surface; ○ neutral. Summary of all known alterations from 1968 to 1977 including laboratory-selected variants (81).

cate by budding. In terms of immune recognition of viral determinants as they are expressed on infected cells, several factors are important to consider. First, it is likely that determinants that can be recognized on the mature virion may not be accessible for recognition as they are expressed on the cell surface and prior to virion maturation. Second, it is likely that some virion determinants are accessible on the surface of infected cells that are not accessible on the intact virion. If such recognition is important in the immunologic defense against viral infection, then immunization with inactivated virion may not be expected to be as effective as the immunization that results from infection (67). Finally, it is possible that a considerable amount of immune recognition is directed towards complex antigenic determinants that may involve both virion and cell surface determinants. Thus, our understanding of immunologic defenses against viral infection may depend on a clear understanding of the recognition of viral determinants as they are expressed on cell surfaces.

Viral Determinants on Infected Cells Recognized by B Cells

Recent studies have demonstrated that serum antibodies directed to viral antigens can cause the lysis of virally infected cells via a complement-mediated process (61). Such findings underline the importance of understanding B cell recognition of virion determinants as they are expressed on infected cells. Recent studies by Gerhard and his co-workers have demonstrated, by the use of monoclonal antibodies directed to the various protein components of the influenza virion, that although both the HA and NA can be readily recognized on the surface of infected cells, little or no representation of the matrix (M) protein can be found on such cells (35, 84). Additionally, a small amount of recognition of the viral nuclear protein (NP) is apparent by low levels of binding of the hybridoma antibody directed to this protein on the surface of infected cells (35, 84). The relevance of this latter recognition remains to be determined. Another viral protein that may be present on infected cells is non-structural protein 1 (NS1) (72).

Recently, this laboratory has carried out an extensive analysis of monoclonal B cell responses to influenza-infected syngeneic murine cell lines (82, 83). To the extent that the analyses have been carried out, it is apparent that vigorous monoclonal responses can be obtained to both the HA and NA of the influenza virion as they are expressed on infected cells. Importantly, the frequency of such responses is somewhat lower than the frequency of responses to the same determinants when the intact virion is used as the immunogen (18). That is, when infected cells are used as the immunogen and antibodies are screened on the intact virion, the frequency of responses by HA-specific B cells is about two thirds of that found in response to the HA as presented on the intact virion. Only approximately one third of B cells stimulated by NA, as expressed on the intact virion, respond when the NA is presented on the infected cell (82). To determine the basis for this lack of responsiveness, monoclonal antibodies derived by stimulation by the intact virion were assessed for their capacity to bind either the HA or the NA as expressed on infected cells. The findings indicate that approximately one third of all virion-stimulated anti-HA antibodies recognize HA determinants not accessible on the infected cell. Similarly, two thirds of the anti-virion NA-specific monoclonal antibodies do not bind to NA as expressed on the infected cell (82). Thus, although both HA and NA are expressed on the surface of infected cells, this expression is apparently not inclusive of all potential responses to these determinants as they are expressed on the virion. Importantly, among the subset missing from the anti-HA response when virus is presented on infected cells is the majority of antibodies that are broadly cross-reactive among closely related viral strains of the H1 subtype. This finding would predict that antibodies derived by stimulation during the course of an infection may be somewhat more specific for the infecting virus than those which may be derived by immunization with high concentrations of the intact virion.

In summary, it would appear that some viral determinants, such as the NP, that are poorly revealed on the intact virion may be recognized on the infected cell. Conversely, it is also apparent that some determinants normally readily recognized on the intact virion are not seen when viral determinants are presented as infected cells. As more information accumulates concerning differences in B cell recognition of the intact virion versus the infected cell, it is likely that a better understanding of the basis and specificity of immune recognition of infected cells will be achieved.

T-Cell Recognition of Viral Determinants on Infected Cells

Precise identification of viral proteins responsible for recognition by cytolytic T lymphocytes (CTL) has proven difficult. Immunologists have approached this task with the reasonable expectations that the viral structures involved should be expressed on the target cell membrane and that the specificity of CTL should, in general, reflect recognition of determinants defined by the structure of viral proteins. Inherent in the latter assumption is the belief that T cell receptors recognize determinants on the foreign antigen. The inability to identify a viral structure that fulfills both of these expectations which, parenthetically, has been the basis for identification of HA as the viral protein involved in MHC-restricted antibody recognition of influenza virus-infected cells (83), has hampered efforts to identify the viral proteins required for T cell recognition of influenza-infected cells.

The basic finding reported by most laboratories that have studied the specificity of influenza-specific CTL is that a CTL population stimulated in response to viral infection with a particular influenza A strain virus cross-reacts on target cells infected with essentially all other A strain viruses (12, 23, 91). This is in marked contrast to the specificity of antibody for which cross-reactivity is restricted to viral strains of the same subtype. The specificity of such CTL populations is defined by the lack of recognition of B strain-infected targets and targets infected with unrelated virus (14). Therefore, CTL demonstrate a broader range of recognition than do antisera. Such cross-reactive recognition is also characteristic of T helper cells and effector cells of delayed-type hypersensitivity specific for influenza (1, 3, 4, 53) and has also been described in T cell recognition of vesicular stomatitis virus (VSV)-infected cells (90).

Several types of evidence suggest that subtype-specific CTL do exist but are obscured by a second population of highly cross-reactive CTL. Ennis and co-workers (24) described such subtype-specific CTL as the only population that could be observed when kidney cells infected with virus were used as CTL targets as opposed to more commonly used tumor cells. With recombinant strains of virus, it was demonstrated that specific recognition required the presence of the viral HA and not NA. It was also possible to demonstrate subtype-specific recognition by cold target competition studies. When cold target inhibitors infected with heterologous A strain virus of a different subtype were added to a cross-reactive effector cell population, the remaining lytic activity was greatest on targets infected with the homologous strain of virus (14, 23, 91). In addition, it was possible to amplify the cross-reactive CTL population preferentially by in vitro stimulation with heterologous A strain virus (14, 23) or specific population by using either UV-inactivated virus (25) or purified HA for in vitro stimulation (13, 91). These data have been interpreted as evidence that at least a portion of the CTL response is subtype specific and is dependent upon expression of the appropriate HA for target recognition. However, there is reason to believe that the precise determinants recognized are not the same as those recognized by HA-specific antisera. First, it is difficult to demonstrate antibody blocking of CTL recognition with HA-specific antisera (5, 14). This is the case even when some subtype-specific CTL are analyzed and is in contrast to the marked inhibition observed with antisera specific for class I MHC molecules (5, 89). Second, when subtype-specific CTL are obtained by stimulation with purified HA, such CTL do not exhibit the same hierarchy of cross-reactive recognition as is observed for HA-specific antibody (13).

If subtype specificity is attributable to HA recognition, then which protein is responsible for cross-reactive recognition? Again, with the basic assumption that the specificity of CTL should reflect the serologically defined antigenic properties of a viral protein, it would be concluded that highly conserved viral proteins are the most likely candidates. This would exclude the two proteins expressed in greatest abundance on infected cells, HA and NA. The recent demonstration that several relatively invarient viral-encoded molecules such as M protein and NS1 may be expressed on the surface of infected cells place these proteins on the list of possible candidates. The only evidence that suggests recognition of M proteins by CTL is the success of one laboratory (68) [in the face of failure by others (14, 40)] to demonstrate CTL inhibition by using anti-M protein antisera. There are no reports as yet on the possible role of NS1.

There is, however, substantial evidence that HA may not only represent the target molecule most often responsible for subtype-specific CTL recognition but may also represent the major target molecule for cross-reactive CTL. First, Askonas & Webster were able to demonstrate antibody inhibition of cross-reactive CTL using a combination of monoclonal antibodies directed against HA and H-2 (5). Second, Koszinowski and co-workers were able to construct target cells for influenza-specific CTL by Sendai-mediated fusion of liposomes containing purified viral glycoproteins (HA and NA) with uninfected H-2 compatible cells (50). The resultant targets were cross-reactively lysed by CTL raised against a different A subtype. Finally, the fact that T helper cells can be stimulated in vitro with purified HA, or HA fragments obtained from a strain heterologous to that used for in vivo priming, indicates that class II MHC-restricted T cells, which are cross-reactive, are unambiguously HA specific (3).

An interesting parallel is the case of VSV recognition by CTL. Nonserologically cross-reactive VSV-G (glycoprotein) have been positively identified as the target structure for recognition by cross-reactive CTL (19, 90). Nevertheless, purified G protein as an immunogen elicits a highly specific CTL population (71) much as does purified HA in the case of influenza-specific CTL. Therefore, for both VSV and influenza, stimulation with viral-infected cells generate both specific and cross-reactive CTL, whereas stimulation with purified viral glycoproteins results in stimulation of the specific population only. Although one interpretation of these data is that a second viral protein is responsible for the cross-reactive CTL population, it is also possible to interpret them based on the difference in the mode of antigen presentation. Whereas the viral HA is expressed as a transmembrane protein on the surface of infected stimulator cells, it is unlikely membrane insertion occurs when purified HA is used as antigen. It is likely that the type of association between H-2 and antigen is qualitatively different when antigen is transmembrane than when it is not. For example, it is possible that association between H-2 and HA in regions proximal to the membrane may be achieved with transmembrane antigen but not solubilized protein, thereby resulting in a greater spectrum of new determinants, including perhaps the majority of those that are cross-reactively recognized. Alternatively, the type of association between H-2 and HA may be significantly affected by the presence of M protein or some other viral protein.

To summarize findings obtained by studies of CTL populations, it would appear that two types of CTL specificities exist: one appears subtype specific and is likely to be HA-directed; the other is cross-reactive and may or may not be HA-specific. The possibility exists that neither type of population reflects recognition of determinants identical to those seen by antibody; however, this could only be resolved at the level of individual receptor specificities by studying CTL clones.

CTL Clones

In the past few years, several laboratories have studied the viral specificity of cloned CTL obtained either by limiting dilution of effector cell populations or cloning after multiple in vitro stimulations of a CTL-containing population. As the findings from each laboratory have contributed different information, each is reviewed separately.

Lu & Askonas described the specificity of a CTL line obtained by multiple in vitro stimulations (54). This line requires interleukin-2 for growth but not stimulator cells. Insofar as clones obtained from this line react with all A strain subtypes, it represents the first demonstration that a single CTL specificity can, indeed, cross-reactively recognize cells infected with distantly related viral strains. It is not possible to estimate the proportion of the original effector cell population that would express such specificity. However, Owen et al (62) have described the specificity of CTL clones obtained by in vitro stimulation under limiting dilution conditions without prior in vitro culture. The frequency of influenza-specific CTL within the spleen of an in vivo primed animal is in the range of 1 in 1700–4700 spleen cells. Of these clones, approximately 80% were cross-reactive on heterologous A strain-infected target cells. Therefore, the vast majority of their clones were of the cross-reactive category.

Another series of CTL clones has been described by Braciale et al (15). These are long-term clones that require both stimulator cells and interleukin-2 for maintenance of activity and growth. Braciale found three categories of specificity in examining 14 such clones. These included clones that demonstrated specificity restricted to the homologous virus used for in vivo stimulation (43%), clones that appeared subtype specific (21%), and 29% that were cross-reactive on all strains tested. One clone demonstrated rather paradoxical recognition insofar as it did not recognize target cells infected with certain heterologous virus of the same subtype as that used for stimulation, yet it did recognize targets infected with virus representing an unrelated subtype.

Our own studies on viral specificity of influenza-specific CTL clones indicate that such paradoxical recognition may indeed be characteristic of a large proportion of PR8-specific CTL clones (A. Vitiello, L. A. Sherman, manuscript in preparation). It was frequently observed that PR8 (H1)-stimulated clones could discriminate among a panel of target cells each infected with different H1 A strain virus and yet would react on target cells infected with an H3-type virus.

How can these various specificities be classified with respect to antigen recognition? Recently, Bennick et al (6) performed an analysis of the specificity of a PR8-specific clone that had the general characteristics of subtype specificity and, therefore, could have been considered HA-specific insofar as the clone reacted with target cells infected with some H1 strains but not heterologous H2 or H3 virus. With defined recombinant viral strains created between PR8 (H1) and Hong Kong (H3), it was found that the requirement for target recognition genetically segregated not with HA but rather with the viral polymerase, P3. Considering that intermediate levels of lysis were observed with different recombinant strains, all of which express the required P3, this would further suggest that the level of expression of the recognized structures is under multigenic control. Although these results do not rule out HA as the requisite target cell structure, they do indicate that in view of the potential for multiple genes to affect expression of the target molecules, it may prove exceedingly difficult to identify precisely the viral proteins required for CTL recognition.

It may be concluded that studies with CTL clones have not yet provided definitive answers with regard to determinant recognition by virus-specific CTL. Clearly, it is also necessary to have well-defined target structures. Perhaps one way this may be achieved is through the use of cloned viral genes as transfection agents. Nevertheless, clones have been quite informative with respect to the demonstration of great diversity within the virus-specific CTL repertoire.

MHC-RESTRICTED RECOGNITION OF VIRAL DETERMINANTS ON INFECTED CELLS

The most striking feature of immune recognition of virally infected cells, particularly by T cells, is the restricted recognition of viral determinants such that antigens are recognized only in the context of MHC-encoded determinants of the infected cell (89). Both the nature of such recognition and the lymphocyte receptor mechanism necessary to accomplish such recognition remain among the most controversial and provocative aspects of the immune system. In general, studies that have attempted to demonstrate a physical association of viral determinants and MHC components have been ambiguous in that, in some instances, such associations may be present (11, 19, 30, 60, 85), whereas in others they appear not to be, or may be of insufficient stability to permit detection (21, 39). Most important, however, is that recognition of viral determinants in the context of given MHC alloantigens is often exclusive for one MHC determinant, and such exclusivity appears to be MHC-haplotype-dependent (22, 87). Thus, recognition would appear to be a function of selection of the receptor repertoire rather than a function of the stability of complexes on the cell surface. In any case, attempts to understand the mechanism of restricted recognition of infected cells will be a central theme of future immunologic studies.

MHC-Restricted Recognition of Virally Infected Cells by T Cells

T cells function through their recognition of cellularly presented antigen. The strategy used by the immune system to restrict T-cell recognition to cells rather than antigen per se would appear to be the requirement that T cells recognize cellular MHC antigens as well as foreign antigen. This is referred to as MHC-restricted recognition. For T helper cells this entails recognition of class II (I region) MHC molecules, whereas CTL has a requirement for recognition of class I molecules (K, D, and L) (89).

MHC restriction is observed as the requirement for MHC sharing between the cell presenting antigen during T cell activation and the target cells recognized by the activated effector cell population. Therefore, MHC determinants as well as antigen select the specificity of the effector cell population. However, available data supporting CTL recognition of determinants on H-2 molecules is far more compelling than available data demonstrating recognition of antigen determinants. Antisera to H-2 blocks target cell recognition by CTL (5, 11, 29, 36, 37, 51, 69). In addition, subtle changes in the structure of the H-2 molecule, such as point mutations, can markedly affect the ability of CTL to recognize target cells. These results are in contrast to the relative insensitivity of CTL reactivity to blocking by antiviral antisera, or to extensive structural changes in viral proteins (see previous section).

The precise H-2 determinants recognized in restricted recognition are not yet known. Some data suggest there may be relatively few H-2 determinants utilized in such a manner. When K^b-restricted CTL are tested for recognition of target cells that bear different mutations in the K^b molecule, the hierarchy of cross-reactive recognition for a variety of different viruses appears to be similar. For example, K^{bm1}-expressing targets are not recognized by K^b-restricted CTL in all viral systems that have been tested (10, 63, 86). This suggests bm1, which differs from wild-type by only two amino acids, may have lost all viral-restricting determinants present on the wild-type molecule. However, that this conclusion may be an oversimplification is suggested by the recent demonstration of the complexity of H-2 determinants recognized in conjunction with minor cell surface histocompatibility antigens (80). As is the case for viral recognition, K^b-restricted and minor antigen-specific CTL do not recognize the K^{bm1} mutant. However, cold target competition studies with other K^b mutants suggest the existence of multiple determinants on the K^b molecule, only some of which are conserved by various K^b mutants. In addition, whereas recognition of the H3.1 minor antigen in association with determinants shared by K^b and K^{bm3} was observed, K^b-restricted H4.1 recognition was not found with bm3 targets. Therefore, a different pattern of restricting elements is seen in conjunction with different minor antigens. In light of these results, a more careful examination of viral-restricting H-2 determinants as may be achieved by studying the specificity of individual clones is required.

The possibility that different antigens may use a different array of H-2 determinants as restricting elements presents the interesting question as to whether or not different H-2 molecules may present a different mosaic of viral determinants. We have recently examined this question by studying the fine specificity of viral recognition by CTL clones restricted to two different H-2 molecules. These results indicate that there is very little overlap between viral determinants recognized by D^b-restricted (B6) versus D^d- and L^d (B10.A5R)-restricted influenza-specific CTL (A. Vitiello, L. A. Sherman, manuscript in preparation). Shared clonotypes represent less than 20% of the repertoires expressed by B6 and B10.A5R. Considering that different clones that share reactivity patterns do not necessarily recognize identical determinants, this represents a maximum estimate of determinant sharing. There are several possible interpretations of these results. First, there is considerable evidence for physical association of H-2 and virus. Physical proximity of H-2 and influenza HA on target cells has been demonstrated by Hackett & Askonas (39). It is possible that the orientation of this association may be different for different H-2 molecules. If so, then different restricting elements may lead to the presentation of different viral determinants. The observation that the same anti-HA monoclonal antibody that is effective in inhibition of D^d-restricted CTL does not inhibit recognition of K^d-restricted CTL may be interpreted as support for this possibility (28). Second, it is possible, that different viral proteins are associated with the MHC molecules of these different murine strains. Third, in view of the evidence that MHC molecules influence CTL repertoire selection (7, 41, 73, 88), it is possible the repertoires are sufficiently different in these two strains as to preclude recognition of comparable viral determinants. Comparison of the allo-specific repertoire of two MHC congenic strains does not necessarily support this view (73). The majority of H-2K^b-specific clonotypes observed in H-2^d mice are also found in H-2^k mice. Rather, it is the frequency of representation of individual clonotypes that is most affected by

H-2 differences. Finally, it is possible that the determinants actually described by these influenza-specific clonotypes are not viral structures per se but instead represent altered H-2 determinants (see below) (74). Regardless of the basis for the lack of correlation between these two repertoires with respect to fine specificity for viral antigens, the finding that the repertoire is so dramatically shifted by the H-2-restricting element makes it less surprising that there is so little correlation between viral determinants recognized by antibodies that are not MHC restricted and determinants recognized by MHC-restricted T cells.

Another dramatic effect that the MHC has on the anti-viral response is the ability of H-2 to determine which MHC molecules are utilized as restricting elements. In general, two types of such immune response deficiencies are attributable to class I MHC molecules. In some situations, lack of responsiveness appears to be inherent in the restriction element itself. For example, B6 mice are low responders to H-2K^b plus influenza (22). This low response phenotype is reflected as a low frequency of CTL activated in the immune animal (2). That this type of defect is inherent in the K^b molecule may be concluded from the fact that the B6-H-2^{bm3} mutant responds well to the K^{bm3}-infected targets (A. Vitiello, L. A. Sherman, manuscript in preparation). This type of nonresponder phenotype is also observed for simian virus 40 (SV40) recognition where B6 responds well in this case to K^b-infected cells but the bm1 mutant is a nonresponder (63). Another example has been described for Moloney leukemia virus (MoLV)-infected cells (76). In this case, mutation in the D^b(bm14) results in lack of D end recognition by the mutant strain.

A second type of immune response defect commonly observed maps to an H-2 molecule other than the one used as the restricting element. For example, D^b-restricted and vaccinia-specific CTL cannot be obtained from mice that bear K^k, whereas B6 mice (K^b D^b) are able to mount a good D^b-restricted vaccinia response (22, 87). As another example, it has been observed that D^d-restricted SV40-specific CTL cannot be raised in mice that bear K^b or K^k antigens, but it can be elicited in K^d and K^q haplotypes (65). The observation that CTL from B10.D2 mice that mount a vigorous D^d-restricted SV40 response can recognize target cells expressing K^b and D^d suggests that the defect is probably not due to association between H-2 and antigen on the cell surface. Similarly, it was demonstrated that whereas H-2^b mice respond to D^b in association with MoLV virus, a mutation in the D^b molecule can result in increased utilization of K^b as a restricting element for this virus (76). Similar observations have been made in other viral systems (89).

The mechanism responsible for these two types of nonresponder phenotypes may or may not be similar. Two types of explanations have been proposed (89). The first evokes lack of association or preferred association between H-2 and antigen as responsible for instances of haplotype preference. Although there is evidence for preferred association of viral protein and H-2 molecules in some systems, for such a model to explain all of the experimental results it would also be necessary to explain why some H-2 molecules associate well enough to act as targets but not well enough to stimulate CTL precursors. In addition, the fact that K^b is a poor restricting element for influenza-specific CTL yet serves as the dominant restricting element for H-2-restricted antibody also makes this less likely (see below) (83). A second model predicts that the CTL repertoire is deleted for recognition of some combinations of H-2 and virus because of elimination of such potential specificities by self tolerance. Recent experiments by Mullbacher (55) support the notion that tolerance to an alloantigen can result in elimination of a particular MHC-restricted antigen response. Finally, it is possible that both types of phenomenon may occur in different situations.

There are numerous studies in which CTL populations that are H-2-restricted and virus or antigen specific have been found to cross-react on alloantigen (8, 16, 27, 44, 64, 78, 79). Since this often represents the specificity of a minor subpopulation, examples of this phenomenon at the level of individual CTL clones were sought and found (16, 44, 78, 79). It is now firmly established that a high proportion of self plus-X-specific T-cell clones can react with alloantigen. Equally impressive is the general trend for specific alloantigens to be recognized more frequently than others. For example, whereas 10% of all CTL clones specific for D^b plus HY reacted on the D^d alloantigen, no clones were found that reacted on H-2^k targets (44). Similar cases of preferred alloreactivity have been reported for I-restricted antigen-specific T-cell clones (70). An extreme example of this type of cross-reactivity has been observed for K^b-restricted influenza recognition. In an analysis of 22 K^b-plus influenza-specific clones many demonstrated lytic activity on uninfected K^{bm10} targets (A. Vitiello, L. A. Sherman, manuscript in preparation). In addition, each of these clones was heteroclytic insofar as lytic activity on bm10 was much greater than on K^b-infected targets. These types of data further support the notion that self tolerance to H-2 antigens may be the most common mechanism responsible for IR phenomena in many of the examples described above. Further, the selectivity as to which alloantigens are recognized suggests that the basis for cross-reactivity is not fortuitous but, rather, may represent truly shared determinants.

MHC-Restricted Recognition of Virally-Infected Cells by B Cells

Recently, several laboratories have demonstrated that in some instances B-cell stimulation and recognition, like T-cell stimulation, can display the need for antigen presentation in the context of an appropriate MHC alloantigen (38, 42, 45, 57, 59, 66, 75, 77). This laboratory has carried out a detailed examination of the role of such recognition by B cells responding to influenza-infected syngeneic cell lines (83). For this purpose, individual B cells were antigenically stimulated by influenza-infected cell lines of two different strains. In both instances, responses were observed wherein monoclonal antibodies recognized influenza antigens expressed on the purified virion. However, in both the responsive H-2^b B cells to influenza-infected EL-4 (H-2^b) cells and H-2^k B cells responsive to influenza-infected L929 (H-2^k) cells, the majority of antibody-forming cell clones produced monoclonal antibodies that recognized the appropriate infected cell but not the purified virion or uninfected cell alone. The majority of these monoclonal antibodies were found to be dependent on the presence of the appropriate influenza HA. The recognition of HA was relatively specific in that little cross-reactivity was observed when the same cell lines were infected with influenza viruses bearing other H1 variants. Most important, a majority of monoclonal antibodies (57%) that responded to influenza-infected cells recognized influenza determinants only in the context of cells expressing either the H-2K or H-2D end antigenic determinants of the syngeneic stimulating cell lines. These findings are summarized in Table 2.

It can be seen that among the monoclonal antibodies derived against influenza-infected L929 (H-2^k) cells, the majority recognized only infected cell lines that bore the H-2D^k class 1 antigen. Similarly, H-2^b B cells that recognized influenza-infected EL-4 (H-2^b) cells recognized primarily influenza-infected cells bearing the H-2K^b molecule. Since panel analyses were carried out with influenza-infected primary kidney fibroblasts, it is

Table 2 MHC restriction of monoclonal antibodies specific for PR8-infected L929 (H–2K) cells

PR8	L929	PR8–L929 (H–2^k)	PR8–BALB.K (H–2^k)	PR8–3T3 (H–2^b)	PR8–B6 (H–2^d)	PR8–C3H.OH (H–2K^d, D^k)	PR8–B10.A(4R) (H–2K^k, D^b)	Percentage of total monoclonal antibodies
+	–	+	NT^a	NT	NT	NT	NT	35
–	+	+	NT	NT	NT	NT	NT	4
–	–	+	–	NT	NT	NT	NT	4
–	–	+	+	+	+	NT	NT	10
–	–	+	+	+	–	NT	NT	5
–	–	+	+	–	+	NT	NT	5
–	–	+	+	–	–	+	+	2
–	–	+	+	–	–	+	–	30
–	–	+	+	–	–	–	+	5

^{a}NT, Not tested. (From 82, 83.)

unlikely that this recognition was dependent on the mode of infection, productive versus nonproductive, or other antigens expressed on the tumor cell lines. Additional evidence that the anti-influenza-infected EL-4 responses were restricted to the H-2K^b locus was obtained by an analysis of binding to influenza-infected fibroblasts obtained from H-2K^b mutants. The majority of these antibodies could discriminate among influenza-infected cells bearing these mutants. Thus, these antibodies were sensitive to both small changes in the influenza HA as well as point mutations in the H-2K^b molecule.

The finding that the majority of antibody responses to infected cells displays restricted recognition reminiscent of T-cell recognition has several important implications for our understanding of immune recognition of infected cells. First, it is clear that the dichotomy previously observed between B-cell recognition and T-cell recognition is not absolute and is based more on the fact that antibodies can be found that bind and neutralize virus particles per se than on the fact that B cells cannot also display restricted recognition as do T cells. Second, it is clear that maximum immune responses by both T and B cells would require the presentation of viral antigens in the context of syngeneic infected cells. Third, the finding that monoclonal antibodies can so readily recognize complex antigen determinants has profound implications for the nature of T-cell receptors, since such recognition obviously does not necessitate the existence of two separate receptors. Fourth, whatever the nature of the complex antigen, it apparently pre-exists antibody interaction with the cell since the infected cells used for binding assays were glutaraldehyde fixed prior to interaction with antibody. Fifth, whatever the mechanism of recognition of complex antigens by immune receptors, such recognition would appear to be extremely specific both with respect to its discriminatory capability for the viral determinants and its discriminatory capability for the H-2 antigen involved. Finally, it is of great interest that the recognition by B cells of influenza-infected cells, like that previously observed for T cells, is markedly skewed towards one end of the MHC locus versus the other in both strains tested. Interestingly, the skewing in both cases is opposite that previously reported for T cell recognition of influenza-infected cells of these haplotypes (22), an observation similar to that of Ohno, et al in a study of anti-HY antisera (59). Thus, although T cells recognizing influenza-infected H-2^b cells are restricted to recognition of viral antigens in the context of D-end antigens, the monoclonal antibodies apparently prefer K end. Similarly, T cells recognizing influenza-infected H-2^k cells are predominantly K end restricted, whereas the monoclonal antibodies appear to be primarily D-end restricted. Whether such endedness in recognition represents a phenomenon similar to the IR phenomenon attributed to T-cell recognition of one end versus another has yet to be determined. Nonetheless, the opposite preference of the two cell systems would seem to eliminate a preference of the virus to complex with one end versus the other as a major determining element in either T-cell or B-cell recognition.

CONCLUSION: COMPARISON OF THE T- AND B-CELL REPERTOIRES SPECIFIC FOR VIRALLY INFECTED CELLS

Comparison of determinant recognition by B and T cell receptors has been an area of great interest to immunologists. Often, such comparisons have led to the conclusion that different epitopes are recognized by these different lymphocyte subsets. However, such results have always been interpreted with the understanding that comparison may not be valid since the B cells that were analyzed were specific for antigen per se whereas T cells are confined to recognition of antigen in the context of MHC molecules. Therefore, the recent demonstration of an additional subset of antibody forming cells which are MHC restricted provides a unique opportunity to study B- and T-cell recognition on common ground.

Of course, the most important conclusion to be drawn from these studies is the fact that a single receptor molecule, immunoglobulin, is sufficient to display restricted recognition. This finding lends credence to the view that a single combining site is sufficient for dual (complex) recognition and, therefore, supports single combining-site models for T-cell recognition as well as providing evidence for H-2-antigen interaction on the cell surface. However, closer inspection of the determinants actually recognized by CTL- and MHC-restricted antibody suggests more differences than similarities between these two repertoires. Certainly, the most striking difference between MHC-restricted antibody and CTL is their choice of different H-2-restricting elements. In H-2^b mice, MHC-restricted influenza-specific antibody is K^b-directed, whereas CTL are D^b-directed. If tolerance is the underlying mechanism responsible for haplotype preference, then this observation implies that T and B cells are differentially tolerized perhaps by different self-determinants, since the CTL and B cell repertoires have been deleted of different self-plus-X specificities. Alternatively, it is possible that the germline repertoires of B cells and CTL are very different to begin with such that few T cell receptor specificities exist that can recognize H-$2K^b$ plus influenza, whereas few B cell specificities exist that can recognize H-$2K^b$ plus influenza. This again would lead us to conclude that T and B cells have different specificity repertoires.

Next, let us consider the nature of viral determinants recognized by MHC-restricted T and B cells. As is the case for antibody recognition of virus per se, MHC-restricted antibodies demonstrate exquisite sensitivity to subtype variation in virus. Indeed, the MHC-restricted component of the anti-HA response represents a subset that is most sensitive to structural variations as measured by the relative lack of cross-reactive recognition of H1 variants. In contrast, the MHC-restricted T cell response is far less discriminatory than antibody responses as measured by the broad range of cross-reactivity.

With respect to MHC recognition, only 57% of cellularly restricted antibodies recognize virus in the context of cells of the immunogenic haplotype (83). This is far less MHC discrimination than is generally observed for CTL. In addition, the K^b determinants described by fine specificity analysis of K^b-restricted antibodies on a mutant panel are a very different set from those observed in allogenic recognition by CTL (83). Taken together, these findings suggest MHC-restricted antibodies see both a different set of H-2 determinants than do CTL and also see a different set of viral determinants than do CTL.

These observations lead us to consider the nature of the determinants actually engaged by MHC-restricted T- and B-cell receptors. There are three possible candidates for immunogenic epitopes. First, there may be structures that are complex determinants composed of sequence information contributed by both the viral molecule and H-2. Second, there may be new determinants that arise on the viral molecule as a result of conformational change induced through interaction with H-2. Finally, there may be new H-2 determinants that result from conformational changes in H-2 that are induced by virus. Responses to all three types of determinants would fulfill the requirements necessary to be considered an MHC-restricted response. It is possible that the differences observed in the T- and B-cell repertoires may be interpreted on the basis of different emphasis by these repertoires as to which of these three types of determinants are immunogenic. For example, the greater specificity of MHC-restricted antibody (versus CTL) to slight variations in viral structure may be interpreted as a greater emphasis by B cells on direct engagement of viral structures whereas the specificity of CTL may be more readily explained as a consequence of T-cell receptor engagement of conformation changes in H-2. That CTL are actually capable of recognition of conformational change in H-2 has been demonstrated by studies of K^b mutants (74), and certainly, there is much evidence that antibody is sensitive to conformational change in antigens.

The concept that the CTL repertoire is predominantly directed against subtle changes in H-2 allows one to explain many experimental observations. As discussed above, CTL clones specific for H-2A plus X often cross-react on specific alloantigen H-2B. This may be explained by considering that H-2A plus X reveals an H-2 determinant already exposed on H-2B.

Also, it provides the best possible explanation as to why a CTL repertoire directed against the alloantigens of the species is indeed the correct choice for recognition of self plus X, since many of the new self H-2 determinants that arise through interaction with antigen are determinants normally exposed on at least some other H-2 haplotypes.

In any case, the resolution of the issues confronted by a comparative assessment of the mode of recognition by B and T cells of viral determinants and virus-infected cells will prove to be central to our ultimate understanding of several major phenomena. As monoclonal B- and T-cell responses continue to be applied to the dissection of determinant recognition and individual MHC and viral antigen expression is manipulated by gene cloning and transfection technology, significant insights should accumulate concerning the mechanism of determinant recognition by the immune system, the mutual effects of viral and immunological evolutionary selective pressures, and the means to maximize prophylactic immunization.

Literature Cited

1. Ada, G. L., Leung, K. N., Ertl, H. 1981. An analysis of effector T cell generation and function in mice exposed to influenza A or sendai virus *Immunol. Rev.* 58:5–24
2. Allouche, M. A., Owen, J. A., Doherty, P. C. 1982. Limit-dilution analysis of weak influenza-immune T cell responses associated with H-2K^b and H-2D^b. *J. Immunol.* 129:689–93.
3. Anders, E. M., Katz, J. M., Brown, L. E., Jackson, D. C., White, D. O. 1981. In *Genetic Variation Among Influenza Viruses,* pp. 547–65. New York: Academic
4. Askonas, B. A., Mullbacher, A., Ashman, R. B. 1982. Cytotoxic T-memory cells in virus infection and the specificity of helper T cells. *Immunology* 45:79–84
5. Askonas, B. A., Webster, R. G. 1980. Monoclonal antibodies to hemagglutinin and to H-2 inhibit the cross reactive cytotoxic T cell population induced by influenza. *Eur. J. Immunol.* 10:151–56
6. Bennick, J. R., Yewdell, J. W., Gerhard, W. 1982. A virus polymerase involved in recognition of influenza virus-infected cells by a cytotoxic T cell clone. *Nature* 296:75–77
7. Bevan, M. J. 1977. In a radiation chimera, host H-2 antigens determine immune responsiveness of donor cytotoxic cells *Nature London New Biol.* 269:417–18
8. Bevan, M. J. 1977. Killer cells reactive to altered-self antigens can also be al-loreactive. *Proc. Natl. Acad. Sci. USA* 74:2094–98
9. Blanden, R. V. 1974. T cell response to viral and bacterial infection. *Transplant. Rev.* 19:56–88
10. Blanden, R. V., Dunlop, M. B. C., Doherty, P. C., Kohn, H. I., McKenzie, I. F. C. 1976. Effects of four H-2K mutations on virus-induced antigens recognized by cytotoxic T cells. *Immunogenetics* 3:541–48
11. Blank, K. J., Lilly, F. 1977. Evidence for an H-2/viral protein complex on the cell surface as the basis for the H-2 restriction of cytotoxicity. *Nature London New Biol.* 269:808–9
12. Braciale, T. J. 1977. Immunologic recognition of influenza virus-infected cells. I. Generation of a virus-strain specific and a cross-reactive subpopulation of cytotoxic T cells in the response to type A influenza viruses of different subtypes. *Cell. Immunol.* 33:423–36
13. Braciale, T. J. 1979. Specificity of cytotoxicity T cells directed to influenza virus hemagglutinin. *J. Exp. Med.* 149:856–69
14. Braciale, T. J., Ada, G. A., Yap, K. L. 1978. Functional and structural considerations in the recognition of virus infected cells by cytotoxic T lymphocytes. In *Contemporay Topics in Molecular Immunology,* ed. R. A. Reisfeld, F. P. Inman, 7:319–64. New York: Plenum
15. Braciale, T. J., Andrew, M. E., Braciale, V. L. 1981. Heterogeneity and specificity of cloned lines of influenza-virus-

specific cytotoxic T lymphocytes. *J. Exp. Med.* 153:910–23

16. Braciale, T. J., Andrew, M. E., Braciale, V. L. 1981. Simultaneous expression of H-2 restricted and alloreactive recognition by a cloned line of influenza virus specific cytotoxic T lymphocytes. *J. Exp. Med.* 153:1371–76

17. Braciale, T. J., Gerhard, W., Klinman, N. R. 1976. The analysis of the *in vitro* humoral immune response to influenza virus. *J. Immunol.* 116:1539–46

18. Cancro, M. P., Gerhard, W., Klinman, N. R. 1978. The diversity of the influenza-specific primary B cell repertoire in Balb/c mice. *J. Exp. Med.* 147:776–87

19. Ciavarra, R., Forman, J. 1981. cell membrane antigens recognized by anti-viral and anti-trinitrophenyl cytotoxic T lymphocytes. *Immunol. Rev.* 58:73–94

20. Cowen, K. M. 1973. Antibody response to viral antigens. *Adv. Immunol.* 17:195–253.

21. Dales, S., Oldstone, M. B. A. 1982. Localization at high resolution of antibody-induced mobilization of vaccinia virus hemagglutinin and the major histocompatibility antigens on the plasma membrane of infected cells. *J. Exp. Med.* 156:1435–47

22. Doherty, P. C., Biddison, W. E., Bennink, J. R., Knowles, B. B. 1978. Cytotoxic T cell responses in mice infected with influenza and vaccinia viruses vary in magnitude with H-2 genotype. *J. Exp. Med.* 148:534–43

23. Effros, R. B., Doherty, P. C., Gerhard, W., Bennink, J. 1977. Generation of both cross-reactive and virus-specific T cell populations after immunization with serologically distinct influenza A viruses. *J. Exp. Med.* 145:557–68

24. Ennis, F. A., Martin, W. J., Verbonitz, M. W. 1977. Hemagglutinin specific cytotoxic T cell response during influenza infection. *J. Exp. Med.* 146:893–98

25. Ertl, H., Ada, G. L. 1981. Roles of influenza virus infectivity and glycosylation of viral antigen for recognition of target cells by cytolytic T lymphocytes. *Immunobiology* 158:239–53

26. Fazekas de St. Groth, S. 1981. The joint evolution of antigens and antibodies. In *The Immune System. A Festschrift in Honor of Niels Kaj Jerne on the Occasion of his 70th Birthday,* ed. C. M. Steinberg, I. Lefkovits, 1:155–68. Basel: Karger

27. Finberg, R., Burakoff, S., Cantor, H., Benacerraf, B. 1978. Biological significance of alloreactivity: T cells stimulated by Sendai virus coated syngeneic cells specifically lyse allogeneic target cells. *Proc. Natl. Acad. Sci. USA* 75:5154–49

28. Frankel, M. E., Effros, R. B., Doherty, P. C., Gerhard, W. 1979. A monoclonal antibody to viral glycoprotein blocks virus-immune effector T cells operating at H-2D^d but not at H-2K^d. *J. Immunol.* 123:2438–40

29. Gardner, I. D., Bowern, N. A., Blanden, R. V. 1974. Cell-mediated cytotoxicity against ectromelia virus-infected target cells. II. Identification of effector cells and analysis of mechanism. *Eur. J. Immunol.* 4:68–72

30. Geiger, B., Rosenthal, K.L., Klein, J., Zinkernagel, R.M., Singer, S. J. 1979. Selective and unindirectional membrane redistribution of an H-2 antigen and an antibody-clustered viral antigen: Relationship to mechanisms of cytotoxic T cell interactions. *Proc. Natl. Acad. Sci. USA* 76:4603–7

31. Gerhard, W. 1977. The delineation of antigenic determinants of the hemagglutinin of influenza A viruses by means of monoclonal antibodies. *Top. Infect. Dis.* 3:15

32. Gerhard, W. 1978. In *The Influenza Virus Hemagglutinin,* ed. W. G. Laver, H. Bachmayer, R. Weil, pp 15–23. Wien: Springer. 259 pp.

33. Gerhard, W., Braciale, T. J., Klinman, N. R. 1975. The analysis of the monoclonal immune response to influenza virus. I. Production of monoclonal antiviral antibodies *in vitro. Eur. J. Immunol.* 5:720–25

34. Gerhard, W., Webster, R. G. 1978. Selection and characterization of antigenic variants of A/PR/8/34(HON1) influenza virus with monoclonal antibodies. *J. Exp. Med.* 148:383–92

35. Gerhard, W., Yewdell, J., Frankel, M. D., Lopes, A. D., Staudt, L. 1980. Monoclonal Antibodies against Influenza Virus. In *Monoclonal Antibodies,* ed. R. Kennett, T. McKearn, K. Bechtol, pp. 317–333. New York: Plenum

36. Germain, R., Dorf, M. E., Benacerraf, B. 1975. Inhibition of T lymphocyte-mediated tumor specific lysis by alloantisera directed against the H-2 serological specificities of the tumor. *J. Exp. Med.* 142:1023–28

37. Gomard, E., Duprez, K., Henin, Y., Levy, J. P. 1976. H-2 region product as determinant in immune cytolysis of syngeneic tumor cells by anti-MSV T lym-

phocytes. *Nature London New Biol.* 260:707–9

38. Gorczynski, R. M., Kennedy, M. J., MacRae, S., Steele, E. J., Cunningham, A. J. 1980. Restriction of antigen recognition in mouse B lymphocytes by genes mapping within the major histocompatibility complex. *J. Immunol.* 124: 590–96

39. Hackett, C. J., Askonas, B. A. 1982. H-2 and viral hemagglutinin expression by influenza infected cells: The proteins are close but do not cocap. *Immunology* 45:431–38

40. Hackett, C. J., Askonas, B. A., Webster, R. G., van Wyke, K. 1980. Quantitation of influenza virus antigens on infected target cells and their recognition by cross reactive cytotoxic T cells. *J. Exp. Med.* 151:1014–25

41. Hunig, T., Bevan, M. J. 1980. Self H-2 antigens influence the specificity of alloreactive cells. *J. Exp. Med.* 151: 1288–98

42. Ivanyi, P., Cornelis, J. M., van Mourik, P., Vlug, A., de Greeve, R. 1979. Lymphocyte antibodies produced by H-2 allo-immunization distinguish between MuLV-positive and -negative substrains of the same haplotype. *Nature* 282: 843–45

43. Joseph, B. S., Oldstone, M. B. A. 1974. Antibody-induced redistribution of measles virus antigens on the cell surface. *J. Immunol.* 113:1205–9

44. Kanagawa, O., Louis, J., Cerottini, J. -C. 1982. Frequency and cross-reactivity of cytolytic T lymphocyte precursors reacting against male alloantigens. *J. Immunol.* 128:2362–6

45. Katz, D. H., Skidmore, B. J., Katz, L. R., Bogowitz, C. B. 1978. Adaptive differentiation of murine lymphocytes. I. Both T and B lymphocytes differentiating in F1-parental chimeras manifest preferential cooperative activity for partner lymphocytes derived from the same parental type corresponding to the chimeric host. *J. Exp. Med.* 148:727–45

46. Kilbourne, E. D., ed. 1975. *The Influenza Viruses and Influenza.* New York: Academic

47. Klinman, N. R., Segal, G. P., Gerhard, W., Braciale, T., Levy, R. 1977. Obtaining homogenous antibody of desired specificity from fragment cultures. In *Antibodies in Human Diagnosis and Therapy,* ed. E. Haber, R. Krause, pp. 225–36. New York: Raven

48. Kohler, G., Milstein, C. 1975. Continuous cultures of fused cells secreting antibody of predefined specificity. *Nature* 256:495–97

49. Koprowski, H., Wiktor, T. 1980. Monoclonal antibodies against rabies virus. In *Monoclonal Antibodies,* ed. R. Kennet, T. McKearn, K. Bechtol, pp. 335–51. New York: Plenum

50. Koszinowski, U. H., Allen, H., Gething, M. J., Waterfield, M.D., Klenk, H. D. 1980. Recognition of viral glycoproteins by influenza A-specific cross reactive cytolytic T lymphocytes. *J. Exp. Med.* 151:945–58

51. Koszinowski, U., Ertl, H. 1975. Lysis mediated by T cells and restricted by H-2 antigen of target cells infected with vaccinia virus. *Nature London New Biol.* 255:552–54

52. Laver, W. G., Air, G. M., Webster, R. G., Gerhard, W., Ward, C. W., Dopheide, T. A. A. 1979. Antigenic drift in type A influenza virus: Sequence differences in the hemagglutinin of Hong Kong (H3N2) variants selected with monoclonal hybridoma antibodies. *Virology* 98:226–37

53. Lin, Y.-L., Askonas, B. A. 1981. Biological properties of an influenza A virus-specific killer T cell clone. *J. Exp. Med.* 154:225–34

54. Lu, L. Y., Askonas, B. A. 1980. Cross reactivity for different type A influenza viruses of a cloned T-killer cell line. *Nature* 288:164–67

55. Mullbacker, A. 1981. Neonatal tolerance to alloantigens alters major histocompatibility complex-restricted response patterns. *Proc. Natl. Acad. Sci. USA* 78:7689–91

56. Niman, H. L., Elder, J. H. 1982. Monoclonal antibodies as probes of protein structure: Molecular diversity among the envelope glycoproteins (gp70s) of the murine retroviruses. In *Monoclonal Antibodies and T cell Products,* ed. D. H. Katz, pp. 23–51 Boca Raton, Fla: CRC

57. Nisbet-Browm, E., Singh, B., Diener, E. 1981. Antigen recognition. V. Requirement for histocompatibility between antigen-presenting cell and B cell in the response to a thymus-dependent antigen, and lack of allogeneic restriction between T and B cells. *J. Exp. Med.* 154:676–87

58. Nowinski, R. C., Stone, M. R., Tam, M. R., Lostrom, M. E., Burnette, W. N., O'Donnell, P. V. 1980. Mapping of viral proteins with monoclonal antibodies. Analysis of the envelope proteins of murine leukemia viruses. In *Monoclonal Antibodies,* ed. R. Kennett, T.

McKearn, K. Bechtol, pp. 295–316. New York: Plenum

59. Ohno, S., Matsunaga, T., Epplen, J. T., Hozumi, T. 1980. Interaction of viruses and lymphocytes in Evolution, Differentiation and Oncogensis. In *Immunology 80-Progress in Immunology IV*, ed. M. Fougereau, J. Dausset, pp. 577–98. London: Academic

60. Oldstone, M. B. A., Fujinami, R. S., Lampert, P. W. 1980. Membrane and cytoplasmic changes in virus-infected cells induced by interactions of antiviral antibody with surface viral antigen. *Prog. Med. Virol.* 26:45–93

61. Oldstone, M. B. A., Sissons, J. G. P., Fujinami, R. S. 1980. Action of Antibody and Complement in Regulating Virus Infection. In *Immunology 80-Progress in Immunology IV*, ed. M. Fougereau, J. Dausett, pp. 599–621. New York: Academic

62. Owen, J. A., Allouche, M., Doherty, P. C. 1982. Limiting dilution analysis of the specificity of influenza-immune cytotoxic T cells. *Cellular. Immunol.* 67:49–59

63. Pan, S.-H., Wettstein, P. J., Knowles, B. B. 1982. H-2K^b mutations limit the CTL response to SV40 TASA. *J. Immunol.* 128:243–46

64. Pfizenmaier, K., Jung, H., Kurrle, R., Rollinghoff, M., Wagner, H. 1980. Anti-H-2D^d alloreactivity mediated by herpes simplex virus specific cytotoxic H-2^k T lymphocytes is associated with H-2D^k. *Immunogenetics* 10:395–404

65. Pfizenmaier, K., Pan, S., Knowles, B. B. 1980. Preferential H-2 association in cytotoxic T cell responses to SV40 tumor-associated specific antigens. *J. Immunol.* 124:1888–97

66. Ramos, T. Moller, G. 1978. Immune response to haptenated syngeneic and allogeneic lymphocytes. *Scand. J. Immunol.* 8:1–7

67. Reiss, C. S., Schulman, J. L. 1980. Cellular immune responses of mice to influenza virus vaccines. *J. Immunol.* 125:2182–88

68. Reiss, C. S., Schulman, J. L. 1980. Influenza type A virus M protein expression on infected cells is responsible for cross-reactive recognition by cytotoxic thymus-derived lymphocytes. *Infect. Immun.* 29:719–23

69. Schrader, J. W., Edelman, G. M. 1976. Partipation of the H-2 antigens of tumor cells in their lysis by syngeneic T cells. *J. Exp. Med.* 143:601–14

70. Schwartz, R. H., Sredni, B. 1982. Alloreactivity of antigen-specific T cell clones. In *Isolation, Characterization, and Utilization of T lymphocyte clones*, ed. C. G. Fathman, F. W. Fitch, pp. 375–84. New York: Academic

71. Sethi, K. K., Brandis, H. 1980. The role of vesicular stomatitis virus major glycoprotein in determining the specificity of virus-specific and H-2 restricted cytolytic T cells. *Eur. J. Immunol.* 10:268–72

72. Shaw, M. W., Lammon, E. W., Compans, R. W. 1981. Surface expression of a non-structural antigen on influenza A virus-infected cells. *Infect. Immun.* 34:1065–67

73. Sherman, L. A. 1982. The influence of the major histocompatibility complex on the repertoire of allospecific cytolytic T lymphocytes. *J. Exp. Med.* 155: 380–89

74. Sherman, L. A. 1982. Evidence for recognition of conformational determinants on H-2 by cytolytic T lymphocytes. *Nature* 297:511–13

75. Singer, A., Hathcock, K. S., Hodes, R. J. 1981. MHC restrictions in the generation of TNP-Ficoll responses: evidence for self-recognition by B cells. *Fed. Proc.* 40:1061 (Abstr.)

76. Stukart, M. J., Vos, A., Boes, J., Melvold, R. W., Bailey, D. W., Melief, C. J. M. 1982. A crucial role of the H-2 D locus in the regulation of both the D- and the K-associated cytotoxic T lymphocyte response against Moloney leukemia virus demonstrated with two D^b mutants. *J. Immunol.* 128:1360–64

77. van Leeuwen, A., Goulmy, E., van Rood, J. J. 1979. Major histocompatibility complex-restricted antibody reactivity mainly, but not exclusively, directed against cells from male donors. *J. Exp. Med.* 150:1075–83

78. von Boehmer, H., Hengartner, H., Nabholz, M., Lenhardt, W., Schreier, M., Haas, W. 1979. Fine specificity of a continuously growing killer cell clone specific for H-Y antigen. *Eur. J. Immunol.* 9:592–97

79. Weiss, A., MacDonald, H. R., Cerottini, J.-C., Brunner, K. T. 1981. Inhibition of cytolytic T lymphocyte clones reactive with Moloney leukemia virus-associated antigens by monoclonal antibodies: A direct approach to the study of H-2 restriction. *J. Immunol.* 126:482–85

80. Wettstein, P. J. 1982. H-2 effects on cell-cell interactions in the response to single non-H-2 alloantigens. V. Effects of H-2K^b mutations on presentation of

H-4 and H-3 alloantigens. *J. Immunol.* 128:2629–33

81. Wyley, D. C., Wilson, I. A., Skehel. J. J. 1981. Structural identification of the antibody binding sites of Hong Kong influenza hemagglutinin and their involvement in antigenic variation. *Nature* 289:373–78

82. Wylie, D. E., Klinman, N. R. 1981. The murine B cell repertoire responsive to an influenza infected syngeneic cell line. *J. Immunol.* 127:194–98

83. Wylie, D. E., Sherman, L. A., Klinman, N. R. 1982. The participation of the major histocompatibility complex in antibody recognition of viral antigens expressed on infected cells. *J. Exp. Med.* 155:403–14

84. Yewdell, J. W., Gerhard, W. 1981. Antigenic Characterization of viruses by monoclonal antibodies. *Ann. Rev. Microbiol.* 35:185–206

85. Zarling, D. A., Miskimen, J. A., Fan, D. P., Fujimoto, E. K., Smith, P. K. 1982. Association of sendai virion envelope and a mouse surface membrane polypeptide on newly infected cells: Lack of association with H-2K/D or alteration of viral immunogenicity. *J. Immunol.* 128:251–57

86. Zinkernagel, R. M. 1976. H-2 compatibility requirement for virus-specific T cell mediated cytolysis. *J. Exp. Med.* 143:437–43

87. Zinkernagel, R. M., Althage, A., Cooper, S., Kreeb, G., Klein, P. A., Sefton, B., Flaherty, L., Stimpfling, J., Shreffler, D., Klein, J. 1978. Ir genes in H-2 regulate generation of antiviral cytotoxic T cells: Mapping to K or D and dominance of unresponsiveness. *J. Exp. Med.* 148:592–606

88. Zinkernagel, R. M., Callahan, G. N., Althage, A., Cooper, J., Klein, P. A., Klein, J. 1978. On the thymus in the differentiation of H-2 self-recognition by T cells: Evidence for dual recognition? *J. Exp. Med.* 147:882–96

89. Zinkernagel, R. M., Doherty, P. C. 1979. MHC-restricted cytotoxic T cells: Studies on the biological role of polymorphic major transplantation antigens determining T cell restriction specificity, function, and responsiveness. *Adv. Immunol.* 27:51–177

90. Zinkernagel, R. M., Rosenthal, K. L. 1981. Experiments and speculation on antiviral specificity of T and B cells. *Immunol. Rev.* 58:131–55

91. Zweerink, H. J., Courtneidge, S. A., Skehel, J. J., Crumpton, M. J., Askonas, B. A. 1977. Cytotoxic T cells kill influenza virus-infected cells but do not distinguish between serologically distinct Type A viruses. *Nature London New Biol.* 267:354–56.

Ann. Rev. Immunol. 1983. 1:87–117
Copyright © 1983 by Annual Reviews Inc. All rights reserved

STRUCTURAL BASIS
OF ANTIBODY FUNCTION

David R. Davies and Henry Metzger

National Institute of Arthritis, Diabetes, Digestive, and Kidney Diseases,
National Institutes of Health, Bethesda, Maryland 20205

INTRODUCTION

It is less than 20 years since the general architecture of antibodies was
elucidated and even less time since the explicit molecular basis of antibody
specificity first became clear. In the subsequent explosion of information
about the immune system, two basic principles have emerged: (*a*) Antibodies remain the only known structures whose diversity is sufficient to explain
the fine specificity exhibited by the immune response; and, (*b*) antibody
function is mediated by a molecule whose structure consists of two distinct
regions—one that carries a recognition site for antigenic determinants, and
a second by which the antibody reacts with receptors of a variety of effector
systems.

   In this review we examine the current information on the structure of
antibodies. We do not describe again the basic four-chain structure of
immunoglobulins nor the division into variable and constant regions, which
are by now well known (e.g. 4, 90, 118). We instead concentrate on the
higher resolution data, much of which is still in the course of refinement.
We discuss the Fabs, in particular with reference to the combining site and
the specificity of binding to hapten; the Fc region with reference to the
binding site of protein A of *Staphylococcus aureus* and of C1q; and the
structure of the hinge with reference to its possible role in separating Fab
and Fc.

   Whereas the characterization of the structures of individual proteins and
of their interactions with small molecules can now be carried out with some

87

0732-0582/83/0410-0087$02.00

sophistication, the interaction of two or more macromolecules still presents considerable difficulties. It is not surprising therefore that our understanding of how antibodies interact with the macromolecular receptors on effector systems (e.g. C1q in the complement pathway, Fc receptors on cell membranes) is much less advanced. Our review of this aspect of antibody structure and function therefore involves more questions than answers.

IMMUNOGLOBULIN STRUCTURE

General Comments

Our knowledge of the three-dimensional structure of antibodies at atomic resolution rests mainly on X-ray diffraction investigations of fragments. Intact proteins for which X-ray analyses have been carried out consist of Kol (106), an IgG1(λ) human myeloma protein, the protein Dob (161), an IgG1(κ) human cryoglobulin, and recently the human myeloma IgG1(λ) protein, Mcg (129). In Kol, and also apparently in another immunoglobulin, Zie (58a), the crystal structure contains an unusual feature: The Fc occupies a number of different positions in the crystal that are not crystallographically related, with the result that no significant electron density occurs in this region of the crystal, there being an abrupt drop in density at the end of the hinge. In both Dob and Mcg there is a 15-amino-acid-residue deletion in the hinge (63, 169), thus, presumably, reducing the flexibility of the molecule and enabling the Fc to be located in the electron density, although in the case of Dob the crystals are disordered and do not diffract to high resolution.

The structures of three Fabs have been published, Newm (148), Kol (106), and McPC603 (152), as well as the structures of a number of V_L dimers and L chain dimers (4). The structure of human IgG Fc has been determined, and also its complex with protein A of *Staphylococcus aureus* (43). These structures have been reviewed most recently by Amzel & Poljak (4) and are not covered comprehensively in this review.

Because of the deficiencies in the crystals of the intact molecules, our knowledge of the whole antibody molecule has to be a sum of its parts. The flexibility of the molecule, in particular in the region between the Fab and the Fc, may preclude for some time visualizing directly by X-ray diffraction an intact molecule with intact hinge at atomic resolution. However, the checks that can be made on this composite three-dimensional picture of the antibody molecule are reassuring. Thus, the Kol Fab in isolated form in the crystal is quite similar to the Fab of the whole molecule in its crystal form. Also, the Fc in Dob has, within the experimental error of the comparison, the same overall structure as does the Fc in the isolated Fc crystals.

Fab Structure and the Antibody Combining Site

McPC603 AND THE PHOSPHOCHOLINE BINDING SITE   The structure of McPC603 Fab, a mouse myeloma IgA (κ) with phosphocholine (PC) binding capability, has been determined at 3.1-Å resolution (41, 42, 125, 152) and is being refined to 2.7 Å (Y. Satow, D. R. Davies, manuscript in preparation). The overall three-dimensional structure of the Fab is illustrated in Figure 1, which demonstrates the strong lateral association between domains of the light (L) and heavy (H) chains, together with the relatively weak longitudinal interactions along each chain. Figure 1 also shows the clustering of six of the seven hypervariable regions at the tip of the Fab, forming the complementarity-determining surface (89, 90). The variable domains have a very similar three-dimensional structure for both the L and H chains and across species (4, 42, 123). The constant domains C_L and C_H1 are also very similar. Both the variable and the constant pairs of domains are related by approximately twofold (rotation about the V axis of 180° will superpose V_L on V_H) and the angle between the two axes has been referred to as the elbow bend of the Fab and has been observed to vary

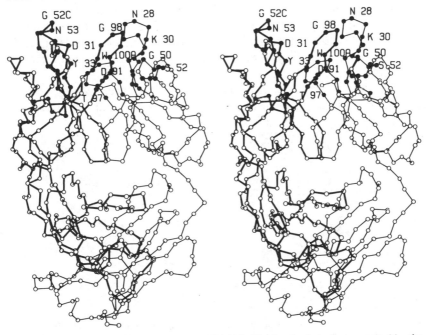

Figure 1   The α-carbon backbone of McPC603 Fab. The heavy chain is represented by the thick line. The two variable domains are at the top and the constant domains are at the bottom of the figure. The complementarity-determining residues (CDR) are shown as filled circles. Two residues in each CDR loop have been labeled.

between approximately 137° for Fab Newm (4), 135° for McPC603 (152), 147° for Dob (161), and approximately 170° for Kol (106).

Figure 2 shows the combining site of McPC603 with PC bound. (Y. Satow, E. A. Padlan, G. H. Cohen, D. R. Davies, manuscript in preparation). The choline is attached at the bottom of a pocket located principally between the hypervariable regions H3 and L3. The phosphate is on the surface and contacts residues from the heavy chain. It is apparent that PC is a small molecule and that the greater part of the hypervariable surface is not directly in contact with it. At the front of the pocket there are two hydrogen bond donors positioned within reasonable hydrogen bonding distance of the phosphate oxygens; these are the hydroxyl group of Tyr 33H and the guanidinium group of Arg 52H (152). The residues lining the inside of the pocket are Tyr 94L on the right side, Asp 91L on the left, Leu 96L at the back, (125, 146), and the side chain of Trp 100aH at the top left. In addition, the backbone of residues 92–94L form the lower rim of the front of the pocket.

CONFORMATIONAL CHANGE   One of the mechanisms proposed for effector function activation involves an allosteric change upon antigen binding (110). Since crystals of immunoglobulins have large solvent channels and can bind to haptens soaked in through these channels, crystallographic investigation offers a direct way for observing conformational changes, when they occur.

When PC binds to McPC603 in the crystal, no significant conformational change occurs in the protein. There is a small movement of Trp 104a away

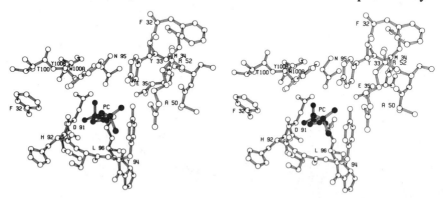

Figure 2   Stereo drawing of the combining site of McPC603 with phosphocholine bound. The lower residues (91–96 and F32) are from the light chain. The remaining residues belong to the three complementarity-determining regions of the heavy chain. The phosphocholine has the phosphate group in front with the choline moiety buried in a pocket.

from the pocket, but no other change of any significance. However, there are several reasons why it cannot be concluded from this observation that antigen-antibody interaction results in no conformation change:

1. The crystals are grown in a concentrated ammonium sulfate solution, and it has been observed that in the absence of PC there is a peak at the phosphate binding site interpreted to be a sulfate ion (126). A conformational change might have been triggered by the presence of this sulfate ion, so that no additional change would be observed upon PC binding. However, in this respect it should be noted that in Fab Newm, (148) no conformational change occurs upon binding of a neutral vitamin K_1 derivative.

2. PC is small and the association constant ($\sim 10^5$ M^{-1}) with McPC603 might be insufficient to trigger a conformational change that could be induced by a larger, more tightly binding antigen. The same consideration applies to Fab Newm.

3. The only two structures at atomic resolution of immunoglobulins with known binding specificity are of Fabs. Although improbable, one cannot rule out the possibility that changes that occur with the intact molecule might not occur with fragments (81).

Thus, although there is no support from X-ray diffraction for a conformational change associated with antigen binding, such a change cannot be rigorously excluded.

McPC603: THE CONTACTING RESIDUES AND THE EFFECT OF CHANGES IN THE COMBINING SITE   The PC molecule is in direct contact with only a limited number of residues. They include side chains from all three heavy chain hypervariable regions and from one (L3) light chain region. The next most distant range of contacts contain many residues that play a role in positioning the directly contacting residues and changes in these might be expected to influence PC binding. An example of such a side chain is Glu 35H, a residue in the interface between V_H and V_L that makes a hydrogen bond with the hydroxyl of Tyr 94L, which is in turn a major contacting residue with hapten. A mutant of S107, a PC-binding myeloma protein, has been observed that has lost the ability to bind PC and also that fails to agglutinate PC-SRBC (144). Amino acid analysis showed that the mutation results in substitution of an alanine for glutamic acid in position 35H. Although a change of this magnitude is likely to produce a significant rearrangement of side chains in its vicinity simply because of the difference in volume of the two side chains, the loss of contact with Tyr 94L reduces an important constraint on a residue in direct contact with hapten.

Another mutant observed by Cook et al (38) is more puzzling. The mutant still bound PC, but it bound less well than S107 to PC coupled to different carriers, and showed a decrease in affinity for a variety of PC-carrier conjugates. The only amino acid change observed was that of Asp → Ala in the fifth position of the heavy chain J region. Changes in the carboxy terminal half of V_L were not entirely ruled out, but they did not appear in a tryptic peptide analysis. The strange aspect of this mutation site is that it is spatially well removed from the PC pocket so that it might not have been expected to be involved in antigen binding. Another curious feature is that diverse carriers coupled to PC were all affected, which implies they all have some contact with this residue, or with a region influenced by it.

SEQUENCE COMPARISON OF PC-BINDING ANTIBODIES   Sequences and binding data are available for a variety of PC-binding myeloma proteins and monoclonal antibodies. They have recently been reviewed by Rudikoff (145) and are only discussed here in relation to the three-dimensional structure of McPC603.

The heavy chain   The sequences of 19 heavy chains of BALB/c immuno-globulins that bind phosphocholine have been analyzed (74, 145). Ten of these are identical and employ the T15 sequence. The remaining nine differ by one to eight residues from the T15 sequence. Gene isolation and analysis have revealed that all 19 of the V_H regions must have arisen from the single germline T15 V_H gene segment (39). M167 is the most divergent protein with eight V_H substitutions. In both M167 and HPCG13 the same change (Thr at position 40) occurs, but all the other substitutions are unique, occurring in only one protein.

The PC-contacting residues Tyr33 and Arg52, together with Glu35, are present in all of these sequences. Similarily, all of the BALB/c sequences with the single exception of M167 contain a tryptophan at 100b. There is considerable variation in the D region for these proteins, accompanied by only relatively small changes in PC affinity, consistent with the fact that, with the exception of Trp100b, CDR3 of the heavy chain does not play a major role in defining the PC pocket.

The light chains   The light chains of PC-binding antibodies can be represented by the three BALB/c myeloma proteins, T15, M603, and M167 (34, 35). These light chains differ considerably in sequence, despite the similarity of their corresponding heavy chain sequences. However, in the PC-binding region, in particular with the contacting residues Tyr94, Pro95, and Leu96, their sequences are the same. They also all employ the same J_L sequence (J5).

The invariant residues in these light and heavy chains provide strong support for the suggestion that these PC-binding antibodies all have the same overall combining site for PC (125), with differences in binding specificity being contributed by the amino acid substitutions.

Fab Newm   The structure of the Fab of Newm, a human IgG (λ) myeloma protein, has been refined to a nominal 2-Å resolution (4, 148). The elbow bend is 137° and the L chain has a seven-residue deletion that includes residues 55 and 56 of CDR2, and 56 to 62 of the framework region FRIII.

The combining site is formed by the association of the remaining five hypervariable regions. The principal feature of the site is a shallow groove 15 × 6 × 6 Å deep, bordered by residues from the H and L chains. Fab Newm binds several haptens at this site with affinity constants ranging from 10^3 to 10^5 M^{-1}. A derivative of vitamin K_1 binds with the higher affinity and a crystallographic investigation has demonstrated that the menadione ring system binds in the shallow groove with the phytyl chain draped along the surface, making contact with a number of residues from the light and heavy chains (5). The number of contacts provided by the phytyl chain can probably account for the difference in binding between menadione and the vitamin K_1 derivative (10^3 vs 1.7×10^5 M^{-1}). As noted above, no conformational change is observed upon the hapten binding.

KoL   The human IgG λ cryoglobulin Kol and its Fab both crystallize and both structures have been solved, the intact molecule at 3.5 Å and the Fab, for which the combining site specificity is unknown, at 1.9 Å resolution (106). The crystals of the intact molecule display an unusual form of disorder, described above, that prevents visualization of the Fc.

The Fab crystallizes well and has provided a detailed high resolution structure. The L chain conformation is quite similar to that of Fab Newm, except for the presence of the seven additional residues around CDR2. However, in the H chain, CDR3 is eight residues longer than Newm (and six longer than McPC603), and these additional residues fold into the combining site and fill it completely, thus obliterating the groove observed in Newm. This combining site is rich in aromatic side chains, being filled largely by Trp (47H, 52H, 90L and 108H), Tyr (35H, 35L and 97L), and His (59H).

In both the crystals of the intact molecule and the Fab, the same contact is made between the hypervariable surface and the hinge segment, Cys221-Cys230, the light chain C terminus Glu212-Ser214, and the residues 133–

138 and 196–199 of C_H1 of a neighboring molecule. This contact involves considerable surface area and it has been estimated that $1314A^2$ of the Kol hypervariable surface is excluded from solvent. The contact is tight and closely packed and involves hydrophobic interactions as well as salt links and hydrogen bonds. Marquart et al (105) suggest that this interaction could be a prototype antibody-antigen interaction and might be responsible for the cryo properties of this molecule.

ANTIBODY COMBINING SITES   Data on antibody combining sites have been comprehensively reviewed by Givol (76). Here, we only highlight a few structural topics.

What common features, if any, will the three-dimensional structures of the combining sites share? Givol (76) notes that the sizes of these sites are comparable to those of some enzymes. Lysozyme, for example, has a groove that will accommodate a hexasaccharide, and this is believed to represent about the upper limit in size for this kind of antigen (88, 102). The binding of antidextrans can be divided into two classes: end-binders like W3129, which bind only the nonreducing end of the dextran; and middle-binders like QUPC52, which bind in the middle of the chain to runs of six glucose units. It has been suggested that the former type of antibody might have a pocketlike site, whereas the latter might be more likely to have a lengthy groove (33, 89).

Since three-dimensional structures are known for but two Fabs with known binding specificities, only a limited picture can be obtained from direct observation. It is nevertheless suggestive that one of these structures, McPC603, has a pocket for binding PC, whereas the other, Newm, has a shallow groove where the menadione binding site is located. Again, if we were to argue from analogy with enzymes we might expect a pocket or groove in most antibodies to provide specificity. However, specificity can also be produced by complementarity between two interacting surfaces without necessarily invoking grooves or pockets, and significant contributions to the free energy of interaction can come from exclusion of hydrophobic groups from contact with water. This kind of specificity, like the interactions between subunits in a multisubunit protein, can be quite precise and can be destroyed by single amino acid changes. For instance, not only is it necessary to maintain complementarity of two interacting surfaces, but there also needs to be a suitable juxtaposition of oppositely charged groups forming salt bridges as well as appropriate disposition of hydrogen bond donors and receptors. Accordingly, although antibodies specific for small antigenic determinants will probably have a groove or pocket, other antibodies specific for an array of amino acids such as epitopes on the surface of a protein need not necessarily have these features. The Kol protein combining site, although similar in some respects to Newm, has its groove partially filled with aromatic amino acid side chains and it is significant that this site interacts strongly with the C-terminal portion of another Fab in what could be a prototype antibody/protein antigen complex.

The availability of monoclonal antibodies to specific antigens through hybridoma technology should provide a significant increase in the diversity of antibodies studies by X-ray diffraction. In particular, they offer the opportunity to study interactions with ligands that are larger than simple haptens and that could completely fill the combining site. One example is an anti-influenza virus neuraminidase that has already been crystallized (37). More of these anti-protein antibodies need to be studied, both alone and complexed with antigen, to obtain structural comparisons with other studies (12, 57, 164). Other monoclonal antibodies to polysaccharides such as $\alpha 1 \rightarrow 6$ linked dextran should clarify the mechanism of binding for linear antigens (90) and the effects of amino acid changes on specificity.

MODEL BUILDING STUDIES OF Fv   Since in the last 10 years high resolution structures have been determined for only three different Fab's, and only two (M603 and Newm) have known binding specificities, it appears that X-ray crystallography can only provide a small fraction of three-dimensional structures for interesting antibody combining sites. There is also the difficulty that only some Fab's can be induced to crystallize. An alternative approach to direct X-ray analysis is to utilize the knowledge available from crystallography together with known amino acid sequences to construct models of the Fv's of interesting antibodies. The general problem of protein folding, i.e. how a polypeptide chain several hundred amino acid residues long folds into its final globular form, is being extensively investigated. However, it is most unlikely that it will soon be possible to predict correctly the final folded protein structure based on sequence. Nevertheless, the forces that contribute to the stability of a protein continue to be studied and are being refined. There is also an increased appreciation of the dynamic aspects of protein structure.

The problem of predicting antibody combining site structures is a special case with some features that make it an attractive candidate for investigation by molecular modeling. The similarity of variable domain structures is quite remarkable (4, 120–122), which indicates a strong conservation of the three-dimensional structure of the framework part of the variable domains. Also, the most variable parts of the V domains occur largely at one end of the domain in the hypervariable loops. As a result, a comparison of the sequence with that of V domains of known structure could, in optimal cases, directly lead to a preliminary model that could then be refined by a suitable energy minimization technique. The problem becomes more complicated when there are amino acid insertions and deletions that change the length of the hypervariable loops. The new loops could be copied wherever possible from similar size loops in proteins whose structures have been determined. The $V_L : V_L$ dimers Rei and Au (60) provide an example of very similar three-dimensional structures associated with loops of the same length, although there are 18 amino acid differences in each domain. A different point of view would be inferred from the L chain dimer of Mcg (58) where corresponding hypervariable loops do not preserve the local twofold symmetry of the V domains but have different conformations as a result of interaction with neighboring molecules in the crystal.

The earliest modeling studies involved the 2,4-dinitrophenol-binding mouse myeloma protein MOPC315. The combining site of MOPC315 has been discussed (135) relative to the structure of Newm. Padlan et al (124) constructed a molecular model for MOPC315 that utilizes the framework structure of M603 and the hypervariable loops derived from other V regions that have CDR's of similar length whose structure had been determined by X-ray diffraction. An extensive nuclear magnetic resonance investigation of MOPC315 (54, 55, 186) and its interaction with ligands has led to a refined model that is similar to the original model, but has a different orientation for the DNP binding site (89). The crystal structure of the Fv of MOPC315 is being investigated and should ultimately provide a basis for evaluating these models (6). Subsequently, Davies & Padlan (40) constructed a model for the homogeneous rabbit antibody (BS5) to type III pneumococcal polysaccharide. Potter et al (136) presented a model for the inulin-binding myeloma protein EPC109. More recently, Stanford & Wu (167) have constructed a backbone model for MOPC325, and Feldmann et al (62) have proposed models for J539 and included a proposal for the binding of hexasaccharide. The crystal structure for J539 has been determined at 4.5 Å and the atomic resolution structure is under investigation so that it should soon be possible to test this model.

None of the models described above has been subjected to any form of energy minimization. They may give an approximate general, low resolution picture of the combining sites particularly where they illustrate some striking insertion or deletion, as in EPC109. However, they are unlikely to be accurate to better than several angstroms for the backbone atoms, and could be quite incorrect in positioning the amino acid side chains. Until, for at least a few cases, they are compared with the results of X-ray diffraction these models should be regarded as being quite hypothetical and should be treated with caution. In the case of MOPC315, the structure investigation by Padlan et al (124) was consistent with the known chemical data from affinity labeling and did lead to the discovery of an error in the sequence determination, but these correlations derive from coarse rather than fine detail in the model. Since single amino acid changes can produce large effects on structure, it is perhaps optimistic at this stage to expect to define an antibody site with reasonable precision. Certainly, some powerful form of energy minimization will be necessary to ensure that the models produced do at least satisfy the basic requirement of stereo-chemistry. However, satisfactory prediction also requires a larger library of known structures of antibodies to a greater variety of antigens.

THE H : L ASSOCIATION   For the combinatorial mechanism for generating antibody diversity to be reasonably effective, most light chains should have the ability to combine with most heavy chains. This requirement has been examined both in vitro and in vivo.

The γ-L interaction has been shown to obey second-order kinetics (7, 13, 22, 69) and has a high affinity with Ka $> 10^{10}$ M^{-1} (13). When the competitive association of autologous and heterologous pairs of chains (i.e. pairs derived or not from the same myeloma) was examined, it was discovered that there was a preferred association with the autologous chain (46, 78, 104, 170). When heterologous Vκ was added to Fd', no significant association took place except in the presence of Cκ. This was not true for the autologous Vκ, which did not require the presence of Cκ (96). Isolated Vκ recombined with the autologous V$_H$, but heterologous pairs did not associate by the criterion of UV difference spectra (80). It is clear that the association of the two complete chains must be significantly aided by the presence of the constant domain of the light chain. Nevertheless, the preferential association for the autologous pair observed in these in vitro experiments does imply some form of prior selection. Whether or not any additional selection is needed other than that for antigen binding per se is not clear. Clearly, if the V$_H$ and V$_L$ do not associate at all or associate very weakly, then their effectiveness in forming a combining site will be greatly reduced.

The source of the variability in the V$_H$:V$_L$ association comes from the involvement of hypervariable residues in the V$_H$:V$_L$ interface, as noted previously (41, 42, 148). Figure 3 shows the residues that make contact across the interface between V$_H$ and V$_L$ in McPC603 (Y. Satow, D.R. Davies, manuscript in preparation). There are many highly conserved residues that interact, particularly those in FRII, such as Glu39. However, at one end of the interface a variety of interactions can be seen to occur between hypervariable and framework residues and between pairs of hypervariable residues. These must contribute to the forces governing the exact positioning of V$_L$ relative to V$_H$. That this is not always the same even in related pairs has been demonstrated by Marquart et al (106) for the proteins

Kol (IgG λ) and Newm (IgG1 λ) where small differences (~9°) occur in the position of V_H relative to V_L that are outside the limit of experimental error.

In vivo results from cell fusion experiments demonstrated the existence of most of the possible "artificial" hybrids, so that it was concluded that most, if not all, H and L chain can randomly recombine to produce new immunoglobulins that do not interfere with cell viability (114). This apparent contradiction raises the possibility that the preferential association is an in vitro artifact, but this would appear to be ruled out by the numerous very carefully controlled experiments. Alternatively, the difference in the affinities between autologous and heterologous pairs of chains may be so small as to be unobservable under cellular conditions.

When a hybrid H/L recombinant is formed, then there is the additional question of how its binding properties will relate to those of the parent molecules. This would appear to depend on how similar the parent chains are, and Sher et al (157) observed that with reconsituted H/L hybrids from

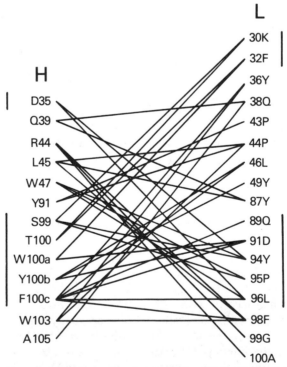

Figure 3   The residues that form the interface between V_H and V_L in McPC603. The vertical bars indicate the complementarity-determining residues. The connecting lines join those residues that have atoms within a distance of 4 Å of one another.

three anti-PC myeloma proteins, those with common idiotypes retained specificity upon recombination, whereas those without common idiotypes did not. Similarily, Manjula et al (103) showed that all the recombinants of anti-galactan myeloma proteins they examined showed anti-galactan activity comparable to the parents. However, as reviewed by Rudikoff (145), these anti-galactan heavy and light chains are so similar to one another that it is not surprising they exhibit the same binding properties. Recently, Kranz & Voss (97) examined the fluorescein binding of all the possible recombinants from six monoclonal antibodies to fluorescein and found that no hybrid has anti-fluorescein activity; the sequence and idiotype of these antibodies was not analyzed.

The Fc

The human IgG Fc has been crystallized and the structure has been determined at 2.9-Å resolution, Figure 4 (43, 44). The structure differs from that of a Fab in that there is no protein/protein contact between the two C_H2 domains. Instead, the complex carbohydrate attached to Asn297 occupies the interface region between the two domains with weak interactions between the two carbohydrate chains. The two C_H3 domains associate in a manner similar to the $C_H1:C_L$ domains of the Fab. The two C_H2 domains would seem to be positioned by the contacts made between C_H2 and C_H3 of the same domain and there is evidence that some flexibility can occur in this region. In the protein Dob, the Fc has a very similar structure to the isolated Fc (161), despite the absence of the hinge disulfide bridge due to a deletion of 15 residues in the heavy chain at this point.

The structure of Fc with a fragment of protein A from *Staphylococcus aureus* attached to it has been determined (43). Two contacts are made between the fragment B of protein A and two neighboring, symmetry-related Fc molecules. One of these, involving primarily hydrophobic interactions and believed to be the interaction occurring in solution, involves residues 251–254, 310–315 of C_H2, and 430–437 of C_H3. Since binding of protein A does not inhibit complement activation (192) and since the binding site of C1q has been reported to be on C_H2 (195), the C1q-Fc interaction site must be located elsewhere on C_H2 (see below and Figure 4).

The Hinge

The hinge region of the heavy chain is encoded for by a separate minigene (147) and varies in length (91) from five residues in IgG to 62 residues in human IgG3, where there appears to be a clear example of gene quadruplication. Marquart et al (106) in the intact Kol crystal observe the structure of the hinge to be double-stranded, disulfide-linked parallel chains, with each chain adopting a three-fold polyproline-like helical structure.

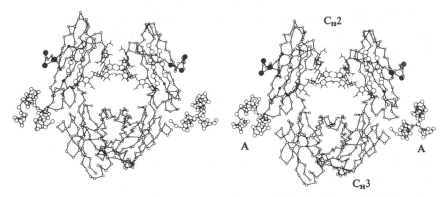

Figure 4 Stereo drawing of the α-carbon backbone of the Fc of human IgG bound to fragment B of protein A from *Staphylococcus aureus.* Based on coordinates obtained from the Brookhaven Protein Data Bank (43). The complex carbohydrate occupies the space between the two C_H2 domains. The protein A is shown with clusters of open circles in the region between C_H2 and C_H3. ● , • , and ○ on the Fc are the α-carbons of those residues, respectively, that have been proposed to bind to C1q in 138, 16, and 24.

Klein et al (91) have investigated the effect of major deletions in the hinge on the expression of biological effector functions. They found that proteins Dob and Lec, both of which have 15 amino acid deletions, leaving them with effectively no hinge peptide, could not express several effector functions. In particular, they showed no ability to bind C1q or to activate complement. However, neither displayed any difference with normal IgG1 in binding to protein A from *Staphylococcus aureus.* The three-dimensional structure for the Dob protein is approximately T shaped (161). In this protein the overall structure of the Fc region is very similar to that observed by Deisenhofer (43) in the crystals of the isolated Fc. Thus, there is no evidence that the lack of ability to perform effector functions is due to a major quaternary change (although minor changes would not be detected). The protein A binding indicates that the junction between Cγ2 and Cγ3 is not significantly changed. However, in the Dob protein the C_L and Cγ2 domains are quite close to each other, and in the conformation observed in the crystal, they would block binding of a large molecule such as the head of C1q to the end of Cγ2 proximal to the Fab. The presence of a somewhat rigid two-stranded hinge, with the strands linked by disulfide bridges and with each strand adopting a polyproline-like conformation (95, 106), would produce a greater separation of the Fab and Fc. Indeed, in IgG3, if the total hinge region adopted an extended rigid polyproline conformation (106), it would be more than 150 Å long.

Membrane Immunoglobulins on B-Lymphocytes

The unique characteristic of B-lymphocytes is the presence of membrane-bound immunoglobulin on their surface. The special nature of this immuno-

Table 1 Extra sequences of membrane immunoglobulin[a]

Segment	Length	Net charge	Homology	Presumed disposition
I	12 (μ) 18 (γ)	−6 to −7	0.08[b] (0.78)[c]	extracellular
II	26	0	0.65 (0.96)	intramembranous
III	3 (μ) 28 (γ)	+2 to +3	1.0 (0.71)	intracellular

[a] Data from IgG1, 2a, 2b, and IgM (142, 182, 194).
[b] Fraction of identical residues at positions shared by μ and γ.
[c] Fraction of identical residues at positions shared by γ.

globulin has recently been elucidated by sequencing the cloned DNA from Balb/c mice. In each case studied so far, the molecules differ from secreted immunoglobulins having a membrane-specific portion coded for by two exons, M1 and M2. M1 is separated from the exon coding for the carboxy terminus of the secreted immunoglobulin by 1.4–1.8 kbases and M2 in turn is separated from M1 by an intron from ~0.1–0.8 kbases. The amino acid sequences show features common to other membrane proteins that have a transmembrane portion: Proceeding in the direction of C-terminus, several negatively charged residues are followed by a hydrophobic stretch that is followed by several positively charged residues (Table 1). It is likely that these represent regions disposed on the outer surface of the membrane, within the bilayer and on the cytoplasmic surface of the membrane, respectively. Tyler et al (182) have suggested that the presence of as many as six negatively charged residues in the stretch of 18 amino acids in the external segment of mIgG may preclude much secondary structure and thus lead to membrane hinge. It contains a cysteine in the middle of the sequence, which may form an interchain disulfide bond (182). In mIgM this region is five residues shorter and contains no cysteine.

A notable feature is the conservation of the hydrophobic sequence of 26 amino acids in the transmembrane portion. The three γ chains show only a single amino acid difference (albeit two base changes) in this segment and nine differences when compared to the membrane μ chains. This is much less divergence than is observed for the other constant region domains of these heavy chains. The conservation is also striking in view of the variability seen both in the hydrophobic leader sequences of various proteins as well as for the intramembranous segments of other (even related) integral membrane proteins (182, 194). That is, the mere requirement that the 26 residue segments are intramembranous would not necessarily lead to the prediction of the marked sequence homology observed. If the 26 hydrophobic segment is an α-helix with 1.5 Å per residue, its length would be ~4 nm, adequate to span the membrane bilayer. Moreover, this would cluster those residues that have diverged on one "side," leaving the other aspect virtually invariant for the four isotypes for which information is available (23, 75). This raises the possibility that these mIg are interacting with another membrane protein (23, 75). It would be interesting to test this possibility by the use of cross-linking reagents or by isolating the mIg under conditions that appear to stabilize interactions between the subunits of other membrane receptors (141).

The hydrophilic segment of mIg presumed to be on the cytoplasmic aspect of the membrane is variable: For mIgM it appears to consist of only three residues (142), whereas for $\gamma 1$, $\gamma 2a$, and $\gamma 2b$ it is 28 residues. Among the latter, the variability increases towards the C-terminus; seven of the final 14 residues show exchanges.

FUNCTIONAL ASPECTS OF IMMUNOGLOBULINS

General Comments

CONSEQUENCES OF ANTIGEN BINDING TO ANTIBODY   The combining sites of most functional proteins such as enzymes reflect a pervasive evolutionary trend. Once a structure capable of effectively cleaving the substrate has evolved, subsequent selective pressures promote its stability. The residues that contact the ligand are most preserved, whereas those that provide simply the scaffolding are less constrained from varying (reviewed in 189). With immunoglobulins the situation is virtually the opposite. It is those residues capable of contacting the antigen that show the greatest diversity, whereas the supporting "framework" residues tend to be more invariant (68, 193).

What is gained with such a system is a diversity of combining sites so great that at least some members of the repertoire are capable of binding to any given cluster of atoms, including epitopes that neither the individual nor even its species are likely to have encountered previously. What is lost is the capacity to catalyze a covalent change in the ligand in the way that the sites on enzymes, finely honed by evolution, can. Instead, the binding of antigen is made productive by other means. In some instances, this occurs simply by the antibody preventing the antigen from adhering to and penetrating vulnerable tissues (107).

In other instances, it is the ability of the antibody complexed with antigen to activate an effector system that is critical. For the most part, the effector systems recruited are cells. The number of different such systems corresponds much more closely to the number of distinctive Fc regions with which they interact than to the number of combining site specificities, but there is no simple one-to-one correspondence of heavy chain isotype to the effector systems it can stimulate. Several different Ig isotypes may activate the same system and any single isotype can initiate several diverse effector reactions.

The effector systems are usefully divided into clonal and nonclonal classes. In clonal systems, the immunoglobulins on a single cell that play a critical role in its activation all bear the same antigen combining site. Within this class are the central elements of the immune response; the antigen receptors (membrane immunoglobulin) on B lymphocytes are a suitable example. In nonclonal systems, an activation event can be mediated by a mixture of immunoglobulins bearing different combining sites. These nonclonal systems constitute the principal mechanisms by which the organism actively rids itself of the offending antigen.

MECHANISM OF ANTIBODY ACTION   The biochemical perturbations activated by the antigen-antibody complexes are many. Nevertheless, the preponderance of evidence suggests some common features. Just as the evolutionary mechanism that generates the diversity of combining sites precludes these sites from having enzymic activity, it seems equally unlikely that in each case the antigen would be capable of acting like a heterotropic allosteric modifier that would change the conformation of the Fc region reproducibly. Similarly, if only those antigens would be effective whose epitopes happened to be just so spaced that a reproducible distortion of the semi-rigid hinge region of the immunoglobulin would ensue, this would substantially limit the number of productive antigen-antibody encounters. These theoretical arguments are supported by considerable experimental data. The latter provide strong evidence against either an allosteric or distortive mechanism of antibody action. The data instead suggest that clustering of two or more Fc regions is the initiating event in most, if not all, of the effector systems activated by antigen-antibody complexes. The older data have been analyzed previously (109, 110, 190); some newer results are discussed in the review of specific systems that follows.

Clonal Systems

B-LYMPHOCYTES   The consequences of antigen binding to the immunoglobulins displayed on the surface of B-lymphocytes are varied. They range from making the cells unresponsive to subsequent challenge (118a) to promoting proliferation, and differentiation into cells that secrete antibodies of different isotypes (80a). Elucidating those factors that are decisive in regulating the outcome is one of the central problems in immunology. It has become apparent that interaction with exogenous molecules and cells, as well as the state of maturation of the B cell itself, critically affect the pathway chosen. Despite this complexity, it is clear that perturbation of the surface immunoglobulin can provide an essential signal without which nothing happens. The most persuasive studies are those in which the perturbation is induced by anti-immunoglobulin. The latter experimental tool permits examination of a population of B-lymphocytes that bears immunoglobulins of diverse specificity (115). With few exceptions (155, 156), those studies show that the critical common feature is the clustering of the surface immunoglobulin by bivalent antibodies or antibody fragments [F(ab'$_2$)]. Such protocols may be somewhat artificial [though see Rajewsky (139a)], but investigations with hapten conjugates have led to similar conclusions (49, 61). The latter more physiological experiments have other complexities associated with them, however.

At present, it is unclear how the size or number of immunoglobulin clusters on the cell affect the eventual outcome. It is also not clear whether there are other events associated with the clustering that are common to all the pathways the stimulated B-lymphocyte can eventually follow. The answers will not come quickly. Current available assessments of the outcome involve measurements temporally distant from the stimulus—often several days. What is sorely needed is to identify early events that presage the ultimate result. Acute changes, apparently resulting from the clustering of the surface immunoglobulin in and of itself, have been reported, e.g. mobilization of bound Ca^{2+} (18), association of the clustered immunoglobulin with the cytoskeletal matrix (17), and changes in the surface distribution of the mIg (177). These cannot yet be correlated with the final fate of the stimulated cells. Other perturbations reported [methylation of phospholipids, activation of a serine esterase, and generation of cytoplasmic factors (94)] appear to require additional exogenous factors (B-cell growth factor and interleukin 2) that are known to affect the cell's future activity. The quantitative role of the clustering of the membrane immunoglobulin has not yet been analyzed in those cases.

Although the immune system as a whole is unlikely to play a significant role in general homeostasis of an organism's internal environment, it obviously does have a variety of regulatory elements within the context of immune responses themselves. The mechanisms by which hormone receptors are controlled (down regulation, desensitization, amplification) are therefore likely to have their counterparts in the control of events mediated by the immunoglobulins on B-lymphocytes also, but appropriate systems for study of the latter are only now emerging.

Two other aspects of the activation of B-lymphocytes mediated by mIg are of interest. Since the mIg are mobile in the plane of the membrane [albeit not as mobile as might be expected from theoretical considerations (56)], it is reasonable to ask why spontaneous clustering of the mIg does not activate the cell if the association is a critical event. One answer is that whereas it may be a critical event, it is likely to be inadequate in the absence of accessory factors, as already discussed. Nevertheless, in systems where such accessory factors are not required, a similar question arises (below). A second question of interest is the following apparent paradox (86). On the one hand, B cells bearing Ig with higher affinities for the antigen are activated preferentially when antigen is limiting (affinity selection) (162); on the other hand, multipoint attachment of antigen to mIg-bearing lymphocytes can be expected on theoretical grounds, to make the amount of antigen bound be quite insensitive to the intrinsic binding affinity of the combining sites on the mIg (86).

The two questions may in fact be related. What is likely to be the significant factor is the size and persistence of the clusters of mIg, whether they arise spontaneously or result from the interaction of antigen with mIg of low or high affinity (111). This subject is discussed further in the context of a simpler system (see below).

T-LYMPHOCYTES   Thymus-processed lymphocytes are capable of mediating a variety of antigen-specific effects apparently without the use of exogenous immunoglobulins. The uncertain nature of the T-cell antigen-specific receptor that must be involved continues to be a major challenge with which immunologists are wrestling (85). Although these receptors may share common features with secreted or membrane-bound Ig, not enough is known to make them an appropriate subject for discussion in a review of antibody structure and function.

Nonclonal Systems: Cells Bearing Fc Receptors

Many cells interact with the Fc region of immunoglobulins (183). In those cases where we have the most information, the interaction is mediated by a plasma membrane glycoprotein usually of molecular weight 50,000–70,000. The binding may be more or less firm to the monomeric Ig and more or less specific for certain heavy chain isotypes and the species of origin. The nature of these "Fc receptors" (130) to which the Ig binds is outside the subject area of this survey; reviews on one or more of them have been published recently (70, 112, 183). Here we focus on two aspects only: (a) the sites on the Ig that interact with the Fc receptors, and (b) the role of the immunoglobulin in mediating the functions in which the receptors participate.

SITES ON Ig TO WHICH THE Fc RECEPTORS BIND

IgG   The literature about the sites in the Fc region that interact with cellular receptors is confusing (reviewed in 2). In part, this arises because various investigators have studied immunoglobulins and cells from different animals. The relationship between isotypes from different species is by no means a simple one nor can one assume that the receptors on alveolar macrophages, peritoneal macrophages, and peripheral blood monocytes will be similar even within a species, let alone from different species. The assays employed may also be important. Those in which multipoint attachment can amplify weak intrinsic association constants can be expected to yield results different from those in which direct (monomeric) binding is examined. Because certain types of fragments of Fc are more readily prepared from the Ig of some species than of others, numerous reports deal with the binding of Ig fragments to cells of a heterologous species. Data derived from such studies may be useful for providing specificity criteria for distinguishing different receptors, but it may confuse more than clarify physiologically significant interactions. We, therefore, focus principally on those studies involving reagents from the same or closely related species. The use of fragments, peptides, or mutant proteins even from homologous species may present additional problems of interpretation because of sequence homologies in the domains of immunoglobulins and related proteins. To cite just two examples, Ciccimarra et al (32) observed binding activity in a decapeptide from $C\gamma3$ that, however, is likely to be buried in the native structure (51). Furthermore, $\beta2$-microglobulin, the light chain of the class I major histocompatibility antigens, was found by Painter et al (128) to bind to macrophage Fc receptors.

In some instances binding of isolated $C\gamma3$ modules to cell receptors can be observed. Human pFc' and $C\gamma3$ fragments interact weakly with human monocytes, although not to human granulocytes (65, 66). A weak interaction has also been seen in an analogous combination with reagents prepared from rabbits (73) and mice (50) but not from guinea pigs (3).

Isolated $C\gamma2$ domains are inactive but many studies demonstrate the involvement of these regions. Cleavage of the interheavy chain disulfides in the hinge region, which increases the segmental flexibility of $C\gamma2$, reduces the binding activity of human IgG1 to human granulocytes (66, 73). Masking the interdomain cleft between $C\gamma2$ and $C\gamma3$ with a fragment of protein A from *Staphylococcus aureus* also interferes with the binding of human IgG1 to human granulocytes, but not to human monocytes (67).

Using variants of the mouse MPC 11 myeloma, Diamond et al (48) showed that the receptor for IgG2b on mouse macrophages recognized a segment in the $C\gamma2$ domain and the receptor for IgG2a, sequences from both $C\gamma2$ and $C\gamma3$ domains. The complexity of these interactions is strikingly illustrated in a report on IgG produced by another variant of this myeloma. It fails to bind to the mouse macrophage receptors for IgG2a even though the variant has completely intact $C\gamma2$ and $C\gamma3$ regions of the 2a isotype (14). This mutant has hybrid heavy chains with the first 220 residues of the 2b isotype and the remaining residues (221 to the carboxy terminus) of the 2a class. Cleavage by papain of the hinge region of this protein occurs one residue away from the normal site of cleavage, after residue 240 versus 239 for both native IgG2a and IgG2b. This shows that this region is conformationally different from that of the normal proteins. The fragment derived from such cleavage of the variant protein binds to the receptors perfectly well. These unusual results indicate that although the principal contacts between the receptors and IgG may be in the $C\gamma2$ and $C\gamma3$ domains, the hinge regions may indirectly affect binding by affecting the relationship of the Fab regions to the $C\gamma2$ domains. Although the human myeloma proteins that have a deleted hinge have so far been tested only on murine macrophages, the results of those studies suggest an analogous phenomenon (95). A study of Facb fragments $((Fab-C\gamma2)_2)$ in a homologous system has been reported by Johanson et al (87). They found that rabbit Facb bound very well to the Fc receptors on membranes of the yolk-sac. Notably, the reduced molecules, which dissociate into monomers of Fab-$C\gamma2$, bound equally well, whereas neither the Fab nor the pFc' $(C\gamma3)$ fragments were active. Activity of Facb fragments albeit with Ig and macrophages from heterologous species has also been observed in certain instances, but not in others (cf. 8, 119).

IgE Several types of cells have one or more different surface receptors capable of binding IgE (26, 165). However, one of the receptors on mast cells and basophils is unique: Its affinity for monomeric IgE is at least 100-fold greater than that observed for any other Fc receptor for any other immunoglobulin. In the homologous human system, the Fc fragments $((C_\epsilon2-C_\epsilon3-C_\epsilon4)_2)$ bind well, but fragments comparable to Facb, i.e. (Fab-$C_\epsilon2)_2$, or pFc', i.e. $(C_\epsilon4)_2$, do not (53). A report that a pentapeptide from the $C_\epsilon2$ domain inhibited binding of the intact IgE (79) could not be corroborated in a detailed study (11). That in the rodent system such a peptide could displace bound IgE (185) is hard to reconcile with data showing that even intact IgE cannot displace bound molecules (187). By exclusion, it has been postulated that $C_\epsilon3$ plays a major role in the interaction (53). This was supported by indirect data concerning the location of irreversible physicochemical changes induced by conditions that lead to irreversible changes in binding. Recently, the rates of tryptic cleavage of unbound IgE and IgE bound to solubilized receptors were compared in the rodent system (132). Whereas cleavage at all sites was inhibited, the most striking (>40-fold) inhibition was observed at a site likely to be close to the $C_\epsilon3$: $C_\epsilon2$ interdomain region. These results also point to the importance of the $C_\epsilon3$ domains. It is interesting that by several criteria the latter domains may be homologous to the $C\gamma2$ domains of IgG (53). There is evidence against the carbohydrate on the IgE contributing significantly to the binding affinity (98).

Overall, the results with IgG and IgE suggest that in most instances the penultimate carboxyterminal C_H domains play a significant, perhaps even the most important, role in the interaction between immunoglobulins and the Fc receptors on cells.

MECHANISM BY WHICH CELLS BEARING Fc RECEPTORS ARE ACTIVATED   The Fc receptors on cells are implicated in a wide variety of immune mechanisms. Some of these can be meaningfully grouped together as involving immunoregulatory functions. Despite the apparent importance of this group, we do not consider it since the mechanism by which Ig mediates these functions is still largely unknown. A useful series of articles on the role of such Fc receptors in regulating the function of T cells has recently been compiled (116).

A second functional group that can be distinguished includes those receptors by which the cell directly or indirectly disposes of antigen. We discuss two of such functions for which the role of the Ig and its mechanism of action have been actively investigated.

Endocytosis   It is likely that engulfment by macrophages of antigens that have successfully penetrated the organisms external barriers is quantitatively the most important mechanism for the disposal of antigens. This engulfment is markedly enhanced if the antigen has antibody bound to it (reviewed in 160). It has been known virtually from the time Fc receptors were discovered that antigen-antibody complexes are readily distinguished by the cell from uncomplexed IgG. This can be observed even when the concentration of the latter exceeds that of the former by four or more orders of magnitude (150). The evidence is overwhelming that the clustering of Fc regions engendered by the antigen-antibody interaction is necessary and sufficient to explain the selectivity. This has been rigorously analyzed theoretically and experimentally by Segal and his colleagues (150, 151, 154). These workers provide evidence that the Fc receptors cycle between the plasma membrane and the interior. They propose that the multipoint attachment that aggregation of Fc promotes can account for the preferential internalization of the complexed immunoglobulin (151, 154). That is, given a moving chain of cycling Fc receptors, the complexes of Ig simply hop aboard more readily. The important corollary is that the complexes need not generate any signal.

Secretion: macrophages   Though endocytosis of immune complexes by macrophages can apparently be accounted for by a simple mechanism, this process seems inadequate to explain the stimulation of synthesis and secretion of various mediators by macrophages when challenged with antigen-antibody complexes (15, 27, 134, 139, 143, 176). In these instances, the cell clearly has been activated to do something new. That the activation occurs on the surface of the cell is indicated by the results of Ragsdale & Arend (139) and Rouzer et al (143). The latter stimulated prostaglandin synthesis and secretion by beads coated with antibodies in the presence of cytochalasin D, which prevents phagocytosis. That immune complexes of guinea pig IgG2 induce superoxide anion production by homologous macrophages threefold more efficiently than IgG1 despite their being equivalently phagocytosed (176) could be similarly interpreted. In this and other studies (134), the greater efficacy of larger complexes was noted. Macrophages also have Fc receptors more or less specific for homologous IgE. Artificially dimerized IgE was sufficient to trigger release of lysosomal enzymes from the cells (47).

Secretion: mast cells   Mast cells have receptors that bind IgE with extraordinarily high affinity. Moreover, triggering such cells via the IgE bound to their surface requires nothing more than something that will react appropriately with the IgE and (in most studies) a 1 mM pinch of Ca^{2+}. These factors have permitted a more complete investigation of this system than has been possible with any other cellular system activated by antibodies (113).

The current consensus is that IgE-mediated triggering of mast cells occurs when antigen, reacting with the cell-bound IgE, clusters the previously unassociated receptors for IgE. Three principal observations form the basis for this model: (*a*) the receptors in situ are univalent (108, 149); (*b*) the receptors are mobile (9, 28, 100, 149, 171); and (*c*) multivalent antibodies to the receptor mimic completely the action of the antigen (83, 84) in the total absence of IgE. The first observation indicates that the process is intermolecular rather than intramolecular; the second shows that the molecules that mediate the reaction need not be closely adjacent prior to triggering; and the third demonstrates that the important molecules are likely to be the receptors and not the IgE.

The mechanism by which clustering of these Fc receptors mediates degranulation is reviewed elsewhere (113). Here we indicate a few salient points only. In this system also, evidence shows that larger and more numerous clusters provoke more complete or rapid activation (64, 92) but that even dimers can provide a unit signal (64, 92, 153). In this system also (see above), the question arises of why, if clustering is all that is required, is spontaneous release not observed?[1] There is as yet no answer to this

---

[1]It could be argued that since these cells have a desensitization mechanism that is also triggered by clustering (92), low levels of clustering are insufficient to cause release because of an unfavorable balance between the activation and deactivation. If this were true, one would then anticipate spontaneous desensitization. Since this has not been observed, the dilemma remains the same.

question based upon experimental data. It is possible that effective clustering cannot occur because of an energy barrier that is only overcome by the free energy provided by the binding of antigen to IgE. It is also possible that the lifetime of spontaneous aggregates is too short to induce a signal; even a ligand that binds relatively weakly to an individual IgE could dramatically increase the time the receptors remain clustered, thereby permitting the necessary molecular changes to take place (45). The interaction of mobile ligands (e.g. in a planar layer of lipids) at a limited interface with the cell-bound IgE is sufficient to cluster receptors (10, 188) and cause mediator release in this system (188), which demonstrates that rather indirect clustering is all that is required.

Other   Although its functional role is undefined, isolation from human leukemic cells of an Fc receptor has been reported by Suzuki and his colleagues (173). On sodium dodecyl sulfate gels, this has a size similar to that reported by Thoenes & Stein (178), but in many of its characteristics it appears to be very different from the Fc receptors isolated by others (183). It is the first receptor for which an intrinsic activity has been claimed (phospholipase A2 activity), an activity stimulated by interaction with IgG in solution, particularly with aggregates of IgG (172, 175). The mechanism by which IgG stimulates this activity remains to be clarified. The same group has reported recently that similar results can be obtained for the receptor specific for IgG2b on the $P388D_1$ line of mouse macrophages (174).

Nonclonal Systems: Complement

CLASSICAL PATHWAY   The activation of the classical pathway of the complement cascade provides a unique opportunity to study an effector system activated by antigen-antibody complexes. The receptor of this system, C1, can be readily isolated in amounts adequate to perform rigorous chemical analyses (140), electron micrographs are yielding detailed pictures of its complex structure (168), and sensitive molecular assays are available to assess its activation rapidly in solution (29). Nevertheless, it has been extraordinarily difficult to define precisely what is going on—a sobering spectacle to those interested in even the simplest cellular system!

Sites on Ig with which C1 interacts   Defining the site of interaction of the globular heads of the C1q subunit of C1 with Ig is complicated by the observation that there may be several peptides derived from or related to immunoglobulins, which in their isolated form can interact with C1q (127, 128). Therefore, studies with fragments generated by procedures that involve even transient exposure to denaturants (e.g. those used to generate an

Facb fragment) must be interpreted cautiously. Nevertheless, the weight of evidence favors interaction of native IgG via the C_H2 domains, although only one such domain in intact IgG appears to be bound by Clq at a time (127). One of the most convincing pieces of evidence for the importance of this region is that mutant IgG's, such as the human proteins Dob and Lec, fail to bind and activate C1 (95). These proteins are missing the hinge region but seem otherwise intact (see above). It appears likely that changes in the disposition of the Fab regions with respect to the $C\gamma2$ domains is responsible (95)—an explanation virtually the same as that used to rationalize the failure of certain mouse IgG mutants to bind to Fc receptors (above), as well as for a variety of observations on the effect of reduction of the hinge disulfides in intact and fragmented IgG (reviewed in 52).

Burton et al investigated the site on rabbit IgG that interacts with Clq by using specific inhibitors, chemical modification, and an analysis of sequence conservation and of accessibility (24). Based on these results, as well as the work of others, they proposed that the site principally involves charged residues in the two carboxyterminal β-strands of the $C\gamma2$ domains (residues 318–322; 332–337). A group in Madrid (16) correlated controlled conjugation of carboxylate groups on human IgG with inactivation and implicated primarily glutamic acid residues 258 and 293. The latter is in a region proposed by Brunhouse and Cebra from analysis of sequence data (20). Work with synthetic peptides led to the proposal that His 285, Lys 288, and Lys 290 were important in human IgG (138). On the other hand, deglycosylation of rabbit IgG also has been reported to decrease Clq binding (191). Results with concanavalin A also suggested that exposure of the carbohydrate might be important in one instance (99) but not in another (184). None of the approaches used distinguishes between direct and indirect effects. The results are depicted in figure 4, which shows the α carbons of the implicated residues. Notably, all these residues are well removed from the binding site for protein A of *Staphylococcus aureus*. Although the residues proposed by the three groups differ, they could all be correct. Firstly, the terminal atoms of the side chains may be topologically considerably closer than the positions of the α carbons of the same residues. Secondly, the site of interaction with Clq could be extensive, involving many amino acid side chains (cf Figure 3). Only procedures giving direct data will yield more precise information: either crystallographic visualization, or the use of cross-linking reagents and isolation of peptides that contain residues from IgG and Clq cross-linked to each other.

Because of some unusual aspects of the activation of C1 by IgM (below), the site(s) on the latter that interact with Clq are of interest. The $C\mu4$ domain is stated to fix C1 efficiently (21, 71), which supports earlier results with fragments of this domain (82). Nevertheless, as noted above with respect to IgG, these results could be quite misleading. Although the findings of Siegel & Cathou (159) do not rule out the possibility that the $C\mu4$ domains are active, these workers interpreted their results on the effect of heating IgM as implicating the $C\mu2$ domains.

Mechanism of activation C1 by Ig   Whereas some data in support of the need for aggregation of IgG (36) have been reinterpreted (166), sufficient other results indicate that Fc clustering is critically important for efficient activation (109, 110, 190). New observations on the lack of effect of antigen alone on the binding IgG to C1q (101, 181) and C1 activation (180) provide further evidence against allosteric effects being critical. That univalent rabbit Fab/c fragments (Fab-($C\gamma2$, $C\gamma3)_2$) efficiently induced complement-mediated lysis of the lymphocytes to which they were bound (77) is a strong argument against a distortive mechanism.

Tschopp has observed that the relative efficacy of artificial oligomers to activate C1 parallels closely their relative affinity for C1q (179). A preparation of monomers of IgG fit the same pattern, but Tschopp could not determine whether or not this was due to contamination with traces of aggregates. It remains possible then that monomeric IgG could induce a signal either solely by its interaction with the globular portion of C1q or perhaps by reacting with the nearby (168) C1r. The role of aggregation could then simply be that of enhancing the binding by multipoint attachment (such as has been proposed for endocytosis by macrophages; see above). It would be useful to know whether or not with monomeric IgG more rigorously freed of aggregates activation was nevertheless observed.

Most studies with IgM suggest that only antigens that engage more than one subunit are capable of stimulating its ability to activate C1 (93, 137, 158), but the work of Koshland's group suggests otherwise (19, 30, 31). Although haptens were ineffective, they found that larger univalent antigens initiated IgM-mediated activation of C1. The mechanism is unclear by which these antigens, which are substantially bigger than what can be accommodated in a combining site, produce this effect. The authors propose several alternatives (31).

A new role for IgG in its interaction with the classical pathway has been described (25). It appears that C4 forms a covalent bond with the $C\gamma1$ domain during the activation of the cascade. Whether or not the IgG plays more than a passive role in this effect is not known.

ALTERNATIVE PATHWAY   Activation of the alternative pathway of complement fixation remains as the single exception to the rule that antibodies stimulate effector mechanisms via their Fc regions. It has been repeatedly demonstrated that $F(ab')_2$ fragments are active—as active as intact IgG—but since this pathway can be initiated in the total absence of immunoglobulins (117), interpretations of the role of the antibody become more uncertain. Several studies suggest that the monovalent Fab' or Fab are inactive or much less active (59, 72, 133, 163), but a recent study reports otherwise (1). Different assays were used and the experts in this field will have to sort this matter out. One aspect about which there does seem to be agreement is that activation requires multiple molecules of antibody attached to the antigen.

ACKNOWLEDGMENT

We wish to thank Enid Silverton, who prepared Figures 1, 2, and 4.

Literature Cited

1. Albar, J. P., Juarez, C., Vivanco-Martinez, F., Brogado, R. B., Ortiz, F. 1981. *Mol. Immunol.* 18:925–35
2. Alexander, M. D., Andrews, J. A., Leslie, R. G. Q., Wood, N. J. 1978. *Immunology* 35:1115–23
3. Alexander, M. D., Leslie, R. G. Q., Cohen, S. 1976. *Eur. J. Immunol.* 6:101–7
4. Amzel, L. M., Poljak, R. J. 1979. *Ann. Rev. Biochem.* 48:961–97
5. Amzel, L. M., Poljak, R. J., Saul, F., Varga, J. M., Richards, F. F. 1974. *Proc. Natl. Acad. Sci. USA* 71:1427–30
6. Aschaffenburg, R., Phillips, D. C., Rose, D. R., Sutton, B. J., Dower, S. K., Dwek, R. A. 1979. *Biochem. J.* 181:497
7. Azuma, T., Isobe, T., Hamaguchi, K. 1975. *J. Biochem.* 77:473
8. Barnett-Foster, D. E., Painter, R. H. 1982. *Mol. Immunol.* 19:247–52
9. Becker, K. E., Ishizaka, T., Metzger, H., Ishizaka, K., Grimley, P. 1973. *J. Exp. Med.* 108:394–409
10. Bell, G. I. 1979. In *Physical Chemical Aspects of Cell Surface Events in Cellular Regulation,* ed. C. DeLisi, R. Blumenthal, pp. 371–92. Heidelberg: Springer
11. Bennich, H., Ragnarsson, U., Johansson, S. G. O., Ishizaka, K., Ishizaka, T., Levy, D. A., Lichtenstein, L. M. 1977. *Int. Arch. Allergy Appl. Immunol.* 53:459–68
12. Berzofsky, J. A., Buckenmeyer, G. K., Hicks, G., Gurd, R. R. N., Feldmann, R. J., Minna, J. 1982. *J. Biol. Chem.* 257:3189–98
13. Bigelow, C. C., Smith, B. R., Dorrington, K. J. 1974. *Biochemistry* 13:4602
14. Birshtein, B. K., Campbell, R., Diamond, B. 1982. *J. Immunol.* 129:610–14
15. Blyden, G., Handschumacher, R. E. 1977. *J. Immunol.* 118:1631–38
16. Boragado, R., Lopez de Castro, J. A., Juarez, C., Albar, J. P., Garcia Pards, A., Ortiz, F., Vivanco-Martinez, F. 1982. *Mol. Immunol.* 19:579–88
17. Braun, J., Hochman, P. S., Unanue, E. R. 1982. *J. Immunol.* 128:1198–204
18. Braun, J., Sha'afi, R. I., Unanue, E. R. 1979. *J. Cell Biol.* 82:755–66
19. Brown, J. C., Koshland, M. E. 1970. *Proc. Natl. Acad. Sci. USA* 74:5682–86
20. Brunhouse, R., Cebra, J. J. 1979. *Mol. Immunol.* 16:907–17
21. Bubb, M. O., Conradie, J. D. 1976. *Immunology* 31:893–902
22. Bunting, P. S., Kells, D. I. C., Kortan, C., Dorrington, K. J. 1977. *Immunochemistry* 14:45
23. Burnstein, Y., Schecter, I. 1978. *Biochemistry* 17:2392–400
24. Burton, D. R., Boyd, J., Braumpton, A. D., Easterbrook-Smith, S. W., Emanuel, E. J., Novotny, J., Rademacher, T. W., Van Schravendijk, M. R., Sternberg, M. J. E., Dwek, R. A. 1980. *Nature* 288:388–444
25. Campbell, R. D., Dodds, A. W., Porter, R. R. 1980. *Biochem. J.* 189:67–80
26. Capron, M., Capron, A., Dessaint, J.-P., Torpier, G., Johansson, S. G. O., Prin, L. 1981. *J. Immunol.* 126: 2087–92
27. Cardella, C. J., Davies, P., Allison, A. C. 1974. *Nature* 242:46–48
28. Carson, D. A., Metzger, H. 1974. *J. Immunol.* 113:1271–77
29. Cooper, N. R., Ziccardi, R. J. 1977. *J. Immunol.* 118:1664–67
30. Chiang, H.-C., Koshland, M. E. 1979. *J. Biol. Chem.* 254:2736–41

31. Chiang, H.-C., Koshland, M. E. 1979. *J. Biol. Chem.* 254:2742–47
32. Ciccimarra, F., Rosen, F. S., Merler, E. 1975. *Proc. Natl. Acad. Sci. USA* 72:2081–83
33. Cisar, J., Kabat, E. A., Dorner, M. M., Liao, J. 1975. *J. Exp. Med.* 142:435–59
34. Claflin, J. L. 1976. *Eur. J. Immunol.* 6:669–74
35. Claflin, J. L., Rudikoff, S. 1976. *Cold Spring Harbor Symp. Quant. Biol.* 41:725–34
36. Cohen, S. 1968. *J. Immunol.* 100: 407–13
37. Colman, P. M., Gough, K. H., Lilley, G. G., Blagrove, R. J., Webster, R. G., Laver, W. G. 1981. *J. Mol. Biol.* 152:609–14
38. Cook, W. D., Rudikoff, S., Giusti, A., Scharff, M. D. 1982. *Proc. Natl. Acad. Sci. USA* 79:1240–44
39. Crews, S., Griffin, J., Huang, H., Calame, K., Hood, L. 1981. *Cell* 25:59–66
40. Davies, D. R., Padlan, E. A. 1977. In *Antibodies in Human Diagnosis and Therapy,* ed. E. Haber, R. Krause, p. 119. New York: Raven
41. Davies, D. R., Padlan, E. A., Segal, D. M. 1975. *Ann. Rev. Biochem.* 44: 639–67
42. Davies, D. R., Padlan, E. A., Segal, D. M. 1975. In *Contemporary Topics in Molecular Immunology,* ed. F. P. Inman, W. J. Mandy, 4:127–55. New York: Plenum
43. Deisenhofer, J. 1981. *Biochemistry* 20:2361–70
44. Deisenhofer, J., Colman, P. M., Epp. O., Huber, R. 1976. *Physiol. Chem.* 359:975
45. DeLisi, C. 1980. *Q. Rev. Biophys.* 13:201–30
46. de Preval, C., Fougereau, M. 1976. *J. Mol. Biol.* 102:657–78
47. Dessaint, J.-P., Waksman, B. H., Metzger, H., Capron, D. 1980. *Cell. Immunol.* 51:280–92
48. Diamond, B., Birshtein, B. K., Scharff, M. D. 1979. *J. Exp. Med.* 150:721–26
49. Dintzis, H. M., Dintzis, R. Z., Vogenstein, B. 1976. *Proc. Natl. Acad. Sci. USA* 73:3671–75
50. Dissanyake, S., Hay, F. C. 1975. *Immunology* 29:1111–18
51. Dorrington, K. J. 1976. *Immunol. Commun.* 5:263–80
52. Dorrington, K. J. 1978. *Can. J. Biochem.* 56:1087–101
53. Dorrington, K. J., Bennich, H. 1978. *Immunol. Rev.* 41:3–25

54. Dower, S. K., Gettins, P., Jackson, R., Dwek, R.A., Givol, D. 1978. *Biochem. J.* 169:179–88
55. Dower, S. K., Wain-Hobson, S., Gettins, P., Givol, D., Roland, W., et al. 1977. *Biochem. J.* 165:207–25
56. Dragsten, P., Henkart, P., Blumenthal, R., Weinstein, J., Schlessinger, J. 1979. *Proc. Natl. Acad. Sci. USA* 76:5163–67
57. East, I., Hurrell, J. G. R., Todd, P. E., Leach, S. J. 1982. *J. Biol. Chem.* 257:3199–202
58. Edmundson, A. B., Ely, K. R., Girling, R. L., Abola, E. E., Schiffer, M., et al. 1974. *Biochemisty* 13:3816-27
58a. Ely, K. R., Colman, P. M. Abola, E. E., Hess, A. C., Peabody, D. S., et al. 1978. *Biochemistry* 17:820-25
59. Ernst, A. 1978. *J. Immunol.* 121: 1206–12
60. Fehlhammer, H., Schiffer, M., Epp, O., Colman, P. M., Lattman, E. E., et al. 1975. *Biophys. Struct. Mech.* 1:139–46
61. Feldmann, M. 1972. *J. Exp. Med.* 135:735–53
62. Feldmann, R. J., Potter, M., Glaudemans, C. P. J. 1981. *Mol. Immunol.* 18:683–98
63. Fett, J. W., Deutsch, H. F., Smithies, O. 1973. *Immunochemistry* 10:115–18
64. Fewtrell, C., Metzger, H. 1980. *J. Immunol.* 125:701–10
65. Foster, D. E. B., Dorrington, K. J., Painter, R. H. 1978. *J. Immunol.* 120:1952–56
66. Foster, D. E. B., Dorrington, K. J., Painter, R. H. 1980. *J. Immunol.* 124:2186–90
67. Foster, D. E. B., Sjoquist, J., Painter, R. H. 1982. *Mol. Immunol.* 19:407–12
68. Franek, F. 1969. In *Developmental Aspects of Antibody Formation and Structure,* ed. J. Sterzl, I. Riha. New York: Academic
69. Friedman, F. K., Chang, M. Y., Beychok, S. 1978. *J. Biol. Chem.* 253:2368
70. Froese, A. 1983. *Prog. Allergy.* In press
71. Fust, G., Csecsi-Nagy, M., Medgyesi, G. A., Kulics, J., Gergely, J. 1976. *Immunochemistry* 13:793–800
72. Gadd, K. J., Reid, K. B. M. 1981. *Immunology* 42:75–82
73. Ganczankowski, M., Leslie, R. G. Q. 1979. *Immunology* 36:487–94
74. Gearhart, P. J., Johnson, N. D., Douglas, R., Hood, L. 1981. *Nature* 291: 29–34
75. Gething, M.-J., Bye, J., Skehel, J., Waterfield, M. 1980. *Nature* 287:301–6
76. Givol, D. 1979. *Int. Rev. Biochem.* 287:301–6

77. Glennie, M. J., Stevenson, G. T. 1982. *Nature* 295:712–14
78. Grey, H. M., Mannik, M. 1965. *J. Exp. Med.* 122:619
79. Hamburger, R. N. 1975. *Science* 189: 389–90
80. Horne, C., Klein, M., Polidoulis, I., Dorrington, K. J. 1982. *J. Immunol.* 129:660–64
80a. Howard, M., Paul, W. E. 1983. *Ann. Rev. Immunol.* 1:00–00
81. Huber, R., Deisenhofer, J., Colman, P. M., Matsushima, M., Palm, W. 1976. *Nature* 264:415–20
82. Hurst, M. M., Volanakis, J., Hester, R. B., Stroud, R. M., Bennett, J. C. 1975. *J. Exp. Med.* 140:1117–21
83. Isersky, C., Taurog, J. D., Poy, G., Metzger, H. 1978. *J. Immunol.* 121: 549–58
84. Ishizaka, T., Chang, T. H., Taggart, M., Ishizaka, K. 1977. *J. Immunol.* 119: 1589–96
85. Janeway, C. A. Jr., Cone, R. E., Rosenstein, R. W. 1982. *Immunol. Today* 3:83–86
86. Jarvis, M. R., Voss, E. W. Jr. 1982. *Mol. Immunol.* 19:1063–69
87. Johanson, R. A., Shaw, A. R., Schlamowitz, M. 1981. *J. Immunol.* 126:194–99
88. Kabat, E. A. 1966. *J. Immunol.* 97:1–11
89. Kabat, E. A. 1978. *Adv. Prot. Chem.* 32:1–75
90. Kabat, E. A. 1982. *Pharm. Rev.* 34:23–38
91. Kabat, E. A., Wu, T. T., Bilofsky, H. 1979. *NIH Publ. No. 80–2008*
92. Kagey-Sobotka, A., Dembo, M., Goldstein, B., Metzger, H., Lichtenstein, L. M. 1981. *J. Immunol.* 127:2285–91
93. Karush, F., Chua, M. M., Rockwell, J. D. 1979. *Biochemistry* 18:2226–31
94. Kishimoto, T., Yoshizaki, K., Okada, M., Miki, Y., Nakagawa, T., et al. 1982. *UCLA Symp. Nature New Biol.* 233: 225–29
95. Klein, M., Haeffner-Cavaillon, N., Isenman, D. E., Rivat, C., Navia, M. A., et al. 1981. *Proc. Natl. Acad. Sci. USA* 78:524–28
96. Klein, M., Kortan, C., Kells, D. I., Dorrington, K. J. 1979. *Biochemistry* 18:1473–81
97. Kranz, D. M., Voss, E. W. Jr. 1981. *Proc. Natl. Acad. Sci. USA* 78:5807–11
98. Kulczycki, A. Jr., Vallina, V. L. 1981. *Mol. Immunol.* 18:723–31
99. Langone, J. J., Boyle, M. D. P., Borsos, T. 1978. *J. Immunol.* 121:327–32
100. Lawson, D., Fewtrell, C., Gomperts, B.,

Raff, M. C. 1975. *J. Exp. Med.* 142: 391–402
101. Liberti, P. A., Bausch, D. M., Schoenberg, L. M. 1982. *Mol. Immunol.* 19:143–49
102. Mage, R. G., Kabat, E. A. 1963. *Biochemistry* 2:1278–88
103. Manjula, B. N., Glaudemans, C. P. J., Mushinski, E. B., Potter, M. 1976. *Proc. Natl. Acad. Sci. USA* 73:932–36
104. Mannik, M. 1967. *Biochemistry* 6:134
105. Marquaret, M., Deisenhofer, J. 1982. *Immunol. Today* 3:160–66
106. Marquart, M., Deisenhofer, J., Huber, R., Palm, W. 1980. *J. Mol. Biol.* 141:369
107. McNabb, P. C., Tomasi, T. B. 1981. *Ann. Rev. Microbiol.* 35:477–96
108. Mendoza, G. R., Metzger, H. 1976. *Nature* 264:548–50
109. Metzger, H. 1974. *Adv. Immunol.* 18:169–207
110. Metzger, H. 1978. *Contemp. Top. Mol. Immunol.* 7:191–224
111. Metzger, H. 1982. *Mol. Immunol.* 19:1071
112. Metzger, H. 1983. *Contemp. Top. Mol. Immunol.* In press
113. Metzger, H., Ishizaka, T. 1982. *Fed. Proc.* 41:7–34
114. Milstein, C., Adetugbo, K., Cowan, N. J., Kohler, G., Secher, D. S., Wilde, C. D. 1976. *Cold Spring Harbor Symp. Quant. Biol.* VI:793–803
115. Moller, G., ed. 1980. *Immunol. Rev.* 52:1–231
116. Moller, G., ed. 1981. *Immunol. Rev.* 56:1–218
117. Muller-Eberhard, H. J., Schrieber, R. D. 1980. *Adv. Immunol.* 29:1–53
118. Nisonoff, A. 1982. In *Introduction to Molecular Immunology.* Sunderland, MA: Sinauer Assoc.
118a. Nossal, G. J. V. 1983. *Ann. Rev. Immunol.* 1:33–62
119. Ovary, Z., Saluk, P. H., Quijada, L., Lamm, M. E. 1976. *J. Immunol.* 116:1265–71
120. Padlan, E. A., 1977. *Q. Rev. Biophys.* 10:35–65
121. Padlan, E. A. 1977. *Proc. Natl. Acad. Sci. USA* 74:2551–55
122. Padlan, E. A. 1979. *Mol. Immunol.* 16:287–96
123. Padlan, E. A., Davies, D. R. 1975. *Proc. Natl. Acad. Sci. USA* 72:819–23
124. Padlan, E. A., Davies, D. R., Pecht, I., Givol, D., Wright, C. 1976. *Cold Spring Harbor Symp. Quant. Biol.* XLI:627
125. Padlan, E. A., Davies, D. R., Rudikoff, S., Potter, M. 1976. *Immunochemistry* 13:945–49

126. Padlan, E., Segal, D., Rudikoff, S., Potter, M., Spande, T., Davies, D. R. 1973. *Nature New Biol.* 245:165–67
127. Painter, R. H., Foster, D. E. B., Gardner, B., Hughes-Jones, N. C. 1982. *Mol. Immunol.* 19:127–31
128. Painter, R. H., Yasmeen, D., Assimeh, S. N., Poulik, M. D. 1974. *Immunol. Commun.* 3:19–34
129. Pajan, S. S., Ely, K. R., Abola, E. E., Wood, M. K., Colman, P. M., et al. 1983. *Biochemistry.* In press
130. Paraskevas, F., Lee S.-T., Orr, K. B., Israels, L. G. 1972. *J. Immunol.* 108: 1319–27
131. Deleted in proof
132. Perez-Montfort, R., Metzger, H. 1982. *Mol. Immunol.* 19:1113–25
133. Perrin, L. H., Joseph, B. S., Cooper, N. R., Oldstone, M. B. A. 1976. *J. Exp. Med.* 143:1027–41
134. Pestel, J., Joseph, M., Santoro, F., Capron, A. 1979. *Ann. Immunol.* 130C: 507–16
135. Poljak, R. J., Amzel, L. M., Chen, B. L., Phizackerley, R. P., Saul, F. 1972. *Proc. Natl. Acad. Sci. USA* 71:3440–44
136. Potter, M., Rudikoff, S., Padlan, E. A., Vrana, M. 1977. In Antibodies in Human Diagnosis and Therapy, ed. E. Haber, R. M. Krause, pp. 9–28. New York: Raven
137. Pruul, H., Leon, M. A. 1978. *Immunochemistry* 15:721–26
138. Prystowsky, M. B., Kehoe, J. M., Erickson, B. 1981. *Biochemistry* 20: 6349–56
139. Ragsdale, C. G., Arend, W. P. 1979. *J. Exp. Med.* 149:954–68
139a. Rajewsky, K. 1983. *Ann. Rev. Immunol.* 1:00–00
140. Reid, K. B. M., Porter, R. R. 1981. *Ann. Rev. Biochem.* 50:433–64
141. Rivnay, B., Wank, S., Poy, G., Metzger, H. 1983. *Biochemistry.* 21: In press
142. Rogers, J., Early, P., Carter, C., Calame, K., Bond, M., et al. 1980. *Cell* 20:303–12
143. Rouzer, C. A., Scott, W. A., Kempe, J., Cohn, Z. A. 1980. *Proc. Natl. Acad. Sci. USA* 77:4279–82
144. Rudikoff, S., Giusti, A. M., Cook, W. D., Scharff, M. D. 1982. *Proc. Natl. Acad. Sci. USA* 79:1979–83
145. Rudikoff, S. 1983. *Contemp. Top. Mol. Immunol.* In press
146. Rudikoff, S., Satow, Y., Padlan, E. A., Davies, D. R., Potter, M. 1981. *Mol. Immunol.* 18:705–11
147. Sakamo, H., Rogers, J. H., Hoppi, K., Brack, C., Trannecker, A., et al. 1979. *Nature* 277:627–33

148. Saul, F., Amzel, L. M., Poljak, R. J. 1978. *J. Biol. Chem.* 253:585–97
149. Schlessinger, J., Webb, W. W., Elson, E. L., Metzger, H. 1976. *Nature* 264: 550–52
150. Segal, D. M., Dower, S. K., Titus, J. A. Submitted for publication
151. Segal, D. M., Dower, S. K., Titus, J. A. 1983. *J. Immunol.* 130:130–37
152. Segal, D. M., Padlan, E. A., Cohen, G. H., Rudikoff, S., Potter, M., Davies, D. R. 1974. *Proc. Natl. Acad. Sci. USA* 71:4298–302
153. Segal, D. M., Taurog, J. D., Metzger, H. 1977. *Proc. Natl. Acad. Sci. USA* 74:2993–97
154. Segal, D. M., Titus, J. A., Dower, S. K. 1983. *J. Immunol.* 130:138–44
155. Sell, S., Mascari, R. A., Hughes, S. J. 1970. *J. Immunol.* 105:1400–5
156. Sell, S., Skaletsky, E., Holdbrook, R., Linthicum, D. S., Raffel, C. 1980. *Immunol. Rev.* 52:141–79
157. Sher, A., Lord, E., Cohn, M. 1971. *J. Immunol.* 107:1226–34
158. Siegel, R. C., Cathou, R. E. 1980. *J. Immunol.* 125:1910–15
159. Siegel, R. C., Cathou, R. E. 1981. *Biochemistry* 20:192–98
160. Silverstein, S. C., Steinman, R. M., Cohn, Z. A. 1977. *Ann. Rev. Biochem.* 46:669–722
161. Silverton, E. W., Navia, M. A., Davies, D. R. 1977. *Proc. Natl. Acad. Sci. USA* 74:5140–44
162. Siskind, G. W., Benacerraf, B. 1969. *Adv. Immunol.* 10:1–50
163. Sisson, J. G. P., Cooper, N. R., Oldstone, M. B. A. 1979. *J. Immunol.* 123:2144–49
164. Smith-Gill, S. J., Wilson, A. C., Potter, M., Prager, E. M., Feldman, R. J., Mainhart, C. R. 1982. *J. Immunol.* 128:314–22
165. Spiegelberg, H. L. 1983. *Fed. Proc.* 42: In press
166. Spouge, J. L., Easterbrook-Smith, S. B. 1982. *Mol. Immunol.* 19:253–56
167. Stanford, J. M., Wu, T. T. 1981. *J. Theor. Biol.* 88:421–39
168. Stang, C. J., Siegel, R. C., Phillips, M. L., Poon, P. H., Schumaker, V. N. 1982. *Proc. Natl. Acad. Sci. USA* 79:586–90
169. Steiner, L. A., Lopes, A. D. 1979. *Biochemistry* 18:4054–67
170. Stevenson, G. T., Mole, L. E. 1974. *Biochem. J.* 139:369
171. Sullivan, A. L., Grimley, P. M., Metzger, H. 1971. *J. Exp. Med.* 134:1403–16
172. Suzuki, T., Sadasivan, R., Taki, T., Stechschulte, D. J., Balentine, L.,

Helmkamp, G. M. Jr. 1980. *Biochemistry* 19:6037–44
173. Suzuki, T., Sadasivan, R., Wood, G., Bayer, W. L. 1980. *Mol. Immunol.* 17:491–503
174. Suzuki, T., Saito-Taki, T., Sadasivan, R., Nitta, T. 1982. *Proc. Natl. Acad. Sci. USA* 79:591–95
175. Suzuki, T., Taki, T., Hachimine, K., Sadasivan, R. 1981. *Mol. Immunol.* 18:55–65
176. Tamoto, K., Koyama, J. 1980. *J. Biochem.* 87:1649–57
177. Taylor, R. B., Duffus, W. P. H., Raff, M. C., dePetris, S. 1971. *Nature New Biol.* 233:225–29
178. Thoenes, J., Stein, H. 1979. *J. Exp. Med.* 150:1049–66
179. Tschopp, J. 1982. *Mol. Immunol.* 19:651–57
180. Tschopp, J., Schulthess, T., Engel, J., Jaton, J.-C. 1980. *FEBS Lett.* 112:152–54
181. Tschopp, J., Villiger, W., Lustig, A., Jaton, J.-C., Engel, J. 1980. *Eur. J. Immunol.* 10:529–34
182. Tyler, B. M., Cowman, A. F., Gerondakis, S. D., Adams, J. M., Bernard, O. 1982. *Proc. Natl. Acad. Sci. USA* 79:2008–12
183. Unkeless, J. P., Fleit, H., Mellman, I. S. 1981. *Adv. Immunol.* 31:247–70
184. Van Schravendijk, M. R., Dwek, R. A. 1981. *Mol. Immunol.* 18:1079–85
185. Vardinon, N., Spirer, Z., Fridkin, M., Schwartz, J., Ben-Ephraim, S. 1977. *Acta Allergol.* 32:291–300
186. Wain-Hobson, S., Dower, S. K., Gettins, P., Givol, D., McLaughlin, A. C., et al. 1977. *Biochem. J.* 165:227–35
187. Wank, S. A., DeLisi, C., Metzger, H. 1983. *Biochemistry.* In press
188. Weis, R. M., Balakarishnan, K., Smith, B. A., McConnell, H. M. 1982. *J. Biol. Chem.* 275:6440–45
189. Wilson, A. C., Carlson, S. S., White, T. J. 1977. *Ann. Rev. Biochem.* 46:573–639
190. Winkelhake, J. L. 1978. *Mol. Immunol.* 15:695–714
191. Winkelhake, J. L., Kunicki, T. J., Elcombe, B. M., Aster, R. H. 1980. *J. Biol. Chem.* 255:2822–28
192. Wright, C., Willan, K. J., Sjodahl, J., Burton, D. R., Dwek, R. A. 1977. *Biochem. J.* 167:661
193. Wu, T. T., Kabat, E. 1970. *J. Exp. Med.* 132:211–50
194. Yamawaki-Kataoka, Y., Nakai, S., Miyata, T., Honjo, T. 1982. *Proc. Natl. Acad. Sci. USA* 79:2623–27
195. Yasmeen, D., Ellerson, J. R., Dorrington, K. J., Painter, R. H. 1976. *J. Immunol.* 116:518

Ann. Rev. Immunol. 1983. 1:119–42
Copyright © 1983 by Annual Reviews Inc. All rights reserved

GENETICS OF THE MAJOR HISTOCOMPATIBILITY COMPLEX: THE FINAL ACT

J. Klein, F. Figueroa, and Z. A. Nagy

Abteilung Immungenetik, Max-Planck-Institut für Biologie, 7400 Tübingen, Federal Republic of Germany

THE MASTER TRICKSTER

Looking back at the 45-year history of the major histocompatibility complex (MHC), one cannot help but be amazed at the tricks the MHC has pulled on its students. The four major tricks which confused immunologists were these:

Trick number one: the MHC is discovered to control the presence of antigens on the surface of red blood cells and is regarded as a blood group locus (26). A whole new dimension is added to serological studies through analysis of the erythrocyte MHC antigens before it is realized that red blood cell expression is the least significant feature of the MHC and has nothing to do with its function (42, 47). It turns out, in fact, that several vertebrate species do not express MHC molecules on erythrocytes at all, and those that do, do so only because of leftover information remaining in the cytoplasm after expulsion of the nucleus (28, 89).

Trick number two: the MHC molecules are recognized as a major obstacle to a free exchange of grafts between individuals of the same species (27), and this recognition becomes a strong impetus for the development of transplantation biology as a separate field of study (42, 89). So firm becomes the tie between MHC and transplantation biology that for years immunologists are unable to rid themselves of the habit of referring to MHC molecules as transplantation antigens. Yet, again, it turns out that its influence on the outcome of grafting is only on the fringe of MHC physiological function.

119

0732-0582/83/0410-0119$02.00

Trick number three: the MHC affords an impression of being the site of mysterious, immune response (Ir)-governing genes distinct from the MHC itself but somehow magically bonded to it (67). The momentum this impression generates is the greatest immunology has known, but it too turns out to be generated by false premises. Eventually, one realizes that Ir genes are nothing but another MHC disguise: The Ir genes turn out to be MHC genes themselves (53, 54). However, the imprint of this trick on immunologists is so profound that as its consequence, immunology may never rid itself of terms such as Ir genes, I region, and Ia antigens.

Trick number four: in a number of vertebrate species, the MHC is found to be intimately associated with genes coding for complement components (7, 58, 69, 87), and this finding alone is enough to convince immunologists that the complement genes are part of the MHC. Yet, a search provides no evidence that ties the two groups of loci together functionally and evolutionarily.

This fourth trick may be the last one of the master trickster, unless it turns out that in our present impression of the true MHC function we have been tricked again. Be that as it may, there is no need to feel foolish because of the tricks. Each of them has had an enormously positive impact on immunology and has led to a series of important discoveries. And in the end they all have brought us closer to an understanding of what the MHC is really about.

TROUBLE WITH DEFINITIONS

In this review we define the MHC as a cluster of loci coding for proteins that provide context for antigen recognition by T lymphocytes. By "context" we mean that a T cell does not recognize an antigen alone; it recognizes the antigen in the context of the MHC molecules of the antigen-presenting cell. We are aware this definition does not meet the approval of other immunologists who traditionally view the MHC as a group of loci that occupy certain arbitrarily delineated chromosomal segments. The latter must be a view based on Montaigne: "Whatever variety of herbs there may be the whole thing is included under the name of salad." Our objection to this traditional view of the MHC is precisely this salad principle, namely, that one lumps together herbs (loci) that may have nothing in common except their occupation of the same bowl (chromosomal segment). The main consequence of our definition is that it excludes the complement loci from the MHC. We justify this exclusion thus: first, the complement components controlled by loci linked to the MHC clearly have a different function than do MHC proteins. The former are involved primarily in humoral immunity; they are plasma proteins that become membrane-bound only secondarily, and they function as typical enzymes. The latter are involved in cellular immunity; they are membrane proteins that may appear in the plasma secondarily, and they have no demonstrable enzymatic activity. Second, there is a considerable degree of homology among the various MHC genes (and proteins), indicating they are all directly related through their origin from common ancestor genes. No such homology could be demonstrated, as yet, between the MHC and the complement genes. We do not wish to go so far as to maintain that no homology exists between the two groups, for it may well be that as more DNA and protein sequences become known, particularly for the complement genes and proteins, some evolutionary relationship between the MHC and complement genes may become apparent. But this relationship, if found, almost certainly would be of a different magnitude than that known to exist among the different MHC genes. Current studies suggest that the MHC genes are related to immunoglobulin (Ig) genes (76); yet, nobody seems willing to consider Ig genes as part of the MHC or vice versa. If homology were found between MHC and complement genes, it would probably not be greater than that between MHC and Ig genes. Hence, if we do not consider the Ig genes as part of the MHC, why should we bestow this privilege on the complement genes? Third, although the *C4* genes are closely linked to the MHC in many species, the *C3* gene is only loosely MHC-linked in some species and apparently is not linked at all in others (1). Since *C3* and *C4* are homologous genes that are almost certainly evolutionarily related, if we consider one (*C4*) to be part of the MHC, we should consider the other (*C3*) as well. However, where do we then place the borders of the MHC? If we include the *C3* locus in the MHC, we must abandon the salad principle anyway because, otherwise, soon half of the genome would become the MHC; and if we abandon the principle, there is no longer any compelling reason to include the complement genes in the MHC. Fourth, recent work indicates that the mouse MHC contains a locus coding for the enzyme neuraminidase (*Neu-1,* see below). According to the salad principle, this locus, too, must be included in the MHC. The absurdity of this inclusion is obvious.

One possible justification for considering the complement loci as part of the MHC is based on the notion of the supergene. Snell (88) and Bodmer (6) have argued that the MHC and loci linked to it may constitute a unit for natural selection and so, from a population genetics point of view, may form a single supergene. However, this notion remains pure speculation, particularly in view of the fact that no effects of natural selection on the MHC have been unequivocally demonstrated—and will be extremely difficult to demonstrate. Moreover, even if the MHC and the linked loci were to form a unit of selection, what would be the advantage of using this observation to define a gene cluster in immunology? A MHC supergene may be a useful concept when discussing one particular aspect of the complex, but to use it as a basis for defining the MHC would only lead to confusion. In summary, we believe it is best to ignore tradition and to define the MHC the way we define virtually all other gene clusters—on the basis of functional similarity and evolutionary homology.

MHC CLASSES

Using the evolutionary definition, one can divide the MHC loci into two classes, which we designate class I and class II (44, 46). Loci belonging to the different classes differ in the properties and function of their respective products.

The characteristic features of class I glycopeptides are a molecular weight of 44,000, a chain length of some 340 to 350 amino acids, the presence of three extramembrane domains (Figure 1), association with β_2 microglobulin, and functional preoccupation with cytolytic T (Tc) cells (25, 42, 73, 106). The characteristics of class II glycopeptides are a molecular weight of 28,000 (β chains) and 34,000 (α chains), a chain length of about 230 (α chain) and 220 (β chain) amino acids, the presence of two extra membrane domains (Figure 1), dimerization of α and β chains, and functional preoccupation with helper T (Th) and suppressor T (Ts) cells (53, 54, 96).

Molecular Weight

The figures given are for glycosylated polypeptide chains. The class I polypeptide may have one or two carbohydrate side chains attached to it,

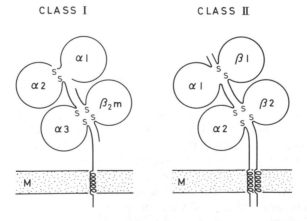

Figure 1 Diagram showing the domain organization of class I and class II MHC molecules. β_2m, β_2 microglobulin; other combinations of Greek letters and numbers indicate individual MHC domains; M, membrane.

depending on the species and probably also on the particular allele. Each side chain consists of some 12 to 15 invariant sugar units, arranged in a fixed order. Each class II polypeptide chain has one side chain attached to it, apparently organized in a manner similar to that of the class I side chains. The difference in the molecular weight of the α and β chains probably arises partially from a difference in the length of the polypeptide chain and partially from a difference in the carbohydrate side chains.

Chain Length

The figures given represent an average length; the actual length may vary by several residues, depending on the particular locus and allele. In some class I or class II genes, several codons or even whole exons can be deleted and the polypeptide chains thus can be shortened correspondingly. Most of the deletions occur in the chain's intracellular region.

Domains

The basic topological unit of the MHC molecule is the domain, a polypeptide some 90 amino-acid residues long and containing cysteines at its opposite ends. The polypeptide probably folds characteristically around a disulfide bridge formed by the cysteines. The compartmentalization of the MHC molecule into domains has its foundation in the segmentation of the MHC gene into exons and introns (Figure 2). Similar domains have also been found in the β_2 microglobulin (one domain), immunoglobulins (from two to six domains per chain), Thy-1 (one domain), and other proteins (102). The presence of this basic structural element in different proteins is usually interpreted as evidence for the common origin of these proteins.

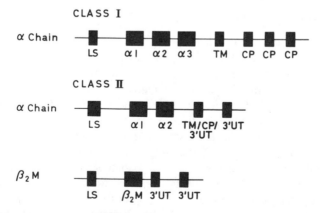

Figure 2 Organization of class I and class II MHC genes. $\alpha1–\alpha3$, exons encoding extramembrane domains of MHC chains; β_2m, β_2 microglobulin; LS, leader sequence; TM, transmembrane sequence; CP, cytoplasmic sequence; 3'UT, 3'untranslated sequence; black boxes, exons; interconnecting lines, introns.

However, since similar domains occur also in proteins such as the superoxide dismutase (84), for which no other evidence is known suggestive of a relationship to MHC or Ig molecules, presumably the presence of these domains does not always signify common ancestry of proteins. The domain may represent an optimal solution to some basic topological problems of protein structure, a solution independently arrived at by different protein families through convergent evolution.

Chain Association

To be stable, apparently, the MHC molecule must contain four extramembrane domains. To achieve this configuration, the three-domain class I chain combines with the one-domain β_2 microglobulin and the two-domain class II α chain associates with the two-domain β chain. The associations are always noncovalent and always between a polymorphic and an oligomorphic chain. (The polymorphism of the β_2 microglobulin and the class II α chains is far lower than that of the class I chain and the class II β chain.) The function of the oligomorphic chains appears to be the stabilization of the MHC molecule. To what degree the chains also contribute to the immunologic functions of the MHC molecule is unclear.

The class II molecules are also associated noncovalently with a third chain that is even less variable than the α chains and, therefore, is referred to as invariant (35). Its function is unknown; however, since it is not controlled by the MHC (57) and is associated with α and β chains only in the cytoplasm but not on the cell surface (57a), its role may be nonimmunologic, perhaps concerned with the integration of the class II chains into the membrane.

Functional Specialization

Although both class I and class II molecules provide context for antigen recognition, a certain division of labor exists between them in that, generally, the former provide the context for antigens recognized by Tc cells, whereas the latter provide context for antigens recognized by Th and Ts cells (54). When MHC molecules themselves serve as antigens (in an allogeneic situation), the same predilection for the stimulation of different T cell sets can often be observed. However, in the latter situation the specialization is not absolute because class I alloantigens can stimulate not only cytolytic T cells, but under certain circumstances also proliferating (presumably Th) cells; similarly, class II alloantigens stimulate not only proliferating T cells (in a mixed lymphocyte culture), but, again under certain circumstances, also cytolytic cells. A corollary to this division of labor is that class I molecules provide context for the recognition of antigens that are an integral component of the plasma membrane (e.g. minor histocompatibility antigens or viral antigens), whereas class II molecules provide recognition context for soluble antigens picked up by macrophage-like antigen-presenting cells. Still another corollary might be that the cells presenting the antigen in the context of class I molecules and those presenting it in the context of class II molecules may not be the same. These three observations, pertaining to the type of the stimulated T cell, the form of the presented antigen, and the type of the antigen-presenting cell, are probably interrelated, but the cause and the effect in this relationship is not clear.

UNITY

We predicted some 10 years ago that the present-day class I and class II loci would turn out to be functionally homologous (43, 52). At that time no biochemical data available could be used to compare the products of these loci, not to mention the loci themselves, and the prediction seemed to go against the then-current concept of the MHC. The general consensus then was that the class II were a special group of loci, in some mysterious way related—through the *Ir* genes—to the T cell receptor, and therefore were more worthy of study than the prosaic transplantation antigen-encoding class I loci. Although some of this *I*-region mystique persists to this day, the prosaic view begins to prevail. There can no longer be any doubt that the class I and class II loci form a single family in terms of their organization and function. They probably are no farther apart than are the Ig light-chain from the heavy-chain genes (61).

A major step in MHC evolution must have been the acquisition by the class I genes of an extra exon coding for the third extramembrane domain. The reorganization of the MHC molecule that followed this step must have been a consequence of this acquisition. The reorganization must have represented an adaptation to a new function—a variant of the function the MHC genes had been carrying out until then. Whether this adaptation relates to T cell function (regulation versus killing), antigen processing (processed versus indigenous antigens), or type of antigen-presenting cell (macrophage-like cell or a lymphocyte) should become apparent once it is resolved why certain antigens are presented in the context of class I and others are presented in the context of class II molecules. Alternatively, the evolution might have proceeded from a three- to two-domain chain, or—and this possibility is the most appealing—the two chain types evolved concurrently from a common ancestor.

EXPRESSION—THE TOLLING OF THE BELLS FOR ALIEN SPECIFICITIES

Another important step in the evolution of the MHC must have been the acquisition of separate controlling elements for the different groups of genes, resulting in a differential expression of these genes in different tissues. Three principal modes of expression have been recognized thus far (47). First, some class I loci (e.g. the K, D, and L loci in the mouse) manifest themselves in most, if not all, somatic tissues, but most strongly in the lymphoid tissues, in particular in the lymphocytes (both T and B). The difference in expression between lymphoid and nonlymphoid tissues is at least 10-fold (2). Second, certain other class I loci (e.g. the Qa and Tla loci in the mouse) preferentially express themselves in the T cell lineage and in cells representing certain stages of this lineage. However, as we discuss later, the restriction to the lineage and stage is not absolute and the differences in tissue distribution may be largely quantitative rather than qualitative. Third, the known class II loci express themselves preferentially in the B cell and monocyte lineages. In these lineages, again, only some cells may express the genes, whereas other cells express them far less or not at all. There may also exist a fourth mode in which certain class II genes express themselves in certain stages of T cell differentiation. This mode of expression has been debated extensively over the last decade and evidence both supporting and disclaiming it has been put forward (29). A new twist to these discussions from a recent report states that there may be a special group of class II loci, perhaps distinct from those already identified, preferentially expressed in T cell subsets (32). Curiously, molecular geneticists thus far have not noticed any sign of these extra loci.

MHC molecules not expressed on all tissues are sometimes referred to as differentiation antigens, but we find this designation misleading. The term differentiation antigen is otherwise used to mean a molecule involved in differentiation, i.e. driving it. We are not aware of any evidence implicating any MHC molecules, including the mouse Qa and Tla molecules, in differentiation. Some of the molecules may serve as markers of differentiation, but this fact does not qualify them as differentiation antigens.

Still another mode of MHC expression has been claimed by the proponents of the alien-specificities hypothesis. According to this hypothesis, each individual possesses most of the MHC genes characterizing a given species (5). Of these, normally only a few become expressed at any given time, but during an aberrant differentiation, such as that characterizing the generation of tumor cells, unexpectedly some of the previously silent genes become activated, which leads to the expression of MHC molecules normally found in other individuals (alien specificities). The hypothesis has been supported by findings from several laboratories (78), all claiming the demonstration of alien specificities in a variety of tumors by a variety of methods. We have criticized these findings by pointing out that the experimental methods used had not been properly controlled, that the purported origin of the tumor lines used had not been satisfactorily demonstrated, and that other explanations of the findings were more likely than those proposed (45). However, the alien-specificities research gained considerable momentum as attested by the number of papers published on this topic and the number of meetings organized. Molecular genetics put a period behind the whole affair: Clearly not enough genes exist in the MHC to code for all the allomorphs (allelic forms of MHC proteins) possessed by a species.

ORGANIZATION

Figure 3 depicts the organization of the MHC in the *H-2^d* mice, as determined by molecular genetics methods (94; L. Hood, personal communication). There are 37 class I and five class II loci on this map—the class II loci all grouped in one cluster, the class I loci split in at least two and perhaps three clusters. Preliminary studies with other *H-2* haplotypes suggest that the number of MHC loci may vary from one inbred strain to another, and that the *H-2^d* strains may, perhaps, carry more loci than do other strains. The *H-2^d* map itself may not be complete, particularly in the class II region, but it is unlikely that large numbers of genes still remain hidden.

How does this map compare to that obtained by classical genetics methods? The first thing one notices is the conspicuous absence of *B*, *C*, and *J* loci, and therefore it is these loci that we discuss now.

The B Locus

The postulate of a separate *B* locus (region, subregion) was originally made by Lieberman and co-workers (62) to explain the genetic control of the antibody response to a myeloma protein MOPC173 (presumably a response against an allotypic determinant on the IgG2a molecule). The critical *H-2* haplotypes involved in the mapping of this locus were *H-2^a*, *H-2^b*, *H-2^h4*, and *H-2^i5*, of which the first was typed as a low responder and the remaining three were typed as high responders to MOPC173. Taking into account the origin of *H-2^h4* and *H-2^i5* from *H-2^a*/*H-2^b* heterozygotes, Lieberman et al proposed that in *H-2^h4* the recombination took place proximally and in *H-2^i5* it took place distally to a hypothetical *B* locus controlling the immune response to MOPC173. The involvement of the *B* locus was later postulated for immune responses to at least five other antigens: lactate dehydrogenase B (LDH$_B$) (68), staphylococcal nuclease

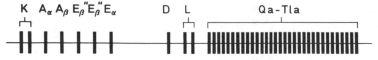

Figure 3   MHC genes present in the *H-2^d* haplotype.

(Nase) (64), oxazolone (12), the male-specific antigen (H-Y, as detected by skin grafting) (33), and trinitrophenylated mouse serum albumin (TNP-MSA) (97). The mapping of the genes controlling these responses revolved in all these cases around the four critical H-2 haplotypes used by Lieberman and her co-workers (Table 1).

We have reanalyzed in detail two of the six responses purportedly controlled by the B locus—the responses to LDH_B and to the myeloma protein MOPC173 (3). The analysis revealed a peculiar interplay of Th and Ts cells: The Th cells of the H-2^a strain recognize the antigen in the context of the A^k molecules, but they cannot respond because they are suppressed by Ts cells that recognize the same antigen in the context of the E^k molecule. As a result of this interplay the H-2^a mice are low responders, not because they carry a low-responder allele at some hypothetical B locus, but because they are inhibited in their response by a simultaneous activation of Ts cells. Hence, the genetic control of the response to LDH_B and to the myeloma protein MOPC173 can be explained without the postulate of a separate B locus, distinct from the previously identified loci coding for the A and E molecules. On the basis of these data, we suggested that the B locus be removed from the genetic map of the H-2 complex.

One might ask, however, what of the other responses? Can they too be explained by similar mechanisms? The response to H-Y and to TNP-MSA probably *can* be, since the responsiveness pattern is identical to that of LDH_B and MOPC173. The responsiveness to oxazolone displays a different pattern, but the responses are highly variable. Not only are there discrepancies among the reports from different laboratories (12, 79), but even in the same laboratory there is a considerable variation from one experiment to another, particularly in mice carrying the H-2^k haplotype (N. Ishii, Z. A. Nagy, unpublished data). Until this variation is brought under control, the mapping of the genes involved in the response to oxazolone remains in

Table 1 Immune responses originally believed to be controlled by a B locus in the H-2 complex

Strain	H-2 haplo-type	Alleles at H-2 loci						Response to antigen[a]					
		K	A_α	A_β	E_β	E_α	D	MOPC173	LDH_B	Nase	H-Y	Oxazolone	TNP–MSA
B10.A	a	k	k	k	k	k	d	−	−	+	−	?	−
C57BL/10	b	b	b	b	b	b	b	+	+	−	+	−	+
B10.A(2R)	$h2$	k	k	k	k	k	b	−	−	+	−	+	−
B10.A(4R)	$h4$	k	k	k	k	b	b	+	+	−	+	−	+
B10.A(5R)	$i5$	b	b	b	b	k	d	+	+	−	+	−	+
B10.BR	k	k	k	k	k	k	k	−	−	+	−	+?	−

[a]+, responder; −, nonresponder. For references see text.

doubt. As for the response to Nase, Berzovsky and his colleagues (4) have established recently that part of the response is controlled by the A loci (A_α and A_β), but they insist that the other part is controlled by the B locus. We suggest as an alternative explanation that Nase is recognized in the context of A molecules in some H-2 haplotypes and in the context of E molecules in other haplotypes and that this switching of recognition context creates the illusion of a B locus. At any rate, the important fact is that the molecular geneticists are unable to find any DNA corresponding to the purportive B locus (L. Hood, personal communication), so there must be some other explanation for all the responses for which the existence of the B locus has been postulated. There is certainly no justification for keeping the locus in the H-2 map.

The C Locus

Evidence of the existence of the C locus is threefold. First, an H-2^{h2} anti-H-2^{h4} antiserum, originally used to define the C locus (9), has recently been reproduced by Sandrin & McKenzie (86). Second, according to Rich and her co-workers (82, 83), antisera purportedly containing C-specific antibodies react with a suppressor factor produced in an allogeneic mixed lymphocyte reaction (MLR). Third, Okuda & David (74) have reported a MLR that occurs in congenic strain combinations presumably differing at the C locus and an inhibition of this reaction by anti-C sera. Mapping by classical methods has suggested a position of the C locus between E_α and the genes coding for the C4 component. Although this chromosomal segment has not yet been fully explored by molecular genetics methods, the data available thus far do not lend any support for the existence of a C locus near the E_α locus. Our own experience agrees with these negative results: We have never been able to obtain any activity in the C-defining H-2^{h2} anti-H-2^{h4} combination, although we have tested this combination repeatedly by serological methods, MLR, graft-versus-host reaction, cell-mediated lymphocytotoxicity (CML), and grafting of tissues (36, 63; J. Klein, unpublished data). If the entire E_α—$C4$ interval indeed does turn out to be devoid of class II loci, then some other explanation will have to be found for the positive results obtained. Molecular geneticists have found an extra E_β-like gene between the E_α and the classical E_β gene (L. Hood, personal communication), and evidence for more than one E molecule has also been presented at the protein level (10, 56, 59, 77). Although this new gene is on the wrong side with respect to the purportive location of the C gene, it might be that some as yet not understood form of interaction between the E genes creates the illusion of the C gene. Another possibility is that the C gene is present only in some H-2 haplotypes. A final possibility is, of course, that the C gene never existed in the first place.

The J Locus

In contrast to the C locus, which has been championed by only three or four laboratories, the existence of the J locus has been accepted, it seems, by every laboratory except ours. Supporting this locus is a mass of data: Anti-J sera have been produced in many laboratories and even are available commercially (71); J-specific monoclonal antibodies have recently been produced by at least two groups (37, 99); functional evidence for the expression on Ts cells has been obtained repeatedly (71); and, recently, an isolation of RNA bearing the J-encoding message has been reported (95). Our reservations have been based on the inability to obtain any biochemical information about the purported J molecule. If it existed, why was it so difficult to isolate? The recent inability to find any difference between the DNAs of the H-2^{i3} and H-2^{i5} haplotypes (L. Hood, personal communication), the two haplotypes originally used to define the J locus (72), of course, puts the scepticism about the J locus on a firm basis. If it can be corroborated that there really is no DNA coding for the J molecule in the class II region, what then is J? Two principal possibilities exist. First, the J gene may be somewhere else than in the class II gene interval: It can be outside the H-2 complex or it can be on a different chromosome altogether. However, if the latter were the case, it would still have to be explained why the anti-J sera have been produced by immunization between congenic lines presumably differing only in a small chromosomal segment within the H-2 complex. Second, the gene could, again, be an illusion arising from an as yet not understood form of interaction between known class II genes. But if this were the explanation, how should one then understand the data on the isolation of J-encoding mRNA? Whatever the correct explanation (and it might be available by the time this article is published), it is clear that J is not what it has been thought to be up to now.

Loci Between K and A_α?

Recently, Monaco & McDevitt (70) have detected a series of 16 spots on two-dimensional gels of cell preparations precipitated by anti-MHC sera. The proteins composing these spots ranged in molecular weights from 15,000 to 30,000. Typing of appropriate H-2 recombinants localized the genes coding for the low-molecular-weight proteins in the chromosomal interval between the K and A_α (A_β) loci. Monaco & McDevitt argue that this interval is occupied by a new class of MHC loci that code for cytoplasmic low-molecular-weight proteins (since the proteins could not be labeled by the lactoperoxidase method, the authors assume they are absent from the membrane) (70). However, even if these loci could be definitely established, what they have to do with the MHC still has to be solved. The fact that neither the class I nor the class II DNA probes have as yet detected them could signify they are unrelated structurally and presumably also functionally to the genuine MHC genes. After all, if there are enzyme-encoding loci entrapped within the MHC, why not also other loci, unrelated to the MHC proper?

The K and D Loci

Iványi & Démant (34) recently have described data suggesting the existence of at least two K and at least three D loci. This suggestion is supported by DNA cloning, which establishes the presence of these loci in the H-2^d haplotype (94). One of the D loci corresponds to Démant's L locus and another, perhaps, to the R locus of Hansen and his co-workers (31).

The existence of multiple loci in the K and D regions has been interpreted occasionally as contradicting the two-locus model for the class I H-2 loci. In fact, however, there is no contradiction because the original hypothesis (55) maintained only that there are two clusters of loci and that all the then-known H-2 recombinants occurred between these two clusters. This is precisely what the methods of both the classical and molecular genetics demonstrate. Moreover, in functional tests the H-2 complex behaves as if it consists of only two class I loci.

The Qa and Tla Loci

The most radical change brought onto the H-2 maps by the DNA cloning methods has occurred in the region occupied by the Tla and Qa loci. Although even classical methods have provided several hints of more than the three or four identified loci, few of us have expected the shower of loci triggered by molecular biologists. Even if 37 loci in the H-2^d haplotype were the maximum number and other haplotypes had considerably fewer loci, important questions concerning the reasons for this multitude of loci would still arise. In a way, deciphering the organization, mode of expression, and function of the Qa-Tla loci is the last major task for immunogeneticists. Because of the uncertainties surrounding them, we now consider these loci in detail.

THE RIDDLES OF THE QA-TLA LOCI

Evidence That Qa and Tla are Class I Loci

In this review, we have classified the Qa and Tla as class I loci and we have done so for the following reasons. First, an antiserum produced in a strain combination differing at the Qa-Tla loci but identical at the K and D loci (A.TL anti-A.TH) cross-reacts with at least two K molecules (K^f and

K^P) (13, 14). We first noticed this cross-reactivity because the antiserum killed 100% of K^f- or K^P-bearing cells in the cytotoxic test under conditions in which only some 30–40% of cells carrying other *H-2* haplotypes were killed. The evidence that the antiserum was K-reactive was provided by testing of *H-2* recombinants and an *H-2* mutant. Second, one can use strain combinations differing at the *Qa-Tla* loci to generate CML, which, unlike that stimulated by non-MHC antigens, is not restricted by MHC loci (Table 2). Several pieces of evidence are now available indicating that the target antigens of this CML are controlled by the *Qa* loci (see below). Third, there is a striking structural homology between the *Qa-Tla* and the *K-D* molecules. This homology is attested to by the association of *Qa* and *Tla* polypeptide chains with the β_2 microglobulin (98), by similar size of the Qa-Tla and K-D polypeptide chains (98), by the peptide composition as determined by biochemical fingerprinting (90), by the similarity in gene organization (93), and by nucleotide sequence similarity (65, 93).

Table 2 Strain combinations in which CML, most likely directed to Qa antigens, has been obtained

Possible target	Strain combination	References
Qa–1^a	C3H (AKR or CBA) anti-B10.BR	17, 38–40
	B10.HTT anti-B10.S (7R)	100
	[or B10.S (9R)]	
	C57BL/6 anti-B6-*Tla*^{a}	I. Vučak, Z. A. Nagy, J. Klein, unpublished data
	A-*Tla*^{b} anti-A	I. Vučak, Z. A. Nagy, J. Klein, unpublished data
	(BALB/c × C57BL/6) F$_1$ anti-B10.A (5R)	18
Qa–1^b	A.TH anti-A.TL	49
	B10.BR anti-C3H (AKR, CBA)	15, 40
	B10.S (9R) anti-B10.HTT	101
	A anti-A-*Tla*^{b}	23
	B6-*Tla*^{a} anti-A.BY/B6	23
	SWR anti-DBA/1	23
	B6-*Tla*^{a} anti-C57BL/6	19
	B6-*Tla*^{a} anti-BALB.B	19
	B6-*Tla*^{a} anti-C3H.SW	19
	B6-*Tla*^{a} anti-129	19
	NZB anti-BALB/c	16, 17, 82, 83
Qa–1^d	C57BL/6 anti-B6-*Tla*^{d}	19
Qa–2^a	BALB/c By anti-BALB/c J	24

Peculiarities of Qa-Tla Loci

Similarities notwithstanding, the *Qa-Tla* and the *K-D* loci differ in several respects. Here we point out three of these differences, pertaining to the expression in different tissues, polymorphism, and function.

The Tla antigens are expressed only on thymocytes at a certain stage of differentiation and on certain leukemias (21). The Qa antigens were originally thought to be present only on a subpopulation of T lymphocytes, but this observation no longer holds. When one uses particularly effective complement, one can demonstrate 100% killing of lymph node or spleen cells with anti-Qa sera, an observation suggesting that the Qa antigens are present not only on virtually all T cells but also on B cells. Their presence on B cells has also been demonstrated with the fluorescence-activated cell sorter (75). However, most of the B cells and some of the T cells must express only very little of these antigens because they are not killed when one uses weak complement that, however, is strong enough to kill all lymphocytes expressing the corresponding K and D antigens.

The polymorphism of the *Qa* and *Tla* loci is significantly lower than that of the *K* and *D* loci. The best evidence in support of this thesis comes from the typing of the B10.W lines carrying *H-2* haplotypes of wild mice on the C57BL/10 (=B10) background. Although we have found 17 different *K* alleles and 20 different *D* alleles in these lines, we could find only three *Qa-1*, two *Qa-2*, and six *Tla* alleles (S. Tewarson, F. Figueroa, J. Klein, unpublished data). Similar paucity of alleles at the *Qa* and *Tla* loci exists also among the standard inbred strains and congenic lines (21).

As for the function of the *Qa-Tla* loci, no evidence has yet been provided for their being able to provide context for antigen recognition. At the same time, however, it must be emphasized that the search for such a function has been hampered by the lack of suitable recombinants separating the *Qa-Tla* from the *K-D* loci and the individual *Qa* loci from one another. In fact, we could find only one report in the literature in which the function of *Qa* molecules as restriction elements has been tested (39).

Organization of the Qa-Tla Loci

Five *Qa* loci (*Qa-1* through *Qa-5*) and one *Tla* locus have been described with classical immunogenetic methods (20, 22, 30, 91). Of the *Qa* loci only two (*Qa-1* and *Qa-2*) have definitely been established as separate loci (they are separated by recombination and they control distinct polypeptide chains) (21). The remaining three loci (*Qa-3, Qa-4,* and *Qa-5*) have been postulated to exist because the antigens they control have a different strain-distribution pattern than those controlled by the *Qa-1* and *Qa-2* loci (22, 30). However, in the past, strain distribution pattern has proved an unreliable criterion for separating loci, and so it seems best to postpone the acceptance of *Qa-3, Qa-4,* and *Qa-5* as separate loci until they can be matched against the genes identified by recombinant DNA methods. Although the Qa-1 and Tla molecules have a different distribution, this fact alone is again insufficient for considering them as controlled by different loci (see below).

As already mentioned, strains differing at *Qa* loci give CML that is not restricted by the MHC loci (24, 49). The *Qa*-associated CML has now been obtained in a number of combinations in the mouse (Table 2) and in the rat (66). The CML loci were originally thought to be distinct from the known *Qa* loci, in part because of an erroneous mapping of the *Qa-1* locus, but also because of certain discrepancies in the strain distribution of the respective antigens. Now that the position of the *Qa-1* locus has been corrected and some of the discrepancies have been resolved, there is no longer any need to postulate separate CML loci. Furthermore, the identity of the *Qa* and CML loci is supported by the observation that the CML can be blocked by Qa-specific antibodies (39, 40). The few discrepancies that remain no doubt are the consequence of this region's genetic complexity and of our inability to separate the individual loci by recombination.

If we accept *Qa-1* and *Tla* as separate loci, we arrive at an order: *D(L)* ... *Qa-2* ... *Tla* ... *Qa-1*. This order is different from that proposed originally and the change has been brought about by a reclassification of certain recombinants in terms of the *Qa-1* locus (92). The chromosome interval containing the *Qa* and *Tla* loci also contains loci coding the C4-binding protein (*C4-bp*) and for the enzymes catalase-2 (*Ce-2*), phosphoglycerate kinase-2 (*Pgk-2*) and urinary pepsinogen-1 (*Upg-1*) (51). The order of these loci with respect to one another and with respect to the *Qa* loci is uncertain; the order given in Figure 4 is based on the typing of B10.W and some other *H-2* congenic lines (41a).

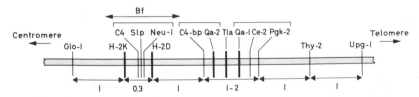

Figure 4 Genetic map of the chromosomal segment occupied by the MHC. Thick vertical lines, MHC loci; thin lines, linked loci (only some MHC loci are shown); numbers indicate length of the particular interval in centimorgans; brackets indicate the order of the bracketted loci is uncertain; *Bf,* complement factor B; *C4,* complement component 4; *C4-bp,* C4-binding protein; *Ce-2,* kidney catalase-2; *Glo-1,* glyoxalase-1; *Neu-1,* neuraminidase-1; *Pgk-2,* phosphoglycerate kinase-2; *Slp,* sex-limited protein; *Thy-2,* thymus cell antigen-2; *Upg-1,* urinary pepsinogen-*1*.

Interpretation

What could be the functional significance of this large cluster of MHC loci? At present, there are two principal ways to view the *Qa-Tla* loci. The first possibility is that these loci carry out an as yet unidentified function that differs from the *K* and *D* loci. The function frequently considered in this context is a guiding role in tissue differentiation. The only reason this role is considered is because of the differential tissue distribution of the *Qa-Tla* molecules. However, we find this proposal unattractive for two reasons. First, because it attaches—unjustifiably, in our opinion—exaggerated importance to a trivial characteristic (tissue distribution), and second, because we have a hard time imagining that genes as closely related as are the *K-D* and *Qa-Tla* clusters could perform so radically different functions.

The second possibility is that the *Qa-Tla* genes either perform the same function as the *K-D* genes but on a more limited scale, or they represent mostly nonfunctional class I genes. The reasons they might be only partially functional, or nonfunctional altogether, could be, first, because many of them might be pseudogenes lacking some important part of the polypeptide chain necessary for carrying out the function (93), and second, because of their reduced expression. With most of the genes, we do not know whether they are expressed at all; but even if all of them were expressed, it is unlikely that the level of expression would reach that of the *K-D* genes (otherwise, one would have detected their products long ago). The conspicuous variation in one's ability to detect the Qa antigens (21) and the apparent influence of the cell cycle on *Tla* gene expression (85), suggest that from most of these genes only minimal quantities of mRNA molecules are copied and translated into membrane proteins. It appears that only cells at a certain favorable stage of differentiation, in certain favorable metabolic conditions, and certain membranous milieu manage to insert enough Qa or Tla molecules into their membranes to reach the sensitivity level of our detection methods. Most of the time, the quantity of the expressed molecules may also be insufficient for their serving as restriction elements in antigen recognition, but under favorable circumstances some of them may "leak through" and become functional.

This view, of course, leads to another important question. If the *Qa-Tla* loci are largely nonfunctional, why do they persist in the genome? To answer this question, it will be necessary to find out first how old these genes are. The low polymorphism of the *Qa-Tla* genes, combined with the suggestion that different *H-2* haplotypes may have different numbers of *Qa-Tla* loci, suggests that the genes are of a relative recent orgin and that most of them arose after the mouse split from its nearest ancestors. The generation of the cluster thus may represent an evolutionary accident and there may be so many genes because there has not been enough time to dispose of them.

Alternatively, one can imagine that a system so dependent on polymorphism as the MHC apparently is requires a great deal of experimentation with the available genetic material to assure this system's optimal adaptability. In this respect, one could view the *Qa-Tla* loci as a "laboratory" in which genes are experimented with before they are put into production. The low and variable expression of the *Qa-Tla* genes would be compatible with this view: The expression would be low enough not to clutter the plasma membrane with unused proteins, yet at the same time high enough to provide an opportunity for selective mechanisms to find suitable candidates for coping with new situations as they arise. The problem with this view is that it does not explain why there is such a "laboratory" only for class I and not for class II genes. There is no compelling evidence that the polymorphism of class II loci is lower than that of the class I loci (50), and so one would expect that the two classes would be treated the same way. Several possibilities may solve this problem: One "laboratory" could serve both the class I and class II loci; a class II "laboratory" may be discovered later in other *H-2* haplotypes; or part of the variation of the class II molecules could be provided by the α chains.

ENTRAPPED LOCI

The mouse neuraminidase-1 (*Neu-1*) locus codes for an enzyme that modifies several other enzymes post-translationally by removing extra sialic acid residues from their carbohydrate units. Thus far, four enzymes have been identified that are influenced by the *Neu-1* locus, all present in liver lysosomes: liver acid phosphatase (60), α-mannosidase (11), acid α-glucosidase (80), and liver arylsulfatase B (8).

Certain mutations in the *Neu-1* genes result in the production of defective neuraminidase that does not manage to remove all the extra sialic acid, and as a consequence the unmodified enzymes have a different mobility in the electric field in comparison to enzymes modified normally. It was this variation in electrophoretic mobility in the liver acid phosphatase and α-mannosidase that was discovered first and was ascribed to the effect of separate loci—*Apl* (60) and *Map-2* (11). However, later it was realized that most likely there are no separate *Apl* and *Map-2* loci but rather the effects seen were a result of the action of the pleiotropic *Neu-1* locus (80, 105). Originally, the *Apl* locus was mapped at a distance of several centimorgans from the *H-2* complex on the telomeric side (104). However, this placement turned out to be wrong and new tests have revealed the *Neu-1* locus to be very closely linked to the *H-2* complex (103). Our discovery of a new form of *Neu-1* variation (41) has allowed us to use several *H-2* recombinants for mapping the locus, and this testing placed the *Neu-1* locus within the *H-2* complex, between the E_α and *D* loci (13). The *Neu-1* locus thus resides in the same chromosomal interval as the *C4*-coding genes. Although it is not certain that *Neu-1* is a structural rather than a regulatory locus, there is no evidence it is in any way related to the MHC loci. We would like to propose, therefore, that *Neu-1* and the *C4* loci occupy a chromosomal segment that has been entrapped accidentally between the MHC genes.

Passing of long chromosomal segments from ancestral forms to present-day species is not an uncommon phenomenon; in fact, it seems to be a rule in the evolutionary process (48). Homologous linkage groups are being discovered all over the genome and in as distant forms as birds and mammals. The preserved chromatin blocks usually contain genes for which no evolutionary or functional relationship is apparent: The preserved linkages seem to be the result of accidental conservation of gene associations rather than any ontological relationship among the genes in the group. The MHC does not seem to be an exception to this rule.

THE FINAL ACT

Although several important questions still remain to be answered with regard to the genetic organization of the MHC, the astonishingly rapid progress of the molecular genetics of this system leaves no doubt that the answers will be provided within the next few years. It is as if molecular genetics has raised the curtain on the final act of the MHC play, and as in every good play this act is turning out to be short and packed with action. It truly was a good play: dead-earnest most of the time, hilarious on occasions, and thrilling all the time. Seeing it nearing its end, one cannot help but feel a sting of nostalgia. For to some of us the play has become part of our lives and after the last curtain call we will realize, undoubtedly, that something more than a good play is over.

ACKNOWLEDGMENTS

We thank Martha Kimmerle and Karina Masur for help in preparation of this manuscript. Part of the support for the experimental work cited in this communication has come from National Institutes of Health and Deutsche Forschungsgemeinschaft grants.

Literature Cited

1. Alper, C. 1981. Complement and the MHC. In *The Role of the Major Histocompatibility Complex in Immunobiology*, ed. M. Dorf, pp. 173–220. New York: Garland STPM
2. Basch, R. S., Stetson, C. A. 1963. Quantitative studies on histocompatibility antigens of the mouse. *Transplantation* 1:469–80
3. Baxevanis, C. N., Nagy, Z. A., Klein, J. 1981. A novel type of T-T cell interaction removes the requirement for *I-B* region in the *H-2* complex. *Proc. Natl. Acad. Sci. USA* 78:3809–13
4. Berzofsky, J. A., Pisetsky, D. S., Killion, D. J., Hicks, G., Sachs, D. H. 1981. Ir genes of different high responder haplotypes for staphylococcal nuclease are not allelic. *J. Immunol.* 127:2453–55
5. Bodmer, W. F. 1973. A new genetic model for allelism at histocompatibility and other complex loci: polymorphism for control of gene expression. *Transplant. Proc.* 5:1471–75
6. Bodmer, W. F. 1976. HLA: A super supergene. *Harvey Lectures Ser.* 72:91–138
7. Curman, B., Östberg, L., Sandberg, L., Malmheden-Eriksson, I., Stalenheim, G., Rask, L., Peterson, P. A. 1975. H-2 linked Ss protein in C4 component of complement. *Nature* 258:243–45
8. Daniel, W. L., Womack, J. E., Henthorn, P. S. 1981. Murine liver arylsulfatase B processing influenced by region on chromosome 17. *Biochem. Genet.* 19:211–55
9. David, C. S., Shreffler, D. C. 1974. Lymphocyte antigens controlled by the *Ir* region of the mouse *H-2* complex. *Transplantation* 17:462–64
10. Delovitch, T., Barber, B. H. 1979. Evidence for two homologous, but nonidentical Ia molecules determined by the *I-EC* subregion. *J. Exp. Med.* 150:100
11. Dizik, M., Elliot, R. W. 1978. A second gene affecting the sialylation of lysosomal α-mannosidase in mouse liver. *Biochem. Genet.* 16:247–60
12. Fachet, J., Andó, I. 1977. Genetic control of contact sensitivity to oxazolone in inbred, H-2 congenic and intra-H-2 recombinant strains of mice. *Eur. J. Immunol.* 7:223–26
13. Figueroa, F., Klein, D., Tewarson, S., Klein, J. 1982. Evidence for placing the *Neu-1* locus within the mouse *H-2* complex. *J. Immunol.* 129:2089–93

14. Figueroa, F., Zaleska-Rutczynska, Z., Kusnierczyk, P., Klein, J. 1983. Cross-reactivity between Qa-1 region and H-2K antigens. *Transplantation.* In press
15. Fischer-Lindahl, K. 1979. Unrestricted killer cells recognize an antigen controlled by a gene linked to *Tla. Immunogenetics* 8:71–76
16. Fischer-Lindahl, K., Bocchieri, M., Riblet, R. 1980. Maternally transmitted target antigen for unrestricted killing by NZB T lymphocytes. *J. Exp. Med.* 152:1583–95
17. Fischer-Lindahl, K., Hausmann, B. 1980. Qed-1—A target for unrestricted killing by T cells. *Eur. J. Immunol.* 10:289–98
18. Fischer-Lindahl, K., Hausmann, B., Flaherty, L. 1982. Polymorphism of a Qa-1-associated antigen defined by cytotoxic T cells. I. Qed-1ᵃ and Qed-1ᵈ. *Eur. J. Immunol.* 12:159–66
19. Fischer-Lindahl, K., Langhorne, J. 1981. Medial histocompatibility antigens. *Scand. J. Immunol.* 14:643–54
20. Flaherty, L. 1976. The Tla region of the mouse: Identification of a new serologically defined locus, *Qa-2. Immunogenetics* 3:533–39
21. Flaherty, L. 1981. Tla-region antigens. In *The Role of the Major Histocompatibility Complex in Immunobiology*, ed. M. Dorf, pp. 33–57. New York: Garland STPM
22. Flaherty, L., Zimmermann, D., Hansen, T. H. 1978. Further serological analysis of the Qa antigens: Analysis of an anti-H-2.28 serum. *Immunogenetics* 6:245–51
23. Forman, J. 1979. *H-2* unrestricted cytotoxic T cell activity against antigens controlled by genes in the Qa/Tla region. *J. Immunol.* 123:2451–55
24. Forman, J., Flaherty, L. 1978. Identification of a new CML target antigen controlled by a gene associated with the *Qa-2* locus. *Immunogenetics* 6:227–33
25. Gachelin, G., Dumas, B., Abastado, J.-P., Cami, B., Papatheakis, J., Kourilsky, P. 1982. Mouse genes coding for the major class I transplantation antigens: a mosaic structure might be related to the antigenic polymorphism. *Ann. Immunol.* 133:3–20
26. Gorer, P. A. 1936. The detection of antigenic differences in mouse erythrocytes by the employment of immune sera. *Br. J. Exp. Pathol.* 17:42–50
27. Gorer, P. A. 1937. The genetic and anti-

genic basis of tumor transplantation. *J. Pathol. Bacteriol.* 44:691–97

28. Götze, D. ed. 1977. *The Major Histocompatibility System in Man and Animals.* Berlin: Springer

29. Hämmerling, G. J. 1976. Tissue distribution of Ia antigens and their expression on lymphocyte subpopulations. *Transplant. Rev.* 30:64–82

30. Hämmerling, G. J., Hämmerling, U., Flaherty, L. 1979. Qa-4 and Qa-5, new murine T-cell antigens governed by the *Tla* region and identified by monoclonal antibodies. *J. Exp. Med.* 150: 108–16

31. Hansen, J. H., Ozato, K., Melino, M. R., Coligan, J. E., Kindt, T. J., Jandinski, J. J., Sachs, D. H. 1981. Immunochemical evidence in two haplotypes for at least three D region-encoded molecules, D, L, and R. *J. Immunol.* 126:1713–16

32. Hiramatsu, K., Ochi, A., Miyatani, S., Segawa, A., Tada, T. 1982. Monoclonal antibodies against unique *I*-region gene products expressed only on mature functional T cells. *Nature* 296:666–68

33. Hurme, M., Chandler, P. R., Heterington, C. M., Simpson, E. 1978. Cytotoxic T-cell responses to H-Y: Correlation with the rejection of syngeneic male skin grafts. *J. Exp. Med.* 147:768–75

34. Iványi, D., Démant, P. 1981. Serological characterization of previously unknown H-2 molecules identified in the products of the K^d and D^k region. *Immunogenetics* 12:397–408

35. Jones, P. P., Murphy, D. B., Hewgill, D., McDevitt, H. O. 1979. Detection of a common polypeptide chain in *I-A* and *I-E* subregion immunoprecipitates. *Mol. Immunol.* 16:51–60

36. Juretić, A., Nagy, Z. A., Klein, J. 1981. Detection of CML determinants associated with *H-2* controlled E_β and E_α chains. *Nature* 298:308–10

37. Kanno, M., Kobayashi, S., Tokuhisa, T., Takei, I., Shinohara, N., Taniguchi, M. 1981. Monoclonal antibodies that recognize the product controlled by a gene in the *I-J* subregion of the mouse *H-2* complex. *J. Exp. Med.* 154:1290–304

38. Kastner, D. L., Rich, R. R. 1979. *H-2*-nonrestricted cytotoxic responses to an antigen encoded telomeric to *H-2D*. *J. Immunol.* 122:196–201

39. Kastner, D. L., Rich, R. R., Chu, L. 1979. *Qa-1*-associated antigens. II. Evidence for functional differentiation from H-2K and H-2D antigens. *J. Immunol.* 123:1239–44

40. Kastner, D. L., Rich, R. R., Shen, F.-W. 1979. *Qa-1*-associated antigens. I. Generation of *H-2*-nonrestricted cytotoxic T lymphocytes specific for determinants of the *Qa-1* region. *J. Immunol.* 123:1232–38

41. Klein, D., Klein, J. 1982. Polymorphism of the *Apl (Neu-1)* locus in the mouse. *Immunogenetics* 16:181–84

41a. Klein, D., Tewarson, S., Figueroa, F., Klein, J. 1982. The minimal length of the differential segment in *H-2* congenic lines. *Immunogenetics* 16:319–28

42. Klein, J. 1975. *Biology of the Mouse Histocompatibility-2 Complex.* New York: Springer

43. Klein, J. 1976. An attempt at an interpretation of the mouse *H-2* complex. *Contemp. Topics Immunobiol.* 5:297–336

44. Klein, J. 1977. Evolution and function of the major histocompatibility system: facts and speculations. In *The Major Histocompatibility System in Man and Animals,* ed. D. Götze, pp. 339–78. Berlin: Springer

45. Klein, J. 1978. *H-2* mutations: Their genetics and effect on immune functions. *Adv. Immunol.* 26:55–146

46. Klein, J. 1979. The major histocompatibility complex of the mouse. *Science* 203:516–21

47. Klein, J. 1981. The histocompatibility-2 (H-2) complex. In *The Mouse in Biomedical Research,* ed. H. L. Foster, J. D. Small, J. G. Fox, 1:119–58. New York: Academic

48. Klein, J. 1982. Evolution and function of the major histocompatibility complex. In *Histocompatibility Antigens: Structure and Function Receptors and Recognition, Series B,* ed. P. Parham, J. S. Strominger, 14:221–39 London: Chapman and Hall

49. Klein, J., Chiang, C.-L. 1978. A new locus (*H-2T*) at the *D* end of the *H-2* complex. *Immunogenetics* 6:235–43

50. Klein, J., Figueroa, F. 1981. Polymorphism of the mouse *H-2* loci. *Immunol. Rev.* 60:23–57

51. Klein, J., Figueroa, F., Klein, D. 1982. *H-2* Haplotypes, genes and antigens: second listing. I. Non-H-2 genes on chromosome 17. *Immunogenetics* 16: 285–317

52. Klein, J., Hauptfeld, V., Hauptfeld, M. 1974. Involvement of H-2 regions in immune reactions. In *Progress in Immunology II,* ed. L. Brent, J. Holborow,

3:197–206. Amsterdam Elsevier: North-Holland
53. Klein, J., Juretić, A., Baxevanis, C. N., Nagy, Z. A. 1981. The traditional and the new version of the mouse *H-2* complex. *Nature* 291:455–60
54. Klein, J., Nagy, Z. A. 1982. MHC restriction and *Ir* genes. *Adv. Cancer Res.* 37:233–317
55. Klein, J., Shreffler, D. C. 1971. The *H-2* model for major histocompatibility systems. *Transplant. Rev.* 6:3–29
56. Koch, N., Hämmerling, G. J. 1981. Ia antigens contain two distinct forms of β chain. *Immunogenetics* 14:437–44
57. Koch, N., Hämmerling, G. J., Szymura, J., Wabl, M. R. 1982. Ia associated I_i chain is not encoded by chromosome 17 of the mouse. *Immunogenetics.* 16:603–6
57a. Koch, N., Koch, S., Hämmerling, G. 1982. Ia invariant chain detected on lymphocyte surfaces by monoclonal antibody. *Nature* 299:645
58. Lachman, P. J., Grennan, D., Martin, A., Démant, P. 1975. Identification of Ss protein as murine C4. *Nature* 258: 242–43
59. Lafuse, W. P., Corser, P. S., David, C. S. 1982. Biochemical evidence for multiple I-E Ia molecules. *Immunogenetics* 15:365–75
60. Lalley, P. A., Shows, T. B. 1977. Lysosomal acid phosphatase deficiency: liver specific variant in the mouse. *Genetics* 87:305–17
61. Larhammar, D., Wiman, K., Schenning, L., Claesson, L., Gustafsson, K., Peterson, P. A., Rask, L. 1981. Evolutionary relationship between HLA-DR antigen β-chains, HLA-A, B, C antigen subunits and immunoglobulin chains. *Scand. J. Immunol.* 14:617–22
62. Lieberman, R., Paul, W. E., Humphrey, W. Jr., Stimpfling, J. H. 1972. *H-2*-linked immune response (*Ir*) genes. Independent loci for *Ir-IgG* and *Ir-IgA* genes. *J. Exp. Med.* 136:1231–40
63. Livnat, S., Klein, J., Bach, F. H. 1973. Graft versus host reaction in strains of mice identical for H-2K and H-2D antigens. *Nature* 243:42–44
64. Lozner, E. C., Sachs, D. H., Shearer, G. M. 1974. Genetic control of the immune response to staphylococcal nuclease. I. Ir-Nase: Control of the antibody response to nuclease by the Ir region of the mouse H-2 complex. *J. Exp. Med.* 139:1204–14
65. Margulies, D. H., Evans, G. A., Flaherty, L., Seidman, J. G. 1982. *H-2*-like

genes in the *Tla* region of mouse chromosome 17. *Nature* 295:168–79
66. Marshak, A., Doherty, P. C., Wilson, D. B. 1977. The control of specificity of cytotoxic T lymphocytes by the major histocompatibility complex (Ag-B) in rats and identification of a new alloantigen system showing no Ag-B restriction. *J. Exp. Med.* 146:1773–90
67. McDevitt, H. O., Chinitz, A. 1969. Genetic control of the antibody response: Relationship between immune response and histocompatibility (*H-2*) type. *Science* 163:1207–8
68. Melchers, I., Rajewsky, K., Shreffler, D. C. 1973. Ir-LDH$_B$: Map position and functional analysis. *Eur. J. Immunol.* 3:754–61
69. Meo, T., Krasteff, T., Shreffler, C. D. 1975. Immunochemical characterization of murine H-2 controlled Ss (serum substance) protein through identification of its human homologue as the fourth component of complement. *Proc. Natl. Acad. Sci. USA* 72:4536–40
70. Monaco, J. J., McDevitt, H. O. 1982. Identification of a fourth class of proteins linked to the murine major histocompatibility complex. *Proc. Natl. Acad. Sci. USA* 79:3001–5
71. Murphy, D. B. 1978. I-J subregion of the murine H-2 gene complex. *Springer Sem. Immunopathol.* 1:111–31
72. Murphy, D. B., Herzenberg, L. A., Okumura, K., Herzenberg, L. A., McDevitt, H. O. 1976. A new I subregion (*I-J*) marked by a locus (*Ia-4*) controlling surface determinants on suppressor T lymphocytes. *J. Exp. Med.* 144:699–712
73. Nathenson, S. G., Uehara, H., Ewenstein, B. M., Kindt, T. J., Coligan, J. E. 1981. Primary structural analysis of the transplantation antigens of the murine H-2 major histocompatibility complex. *Ann. Rev. Biochem.* 50:1025–51
74. Okuda, K., David, C. S. 1978. A new lymphocyte-activating determinant locus expressed on T cells, and mapping in I-C subregions. *J. Exp. Med.* 147: 1028–36
75. Okuda, M., Stanton, T. H., Kuppers, R. C., Henney, C. S. 1981. The differentiation of cytotoxic T cells in vitro. IV. Interleukin-2 production in primary mixed lymphocyte cultures involves cooperation between Qa-1$^+$ and Qa-1$^-$ "helper" T cells. *J. Immunol.* 126: 1635–39
76. Orr, H. T., Lancet, D., Robb, R. J., Lopez de Castro, J. A., Strominger, J. L. 1979. The heavy chain of human his-

tocompatibility antigen HLA-B7 contains an immunoglobulin-like region. *Nature* 282:266–70

77. Ozato, K., Sachs, D. H. 1982. Detection of at least two distinct mouse I-E antigen molecules by the use of a monoclonal antibody. *J. Immunol.* 128: 807–10

78. Parmiani, G., Carbone, G., Invernizzi, G., Pierotti, M. A., Sensi, M. L., Rogers, M. J., Apella, E. 1979. Alien histocompatibility antigens on tumor cells. *Immunogenetics* 9:1–24

79. Pelkonen, J. L. T., Kaartinen, M., Karjalainen, K., Mäkele, O. 1979. A hapten-specific Ir gene. *J. Immunol.* 123: 1558–64

80. Peters, J., Swallow, D. M., Andrews, S. J., Evans, L. 1981. A gene (*Neu-1*) on chromosome 17 of the mouse affects acid α-glucosidase and codes for neuraminidase. *Genet. Res.* 38:47–55

81. Porter, R. R., Reid, K. B. M. 1979. Activation of the complement system by antibody-antigen complexes: the classical pathway. *Adv. Prot. Chem.* 33:1–71

82. Rich, R. R., Sedberry, D. A., Kastner, D. L., Chu, L. 1979. Primary in vitro cytotoxic response of NZB spleen cells to *Qa-1ᵇ*-associated antigenic determinants. *J. Exp. Med.* 150:1555–60

83. Rich, S. S., David, C. S., Rich, R. R. 1979. Regulatory mechanisms in cell-mediated immune responses. VII. Presence of I-C subregion determinants on mixed leukocytes reaction suppressor factor. *J. Exp. Med.* 149:114–26

84. Richardson, J. S., Richardson, D. C., Thomas, K. A., Silverton, E. W., Davies, D. R. 1976. Similarity of three-dimensional structure between the immunoglobulin domain and the copper, zinc superoxide dismutase subunit. *J. Mol. Biol.* 102:221–35

85. Rothenberg, E. 1982. A specific biosynthetic marker for immature thymic lymphoblasts. Active synthesis of thymus-leukemia antigen restricted to proliferating cells. *J. Exp. Med.* 155: 140–54

86. Sandrin, M. S., McKenzie, I. F. C. 1981. Production of a cytotoxic anti-Ia.6 antibody. *Immunogenetics* 14: 345–50

87. Shreffler, D. C., Owen, R. D. 1963. A serologically detected variant in mouse serum: inheritance and association with the histocompatibility-2 locus. *Genetics* 48:9–25

88. Snell, G. D. 1968. The H-2 locus of the mouse: Observations and speculations concerning its comparative genetics and

its polymorphism. *Folia Biol.* 14: 335–58

89. Snell, G. D., Dausset, J., Nathenson, S. 1976. *Histocompatibility.* New York: Academic

90. Soloski, M. J., Uhr, J. W., Flaherty, L., Vitetta, E. S. 1981. Qa-2, H-2K, and H-2D alloantigens evolved from a common ancestral gene. *J. Exp. Med.* 153:1080–93

91. Stanton, T. H., Boyse, E. A. 1976. A new serologically defined locus, *Qa-1*, in the Tla-region of the mouse. *Immunogenetics* 3:525–31

92. Stanton, T. H., Carbon, S., Maynard, M. 1981. Recognition of alternate alleles and mapping of the *Qa-1* locus. *J. Immunol.* 127:1640–43

93. Steinmetz, M., Moore, K. W., Frelinger, J. G., Sher, B. T., Shen, F.-W., Boyse, E. A., Hood, L. 1981. A pseudogene homologous to mouse transplantation antigens: transplantation antigens are encoded by eight exons that correlate with protein domains. *Cell* 25: 683–92

94. Steinmetz, M., Winoto, A., Minard, K., Hood, L. 1982. Clusters of genes encoding mouse transplantation antigens. *Cell* 28:489–98

95. Taniguchi, M., Tokuhisa, T., Kanno, M., Yaoita, Y., Shimizu, A., Honjo, T. 1982. Reconstitution of antigen-specific suppressor activity with translation products of mRNA. *Nature* 298: 172–74

96. Uhr, J. W., Capra, J. D., Vitetta, E. S., Cook, R. G. 1979. Organization of the immune response genes. Both subunits of murine I-A and I-E/C molecules are encoded within the I region. *Science* 206:292–97

97. Urba, W. J., Hildemann, W. H. 1978. *H-2*-linked recessive *Ir* gene regulation of high antibody responsiveness to TNP hapten conjugated to autologous albumin. *Immunogenetics* 6:433–45

98. Vitetta, E. S., Capra, J. D. 1978. The protein products of the murine 17th chromosome: genetics and structure. *Adv. Immunol.* 26:147–93

99. Waltenbaugh, C. 1981. Regulation of immune response by *I-J* gene products. I. Production and characterization of anti-I-J monoclonal antibodies. *J. Exp. Med.* 154:1570–83

100. Wernet, D., Klein, J. 1979. Unrestricted cell-mediated lympholysis to antigens linked to the Tla locus in the mouse. *Immunogenetics* 8:361–65

101. Wernet, D., Klein, J. 1981. Cell-mediated lympholysis in H-2K/D iden-

tical congenic strain combinations. In *Control of Cellular Division and Development, Part A,* ed. D. Cunningham, E. Goldwasser, J. Watson, C. F. Fox, pp. 573–77. New York: Liss

102. Williams, A. F., Gagnon, J. 1982. Neuronal cell Thy-1 glycoprotein: homology with immunoglobulin. *Science* 216:696–703

103. Womack, J. E., David, C. S. 1982. Mouse gene for neuraminidase activity (*Neu-1*) maps to the *D* end of *H-2. Immunogenetics* 16:177–80

104. Womack, J. E., Eicher, E. M. 1977. Liver-specific lysosomal acid phosphatase deficiency (*Apl*) on mouse chromosome 17. *Mol. Gen. Genet.* 155:315–17

105. Womack, J. E., Yan, D. L. S., Potier, M. 1981. Gene for neuraminidase activity on mouse chromosome 17 near H-2: Pleiotropic effects on multiple hydrolases. *Science* 212:63–65

106. Zinkernagel, R. P., Doherty, P. C. 1979. MHC restricted cytotoxic T-cells: Studies on the biological role of polymorphic major transplantation antigens determining T-cell restriction specificity. *Adv. Immunol.* 27:51–177

Ann. Rev. Immunol. 1983. 1:143–73

IMMUNOBIOLOGY OF TISSUE TRANSPLANTATION: A RETURN TO THE PASSENGER LEUKOCYTE CONCEPT

Kevin J. Lafferty, Stephen J. Prowse, and Charmaine J. Simeonovic

Transplantation Biology Unit, John Curtin School of Medical Research, PO Box 334, Canberra City, ACT 2601, Australia

Hilary S. Warren

Cancer Research Unit, Woden Valley Hospital, Canberra, Australia

INTRODUCTION

Once Medawar (76) had established the immunological basis of the allograft reaction, tissue antigen was seen to be the barrier to the grafting of tissues and organs. At the root of this notion was the idea that antigen recognition caused the activation of the potentially responsive lymphocyte. Antigen, like an embryonic inducer, was thought to direct the final differentiation of specific immunocyte clones (77). There is no doubt that antigen recognition is involved in the processes of tissue rejection. However, although antigen recognition is responsible for graft rejection, it may not be correct to deduce that transplantation antigen itself constitutes the major barrier to grafting. This is because antigen recognition alone may not be sufficient for lymphocyte activation.

The idea that tissue antigen was the obstacle to grafting has had a profound effect on clinical transplantation. If antigen is the barrier, then we must either match the tissues to the recipient or treat the recipient in a way that renders it no longer responsive to the antigenic challenge of the graft. Tissue matching plus recipient immunosuppression have been used with

143

0732-0582/83/0410-0143$02.00

considerable success, more so the latter, in the clinical development of organ and tissue transplantation (74). In this article, we first review recent developments concerning the mechanism of tissue rejection. We explore in some depth the idea that antigen recognition alone is not sufficient for lymphocyte activation. This leads to the conclusion that transplantation antigen, whether it be class I or class II major histocompatibility complex (MHC) antigen, is not the primary barrier to grafting, and we return to the concept initially proposed by Snell (90), that "passenger" leukocytes carried in the donor tissue provide the major immunogenic stimulus for the host. It therefore makes good theoretical and practical sense to attempt to alter tissue immunogenicity by treating the tissue to be grafted rather than the recipient. We review the results of this approach to the transplantation problem.

EFFECTOR MECHANISMS IN GRAFT REJECTION

Antibody was initially thought to play little part in the process of graft rejection (13, 78, 79). It is now clear that this is not the case. The failure of initial attempts to transfer antibody passively and induce skin graft rejection can be attributed to deficiencies in the complement system of rodents (35); antibody-mediated hyperacute rejection of skin allografts and xenografts has now been demonstrated in mice, but only when an exogeneous source of complement is transferred along with the specific antibody (21, 75). Different tissues also vary in their susceptibility to antibody-mediated rejection. Allografts prepared by primary vascular anastomosis are highly sensitive to antibody and complement and retain this sensitivity for a prolonged period (44).

In contrast, skin grafts show marked variations with time in their susceptibility to attack by antibody and complement (36, 50, 51). This effect is related to the replacement of donor endothelium in the skin by host cells (51). Jooste et al (50, 51) showed that soon after grafting mice with rat skin, rat cell surface antigens could be demonstrated on the graft vasculature. As the graft became less sensitive to anti-rat antibody-mediated damage, mouse cell surface antigens were detected on the endothelium. Eventually, all the graft vasculature was of recipient origin, and the graft was resistant to passively transferred antibody (51).

Administration of immune serum causes rapid rejection of islet allografts in tolerant rats (83). However, when endothelium is destroyed by a period of organ culture prior to transplantation (88), grafts are resistant to the action of antibody and complement given either at the time of transplantation (M. Agostino, unpublished observations) or approximately 100 days after grafting (1). These findings suggest that antibody-mediated damage is a vascular phenomenon and that tissue parenchymal cells may be less sensitive to this form of damage. Endothelial cells are readily lysed by alloantiserum or xenoantiserum and complement (26), and the target antigens for this antibody-mediated damage are likely to be any expressed on the surface of vascular endothelium. In man, both class I and class II MHC antigens as well as blood group antigens are expressed on endothelial cells and could provide a target for antibody-mediated damage (46). Mouse skin can be acutely rejected by antibody directed against class I alloantigens (27, 75), but not by antisera against class II antigens (75). This probably reflects the lack of Ia antigens expression on mouse endothelium (89).

A mechanism for antibody-mediated graft rejection has been described by Bogman et al (15). Small amounts of antibody binding to the vessel wall fix complement, resulting in endothelial damage and the release of chemotactic factors. This results in the attraction and binding, through complement component receptors, of polymorphonuclear cells, which cause further endothelial damage. Following the exposure of subendothelial tissue, thrombosis occurs. In the presence of large amounts of antibody, the endothelium is likely to be rapidly destroyed exposing thrombogenic tissue and thus causing thrombosis without polymorphonuclear cell involvement. Where animals are deficient in complement components, hyperacute rejection will not naturally occur.

Cellular Rejection Mechanisms

The idea that allograft rejection was a cell-mediated phenomenon came from early studies of Mitchison (79) and Billingham et al (13). Accelerated rejection of an allogeneic tumor graft could be transferred to a naive animal by cells from the draining lymph nodes of tumor-bearing animals (79). Billingham et al (14) showed that immunity to a skin allograft could also be adoptively transferred in a similar manner. It was later shown that neonatally tolerant mice carrying skin allografts of the tolerated genotype would reject these grafts when injected with normal lymphoid cells, syngeneic with the recipient; if the lymphoid cells came from a sensitized animal, then a "second set" rejection occurred (14).

There is now good evidence that graft rejection can be triggered by the interaction of sensitized T lymphocytes and graft antigens. Transfer of cell populations to sublethally irradiated rats carrying heart allografts demonstrated that long-lived, recirculating, Ig-negative cells were responsible for initiating rejection (39); the transfer of cells from sensitized animals showed that memory was carried by long-lived, non-recirculating, Ig-negative cells (40). Loveland and co-workers studied the rejection of skin grafts from adult thymectomized, irradiated, bone-marrow reconstituted (ATXBM) mice by transfer of cells from sensitized donors and showed that the Lyt 1^+2^- T cell could trigger rejection (70, 71). These authors went on to conclude that Lyt 1^+2^+ T cells were not involved in the rejection process (70). Talmage's group (112), working with cultured thyroid allografts, observed a long delay in rejection following adoptive immunization and a lack of correlation with cytotoxic activity in the transferred cells. They also suggested that cytotoxic T cells may not be required for tissue rejection.

Although Lyt 1^+2^- cells may mediate skin graft and possibly thyroid rejection, it would seem unwise on the basis of the above evidence to exclude completely a role for Lyt 1^+2^+ cells. Pancreatic islets can be transplanted to normal allogeneic mice if the islet tissue is first cultured in an oxygen-rich atmosphere (16–18). This tissue is functional and will reverse chemically induced diabetes (17, 18). In this system, the transfer of sensitized Lyt 1^+2^+ T cells is required to initiate acute graft rejection (Figure 1).

The subclass of T lymphocyte responsible for graft rejection may depend on the type of tissue transplanted, which in turn reflects the particular class

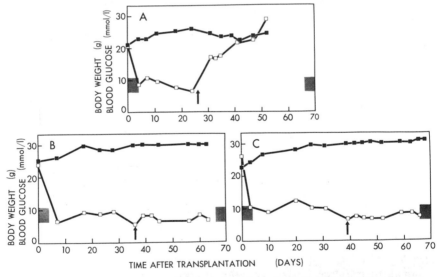

Figure 1 Streptozotocin-induced diabetic CBA mice were grafted with cultured islet clusters. Blood glucose levels (□) and body weight (■) were monitored. Mice became normoglycemic within 12 days. 5×10^7 CBA anti-P815 immune spleen cells or remaining cells after treatment with antibody and complement were injected i.v. (indicated by arrow). The immune cells were untreated (*A*), treated with anti-Thy-1.2 and complement (*B*), or treated with anti-Lyt 2.1 and complement (*C*). The untreated cells had a total cytotoxic activity of 5.2 $\log_{10}$ cytotoxic units. The treated cells had a cytotoxic activity which was less than 3.9 $\log_{10}$ cytotoxic units. The results presented are for one mouse in each group and are representative of the four mice in each treatment group. The normal range for blood glucose for CBA male mice (mean ±2 SD) is indicated by the darkly hatched area (see 87).

of alloantigen expressed on the graft. Cultured islet allografts express only class I antigens (89), whereas cells present in skin grafts express both class I and class II antigens (41). Class I and class II antigens cause the activation of Lyt 1^+2^+ and Lyt 1^+2^- cells, respectively (24). Lyt 1^+2^+ could mediate graft rejection by a cytotoxic mechanism (24, 68). Also, both Lyt 1^+2^+ and Lyt 1^+2^- T cells, and their human equivalents, have been shown to produce lymphokine when appropriately triggered (3, 72). Thus, cells of either subclass could mediate graft rejection via lymphokine release and an associated inflammatory response. The protracted rejection of thyroid allografts (112) by adoptively immunized animals would be consistent with such a mechanism. The acute rejection of islet allografts, mediated by Lyt 1^+2^+ T cells, would be more consistent with a direct cytotoxic effect of the immune cells. We should stress, however, that at this stage we do not know the actual mechanism by which Lyt 1^+2^+ T cells mediate pancreatic islet rejection. The differential sensitivity of skin, thyroid, and islet tissue to rejection by transferred immune T cells requires further investigation.

Although T cells may initiate graft rejection, other inflammatory cells enter the rejection site and greatly outnumber T cells as the process of rejection develops (5, 48, 106). Cytotoxic activity in rejecting allografts has been shown to be mediated by both T cells and other non-thymus-derived leukocytes (4, 5). Cells involved in antibody-dependent cell-mediated cytotoxicity have also been found in rejecting grafts. The general picture that emerges is one of antigen-specific, T cell-mediated triggering of a rejection process that is probably mediated by multiple specific and nonspecific effector mechanisms.

THE TRANSPLANTATION BARRIER

When Snell in 1957 (99) suggested that leukocytes associated with the transplanted tissue were a major source of tissue immunogenicity, he was formally stating the common observation that immunization of a recipient animal with donor spleen or lymph node cells would sensitize to a tumor allograft, whereas antigen extracts of the tumor were only weakly immunogenic and, under certain conditions (52), would promote the growth of the tumor graft. It was thought, at the time, that antigen density or the form of antigen presentation on the surface of viable leukocytes rendered these transplantation antigens more immunogenic.

Studies of the graft-versus-host reaction (GVHR), carried out in the late 1960s, rekindled interest in the role leukocytes in the tissues play in alloreactivity (29–31, 63). Elkins' investigations in rats implied a major role for leukocytes in the inflammatory reaction produced when parental strain lymphocytes were introduced under the kidney capsule of F1 recipients

(29–31). Steinmuller showed that allogeneic leukocytes carried within a skin graft were highly immunogenic. Skin isografts taken from strain A mice rendered neonatally tolerant of strain B, and transplanted to naive A mice, immunized the latter against strain B skin allografts (102). Since the tolerant donors were hemopoetic chimeras, the phenomenon was explained by the passive transfer of allogeneic leukocytes via the primary skin isograft. In support of this proposal, the immunizing capacity of such skin isografts was shown to be a function of the number of allogeneic leukocytes present and the degree of leukocyte chimerism of the donors (103). Our own studies in the chicken embryo lead us to conclude that the interaction of lymphocytes with lymphoreticular components in the graft was required for the activation of the allograft response (63). Thus, the transfer of adult lymphoid cells to syngeneic embryos grafted with allogeneic tissues would confer on the embryo the capacity to reject the allograft. Strong reactions were seen with grafts of bone, with its associated marrow, spleen, and fetal liver. However, heart muscle allografts, cleared of blood cells by perfusion prior to transplantation, were not destroyed. In contrast, perfusion of the embryonic heart with leukocytes prior to grafting resulted in the activation of a violent allogeneic reaction that resulted in total allograft destruction. The intensity of the allograft reaction was correlated with the lymphoreticular cell content of the tissue transplanted.

The idea that something over and above simple antigen recognition was involved in the triggering of the allograft response came from a study of the species specificity of the GVHR in the chicken embryo. Simonsen (98) had clearly shown this reaction to be an immune phenomenon initiated when lymphoid cells from adult birds were introduced into genetically different embryos that were unable, because of their immaturity, to reject the grafted cells. However, these reactions proved to be species specific. For example, pigeon spleen cells failed to induce a GVHR when injected into the developing chicken embryo (63). This failure to react was not due to a failure of the xenogeneic cells to survive in the embryonic host; pigeon spleen fragments survived quite well when grafted to the chorioallantoic membrane of the chicken embryo, yet caused no damage to the undoubtedly foreign host (63). Moreover, pigeon cells were capable of expressing immunologic function in a xenogeneic (chicken) environment. Pigeon spleen cells rejected allogeneic pigeon bone grafts transplanted to the same chicken embryo host (63). We concluded from this evidence that the failure of the xenogeneic pigeon cells to cause damage when introduced into the chicken embryo meant that contact with foreign antigen was not a sufficient requirement for the initiation of transplantation reactions.

Similar conclusions can be derived from the study of T cell activation in vitro. When normal T cells are cultured with allogeneic leukocytes, some are activated to proliferate and a proportion of the activated cells differentiate into cytotoxic cells reactive to MHC antigens of the stimulating cell. Early attempts to explain this phenomenon assumed that antigen binding by receptors on the T cells caused their activation (77). According to such a model, one signal (antigen binding) was sufficient for lymphocyte activation, and the presentation of alloantigen alone should induce the differentiation of lymphocyte clones. This prediction of the one-signal model was not supported by experimental observation. The capacity of allogeneic cells to stimulate T cell activation is a function of metabolically active cells (42, 45, 64, 69, 92). Cells killed by any of a variety of procedures do not stimulate, although they can be shown to express antigen (Table 1). Furthermore, not all viable cells have the capacity to stimulate allogeneic T cells in vitro. Antigen-bearing non-lymphoid cells such as fibroblasts, erythrocytes, or platelets do not stimulate normal allogeneic lymphocytes in vitro (43). Also, the capacity to stimulate is phylogenetically restricted and the response to nonconcordant xenogeneic leucocytes is generally poor or nonexistent (113). Clearly, we are faced, once again, with evidence that something other than simple antigen recognition is involved in T cell responses to MHC antigen.

The Two-Signal Model For T-Cell Activation

In an attempt to explain these phenomena, Lafferty & Cunningham (62) took up the Bretscher-Cohn suggestion that two signals were required for lymphocyte activation (19) and developed a two-signal model for the initial step in T cell activation (Figure 2). This model postulated that a stimulator cell (S^+) was required for the presentation of antigen to the potentially responsive T cell. Antigen binding by the T cell provides signal one for T cell activation. The second signal was postulated to be an inductive mole-

Table 1   Two signals, antigen and a source of CoS activity, are required to activate the in vitro T-cell response to alloantigen

Cells used for in vitro stimulation of C57B1 lymphocytes ($H–2^b$)	Cytotoxic activity ($\log_{10}$ CU/culture)	
	No CoS[a] activity added	CoS activity added
γ-Irradiated P815^b (S^+)	5.4	6.0
UV-irradiated P815 (S^-)	< 2.5	6.1
γ-Irradiated CaD2^b (S^-)	< 3.7	6.0
None	< 3.7	< 3.7

[a] CoS activity provided by the supernatant of Con A-activated spleen cells (87, 111). Reproduced with permission (64a).
[b] P815 is a lymphoreticular tumor and CaD2 is an epithelial tumor of the DBA/2 mouse strain.

cule, said to express co-stimulator (CoS) activity. Once the responsive cell was activated, factors may be released that greatly amplify the overall level of proliferation in the population; this model was only concerned with the events leading to the activation of the responsive cell (25, 62). The species specificity of mixed leukocyte reactions was seen to be expressed at the level of the second signal. Provision of signal one alone does not cause T-cell activation and may have a negative, i.e. suppressive, effect (Figure 2) (see Appendix).

Two Signals Required for Cytotoxic T-Cell Activation

There is now clear evidence in support of the two-signal model for T-cell activation by alloantigen (see Table 1) and it has been shown that the species specificity of alloreactivity is expressed at the level of the second signal, CoS activity (113). Using cloned tumor lines as the source of allogeneic stimulator cells, Talmage et al (111) demonstrated the existence of stimulating (S^+) and nonstimulating (S^-) tumor lines. Tumors of lymphoreticular origin expressed the S^+ phenotype, whereas the S^- tumor was an epithelial cell line. The stimulator cells must be metabolically active to express their S^+ function (111); metabolic inactivation of the S^+ population with UV irradiation destroys its capacity to activate a cytotoxic T-cell response. Table 1 shows the experimental evidence for the two-signal model of cytotoxic T cell activation. The lymphokine used as a source of the second signal is the

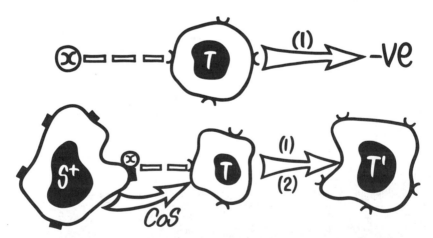

Figure 2 Diagrammatic representation of the requirement for T-cell activation. Binding of antigen, *x,* by the T-cell receptor does not cause T-cell activation. T-cell activation occurs when two signals are provided for the responsive T cell. Antigen binding by the T-cell receptor provides signal 1 (1), and a stimulator cell (S^+) provides an inductive molecule, CoS activity, which is the second signal (2) for T-cell activation. Interaction with the control structure (dark blocks) on the surface of the stimulator cell triggers CoS release.

supernatant of Con A-activated spleen cells (87, 111), which provides a source of CoS activity. Neither antigen provided on the S⁻ tumor cells nor CoS activity alone provided a sufficient requirement for cytotoxic T cell activation. However, a strong cytotoxic T cell response was activated when both antigen and CoS activity were provided for the potentially responsive T cell. Clearly, the S⁻ cells provide all the antigenic requirements for T cell activation, but alloantigen alone does not activate the cytotoxic cell.

A careful kinetic analysis of cytotoxic T cell activation has demonstrated two phases in the activation process (59).

$$T \xrightarrow[2]{1} T' \qquad \text{(a)}.$$

$$T' \xrightarrow[2]{} nT' \qquad \text{(b)}.$$

Resting T cells (T) require two signals for their conversion to the differentiated state (T'); T' cells can express cytotoxic activity. T' cells proliferate in the presence of lymphokine (clonal expansion). Interleukin 1 (IL1) has been shown to provide the second signal for reaction (a) and interleukin 2 (IL2 or T cell growth factor) will induce clonal expansion of the activated T cell, reaction (b) (2a); see appendix.

Mitogen-induced activation of T cells has also been shown to be a two-signal process in which mitogen (Con A) provides signal one and Ia⁺ spleen cells are required for the generation of the second signal (2, 66), which also appears to be provided by IL1 (66).

Role of Class I and Class II MHC Antigens

Bach et al (7), arguing from an immunogenetic viewpoint, developed an alternative two-signal model for allogeneic T cell activation in which recognition of the Ia epitope on the allogeneic cell activated the helper T cell subset, which then provided the second signal for cytotoxic T cell activation. There is now evidence, however, that recognition of the class II antigen is not required for cytotoxic T cell activation, the clearest being the demonstration of cytotoxic T cell activation in the situation where the stimulator and responder strains differ only in the K region of MHC (6–8, 82). Also, Batchelor's group (10) has examined the response of lymphocytes to the class II MHC antigen and shown that recognition of the Ia epitope alone is not a sufficient requirement for activation of the helper lymphocyte subset.

T cells responsive to class I MHC antigens belong to the Lyt 1⁺2⁺3⁺ cytotoxic subset whereas those responsive to class II MHC antigens are part of the Lyt 1⁺2⁻3⁻ helper subset (24). Similar signalling requirements are required for the activation of cells of either subset (59). That is, reactions

(a) and (b) describe the activation of both cytotoxic and helper T cells. We now also know that cells of both the cytotoxic and the helper subsets produce equivalent amounts of lymphokine when triggered by antigen alone (3, 72). The clearly established synergy between class I and class II reactive T cells for the in vitro generation of cytotoxic T cell response can be attributed to the augmentation of IL2 production in the culture system. However, such synergistic effects have not been clearly demonstrated in vivo and allografts are promptly rejected when class I antigenic differences only are recognized (100, 105).

This does not mean that Ia$^+$ cells are unimportant in the process of allogeneic T cell activation. Inactivation of these cells with anti-I-region antibody and complement removes the stimulating capacity of the cell population (93, 114). However, not all Ia$^+$ cells stimulate (37, 67, 115), and some Ia$^-$ cells, such as P815, do stimulate (Table 1). That is, although there is a strong correlation between Ia positivity and the expression of the S$^+$ phenotype, exceptions to the correlation show that recognition of Ia antigen itself cannot provide the source of the second signal required for cytotoxic T cell activation. Nevertheless, the Ia$^+$ cell is probably the cell responsible for the stimulation of both class I and class II MHC-reactive T cells in vivo. The most likely candidate for the physiologically relevant stimulator cell is the Ia-rich dendritic cell (67).

Such a model for immunocyte induction has a profound influence on the way we view allogeneic interactions. In the case of allogeneic T-cell activation, the antigen is built into the surface of the allogeneic cell. Cells of the S$^+$ phenotype have the capacity to produce CoS activity; other cells are S$^-$. Thus, only antigen-bearing S$^+$ cells will stimulate the allogeneic T cells, since it is only in this situation that the two signals required for T cell activation can be provided. Thus, alloantigen on the surface of metabolically active S$^+$ cells will be highly immunogenic for T cells in vitro, whereas the same antigen presented on the surface of cells of the S$^-$ phenotype will not be immunogenic (see Appendix for detailed treatment). Since the S$^+$ phenotype is only expressed by cells of lymphoreticular origin (42, 43, 63, 111), we can see why leukocytes are strong stimulators of allogeneic T cells in vitro.

Passenger Leukocyte and Graft Rejection

It follows from the above discussion that leukocytes carried within the grafted tissue play a major role in regulation of tissue immunogenicity. Stimulator cells (dendritic cells?) will activate recipient T cells to generate a graft-specific response (see Appendix). Antigen carried on graft parenchymal cells will not activate T cells directly because the parenchymal cells cannot provide a source of CoS activity. Free antigen, shed from the graft, could be carried to the immune system of the recipient and there presented on recipient stimulator cells to recipient T cells. Such indirect antigen presentation would not lead to the generation of graft-specific T cell response; T cells activated by indirect antigen presentation would be specific for graft antigen in association with the MHC of the recipient (see Appendix). Thus, it is stimulator cells in the graft that constitute the major immunogenic stimulus of the graft and therefore represent the major barrier to tissue grafting. Since stimulator cells of the graft, such as dendritic cells, are Ia$^+$, it follows that Ia$^+$ cells carried within the grafted tissue will constitute a major immunogenic component of the graft. This is not because Ia antigen itself is highly immunogenic (9, 10), but rather that the class II antigen is a marker for the cell that expresses the S$^+$ phenotype (93, 114), i.e. the cell that can provide a source of CoS activity. Against this theoretical background it makes practical sense to attempt a reduction of tissue immunogenicity by removal of leukocytes from the graft prior to transplantation (see Appendix).

MODULATION OF TISSUE IMMUNOGENICITY

Organ Culture Prior to Transplantation

In the early 1970s, several groups investigated the effect of organ culture on the immunogenicity of endocrine organs and certain tumor tissue (65, 79). The idea that organ culture might reduce tissue immunogenicity was not new. There were several reports in the 1930s suggesting clinical benefit when parathyroid tissue was held in organ culture for a period prior to transplantation to patients with hypoparathyroidism (104). However, these studies were not genetically controlled, and without the support of an adequate theoretical base, enthusiasm for the experiments quickly waned. Our interest in organ culture was stimulated by a report from Summerlin et al—subsequently not confirmed—that organ culture prior to transplantation could facilitate the grafting of skin to normal allogeneic recipients (108). Jacobs (49) suggested that antigen expression was modified during the period in organ culture. Such an explanation did not seem very likely, and against the above theoretical background, the effect of organ culture could be readily explained if blood cells in the tissue died or were inactivated during the culture period.

The organ culture technique has proved spectacularly successful in the case of endocrine organ transplantation. Mouse thyroid can be transplanted under the kidney capsule of isogeneic thyroidectomized recipients, where its ability to concentrate radioactive iodine can be used as a measure of graft function (60). Organ culture of thyroid tissue in an atmosphere of 95%

O_2 and 5% CO_2 for 14 days extends allograft survival (61), and after a culture period of 3–4 weeks, allografts show no evidence of rejection over an observation period of 100 days (60). Pretreatment of the tissue donor with cyclophosphamide (300 mg/Kg) 4 and 2 days prior to the harvest of tissue reduced the period of organ culture required to facilitate allograft survival; after this treatment, tissue cultured for approximately 2 weeks could be transplanted to normal allogeneic recipients where it functioned indefinitely (65). Under these conditions both thyroid and parathyroid show indefinite survival in allogeneic recipient mice (65). Similar results were reported with other species (84) and the procedure has been shown to work for xenotransplantation of thyroid and islet tissue from rat to mouse (58, 101).

What is the explanation for this phenomenon? Does antigen modulation occur in organ culture as suggested by Jacobs (49), or does the effect result from the postulated loss of passenger leukocytes? During organ culture there is a rapid degeneration of the vascular bed and blood elements within the cultured tissue (88). The tissue retains antigen, recognizable by immunoferritin labeling (88), and the tissue is rejected when as few as 10^3 viable peritoneal cells of donor origin are injected into the recipient animal at the time the cultured tissue is transplanted (110). Clearly, the cultured allograft carries functionally recognizable antigen. Established thyroid allografts are also rejected when a second uncultured thyroid of donor origin is transplanted to the recipient (65). This effect is antigen-specific since an uncultured third party graft can be rejected, but its rejection does not affect the integrity of the established cultured allograft (65).

Hirschberg & Thorsby (46) suggested that vascular endothelium may provide a major stimulus of allograft immunity. Vascular endothelium degenerates during organ culture (88) and its destruction could account for the reduction in tissue immunogenicity achieved by organ culture. Endothelial cells express MHC antigens and would, therefore, provide a primary target for the attack of activated graft-specific T cells or graft-specific antibody (see above). The question to be addressed is whether or not endothelium provides a major source of stimulating activity for recipient T cells.

Cyclophosphamide treatment of animals for 2 days prior to the harvest of tissues causes a profound drop in the capacity of spleen cells to stimulate allogeneic T cells in culture (60). This treatment of the donor with cyclophosphamide has no obvious effect on thyroid endothelium. However, after this treatment alone, approximately 30% of thyroid allografts function normally over a prolonged observation period (65). If donor endothelium were a major source of tissue immunogenicity we would not expect to observe such an effect. Batchelor's studies using kidney transplantation (9, 11) led to similar conclusions (see below).

Pancreatic Islet Transplantation

Initial attempts to apply the organ culture pretreatment to pancreatic islet transplantation were frustrated by technical difficulties associated with this procedure (16, 56). The loss of tissue immunogenicity during organ culture is an oxygen-dependent phenomenon, thought to result from the sensitivity of leukocytes to oxygen toxicity (109). Pancreatic islet tissue cannot be treated in the same way. Single islets, isolated from the adult pancreas, are extremely sensitive to oxygen toxicity and rapidly degenerate when cultured in 95% oxygen. This toxicity problem can be overcome by allowing groups of approximately 50 islets to aggregate and fuse together (16, 18, 58).

The islet clusters are more resistant to oxygen toxicity and can be successfully allotransplanted in mice after 1 week of culture under these conditions (16, 18, 58). Following organ culture in 95% oxygen, islet allografts of 6 to 7 islet clusters have been shown to reverse streptozotocin-induced diabetes in mice. Blood sugar levels of transplanted animals rapidly return to normal, and the animals become aglycosuric and respond normally to a glucose challenge (17, 18). Uncultured islets, on the other hand, temporarily reverse diabetes, but recipient animals return to the diabetic state within 4 weeks of transplantation (17). Lacy's group (56) achieved similar results in the rat. However, in their initial studies, islets were not cultured in 95% O_2 and allograft acceptance was only achieved when recipient animals were treated with a single dose of ALS at the time of transplantation. Similar results have now been reported in the case of rat islet xenotransplantation to mice (57, 58). Faustman et al (32) demonstrated a dependence of islet immunogenicity on the presence of Ia^+ cells in the transplanted tissue, by achieving allograft survival following anti-Ia serum and complement treatment of the donor tissue prior to transplantation. However, Ia antigen recognition is not required; Morrow et al (80) have shown that islet allograft rejection occurs when there is I region compatibility between donor and recipient. The fact that rejection is dependent on the presence of an Ia^+ cell in the tissues is consistent with the notion that the S^+ stimulator cell carries Ia antigen on its surface (93, 114).

Fetal Pancreas Transplantation

Fetal pancreas can be used as a source of islet tissue for transplantation. Following organ culture or transplantation, the exocrine pancreas degenerates whereas the endocrine tissue continues to differentiate and develop endocrine function. In rodents, transplantation of one fetal pancreas to the kidney capsule has been shown to reverse diabetes (20, 47, 73, 95). However, early attempts to condition fetal pancreas for allotransplantation by organ culture prior to transplantation met with little or no success (73). When we compared the immunogenicity of fetal pancreas and islets isolated from the adult pancreas, in the same strain combinations, we found the fetal pancreas much more difficult to prepare for allotransplantation (94). This difference in behavior of different tissues was related to the large amount of lymphoid tissue associated with the fetal pancreas (94). It is possible to reverse diabetes in CBA mice with tissue from two BALB/c fetal pancreases cultured for 20 days in 95% O_2 (Figure 3). However, the success rate of allografts is low; approximately 40% of recipients reverse their diabetes. Tissue cultured in this way is slow to reverse diabetes even in an isogeneic situation (95). However, the fact that cultured tissue can reverse diabetes in isogeneic animals, sometimes as late as 6 months post-transplantation, provides evidence that the cultured fetal pancreas has a capacity for continued growth and differentiation after transplantation (95).

Immature precursors of islet tissue, pro-islets, can be isolated from the pancreas by collagenase digestion and culture in air for 4 days (96). Following transplantation, this tissue has the capacity to grow and differentiate into functional islet tissue that will reverse streptozotocin-induced diabetes (97). In the digestion procedure the fetal pro-islets appear to be partially separated from lymphoid elements associated with the fetal pancreas. Pro-

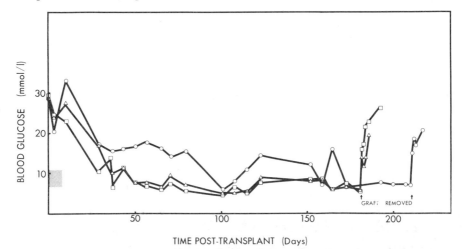

TIME POST-TRANSPLANT (Days)

Figure 3 Nonfasting blood glucose response of diabetic CBA mice following allotransplantation of two 20-day cultured BALB/c fetal pancreases (six 20-day cultured fetal pancreas segments) under the kidney capsule. The shaded region defines the nonfasting blood glucose range of normal CBA mice (mean ±95% confidence interval). Grafts were removed by excision of the graft-bearing kidney. Approximately 40% of 20-day cultured fetal pancreas allografts are successful in reversing streptozotocin-induced diabetes.

islets are less immunogenic than the whole fetal pancreas and may provide a more suitable source of tissue for clinical transplantation (97).

Reversal of Juvenile-Onset Diabetes

There are suggestions that spontaneous juvenile-onset diabetes may have an autoimmune etiology (23, 86) and, if this is the case, a transplant could succumb to the same fate as the animal's own islet tissue. Naji et al (85), working with the spontaneous diabetic BB rat, found that MHC-compatible transplants failed when animals were grafted soon after the onset of overt diabetes. Transplants appeared to fare better when animals were grafted some months after the onset of disease. These findings suggest that the transplant may be more vulnerable to attack when grafted during the acute phase of the disease.

Inflammatory damage of islet tissue may be virally mediated (86). Thus, if a virus grows specifically in the β cells of the pancreas, the immune response to the virus-infected cells could irreversibly damage the infected β cells. If such a process was involved in the induction of spontaneous diabetes, one would expect an allogeneic graft to be less susceptible to damage than the MHC-compatible graft. MHC restriction of the T cell response to virally infected host cells would protect the allograft from a cross-reactive attack mediated by activated T cells of the diabetic animals.

In our laboratory we have studied a case of spontaneous juvenile-onset diabetes that occurred in a young CBA mouse (Figure 4). This animal was transplanted with a cultured BALB/c allograft soon after the onset of diabetes. Within a week of transplantation, the blood sugar level returned to normal and remained normal for a further 180-odd days (Figure 4). During this period, the animal underwent two normal pregnancies with no evidence of abnormal blood sugar responses. This animal was killed at the completion of the study and both its own pancreas and the transplanted tissue were examined histologically. The animal's own pancreas contained few islets of Langerhans and those islets that were seen were grossly abnormal with few granulated β cells (Figure 5). The transplanted tissue, on the other hand, showed no evidence of damage (Figure 5). These findings are very promising from the clinical viewpoint; although there may be technical problems associated with the transfer of this technology to the treatment of juvenile-onset diabetes, there are no conceptual problems.

Organ Transplantation

Initial attempts to remove passenger leukocytes from tissues prior to transplantation were not successful. These attempts (56, 78, 107) involved the induction of leukopenia in the tissue donor by procedures such as whole

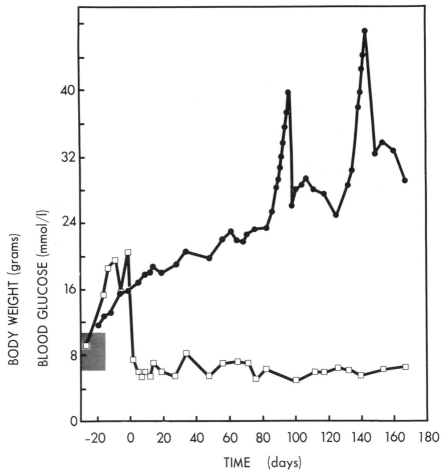

Figure 4 Nonfasting blood glucose concentrations (open squares) and body weight (closed circles) of a spontaneously diabetic CBA/H female mouse transplanted on day 0 with eight clusters of cultured BALB/c islets. Pregnancy is indicated by the sharp body weight increase. The shaded area shows the range of normal blood glucose concentrations (mean ±2 SD).

body irradiation, cyclophosphamide treatment, or treatment with anti-lymphocyte serum. At best, only marginal effects were observed for tissues transplanted across a major histocompatibility barrier (38, 55, 107), and rat heart allografts transplanted across multiple minor differences (34). Stuart et al (107) attempted to remove passenger leukocytes by first allografting rat kidneys to passively enhanced intermediate hosts. At 60 to 300 days post-transplantation, the kidney grafts were retransplanted to naive recipient rats isologous to the intermediate host. Only a delay in the onset of rejection was achieved. These marginal or weak effects led to some confu-

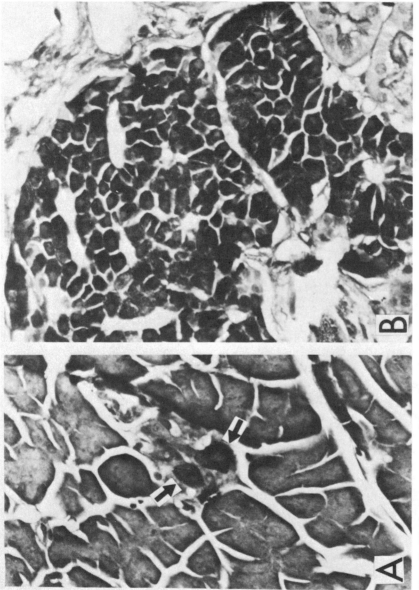

Figure 5 Histological appearance of tissues from the spontaneous juvenile-onset diabetic mouse. (*A*) Damaged islet from the animal's pancreas. Remaining granulated β cells are marked with arrows. (*B*) Well-granulated β cells in allograft under the kidney capsule. Tissue examined 180 days after transplantation. Aldehyde fuchsin, ×450.

sion over the extent to which passenger leukocytes contributed to tissue immunogenicity (12). Nevertheless, the clinical success reported with renal allografts taken from cadaver donors pretreated with cyclophosphamide and methylprednisolone helped maintain some interest in the passenger leukocyte concept (37a).

Studies from Batchelor's laboratory have now clarified the role passenger leukocytes play in renal allografting (9–11, 67). This group demonstrated that long-surviving, immunologically enhanced MHC-incompatible rat kidney grafts, when transplanted from a primary to a secondary recipient of the same genotype, do not elicit T-cell alloimmunity in the secondary recipient. In contrast, normal primary kidney allografts in the relevant donor/recipient combination are regularly rejected in 12 days. The failure of long-surviving kidney grafts to activate a T-cell response could not be attributed to a lack of either class I or II MHC antigens (67), and Batchelor's group suggested that the effect resulted from a lack of passenger leukocytes. They then went on to show, quite convincingly, that donor strain dendritic cells in very low numbers would trigger rejection of kidneys taken from an intermediate enhanced recipient (67). The earlier failure to see such a dramatic effect probably resulted from the strain combinations used in these studies (107). When discussing this strain-dependent variation in survival time, Lechler & Batchelor (67) emphasize, and we would agree, that acute graft destruction is not a precise measure of a recipient's response. In those strain combinations where acute graft destruction occurs, no quantitative comparison has been made to determine the relative strength of responses induced by normal allografts and those depleted of passenger leukocytes. They go on to predict that minimal immunosuppression at levels ineffective for normal grafts would maintain passenger cell-depleted allografts in good function. This proposal deserves careful investigation.

Development of Tolerance

Cultured allografts of both thyroid and pancreatic islet tissue carry recognizable antigen and are promptly rejected if recipient animals are challenged with donor leukocytes at the time of transplantation. However, animals carrying allografts for a prolonged period (≥100 days) become progressively more resistant to challenge with donor cells (18, 28, 116). In the case of the thyroid we have demonstrated that the grafted tissue retains antigen, and that the adaptation of the graft to its host results from the development of tolerance in the adult recipient (28). We argued that this phenomenon might develop as the result of the slow leakage of free antigen into the immune system of the recipient, possibly leading to tolerance induction in a manner akin to active enhancement. Some support for this notion has been obtained in the pancreatic islet system. When transplanted animals are injected with UV-killed donor spleen cells, a source of alloantigen on nonstimulating cells, around 30 days post-transplantation, graft rejection is not stimulated; administration of viable donor cells at this time promptly activates a rejection reaction. That is, the UV-irradiated cells are not immunogenic. When animals that received UV-irradiated cells were subsequently challenged with viable donor spleen cells, the allografts were not rejected. The UV-irradiated cells, which provide a source of antigen alone, have stimulated the development of tolerance in the allografted animals. The specificity of this tolerance was demonstrated by the fact that these animals would accept an uncultured thyroid allograft of donor origin but would reject a third-party thyroid grafted to the same recipient. We have shown, in the case of both thyroid- and islet-transplanted animals, that this phenomenon is not a deletion form of tolerance. Lymphocytes from "tolerant" animals respond normally to donor alloantigen in vitro. More recently, Faustman et al (33) have suggested such unresponsiveness may be due to the action of suppressor cells; the role of suppressor cells in allograft tolerance has been reviewed elsewhere (91). A similar phenomenon has been observed following heart allografting under cover of immune suppression with cyclosporin A (81). When the cyclosporin A is withdrawn, the heart continues to function but can be rejected if the recipient is appropriately challenged with donor tissue; at this stage, the heart allograft is in a metastable condition. With the further passage of time, a stable form of donor tolerance gradually develops (81).

In summary, we can now say it is possible to either eliminate or reduce tissue immunogenicity by removal of leukocytes (dendritic cells) from the transplant prior to grafting. This phenomenon has been demonstrated to be effective with both endocrine tissue and larger organs, such as the kidney. There is some variation in the effectiveness of the technique with different tissues, which can be related to the degree of leukocyte contamination. Long-standing allografts of pretreated tissue induce a state of specific unresponsiveness in the adult recipient. Although the mechanism of this "tolerance" is not fully understood, the existence of the phenomenon is one of the most encouraging recent developments in transplantation biology.

THEORETICAL APPENDIX

Immunobiology requires a theoretical framework on which to arrange the experimental findings of cellular immunology. Such a framework must comfortably explain why it is the MHC complex antigens play the dual role of immune regulation and control of allograft rejection. In the past, this relationship has been explained by the "immune surveillance" concept, which postulated that the immune system developed to recognize components closely related to "self," including foreign MHC antigens, so that tumors would be readily detected and eliminated from the system. The experimental observation that tissues expressing foreign MHC antigens survive in non-immunosuppressed recipients and show no sign of lymphocytic infiltration (65) points out the inadequacy of the surveillance idea. The following analysis attempts a solution to this problem.

Theory

Our theoretical development accepts the postulates of "clonal selection" (22), namely, that lymphocyte receptor diversity is generated by a random process, and that self-reactivity is forbidden by a mechanism learned during development and enforced throughout the life of the animal. We develop a theoretical system based on a further two postulates that govern the process of clonal selection and expansion. We then use a symbolic terminology, which is abstract yet precise, to derive the characteristics of the immune system that follow from these postulates.

Terminology

In the following development the accessory cell required for immunocyte activation is called the stimulator cell, S.

$S_{(x)}^{c}$ is a stimulator cell that carries on its surface a control structure, c, and a receptor for antigen, x; the term (x) is read as: receptor for x. This receptor may be a product of the stimulator cell or could be a cytophylic product of some other cell. When $S_{(x)}^{c}$ interacts with x, it can behave as an antigen-presenting cell S_{x}^{c}, which presents x to the potentially responsive immunocyte. We can write this reaction as

$$S_{(x)}^{c} + x \longrightarrow S_{x}^{c}.$$

The control structure is a genetically determined product and can be used to define the genotype of the animal producing the cell; we use c to define a genotype in the same way that blood group antigens are used for this purpose.

$T_{(x)}$ is a resting T cell that carries a receptor for x. $T'_{(x)}$ is an activated cell that carries a receptor for x. The genotype of the resting or activated T cell can be represented as $T_{(x)}^{c}$ and $T'_{(x)}^{c}$, respectively; these are resting and activated cells of x specificity, derived from animals of c genotype. Activated T cells express effector functions such as the expression of cellular cytotoxicity or lymphokine production. Activated T cells release a number of lymphokine activities (3, 53), and lymphokine release is triggered by signal 1 alone (59). That is,

$$T'_{(x)} + x \xrightarrow{1} T'_x + l.$$

Here l is used as a generic term to cover a number of lymphokine activities.

Cell surface antigens other than those that form the control structure, c, are represented by y. That is, $(c + y)$ represents the sum of all cell surface antigens.

The terms c, x, and y can be used in a generic sense. Particular examples of each are represented as c, c_1, c_2, c_3 ... etc.; x, x_1, x_2, x_3 ... etc.; y, y_1, y_2, y_3 ... etc.

Postulates of the theory

The first postulate maintains that T cells do not respond to contact with antigen alone, but respond to antigen in conjunction with a second signal provided by an inductive molecule, said to express CoS activity (Figure 2).

FIRST POSTULATE   Two signals—antigen activity and CoS activity—are required for initiation of T cell activation.

$$T \xrightarrow[2]{1} T'.$$

T, the resting T cell, is converted to the activated state T' following antigen binding by the T-cell receptor (delivery of signal 1) in the presence of CoS activity, which provides signal 2. Since CoS activity is a cellular product, a corollary of the first postulate is the idea that a stimulator cell is required for T cell activation (Figure 2).

The second postulate concerns the regulation of stimulator cell function.

SECOND POSTULATE: A control structure on the surface of the S^+ (stimulator) cell regulates the release of CoS activity; CoS activity is only released when this structure is engaged by the potentially responsive T cell.

Initiation of the Immune Response

In the resting immune system, spontaneous T cell activation does not occur because T cells do not express a functional receptor for self-antigens, and c, the control structure, is a self-antigen. Thus, the resting situation in the immune system can be written

$$S^c_{(x)} + T^{\,c}_{(\,)} \xrightarrow{} -Ve,$$

where $S^c_{(x)}$ is a stimulator cell that carries a control structure c and a receptor for x. $T^c_{(x)}$ is a T cell of the same genotype that has a receptor specificity defined at this stage only as being unresponsive to self-antigens such as c. That is, $T^c_{(x)}$ is unable to interact with the control structure of $S^c_{(x)}$ and so trigger CoS release.

Consider, now, the entry of foreign antigen, x, into the system. Antigen x can interact with $S^c_{(x)}$, the antigen-presenting cell:

$$S^c_{(x)} + x \longrightarrow S^c_x,$$

if $S^c_x \neq S^c$.

That is, if x interacts with c in such a way that $c \cdot x$ (the interaction between c and x on the antigen-presenting cell) is no longer equivalent to c, then the following situation will hold:

$$S^c_x + T^c_{(c \cdot x)} \xrightarrow{\quad \frac{1}{2} \quad} T'^c_{(c \cdot x)}.$$

That is, T cells of $c \cdot x$ specificity will interact with $c \cdot x$ on the surface of the antigen-presenting cell. This will provide signal 1 for the T cell. Signal 2, CoS activity, is released from the stimulator cell because the control structure is now engaged in the interaction with the T cell of $c \cdot x$ specificity.

FIRST INFERENCE   The first inference we draw from our postulates is that the specificity of the T cell response to exogenous antigen, x, is defined by both x and the control structure, c. That is, T cell responses to x are restricted by c.

Alloreactivity

An analysis of alloreactivity defines the nature of c, the control structure.

Consider in vitro allogeneic interactions in which T cells of one genotype interact with allogeneic stimulator cells (Figure 6). There are basically two different classes of such allogeneic interactions: Those that involve differences in the control structure (differences in c), and those that involve differences in other cell surface antigens (differences in y) (Figure 6).

Allogeneic interactions between cells of different c-type have the form

$$^yS^c + {}^yT^{c1}_{(c)} \xrightarrow{\quad \frac{1}{2} \quad} {}^yT'^{c1}_{(c)},$$

and T cells specific for the control structure on the stimulator cell, c, are generated. On the other hand, allogeneic interactions between cells of the same c-type, but differing in y-type antigens, do not proceed because the responsive T cell is unable to interact with the control structure on the stimulator cell (Figure 6). That is,

$$^{y}S^{c} + {}^{yl}T^{c}_{(y)} \xrightarrow{\quad 1 \quad} -Ve.$$

The potentially responsive T cell may carry a functional receptor for y, the alloantigen on the stimulator cell, but an interaction with the stimulator cell via y will not involve the control structure c. As a result, CoS activity will not be released from the S^{+} cell, thus depriving the system of a source of signal 2 (see Figure 6).

There is another important characteristic of c-type antigens. Allogeneic T cells respond to these antigens in an unrestricted manner:

$$S^{c} + T^{c}{}^{1}_{(c)} \xrightarrow[2]{\quad 1 \quad} T'^{c}{}^{1}_{(c)}.$$

Responses to other antigens, whether they be minor histocompatibility antigens (y-type) or other foreign antigens, x, will always be restricted by the control structure of the antigen-presenting cell:

$$S^{c}_{(y)} + y \longrightarrow S^{c}_{y}.$$

If $S^{c}_{y} \neq S^{c}$,

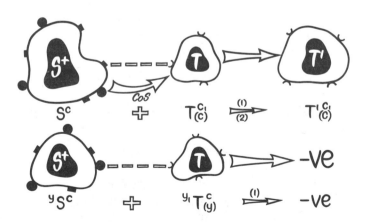

Figure 6   Alloreactivity. T cells are activated when they interact with c-type antigens (closed blocks) on the surface of viable S^{+} cells because this interaction provides signals 1 and 2 required for lymphocyte activation. y-Type antigens (closed circles) on the surface of metabolically active S^{+} cells are not immunogenic under these conditions because CoS activity is not released from the S^{+} cell. The symbolic representation of each interaction is set out below the diagrammatic sequence.

$$\text{then } S^c_y + T^c_{(c \cdot y)} \xrightarrow{\quad \frac{1}{2} \quad} T'^c_{(c \cdot x)}.$$

Similarly, T cell responses to x antigens are of the type $T'^c_{(c \cdot x)}$.

SECOND INFERENCE The second inference we derive from the postulates is that there are two classes of alloantigen, c-type and y-type antigens. c-Type antigens are highly immunogenic for allogeneic T cells in vitro, when they are carried on the surface of metabolically active stimulator cells. y-Type antigens can be seen by T cells in an unrestricted manner, whereas y-type antigens are not immunogenic under these conditions. c-Type antigens are seen in association with the control structure of the antigen presenting cell.

Thus, we can use alloreactivity to define experimentally the nature of c, the control structure. c-Type structures will be that group of alloantigens that are highly immunogenic for T cells in vitro. Experimentally we know that the MHC antigens behave in this way. That is, MHC antigens constitute the control structures of the immune system.

Complexity of the Control System

The MHC of the species is complex and consists of at least two classes of cell surface antigens that show strong homologies within the class and also display a characteristic tissue distribution. Klein (54) has called these antigens class I and class II MHC antigens. In this analysis we refer to class I antigens as K-type antigens and class II antigens as I-type antigens.

We know from the analysis of alloreactivity among cell populations that differ only in K-type or I-type antigens that K-type and I-type antigens control the activation of different T cell subsets (24):

$$S^K + T^c_{(K)}{}^1 \xrightarrow{\quad \frac{1}{2} \quad} T'^c_{(K)}{}^1.$$

In the mouse, $T'^c_{(K)}{}^1$ is an Lyt $1^+2^+3^+$ T cell of the cytotoxic subset (24). Similarly:

$$S^I + T^c_{(I)}{}^1 \xrightarrow{\quad \frac{1}{2} \quad} T'^c_{(I)}{}^1,$$

where in the mouse $T'^c_{(I)}{}^1$ is an Lyt $1^+2^-3^-$ T cell of the helper subset (24). Both $T'^c_{(K)}{}^1$ and $T'^c_{(I)}{}^1$ release lymphokine (l) when they are triggered by antigen alone (3, 53, 72). That is:

$$T'^c_{(K)}{}^1 + K \xrightarrow{\quad 1 \quad} T'^c_K{}^1 + l,$$

and

$$T'^c_{(I)}{}^1 + I \xrightarrow{\quad 1 \quad} T'^c_I{}^1 + l.$$

Response to MHC-Incompatible Tissue Grafts

The above analysis shows that the control structures of the immune system behave as MHC antigens in the mixed leukocyte reaction. Let us now consider how these antigens behave in the case of tissue transplantation. The in vitro response to a tissue graft is more complex than the in vitro unidirectional mixed leukocyte interactions. The transplanted tissue will contain antigen-bearing parenchymal cells (Figure 7). The tissue will also carry lymphoreticular cells, either as fixed tissue elements or as passenger leukocytes. These cells provide a source of antigen-bearing S^+ cells (Figure 7).

The simplest way to analyze the allograft situation is to consider separately the types of interaction that may occur in the recipient's immune system and in the transplanted tissue itself.

Interactions in the Recipient's Immune System

Antigen-bearing S^+ cells from the grafted tissue will be carried to lymph nodes draining the transplantation site by way of the lymphatic system. In the node, the control structure (i.e. MHC antigens) of the donor (c genotype) will be presented directly to the recipient's T cells (c_1 genotype):

$$S^c + T^{c\,1}_{(c)} \xrightarrow{\quad \frac{1}{2} \quad} T'^{c\,1}_{(c)}.$$

If the tissue donor differs from the recipient in both K- and I-type control structures, the activated T cells will be specified as $T'^{c\,1}_{(K)}$ and $T'^{c\,1}_{(I)}$. This direct stimulation of the recipient's T cells will therefore lead to the generation of T cells reactive to both the K- and I-type antigens of the donor.

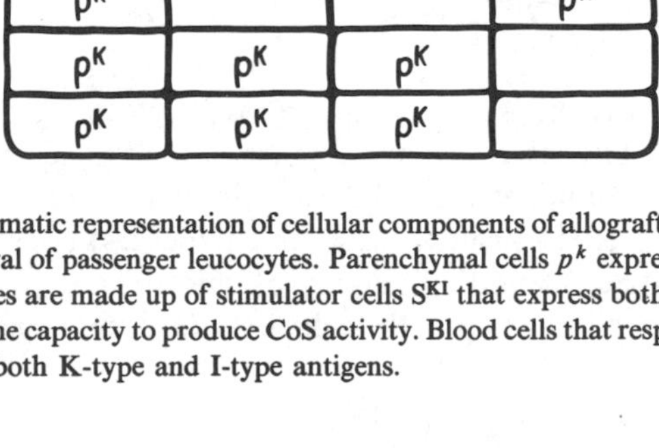

Figure 7 Diagrammatic representation of cellular components of allograft before (above) and after (below) removal of passenger leucocytes. Parenchymal cells p^k express K-type antigens. Passenger leukocytes are made up of stimulator cells S^{KI} that express both K-type and I-type antigens and have the capacity to produce CoS activity. Blood cells that respond to lymphokine (R^{KI}) also express both K-type and I-type antigens.

A second type of recipient response will occur when tissue antigens, designated collectively as x, reach the lymph node either as membrane fragments or as free antigen. In the lymph node, these antigens will be presented to recipient (c genotype) antigen-presenting cells. This process of indirect antigen presentation has the form:

$$S^{c\,1}_{(x)} + x \longrightarrow S^{c1}_x.$$

If $S^{c1}_x \neq S^{c\,1}$,

$$\text{then } S^{c1}_x + T^{c\,1}_{(c_1 \cdot x)} \longrightarrow T'^{c\,1}_{(c_1 \cdot x)}.$$

The activated T cells generated are of specificity $(c_1 \cdot x)$ and are not graft specific; graft antigens are x alone. That is, indirect antigen presentation does not lead to a graft-specific T cell response.

Reactions in the Transplanted Tissue

In the case of a solid tissue transplant, the graft will consist of tissue parenchymal cells, including vascular endothelium, that carry K-type antigens, and lymphoreticular cells that carry both K-type and I-type antigens. Reactions in the transplanted tissue will be regulated primarily by the interaction of recipient T cells with graft antigens. Activated T cells specific for graft K-type antigens have the potential to express cytotoxic activity and in this way damage graft endothelium and parenchymal cells. In addition, binding of these cells to K̇-type antigens on target cells will trigger lymphokine release and so activate a nonspecific inflammatory process in the transplant. T cells specific for I-type antigens of the graft are unable to interact directly with graft parenchymal cells which do not express I-type antigens. However, these cells could play an important role in activating donor-type lymphoreticular cells in the transplant by way of lymphokine release (Figure 7):

$$T'^{c\,1}_{(I)} + R^I \xrightarrow{\;\;2\;\;} R'^I.$$

Thus, graft damage could result from a cytotoxic effect directed against cells of the transplant and/or a severe nonspecific inflammatory response. For this reason c-type antigens carried on stimulator cells and other lymphoreticular cells of the graft (passenger leukocytes) constitute the major barrier to tissue transplantation (Figure 6), and their removal from a graft prior to transplantation will greatly reduce the immunogenic strength of the graft (Figure 7).

ACKNOWLEDGMENTS

We are most grateful to Jane Dixon for her assistance. This work was supported by Grant No. AM28126 from the National Institutes of Health, United States Public Health Service, and grants from the Kroc Foundation and the Kellion Foundation.

Literature Cited

1. Agostino, M., Prowse, S. J., Lafferty, K. J. 1981. Resistance of established islet allografts to rejection by antibody and complement. *Aust. J. Exp. Biol. Med. Sci.* 60:219–22
2. Andersson, J., Gronvik, K., Larsson, E., Coutinho, A. 1979. Studies on T lymphocyte activation. I. Requirements for the mitogen-dependent production of T cell growth factors. *Eur. J. Immunol.* 9:581–87
2a. Andrus, L., Lafferty, K. J. 1982. Inhibition of T cell activity by cyclosporin A. *Scand. J. Immunol.* 15:449–58
3. Andrus, L., Prowse, S. J., Lafferty, K. J. 1981. Interleukin 2 production by both Ly2+ and Ly2- T cell subsets. *Scand. J. Immunol.* 13:297
4. Ascher, N. L., Ferguson, R. M., Hoffman, R., Simmons, R. L. 1979. Partial characterisation of cytotoxic cells infiltrating sponge matrix allografts. *Transplantation* 27:254–59
5. Ascher, N. L., Hoffman, R., Chen, S., Simmons, R. L. 1980. Specific and nonspecific infiltration of sponge matrix allografts by specifically sensitized cytotoxic lymphocytes. *Cell Immunol.* 52:38–47
6. Bach, F. H., Alter, B. J. 1978. Alternative pathways of T-lymphocyte activation. *J. Exp. Med.* 148:829–34
7. Bach, F. H., Bach, M. L., Sondel, P. M. 1976. Differential function of major histocompatibility complex antigens in T-lymphocyte activation. *Nature* 259:273–81
8. Bach, F., Widmer, M. B., Segall, M., Bach, M. L., Klein, J. 1972. Genetic and immunological complexity of major histocompatibility regions. *Science* 176:1024–37
9. Batchelor, J. R. 1978. The riddle of kidney graft enhancement. *Transplantation* 26:139–41
10. Batchelor, J. R., Welsh, K. I., Burgos, H. 1978. Transplantation antigens *per se* are poor immunogens within a species. *Nature* 273:54–56
11. Batchelor, J. R., Welsh, K. I., Maynard, A., Burgers, H. 1979. Failure of long surviving, passively enhanced allografts to provoke T-dependent alloimmunity. I. Retransplantation of (ASx AUG)F$_1$ kidneys into secondary AS recipients. *J. Exp. Med.* 150:455–64
12. Billingham, R. E. 1971. The passenger cell concept in transplantation immunology. *Cell. Immunol.* 2:1–12
13. Billingham, R. E., Brent, L., Medawar, P. B. 1954. Quantitative studies on tissue transplantation immunity. II. The origin, strength and duration of actively and adoptively acquired immunity. *Proc. R. Soc. Lond. Biol.* 143:58–80
14. Billingham, R. E., Silvers, W. K., Wilson, D. B. 1963. Further studies on adoptive transfer of sensitivity to skin homografts. *J. Exp. Med.* 118:397–419
15. Bogman, M. J., Berden, J. H., Hagemann, J. F., Marass, C. N., Koene, R. A. 1980. Patterns of vascular damage in the antibody-mediated rejection of skin xenograft in the mouse. *Am. J. Pathol.* 100:727–37
16. Bowen, K. M., Andrus, L., Lafferty, K. J. 1980. Successful allotransplantation of mouse pancreatic islets to nonimmunosuppressed recipients. *Diabetes* 29:98–104
17. Bowen, K. M., Lafferty, K. J. 1980. Reversal of diabetes by allogeneic islet transplantation without immunosuppression. *Aust. J. Exp. Biol. Med. Sci.* 58:441–47
18. Bowen, K. M., Prowse, S. J., Lafferty, K. J. 1981. Reversal of diabetes by islet transplantation: vulnerability of the established allograft. *Science* 213:1261–62
19. Bretscher, P., Cohn, M. 1970. A theory of self-nonself discrimination. *Science* 169:1042
20. Brown, J., Heininger, D., Kuret, J., Mullen, Y. 1981. Islet cells grow after transplantation of fetal pancreas and control of diabetes. *Diabetes* 30:9–13
21. Burdick, J. F., Russell, P. S., Winn, H. J. 1979. Sensitivity of long standing xenografts of rat hearts to humoral antibodies. *J. Immunol.* 123:1732.

22. Burnet, F. M. 1959. *The Clonal Selection Theory of Acquired Immunity.* Cambridge: Cambridge Univ.

23. Cahill, G. F., McDevitt, H. O. 1981. Insulin dependent diabetes mellitus: The initial lesion. *New Engl. J. Med.* 304:1454–65

24. Cantor, H., Boyse, E. A. 1975. Functional subclasses of T lymphocytes bearing different Ly antigens. II. Cooperation between subclasses of Ly⁺ cells in the generation of killer activity. *J. Exp. Med.* 141:1390

25. Cunningham, A. J., Lafferty, K. J. 1977. A simple conservative explanation of the H-2 restriction of interactions between lymphocytes. *Scand. J. Immunol.* 6:1

26. De Bano, D. 1964. Effects of cytotoxic sera on endothelium *in vitro. Nature* 252:83–84

27. De Waal, R. M. W., Capel, P. J. A., Koene, R. A. P. 1980. The role of anti-H-2K and H-2D alloantibodies in enhancement and acute antibody-mediated rejection of mouse skin allografts. *J. Immunol.* 124:719–23

28. Donohoe, J. A., Andrus, L., Bowen, K. M., Simeonovic, C., Prowse, S. J., Lafferty, K. J. 1983. Cultured thyroid allografts induce a state of partial tolerance in adult recipient mice. *Transplantation* In press.

29. Elkins, W. L. 1966. The interaction of donor and host lymphoid cells in the pathogenesis of renal cortical destruction induced by a local graft versus host reaction. *J. Exp. Med.* 123:103–18

30. Elkins, W. L. 1971. The sources of immunogenic stimulation of lymphoid cells mediating a local graft-versus-host reaction in rat kidney. *Transplantation* 11:551–60

31. Elkins, W. L., Guttmann, R. D. 1968. Pathogenesis of a local graft versus host reaction: immunogenicity of circulating host leucocytes. *Science* 159:1250–51

32. Faustman, D., Hauptfeld, V., Lacy, P., Davie, J. 1981. Prolongation of murine islet allograft survival by pretreatment of islets with antibody directed to Ia determinants. *Proc. Natl. Acad. Sci. USA* 78:5156–59

33. Faustman, D., Lacy, P., Davie, J., Hauptfeld, V. 1982. Prevention of allograft rejection by immunization with donor blood depleted of Ia-bearing cells. *Science* 217:157–58

34. Freeman, J. S., Chamberlain, E. C., Reemtsma, K., Steinmuller, D. 1971. Prolongation of rat heart allografts by donor pretreatment with immunosuppressive agents. *Transplant. Proc.* 3:580–82

35. French, M. E. 1972. The early effects of alloantibody and complement on rat kidney allografts. *Transplantation* 13:447–51

36. Gerlag, P. G., Capel, P. J., Hageman, J. F., Koene, R. A. 1980. Adaption of skin grafts in the mouse to antibody-mediated rejection. *J. Immunol.* 125:583–86

37. Glimcher, L. H., Kim, K. J., Green, I., Paul, W. E. 1982. Ia antigen-bearing B cell tumour lines can present protein antigen and allo-antigen in a major histocompatibility complex-restricted fashion to antigen-reactive T cells. *J. Exp. Med.* 155:445–49

37a. Guttman, R. D., Beadoin, J. G., Moorehouse, P. D., Klassen, J., Knaack, J., Jeffery, J., Chassot, P. G., Abbou, C. C. 1975. Donor pretreatment as an adjunct to cadaver renal allotransplantation. *Transplant. Proc.* 7:117–21

38. Guttman, R. D., Carpenter, C. B., Lindquist, R. R., Merrill, J. P. 1967. An immunosuppressive site of action of heterologous antilymphocyte serum. *Lancet* 1:248–49

39. Hall, B. M., Dorsch, S., Roser, B. 1978. The cellular basis of allograft rejection in vivo I. The cellular requirements for first-set rejection of heart grafts. *J. Exp. Med.* 148:878–89

40. Hall, B. M., Dorsch, S., Roser, B. 1978. The cellular basis of allograft rejection in vivo II. The nature of memory cells mediating second set heart graft rejection. *J. Exp. Med.* 148:890–902

41. Hammerling, G. J. 1976. Tissue distribution of Ia antigens and their expression on lymphocyte subpopulations. *Transplant. Rev.* 30:64

42. Hardy, D. A., Knight, S., Ling, N. R. 1970. The interaction of normal lymphocytes and cells from lymphoid cell lines. *Immunology* 19:329–42

43. Hardy, D. A., Ling, N. R. 1969. Effects of some cellular antigens on lymphocytes and the nature of the mixed lymphocyte reaction. *Nature* 221:545–48

44. Hart, D. N. J., Winnearls, C. G., Fabre, J. W. 1980. Graft adaption: studies on possible mechanisms in long-term surviving rat renal allografts. *Transplantation* 30:73–80

45. Hayrey, P., Andersson, L. C. 1976. Generation of T memory cells in one-way mixed lymphocyte culture. IV. Primary and secondary responses to solu-

ble and insoluble membrane preparations and to ultraviolet light inactivated stimulator cells. *Scand. J. Immunol.* 5:391–99

46. Hirschberg, H., Thorsby, E. 1981. Immunogenicity of foreign tissues. *Transplantation* 31:96–97

47. Hoffman, L., Martin, F. I. R., Carter, W., Mandel, T. E., Campbell, D. G. 1981. Effect of short term insulin treatment on the reversal of diabetes after transplantation of syngeneic cultured or uncultured foetal mouse pancreas. *Transplantation* 32:342–45

48. Hopt, U. T., Sullivan, W., Hoffman, R., Simmons, R. L. 1980. Migration and cell recruiting activity of specifically sensitized lymphocytes in sponge matrix allografts. *Transplantation* 30:411–16

49. Jacobs, B. B., Huseby, R. A. 1967. Growth of tumors in allogeneic hosts following organ culture explantation. *Transplantation* 5:410–19

50. Jooste, S. V., Colvin, R. B., Soper, W. D., Winn, H. J. 1981. The vascular bed as the primary target in the destruction of skin grafts by antiserum. I. Resistance of freshly placed xenografts of skin to antiserum. *J. Exp. Med.* 154:1319–31

51. Jooste, S. V., Colvin, R. B., Winn, H. J. 1981. The vascular bed as the primary target in the destruction of skin grafts by antiserum. II. Loss of sensitivity to antiserum in long-term xenografts of skin. *J. Exp. Med.* 154:1332–41

52. Kaliss, N., Kandutsch, A. A. 1956. Acceptance of tumour homografts by mice injected with antiserum. I. Activity of serum fractions. *Proc. Soc. Exp. Biol. Med.* 91:118–21

53. Kelso, A., Glasebrook, A. L., Kanagawa, O., Brunner, K. T. 1982. Production of macrophage activating factor by T lymphocyte clones and correlation with other lymphokine activities. *J. Immunol.* 129:550–56

54. Klein, J. 1979. The major histocompatibility complex of the mouse. *Science* 203:516

55. Kyger, E. R., Salyer, K. E. 1973. The role of donor passenger leukocytes in rat skin allograft rejection. *Transplantation* 16:537–43

56. Lacy, P. E., Davie, J. M., Finke, E. H. 1979. Prolongation of islet allograft survival following *in vitro* culture (24°C) and a single injection of ALS. *Science* 204:312–13

57. Lacy, P. E., Davie, J. M., Finke, E. H. 1980. Prolongation of islet xenograft survival without continuous immunosuppression. *Science* 209:283–85

58. Lacy, P. E., Finke, E. H., Janney, C. G., Davie, J. M. 1982. Prolongation of islet xenograft survival by *in vitro* culture of rat megaislets in 95%0$_2$. *Transplantation* 33:588–92

59. Lafferty, K. J., Andrus, L., Prowse, S. J. 1980. Role of lymphokine and antigen in the control of specific T cell responses. *Immunol. Rev.* 51:279–314

60. Lafferty, K. J., Bootes, A., Kilby, V. A. A., Burch, W. 1976. Mechanism of thyroid allograft rejection. *Aust. J. Exp. Biol.* 54:573–86

61. Lafferty, K. J., Cooley, M. A., Woolnough, J., Walker, K. Z. 1975. Thyroid allograft immunogenicity is reduced after a period in organ culture. *Science* 188:259–61

62. Lafferty, K. J., Cunningham, A. J. 1975. A new analysis of allogeneic interactions. *Aust. J. Exp. Biol. Med. Sci.* 53:27

63. Lafferty, K. J., Jones, M. A. S. 1969. Reactions of the graft versus host (GVH) type. *Aust. J. Exp. Biol. Med. Sci.* 12:198

64. Lafferty, K. J., Misko, L. S., Cooley, M. A. 1974. Allogeneic stimulation modulates the *in vitro* response of T cells to transplantation antigen. *Nature* 249:275–76

64a. Lafferty, K. J., Warren, H. S., Woolnough, J. A., Talmage, D. W. 1978. Immunological induction of T lymphocytes: role of antigen and the lymphocyte costimulator. *Blood Cells* 4:395

65. Lafferty, K. J., Woolnough, J. A. 1977. The origin and mechanism of the allograft reaction. *Immunol. Rev.* 35:231

66. Larsson, E. L., Iscove, N., Coutinho, A. 1980. Two distinct factors are required for induction of T cell growth. *Nature* 283:664

67. Lechler, R. I., Batchelor, J. R. 1982. Restoration of immunogenicity to passenger cell-depleted kidney allografts by the addition of donor strain dendritic cells. *J. Exp. Med.* 155:31–41

68. Ledbetter, J. A., Rouse, R. V., Micklem, H. S., Herzenberg, L. A. 1980. T cell subsets defined by expression of Lyt-1,2,3 and Thy-1 antigens. Two-parameter immunofluorescence and cytotoxicity analysis with monoclonal antibodies modifies current views. *J. Exp. Med.* 152:280

69. Lindahl-Kiessling, K., Safwenberg, J. 1971. Inability of UV-irradiated lymphocytes to stimulate allogeneic cells in

mixed lymphocyte culture. *Int. Arch. Allergy* 41:670–78

70. Loveland, B. E., Hogarth, P. M., Ceredig, R. H., McKenzie, I. F. C. 1981. Cells mediating graft rejection in the mouse I. Lyt-1 cells mediate skin graft rejection. *J. Exp. Med.* 153:1044–57

71. Loveland, B. E., McKenzie, I. F. C. 1982. Which T cells cause graft rejection? *Transplantation* 33:217–21

72. Luger, T. A., Smolen, J. S., Chused, T. M., Steinberg, A. D., Oppenheim, J. J. 1982. Human lymphocytes with either the OKT4 or OKT8 phenotype produce interleukin 2 in culture. *J. Clin Invest.* 70:470–73

73. Mandel, T. E., Higginbotham, L. 1979. Organ culture and transplantation of fetal mouse pancreatic islets. *Transp. Proc.* 11:1505–6

74. Masshoff, J. W., ed. 1977. *Transplantation.* Berlin: Springer

75. McKenzie, I. F. C., Henning, M. M. 1978. The differential destructive and enhancing effects of anti-H-2K, H-2D and anti-Ia antisera on murine skin allografts. *J. Exp. Med.* 147:611–16

76. Medawar, P. B. 1944. The behaviour and fate of skin autografts and skin homografts in rabbits. *J. Anat.* 78: 176–99

77. Medawar, P. B. 1963. Introduction: Definition of the immunologically competent cell. In *The Immunologically Competent Cell: Its Nature and Origin, Ciba Found. Study Group No. 16,* ed. G. E. W. Wolstenholme, J. Knight, pp. 1–3. London: Churchill

78. Mitchison, N. A. 1953. Passive transfer of transplantation immunity. *Nature* 171:267–68

79. Mitchison, N. A. 1954. Passive transfer of transplantation immunity. *Proc. R. Soc. London B.* 142:72–87

80. Morrow, C. E., Sutherland, D. E. R., Steffes, M. W., Najarian, J. S., Bach, F. H. 1983. The effect of isolated H-2 K,D and I region encoded histocompatibility antigen differences on mouse pancreatic islet allograft rejection. *Science.* In press.

81. Nagao, T., White, D. J. G., Calne, R. Y. 1982. Kinetics of unresponsiveness induced by a short course of cyclosporin A. *Transplantation* 33:31–35

82. Nairn, R., Yamaga, K., Nathenson, S. G. 1980. Biochemistry of the gene products from murine MHC mutants. *Ann. Rev. Genet.* 14:241–77

83. Naji, A., Barker, C. F., Silvers, W. K. 1979. Relative vulnerability of isolated pancreatic islets, parathyroid and skin allografts to cellular and humoral immunity. *Transp. Proc.* XI:560–62

84. Naji, A., Silvers, W. K., Barker, C. F. 1979. Effect of culture in 95%O₂ on the survival of parathyroid allografts. *Surgical Forum* 30:109–11

85. Naji, A., Silvers, W. K., Bellgrau, D., Barker, C. F. 1981. Spontaneous diabetes in rats: destruction of islets is prevented by immunological tolerance. *Science* 213:1390–92

86. Onoderra, T., Ray, U. R., Melez, K. A., Suzuki, H., Toniolo, A., Notkins, A. L. 1982. Virus induced diabetes mellitus. Autoimmunity and polyendocrine disease prevented by immunosuppression. *Nature* 297:66–68

87. Paetkau, V., Mills, G., Gerhart, S., Monticone, V. 1976. Proliferation of murine thymic lymphocytes in vitro is mediated by the concanavalin A-induced release of a lymphokine (costimulator). *J. Immunol.* 117:1320

88. Parr, E. L., Bowen, K. M., Lafferty, K. J. 1980. Cellular changes in cultured mouse thyroid glands and islets of Langerhans. *Transplantation* 30: 135–41

89. Parr, E. L., Lafferty, K. J., Bowen, K. M., McKenzie, I. F. C. 1980. H-2 complex and Ia antigens on cells dissociated from mouse thyroid glands and islets of Langerhans. *Transplantation* 30:142

90. Prowse, S. J., Warren, H. S., Agostino, M., Lafferty, K. J. 1983. Transfer of sensitised Lyt 2⁺ cells triggers acute rejection of pancreatic islet allografts. *Aust. J. Exp. Biol. Med. Sci.* In press.

91. Roser, B., Dorsch, S. 1979. The cellular basis of transplantation tolerance in the rat. *Immunol. Rev.* 46:55–86

92. Schellekans, P. T. A., Eijsvoogel, V. P. 1970. Lymphocyte transformation in vitro III. Mechanism of stimulation in the mixed lymphocyte culture. *Clin. Exp. Immunol.* 7:229–39

93. Silberberg-Sinakin, I., Gigli, I., Baer, R., Thorbecke, G. 1980. Langerhans cells: role in contact hypersensitivity and relationship to lymphoid dendritic cells and to macrophages. *Immunol. Rev.* 53:203–32

94. Simeonovic, C. J., Bowen, K. M., Kotlarski, I., Lafferty, K. J. 1980. Modulation of tissue immunogenicity by organ culture. Comparison of adult islets and fetal pancreas. *Transplantation* 30: 174–79

95. Simeonovic, C. J., Lafferty, K. J. 1981. Effect of organ culture on function of transplanted foetal pancreas. *Aust. J. Exp. Biol. Med. Sci.* 59:707–12

96. Simeonovic, C. J., Lafferty, K. J. 1982. The isolation and transplantation of foetal mouse proislets. *Aust. J. Exp. Biol. Med. Sci.* 60:383–90
97. Simeonovic, C. J., Lafferty, K. J. 1982. Immunogenicity of isolated foetal mouse proislets. *Aust. J. Exp. Biol. Med. Sci.* 60:391–95
98. Simonsen, M. 1962. Graft versus host reactions. Their natural history and applicability as tools of research. *Prog. Allergy.* 6:349–467
99. Snell, G. D. 1957. The homograft reaction. *Ann. Rev. Microbiol.* 11:439
100. Sollinger, H. W., Bach, F. H. 1976. Collaboration between *in vivo* responses to LD and SD antigens of major histocompatibility complex. *Nature* 259:487–88
101. Sollinger, H. W., Burkholder, P. M., Rasmus, W. R., Bach, F. H. 1976. Prolonged survival of xenografts after organ culture. *Surgery* 81:74
102. Steinmuller, D. 1967. Immunisation with skin isografts taken from tolerant mice. *Science* 158:127–29
103. Steinmuller, D. 1969. Allograft immunity produced with skin isografts from immunologically tolerant mice. *Transp. Proc.* 1:593–96
104. Stone, H. B., Owings, J. C., Grey, G. O. 1934. Transplantation of living grafts of thyroid and parathyroid glands. *Ann. Surg.* 100:613
105. Streilein, J. W., Toews, G. B., Bergstresser, P. R. 1979. Corneal allografts fail to express Ia antigens. *Nature* 282:326–27
106. Strom, T. B., Tilney, N. L., Paradysz, J. M., Bancewicz, J., Carpenter, C. B. 1977. Cellular components of allograft rejection: identity, specificity and cytotoxic function of cells infiltrating acutely rejecting allografts. *J. Immunol.* 118:2020–26
107. Stuart, F. P., Bastien, E., Holter, A., Fitch, F. W., Elkins, W. L. 1971. Role of passenger leukocytes in the rejection of renal allografts. *Transplant. Proc.* 3:461–64
108. Summerlin, W. T., Broutbar, C., Foanes, R. B., Payne, R., Stutman, O., Hayflick, L., Good, R. A. 1973. Acceptance of phenotypically differing cultured skin in man and mice. *Transp. Proc.* 5:707–10
109. Talmage, D. W., Dart, G. A. 1978. Effect of oxygen pressure during culture on survival of mouse thyroid allografts. *Science* 200:1066–67
110. Talmage, D. W., Dart, G., Radovich, J., Lafferty, K. J. 1976. Activation of Transplant Immunity: Effect of donor leukocytes on thyroid allograft rejection. *Science* 191:385
111. Talmage, D. W., Woolnough, J. A., Hemmingsen, H., Lopez, L., Lafferty, K. J. 1977. Activation of cytotoxic T cells by nonstimulating tumor cells and spleen cell factor(s). *Proc. Natl. Acad. Sci. USA* 74:4610–14
112. Vesole, D. H., Dart, G. A., Talmage, D. W. 1982. Rejection of stable cultured allografts by active or passive (adoptive) immunization. *Proc. Natl. Acad. Sci. USA* 79:1626–28
113. Woolnough, J. A., Misko, I. S., Lafferty, K. J. 1979. Cytotoxic and proliferative lymphocyte responses to allogeneic and xenogeneic antigens *in vitro*. *Aust. J. Exp. Biol. Med. Sci.* 57:467–77
114. Yamashita, U., Shevach, E. 1977. The expression of Ia antigens on immunocompetent cells in the guinea pig. II Ia antigens on macrophages. *J. Immunol.* 119:1584–88
115. Zitron, I. M., Ono, J., Lacy, P. E., Davie, J. M. 1981. The cellular stimuli for the rejection of established islet allografts. *Diabetes* 30:242–46
116. Zitron, I. M., Ono, J., Lacy, P. E., Davie, J. M. 1981. Active suppression in the maintenance of pancreatic islet allografts. *Transplantation* 32:156–58

Ann. Rev. Immunol. 1983. 1:175-210

AUTOIMMUNITY – A PERSPECTIVE

Howard R. Smith and Alfred D. Steinberg

Section on Cellular Immunology, Arthritis Branch, National Institute of
Arthritis, Diabetes, and Digestive and Kidney Diseases, National Institutes of
Health, Bethesda, Maryland 20205

AUTOIMMUNITY

The conceptual framework for understanding the basis for autoimmune
diseases has changed drastically in the past few years (Table 1). This change
has allowed what were previously thought to be mysterious disorders to
take their place alongside those with less complexity.

Paul Erlich viewed autoimmunity as the "horror autotoxicus," which the
organism was to avoid at all costs. Dysfunction of Burnet's thymic censor,
which was supposed to prevent self-reactive cells from emerging, provided
the primitive conceptual basis for disorders characterized by self-reactivity.
In addition, immunology was deeply immersed in departments of microbiology, which held Koch's postulates as sacred long after their usefulness
had passed. As a result, autoimmune disorders were thought to be caused
by single agents, which one day would be uncovered. The discovery of slow
viruses added fuel to this line of reasoning. Nevertheless, this conceptual
framework left autoimmune disorders shrouded in a cloak of mystery.

In the past 10–15 years there has emerged a new conceptual framework
for viewing autoimmune diseases. Although never explicitly stated, the
framework can be readily outlined (Table 1). The theoretical considerations
have not yet produced a complete picture. Moreover, the precise details of
the pathogenesis and etiology of most autoimmune disorders have not been
elucidated. Nevertheless, there is sufficient information, coupled with the
new conceptual framework, to move such diseases from the realm of the
mysterious to the realm of the comprehensible. The shift from mystery to
understanding has been complex. At least one important factor is the advance of information in immunology. Another is the recognition that such
diseases may not have a single etiology. Since immunology emerged from

175

0732-0582/83/0410-0175$02.00

Table 1　Autoimmune diseases: conceptual framework

Era	Beliefs
I. DARK AGES	Horror autotoxicus of Ehrlich Thymic censor of Burnet prevents emergence of forbidden clones Qualitative abnormalities lead to disease Single cause (especially viral)
II. ENLIGHTENMENT	Self-reactivity is normal Self-self reactions form the bases for both normal immune reactivity and normal immune regulation Disease results from quantitative abnormalities (e.g. amount of stem cell proliferation, amount of autoantibody, quantity of immune complexes) Multifactorial etiology 　Genetic 　　Individual genes predispose to particular abnormalities 　　Disease is polygenic 　Environmental factors 　　Stimulate immune system 　　Interfere with normal immune regulation 　Hormonal factors (modify disease manifestations) 　Abnormal immune regulation (may result, in part, from the above)

microbiology, there was a strong prejudice in favor of a single causative agent, or at least a unifying cause, for a disease. That view may be fine for many infectious diseases; however, Koch's postulates are useless for a disease in which host susceptibility is the most important factor. They are further compromised by the need for two or three additional factors for disease to become manifest.

The role of individual infectious agents in many autoimmune diseases is still uncertain. The experiences with syphilis and Whipple's disease still render us sufficiently humble to acknowledge that an organism may yet be found for each autoimmune disease. In fact, the observation that Lyme arthritis is caused by an infectious agent has recently underlined this possibility. Nevertheless, the best evidence points strongly to a multifactorial etiology for many autoimmune diseases. If an organism is one of the factors, it is only one. In this paper, we try to define and discuss autoimmune disease in the light of modern immunologic concepts and on the basis of the most recent studies.

The immune system displays a complexity, which includes a requirement for self-recognition. Internal regulation of the system occurs through that self-recognition and constitutes the most basic form of autoimmunity. Regulation may involve cells, antibodies, amplification systems (e.g. complement), and combinations of these elements. Thus, certain cell-cell interactions do not occur efficiently unless there is a recognition by one cell of specific self-determinants on the other. This self recognition is a form of autoimmunity. In addition, production of antibodies (Ab_1) to an antigen may lead to regulation of that response through (auto)antibodies (Ab_2) specific for Ab_1. Such anti-idiotype antibody (Ab_2) production represents another form of autoimmunity. In health, the regulatory circuits flow smoothly. They regulate the production of antibodies to foreign organisms and play an important role in normal homeostasis. Thus, the horror autotoxicus of Paul Erlich does not necessarily follow from anti-self immunity. However, alterations in the normal functioning of the immune regulatory circuits may cause dramatic and sometimes deleterious results. Whereas the regulated production of autoantibodies appears to be consistent with health, uncontrolled production of autoantibodies to certain antigens may lead to disease. The same is true of cells, which recognize self determinants on other cells. Diseases characterized by abnormal and/or excessive anti-self immune responses have been termed autoimmune diseases.

In this paper we discuss mechanisms by which autoimmune diseases may occur. A number of diseases serve to illustrate particular principles. Thereafter, we discuss two in greater depth: autoimmune thyroiditis (AIT) and systemic lupus erythematosus (SLE). Although we concentrate on human disease, we rely upon animal models where necessary.

The Immunological Basis for Disease

Before embarking, we wish to classify autoimmune diseases. These are summarized in Table 2. A defect in the afferent limb of the immune system

Table 2   Classification of autoimmune diseases

Class	Description
A	A defect in the afferent limb of the immune system initiated without a requirement for a specific external agent (SLE) With important genetic requirements Without important genetic requirements
B	A defect in the afferent limb of the immune system initiated by a specific external agent (acute rheumatic fever) With important genetic requirements Without important genetic requirements
C	A defect in effector mechanisms of immunity initiated without a requirement for a specific external agent (hereditary angioedema)
D	A defect in effector mechanisms of immunity initiated by a specific external agent (certain central nervous system viral infections)
E	Combinations of the above

initiated without a requirement for a specific external agent (class A) characterizes the classic autoimmune diseases such as SLE, AIT, and rheumatoid arthritis. In them, the primary events are presumed to be initiated by immune processes directed against self antigens (Table 3). The abnormality may lie in the self antigen, the immune process, or both. Thus, one may have (a) an abnormal self antigen with a normal immune response to it, (b) an abnormal response to a normal self antigen, or (c) an abnormal response to an abnormal antigen.

Antibodies reactive with self determinants can be caused by an exogenous agent. Acute rheumatic fever is an example of a class B disease. The initiating agent is a group A beta-hemolytic streptococcus. In susceptible individuals, a severe infection with the organism leads to an immune response, which includes responsiveness to certain self antigens. In this case the self antigens include heart antigens, which cross-react with those of the infecting organism; thus, the disease process ensues (50).

For a variety of formal and historical reasons, diseases of class C tend not to be included among autoimmune diseases. In hereditary angioedema, a defect in the functional capacity of the inhibitor of the first component of complement (because of either a structural aberration or a quantitative reduction) leads to edema (initiated by trauma, endotoxin release, or other stimuli), which can be life threatening (55). Thus, the disease is immune-mediated (by components of the immune system); moreover, it can be lethal. However, most textbooks and teachers do not include it among the autoimmune diseases. We believe there is a bias in favor of including among autoimmune diseases those in which there is apparent specificity. This bias comes from the historical bias among immunologists that specificity is the most important aspect of the immune system and that disorders of the immune system should be manifested in terms of alteration in specificity, if not qualitatively, at least quantitatively. In other words, there should be specific antibody or cell-mediated immunity that is heightened to account for the disease process. Unfortunately, not all diseases fit into these preconceived notions. It has been pointed out that the most nonspecific aspects of immunity may account for disease (112). Thus, in a strictly formal sense, class C diseases could be regarded as perfectly acceptable autoimmune disorders.

The class D disorders are clearly immune mediated; however, they are initiated by known exogenous agents. It is sometimes difficult to determine the extent to which the inflammatory response to exogenous agents is beneficial and the extent to which it is detrimental. If more detrimental than beneficial, it could be termed pathologic. Such a designation would be very broad, probably too broad to be practical. Nevertheless, it is useful to recognize that many toxins (e.g. from an insect) induce inflammatory reac-

Table 3 Self-antigens in autoimmune diseases of class A

Self-antigen	Disease
Intracellular	
Nucleolus	SLE, Sjogren's syndrome, scleroderma, polymyositis
DNA	SLE, chronic active hepatitis, Sjogren's syndrome
RNA	SLE, Sjogren's syndrome
Sm-RNA	SLE, mixed connective tissue disease, scleroderma, poly- myositis
Histones	SLE
Ribosomes	SLE
Cytoplasm	Insulin-dependent diabetes, autoimmune thyroid disease, idiopathic Addison's disease, pernicious anemia, SLE
Mitochondria	SLE, primary biliary cirrhosis, Sjogren's syndrome, chronic active hepatitis
Lysosomes	SLE
Microsomes	Autoimmune thyroiditis, idiopathic Addison's disease, pernicious anemia
Melanosomes	Vitiligo
Filaments/tubules	Chronic active hepatitis, gluten-sensitive enteropathy, SLE, mixed connective tissue disease
Cell membrane receptors	
Thyroid receptor	Graves' disease (stimulates), AIT (blocks)
Acetylcholine receptor	Myasthenia gravis
Insulin receptor	Insulin-dependent diabetes
Cell membranes	
Red blood cells	Autoimmune hemolytic anemia, SLE
Lymphocytes	SLE
Neutrophils	Autoimmune neutropenia, SLE
Platelets	Idiopathic thrombocytopenic purpura, SLE
Muscle	Rheumatoid arthritis, SLE, cirrhosis, polymyositis
Neurons	SLE
Myelin components	Multiple sclerosis
Epidermal cells	Pemphigus vulgaris
Spermatazoa	Autoimmune (male and female) infertility
Ovum	Female infertility
Extracellular	
Basement membranes	Goodpasture's disease, bullous pemphigoid, discoid lupus
Intercellular substance	Pemphigus vulgaris
Plasma proteins	
Immunoglobulins	Rheumatoid arthritis, SLE, Sjogren's syndrome, chronic active hepatitis
Complement components	SLE, rheumatoid arthritis, dense deposit disease
Clotting factors	SLE
Hormones	
Insulin	Insulin-dependent diabetes
Thyroid hormones	AIT
Glucagon	Insulin-dependent diabetes
Intrinsic factor	Pernicious anemia

tions; it could be argued that in the absence of the body's reaction, morbidity would be minimal. An extreme example of an apparently misplaced response to an exogenous agent is the inflammatory response of the central nervous system to certain viral infections. The viruses may not be limited in any important way by the response and they may not be very pathologic by themselves; however, the response to the infection may be disabling or fatal. Is this an autoimmune response? In the absence of knowledge about the virus and with the presence of hypergammaglobulinemia, antiglobulins, anti-nuclear or antilymphocyte antibodies, and/or circulating immune complexes, such diseases would definitely be labeled autoimmune. Our knowledge of the role of the virus allows us to philosophize about the relative merits of including such immune diseases among the autoimmune ones.

Unfortunately, the classification shown in Table 2 is an oversimplification. Many diseases that are immune-mediated (all diseases?) are influenced to a greater or lesser extent by the genetic background of the individual and the state of the immune system. These and other factors may be critical to the expression of illness or the degree of abnormality. In addition, there may be more than one class of defects or mechanisms that may operate in a given person with a particular disease. Therefore, the multifactorial nature of diseases, especially of autoimmune diseases, must be appreciated. Although this adds a degree of complexity, it helps to explain the great variability among patients with the same syndrome. It also helps to explain variability in the expression of diseases among closely related individuals, identical twins, or even highly inbred rodents.

Disease Mechanisms

Many autoimmune diseases have common features such as arthritis or glomerulonephritis. In part, this may be due to the limited outcomes permissible in the affected tissues. It may also be due to the relatively limited mechanisms that mediate autoimmune disorders. The immune mechanisms involved with the pathogenesis of autoimmunity are those as described by Coombs & Gell (18): type I, anaphylactic; type II, cytotoxic; type III, antigen-antibody complexes; and type IV, cell mediated. These specific mechanisms can be augmented by nonspecific measures: amplification systems involving either cellular (via lymphokines) or humoral (i.e. complement, coagulation, kinin, and fibrinolytic) systems. The overall interaction of these complex processes leads to the specific outcomes of each disease.

The cytotoxic mechanism type II encompasses the reaction of circulating autoantibody with antigen fixed either to the surface of cells or to tissue. The antigen may be the cell membrane, a receptor of that membrane, or an antigen such as a virus or drug that has become attached to the cell/tissue surface. The pathology of many autoimmune diseases is initiated by this mechanism.

The final outcome of the cytotoxic process may be mediated by complement as in the reaction of C'-fixing antibody with cell-bound antigen, or may be mediated by cells as in antibody-dependent cell-mediated cytotoxicity (ADCC). ADCC may be the most important mechanism in most autoimmune diseases. The final effect, irrespective of mediation, is usually cell lysis, elimination, or inactivation. This is the mechanism of many autoimmune hematologic disorders. However, antibodies to receptors can lead to stimulation. In Graves' disease, long-acting thyroid stimulator, an autoantibody [immunoglobulin G (IgG)] to the thyroid stimulating hormone receptor, stimulates the receptor and thus mimics the function of the hormone intended for that receptor (3, 23). An opposite effect is seen in myasthenia gravis. An autoantibody to the acetylcholine receptor at the neuromuscular junction combines with that receptor and thus blocks the neurotransmitter, acetylcholine. Besides blocking the receptors, these anti-receptor antibodies accelerate the degradation of the receptor (22). The autoimmune endocrine diseases commonly have antibodies to receptors. Insulin-dependent diabetes may have autoantibodies to the islet cell surface or cytoplasm, insulin receptors, or insulin (48). Insulin may thus be bound by anti-insulin antibody leading to decreased availability of insulin. (This bound insulin may form immune complexes that lead to additional disease manifestations.) Antibodies to the insulin cell surface receptor as seen in insulin-dependent diabetes and acanthosis nigricans can block responsiveness or occasionally can lead to hypoglycemia by mimicking the insulin effect.

Immune complexes comprise the type III mechanism. The formation of the complex of antigen plus antibody may fix complement and thereby initiate inflammatory processes. Any organ in which such complexes are deposited may be subjected to inflammation and ultimately destruction. The original antigen can be either exogenous, as in microbial diseases, or endogenous, as in many autoimmune diseases. Nucleic acids often serve as the antigen in SLE. The vasculitis of polyarteritis nodosa results, at least in some patients, from immune complex deposition of antibody and hepatitis B surface antigen (39). Often, deposition occurs in the glomerulus and accounts for the immune complex glomerulonephritis of many autoimmune disorders.

Localization of the immune complex is not limited to the vessels and glomerulus. The uveal tract can be involved with deposition in the ocular basement membrane, thus leading to uveitis. The choroid plexus has been demonstrated to be a site for a type III mechanism in SLE. Many autoimmune skin diseases result from this process, including cutaneous vasculitis, erythema nodosum, and discoid lupus.

Type IV reactions comprise immune mechanisms that are mediated by interactions of cells or their soluble produces with antigen rather than by antibody and complement. The classical example, delayed hypersensitivity, is characterized by a reaction that is time-dependent, has a specific histologic sequence in terms of cellular infiltration and inflammation, and can only be transferred by cells and not by serum. The effector mechanism of cytotoxicity can include direct cell interaction with antigen or elaboration of lymphokines. The latter include a number of mediators that have diverse functions. They primarily amplify the initial reaction by nonspecifically recruiting inflammatory cells such as neutrophils and macrophages to the reaction area. At that inflammatory site, cells become activated, undergo transformation and proliferation, and further secrete lymphokines, thus generating a cascade effect. Accordingly, in vitro assessment of T cell responses include both quantitative and qualitative measurements of lymphokine production as well as proliferation of effector cells.

In a similar manner, the humoral system has amplification systems that intensify the type I, II, and III reactions. This system involves the cascades of coagulation, complement, kinin, and fibrinolysis, all of which primarily interact intravascularly to maximize the effect of antibody. Although the cellular, humoral, and amplification mechanisms can be dissected, in vivo these reactions tend to act in concert.

SPONTANEOUS THYROIDITIS

Spontaneous thyroiditis was chosen to represent organ-specific diseases. It has been studied extensively in chickens, rats, and humans.

OS Model

The obese chicken (OS) represents a model for the spontaneous development of thyroiditis. The disease resembles the human disease AIT (Hashimoto's) with regard to histopathology, serological abnormalities, and phenotypic corollaries. Both the distinct anatomical separation of the B and T cell generative organs and well-characterized immunogenetics of *Gallus gallus* has allowed otherwise inaccessible exploration into the pathogenesis of this disorder. Cellular manipulations of T or B cells in the OS model are made possible by thymectomy or bursectomy, respectively. It has been repeatedly demonstrated that neonatal thymectomy accelerates thyroiditis whereas neonatal bursectomy prevents disease (128). Transfer experiments have subsequently been performed to elucidate further the site of the cellular defect. Bursal cells from OS chickens or their normal counterparts (CS chickens) were transferred to neonatally, chemically (cyclophosphamide) bursectomized OS and CS chicks (85). The reconstituted normal CS chicks did not develop disease or autoantibodies, whereas the reconstituted OS chicks still developed thyroiditis and autoantibodies. In contrast, T cell transfers from OS chickens to normal T cell-depleted (irradiated) CS chickens caused thyroiditis, whereas the reverse (CS T cells into irradiated OS chickens) did not cause disease (95). These cellular studies indicate that although B cells are required for disease expression, they are probably normal; however, the T cells appear to be defective. Further investigations have characterized the T cell defect to reside in the T suppressor cell population (95). It has been postulated that T suppressor cells fail to migrate adequately to the peripheral lymphoid tissue secondary to delayed thymic maturation, and/or that relative unresponsiveness to T suppressive actions exists at the target tissue (126).

The OS chicken model has a genetic predisposition to the development of disease. This appears to be under the polygenic control of at least three loci (127): the major histocompatibility complex (B locus), a non-B (minor histocompatibility) locus, and a thyroid locus.

Other models of spontaneous autoimmune thyroiditis (SAT) exist (beagles, marmosets), but most notable is that of Buffalo rats (105). Results in that model are quite similar to those found in OS chickens; again, neonatal (but not adult) thymectomy promotes thyroiditis, which suggests significant T suppressor restraint of disease. In contrast, experimentally induced autoimmune thyroiditis (EAT) of rabbits, guinea pigs, and mice appears to have a different cellular basis for disease. The induction of disease with homologous thyroglobulin has been abrogated by neonatal thymectomy, thus indicating that the induced disease is thymic (helper) dependent. Further differences exist between EAT and SAT: thyroid destruction is largely T-cell mediated in EAT as opposed to important antibody-mediated mechanisms in SAT and human AIT. Although it can be helpful for elucidation of immunogenetic details, the usefulness of EAT as a model for human AIT may be limited.

HUMAN AUTOIMMUNE THYROIDITIS (AIT)

Antibodies

The finding of antibodies to thyroglobulin in the serum of patients with AIT by Roitt et al (93) in 1956 marked the beginning of the investigations into the autoimmune nature of this disease. Rose & Witebsky (96) were concurrently able to demonstrate that similar lesions were produced in animals by immunizations with homologous thyroid antigens mixed with Freund's adjuvant. Subsequently, a large number of antibodies to normal thyroid tissue components were discovered in patients' sera. The major antibody is to thyroglobulin. Antibodies against other thyroid antigens include those to the second component of colloid, microsomal antigen, nuclear component, thyroxine, triiodothyronine, and thyroid-stimulating hormone receptor (26).

The antibody to thyroglobulin is usually found in high titers in patients with AIT, but it may also be found in Graves' disease (70), in low titers in other autoimmune disorders, and even in normal persons. As with the other autoantibodies against thyroid antigens, thyroglobulin-antibody is usually of the IgG class.

The exact role of these various antibodies in the pathogenesis of the disease is unclear. By themselves, they appear unable to cause thyroid lesions. Studies have shown that transfer of serum from individuals with AIT to monkeys does not cause disease and that newborns with high titers of antibodies obtained by transplacental transfer from mothers with AIT also do not develop disease. Even though complement-dependent cytotoxicity by thyroid microsomal antibody against thyroid cells occurs in tissue culture, the role in vivo may follow initial cell damage by another mechanism (52). However, there is a relationship between the levels of serum anti-thyroid antibodies and lymphocytic infiltration of the thyroid gland (131). The importance of these autoantibodies may lie in their involvement in immune complex formation (49) or ADCC (17), mechanisms that may be involved in thyroid tissue damage. In AIT, both soluble immune complexes in serum and electron-dense deposits in thyroid follicular basement membrane occur and presumably represent thyroglobulin-anti-thyroglobulin antibody (49). Recently, ADCC activity and K cell numbers have been shown to be increased in AIT (7); their increase corresponded to the appearance of the thyroid lesion (56). Although several mechanisms for the induction of inflammation appear to be responsible for AIT, ADCC appears to be a critical component.

Cells

Because a pathogenetic role could not readily be demonstrated for serum autoantibodies in AIT, investigators turned their attention to the cellular basis for the disease. By use of monoclonal antibodies against cell (sub)-populations, defects in T cell numbers and in suppressor T cells have been found in the peripheral blood of patients with AIT (81). Nonspecific suppressor T cell function appears to be normal, whereas antigen-specific T suppressor dysfunction recently has been demonstrated (52). The latter points to the possibility that altered immunoregulation via loss of and/or defective T suppressor cells may contribute to AIT by allowing unhindered B cell autoantibody production.

Further defects in tests of cell-mediated immunity in AIT raise the possibility that other mechanisms participate in the pathogenetic events. T lymphocytes and monocyte-macrophages may be involved. Lymphokines are generated by sensitized T lymphocytes in AIT, indicating specific recognition of thyroid antigens. T lymphocytes from AIT patients can be directly cytotoxic for thyroid cells in tissue culture, for mouse mastocytoma cells coated with thyroid microsomes, or for heterologous red blood cells coated with thyroglobulin.

Monocyte-macrophage function in AIT is not well characterized: however, it may be important for tissue inflammation. Monocytes presented with cytophilic anti-thyroglobulin-antibody become armed. If presented with thyroglobulin-coated red blood cells they are able to lyse the target cells by an extracellular mechanism (68). Anti-microsomal antibodies can also be cytophilic for monocytes (68). In addition, the interaction of macrophages with lymphocytes differs in autoimmune thyroid disorders as compared to normals: There is an increase in macrophage-lymphocyte rosettes. Thus, the defective interaction of macrophages with both monocytes and lymphocytes may contribute to disease.

Summary of AIT

Spontaneously occurring autoimmune thyroiditis in humans may have more than one etiology and more than one mechanism of inflammation and glandular destruction. The animal studies indicate that at least three genes are necessary for disease production. The human disease is significantly associated with HLA-D3, however with only a moderate relative risk, which suggests that additional genes may be involved. In some susceptible individuals, various initiating events may trigger the disease process. These have not yet been elucidated. In other individuals, an inherited thyroid defect may be sufficient to trigger the disease in the presence of the appropriate genes for the immune abnormalities.

SYSTEMIC LUPUS ERYTHEMATOSUS (SLE)

Introduction

In contrast to AIT, SLE is an example of a widespread organ nonspecific disorder. This autoimmune disease is characterized by B-cell hyperactivity, which results in hypergammaglobulinemia and in the production of a variety of antibodies reactive with organ-nonspecific antigens such as DNA, RNA, and cell membrane structures. Disease signs and symptoms may occur in virtually any organ in the body. It is widely held that immune complexes of antigen and antibody and complement components initiate the inflammatory process in many organs; however, alternative or additional mechanisms are likely. A major problem in the study of SLE is determining which immune abnormalities result from the illness and which lead to disease. This aspect has been especially difficult to dissect in patients who usually appear with the syndrome already present. However, the availability of mice known to develop the illness allows for study prior to the onset of overt symptoms or signs of disease.

In the past few years, work with the experimental animals that develop an illness resembling human SLE has contributed substantially to our knowledge. In addition, important advances have been made in the study of the human disease. Most of these advances have emphasized pathogenetic factors in the induction of the syndrome. They suggest that multiple factors play a role in the development and expression of illness, including genetic factors, immune factors, environmental factors, and hormones. Although the exact mechanisms have not been completely elucidated, broad outlines are now possible.

Murine Lupus

A number of different strains of mice develop spontaneously an autoimmune disorder with features of human SLE. However, the clinical syndromes vary substantially among the various murine models of SLE. Some of the features of the most widely studied strains are shown in Table 4. NZB mice produce anti-T cell antibodies from early in life. Thereafter, the pace of their illness is very slow. These mice do not die until 15–20 months of age. In contrast, MRL-*lpr/lpr* mice have a rapidly progressive illness characterized by massive lymphadenopathy and death by 7 months of age. BXSB males die between 6 and 10 months of life, whereas their sisters live well into year 2. The opposite occurs in the (NZB X NZW) F_1: The females die of immune complex renal disease at about 10 months whereas the males live until 15–20 months. Therefore, although any one mouse may be a model of human SLE, they, like the patients, vary tremendously among themselves.

B CELL HYPERACTIVITY   All of the strains mentioned have B cell hyperactivity: hypergammaglobulinemia and increased numbers of antibody-forming cells. This hyperactivity appears to be polyclonal; it is not limited to the production of autoantibodies. As a result, murine SLE could be viewed as merely a manifestation of polyclonal B cell activation, with some of the B cells producing autoantibodies that lead to injury. The cause of the B cell activation is uncertain; moreover, it may vary from strain to strain. NZB mice produce enormous amounts of IgM; MRL-*lpr/lpr* mice produce IgG1 and IgG2a in great excess; BXSB mice produce large amounts of IgG1 and IgG2b. Nevertheless, the information for the polyclonal B cell activation resides in the bone marrow stem cells. In all of the strains, marrow stem cells can transfer the characteristic disease to irradiated

Table 4 Abnormalities in murine lupus

Feature	NZB	(NZB × NZW) F_1	MRL-MP/*lpr/lpr*	BXSB
Genetic	At least 6 autosomal genes	Multiple genes, some from NZW	Multiple background genes *lpr* major accelerator	Multiple background genes Y chromosome gene major accelerator
Effect of *xid*	Prevents disease	Prevents disease	Reduces anti-ssDNA but does not prevent disease	Prevents disease
Sex	Little effect Recessive gene for androgen insensitivity	Marked effect Androgens protect Estrogens worsen	Androgens protect	Marked acceleration in males—not hormonal
Effect of neonatal thymectomy	Worsens	Worsens	Cures	Worsens
Tolerance to BGG	No	No	No	No
Immunoglobulin	↑ IgM	↑ IgM, ↑ IgG2	↑ IgG1, ↑ IgG2a	↑ IgG1, ↑ IgG2b
Lymphoid organs	Lymphoid hyperplasia	Lymphoid hyperplasia	Marked ↑ T cells	Moderate ↑ B cells
Disease	Marked IgM production	Anti-DNA	Marked lymphadenopathy	Immune complex glomerulonephritis
Manifestations	Anti-T cell antibodies Coombs-positive hemolytic anemia Membranous glomerulonephritis Vasculitis Splenic hyperdiploidy resulting in malignancy	Membranoproliferative glomerulonephritis Sjogren's syndrome	Anti-DNA Arthritis and anti-Ig Membranoproliferative glomerulonephritis Anti-Sm	Arteritis with coronary artery disease Serologically less abnormal than others

histocompatible recipients (5, 25, 74, 121). Therefore, the problem is, at least in part, proximal to the mature B cell.

T cells normally serve to regulate the magnitude of B cell responses. In the autoimmune mouse strains, it is possible that B cell hyperactivity might result, in part, from T cell regulatory dysfunctions. It appears that in different strains different T cell functions are prominent. Therefore, it is necessary to examine T cell function before conclusions regarding the B cell hyperactivity are possible.

T CELL FUNCTIONS IN MURINE LUPUS   The most glaring examples of T cell abnormalities are found in the MRL-*lpr/lpr* mouse. This mouse develops massive lymphadenopathy (77). The predominant cellular increase consists of Ly $1^+,2^-$,Thy 1^+ (helper phenotype) T cells (63). However, the intensity of staining of these T cells with anti-Ly 1 is low; therefore, the cells may not be representative of the normal distribution of Ly 1^+ T cells. The presence of such enormous numbers of Ly 1^+ T cells suggests the possibility that such T cells might drive B cells to produce autoantibodies. In the absence of the neonatal thymus, lymphadenopathy does not occur; moreover, anti-DNA, renal disease, and death are markedly retarded (116). If a thymus graft is provided, the autoimmune syndrome can occur (121). Thus, the stem cells of the MRL-*lpr/lpr* mouse give rise to abnormal T cells. Such bone marrow stem cells give rise to pre-T cells that must mature in a thymus. Thus, the defect that leads to autoimmunity in MRL-*lpr/lpr* mice is not in the thymus, it is merely thymic dependent. The abnormal stem cell needs a proper environment in which to mature.

NZB mice also have both B cell and T cell abnormalities. Although nu/nu NZB mice develop disease (32) and stem cells can transfer it to non-autoimmune-prone recipients (74), the addition of NZB T cells to NZB marrow cells in neonatally thymectomized radiation chimeras increases the autoimmune disease (120). Recent studies have demonstrated that a marrow pre-T cell contributes to the abnormality in experimental tolerance (61). This cell can mature in a non-NZB thymus. Therefore, the T cell abnormality of NZB mice, like that of MRL-*lpr/lpr* mice, is based upon an abnormal stem cell that requires a thymus for maturation.

The male BXSB mouse has not only polyclonal B cell activation, but also B cell hyperplasia (121). Unlike the T cell adenopathy of the MRL-*lpr/lpr* mouse, the lymphadenopathy of the BXSB mouse consists largely of B cells. Recently, the B cell hyperplasia of the male BXSB mouse has been found to be under T cell regulation (106a). Neonatal thymectomy of male, but not female, BXSB mice leads to a marked increase in lymphadenopathy; the enormously enlarged nodes contain large numbers of B cells. Of interest, these neonatally thymectomized BXSB mice contain very few Lyt2$^+$ cells, suggesting that T cells of the suppressor circuits normally serve to regulate the B cell hyperactivity.

(NZB X NZW) F_1 T cells also have been found to regulate the pace of illness. Their effect is more like that seen in BXSB mice than in MRL-*lpr/lpr* mice—neonatal thymectomy tends to accelerate disease (113, 114, 116). However, thymectomy at 6 or 10 weeks of life does not (114). Therefore, whatever serves to regulate the pace of illness must work within the first few weeks of life for it is no longer present in older (NZB X NZW) F_1 mice. Consistent with these cellular studies is suppressor factor production deficiency in these mice (86), which may be replaced with some benefit.

STUDIES OF MARROW STEM CELLS   In view of the ability of marrow from all of the autoimmune strains to transfer disease, it appears that the information necessary to produce abnormal lymphocytes is contained within the marrow stem cells. Stem cells from NZB mice were found to produce large numbers of endogenous granulocyte and monocyte spleen colonies (75, 122, 125). However, lethally irradiated NZB mice given a fixed number of syngeneic marrow cells did not have an increase in colonies (125). Two explanations are possible: Spleen stem cells manifest excessive endogenous colony formation; marrow stem cells, which give rise to colonies, are increased in absolute number but not percentage in NZB mice. The latter possibility would explain why studies of a fixed number of marrow cells leads to fewer abnormalities than study of the whole marrow potential. Nevertheless, NZB marrow cells were able to generate increased numbers of B cell colonies in vitro (53), which were resistant to suppressive factors such as prostaglandins or IgM (54). The increase in B cell colonies is observed very early in life but becomes subnormal by 15 weeks of age (47). These studies suggest that the NZB marrow stem cells are capable of excessive production of B cell offspring; however, if that process plays a role in disease, there must be a long lag between the peripheralization of the cells and their ultimate contribution to disease. Perhaps the increased marrow production of B cell colonies gives rise to an abnormal distribution of peripheral B cells, which ultimately explains the unusual IgM production, the increase in Ly 1^+, IgM$^+$ cells, hyperdiploid cells, and the autoantibody production characteristic of NZB mice. It may also explain the splenic localization of the abnormal cells after the first few months of life.

GENETIC STUDIES   A number of genetic factors have been found to influence the expression of autoimmunity in various strains of mice. The *lpr/lpr* gene is a single gene locus that causes lymphoproliferation. In homozygous form, this gene is responsible for the lymphoproliferation seen in MRL-*lpr/lpr* mice. It has been suggested that this is a single gene model for autoimmunity (77) and that a specific defect might be the cause (130). However, the MRL +/+ mice are not normal. They spontaneously produce antinuclear antibodies. In addition, MRL mice have phenomenal hyper-maturity. We have observed that they are born more than twice as big as mice of other inbred strains and are able to reproduce as early as 5–6 weeks of age (which contrasts with the 8–12 weeks in other inbred strains). Thus, there appear to be background genes unrelated to *lpr* that may predispose to hypermaturity, accelerated aging, and autoimmunity. This is perhaps best demonstrated by examination of other strains of mice with the *lpr/lpr* gene. In the absence of the MRL background genes, C57BL/6-*lpr/lpr* mice manifest much less vigorous autoimmune disease (77, 130). However, in the presence of appropriate background genes from another source (e.g. NZB), the pace of illness induced by *lpr* is markedly accelerated. In other words, *lpr/lpr* has its major effect in accelerating disease when background genes for autoimmunity are present.

The background genes for autoimmunity have been studied best in the NZB strain. Here, at least six genes are responsible for the full-blown illness (87). Some of these genes appear to code for individual abnormalities in the mice. Thus, a single gene is responsible for the predisposition to anti-DNA production and another unlinked gene allows anti-T cell antibodies to be produced in large amounts (87, 90). There appears to be no linkage to H-2 or allotype in the RI lines (13). In the NZB mice there is no single gene that predisposes to, or causes, autoimmunity; rather, a number of genes contribute (113).

The BXSB mouse initially appeared to have a single gene for autoimmunity (78). This was a Y chromosome-linked factor. Recent studies have clearly demonstrated that the Y-linked factor cannot, by itself, induce autoimmunity. For example, crossing a BXSB male with a non-autoimmune female leads to male offspring that do not manifest autoimmunity (113); however, crossing a BXSB male with an autoimmune-prone female leads to male offspring with disease that is accelerated relative to their sisters (121). Thus, the Y chromosome-linked factor operates in concert with background genes to induce accelerated autoimmunity. Thus, like the MRL and NZB mice, BXSB mice have background genes for autoimmunity. In all of these mice, the *lpr/lpr* or BXSB Y chromosome can cause marked acceleration of disease. The exact nature of the background genes and the precise modes of action of *lpr/lpr* or the BXSB Y remain to be determined. Characterization of the gene products would be a major advance in our understanding of the basis for autoimmunity.

B CELL SUBSETS RESPONSIBLE FOR AUTOANTIBODIES   More extensive studies of B cell subsets responsible for autoantibody production have been performed with NZB mice and their hybrids. Congenic NZB·*xid* mice are mice that are largely NZB but bear the *xid* gene, a gene that originated from the CBA/N X chromosome. Mice that bear *xid* in homozygous or hemizygous form are deficient in a subset of mature B cells (Lyb 3⁺, Lyb 5⁺), which is responsible for immune responses to polysaccharide antigens with repeating structures (103). NZB·*xid* mice fail to manifest the autoimmune syndrome characteristic of NZB mice; they have a much reduced quantity of autoantibodies and live almost a normal life span (119). The same is true for (NZB X NZW) F_1·*xid* (117), BXSB·*xid* and (NZB X BXSB) F_1·*xid* mice (H. R. Smith, T. M. Chused, A.D. Steinberg, submitted for publication). Therefore, the *xid* gene appears to inhibit markedly the development of autoantibodies in NZB and BXSB mice and their hybrids. The simplest explanation is that the Lyb 3⁺,5⁺ B cells are necessary for autoantibody production (they make the autoantibodies) and that in their absence fewer antibodies are made. One caveat remains, however. It is possible that *xid* has additional effects besides the deletion of Lyb 3⁺, 5⁺ cells. In fact, it appears that it also alters macrophage markers (97). Therefore, it is possible the *xid* gene more indirectly reduces autoantibody production by eliminating an important macrophage-B cell interaction. Nevertheless, it is clear that the Lyb 3⁺,5⁺ B cell subset limb of the immune response is very important for autoantibody production and that in its absence autoantibody production is reduced markedly. The reduction is, however, not absolute. Low levels of autoantibodies occur spontaneously in *xid*-bearing New Zealand mice (80, 94), and prolonged administration of polyclonal immune activators can stimulate such mice to produce autoantibodies (106). Thus, although the Lyb 3⁺,5⁺ B cell subset appears to be largely responsible for autoantibody production, other B cells can produce autoantibodies if stimulated sufficiently.

The B cell subset responsible for autoantibody production in MRL-*lpr/lpr* and BXSB mice has not been completely elucidated. We have now developed fully congenic mice bearing *xid* and such studies are in progress. Studies with more limited inbreeding suggest that the hyperactivity of MRL-*lpr/lpr* mice is not as completely inhibited by *xid* as is the case with NZB mice. Thus, the Lyb 3⁻,5⁻ cells may play a larger role in autoantibody production in the MRL mice (51). However, it is possible that the abnormal T cells of the MRL-*lpr/lpr* mouse drive the B cells to produce antibody very much as the exogenously administered polyclonal immune activators drive the B cells in the NZB·*xid* mice (106).

A subset of B cells has been described with an unusual phenotype. This Ly 1⁺ B cell is quantitatively increased in NZB mice (66). In addition, it is increased in neonatally thymectomized BXSB mice with accelerated autoimmunity (106a). It is possible that this B cell subset is responsible for autoantibody production. The relationship between this B cell subset and the Lyb $3^+,5^+$ subset remains to be determined.

DEFECTS IN TOLERANCE   The development of widespread autoimmune disease in NZB, (NZB X NZW) F_1, MRL-*lpr/lpr*, BXSB, and (NZB X BXSB) F_1 mice may be viewed as a loss of self tolerance. This loss has been studied not only by assessing the development of autoantibodies, but also by assessing experimentally induced tolerance. Tolerance to heterologous gamma globulins has been studied most extensively. These mice have been found to be defective in the development of experimental tolerance to both bovine gamma globulin (BGG) and human gamma globulin (41, 62, 113). The defect can be transferred to lethally irradiated recipients in both the NZB and BXSB by marrow stem cells (25, 41, 61, 62, 120). In the NZB mouse, the T cells and non-T cells are both abnormal; however, the marrow pre-T cell was found to be sufficient to produce the defect (61, 62). In contrast, studies in the BXSB mice suggest that a non-T B cell might be most abnormal (41). Earlier studies in (NZB X NZW) F_1 mice indicated that marrow stem cells contained the information for intolerance to BGG (111). Of interest, the young (NZB X NZW) F_1 thymus contains cells that promote tolerance or prevent the escape from tolerance to BGG; these cells also retard spontaneous anti-DNA production (114). Therefore, defects in experimental tolerance, like the disease, can be transferred by stem cells. The exact contributions of different mature lymphoid populations to the defect may vary from strain to strain or may be dependent upon the details of experimental design.

Other forms of tolerance have been less completely studied. Recently NZB mice were found to have a defect in the development of systemic tolerance following gastrointestinal exposure to antigen (18a). Antibody-mediated suppression of immunity has been found to be defective in NZB and MRL-*lpr/lpr* mice, but not in BXSB or (NZB X NZW) F_1 (18b). Therefore, these functions may be irrelevant to the development of autoimmunity; alternatively, the differences among the strains with regard to certain immune functions may underlie the differences in their expression of the autoimmune syndrome.

SEX   Most patients with SLE are females. In fact, during the childbearing years of females (approximately ages 12–40 years) more than 90% of the patients developing SLE are females. As a result, there has been an attempt to understand the female predominance in human SLE by studying sex hormones in mice. Initial studies of (NZB X NZW) F_1 mice demonstrated only partial effects of sex hormone reversal. Such studies involved mice that were 5 weeks of age. Subsequent studies demonstrated that the age at which sexual alteration was carried out was critical to the outcome. Castration of females and males at 2–3 weeks of age and treatment with the opposite sex hormone led to accelerated disease in males and retarded disease in females (98, 115). In these studies it appeared that androgens retarded disease to a greater extent than estrogens accelerated disease; however, both processes were operative. To the extent that patients were like (NZB X NZW) F_1 mice, a hormonal explanation for the female predominance of SLE was possible. The cellular basis for the result is not completely understood. It has been found that androgens tend to favor the development of thymic suppressor cells and that estrogens favor helper cell differentiation (E. A. Novotny, E. S. Raveche, M. Ottinger, A. D. Steinberg, submitted for publication). Studies of experimental tolerance have further shown that in the presence of estrogens a single cellular defect is sufficient to transmit a defect in tolerance; in the presence of androgens, two cellular defects are necessary for a tolerance defect (62). Therefore, females, by virtue of their estrogens, might be expected to have a predisposition to SLE with more modest cellular abnormalities and males would be protected by their androgens unless they had profound cellular abnormalities.

There are much less marked sex differences in the severity of disease in MRL-*lpr/lpr* and NZB mice than in NZB hybrids. Nevertheless, androgens tend to retard and estrogens accelerate disease in these strains, even if it is quantitatively less impressive than in the NZB hybrids. A notable exception is a gene for androgen insensitivity in NZB mice that regulates anti-T cell production (113).

BXSB mice, superficially, represent an exception to the above. The male BXSB mice have markedly accelerated disease relative to their female littermates. This effect is unrelated to sex hormones (24) and appears to be caused by the stem cell defect that requires the BXSB Y chromosome (25). However, in crosses of CBA/J mice with BXSB, the female offspring have more severe disease than do the males whether the BXSB is the mother or the father (113). Thus, only when the BXSB Y chromosome can act upon other suitable background genes does the male predominance become manifest.

VIRUSES IN MURINE SLE   There has been considerable interest in the search for environmental agents, such as viruses, to trigger the immune system and cause SLE. Both high titers of xenotropic type C RNA virus and circulating levels of the murine retroviral envelope glycoprotein, gp 70, are found in murine SLE strains (121). However, features of autoimmunity have been dissociated from viral expression in F_2 and RI mice (19). Some viruses, such as lymphocytic choriomeningitis, can accelerate disease (123) whereas others, such as lactate dehydrogenase virus, can ameliorate disease and prolong survival (83). Although available evidence does not support an etiologic role for viruses, it appears that antibody formation to virus and/or viral antigens causes immune complexes which can deposit in the kidney and contribute importantly to the development of glomerulonephritis.

PRIMARY VERSUS SECONDARY DEFECTS   Despite an intensive effort over the course of a quarter century, it remains unclear as to whether individual cellular abnormalities are primary or secondary to other defects. Many abnormalities appear to result from the illness rather than induce the illness. These include reduced primary antibody responses in vivo, reduced responses of T cells to mitogens in vitro, reduced IL-2 production, and impaired-cell mediated immunity (130). Many of these are also observed in patients with active SLE (see below). These defects may provide insight into the effects of ongoing autoimmune disease. They may also provide a basis for therapy; however, they are unlikely to provide insights into the early stages that precede the development of the autoimmune syndrome.

SUMMARY OF MURINE SLE   There are a number of different mouse strains with spontaneous autoimmunity. They all have defects in self-tolerance (they produce pathogenic autoantibodies) and in experimental tolerance. However, the details of their clinical syndromes and the basis for illness in the mature lymphocytes vary among them. The major common feature is the ability of lymphoid stem cells from the mice to carry the information necessary to transmit the disease. Nevertheless, a variety of environmental factors can modify the expression of illness, as can a variety of background genes. Finally, once the illness is fullblown, multiple immune defects may be measured that bear little relationship to the primary (preclinical) underlying process. The relationship to human SLE is discussed below.

Human SLE

Human SLE varies from a mild disorder that may require little treatment to a rapidly progressive and possibly fatal disorder. If one examines the spectrum of diseases that characterize the murine models of SLE and adds to them the offspring of matings among themselves and with another 20 inbred strains, one begins to develop a picture of the spectrum of human SLE. Nevertheless, certain differences may characterize humans. Disease manifestations occur in many humans early in life, especially in the first half of the reproductive life span of females. In contrast, many mice with SLE do not develop illness (as opposed to serological abnormalities) until a later time. Only *lpr/lpr* females die during their reproductive period and share this feature with humans. Rashes are more common in humans, as are serositis and alopecia. These are not characteristic of the mice. Joint disease is a feature only of MRL-*lpr/lpr* mice, but it occurs in the majority of patients. One very important difference between mice and people is that the studies in the mice usually are performed in the absence of therapy, whereas the people often have been treated previously or are currently being treated. A second difference is the ability to study the mice prior to the onset of disease. Finally, a variety of manipulations are possible in mice, but not in humans.

B CELLS AND B CELL FUNCTIONS IN SLE   Patients with active SLE have reduced absolute numbers of circulating B cells (37) but increased percentages of cells staining with anti-B1 (72). Polyclonal B cell hyperactivity characterizes the great majority of SLE patients with active disease. They have markedly increased numbers of antibody-forming cells as determined by reverse plaque assay, especially those of the IgG and IgA classes (10, 35). This increase in antibody-forming cells is highly correlated with the degree of disease activity (10). These SLE B cells produce large quantities of autoantibodies, such as to DNA; however, they also produce increased numbers of antibody-forming cells that produce antibodies to irrelevant chemical haptens (16, 71). Although patients with SLE may have generalized antibody production, it may be especially directed toward a limited number of antigens. A stem cell defect could account for the increase in B cell function as SLE patients have increased numbers of B cell colonies generated by peripheral blood non-T cells (57).

Paradoxically, when SLE peripheral blood B cells are stimulated with B cell mitogens, they do not necessarily produce more immunoglobulin than normal, and they actually may have a markedly depressed in vitro response to pokeweed mitogen (27). However, in short term unstimulated cultures of fresh peripheral blood lymphocytes, immunoglobulin synthesis by SLE cells has been found to be much greater than that by normal cells (35, 45, 79) and parallels SLE disease activity (35). It has recently been observed that SLE blood gives rise to greater numbers of B cell colonies, a defect that maps to B but not T cells (57).

T CELLS AND T CELL FUNCTIONS IN SLE   A considerable number of studies have been performed on this subject (88). Lymphopenia, including reductions in circulating T cells, characterize most patients with active SLE. The abnormality is much less marked when the disease is quiescent. Various subsets of T cells have been reported to be reduced in patients with SLE. These include active or early rosette-forming cells, T_G cells, and more recently T cells that rosette with human erythrocytes (T_{ar}). These cells may all overlap to a substantial degree. The T_{ar} subpopulation of T cells has been found to have the surface and functional characteristics of post-thymic precursor T cells, can generate suppression function and natural killer cell activity, and can act as the responding cell in the autologous mixed lymphocyte reaction (AMLR). The decrease in T_{ar} cells in active disease may be due, in part, to antilymphocyte antibodies, as antibodies specific for such cells have been found; of interest, partial correction of the defect occurred when such cells were incubated with fresh human serum (or serum thymic factor) (84). This result suggests that the T_{ar} precursor cell was present or that the erythrocyte receptor was blocked or not expressed.

The relationship between circulating T cells and the human total body T cells is unknown. Although a reduction in peripheral blood T cells is found in SLE, it is not known whether this represents a migration of cells from the circulation to the tissue or a decrease in absolute total body T cells. Many patients with SLE have impaired delayed hypersensitivity responses to a variety of antigens (11). In addition, they have impaired ability to be sensitized (28). At least part of this defect may result from the lymphopenia. An additional component derives from suppression, perhaps by macrophages and their products. We have found that treatment of patients with azathioprine or cyclophosphamide in moderate dosage often leads to normalization of skin test responses.

Variable results have been reported regarding the ability of SLE T cells to respond to mitogens. Some of the variability stems from use of whole peripheral blood mononuclear cells as opposed to more purified cell populations. However, it is often found that responsiveness to the T cell mitogen concanavalin A (ConA) is reduced in patients with active SLE and that there is a return toward normal in less active disease. It is thought that there is a defect in cell-cell communication in patients with active SLE. This is perhaps best exemplified by the impairment in the AMLR (100). However, that reaction also returns toward normal in disease quiescence (101). At a time when B cell hyperactivity is most abnormal, helper T cell function is most deficient in SLE (21). Therefore, the maintenance of B cell hyperactivity does not appear to be dependent upon helper T cells.

In the absence of excessive helper function, a defect in suppressor function has been sought. A number of papers have provided evidence for a defect in suppressor function in patients with SLE (1, 14, 99). However, even when such a defect has been described, some patients did not have a defect and others had a defect with regard to some suppressor functions but not others (99). Finally, a defect in suppressor function has also been denied. An important serial study demonstrated that both the defect in the AMLR and the defect in suppressor function became normalized when the disease became less active (101). As a result, most of the studies of T cell functions appeared to be analyses of defects that resulted from the disease rather than those that lead to the disease.

The value of ConA-induced suppressor cells has been questioned. However, a defect in suppressor function has been demonstrated even in the absence of ConA (58a). Moreover, antibodies reactive with T cells are produced spontaneously by many patients with SLE (and by NZB mice). Such antibodies are capable of interfering with T cell functions. Both IgM (129) and IgG (58, 82) antibodies may be capable of eliminating T cells. Suppressor function has been reduced or abolished by such antibodies in vitro (102). Therefore, it is possible that defects in T cell functions may not lead to disease, but, rather, result from the disease process. Nevertheless, such antibodies may play an important role in perpetuating a process that otherwise might be self-limited.

GENETIC STUDIES   A genetic predisposition to the development of SLE has been established both by familial studies and by analysis of major histocompatibility complex associations. Although relatives of patients with SLE have an increased frequency of developing the disease (approximately 5%) (12), concordance of SLE in identical twin pairs is about 70% (12). There is some evidence to suggest that environmental effects contribute to the familial occurrence of SLE (8, 15); however, the concordance of SLE in dizygotic twins is no more than that of first degree relatives (12). Although a high incidence of serologic and immunoregulatory abnormalities occurs in clinically unaffected relatives of patients with SLE, including diminished suppressor cell function (67), reduced numbers of erythrocyte C3b receptors (43, 69, 128a), antinuclear antibodies (65), lymphocytotoxic antibodies, and anti-RNA antibodies (20), the possible role of environmental factors in some of these defects has not been fully elucidated. Recent studies suggest that a genetic deficiency of C3b receptors may be responsible for impaired clearance of immune complexes. Previously it was thought that circulating immune complexes were responsible for decreased C3b receptors. Population studies indicate that particular groups such as American blacks (104) and American Sioux Indians (76) have an increased incidence of SLE.

Analysis of the relationship between the major histocompatibility complex and SLE discloses an association between the disease and genes that are grouped in the HLA (human leukocyte antigen) region of the sixth chromosome. In animal species, genes of this region play a fundamental role in the control of normal immune responses (9). Initial investigations into the associations between SLE and certain HLA A or B locus antigens failed to highlight any strong ones. However, the study of the D region antigens showed a high incidence of two alloantigens, DRw2 and DRw3 (34, 92).

An even greater association was found with an Ia-like antibody: 715. In family studies, inheritance of two of the markers DR2, DR3, or 715 was highly associated with features of SLE (91).

An interesting association exists between SLE (and similar disorders) and a variety of hereditary deficiencies of complement components, primarily C2. A substantial percentage of women with homozygous C2 deficiency, perhaps as many as 60%, develop a lupus-like illness (4). It has been speculated that the C2-deficient state may play a role in the development of the rheumatic disease, possibly by allowing a predisposition to certain infectious agents that stimulate the immune system and thus favor the development of SLE. However, a strong linkage is found between the hereditary C2 deficiency and the DR2 antigen, which itself is associated with SLE. Thus, the C2 defect may be a marker for a disease susceptibility gene. A related preliminary finding is a strong association between certain C4 alleles and susceptibility to SLE (124).

SEX   Hormonal factors play a role in certain murine SLE models (see above) and, similarly, it is becoming clear that sex hormones are important in both normal and abnormal human immunoregulation. Human SLE is predominantly a female disease, especially between the ages of 12 and 40. Exacerbations of human SLE are often seen with hormonal changes, such as during or after pregnancy (109), at menarche, or with oral estrogen-containing contraceptive use (46). It has been shown that both male and female patients with SLE (60), SLE-Klinefelter's patients (118), and first degree relatives of SLE patients (59) all have an increase in hydroxylation of estrogen to the potent estrogenic urinary 16 alpha metabolites. In addition, male SLE patients have elevated ratios of estradiol/testosterone (44). Thus, some patients with SLE may have a metabolic abnormality that leads to excessive production of certain estrogenic components. The mechanisms whereby estrogens exacerbate disease have not been entirely elucidated. Nevertheless, sex hormones and their metabolism provide yet another factor that modifies disease expression in SLE.

HETEROGENEITY IN SLE   Among the various murine models of SLE, there is heterogeneity with regard to the genetics, cellular basis, and expression of disease. These animal models have revealed that similar outcomes may be the sequelae of different perturbations of immunoregulatory pathways (113). Therefore it would not be surprising to find that human SLE was similarly heterogeneous. Although the wide spectrum of varied clinical presentations and serologic abnormalities of patients with SLE has long been appreciated, most physicians generally consider SLE a single illness. However, like the mouse models, human SLE demonstrates heterogeneity for (*a*) genetics (HLA types, C receptors, immune responses), (*b*) cellular basis (variations in helper functions, suppressor function, AMLR, antibody-forming cells), and (*c*) expression of disease (types of autoantibodies, organ system involvement). Recently, patients have been divided into clinical subgroups based upon differences in their peripheral blood T cell subset composition (108). Certain clinical characteristics of patients with SLE were associated with the ratio of these regulatory T cells. These studies provide a clue to relationships between clinical subsets of disease and differences in cellular abnormalities. Thus, SLE may not be one illness, but rather a symptom complex with common manifestations.

MECHANISMS OF DISEASE IN SLE   The various immune abnormalities-B cell hyperactivity with autoantibody formation, concomitant T cell aberrations, and defective amplification systems (i.e. complement)-all interact to culminate in tissue pathology and disease manifestations. In both murine and human SLE, antibody appears to be the primary initiator of tissue damage. Immune complexes (IC), which are found in various sites including kidney, blood vessels, and skin, are both locally deposited from the circulation and formed in situ. The sites are determined by local tissue factors, such as vascular permeability, whereas multiple properties of IC account for their pathogenicity (including size, avidity for antibody, ability to fix complement, and clearance of IC by the mononuclear phagocytic system). The mononuclear phagocytic system in SLE has been demonstrated to be defective in the immunospecific clearance of IgG-coated autologous erythrocytes, which would allow increased levels of circulating immune complexes (30, 84a). The composition of IC includes antibody to DNA/anti-DNA. Any superimposed inflammatory process or viral or bacterial infection would further stress the system.

Recently, low-molecular-weight DNA fragments similar to those found in DNA/anti-DNA antibody IC were found to accumulate in peripheral blood lymphocytes in SLE patients, suggesting that they may serve as a primary source of autoantigen for anti-DNA production and/or pathogenic IC (73). Not all anti-DNA may deposit in the form of IC. DNA may first bind to basement membranes and circulating antibody secondarily bind to the DNA to initiate the inflammatory process. Other autoantibodies, such as to platelets, red blood cells, or lymphocytes, can account for disease manifestations such as thrombocytopenia, hemolytic anemia, and lymphopenia, respectively. In addition to specific anti-lymphocyte antibody, anti-ribonucleoprotein antibody from SLE sera can penetrate into suppressor T_G cells via their Fc receptors, bind to their nuclei, cause their deletion, and abrogate their suppressor function (6). Furthermore, disruption of normal amplification systems (e.g. prostaglandins) by autoantibodies occurs. SLE patients produce antibodies to an inhibitor of phospholipase A_2, lipomodulin, which interferes with lipomodulin's inhibition of cell triggering. Thus, such an autoantibody could lead to abnormal cellular activation characteristic of patients with SLE (42).

Cellular mechanisms may be involved in the formation of SLE disease. Damage to human suppressor T cells by anti-T cell antibodies can be effected in the absence of complement by ADCC (58), whereas anti-DNA antibodies in SLE serum have been shown to act as specific antibodies with a model target cell (i.e. cell-mediated lysis of DNA-coated cells) (33). Although not as well characterized, NK activity is defective in SLE (40), as is macrophage function. Titers of alpha interferon, which can be produced by NK cells, are increased in active SLE (31) and other interferons may cause increases in macrophage activation and Ia expression as well as suppressor T cell defects.

SUMMARY OF THE PATHOGENETIC FACTORS IN SLE   Humans with SLE have B cell hyperactivity. In most individuals, such hyperactivity is measurable as hypergammaglobulinemia, autoantibody production, and increased numbers of antibody-forming cells in the blood. Similarly, mice with SLE have hypergammaglobulinemia, autoantibody production, and increased numbers of antibody-forming cells (usually measured in the spleen). The basis for the B cell hyperactivity is difficult to study in vitro. In particular, the cells of the mice and the patients do not excessively respond to B or T cell signals in culture; rather, they are more likely to hyporespond. Therefore, an in vivo stimulus to B cell hyperactivity is lost in culture, but is present in the body. This stimulus may be a T cell in the case of MRL-*lpr/lpr* mice, as in graft-versus-host disease in mice. It may be a T cell in a small number of patients with SLE. However, in most mice and patients, it is difficult to demonstrate excessive helper T cell function. In addition, defective suppressor T cell function is not a uniform finding. That is not to say that either or both are not important in perpetuating a process. However, it suggests that another cause of B cell hyperactivity is operative. Perhaps the antigen-presenting cell is defective, providing excessive stimulation to B cells. There is no doubt that macrophage defects are present in SLE. All of the murine models have defects in macrophage function. It is possible that factors are produced by some cell that drives the B cells in SLE. Alternatively, the B cells may be endogenously hyperactive. This could be a result of a primary B cell defect. Alternatively, it could be a result of excessive production of B cells (individual B cells would be normal).

SLE patients, even when the SLE is inactive, have excessive proliferation in their B cell-enriched fraction of leukocytes (37). Most of the autoimmune mice also have excessive spontaneous lymphoid proliferation (89). Therefore, SLE could be viewed as a disorder of proliferation. That would put it in the class of diseases in which malignancies are placed; however, the proliferation is polyclonal rather than monoclonal. In the mice, the increased proliferation is found at the precursor stem cell level. At least one study of humans has found evidence for abnormal pre-B stem cells in SLE (57). Moreover, in that study T cells were not necessary for expression of the defect. Therefore, excessive stem cell turnover may be sufficient to lead to excessive proliferation. A second factor may be necessary to result in differentiation. This factor is what turns a patient with inactive SLE into one with manifested disease activity. Any of a variety of triggers may lead to the signals that allow the excessively proliferating B cells to differentiate into antibody-producing cells.

This formulation may allow for a synthesis of diverse SLE-like diseases. Graft-versus-host disease may provide such an extreme stimulus to proliferation and differentiation that autoimmunity may occur in the absence of a genetic predisposition in either mice (36, 64) or people (scleroderma-like illness). In addition, severe chronic stimulation of the immune system by polyclonal immune activators such as nucleic acids or endotoxin (29, 107) or by certain viruses (Epstein-Barr, aleutian mink) may be sufficient to produce disease. Finally, a variety of genetic diseases may lead to an individual with a predisposition to the development of disease by virtue of excessive stem cell activity. Additional signals would lead to differentiation and the disease. Genes for high responsiveness to certain autoantigens would allow specific autoantibodies to be produced. However, a sufficient disruption of normal immune regulatory processes could lead to a variety of autoantibodies despite no strong genetic susceptibility.

Viewed in this way, SLE represents a syndrome that may be brought about by a great number of immune perturbations. Some patients may inherit a very strong predisposition such that trivial environmental factors (or even endogenous factors) may be sufficient to induce the syndrome. At the other end of the spectrum, a person who is not particularly susceptible may receive a sufficiently strong stimulus to induce disease. There are probably several genes that particularly predispose to, or protect against, the development of the syndrome. However, their predisposing or protective ability may be influenced by other genes or the environment. The *xid* gene in NZB mice is a good example. NZB mice are protected against autoimmunity by virtue of the imposition of the *xid* gene (119). However, removal of normal T cell regulatory function by neonatal thymectomy will allow disease to occur. Alternatively, stimulation of the immune system with polyclonal immune activators will allow the syndrome to occur. Thus it is possible that any of several perturbations of the immune system will allow protective functions to be overwhelmed.

Once the syndrome of polyclonal B cell activation and hypergammaglobulinemia is induced, a great number of secondary factors operate to perpetuate the disease process and to produce disease symptoms. Activated T cells and other cells produce interferon. Interferon may lead to increased Ia or DR expression on macrophages and more efficient macrophage function. Initially, this would be expected to heighten immune responses. Ordinarily, interferon might be expected to suppress antibody production; however, autoimmune NZB mice, and perhaps patients with SLE, appear to be relatively resistant to those suppressive actions of interferon. As a result, only the stimulatory activities are observed. Ultimately, excessive macrophage function suppresses new immune responses whereas the autoimmune process continues unabated. Some of the autoantibodies produced perpetuate disease. Antibodies reactive with T lymphocytes tend to impair regulatory function; antibodies reactive with B lymphocytes may stimulate further B cell proliferation.

Why is the autoantibody response of SLE patients not regulated as are normal immune responses? In periods of remission it may be. Normal suppressor functions may serve to limit the extent of autoantibody production. Reduction in suppressor function in BXSB mice by neonatal thymectomy or in normal mice given polyclonal B cell activators tends to allow accelerated disease. In addition, anti-idiotype antibodies may be produced during remission but may be ineffective during disease activity (2). It appears that the degree of immune stimulation that triggers the active disease process may be sufficiently strong to overcome any normal regulatory processes. In individuals with impaired regulatory processes, it is even easier to trigger disease. Moreover, as indicated above, the process tends to be self-perpetuating. Once under way, the autoimmune process tends to favor continued stimulation of autoantibody production and thereby continued inflammation, which tends to favor stimulation of the process. Furthermore, if B cells already are excessively proliferating and require only a signal for differentiation to initiate disease, many regulatory processes that operate on proliferation have already been circumvented. Therefore, once the disease is full-blown, a very strong immunosuppressive intervention may be necessary to bring the entire system back toward normal.

AUTOIMMUNITY AND AUTOIMMUNE DISEASES

What can the analysis of AIT and SLE tell us about autoimmune diseases? They suggest that whereas autoimmunity may be normal, autoimmune diseases have a variety of features that lead to pathology. A genetic basis for disease may be a unique gene; this could be true of certain HLA-B27-associated diseases. Even there, additional genes and an environmental trigger may be required. In AIT and SLE, a greater number of genes appear to be necessary to predispose to disease. Some genes may code for cell surface determinants (e.g. a thyroid structural determinant in AIT or reduced or abnormal surface C receptors in SLE). Others may code for high responsiveness to certain antigens (thyroglobulin or the thyroid stimulating hormone receptor in AIT, nucleic acids or circulating cells in SLE). Still other genes may predispose to immune defects such as B-cell hyperactivity or impaired immune regulatory processes. In view of the great complexity of the immune regulatory circuits (110), a number of different regulatory defects could all predispose to disease. A variety of different genes may be very important in modulating the extent of inflammatory processes through the regulation of a variety of molecules that are so important in those processes.

In addition to genetic factors, environmental factors may serve to promote or retard disease. Environmental immune activators may be critical to the triggering of disease in certain patients with SLE. Female sex hormones may predispose to, and male sex hormones may retard, disease. Nutritional factors may determine the degree of inflammation or autoantibody production (110). Different factors may be more important in different individuals. In AIT and SLE, and especially in other diseases, such as some of the arthritides, chronic immune stimulation may initiate and/or perpetuate illness. Such stimulation may be viral (e.g. Epstein-Barr) or bacterial. Bacterial stimulation can occur through chronic release of stimulatory factors (e.g. endotoxin) or the persistence of nondegradable cell wall components.

Does any of this help us to plan for the treatment or prevention of disease? It is possible that a complete understanding of the pathogenetic and etiologic mechanisms of disease might not immediately lead to prevention or specific therapy. On the other hand, it is possible that a direct approach to specific therapy or prevention might be apparent or at least theoretically possible.

Appreciation that a subset of B cells is responsible for autoantibody production, and that elimination of that subset prevents disease in NZB, BXSB, (NZB X NZW) F_1, and (NZB X BXSB) F_1 mice with SLE-like illness, suggests that a related approach might be possible for patients with SLE. If a subset of human B cells analogous to the Lyb $3^+,5^+$ cells of mice is found to be responsible for disease, that subset could be specifically eliminated by a monoclonal antibody specific for that subset. Such an appproach might be entirely nontoxic and might not interfere with most other normal immune responses.

Recognition of the importance of antireceptor antibodies in myasthenia gravis could lead to specific prevention of the synthesis of those autoantibodies.

An appreciation that certain viruses or bacteria act to stimulate the immune systems of many people could lead to a vaccination program. We believe that the Epstein-Barr virus may be a very important underlying stimulus to excessive immune activity in a variety of situations. Vaccination against that virus alone might substantially reduce the number of people afflicted with a variety of autoimmune diseases. A similar attack on other stimulatory viruses, or bacteria with nondegradable but stimulatory cell wall components, could be carried out. In addition to preventing a variety of autoimmune diseases in many people, such an approach might reduce the severity of disease in others.

Despite the possibility of specific intervention, study of autoimmune diseases should provide a deep respect for the variety of mechanisms responsible for normal immune regulation and therefore the multiple mechanisms by which those may be rendered abnormal. Nevertheless, the autoimmune diseases appear to be losing their mystery. Some are relatively simple; others are less so. They are complicated because of the multifactorial nature of the disorders, the complexity of the immune system, the multigenetic bases for disease, and our imperfect understanding of these and their interrelationships. Nevertheless, substantial insights are being gained into all of these. Knowledge of the bases for individual autoimmune diseases is increasing rapidly. As more is learned, a greater number of details will fit into an even more understandable whole.

Literature Cited

1. Abdou, N. I., Sagawa, A., Pascual, E., Hebert, J., Sadeghee, S. 1976. Suppressor T cell abnormality in idiopathic systemic lupus erythematosus. *Clin. Immunol. Immunopathol.* 6:192–99

2. Abdou, N. I., Wall, H., Lindsley, H. B., Halsey, J. F., Suzuki, T. 1981. Network theory in autoimmunity. In vitro suppression of serum anti-DNA antibody binding to DNA by anti-idiotypic antibody in systemic lupus erythematosus. *J. Clin. Invest.* 67:1297–304

3. Adams, D. D., Kennedy, T. H., Stewart, R. 1974. Correlation between long-acting thyroid stimulator protector level and thyroid ^{131}I uptake in thyrotoxicosis. *Brit. Med. J.* 2:199–201

4. Agnello, V. 1978. Association of systemic lupus erythematosus and systemic lupus erythematosus-like syndromes with hereditary and acquired complement deficient states. *Arthritis Rheum.* 21:S146–S60

5. Akizuki, M., Reeves, J. P., Steinberg, A. D. 1978. Expression of autoimmunity by NZB/NZW marrow. *Clin. Immunol. Immunopathol.* 10:247–50

6. Alarcon-Segovia, D. 1982. Antibody penetration into living cells. III. Effect of antiribonucleoprotein IgG on the cell cycle of human peripheral blood mononuclear cells. *Clin. Immunol. Immunopathol.* 23:22–33

7. Amino, N., Hidemitsu, M., Yoshinori, I., Seishi, A., Izumiguchi, Y., et al. 1982. Peripheral K lymphocytes in autoimmune thyroid disease: Decrease in Graves' disease and increase in Hashimoto's disease. *J. Clin. Endo. Met.* 54:587–91

8. Arnett, G. C., Shulman, L. E. 1976. Studies in familial systemic lupus erythematosus. *Medicine* 55:313–22

9. Benacerraf, B. 1981. Role of MHC gene products in immune regulation. *Science* 212:1229–38

10. Blaese, R. M., Grayson, J., Steinberg, A. D. 1980. Elevated immunoglobulin secreting cells in the blood of patients with active system lupus erythematosis: correlation of laboratory and clinical as-

sessment of disease activity. *Am. J. Med.* 69:345–50

11. Block, S. R., Gibbs, C. B., Stevens, M. B., Shulman, L. E. 1968. Delayed hypersensitivity in sytemic lupus erythematosus. *Ann. Rheum. Dis.* 27: 311–18

12. Block, S. R., Lockshin, M. D., Winfield, J. B., Wecksler, M. E., Iamura, M., et al. 1976. Immunologic observations on 9 sets of twins either concordant or discordant for systemic lupus erythematosus. *Arthritis Rheum.* 19:545–54

13. Bocchieri, M. H., Cooke, A., Smith, J. B., Weigert, M., Riblet, R. J. 1982. Independent segregation of NZB autoimmune abnormalities in NZB X C58 recombinant inbred mice. *Eur. J. Immunol.* 12:349–54

14. Bresnihan, B., Jasin, H. E. 1977. Suppressor function of peripheral blood mononuclear cells in normal individuals and patients with systemic lupus erythematosus. *J. Clin. Invest.* 59:106–16

15. Buckman, K. J., Moore, S. K., Ebbin, A. J., Mavis, B. C., Dubois, E. L. 1978. Familial systemic lupus erythematosus. *Arch. Intern. Med.* 138:1674–80

16. Budman, D. R., Merchant, E. B., Steinberg, A. D., Doft, B., Gershwin, M. E., et al. 1977. Increased spontaneous activity of antibody forming cells in the peripheral blood of patients with active systemic lupus erythematosus. *Arthritis Rheum.* 20:829–33

17. Calder, E. A., Penhale, W. J., McLennan, D. 1973. Lymphocyte-dependent antibody-mediated cytotoxicity in Hashimoto's thyroiditis. *Clin. Exp. Immunol.* 14:153–58

18. Coombs, R. R. A., Gell, P. G. H. 1975. Classification of allergic reactions responsible for clinical hypersensitivity and disease. In *Clinical Aspects of Immunology,* ed. P. G. H. Gell, R. R. A. Coombs, P. J. Laehmann, p. 761. Oxford: Blackwell

18a. Cowdery, J. S., Curtin, M. F. Jr., Steinberg, A. D. 1982. The effect of prior intragastric antigen administration on primary and secondary anti-ovalbumin responses of C57BL/6 and NZB mice. *J. Exp. Med.* 156:1256–61

18b. Cowdery, J. S., Steinberg, A. D. 1982. Regulation of primary, thymus-independent, anti-hapten responses of normal and autoimmune mice by syngeneic antibody. *J. Immunol.* 129: 1250–55

19. Datta, S. D., Owen, F. L., Womack, J. E., Riblet, R. J. 1982. Analysis of recombinant inbred lines derived from "autoimmune" (NZB) and "high leukemia" (C58) strains: independent multigenic systems control B cell hyperactivity, retrovirus expression, and autoimmunity. *J. Immunol.* 129:1539–44

20. DeHoratius, R. J., Messner, R. P. 1975. Lymphocytotoxic antibodies in family members of patients with systemic lupus erythematosus. *J. Clin. Invest.* 55:1254–58

21. Delfraissy, J. F., Segond, P., Galanaud, P., Wallon, C., Massias, P., et al. 1980. Depressed primary in vitro antibody response in untreated systemic lupus erythematosus: T helper cell defect and lack of defective suppressor cell function. *J. Clin. Invest.* 66:141–48

22. Drachman, D. B., Adams, R. N., Josifek, L. F., Self, S. G. 1982. Functional activities of autoantibodies to acetylcholine receptors and the clinical severity of myasthenia gravis. *N. Engl. J. Med.* 307:769–75

23. Drexhage, H. A., Bottazo, G. F., Bitensky, L., Chayen, J., Doniach, D. 1981. Thyroid growth-blocking antibodies in primary myxoedema. *Nature* 289:594–98

24. Eisenberg, R. A., Dixon, F. J. 1980. Effect of castration on male-determined acceleration of autoimmune disease in BXSB mice. *J. Immunol.* 125:1959–61

25. Eisenberg, R. A., Izui, S., McConahey, P. J., Hang, L., Peters, C. J., et al. 1980. Male determined accelerated autoimmune disease in BXSB mice: transfer by bone marrow and spleen cells. *J. Immunol.* 125:1032–36

26. Endo, K., Kasagi, K., Konishi, J., Ikekubo, K., Okuno, K., et al. 1978. Detection and properties of TSH-binding inhibitor immunoglobulins in patients with Graves' disease and Hashimoto's thyroiditis. *J. Clin. Endocrinol. Metab.* 46:734–39

27. Fauci, A. S., Steinberg, A. D., Haynes, B. F., Whalen, G. 1978. Immunoregulatory aberrations in systemic lupus erythematosus. *J. Immunol.* 121:1473–79

28. Foad, B. S., Khullar, S., Freimer, E. H., Kirsner, A. B., Sheon, R. P. 1975. Cell-mediated immunity in systemic lupus erythematosus: alterations with advancing age. *J. Lab. Clin. Med.* 85:133–39

29. Fournie, G. J., Lambert, P. H., Miescher, P. A. 1974. Release of DNA in circulating blood and induction of anti-DNA antibodies after injection of bacterial lipopolysaccharides. *J. Exp. Med.* 140:1189–206

30. Frank, M. M., Hamburger, M. I., Lawley, T. J., Kimberly, R. P., Plotz, P. H. 1979. Defective reticuloendothelial system Fc-receptor function in systemic lupus erythematosus. *N. Engl. J. Med.* 300:518–23

31. Friedman, R. M., Preble, O., Black, R., Harrell, S. 1982. Interferon production in patients with systemic lupus erythematosus. *Arthritis Rheum.* 25:802–3

32. Gershwin, M. E., Castles, J. J., Erickson, K., Ahmed, A. 1979. Studies of congenitally immunologic mutant New Zealand mice. II. Absence of T cell progenitor populations and B cell defects of congenitally athymic (nude) New Zealand Black (NZB) mice. *J. Immunol.* 122:2020–25

33. Gershwin, M. D., Glinski, W., Chused, T. M., Steinberg, A. D. 1977. Lymphocyte dependent antibody mediated cytolysis of DNA-anti-DNA coated target cells using human and murine effector populations. *Clin. Immunol. Immunopathol.* 8:280–91

34. Gibofsky, A., Winchester, R. J., Hansen, J., Patarroyo, M., Dupont, D., et al. 1978. Contrasting patterns of newer histocompatibility determinants in patients with rheumatoid arthritis and systemic lupus erythematosus. *Arthritis Rheum.* 21:134s–38s

35. Ginsburg, W. W., Finkelman, F. D., Lipsky, P. E. 1979. Circulating and pokeweed mitogen-induced immunoglobulin-secreting cells in system lupus erythematosus. *Clin. Exp. Immunol* 35:76–88

36. Gleichmann, E., Issa, P., Van Elven, E. H., Lamers, M. C. 1978. The chronic graft-versus-host reaction: A lupus erythematosus-like syndrome caused by abnormal T-B cell interactions. *Clin. Rheum. Dis.* 4:587–602

37. Glinski, W., Gershwin, M. E., Budman, D. R., Steinberg, A. D. 1976. Study of lymphocyte subpopulations in normal humans and patients with systemic lupus erythematosus by fractionation of peripheral blood lymphocytes on a discontinuous ficoll gradient. *Clin. Exp. Immunol.* 26:228–38

38. Glinski, W., Gershwin, M. E., Steinberg, A. D. 1976. Fractionation of cells on a discontinuous ficoll gradient. Study of subpopulations of human T-cells using anti-T-cell antibodies from patients with systemic lupus erythematosus. *J. Clin. Invest.* 57:604–14

39. Gocke, D. J., Hsu, K., Morgan, C., Bombardiere, S., Lockshin, M., et al. 1970. Association between polyarteritis and Australia antigen. *Lancet* 2:1149–53

40. Goto, M., Tanimoto, K., Horiuchi, Y. 1980. Natural cell mediated cytotoxicity in systemic lupus erythematosus. Suppression by antilymphocyte antibody. *Arthritis Rheum.* 23:1274–81

41. Hang, L., Izui, S., Slack, J. H., Dixon, F. J. 1982. The cellular basis for resistance to induction of tolerance in BXSB systemic lupus erythematosus male mice. *J. Immunol.* 129:787–89

42. Hirata, F., del Carmine, R., Nelson, C. A., Axelrod, J., Schiffman, E., et al. 1981. Presence of autoantibody for phospholipase inhibitory protein, lipomodulin, in patients with rheumatic diseases. *Proc. Natl. Acad. Sci. USA* 78:3190–94

43. Iida, K., Mornaghi, R., Nussenzweig, V. 1982. Complement receptor deficiency in erythrocytes from patients with systemic lupus erythematosus. *J. Exp. Med.* 155:1427–38

44. Inman, R. D., Jovanovic, L., Dawood, M. Y., Longcope, C. 1979. Systemic lupus erythematosus in the male: A genetic and endocrine study. *Arthritis Rheum.* 22:624

45. Jasin, H. E., Ziff, M. 1975. Immunoglobulin synthesis by peripheral blood cells in systemic lupus erythematosus. *Arthritis Rheum.* 18:219–28

46. Jungers, P. J., Dougados, M., Pelissier, C., Kuttern, F., Tron, F., et al. 1982. Influence of oral contraceptive therapy on the activity of systemic lupus erythematosus. *Arthritis Rheum.* 25:618–23

47. Jyonouchi, H., Kincade, P. W., Landreth, K. S., Lee, G., Good, R. A., et al. 1982. Age-dependent deficiency of B lymphocyte lineage precursors in NZB mice. *J. Exp. Med.* 155:1665–78

48. Kahn, C. R. 1980. Role of insulin receptors in insulin-resistant states. *Metabolism* 29:455–66

49. Kalderon, A. E. 1980. Emerging role of immune complexes in autoimmune thyroiditis. *Pathol. Ann.* 15:23–35

50. Kaplan, M. H., Svec, K. H. 1964. Immunologic relation of streptococcal and tissue antigens. III. Presence in human sera of streptococcal antibody cross-reactive with heart tissue. Association with streptococcal infection, rheumatic fever, and glomerulonephritis. *J. Exp. Med.* 119:651–65

51. Kemp, J. D., Cowdery, J. S., Steinberg, A. D., Gershon, R. K. 1982. Genetic controls of autoimmune diseases: In-

teractions between *xid* and *lpr. J. Immunol.* 128:388–92

52. Kidd, A., Okita, N., Row, V. V., Volpe, R. 1980. Immunologic aspects of Graves' and Hashimoto's diseases. *Metabolism* 29:80–99

53. Kincade, P. W., Lee, G., Fernandes, G., Moore, M. A. S., Williams, N., et al. 1979. Abnormalities in clonable B lymphocytes and myeloid progenitors in autoimmune NZB mice. *Proc. Natl. Acad. Sci. USA* 76:3464–68

54. Kincade, P. W., Paige, C. J., Parkhouse, R. M. E., Lee, G. 1978. Characterization of murine colony-forming B cells. I. Distribution, resistance to anti-immunoglobulin antibodies, and expression of Ia antigens. *J. Immunol.* 120:1289–96

55. Kohler, P. F., Percy, J., Campion, W. M., Smyth, C. J. 1974. Hereditary angiooedema and "familial" lupus erythematosus in identical twin boys. *Am. J. Med.* 56:406–11

56. Kotani, T., Komuro, K., Yoshiki, T., Itoh, T., Aizawa, M. 1981. Spontaneous autoimmune thyroiditis in the rat accelerated by thymectomy and low doses of irradiation: mechanisms implicated in the pathogenesis. *Clin. Exp. Immunol.* 45:329–37

57. Kumagai, S., Sredni, B., House, S., Steinberg, A. D., Green, I. 1982. Defective regulation of B lymphocyte colony formation in patients with systemic lupus erythematosus. *J. Immunol.* 128:258–62

58. Kumagai, S., Steinberg, A. D., Green, I. 1981. Antibodies to T cells in patients with systemic lupus erythematosus mediated ADCC against human T cell. *J. Clin. Invest.* 67:605–14

58a. Kumagai, S., Steinberg, A. D., Green, I. 1981. Immune responses to hapten-modified self and their regulation in normal individuals and patients with SLE. *J. Immunol.* 127:1643–52

59. Lahita, R. G., Bradlow, L., Fishman, J., Kunkel, H. G. 1982. Estrogen metabolism in systemic lupus erythematosus: patients and family members. *Arthritis Rheum.* 25:843–46

60. Lahita, R. G., Bradlow, H. L., Kunkel, H. G., Fishman, J. 1981. Increased 16 α-hydroxylation of estradiol in systemic lupus erythematosus. *J. Clin. Endocrinol. Metab.* 53:174–78

61. Laskin, C. A., Smathers, P. A., Reeves, J. P., Steinberg, A. D. 1982. Studies of defective tolerance induction in NZB mice: Evidence for a marrow pre-T cell defect. *J. Exp. Med.* 155:1025–36

62. Laskin, C. A., Taurog, J. D., Smathers, P. A., Steinberg, A. D. 1981. Studies of defective tolerance in murine lupus. *J. Immunol.* 127:1743–47

63. Lewis, D. E., Giorgi, J. V., Warner, N. L. 1981. Flow cytometry analysis of T cells and continuous T-cell lines from autoimmune MRL/1 mice. *Nature* 289:298–300

64. Lewis, R. M., Armstrong, M. Y. K., Andre-Schwartz, J., Muftooglu, A., Beldotti, L., et al. 1968. Chronic allogeneic disease. I. Development of glomerulonephritis. *J. Exp. Med.* 128:653–67

65. Lowenstein, M. B., Rothfield, N. F. 1977. Family study of systemic lupus erythematosus: analysis of the clinical history, skin immunofluorescence, and serologic parameters. *Arthritis Rheum.* 20:1293–303

66. Manohar, V., Brown, E., Leiserson, W. M., Chused, T. M. 1982. Expression of Ly-1 by a subset of B lymphocytes. *J. Immunol.* 129:532–38

67. Miller, K. B., Schwartz, R. S. 1979. Familial abnormalities of suppressor cell function in systemic lupus erythematosus. *N. Engl. J. Med.* 301:803–9

68. Mitsunaga, M., Suzuki, S., Natagawo, O., Hirakawa, S., Miura, H., et al. 1980. Cytophilic antibodies and monocyte-mediated cytotoxicity in Hashimoto's thyroiditis. In *Thyroid Research VIII,* ed. J. R. Stockigt, T. Nagataki, pp. 789–92. New York: Pergamon

69. Miyakawa, Y., Yamada, A., Kosaka, K., Tsuda, F., Kosugi, E., et al. 1981. Defective immune-adherence (C3b) receptor on erythocytes from patients with systemic lupus erythematosus. *Lancet* 2:493–97

70. Mori, T., Kriss, J. P. 1971. Measurements by competitive binding radioassay of serum anti-microsomal and anti-thyroglobulin antibodies in Graves' disease and other thyroid disorders. *J. Clin. Endocrinol. Metab.* 33:688–98

71. Morimoto, C., Abe, T., Hara, M., Homma, M. 1977. In vitro TNP-specific antibody formation by peripheral lymphocytes from patients with systemic lupus erythematosus. *Scand. J. Immunol.* 6:575–79

72. Morimoto, C., Reinherz, E. L., Nadler, L. M., Distaso, J. A., Steinberg, A. D., et al. 1982. Comparison in T- and B-cell markers in patients with Sjogren's syndrome and systemic lupus erythematosus. *Clin. Immunol. Immunopathol.* 22:270–78

73. Morimoto, C., Sano, H., Abe, T., Homma, M., Steinberg, A. D. 1982. Correlation between clinical activity of systemic lupus erythematosus and the amounts of DNA in DNA/anti-DNA antibody immune complexes. *J. Immunol.* 129:1960–65

74. Morton, J. I., Siegel, B. V. 1974. Transplantation of autoimmune potential. I. Development of antinuclear antibodies in H-2 histocompatible recipients of bone marrow from New Zealand black mice. *Proc. Natl. Acad. Sci. USA* 71: 2162–65

75. Morton, J. I., Siegal, B. V. 1978. Transplantation of autoimmune potential. III. Immunological hyper-responsiveness and elevated endogenous spleen colony formation in lethally irradiated recipients of NZB bone marrow cells. *Immunology* 34:863–68

76. Morton, R. O., Gershwin, M. E., Brady, C., Steinberg, A. D. 1976. Incidence of systemic lupus erythematosus (SLE) in North American Indians. *J. Rheum.* 3:186–90

77. Murphy, E. D. 1981. Lymphoproliferation (LPR) and other single-locus models for murine lupus. In *Immunologic Defects in Laboratory Animals,* ed. M. E. Gershwin, B. Merchant, 2:143–73. New York: Plenum

78. Murphy, E. D., Roths, J. B. 1979. A Y-chromosome associated factor in strain BXSB producing accelerated autoimmunity and lymphoproliferation. *Arthritis Rheum.* 22:1188–94

79. Nies, K. M., Louie, J. S. 1978. Imparied immunoglobulin synthesis by peripheral blood lymphocytes in systemic lupus erythematosus. *Arthritis Rheum.* 21:51–57

80. Ohsugi, Y., Gershwin, M. E., Ahmed, A., Skelly, R., Milich, D. R. 1982. Studies of congenitally immunologic mutant New Zealand mice IV. Spontaneous and induced autoantibodies to red cells and DNA occur in New Zealand X-linked immunodeficient mice without phenotypic alterations of the X^{id} gene or generalized polyclonal B cell activation. *J. Immunol.* 128:2220–27

81. Okita, N., Kidd, A., Row, V. V., Volpe, R. 1981. Suppressor T lymphocyte deficiency in Graves' disease and Hashimoto's thyroiditis. *J. Clin. Endocrinol. Metab.* 52:528–33

82. Okudaira, K., Searles, R. P., Tanimoto, K., Horiuchi, Y., Williams, R. C. Jr. 1982. T lymphocyte interaction with immunoglobulin G antibody in systemic lupus erythematosus. *J. Clin. Invest.* 69:1026–38

83. Oldstone, M. B. A., Dixon, F. J. 1972. Inhibition of antibodies to nuclear antigen and to DNA in New Zealand mice infected with lactate dehydrogenase virus. *Science* 175:784–88

84. Palacios, R., Alarcon-Segovia, D., Llorente, L., Ruiz-Arguelles, A., Diaz-Jouanen, E. 1981. Human postthymic precursor cells in health and disease. II. Their loss and dysfunction in systemic lupus erythematosus and their partial correction with serum thymic factor. *J. Clin. Lab. Immunol.* 5:71–80

84a. Parris, T. M., Kimberly, R. P., Inman, R. D., McDougal, J. S., Gibofsky, M. D., Christian, C. L. 1982. Defective Fc receptor-mediated function of the mononuclear phagocyte system in lupus nephritis. *Ann. Int. Med.* 97:526–32

85. Polley, C. R., Bacon, L. D., Rose, N. R. 1977. Failure of functional bursa cells transplanted between susceptible OS and resistant CS strains of chickens to influence spontaneous autoimmune thyroiditis. *Fed. Proc.* 36:1207

86. Ranney, D. F., Steinberg, A. D. 1976. Differences in the age-dependent release of a low molecular weight suppressor (LMWS) and stimulators by normal and NZB/W lymphoid organs. *J. Immunol.* 117:1219–25

87. Raveche, E. S., Novotny, E. A., Hansen, C. T., Tjio, J. H., Steinberg, A. D. 1981. Genetic studies in NZB mice. V. Recombinant inbred lines demonstrate that separate genes control autoimmune phenotype. *J. Exp. Med.* 153:1187–97

88. Raveche, E. S., Steinberg, A. D. 1979. Lymphocytes and lymphocyte functions in systemic lupus erythematosus. *Clinics Haematol.* 16:344–70

89. Raveche, E. S., Steinberg, A. D., DeFranco, A. L., Tjio, J. H. 1982. Cell cycle analysis of lymphocyte activation in normal and autoimmune strains. *J. Immunol.* 129:1219–26

90. Raveche, E. S., Steinberg, A. D., Klassen, L. W., Tjio, J. H. 1978. Genetic studies in NZB mice. I. Spontaneous autoantibody production. *J. Exp. Med.* 147:1487–502

91. Reinertsen, J. L., Klippel, J. H., Johnson, A. H., Steinberg, A. D., Decker, J. L. 1982. Family studies of B lymphocyte alloantigens in systemic lupus erythematosus. *J. Rheumatol.* 9:253–62

92. Reinertsen, J. L., Klippel, J. H., Johnson, A. H., Steinberg, A. D., Decker, J. L. et al. 1978. B-lymphocyte alloanti-

gens associated with systemic lupus erythematosus. *N. Engl. J. Med.* 299: 515–58

93. Roitt, I. M., Doniach, D., Campbell, R. N., Hudson, R. V. 1956. Autoantibodies in Hashimoto's disease. *Lancet* 2:820–21

94. Romain, P. L., Cohen, P. L., Fish, F., Ziff, M., Vitetta, E. S. 1980. The specific B cell subset lacking in the CBA/N mouse is not required for the production of autoantibody in (CBA/N X NBZ)F₁ male mice. *J. Immunol.* 125: 246–51

95. Rose, N. R., Kang, Y. M., Okayasu, I., Giraldo, A. A., Beisel, K., et al. 1981. T-cell regulation in autoimmune thyroiditis. *Immunol. Rev.* 55:299–314

96. Rose, N. R., Witebsky, E. 1956. Studies on organ specificity V. changes in the thyroid gland of rabbits following active immunization with rabbit thyroid extracts. *J. Immunol.* 76:417–27

97. Rosenwasser, L. J., Huber, B. T. 1981. The *xid* gene controls Ia. W39-associated immune response gene function. *J. Exp. Med.* 153:1113–23

98. Roubinian, J. R., Papoian, R., Talal, N. 1977. Androgenic hormones moderate autoantibody responses and improve survival in murine lupus. *J. Clin. Invest.* 59:1066–70

99. Sakane, T., Steinberg, A. D., Green, I. 1978. Studies of immune functions of patients with systemic lupus erythematosus. I. Failure of suppressor T cell activity related to impaired generation of, rather than response to, suppressor cells. *Arthritis Rheum.* 21: 657–64

100. Sakane, T., Steinberg, A. D., Green, I. 1978. Failure of autologous mixed lymphocyte reactions between T and non-T cells in patients with systemic lupus erythematosus. *Proc. Natl. Acad. Sci. USA* 75:3464–68

101. Sakane, T., Steinberg, A. J., Green, I. 1980. Studies of immune functions of patients with systemic lupus erythematosus. V. T-cell suppressor function and autologous MLR during active and inactive phases of disease. *Arthritis Rheum.* 23:225–31

102. Sakane, T., Steinberg, A. D., Reeves, J. P., Green, I. 1979. Studies of immune functions of patients with systemic lupus erythematosus. Complement dependent immunoglobulin M antithymus-derived cell antibodies preferentially inactivate suppressor cells. *J. Clin. Invest.* 63:954–65

103. Scher, I., Steinberg, A. D., Berning, A. D., Paul, W. E. 1975. X-linked B-lymphocyte immune defect in CBA/N mice. II. Studies of mechanisms underlying the immune defect. *J. Exp. Med.* 142:637–50

104. Siegel, M., Holley, H. L., Lee, S. L. 1970. Epidemiologic studies on systemic lupus erythematosus. *Arthritis Rheum.* 13:802–11

105. Silverman, D. A., Rose, N. R. 1974. Neonatal thymectomy increases the incidence of spontaneous and methylcholanthrene-enhanced thyroiditis in rats. *Science* 184:162–63

106. Smathers, P. A., Steinberg, B. J., Reeves, J. P., Steinberg, A. D. 1982. Effects of polyclonal immune stimulators upon NZB·*xid* congenic mice. *J. Immunol.* 128:1414–19

106a. Smith, H. R., Chused, T. M., Smathers, P., Steinberg, A. D. 1983. Evidence for thymic regulation of autoimmunity in BXSB mice: Acceleration of disease by neonatal thymectomy. *J. Immunol.* 130:In press

107. Smith, H. R., Green, D. R., Raveche, E. S., Smathers, P. A., Gershon, R. K., et al. 1982. Studies on the induction of anti-DNA in normal mice. *J. Immunol.* 129:2332–34

108. Smolen, J. S., Chused, T. M., Leiserson, W. M., Reeves, J. P., Alling, D. W. 1982. Heterogeneity of immunoregulatory T cell subsets in systemic lupus erythematosus. Correlation with clinical features. *Am. J. Med.* 72:783–90

109. Smolen, J. S., Steinberg, A. D. 1981. Systemic lupus erythematosus and pregnancy. Clinical, immunological and theoretical aspects. In *Reproductive Immunology,* ed. N. Gleicher, pp. 283–302. New York: Liss

110. Smolen, J. S., Steinberg, A. D. 1982. Disorders of immune regulation. In *The Pathophysiology of Human Immunologic Disorders,* ed. J. Twomey, 10:173–98. Baltimore: Urban & Schwarzenberg

111. Staples, P. J., Steinberg, A. D., Talal, N. 1970. Induction of immunological tolerance in older New Zealand mice repopulated with young spleen, bone marrow or thymus. *J. Exp. Med.* 131:1223–38

112. Steinberg, A. D. 1980. Autoimmunity. In *Strategies of Immune Regulation,* ed. E. E. Sercarz, A. J. Cunningham, p. 503. New York: Academic

113. Steinberg, A. D., Huston, D. P., Taurog, J. D., Cowdery, J. S., Raveche, E. S. 1981. The cellular and genetic basis

of murine lupus. *Immunol. Rev.* 55:121–54

114. Steinberg, A. D., Law, L. W., Talal, N. 1970. The role of the NZB/NZW F_1 thymus in experimental tolerance and autoimmunity. *Arthritis Rheum.* 13:369–77

115. Steinberg, A. D., Melez, K. A., Raveche, E. S., Reeves, J. P., Boegel, W. A., et al. 1979. Approach to the study of the role of sex hormones in autoimmunity. *Arthritis Rheum.* 22:1170–76

116. Steinberg, A. D., Roths, J. B., Murphy, E. D., Steinberg, R. T., Raveche, E. S. 1980. Effects of thymectomy or androgen administration upon the autoimmune disease of MRL/MP-*lpr/lpr* mice. *J. Immunol.* 125:871–73

117. Steinberg, B. J., Smathers, P. A., Frederiksen, K., Steinberg, A. D. 1982. Ability of the *xid* gene to prevent autoimmunity in (NZB X NZW) F_1 mice during the course of their natural history after polyclonal stimulation or following immunization with DNA. *J. Clin. Invest.* 70:587–97

118. Stern, R., Fishman, J., Brusman, H., Kunkel, H. G. 1977. Systemic lupus erythematosus associated with Klinefelter's syndrome. *Arthritis Rheum.* 20:18–22

119. Taurog, J. D., Raveche, E. S., Smathers, P. A., Glimcher, L. H., Huston, D. P., et al. 1981. T cell abnormalities in NZB mice occur independently of autoantibody production. *J. Exp. Med.* 153:221–34

120. Taurog, J. D., Smathers, P. A., Steinberg, A. D. 1980. Evidence for abnormalities in separate lymphocyte populations in NZB mice. *J. Immunol.* 125:485–90

121. Theofilopolous, A. N., Dixon, F. J. 1981. Etiopathogenesis of murine systemic lupus erythematosus. *Immunol. Rev.* 55:179–216

122. Till, J. E., McCulloch, E. A. 1964. Repair processes in irradiated mouse hematopoietic tissue. *Ann. NY Acad. Sci.* 114:115–25

123. Tonietti, G., Oldstone, M. B., Dixon, F. J. 1970. The effect of induced chronic viral infections on the immunologic diseases of New Zealand mice. *J. Exp. Med.* 132:89–109

124. Walport, M. J., Fielder, A. H. L., Batchelor, J. R., Black, C. M., Rynes, R. I., et al. 1982. HLA linked complement allotypes and genetic susceptibility to systemic lupus erythematosus. *Arthritis Rheum.* 24:s41

125. Warner, N. W., Moore, M. A. S. 1971. Defects in hematopoietic differentiation in NZB and NZC mice. *J. Exp. Med.* 134:313–34

126. Wick, G., Boyd, R., Hala, K., Thunold, S., Kofler, H. 1982. Pathogenesis of spontaneous autoimmune thyroiditis in obese strain (OS) chickens. *Clin. Exp. Immunol.* 47:1–18

127. Wick, G., Gundolf, R., Hala, K. 1979. Genetic factors in spontaneous autoimmune thyroiditis in OS chickens. *J. Immuno. Genet.* 6:177–83

128. Wick, G., Kite, J. H. Jr., Witebsky, E. 1970. Spontaneous thyroiditis in the obese strain of chickens. IV. The effect of thymectomy and thymo-bursectomy on the development of the disease. *J. Immunol.* 104:54–62

128a. Wilson, J. G., Wong, W. W., Schur, P. H., Fearon, D. T. 1982. Mode of inheritance of decreased C3b receptors on erythrocytes of patients with systemic lupus erythematosus. *N. Engl. J. Med.* 307:981–86

129. Winfield, J. B., Winchester, R. J., Wernet, P., Fu, S. M., Kunkel, H. G. 1975. Lymphocytotoxic antibodies to lymphocyte surface determinants in systemic lupus erythematosus. *Arthritis Rheum.* 18:1–8

130. Wofsky, D., Murphy, E. D., Roths, J. B., Dauphinee, M. J., Kipper, S. B., et al. 1981. Deficient interleukin 2 activity in MRL/Mp and C57B1/6J mice bearing the *lpr* gene. *J. Exp. Med.* 154:1671–80

131. Yoshida, H., Amino, N., Yagawa, K., Uemura, K., Satoh, M., et al. 1978. Association of serum antithyroid antibodies with lymphocytic infiltration of the thyroid gland. Study of 70 autopsied cases. *J. Clin. Endocrinol. Metab.* 46:859–62

Ann. Rev. Immunol. 1983. 1:211–41

MECHANISMS OF
T CELL-B CELL INTERACTION

Alfred Singer and Richard J. Hodes

Immunology Branch, National Cancer Institute, National Institutes of Health, Bethesda, Maryland 20205

INTRODUCTION

The demonstration that interactions among distinct immunocompetent cell types were required for the generation of most immune responses signalled the beginning of a new era in cellular immunology. The subsequent demonstrations that these cellular interactions were regulated by cell surface gene products encoded by the major histocompatibility complex (MHC) provided investigators with a precise and potent tool for the dissection of immune cell interactions. In addition, these observations provided the basis for truly dramatic insights into the reasons for the existence of an MHC. Thus, modern day cellular immunology is the offspring of the union between cell biology and immunogenetics. In this review, we focus on the genetic regulation of the cellular interactions required for the T helper cell-dependent activation of B cells to secrete immunoglobulin. However, it is becoming increasingly clear that the general principles that apply for the cellular interactions involved in B cell activation also apply for the cell interactions involved in a broad variety of immune responses.

ACTIVATION OF ANTIGEN-SPECIFIC B CELLS REQUIRES INTERACTIONS WITH T HELPER CELLS

The first demonstration that cell interactions were required for the generation of immune responses came from the seminal experiments of Claman et al (21), which were soon followed by the experiments of Davies et al (22) and of Mitchell & Miller (64, 66, 67). Claman and co-workers demonstrated that the antibody response to sheep erythrocyte (SRBC) antigen by lethally irradiated mice could be fully reconstituted by the adoptive transfer of spleen cells, but could not be reconstituted by the adoptive transfer of either

211

0732-0582/83/0410-0211$02.00

bone marrow or thymus cells. However, the antibody response to SRBC antigen in lethally irradiated mice could be fully reconstituted by the simultaneous transfer of both bone marrow cells and thymus cells. Thus, the transferred bone marrow and thymus cells cooperated synergistically, resulting in the generation of an anti-SRBC response. These investigators suggested that the bone marrow was the source of antibody-secreting cells and that the thymus was the source of an "auxilliary" cell.

Definitive evidence that the precursors of antibody-producing cells were contained within the bone marrow population and not within the thymus population was provided by the experiments of Miller & Mitchell (64, 66, 67). Using an adoptive transfer system in which the bone marrow cells and the thymus cells were derived from H-2 different strains, these investigators observed that all antibody-secreting cells expressed the H-2 type of the bone marrow. Three conclusions could be drawn from these experiments. First, antibody-secreting cells were derived from the bone marrow (referred to as B cells). Second, thymus cells (referred to as T cells) were required as "helper" cells for the activation of B cells. And third, histoincompatible T and B cells could cooperate with each other.

EFFECTIVE T_H-B CELL INTERACTION REQUIRES T_H CELL RECOGNITION OF CARRIER DETERMINANTS AND B CELL RECOGNITION OF HAPTENIC DETERMINANTS

The experiments that provided the first understanding of the mechanism by which T helper (T_H) cells and B cells interacted were performed by Mitchison (68), by Rajewsky et al (82), and by Raff (81). Assessing the anti-hapten antibody response stimulated by carrier-hapten conjugates, these investigators observed that optimal IgG anti-hapten secondary responses required that the T cells be previously primed to carrier and the B cells be previously primed to the hapten. They concluded that the T cells recognized carrier determinants whereas the B cells recognized haptenic determinants. Furthermore, they noted that optimal anti-hapten responses were only elicited when the carrier and hapten were physically linked on the same molecule. Thus, these investigators proposed the key concept that effective T_H-B cell collaboration involved the physical joining of a T_H cell with its partner B cell via an antigen bridge. It was clear from this model that the antigenic determinants recognized by the T_H cell could be (must be?) quite distinct from the determinants recognized by its partner B cell (i.e. the haptenic determinants).

INVOLVEMENT OF MHC ENCODED GENE PRODUCTS IN T$_H$-B CELL COLLABORATION

Next, the possibility that the interaction between T$_H$ cells and their partner B cells might be genetically regulated was systematically addressed. In contrast to the repeated observation from a number of laboratories that histoincompatible T$_H$ cells and B cells could cooperate with each other, experiments reported by Kindred & Shreffler suggested that thymus-derived cells could not effectively cooperate with histoincompatible nude B cells (54). Katz et al then provided critical insights into this issue by demonstrating that the mechanism by which histoincompatible T$_H$ and B cells collaborated was distinct from the mechanism by which histocompatible T$_H$ and B cells collaborated (30, 50).

Using a carrier-hapten response, they observed that anti-hapten B cell responses resulting from the interaction of histoincompatible T$_H$ and B cells did not require carrier priming and did not require carrier-hapten linkage. In contrast, they observed that anti-hapten responses resulting from the interaction of histocompatible T$_H$ cells and B cells did require carrier-primed T$_H$ cells and did require carrier-hapten linkage. These experiments were interpreted as demonstrating that the activation of B cells required two distinct signals, one resulting from the binding of antigen by the B cell and one derived from the T$_H$ cells. Thus, the activation of antigen-specific B cells by either histocompatible or histoincompatible T$_H$ cells involved B cell binding of the hapten as well as B cell triggering by a T$_H$ cell-derived B cell activation signal. It was in the mechanism by which the T$_H$ cell-derived activation signal was elicited from the T$_H$ cells that the two responses fundamentally differed. In the case of histoincompatible T$_H$ cells, the activation signal was elicited from the T$_H$ cells by their recognition of the allogeneic MHC determinants expressed by the B cells. This mechanism was termed the allogeneic effect and did not require T$_H$ cell recognition of any carrier determinants. In contrast, in the case of histocompatible T$_H$ cells, the activation signal was elicited from the T$_H$ cells by their recognition of carrier determinants simultaneously with their recognition of syngeneic MHC determinants expressed by the B cells. This latter mechanism was termed physiologic cooperation because it was presumably the mechanism by which T$_H$-B cell interactions occured in a normal animal.

It should be emphasized that the experiments by Katz et al did not refute the ability of T$_H$ cells to activate histoincompatible B cells, but rather demonstrated that such an interaction resulted from an entirely different mechanism than that involving histocompatible T$_H$ and B cells. Katz et al went on to show that physiologic interactions between T$_H$ and B cells did not require complete MHC compatibility, but only required I-region compatibility (49). Since it was uncertain that the I-region-encoded structures involved in T_H-B cell interactions were identical with the serologically defined Ia antigens, they referred to these structures as cell interaction molecules.

Nearly simultaneously with the reports of Katz et al, Rosenthal & Shevach reported that T-accessory cell interactions were similarly regulated by MHC encoded genes in that only MHC compatible accessory cells could present antigen to antigen-primed T cells (84). In addition, these investigators demonstrated that anti-MHC antisera effectively blocked T-accessory cell interactions, formally demonstrating that this cellular interaction was mediated by MHC encoded gene products that were expressed on the cell surface, and strongly suggesting that the cellular interaction molecules were likely to be the serologically defined MHC antigens themselves.

MHC REGULATION OF T_H-B CELL INTERACTION IS MEDIATED BY A RECEPTOR-LIGAND INTERACTION

The requirement for Ia compatibility for the physiologic cooperation between T_H cells and B cells still needed to be explained. One possibility was that a like-like interaction between Ia encoded cell surface molecules expressed on T_H cells and B cells was required to activate B cells. A second possibility was that carrier-specific T_H cells needed to recognize via their cell interaction structures the Ia-encoded determinants expressed by B cells. Thus, the requirement for MHC compatibility between T_H and B cells could have reflected either a like-like interaction or a receptor-ligand interaction in which carrier-specific T_H cells could only recognize syngeneic, but not allogeneic, Ia encoded B cell interaction structures.

The answer was provided quite clearly by the experiments of von Boehmer et al (111). These investigators constructed double-parental radiation bone marrow chimeras of the type $P_1+P_2 \rightarrow (P_1 \times P_2)F_1$. The lethally irradiated F_1 mice were completely repopulated by lymphoid elements derived from bone marrow stem cells of both parents. These chimeric mice were then primed in vivo to SRBC, and the SRBC-primed chimeric T cells were assayed for their ability to cooperate with SRBC-specific B cells of P_1 or P_2 type. Since some of the chimeric T cells were of P_1 origin and some of P_2 origin, these investigators used strain-specific alloantisera plus complement (C) to eliminate T cells of either P_1 or P_2 origin selectively. Upon isolating the P_1 T cells from the chimeras, it was found that, as a result of the chimerization, P_1 T cells were tolerant to both parental haplotypes so that these T cells could not activate either parental B cell population via an allogeneic effect. When SRBC-primed P_1 T cells were assayed for their ability to cooperate with B cells of either P_1 or P_2 origin, it was observed that they could effectively, and physiologically, cooperate with either parental B cell population for the generation of anti-SRBC responses. Thus, since chimeric P_1 T cells could cooperate with P_2 B cells, these experiments demonstrated that physiologic cooperation between T_H cells and B cells did not require MHC identity and, therefore, that T_H-B cell interactions could not be mediated by a like-like mechanism.

Although the experiments of von Boehmer et al effectively ruled out a like-like interaction mechanism, their results were also apparently discrepant with the experiments of Katz et al in that they failed to show any role for MHC at all. Thus, these experiments provoked renewed skepticism about the central role of the MHC in the regulation of physiologic cooperation between T_H cells and B cells. After all, Katz et al had demonstrated that, in the absence of allogeneic effects, histoincompatible T_H and B cells could not cooperate. The field was at an impasse.

The way out of the impasse was provided by Katz, who proposed that the conflict between his and von Boehmer's experimental results derived from the fact that the chimeric P_1 T cells that cooperated with allogeneic P_2 B cells were not simply tolerant to the allogeneic MHC determinants expressed by the P_2 B cells. Rather, because these T cells had differentiated from stem cell precursors in a P_2 MHC environment, they had "learned" to cooperate with P_2 B cells, a process he termed adaptive differentiation (47). This hypothesis accepted the conclusion that like-like interactions were not the mechanism of T_H-B cell interaction. More importantly, however, it proposed that the physiologic interactive capacity of T_H cells was "plastic" and could be changed by the environment in which the T_H cells had differentiated.

Renewed support for a central role of the MHC in regulating T_H-B cell interaction was then provided by the elegant experiments of Sprent (97). Sprent utilized as a T_H cell source normal strain$_A$ adult T cells that had been in vivo primed with SRBC and then depleted of T cells that were alloreactive to strain B MHC determinants by filtration through an $(A \times B)F_1$. Such SRBC-specific T_H cell populations were non-alloreactive to either strain$_A$ or strain$_B$ B cells and could not mediate allogeneic effects in response to either haplotype. However, the strain$_A$ T_H cells (which had been primed in a strain$_A$ environment) only cooperated with strain$_A$ B cells. Thus, these results confirmed the earlier studies of Katz et al and reaffirmed that MHC played a central role in regulating physiologic T_H-B cell interaction.

The central role of the MHC in regulating T_H-B cell interactions was now firmly established. Furthermore, the role of the MHC in these interactions could not be explained as a like-like interaction between T_H cells and B cells, but was instead most likely a reflection of a receptor-ligand interaction with T_H cells expressing receptors specific for the recognition of self-MHC determinants expressed by B cells. In contrast to allorecognition, this sort of "physiologic" recognition of MHC maintained the requirement for recognition by T_H cells of carrier determinants, and has been termed self-recognition.

Compelling evidence for the receptor-ligand-mediated self-recognition mechanism of T_H-B cell interaction was provided by several different groups and was based upon Burnet's clonal selection theory (19). The fundamental tenet of Burnet's clonal selection theory was that receptor expression by immunocompetent and antigen-specific lymphocytes is clonally excluded. That is, some lymphocyte clones express one receptor specificity whereas other clones express other receptor specificities. In contrast, non-receptor cell surface components such as MHC determinants are not clonally excluded and are expressed by all members of a given cell type. For example, essentially all B cells from a given individual express the same genetically encoded Ia determinants. Thus, if the interactive capacity of T_H cells were clonally excluded, it would be direct evidence that the interactive capacity of the T_H cells was due to a receptor mechanism. Paul and co-workers demonstrated that antigen-specific $(A \times B)F_1$ T cells consisted of at least two distinct subpopulations, one able to respond to antigens presented by parent$_A$ accessory cells and one able to respond to the same antigens presented by parent$_B$ accessory cells (77). Thus, their results demonstrated that the ability of antigen-specific F_1 T cells to cooperate with parental accessory cells was clonally excluded and, hence, was mediated by receptors specific for parental (i.e. self) MHC determinants.

Sprent (98, 99) and Swierkiosz et al (102) made use of this observation to select positively for $(A \times B)F_1$ T_H cells specific for the self-recognition of the Ia determinants of one or the other parent. In the case of Swierkiosz et al, $(A \times B)F_1$ T_H cells were precultured on antigen-pulsed parent$_A$ accessory cell monolayers. The T cells that adhered to the monolayers were then assayed for their ability to cooperate with B cells of either parental haplotype. It was found that the adherent T cells were markedly enriched in their ability to cooperate with parental B cells that were MHC identical to the accessory cell monolayer. Thus, these experiments provided strong direct evidence that the ability of antigen-specific T_H cells to cooperate with B cells was clonally excluded and, therefore, was mediated by an MHC-specific receptor mechanism.

From this perspective, the original experiments of Katz and Sprent in which in vivo primed strain$_A$ T_H cells had failed to cooperate physiologically with allogeneic strain$_B$ B cells did not, in retrospect, necessarily reflect an intrinsic inability of strain$_A$ T_H cells to cooperate physiologically with allogeneic strain$_B$ B cells. Rather, it is possible that their results with strain$_A$ T$_H$ cells, like those with (A × B)F$_1$ T$_H$ cells, specifically reflected the failure of T$_H$ cells that had been primed on strain$_A$ accessory cells to cooperate with strain$_B$ B cells. Thus, if it were possible to prime strain$_A$ T$_H$ cells on allogeneic strain$_B$ accessory cells in the absence of allogeneic effects, such T$_H$ cells might be able to cooperate physiologically with allogeneic strain$_B$ B cells. Such an experimental outcome would be consistent with the concept that MHC-restricted T$_H$-B cell interaction never reflects a requirement for MHC identity between T$_H$ cells and B cells, but instead reflects the existence of a receptor-ligand interaction in which T$_H$ cells are required to recognize the MHC determinants expressed by their partner B cells.

Additional evidence that the antigen-dependent T$_H$-B cell interaction was mediated by T$_H$ cell receptors that recognized B cell Ia determinants came from chimera experiments that had been based on the exciting findings of Zinkernagel et al (114, 115) and Bevan et al (15, 25). These investigators had shown that for cytotoxic T lymphocyte (CTL) responses in (A × B)-F$_1$ → A radiation bone marrow chimeras, the only F$_1$ T cells that differentiated into functional competence were those capable of recognizing antigens in conjunction with host-type strain$_A$ MHC determinants. In addition, these investigators had shown that the specific host element responsible for determining the self-MHC specificity of the T cells in (A × B)F$_1$ → A chimeras was the thymus in which the T cells had differentiated. These studies, together with later studies in A → B fully allogeneic radiation bone marrow chimeras (27, 56, 63, 93, 94), confirmed the plasticity of the T cell self-MHC repertoire proposed by the adaptive differentiation concept of Katz. Sprent was the first to assess the ability of T$_H$ cells from such (A × B)F$_1$ → A chimeras to cooperate with B cells from each parental strain (100). He found that even though the chimeric T cells were genotypically (A × B)F$_1$ and tolerant to both parental haplotypes, they were only able to cooperate with parent$_A$ B cells for the generation of anti-SRBC responses. These results were soon confirmed by two other groups (45, 91).

Thus, it was quite clear that T$_H$ cell self-recognition of MHC determinants was a critical event in the immune cell interactions required for the generation of antigen-specific B cell responses. The precise identification of those cell interactions that required T$_H$ cell recognition of self-MHC was to be the next area of controversy.

T$_H$ CELL RECOGNITION OF MHC DETERMINANTS ON ACCESSORY CELLS VERSUS B CELLS

The experiments reviewed above demonstrated unequivocally that T$_H$ cells were required to recognize the self-MHC determinants expressed by cells contained within the B cell population. However, these previous experiments had not attempted to distinguish whether this self-recognition requirement actually existed between the T_H cells and the responding B cells, between the T_H cells and the accessory cells contained within the B cell population, or both. Indeed, the existence of MHC restrictions for the activation of T cells by accessory cells had been described virtually simultaneously with the existence of MHC restrictions for the activation of B cells. As a result, it was possible that the only cellular interaction in which T_H cell recognition of self-MHC determinants was required was that between T_H cells and antigen-presenting accessory cells.

Katz et al reasoned that in their experiments, T_H cells were indeed required to directly recognize B cell MHC determinants since their experiments had been performed in lethally irradiated F_1 recipient mice, which should have provided F_1 accessory cells that would have fulfilled any MHC requirements for the activation of the T_H cell populations (48). Furthermore, the finding of a requirement for carrier-hapten linkage in their responses was interpreted as providing additional supportive evidence for a direct interaction between T_H cells and B cells. Similarly, Sprent also reported (98, 99) that secondary adoptive transfer responses to SRBC involved MHC-restricted T_H-B cell interactions since he found that this restriction could not be fulfilled by the intentional addition of accessory cells that were appropriate for the activation of the T_H cells. Similar observations were reported by Swierkosz et al (102). Thus, these results suggested a rather elegant model for B cell activation in which T_H cells were initially activated by the recognition of antigen in the context of accessory cell MHC determinants and were then triggered to provide their helper signals by the subsequent recognition of the identical antigen and the identical MHC determinants expressed by B cells.

However, conflicting data were simultaneously reported from a number of other laboratories. Erb & Feldmann (24) observed that T_H cells were MHC restricted in their interaction with accessory cells, but, once activated, were not MHC restricted in their interaction with B cells. Similar results were reported by McDougal & Cort (61) and Singer et al (91–93) for antigen-primed T_H cells and unprimed T_H cells, respectively. Finally, Shih et al (90) extended these previous reports by demonstrating that MHC unrestricted T_H-B cell interactions for IgG secondary antibody responses could still require carrier-hapten linkage. Thus, these investigators suggested that the activation of T_H cells required the recognition of antigen and accessory cell MHC determinants, but that, as a result of this interaction, the T_H cells were already triggered to provide their helper signals and did not need to recognize again the identical MHC determinants on B cells.

Both sets of investigators agreed that the initial activation of T_H cells required T_H cell recognition of antigen and accessory cell MHC determinants. However, they were clearly divided over whether a similar require-ment existed for T_H cell recognition of antigen and B cell MHC determinants. This conflict did not appear to correlate easily with experi-mental variables such as the use of primed versus unprimed T_H cells or B cells, the use of in vitro or in vivo response conditions, or the use of soluble versus particulate antigens. The field was again at an impasse.

In an attempt to understand why T_H-B cell interactions were clearly MHC restricted in some experimental systems but not others, Shreier, Andersson, Melchers and colleagues (3, 62, 85) attempted to study the early events involved in the T cell-dependent activation of B cells. These investi-gators found that the transformation of small resting B cells into B cell blasts required an antigen-specific and MHC-restricted interaction with T_H cells. In contrast, B cell blasts could then be further driven into antibody secretion by T_H cells by an antigen-nonspecific and MHC unrestricted mechanism. Thus, these results suggested that the conflict in the field re-sulted from the fact that small resting B cells had been inadvertently utilized in those studies demonstrating MHC restricted T_H-B cell interactions, whereas partially activated B cell blasts had been inadvertently utilized in those studies that failed to observe MHC restricted T_H-B cell interactions.

Although these studies provided important insights into some of the mechanisms of T_H cell activation of B cells, they did not appear to be able to resolve the discrepant results from the various laboratories. For example, if the discrepant results were due to the use of small, resting B cells versus partially activated B cell blasts, it would be expected that those studies that utilized B cells recently immunized with TNP-LPS would have failed to observe MHC restricted T_H-B cell interactions and those laboratories that had utilized unimmunized and unmanipulated B cells would have success-fully observed MHC restricted T_H-B cell interactions. Contrary to these expectations, experiments reported by the Kappler-Marrack laboratory fre-quently utilized B cells that had been recently primed with TNP-LPS (102), yet the investigators consistently observed MHC restricted T_H-B cell in-teractions, whereas Singer and Hodes had consistently failed to observe MHC restricted T_H-B cell interactions even though they had utilized un-primed, unmanipulated B cells (91, 92).

Another possible resolution to this conflict was suggested by experiments demonstrating that optimal T_H cell function required two distinct interact-ing T cell subsets, which could be distinguished by their sensitivity to adult thymectomy or to anti-lymphocyte serum (5, 43). The results of experi-ments by Keller et al (46) and by Tada et al (103) further suggested that one T_H cell subset was MHC restricted in its interaction with B cells whereas the other one was not, a concept extended by the work of Bottomly & Mosier (18). However, in these experiments there existed a strict requirement for the function of the B cell-restricted T_H cell subset; the second T_H cell subset appeared to augment, rather than initiate, B cell responses. Thus, these experiments could not readily explain the failure of a number of laboratories to observe MHC restricted T_H cell activation of B cells. Furthermore, it has more recently been shown that monoclonal T_H cell populations, which only consist of the progeny of a single T cell, are sufficient to activate B cells and can trigger B cells without recognizing the MHC determinants expressed by the responding B cells even though they are required to recognize the MHC determinants expressed by antigen-presenting accessory cells (34). It should be emphasized that even though heterogeneity of T_H cell subsets appears insufficient to fully resolve the conflict over the role of MHC in T_H-B cell interactions, it is likely that such heterogeneity does exist.

DISTINCT B CELL SUBPOPULATIONS DIFFER IN THEIR GENETIC REQUIREMENTS FOR ACTIVATION BY T_H CELLS

One facet of the problem of T_H cell-dependent B cell activation that had not been examined was the fact that B cells consisted of distinct B cell subpopulations that fundamentally differed in their requirements for activation. It had been shown for T-independent B cell responses that such responses were mediated by two distinct B cell subsets that differed in their ontogeny as well as in their expression of a number of B cell differentiation antigens, such as Lyb3,5, and 7 (1, 2, 39, 101). Thus, Lyb5$^-$ B cells were present from birth in normal mice and responded to T-independent antigens such as TNP-LPS and TNP-*Brucella abortus* (TNP-BA), which were grouped together as TI-1 antigens (70, 72). In contrast, Lyb5$^+$ B cells appeared at 2–3 wks of age and responded to T-independent antigens such as TNP-Ficoll and TNP-Levan, which were grouped together as TI-2 antigens (70, 72). In addition, it had been shown for these TI responses that Lyb5$^-$ and Lyb5$^+$ B cell subpopulations differed in their cellular interaction requirements for activation such that Lyb5$^+$ B cell responses to TI-2 antigens required Ia$^+$ antigen-presenting accessory cells (16, 17, 20, 71), whereas Lyb5$^-$ B cell responses to TI-1 antigens proceeded without accessory cells (16, 17).

An early suggestion that alternate pathways of B cell activation might also exist for T-dependent responses came from the work of Pierce & Klinman (78). In their work, it was observed that I-region syngeny between T_H cells and B cells was necessary for IgG$_1$ responses by unprimed B cells, but was not necessary for IgM responses by unprimed B cells. In addition it was observed that I-region syngeny between T_H cells and B cells was also not required for the IgG_1 responses of primed B cells in spite of a strict requirement for carrier-hapten linkage in these responses. The possibility that MHC-restricted and MHC-unrestricted T-dependent activation of B cells might be mediated by different B cell subpopulations was first suggested by the work of Lewis et al (57). These authors separated hapten-primed B cells into complement receptor positive (CR^+) and complement receptor negative (CR^-) fractions and assessed the presence or absence of MHC restrictions in the activation of each subpopulation by carrier-primed T_H cells. It was observed that CR^+ B cells were activated by T cells across allogeneic barriers, whereas CR^- B cells were activated only by I-region-compatible T cells, findings interpreted by the authors as suggesting the existence of alternative cooperative pathways between T cells and distinct B cell subpopulations.

Compelling evidence that distinct B cell subpopulations existed that fundamentally differed in their requirement for MHC-restricted interaction with T_H cells came from experiments that assessed the MHC restriction requirements of $Lyb5^+$ and $Lyb5^-$ B cells, the two B cell subsets that had been defined by differences in their responsiveness to T-independent antigens. A primary T_H cell-dependent in vivo adoptive transfer response to SRBC was utilized (96) in which the unprimed T_H cell population came from $(A \times B)F_1 \to A$ radiation bone marrow chimeras so that these T cells were tolerant to both parental haplotypes but were only able to recognize the self-MHC determinants of parent$_A$; also utilized were unprimed B cells from the "inappropriate" parent (parent$_B$), which were selected such that they contained either $Lyb5^-$ B cells exclusively or contained both $Lyb5^-$ and $Lyb5^+$ B cells. It was found that upon the addition of parent$_A$ accessory cells, the $(A \times B)F_1 \to A$ chimeric T_H cells could collaborate with the parent$_B$ B cell populations that contained $Lyb5^+$ B cells but could not collaborate with parent$_B$ B cells that were devoid of $Lyb5^+$ B cells. In contrast, normal $(A \times B)F_1$ T_H cells, which could recognize the self-MHC determinants expressed by parent$_B$ B cells, could collaborate with either parent$_B$ B cell subpopulation. Thus, in a single system of immune response to SRBC, these experiments demonstrated that the interaction of T_H cells with $Lyb5^+$ B cells did not require T_H cell recognition of B cell MHC determinants, whereas the interaction of T_H cells with $Lyb5^-$ B cells did require T_H cell recognition of B cell MHC determinants. Similar results have also recently been observed in vitro for the generation of anti-SRBC responses (Ono, Yaffe, Singer, manuscript in preparation) as well as for the generation of anti-TNP-KLH responses (10). It has also been observed (9) that a single antigen-specific monoclonal T_H cell population that is MHC restricted in its interaction with accessory cells could, under different conditions, activate both $Lyb5^-$ and $Lyb5^+$ B cells, but that its activation of

Lyb5⁻ B cells requires recognition of the MHC determinants expressed by Lyb5⁻ B cells, whereas its activation of Lyb5⁺ B cells does not require recognition of the MHC determinants expressed by Lyb5⁺ B cells.

These experiments provided a novel resolution for the long-standing dispute over the requirements for T_H cell recognition of B cell MHC determinants. Essentially, they demonstrated that the same T_H cells could interact with B cells in either an MHC restricted or MHC unrestricted way, but that the B cells which were triggered in each case were different (Figure 1).

However, since MHC unrestricted Lyb5⁺ B cells were present in all unfractionated normal B cell populations, it remained to be explained why many of the experiments had failed to observe MHC unrestricted activation of this B cell subset. The answer apparently involves the existence of still greater complexities in the cell interactions involved in the activation of B cells by T_H cells. In most experimental systems, it appears that the conditions utilized predominantly, if not exclusively, favor the activation of one or the other B cell subset, but rarely result in the simultaneous activation of both B cell subsets. Thus, in vitro primary responses to T-dependent antigens that utilize unprimed T_H and unprimed B cell populations are almost exclusively mediated by Lyb5⁺ B cells and do not require an MHC restricted T_H-B cell interaction (16, 96); secondary in vitro responses to low concentrations of T-dependent antigens that utilize primed T_H and primed B cells appear to be almost exclusively mediated by Lyb5⁻ B cells and require an MHC-restricted T_H-B cell interaction (10); in contrast, augmented primary in vitro responses to T-dependent antigens that utilize primed T_H but unprimed B cells are mediated by both Lyb5⁺ and Lyb5⁻ B cells, at least to SRBC (Ono, Yaffe, Singer, manuscript in preparation). These differences are likely to result from several different factors, at least one of which involves the participation of regulatory T-suppressor (T_S)

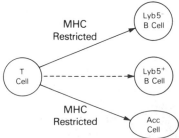

Figure 1 Existence of distinct B cell subpopulations that differ in their genetic requirements for activation by T_H cells. The interaction of T_H cells with accessory cells is MHC restricted, as indicated by a solid arrow. Once activated, the interaction of T_H cells with Lyb5⁻ B cells is also MHC restricted. In contrast, the interaction of activated T_H cells with Lyb5⁺ B cells need not be MHC restricted and can be mediated by soluble T_H cell-derived B cell activation factors such as TRF (indicated by broken arrow).

cells, since it has been shown (8) that heterogeneous T_H cell populations contain T_S cells able to differentially suppress the activation of Lyb5$^-$ but not Lyb5$^+$ B cell responses. Thus, the B cell subset that is actually triggered in a given experimental situation must be directly assessed, since it cannot be assumed that a given response will be mediated by both B cell subsets.

For the purposes of this review, the complexities involved in T_H-B cell interactions can be reduced with reasonable accuracy to the fact that distinct B cell subpopulations exist that differ in their requirements for an MHC restricted interaction with T_H cells. This difference is precisely correlated with the ability of Lyb5$^+$ but not Lyb5$^-$ B cells to respond to antigen-nonspecific T_H cell factors, a feature of B cell activation considered in greater detail below.

ROLE OF MHC-LINKED IMMUNE RESPONSE GENES IN T_H-B CELL INTERACTIONS

In parallel with studies examining the general role of the MHC in T_H-B cell interactions were a large number of studies examining the role of MHC-linked immune response (Ir) genes in T_H-B cell interactions. The earliest studies investigating mechanisms of Ir gene function attempted to identify those cell types whose immune functions were regulated by Ir genes. In T-dependent antibody responses, the first studies of this question variously concluded that Ir genes controlled the function of T_H cells, B cells, or both (12, 13, 44, 65, 73–75, 80, 87, 88). Appreciation of the role of accessory cells in T-dependent responses further complicated this issue by requiring the potentially difficult distinction of Ir gene control of accessory cell function versus B cell function. More recently, a large body of experimental data has suggested that regulation of immune responses by MHC-linked Ir genes is an antigen-specific facet of the antigen-nonspecific regulation of immune cell interactions by MHC genes. Thus, the fact that MHC restrictions result from a receptor-ligand interaction in which T_H cells, regardless of their antigen specificity, are required to recognize self-Ia determinants expressed by the cells with which the T_H cells directly interact implied that Ir gene function might better be understood as regulating the interaction with B and accessory cells of those T_H cells involved in the generation of Ir gene regulated immune responses.

A number of studies concluded that T_H-B cell interactions are regulated by MHC-linked Ir genes. For example, it was observed that (Responder X Nonresponder)F_1 $[(R \times NR) F_1]$ T_H cells cooperated with R but not NR parental B cells for the generation of antibody responses (51, 59, 113). Furthermore, NR T_H cells that had matured in a responder environment, i.e. NR → (R X NR)F_1 chimera, could in most (45) but not all (60) cases also cooperate with R but not NR B cells. Since it was found in these studies that R strain accessory cells did not reverse the unresponsiveness of NR B cells, it was concluded that T_H-B cell interaction for these Ir gene-controlled responses was regulated by Ir genes.

In contrast, analogous studies were performed by other investigators in which primary Ir gene-controlled responses were analyzed in vitro and in which T_H cells, B cells, and accessory cells were separated into independent cell pools and mixed together in three-cell mixing experiments (31). It was found that T_H cells of R, (RxNR)F_1, or NR→ (RxNR)F_1 origin were activated by accessory cells that expressed R Ia determinants but were not activated by accessory cells that only expressed NR Ia determinants. However, once activated by accessory cells of R phenotype, the T_H cells could collaborate equally well with either R or NR B cells. Thus, these studies concluded that T_H-accessory cell interactions were regulated by MHC-linked Ir genes but that, in contrast, T_H-B cell interactions were not regulated by MHC-linked Ir genes.

There are obvious parallels between this conflict over Ir gene regulation of T_H-B cell interactions and the conflict over MHC regulation of T_H-B cell interactions in general. Essentially, all investigators agreed that Ir genes did regulate T_H-accessory cell interactions but strongly disagreed whether or not Ir genes also regulated T_H-B cell interactions. In fact, those laboratories that found Ir gene regulation of T_H-B cell interactions were precisely the same laboratories that had also found MHC regulation of T_H-B cell interaction; whereas those laboratories that did not observe Ir gene regulation of T_H-B cell interaction were precisely the same laboratories that had also not found MHC regulation of T_H-B cell interaction. As was the case for MHC regulation of T_H-B cell interactions, the way out of this impasse is probably also provided by an appreciation of B cell heterogeneity.

Experiments were performed in which carrier-primed T_H cells and hapten-primed B cells were challenged in vitro with either low or high concentrations of Ir gene-controlled antigens. As for non-Ir gene-controlled antigen responses, low in vitro antigen concentrations predominantly resulted in IgG responses mediated by Lyb5$^-$ B cells, whereas high antigen concentrations predominantly resulted in IgM responses mediated by Lyb5$^+$ B cells (11). Responses mediated by either Lyb5$^-$ or Lyb5$^+$ B cells required the presence of R accessory cells, demonstrating again that T_H-accessory cell interactions were regulated by Ir genes. However, following their activation by R accessory cells, the T_H cells could only collaborate with R Lyb5$^-$ B cells and could not collaborate with NR Lyb5$^-$ B cells. Thus, Ir genes regulate the interaction of T_H cells with Lyb5$^-$ B cells. In contrast, once they were activated by R accessory cells, the T_H cells could collaborate equally well with either R or NR Lyb5$^+$ B cells, demonstrating that Ir genes do not regulate the interaction of T_H cells with Lyb5$^+$ B cells.

These latest findings dramatize the parallel between antigen-specific Ir gene regulation of cell interactions and antigen-nonspecific MHC regulation of cell interactions. Specifically, both regulate T_H-accessory cell interactions, both regulate T_H-Lyb5$^-$ B interactions, and neither regulates T_H-Lyb5$^+$ B cell interactions.

Although it is now generally accepted that Ir gene regulation represents an antigen-specific regulation of those cell interactions regulated by MHC genes, it remains controversial how the antigen-specificity of Ir gene regulation is derived. Three schools of thought currently exist that can be characterized in the following way. One viewpoint is that T_H cells only recognize a molecular association of Ia and antigen and that Ir gene-controlled antigens are unable to form an immunogenic complex with NR Ia determinants even though there exist NR T cells that would be capable of recognizing such a complex, were it formed. This viewpoint has been referred to as determinant selection (14, 83). A second viewpoint is that the problem in the NR does not lie in the failure to form an Ia-antigen complex, but rather that there do not exist in the repertoire of the species T_H cells capable of recognizing the complex created between NR Ia and antigen (89). The third viewpoint represents a middle ground and suggests that antigen can form a complex with NR-Ia molecules but that the particular complex formed cannot be recognized by T_H cells that had matured in an environment that would have tolerized them to NR Ia determinants (42, 79, 86, 93). This latter viewpoint has been referred to as clonal deletion (86).

Which if any of these perspectives is correct remains unclear, but each is fundamentally compatible with the data that presently exist. However, whichever is correct, it is now clear that antigen-specific Ir genes only regulate those cell interactions that are regulated for all antigens by MHC genes. Indeed, MHC regulation of cell interactions can be viewed as identical with Ir gene regulation for all antigen responses, even those for which a nonresponder strain does not exist.

ROLE OF ANTIGEN-SPECIFIC T_H FACTORS IN B CELL ACTIVATION

An alternative and potentially powerful approach to the study of T_H-B cell interactions is the study of soluble T_H cell products that possess some or all of the functional properties of T_H cells. Indeed, both antigen-specific and antigen-nonspecific T_H cell factors have been described that are either partially or completely T_H cell-replacing.

The study of T_H cell factors in general and of antigen-specific T_H cell factors in particular has proceeded through a relatively stereotyped general experimental approach. It has been necessary to generate soluble T cell products, to establish a T-dependent B cell response system, and then to assess the functional activity of soluble T cell products in that response system. Among the first studies of antigen-specific T_H factors were those of Taussig and co-workers and Mozes et al (74, 75, 108), who characterized factors specific for synthetic polypeptides such as TGAL. These helper factors were generated by the education of thymocytes through transfer with antigen in irradiated recipients, and the subsequent in vitro stimulation of these educated T cells with specific antigens. The soluble supernatant from such cultures was then assayed for its ability to trigger, in the absence of T_H cells but in the presence of antigen, antibody production by bone marrow cells adoptively transferred into irradiated hosts. The results of such studies demonstrated that soluble products derived from educated T_H cells could help in an antigen-specific manner the production of antibody by bone marrow-derived cells. It was also shown that MHC-linked Ir genes influenced these responses both by determining the ability of T cells to respond to antigen for the generation of active factor, and by determining the ability of antigen-specific B cells to be triggered by the active factor. No MHC restriction was observed, however, between the strain of origin of active T cell factor and the strain of high responder B cell. The conclusion that such antigen-specific T cell factors acted directly on B cells was based upon the additional observation that B cell-containing populations were uniquely able to absorb out the functional activity from these factor preparations.

These early studies opened the way for the pursuit of a number of critical issues. First, there was a need for a more rigorous identification of the cells producing active T_H cell factors, and of the activation requirements for its production. Second, the nature of the target cells directly triggered by T_H factors required more precise identification. Finally, the T_H factor itself required characterization at the biochemical and serologic, as well as functional, level. In particular, the demonstration of soluble, antigen-specific T cell factors offered a promising system in which to evaluate the character of the T cell receptor, of MHC restricted recognition in general, and of T_H-B cell interactions in particular.

A more precise identification of the T cell origin of helper factors was first facilitated by the work of Feldmann and colleagues, who were able both to prime murine spleen cells in vitro and stimulate the release of helper factor in vitro (36). The most definitive approach to this question came, however, from the work of Mozes, Lonai, Feldmann and their colleagues, who established cloned T cell hybridomas that produced active helper factors (4, 26,

58). Although these findings established the ability of hybridoma T cells to synthesize active factors, they did not resolve the question of what cell interactions might normally result in the triggering of factor production by T_H cells. Other investigators have suggested that accessory cells and even B cells may be involved in the production of helper factors by T cells (55).

It was also unclear from the initial studies in this area whether T helper factors could function to activate B cells directly, or whether additional cell interactions were required for B cell activation. Howie & Feldmann demonstrated that antigen-specific helper factors produced by heterogeneous T cell populations could help T cell-depleted spleen cells, suggesting that T cells did not play an essential intermediary role in the activation of antigen-specific B cells by antigen-specific T_H factor (36). However, it has been suggested by Howie & Feldmann (37) that adherent accessory cells are required for B cell activation in the presence of antigen-specific T helper factors, and indeed that genetic restrictions between B cells and accessory cells can be observed under these conditions. Thus, although T_H factors may act in the absence of additional T cells, the activation of B cells may require additional signals provided by accessory cells. The question of whether or not genetic restrictions exist in the ability of T cell factors to help B cells was approached in part in the early studies by Taussig and colleagues described above. More recently, this issue was analyzed by Lonai et al employing the soluble products of monoclonal T cell hybridomas (58). It was found in these studies that there existed a strict H-2 restriction (mapping to the K or I-A region) between a factor of $H-2^b$ origin and the B cells with which the factor cooperated. A similar restriction was observed in the ability of B cells to absorb factor. These studies did not formally distinguish whether the observed restrictions were between T cell factor and B cells, or between factor and accessory cells contained in the B cell population.

Biochemical and immunochemical analyses of T helper factors have been carried out in recent years both for the products of heterogeneous T cell populations and for the products of monoclonal hybridomas. Howie et al reported that factors derived from conventional T_H cell populations bound antigen specifically, expressed I-A- or I-J-encoded (Ia) determinants, and reacted with a chicken anti-mouse Ig reagent, raising the possibility that T cell factors express Ig-like determinants (38). Apte et al found that a TGAL-specific factor bound antigen, although with a fine specificity distinct from that of antibody (4). This T cell factor expressed idiotypic determinants cross-reactive with those expressed by anti-TGAL antibodies, reacted with an anti-V_H reagent, and consisted of two chains, one of which was V_H positive and the other of which was Ia positive. Lonai et al similarly reported that a CGG-specific hybridoma factor expressed both V_H and I region products (58).

Thus, antigen-specific T cell factors, including those obtained from monoclonal T cell populations, appear to interact with B cells in a manner analogous to that demonstrated for T_H-B cell interactions. The behavior of antigen-specific T_H cell factors for activating distinct B cell subpopulations such as those represented by Lyb5$^-$ and Lyb5$^+$ B cells remains to be examined. The "physiologic" role of such factors in the cooperation of T_H cells and B cells remains a reasonable but unconfirmed hypothesis.

ROLE OF ANTIGEN-NONSPECIFIC T_H FACTORS IN B CELL ACTIVATION

A great deal of experimental data has recently been accumulated assessing the role of antigen-nonspecific factors in B cell triggering. These studies have attempted to identify the pertinent lymphokines elaborated as a result of T_H-accessory and T_H-B cell interactions involved in activating responding B cells. A variety of such factors have been identified and include interleukin 1 (IL-1), interleukin 2 (IL-2), B cell growth factor, and T cell replacing factor (TRF) (69). It remains unclear at this time whether some or all of these factors are involved in B cell triggering, and it remains controversial in what sequence these factors act on responding B cells. An interesting model has been proposed by Hoffman (35), who suggested that the accessory-cell-derived lymphokine IL-1 first binds to B cells, resulting in the B cells becoming receptive to T cell-derived lymphokines.

Although the identification of the precise lymphokines involved in B cell activation remains uncertain, studies with soluble supernatants that probably contain multiple lymphokines have provided a number of insights into the mechanism of T_H-B cell interactions. One of the earliest antigen-nonspecific factors studied was derived from the supernatants of mixed lymphocyte reactions directed against allogeneic stimulator cells (51). These supernatants were designated as allogeneic effect factor (AEF), and were able to trigger B cells in the absence of T_H cells but still required the presence of antigen (6, 7). As originally described, AEF was capable of activating B cells of any MHC haplotype as well as B cells of any antigen specificity. Thus, activation of B cells by AEF was remarkably similar to the activation of B cells by histoincompatible T_H cells. Indeed, it seemed likely that secretion of AEF by alloreactive T_H cells was the molecular mechanism by which B cells were triggered by histoincompatible T_H cells.

More recent studies with AEF by Delovitch et al have shown that it is possible to obtain "restricted" AEF capable of activating B cells of a specific MHC haplotype (23). If the MLR from which the AEF supernatants were obtained consists of T cells depleted of Ia^+ cells and stimulator cells devoid of T cells, the resulting AEF specifically activated B cells of stimulator type. It was suggested that the restricted AEF consists in part of molecules secreted by the responding T cells, which recognize the Ia determinants of the stimulator cells. Thus, the ability of restricted AEF to trigger B cells depended in part on its anti-Ia specificity. These studies provided a molecular model for T_H-B cell activation and confirmed the importance of Ia recognition in the cell interactions required for B cell triggering.

In light of the controversy over T_H cell recognition of Ia determinants expressed by accessory cells versus B cells, it was clearly of interest to determine whether the specificity of restricted AEF was for accessory cell Ia determinants, B cell Ia determinants, or both. It has recently been observed (Delovitch et al, submitted for publication) that restricted AEF is in fact specific for the Ia determinants expressed by accessory cells but. is not specific for the Ia determinants expressed by the responding B cells. As a result, even though restricted AEF is Ia specific and is able to trigger B cells, these studies strongly support the existence of T_H-B cell interaction pathways that do not require T_H cell recognition of B cell Ia determinants. Given the recent awareness of B cell hetcrogeneity in T_H cell-dependent antigen responses, it would be informative to know which B cell subset is activated by restricted AEF. Indeed, Delovitch et al have suggested (submitted for publication) that restricted AEF activates $Lyb5^+$ but not $Lyb5^-$ B cells. Thus, restricted AEF behaves in a manner consistent with the observations that T_H cell recognition of B cell Ia determinants is not required for T_H cell activation of $Lyb5^+$ B cells but is required for T_H cell activation of $Lyb5^-$ B cells.

A second approach to the study of B cell activation is illustrated by the studies of Huber et al, who utilized Lyb3-specific antisera in T_H cell-dependent antibody responses (39). The Lyb3 determinant is non-allelic and is only expressed on a late-appearing B cell population that is essentially identical with the B cells that express Lyb5. These investigators demonstrated that Lyb3 antisera significantly augmented the T_H cell-dependent B cell response to suboptimal doses of SRBC. As would be expected, the augmentation of anti-SRBC responses by Lyb3 antisera was only observed in B cell populations that contained $Lyb3^+5^+$ B cells. Thus, these investigators suggested the possibility that the augmentation of anti-SRBC B cell responses by Lyb3 antisera was due to the fact that the Lyb3 determinant was part of the B cell membrane receptor for T_H cell-derived B cell activation signals.

A third and exceedingly informative approach has been the study of antigen-nonspecific supernatant factors that replace T_H cell function in "physiologic," rather than allogeneic, T_H-B cell interactions. Such supernatant factors have been termed T cell replacing factors (TRF), and are both antigen-nonspecific and haplotype-nonspecific (84a). TRF has been shown to activate B cells via a mechanism that is MHC unrestricted and that does not require carrier-hapten linkage (40, 41). Among the most elegant studies with TRF were those performed by Takatsu, Hamaoka, and co-workers, who utilized a monoclonal TRF produced by a T cell hybridoma (104–106). In addition, these investigators have produced antisera specific for the TRF acceptor molecule expressed by B cells (104). Thus, these investigators have been able to identify the distinct B cell subpopulations that are able to bind TRF as well as to functionally respond to TRF. Their studies demonstrated that the expression of the TRF-acceptor molecule was genetically controlled and that there exists a mouse strain, DBA/2Ha, which is both specifically deficient in the expression of this receptor molecule and also functionally unresponsive to TRF (109). Moreover, these investigators further demonstrated that there exist in normal mouse strains two distinct B cell subpopulations, one of which is TRF-acceptor positive and one of which is TRF-acceptor negative. The TRF-acceptor positive B cells are responsive to TRF and can be triggered by an unlinked carrier-hapten mechanism. In contrast, TRF-acceptor negative B cells are unresponsive to TRF and require a linked carrier-hapten mechanism for their activation. Interestingly, the TRF-acceptor positive B cells appear to be predominantly $Lyb3^+5^+$, whereas the TRF-acceptor negative B cells appear to be predominantly $Lyb3^-5^-$. The relationship between the TRF-acceptor molecule and the Lyb3 determinant is unknown, but anti-TRF acceptor antibody, like anti-Lyb3 antibody, augments the response to suboptimal doses of SRBC (107). Thus, the functional parallels between anti-TRF-acceptor and anti-Lyb3 antisera are quite striking, and it might be suspected that the Lyb3 determinant is a component of the TRF acceptor molecule.

The effect of TRF-containing supernatants on MHC-restricted and MHC-unrestricted T_H-B cell interactions has also been directly examined. Studies have been performed utilizing supernatants from either Con A-stimulated spleen cells (29, 96) or from antigen-specific, Ia-restricted T_H cell clones (33, 34). In both cases it was observed that such supernatants in the presence of antigen triggered $Lyb5^+$ B cells but did not trigger $Lyb5^-$ B cells. The studies with TRF derived from antigen-specific and Ia restricted T_H cell clones were especially informative in sorting out some of the complexities of T_H-B cell interactions and the potential role performed by soluble TRF in these interactions. As mentioned previously, the T_H cell clones, in response to low in vitro concentrations of antigen, activated $Lyb5^+$ B cells via an MHC restricted mechanism and the same T_H cell clones, in response to high in vitro concentrations of antigen, activated $Lyb5^+$ B cells via an MHC unrestricted mechanism. Under high antigen dose conditions, the T_H cell clone was able to secrete soluble TRF, which functioned in a manner analogous to the clone itself under high antigen dose conditions in that the TRF only activated $Lyb5^+$ B cells, and activated these B cells in an MHC unrestricted manner (90a). Thus, these studies provided direct evidence that the secretion of TRF by T_H cells was the probable mechanism by which such T_H cells activated $Lyb5^+$ B cells in an MHC unrestricted manner. In addition, these studies suggested one further complexity that had not been previously appreciated. Specifically, they demonstrated that in response to low concentrations of antigen, the T_H cell clones were activated to function as helper cells but were not activated to secrete detectable quantities of soluble TRF, suggesting that a monoclonal T_H cell population could potentially exist in two distinct activation states, one in which it is able to activate B cells via an MHC restricted T_H-B cell interaction, and one in which it secretes TRF and is able to activate B cells via an MHC unrestricted T_H-B cell interaction.

Taken together, the studies with antigen-nonspecific factors have provided important insights into the possible molecular mechanisms underlying MHC-unrestricted T_H-B cell interactions. Specifically, they have provided strong evidence that the MHC-unrestricted activation of $Lyb5^+$ B cells by T_H cells is mediated by soluble TRF secreted by T_H cells which is then bound by receptors on the surface of $Lyb5^+$ B cells. In contrast, these studies have not provided any insights into the mechanisms of MHC-restricted activation of $Lyb5^-$ B cells by T_H cells. A priori, it would seem a reasonable presumption that the MHC-restricted activation of $Lyb5^-$ B cells by T_H cells would also be mediated by the local release of small quantities of TRF. However, such B cells appear not to express detectable quantities of TRF-acceptor molecules and fail to be triggered by soluble TRF. Consequently, the possibility needs to be considered that contrary to current presumptions, the T_H signals involved in the MHC-restricted activation of $Lyb5^-$ B cells may not only be quantitatively distinct but may also be qualitatively distinct from the T_H signals involved in the MHC-unrestricted activation of $Lyb5^+$ B cells.

A recent study has examined the consequences, but not the mechanism, of MHC restricted T_H-$Lyb5^-$ B cell interactions (Ono, Yaffe, Singer, manuscript in preparation). This study assessed the TRF responsiveness and MHC restriction requirements of SRBC-specific $Lyb5^-$ B cells after they had been initially activated by an MHC-restricted T_H-B cell interaction. Unprimed mutant CBA/N (i.e. $Lyb5^-$) B cells specific for SRBC were unresponsive to TRF and were triggered in vitro by T_H cells only via an

MHC restricted T_H-B cell interaction. In contrast, CBA/N B cells that had been previously primed in vivo with SRBC and so had potentially undergone an MHC-restricted T_H-B cell interaction in vivo were responsive to TRF and were triggered in vitro by T_H cells via an MHC-unrestricted T_H-B cell interaction. These results are concordant with the observation that SRBC-primed CBA/N B cells express the TRF-acceptor molecule (104). These results suggest that TRF-acceptor negative Lyb5$^-$ B cells can be induced by an MHC-restricted interaction with T_H cells to express the TRF-acceptor molecule and become functionally responsive to TRF. However, the molecular mechanism involved in the initial activation of Lyb5$^-$ B cells via an MHC-restricted interaction with T_H cells remains obscure.

It now seems unlikely that insights into the molecular mechanisms involved in MHC-restricted T_H-B cell interactions will come from studies of antigen-nonspecific T_H factors. Rather, it seems more likely that such insights will come from the isolation and characterization of T_H cell factors that are both specific for antigen and specific for B cell Ia determinants.

DO ANTIGEN-SPECIFIC B CELLS RECOGNIZE SELF-MHC DETERMINANTS?

Thus far, the role of the MHC in regulating T_H-B cell interactions has been discussed from the perspective of a receptor-ligand interaction in which antigen-specific T_H cells express receptors specific for the recognition of self-MHC determinants expressed by the cells with which the T_H cells directly interact, i.e. accessory cells and some B cells. However, the possibility can also be considered that antigen-specific B cells also express receptors specific for self-MHC determinants expressed by cells with which the B cells might directly interact, i.e. accessory cells and T_H cells. It has generally been presumed that an MHC-restricted self-repertoire is not expressed by B cells because B cells express on their surface, as well as secrete, Ig, which is capable of binding free antigen directly. However, several groups of investigators have reported that it is possible to devise experimental systems in which MHC-restricted self-recognition by B cells can be observed.

The first suggestion that the interactive phenotype of B cells can be MHC-restricted came from experiments by Katz et al (53). These investigators assessed the ability of B cells from $F_1 \rightarrow$ parent radiation bone marrow chimeras to accept help from T_H cells. It was observed that such B cells preferentially interacted with T_H cells syngeneic to the chimeric host rather than with T_H cells syngeneic to the chimeric donor. Thus, the chimeric F_1 B cells appeared to distinguish T_H cells according to the MHC determinants the T_H expressed.

Subsequently, Gorczynski et al utilized an anti-SRBC response system, which required accessory cells but in which the usual requirement for T_H

cells could apparently be replaced by low concentrations of LPS (28). In this response system these investigators observed that the responding B cells that had been previously primed in vivo to SRBC could distinguish among different populations of accessory cells such that the B cells only interacted with accessory cells that were syngeneic to the haplotype of the environment in which the B cells had originally been primed. In a more conventional anti-SRBC response utilizing a short-term adoptive transfer system, Nisbet-Brown et al observed that optimal T_H cell-dependent responses were obtained when the irradiated host, which presumably provided accessory cells, was syngeneic to the responding B cells (76). Again, these results were consistent with the concept that antigen-specific B cells could distinguish the MHC determinants expressed by different accessory cells.

In an effort to more simply analyze MHC restrictions between a specific subpopulation of antigen-specific B cells and accessory cells, Singer and Hodes analyzed the response stimulated by the polysaccharide antigen TNP-Ficoll (32, 95). Antibody responses stimulated by TNP-Ficoll are mediated exclusively by $Lyb5^+$ B cells, require Ia^+ accessory cells, and exhibit both a T-independent and a T-dependent component. For the T-independent component of the TNP-Ficoll response (95), it was observed that spleen cells from $(A \times B)F_1 \rightarrow parent_A$ and from full allogeneic $A \rightarrow B$ radiation bone marrow chimeras were MHC-restricted in their interactions with accessory cells such that these chimeric B cells only cooperated with host-type, but not donor-type, accessory cells. Such restrictions could not fully be accounted for by restricted T_H cell recognition of accessory cell Ia determinants, because the intentional addition of appropriate T_H cells consistently failed to alter the observed MHC restrictions. More direct evidence that such MHC restrictions actually resulted from the restricted recognition by B cells of accessory cell MHC determinants came from experiments in which "unrestricted" $strain_A$ spleen cells from normal mice were cultured together with "restricted" $strain_A$ spleen cells from $A \rightarrow B$ chimeras and stimulated with TNP-Ficoll. Thus, in addition to TNP-Ficoll, these co-cultures contained $strain_A$ T_H cells, $strain_A$ accessory cells, and $strain_A$ B cells derived from both normal $strain_A$ mice and $A \rightarrow B$ chimeric mice. If the MHC restrictions observed in TNP-Ficoll responses only reflected the MHC requirements for triggering T_H cells, despite the fact such responses appeared to be T-independent, then the normal $strain_A$ T_H cells in these cultures would be triggered by the $strain_A$ accessory cells and in turn should activate both the normal as well as the chimeric $strain_A$ B cells. Alternatively, if the MHC restrictions had in fact resulted from the MHC recognition requirements of the TNP-Ficoll responsive B cells, then the $strain_A$ accessory cells present in these co-cultures would only be recognized by the normal $strain_A$ B cells but would not be recognized by the chimeric $strain_A$ B cells; hence, only the normal $strain_A$ B cells would be activated in these co-cultures. It was observed that in these co-cultures only the normal strain$_A$ B cells, but not the chimeric A → B strain$_A$ B cells, responded to TNP-Ficoll. These experiments provided strong evidence that TNP-Ficoll-responsive Lyb5$^+$ B cells, at least Lyb5$^+$ B cells from fully allogeneic chimeras, recognize accessory cell MHC determinants.

For the T-dependent component of the TNP-Ficoll response, it was observed that both the T$_H$ cells as well as the TNP-Ficoll-responsive B cells were required to recognize self-MHC determinants expressed on accessory cells, but were not required to recognize each other (32). Thus, for T-dependent responses to TNP-Ficoll, the T$_H$-accessory cell interaction was MHC restricted, the B-accessory cell interaction was MHC-restricted, but the T$_H$-B cell interaction was not MHC-restricted. Indeed, it seemed likely that the T$_H$-B cell interaction actually occurred via an accessory cell intermediary. It was also observed in these experiments that the addition of soluble TRF to the responding cultures obviated any requirement for accessory cells and obscured any requirement for B cell recognition of accessory cell MHC determinants. As a result, it was concluded that T-dependent responses to TNP-Ficoll differed from classical T-dependent responses such as those to TNP-KLH in that the participating T$_H$ cells did not secrete detectable quantities of TRF. Thus, the existence of a cell interaction pathway involving self-recognition of accessory cell MHC determinants by both T$_H$ cells and Lyb5$^+$ B cells (32a) would be obscured in classical T-dependent responses which result in the elaboration of TRF (Figure 2).

Taken together, all of these different experiments suggested that antigen-specific B cells, like antigen-specific T$_H$ cells, might express a requirement

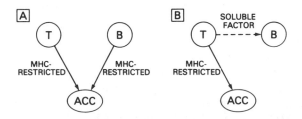

Figure 2 T-dependent responses to TNP-Ficoll reveal the existence of a cell interaction pathway requiring MHC-restricted recognition of accessory cells by both T$_H$ cells and B cells. (*A*) T-dependent responses to TNP-Ficoll require the self-recognition of accessory cell MHC determinants by both T$_H$ cells and TNP-Ficoll responsive Lyb5$^+$ B cells. Since soluble T$_H$ cell-derived factors can overcome any requirement for B-accessory cell interactions, the existence of this cell interaction pathway in TNP-Ficoll responses suggests that TNP-Ficoll, unlike classical T-dependent antigens, does not induce the release of TRF from T$_H$ cells. (*B*) For classical T-dependent antigen responses that stimulate T$_H$ cells to secrete TRF, it is likely that any genetic restrictions between Lyb5$^+$ B cells and accessory cells would be obscured.

for recognition of MHC determinants expressed by the cells with which they directly interact. It would be simplest to imagine that a requirement for MHC recognition by B cells would be mediated, not by a separate receptor, but by its sIg. Indeed, it might be conceived that, in the absence of T_H cell-derived B cell activation signals, Lyb5$^+$ B cells such as those responding to TNP-Ficoll elicit activation signals from the accessory cells with which they interact by expressing sIg that recognizes TNP in conjunction with accessory cell Ia determinants. Evidence consistent with this possibility has recently been reported (110, 112). For example, Wylie et al observed in the splenic focus assay that the antibodies elicited in response to virally infected fibroblasts were to a large extent not specific for free virus, but were specific for virus in conjunction with the MHC determinants expressed by the infected fibroblasts. Although it is uncertain that such a result would have been obtained in response to immunization with free virus, this study demonstrates that it is at least possible to manipulate B cell responses to reveal the existence of MHC-restricted antibodies.

Recognition of self-MHC determinants by antigen-specific B cells increases enormously the complexities of T_H-B cell interactions. Indeed, it raises the provocative possibility that each MHC-restricted cell interaction involved in the activation of B cells might simultaneously involve two discrete sets of receptor-ligand interactions in which each partner cell expresses MHC determinants as well as anti-MHC receptors.

CONCLUSIONS

The enormous complexities involved in T_H-B cell interactions are still emerging. Indeed, the history of this research area consistently demonstrates that areas of controversy and disagreement reflected the existence of greater complexities than were appreciated at the time. Thus, B cell activation required cell interactions among distinct immunocompetent cell populations. These cell interactions were regulated by MHC-encoded gene products, reflecting a requirement for a receptor-ligand interaction in which the T_H cells recognized MHC determinants expressed on the surface of cells with which they directly interacted. The heated controversy over T_H cell recognition of B cell MHC determinants now appears to have been due to the existence of distinct B cell subpopulations that differed in their ability to respond to antigen-nonspecific and MHC-unrestricted soluble factors secreted by activated T_H cells. The major area of controversy that is developing at this time involves the ability of antigen-specific B cells to recognize self-MHC determinants on the cells with which the B cells directly interact. Whatever the final resolution to this issue, it is likely to reveal still greater levels of complexity than are now appreciated.

236   SINGER & HODES

Literature Cited

1. Ahmed, A., Scher, I. 1979. Murine B cell heterogeneity defined by anti-Lyb5, an alloantiserum specific for a late-appearing B lymphocyte subpopulation. In *B Lymphocyte and Immune Responses,* ed. M. Cooper, D. E. Mosier, E. S. Vitetta, I. Scher, pp. 117–24. New York: Elsevier-North Holland
2. Ahmed, A., Scher, I., Sharrow, S. O., Smith, A. H., Paul, W. E., Sachs, D. H., Sell, K. W. 1977. B lymphocyte heterogeneity: development of an alloantiserum which distinguishes B lymphocyte differentiation alloantigens. *J. Exp. Med.* 145:101
3. Andersson, J., Schreier, M. H., Melchers, F. 1980. T-cell-dependent B-cell stimulation is H-2 restricted and antigen dependent only at the resting B-cell level. *Proc. Natl. Acad. Sci. USA* 77:1612–16
4. Apte, R. N., Eshhar, Z., Lowy, I., Zinger, H., Mozes, E. 1981. Characteristics of a poly(Ltyr,LGlu)-poly(D-LAla)—poly(LLys)-specific helper factor derived from a T cell hybridoma. *Eur. J. Immunol.* 11:931–36
5. Araneo, B. A., Marrack (Hunter), P. C., Kappler, J. W. 1975. Functional heterogeneity among the T-derived lymphocytes of the mouse. II. Sensitivity of subpopulations to antithymocyte serum. *J. Immunol.* 114:747–51
6. Armerding, D., Katz, D. H. 1974. Activation of T and B lymphocytes in vitro. II. Biological and biochemical properties of an allogeneic effect factor (AEF) active in triggering specific B lymphocytes. *J. Exp. Med.* 140:19–37
7. Armerding, D., Sachs, D. H., Katz, D. H. 1974. Activation of T and B lymphocytes in vitro. III. Presence of Ia determinants on allogeneic effect factor. *J. Exp. Med.* 140:1717–24
8. Asano, Y., Hodes, R. J. 1982. T-cell regulation of B cell activation. T cells independently regulate the responses mediated by distinct B cell subpopulations. *J. Exp. Med.* 155:1267–76
9. Asano, Y., Shigeta, M., Fathman, C. G., Singer, A., Hodes, R. J. 1982. Role of the major histocompatibility complex in T cell activation of B cell subpopulations. A single monoclonal T helper cell population activates different B cell subpopulations by distinct pathways. *J. Exp. Med.* 156:350–60
10. Asano, Y., Singer, A., Hodes, R. J. 1981. Role of the major histocompatibility complex (MHC) in T cell activa-

tion of B cell subpopulations. MHC restricted and unrestricted B cell responses are mediated by distinct B cell subpopulations. *J. Exp. Med.* 154:1100–15
11. Asano, Y., Singer, A., Hodes, R. J. 1983. Role of the major histocompatibility complex in T cell activation of B cell subpopulations. Ir gene regulation of the T cell dependent activation of distinct B cell subpopulations. *J. Immunol.* 130:67–72
12. Bechtol, K. B., Freed, J. H., Herzenberg, L. A., McDevitt, H. O. 1974. Genetic control of the antibody response to poly-L(TYR,GLU)-poly-D,L-ALA—poly-L-LYS in C3H ↔ CWB tetraparental mice. *J. Exp. Med.* 140:1660–75
13. Bechtol, K. B., McDevitt, H. O. 1976. Antibody response of C3H ↔ (CKB X CWB)F₁ tetraparental mice to poly-L(TYR,Glu)-poly-D,L-ALA—poly-L-LYS immunization. *J. Exp. Med.* 144:123–44
14. Benacerraf, B. 1978. A hypothesis to relate the specificity of T lymphocytes and the activity of I region-specific Ir genes in macrophages and B lymphocytes. *J. Immunol.* 120:1809–12
15. Bevan, M. J. 1977. In a radiation chimera, host H-2 antigens determine the immune responsiveness of donor cytotoxic cells. *Nature* 269:417
16. Boswell, H. S., Ahmed, A., Scher, I., Singer, A. 1980. Role of accessory cells in B cell activation. II. The interaction of B cells with accessory cells results in the exclusive activation of an Lyb5+ B cell subpopulation. *J. Immunol.* 125:1340–48
17. Boswell, H. S., Nerenberg, M. I., Scher, I., Singer, A. 1980. Role of accessory cells in B cell activation. III. Cellular analysis of primary immune response deficits in CBA/N mice: presence of an accessory cell-B cell interaction defect. *J. Exp. Med.* 152:1194–209
18. Bottomly, K., Mosier, D. E. 1981. Antigen-specific helper T cells required for dominant idiotype expression are not H-2 restricted. *J. Exp. Med.* 154:411–21
19. Burnet, F. M. 1959. *The Clonal Selection Theory of Acquired Immunity.* Cambridge: Cambridge Univ.
20. Chused, T. M., Kassan, S. S., Mosier, D. E. 1976. Macrophage requirement for the in vitro response to TNP-Ficoll: a thymic independent antigen. *J. Immunol.* 116:1579

21. Claman, H. N., Chaperon, E. A., Triplett, R. F. 1966. Thymus-marrow cell combinations. Synergism in antibody production. *Proc. Soc. Exp. Biol. Med.* 122:1167–71

22. Davies, A. J. S., Leuchers, E., Wallis, V., Marchant, R., Elliot, E. V. 1967. The failure of thymus derived cells to produce antibody. *Transplantation* 5: 222

23. Delovitch, T. L., Phillips, M. L. 1982. The biological and biochemical basis of allogeneic effect factor (AEF) activity: relationship to T cell alloreactivity. *Immunobiology* 161:51

24. Erb, P., Feldmann, M. 1975. The role of macrophages in the generation of T helper cells. II. The genetic control of macrophage-T-cell interaction for helper cell induction with soluble antigen. *J. Exp. Med.* 142:460

25. Fink, P. J., Bevan, M. J. 1978. H-2 antigens of the thymus determine lymphocyte specificity. *J. Exp. Med.* 148: 766–75

26. Fischer, A., Beverly, P. C. L., Feldmann, M. 1981. Long-term human T-helper lines producing specific helper factor reactive to influenza virus. *Nature.* 294:166–68

27. Glimcher, L. H., Longo, D. L., Green, I., Schwartz, R. H. 1981. The murine syngeneic mixed lymphocyte response. I. The target antigens are self Ia molecules. *J. Exp. Med.* 154:1652–70

28. Gorczynski, R. M., Kennedy, M. J., MacRae, S., Steele, E. J., Cunningham, A. J. 1980. Restriction of antigen recognition in mouse B lymphocyte by genes mapping within the major histocompatibility complex. *J. Immunol.* 124: 590–96

29. Greenstein, J. L., Lord, E., Kappler, J., Marrack, P. C. 1981. Analysis of the response of B cells from CBA/N-defective mice to nonspecific T cell help. *J. Exp. Med.* 154:1608–17

30. Hamaoka, T., Osborne, D. P. Jr., Katz, D. H. 1973. Cell interactions between histoincompatible T and B lymphocytes. I. Allogeneic effect by irradiated host T cells on adoptively transferred histoincompatible B lymphocytes. *J. Exp. Med.* 137:1393–404

31. Hodes, R. J., Hathcock, K. S., Singer, A. 1979. Cellular and genetic control of antibody responses. VI. Expression of *Ir* gene function by H-2ᵃ accessory cells, but not H-2ᵃ T or B cells in responses to TNP-(T,G)-A—L. *J. Immunol.* 123: 2823–29

32. Hodes, R. J., Hathcock, K. S., Singer, A. 1983. Major histocompatibility complex restricted self-recognition in responses to trinitrophenyl-ficoll. A novel cell interaction pathway requiring self-recognition of accessory cell H-2 determinants by both T cells and B cells. *J. Exp. Med.* In press

32a. Hodes, R. J., Hathcock, K. S., Singer, A. 1983. Major histocompatibility complex-restricted self-recognition by B cells and T cells in responses to TNP-Ficoll. *Immunol. Rev.* In press

33. Hodes, R. J., Kimoto, M., Hathcock, K. S., Fathman, C. G., Singer, A. 1981. Functional helper activity of monoclonal T cell populations. Antigen-specific and *H-2* restricted cloned T cells provide help for in vitro antibody responses to trinitrophenyl-poly-L-(Tyr,Glu)-poly-D,L-Ala—poly-L-Lys. *Proc. Natl. Acad. Sci. USA* 78:6431–35

34. Hodes, R. J., Shigeta, M., Hathcock, K. S., Fathman, C. G., Singer, A. 1982. Role of the major histocompatibility complex in T cell activation of B cell subpopulations. Antigen-specific and *H-2* restricted monoclonal Tᴴ cells activate Lyb5+ B cells through an antigen-non-specific and *H-2* unrestricted effector pathway. *J. Immunol.* 129:267–71

35. Hoffman, M. K. 1980. Macrophages and T cells control distinct phases of cell differentiation in the humoral response in vitro. *J. Immunol.* 125: 2076–81

36. Howie, S., Feldmann, M. 1977. In vitro studies on H-2-linked unresponsiveness to synthetic polypeptides. III. Production of an antigen-specific T helper cell factor to (T,G)-A—L. *Eur. J. Immunol.* 7:417–21

37. Howie, S., Feldmann, M. 1978. Immune response (Ir) genes expressed at macrophage-B lymphocyte interactions. *Nature* 273:664–66

38. Howie, S., Parish, C. R., David, C. S., McKenzie, I. F. C., Maurer, P. H., Feldmann, M. 1979. Serological analysis of antigen-specific helper factors specific for poly-L(Tyr,Glu)-poly-DLAla-poly-LLys ((T,G)-A—L) and LGlu⁶⁰-LAla³⁰-LTyr¹⁰(GAT). *Eur. J. Immunol.* 9:501–6

39. Huber, B., Gershon, R. K., Cantor, H. 1977. Identification of a B cell surface structure involved in antigen-dependent triggering: absence of the structure on B cells from CBA/N mutant mice. *J. Exp. Med.* 145:10

40. Hunig, T., Schimpl, A., Wecker, E. 1977. Mechanism of T-cell help in the

immune response to soluble protein antigens. I. Evidence for in situ generation and action of T-cell-replacing factor during the anamnestic response to dinitrophenyl keyhole limpet hemocyanin in vitro. *J. Exp. Med.* 145:1216–27

41. Hunig, T., Schimpl, A., Wecker, E. 1977. Mechanism of T-cell help in the immune response to soluble protein antigens. II. Reconstitution of primary and secondary in vitro immune responses to dinitrophenyl-carrier conjugates by T-cell-replacing factor. *J. Exp. Med.* 145:1228–36

42. Ishii, N., Baxevanis, C. N., Nagy, Z. A., Klein, J. 1981. Responder T cells depleted of alloreactive cells react to antigen presented on allogeneic macrophages from nonresponder strains. *J. Exp. Med.* 154:978–82

43. Kappler, J. W., Hunter, P. C., Jacobs, D., Lord, E. 1974. Functional heterogeneity among the T-derived lymphocytes of the mouse. I. Analysis by adult thymectomy. *J. Immunol.* 113:27–38

44. Kappler, J. W., Marrack, P. 1977. The role of H-2-linked genes in helper T-cell function. I. In vitro expression in B cells of immune response genes controlling helper T-cell activity. *J. Exp. Med.* 146:1748–63

45. Kappler, J. W., Marrack, P. 1978. The role of H-2 linked genes in helper T-cell function. IV. Importance of T-cell genotype and host environment in I-region and Ir gene expression. *J. Exp. Med.* 148:1510–22

46. Keller, D. M., Swierkosz, J. E., Marrack, P., Kappler, J. W. 1980. Two types of functionally distinct, synergizing helper T cells. *J. Immunol.* 124:1350–59

47. Katz, D. H. 1977. The role of the histocompatibility gene complex in lymphocyte differentiation. *Cold Spring Harbor Symp. Quant. Biol.* 41:611

48. Katz, D. H., Benacerraf, B. 1975. The function and interrelationship of T cell receptors, Ir genes, and other histocompatibility gene products. *Transplant. Rev.* 22:175

49. Katz, D. H., Graves, M., Dorf, M. E., Dimuzio, H., Benacerraf, B. 1975. Cell interactions between histoincompatible T and B lymphocytes. VII. Cooperative responses between lymphocytes are controlled by genes in the I region of the H-2 complex. *J. Exp. Med.* 141:263–68

50. Katz, D. H., Hamaoka, T., Benacerraf, B. 1973. Cell interactions between histoincompatible T and B lymphocytes.

II. Failure of physiologic cooperative interactions between T and B lymphocytes from allogeneic donor strains in humoral response to hapten-protein conjugates. *J. Exp. Med.* 137:1405

51. Katz, D. H., Hamaoka, T., Dorf, M. E., Maurer, P. H., Benacerraf, B. 1973. Cell interactions between histoincompatible T and B lymphocytes. IV. Involvement of the immune response (IR) gene in the control of lymphocyte interactions in responses controlled by the gene. *J. Exp. Med.* 138:734–39

52. Katz, D. H., Paul, W. E., Goidl, E. A., Benacerraf, B. 1971. Carrier function in anti-hapten immune responses. III. Stimulation of antibody synthesis and facilitation of hapten-specific antibody responses by graft versus host reaction. *J. Exp. Med.* 133:169

53. Katz, D. H., Skidmore, B. J., Katz, L. R., Bogowitz, C. A. 1978. Adaptive differentiation of murine lymphocytes. I. Both T and B lymphocytes differentiating in $F_1 \rightarrow$ parental chimeras manifest preferential cooperative activity for partner lymphocytes derived from the same parental type corresponding to the chimeric host. *J. Exp. Med.* 148:727–45

54. Kindred, B., Shreffler, D. C. 1972. H-2 dependence of cooperation between T and B cells. *J. Immunol.* 109:940

55. Kirov, S. M., Parish, C. R. 1976. Carrier-specific B cells play a role in the production of an antigen-specific T-cell-replacing factor. *Scand. J. Immunol.* 5:1155–62

56. Kruisbeek, A. M., Hathcock, K. S., Hodes, R. J., Singer, A. 1982. T cells from fully H-2 allogeneic (A → B) radiation bone marrow chimeras are functionally competent and host restricted, but are alloreactive against hybrid Ia determinants expressed on (AXB)F_1 cells. *J. Exp. Med.* 155:1864–69

57. Lewis, G. K., Goodman, J. W., Ranken, R. 1978. Activation of B cell subsets by T-dependent and T-independent antigens. In *Advances in Experimental Medicine and Biology*, ed. Z. Atassi, A. Staritsky, 98:339–56. New York: Plenum

58. Lonai, P., Puri, J., Bitton, S., Benneriah, Y., Givol, D., Hammerling, G. 1981. H-2-restricted helper factor secreted by cloned hybridoma cells. *J. Exp. Med.* 154:942–51

59. Marrack, P., Kappler, J. W. 1978. The role of H-2-linked genes in helper T-cell function. III. Expression of immune response genes for trinitrophenyl conjugates of Poly-L(Tyr,Glu)-Poly-D,L-Ala

—Poly-L-Lys in B cells and macrophages. *J. Exp. Med.* 147:1596–610

60. Marrack, P., Kappler, J. W. 1979. The role of H-2-linked genes in helper T-cell function. VI. Expression of Ir genes by helper T cells. *J. Exp. Med.* 149:780–85

61. McDougal, J. S., Cort, S. P. 1978. Generation of T helper cells in vitro. IV. F_1 T helper cells primed with antigen-pulsed parental macrophages are genetically restricted in their antigen-specific activity. *J. Immunol.* 120:445–51

62. Melchers, F., Andersson, J., Lernhardt, W., Schreier, M. H. 1980. H-2-restricted polyclonal maturation without replication of small B cells induced by antigen-activated T cell help factors. *Eur. J. Immunol.* 10:679–85

63. Miller, J. F. A. P., Gamble, J., Mottran, D., Smith, F. I. 1979. Influence of thymus genotype on acquisition of responsiveness in delayed-type hypersensitivity. *Scand. J. Immunol.* 9:29

64. Miller, J. F. A. P., Mitchell, G. F. 1968. Cell to cell interaction in the immune response. I. Hemolysis forming cells in neonatally thymectomized mice reconstituted with thymus or thoracic duct lymphocytes. *J. Exp. Med.* 128:801

65. Mitchell, G. F., Grumet, C., McDevitt, H. O. 1972. Genetic control of the immune response. The effect of thymectomy on primary and secondary antibody response of mice to poly-L-(Tyr,-Glu)-poly-D,L-Ala-poly-L-Lys. *J. Exp. Med.* 135:126

66. Mitchell, G. F., Miller, J. F. A. P. 1968. Cell to cell interaction in the immune response II. The source of hemolysin forming cells in irradiated mice given bone marrow and thymus or thoracic duct lymphocytes. *J. Exp. Med.* 128:821

67. Mitchell, G. F., Miller, J. F. A. P. 1968. Immunological activity of thymus and thoracic duct lymphocytes. *Proc. Natl. Acad. Sci. USA* 59:296

68. Mitchison, N. A. 1971. The carrier effect in the secondary response to hapten-protein conjugates. II. Cellular cooperation. *Eur. J. Immunol.* 1:18

69. Moller, G., ed. 1982. Interleukins and Lymphocyte Activation. *Immunol. Rev.* 63

70. Mond, J. J., Scher, I., Mosier, D. E., Blaese, M., Paul, W. E. 1978. T-independent responses in B cell defective CBA/N mice to Brucella abortus and to trinitrophenyl (TNP) conjugates of Brucella abortus. *Eur. J. Immunol.* 8:459

71. Morrissey, P. J., Boswell, H. S., Scher, I., Singer, A. 1981. Role of accessory cells in B cell activation. IV. Ia+ accessory cells are required for the in vitro generation of thymic independent type 2 antibody responses to polysaccharide antigens. *J. Immunol.* 127:1345–47

72. Mosier, D. E., Zitron, I. M., Mond, J. J., Ahmed, A., Scher, I., Paul, W. E. 1977. Surface immunoglobulin D as a functional receptor for a subclass of B lymphocytes. *Immunol. Rev.* 37:89

73. Mozes, E. 1974. The role of thymocytes and bone marrow cells in the genetic control of immune responsiveness to synthetic polypeptides. *Ann. Immunol.* 125:185–87

74. Mozes, E., Isac, R., Taussig, M. J. 1975. Antigen-specific T-cell factors in the genetic control of the immune response to poly(TYR,GLU)-Poly DL ALA—polyLYS. Evidence for T- and B-cell defects in SJL mice. *J. Exp. Med.* 141:703–7

75. Munro, A. J., Taussig, M. J. 1975. Two genes in the major histocompatibility complex control immune response. *Nature* 256:103–6

76. Nisbet-Brown, E., Singh, B., Diener, E. 1981. Antigen recognition. V. Requirement for histocompatibility between antigen-presenting cells and B cells in the response to a thymus-dependent antigen, and lack of allogeneic restriction between T and B cells. *J. Exp. Med.* 154:676–87

77. Paul, W. E., Shevach, E. M., Pickeral, S., Thomas, D. W., Rosenthal, A. S. 1977. Independent populations of primed F_1 guinea pig T lymphocytes respond to antigen-pulsed parental peritoneal exudate cells. *J. Exp. Med.* 145:618–30

78. Pierce, S. K., Klinman, N. R. 1976. Allogeneic carrier-specific enhancement of hapten-specific secondary B-cell responses. *J. Exp. Med.* 144:1254–62

79. Pierce, S. K., Klinman, N. R., Maurer, P. H., Merryman, C. F. 1980. Role of the major histocompatibility gene products in regulating the antibody response to dinitrophenyl poly(L-Gly^{55},L-Ala^{35}, L-Phe^9). *J. Exp. Med.* 152:336–49

80. Press, J. L., McDevitt, H. O. 1977. Allotype-specific analysis of anti-(TYR,-GLU)-ALA—LYS antibodies produced by Ir-1A high and low responder chimeric mice. *J. Exp. Med.* 146:1815–20

81. Raff, M. C. 1970. Role of thymus derived lymphocytes in the secondary

humoral immune response in mice. *Nature* 226:1257

82. Rajewsky, K., Schirrmacher, V., Nase, S., Jerne, N. K. 1969. The requirement of more than one antigenic determinant for immunogenicity. *J. Exp. Med.* 129:1131

83. Rosenthal, A. S., Barcinski, M. A., Blake, J. T. 1977. Determinant selection is a macrophage dependent immune response gene function. *Nature* 267:156–58

84. Rosenthal, A. S., Shevach, E. M. 1973. Function of macrophages in antigen recognition by guinea pig T-lymphocytes. *J. Exp. Med.* 138:1194

84a. Schimpl, A., Wecker, A. 1973. Stimulation of IgG antibody response in vitro by T-cell-replacing factor. *J. Exp. Med.* 137:547

85. Schreier, M. H., Andersson, J., Lernhardt, W., Melchers, F. 1980. Antigen-specific T-helper cells stimulated H-2 compatible and H-2 incompatible B-cell blasts polyclonally. *J. Exp. Med.* 151:194–203

86. Schwartz, R. H. 1978. A clonal deletion model for Ir gene control of the immune response. *Scand. J. Immunol.* 7:3–10

87. Shearer, G. M., Mozes, E., Sela, M. 1971. Cellular basis of the genetic control of immune responses. *Prog. Immunol.* 1:509–28

88. Shearer, G. M., Mozes, E., Sela, M. 1972. Contribution of different cell types to the genetic control of immune responses as a function of the chemical nature of the polymeric side chains (poly-L-PROLYL and poly-DL-ALA-NYL) of synthetic immunogens. *J. Exp. Med.* 135:1009–27

89. Shevach, E. M., Rosenthal, A. S. 1973. Function of macrophages in antigen recognition by guinea pig T lymphocyte. II. Role of the macrophage in the regulation of genetic control of the murine response. *J. Exp. Med.* 145:726

90. Shih, W. H., Matzinger, P. C., Swain, S. L., Dutton, R. W. 1980. Analysis of histocompatibility requirements for proliferative and helper T cell activity. T cell population depleted of alloreactive cells by negative selection. *J. Exp. Med.* 152:1311–28

90a. Singer, A., Asano, Y., Shigeta, M., Hathcock, K. S., Ahmed, A., Fathman, C. G., Hodes, R. J. 1982. Distinct B cell subpopulations differ in their genetic requirements for activation by T helper cells. *Immunol. Rev.* 64:137–60

91. Singer, A., Hathcock, K. S., Hodes, R. J. 1979. Cellular and genetic control of antibody responses. V. Helper T cell recognition of H-2 determinants on accessory cells but not B cells. *J. Exp. Med.* 149:1208–26

92. Singer, A., Hathcock, K. S., Hodes, R. J. 1980. Cellular and genetic control of antibody responses. VIII. MHC restricted recognition of accessory cells, not B cells, by parent-specific subpopulations of normal F_1 T helper cells. *J. Immunol.* 125:914–20

93. Singer, A., Hathcock, K. S., Hodes, R. J. 1981. Self-recognition in allogeneic radiation bone marrow chimeras. A radiation-resistant host element dictates the self specificity and immune response gene phenotype of T-helper cells. *J. Exp. Med.* 153:1286–301

94. Singer, A., Hathcock, K. S., Hodes, R. J. 1982. Self-recognition in allogeneic thymic chimeras. Self-recognition by T helper cells from thymus engrafted nude mice is restricted to the thymic H-2 haplotype. *J. Exp. Med.* 155:339–44

95. Singer, A., Hodes, R. J. 1982. Major histocompatibility complex restricted self-recognition in responses to trinitrophenyl-Ficoll. Adaptive differentiation and self-recognition by B cells. *J. Exp. Med.* 156:1415-34

96. Singer, A., Morrissey, P. J., Hathcock, K. S., Ahmed, A., Scher, I., Hodes, R. J. 1981. Role of the major histocompatibility complex (MHC) in T cell activation of B cell subpopulations. Lyb5⁺ and Lyb5⁻ B cell subpopulations differ in their requirement for MHC restricted T cell recognition. *J. Exp. Med.* 154:501–16

97. Sprent, J. 1976. Helper function of T cells depleted of alloantigen-reactive lymphocytes by filtration through irradiated F_1 hybrids. I. Failure to collaborate with allogeneic B cells in a secondary response to sheep erythrocytes measured in vivo. *J. Exp. Med.* 144:617

98. Sprent, J. 1978. Restricted helper function of F_1 hybrid T cells positively selected to heterologous erythrocytes in irradiated parental strain mice. I. Failure to collaborate with B cells of the opposite strain not associated with active suppression. *J. Exp. Med.* 147:1142–58

99. Sprent, J. 1978. Restricted helper function of F_1 hybrid T cells positively selected to heterologous erythrocytes in irradiated parental strain mice. II. Evidence for restrictions affecting helper cell induction and T-B collaboration,

both mapping to the K-end of the H-2 complex. *J. Exp. Med.* 147:1159–74

100. Sprent, J. 1978. Restricted helper function of $F_1 \to$ parent bone marrow chimeras controlled by K-end of H-2 complex. *J. Exp. Med.* 147:1838–42

101. Subbarao, B., Ahmed, A., Paul, W. E., Scher, I., Lieberman, R., Mosier, D. E. 1979. Lyb-7, a new B cell alloantigen controlled by genes linked to the Ig C_H locus. *J. Immunol.* 122:2279

102. Swierkosz, J. E., Rock, K., Marrack, P., Kappler, J. W. 1978. The role of H-2-linked genes in helper T-cell function. II. Isolation on antigen-pulsed macrophages of two separate populations of F_1 helper T cells each specific for antigen and one set of parental H-2 products. *J. Exp. Med.* 147:554–70

103. Tada, T., Takemori, T., Okumura, K., Nonaka, M., Tokuhisa, T. 1978. Two distinct types of helper T cells involved in the secondary antibody response: independent and synergistic effects of Ia⁻ and Ia⁺ helper T cells. *J. Exp. Med.* 147:446–58

104. Takatsu, K., Hamaoka, T. 1982. DBA/2Ha mice as a model of an X-linked immunodeficiency which is defective in the expression of TRF-acceptor site(s) on B lymphocytes. *Immunol. Rev.* 64:25

105. Takatsu, K., Tominaga, A., Hamaoka, T. 1980. Antigen-induced T cell-replacing factor (TRF). I. Functional characterization of a TRF-producing helper T cell subset and genetic studies on TRF production. *J. Immunol.* 124:2414–22

106. Takatsu, K., Tanaka, K., Tominaga, A., Kumahara, Y., Hamaoka, T. 1980. Antigen-induced T cell-replacing factor (TRF). III. Establishment of T cell hybrid clone continuously producing TRF and functional analysis of released TRF. *J. Immunol.* 125:264–65

107. Takatsu, K., Sano, Y., Tomita, S., Hashimoto, H., Hamaoka, T. 1981. Antibody against T cell-replacing factor acceptor site(s) augments in vivo primary IgM responses to suboptimal doses of

heterologous erythrocytes. *Nature* 292:360–62

108. Taussig, M. J., Mozes, E., Isac, R. 1974. Antigen-specific thymus cell factors in the genetic control of the immune response to poly-(TYROSYL,GLUTAMYL)-POLY-D, L-ALANYL—POLY-LYSYL. *J. Exp. Med.* 140:301–12

109. Tominaga, A., Takatsu, K., Hamaoka, T. 1980. Antigen-induced T cell-replacing factor (TRF). II. X-linked gene control for the expression of TRF-acceptor sites(s) on B lymphocytes and preparation of specific antiserum to that acceptor. *J. Immunol.* 124:2423–29

110. van Leeuwen, A., Goulmy, E., van Rood, J. J. 1979. Major histocompatibility complex-restricted antibody reactivity mainly, but not exclusively directed against cells from male donors. *J. Exp. Med.* 150:1075–83

111. von Boehmer, H., Hudson, L., Sprent, J. 1975. Collaboration of histocompatible T and B lymphocytes using cells from tetraparental bone marrow chimeras. *J. Exp. Med.* 142:989

112. Wylie, D. E., Sherman, L. A., Klinman, N. R. 1982. Participation of the major histocompatibility complex in antibody recognition of viral antigens expressed on infected cells. *J. Exp. Med.* 155:403

113. Yamashita, U., Shevach, E. M. 1978. The histocompatibility restrictions on macrophage T-helper cell interaction determine the histocompatibility restrictions on T-helper cell B-cell interaction. *J. Exp. Med.* 148:1171–85

114. Zinkernagel, R. M., Callahan, G. N., Althage, A., Cooper, S., Klein, P. A., Klein, J. 1978. On the thymus in the differentiation of "H-2 self-recognition" by T cells: evidence for dual recognition? *J. Exp. Med.* 147:882–96

115. Zinkernagel, R. M., Callahan, G. N., Althage, A., Cooper, S., Streilein, J. W., Klein, J. 1978. The lymphoreticular system in triggering virus plus self-specific cytotoxic T cells: evidence for T help. *J. Exp. Med.* 147:897–911

Ann. Rev. Immunol. 1983. 1:243–71

COMPLEMENT LIGAND-RECEPTOR INTERACTIONS THAT MEDIATE BIOLOGICAL RESPONSES

Douglas T. Fearon and Winnie W. Wong

Department of Medicine, Harvard Medical School, and the Department of Rheumatology and Immunology, Brigham and Women's Hospital, Boston, Massachusetts 02115

INTRODUCTION

Complement is the principal humoral effector system of inflammation. It is comprised of 18 proteins that are present in plasma and, at lesser concentrations, in extravascular fluids. The system can be activated through the classical pathway by immune complexes containing immunoglobulin IgG or IgM and through the alternative pathway by particulate material having certain chemical characteristics, such as the absence of sialic acid (23, 78, 93). During activation of these pathways, proteases that hydrolyze complement proteins are assembled and produce cleavage fragments that then serve as ligands for specific receptors on certain cells, including polymorphonuclear leukocytes, eosinophils, monocytes and macrophages, mast cells, and lymphocytes. These ligand-receptor interactions elicit cellular responses that may culminate in an inflammatory reaction.

The important role of the complement system in host resistance to microbial infection relates to its capacity to induce inflammation at a tissue site. Two types of ligands are generated during complement activation to serve this function: those bound to the microorganism, and those freely diffusible in the fluid phase. Examples of the former are C1q, C4b, and C3b, which may be considered bifunctional since they can attach both to the target of complement activation and to effector cells bearing the appropriate receptors. Ligand-receptor interactions of this type directly promote the

243

0732-0582/83/0410-0243$02.00

removal of microorganisms through phagocytosis by leukocytes. The diffusible ligands, which are low-molecular-weight cleavage peptides such as C3a and C5a, act as local hormones that cause receptor-mediated cellular responses of secretion, altered metabolism, and migration in the microenvironment. These ligand-receptor interactions indirectly promote the clearance of infectious agents by augmenting the accumulation of large numbers of phagocytic cells, thereby synergistically enhancing the effects of the complement ligands bound to microorganisms.

This review considers the molecular reactions that lead to formation of these two classes of ligands, their cellular receptors, and the cellular reactions induced by occupancy of the receptors.

MOLECULAR REACTIONS LEADING TO THE GENERATION OF COMPLEMENT LIGANDS FOR CELLULAR RECEPTORS

Ligands Attached to the Target of Complement Activation

C3   C3 is comprised of two polypeptide chains, the molecular weights of which are 120,000 (α-chain) and 75,000 (β-chain) (Figure 1). Cleavage of the α-chain at several sites yields six fragments that can interact with cellular receptors. All of these cleavage products except C3a and C3e are covalently bound to the target of complement activation and are therefore bifunctional ligands.

Activation of the classical or alternative complement pathways leads to the assembly of their respective C3-cleaving enzymes, C4b,2a and C3b,Bb, which hydrolyze the Arg 77-Ser 78 bond in the α-polypeptide of C3 to generate the M_r 9000 C3a peptide and the M_r 186,000 C3b molecule (Figure 1). This proteolytic reaction causes a rapid conformational change in C3b that is associated with an increase in its surface hydrophobicity (55) and exposure of the internal thioester (113) between cysteinyl and glutamyl residues in the α-polypeptide. This structure mediates the covalent attachment of C3b by transesterifying to nucleophilic sites on the surface bearing the C3 convertase, such as hydroxyl groups of carbohydrates (68). Covalent binding of C3b is relatively nonspecific as exemplified by the capacity of the molecule to fix on many different types of cell surfaces and on immune complexes.

The C3b is subject to conversion to iC3b by the serine protease, factor I, which cleaves a 3000-M_r peptide from the α-polypeptide, leaving segments of 68,000 and 43,000 M_r that are disulfide-bonded to the β-polypeptide (45, 67, 98). This reaction generally requires the presence of factor H, which binds to C3b and presumably induces the conformation that is suit-

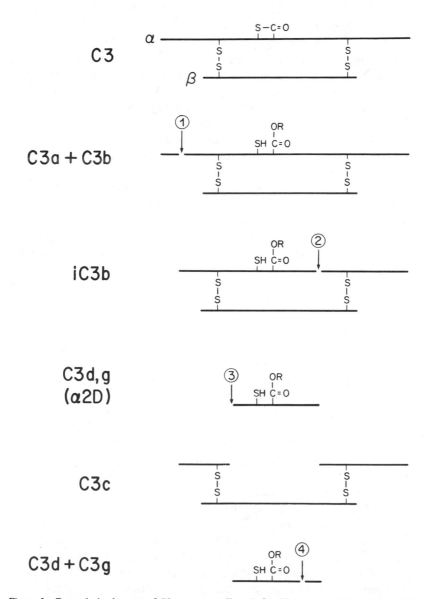

Figure 1 Proteolytic cleavage of C3 to generate ligands for C3 receptors. The position of the internal thioester capable of forming an ester linkage for covalent deposition is shown on the α-chain. The sequential degradation of native C3 to its biologically active fragments occurs in four steps. Step ① involves the cleavage of the α-chain at the designated site by the classical or alternative pathway convertases to form C3a and C3b. Step ② cleavage is mediated by factor I in the presence of cofactors H or C3b receptor and results in formation of iC3b. The source of the enzymes responsible for Step ③ cleavage to form C3d,g and releasable C3c is not well defined. Step ④ involves the cleavage of C3d,g by tryptic enzymes to form C3d and C3g. Not shown in this figure is the production of C3e, which is thought to be derived from C3c by unknown proteolytic reactions.

able for interaction with factor I (88, 125). Since there is impaired binding of factor H to C3b that is covalently attached to activators of the alternative pathway (23, 78), C3b at these sites is relatively resistant to factor I. The cofactor function of factor H can also be performed by purified, solubilized C3b receptor (21) and probably by membrane-associated receptor (73, 74). Since conversion of C3b to iC3b greatly reduces the capacity of the protein to bind to the C3b receptor, this receptor, in concert with factor I, has a mechanism for releasing bound ligand.

The iC3b is cleaved at another site in the α-polypeptide that is on the opposite side of the region containing the covalent binding site. This reaction leads to formation of two new fragments, C3d,g (also termed α2D) of 41,000 M_r, that remains bound to the complement-activating target, and C3c of 143,000 M_r, that is released (61, 102, 124). The enzyme responsible for the physiologic cleavage of iC3b has not been conclusively identified. Incubation of iC3b in serum at 37°C for up to 24 hr causes slow conversion of the protein to the C3d,g and C3c fragments (61). Although plasmin can hydrolyze the C3d,g fragment from iC3b (61, 80), it also splits C3d,g into the C3d and C3g fragments, which are not observed after treatment of iC3b with serum. Moreover, removal of plasmin/plasminogen from serum does not diminish the rate of cleavage of iC3b (61). The most likely candidate for the enzyme effecting the appropriate scission of iC3b is factor I by analogy to its capacity to generate the C4d and C4c fragments from C4b in the presence of C4-binding protein (27, 79). Factor H and the C3b receptor may be the necessary fluid phase and solid phase cofactors, respectively, since they have been reported to have weak affinity for iC3b (98), which is much less than their respective affinities for C3b. In support of this possibility is the recent demonstration of cleavage of C3c from cell-bound iC3b by factor I in the presence of membrane-associated C3b receptor (73). If this finding is confirmed, the C3b receptor has the unusual role of participating in the formation of two ligands, iC3b and C3d,g, for other receptors.

The final cleavage reaction involving C3 is that which generates the C3d fragment (M_r 33,000) by release of C3g (M_r 8000) from C3d,g (61). This can be effected by plasmin, trypsin, leukocyte elastase, and cathepsin G. Only recently have investigators distinguished between this fragment and the α2D or C3d,g fragment, so that much of the literature is confusing in this respect. In general, if the "C3d" was generated by serum enzymes, the fragment probably is C3d,g, whereas in those instances in which trypsin was utilized, the 33,000-M_r C3d fragment was formed. Since C3d,g is the final C3 fragment found in plasma of patients with complement activation and on erythrocytes of patients with autoimmune hemolytic anemia, generation of C3d may occur only in tissues where there is an influx of granulocytes secreting elastase and cathepsin G.

Clq   The first component of the classical pathway of complement activation is comprised of three subcomponents, Clq, Clr, and Cls, in a molar ratio of $1:2:2$, held together in a calcium-dependent complex. The globular heads of the Clq subcomponent bind to the Fc region of IgM and IgG in immune complexes, and the collagen-like region binds the Clr and Cls subcomponents. Following binding and activation of C1 by immune complexes, viruses or lipopolysaccharides, the Clr and Cls subcomponents are dissociated from Clq by the C1 inhibitor that forms the complex, Clr-Cls-(C1 inhibitor)$_2$ (108, 129). The Clq remains bound to the activator, and its collagen-like region is now exposed and capable of interacting with cellular Clq receptors.

FACTOR H   Factor H is the regulatory protein of the alternative complement pathway that binds to C3b and promotes its cleavage by factor I (23, 45, 67, 78, 88, 125). Recently, a receptor for this protein has been identified on leukocytes (63, 66, 87, 104). Since factor H is present in plasma at concentrations of 300 to 500 $\mu g/ml$ and factor H receptors are presumably not continuously occupied by this ligand, physical alteration in the molecule must occur that confers on it receptor-binding activity. Hexamers of factor H produced by concentrating the purified protein have been reported to bind to factor H receptors on Raji cells (63), suggesting that conversion of the protein into a multivalent ligand capable of multipoint attachment may be critical for its binding activity. The occurrence of soluble multimers of factor H in vivo has not been demonstrated, but a possible mechanism for polymerization may involve the binding of factor H to immune complexes bearing several molecules of C3b. In support of this possibility is the finding of deposition of factor H at tissue sites corresponding to the location of C3 antigen (10). Since the interaction between immune complex-bound C3b molecules and factor H probably represents the means for production of an effective ligand, factor H is considered as a bound ligand rather than a freely diffusible ligand.

Diffusible Ligands

Release of C5a from its parent molecule, C5, occurs by a mechanism similar to that already described for C3a. C5 is comprised of two polypeptides of M_r 112,000 (α-chain) and M_r 74,000 (β-chain). The classical and alternative pathway C5 convertases, C4b,2a,3b and C3b$_{(n)}$,Bb, respectively, cleave the α-chain at Arg 74- x 75 to release the M_r 11,000 glycopeptide, C5a (23, 78, 93). The C-terminal arginines of C3a and C4a are removed by serum carboxypeptidase N to produce the desArg derivatives of these peptides that lack biologic activity. In serum, this reaction occurs so rapidly that the time during which C3a and C4a can diffuse before losing their capacity to interact with their receptors on mast cells is probably brief. Serum carboxypeptidase N also releases the C-terminal arginyl residue of C5a to diminish greatly its capacity to interact with receptors on mast cells, but C5a desArg retains the capacity to bind to receptors on polymorphonuclear leukocytes and monocytes. Therefore, C5a desArg can diffuse from the site of its generation and can establish the concentration gradient necessary for its leukoattracting function.

A fragment of C3 termed C3e has been identified by its leukocytosis-inducing activity when injected into rabbits (35, 100). It can be generated by prolonged incubation of serum or by treatment of C3 with trypsin. This M_r 12,000 peptide is thought to be derived from the α-chain of C3c and thus represents a relatively late degradation product.

MEMBRANE RECEPTORS FOR COMPLEMENT-DERIVED LIGANDS

Specific cellular receptors for ligands derived from complement proteins (Table 1) have been identified in studies with assays that vary in their reliability. For example, the binding to a cell of antibody monospecific for a receptor provides strong evidence for the presence of that receptor. However, some otherwise distinct membrane proteins may share antigenic determinants (128). The saturable binding of a purified ligand might also be considered proof of the existence of a particular receptor on a cell, but some proteins, such as iC3b and C3d,g, apparently can bind to two different receptors on lymphocytes (97). The binding to cells of indicator particles, such as erythrocytes, coated with certain complement proteins is a frequently used assay for complement receptors, but it may yield misleading results: The ligand may be modified during interaction with cells (63, 97); more than one potential ligand may be present on the indicator particle (97); and classical studies of saturability and reversibility of binding are not possible. The most indirect assay involves the induction by a putative ligand of a biologic response from a cell or mixture of cells. For example, the capacity of C3a to cause contraction of the guinea pig ileum is considered indicative of C3a receptors on mast cells contained within this tissue. Other problems in experiments designed to demonstrate the presence of particular receptors on certain cell types are related to impurity of the ligand and inadequate identification of cells when mixed populations, such as lymphocytes, are utilized. For these reasons, this section reviews in some detail the methodology used for studies relating to the cellular distribution of receptors.

Receptors for Ligands Attached to the Target of Complement Activation

Clq RECEPTOR   The capacity of certain cells to bind soluble Clq which was labeled either with ^{125}I or indirectly with fluorescein (115, 116), and to form rosettes with erythrocytes bearing Clq (28), has indicated the presence of cellular Clq receptors. Human peripheral blood granulocytes, monocytes, most B lymphocytes, less than 8% of T cells (identified by formation of rosettes with sheep erythrocytes), and a small proportion of null cells bound fluoresceinated Clq (116). Inexplicably, in one study Clq-bearing erythrocytes formed rosettes only with B cells but not with monocytes or granulocytes (28). Scatchard analysis of the specific binding of ^{125}I-labeled Clq to granulocytes indicated the presence of 150,000–600,000 receptors per cell (116). Binding of radiolabeled Clq to mononuclear cells and of Clq-coated particles was inhibited by type I collagen and by pepsin-resistant Clq fragments; the intact molecular complex of Clqrs was not taken

Table 1   Cell types bearing complement receptors

Receptor	Cell type	Cellular response
Clq	Neutrophil	Respiratory burst
	Monocyte	
	Null cells	Enhanced ADCC
	B lymphocytes	
C4a, C3a	Mast cells	Secretion
C3b	Erythrocyte	Immune complex clearance? production of iC3b and C3d,g?
	Neutrophil	Enhanced phagocytosis; adsorptive endocytosis
	Monocyte/Macrophage	Same as neutrophil
	Eosinophil	Enhanced phagocytosis
	B lymphocyte	
	T lymphocyte	
	Glomerular podocyte	
CR2	B lymphocyte	
	Null cells?	Enhanced ADCC
CR3	Same as C3b receptor	Same as C3b receptor?
C3e	Neutrophils	Release from bone marrow
C5a	Mast cells	Secretion; leukotriene synthesis?
	Neutrophil	Chemotaxis; secretion increased stickiness; increased C3b receptor expression
	Monocyte/Macrophage	Chemotaxis; secretion; spreading leukotriene synthesis?
Factor H	B lymphocytes	Secretion of factor I; mitogenesis
	Monocytes	Respiratory burst
	Neutrophils	

up by these cells. Thus, the portion of the C1q molecule interacting with receptors was identified as the collagen-like tail region. The calcium-dependent association of C1q with platelets is reported to occur through C1s bound to the platelet and thus is distinct from binding mediated by C1q receptors (121).

The affinity of the C1q receptor on granulocytes and monocytes for monomeric C1q at physiological ionicity was so low that binding was reproducibly detected only in hypotonic media (115, 116). In the presence of a relative salt concentration of 0.09, the average equilibrium binding constant of the receptor for monomeric C1q on granulocytes was 7.6×10^6 M^{-1} (116). Since C1q would usually be presented to cells as an aggregate or array on immune complexes, multipoint attachment to C1q receptors would greatly increase the effective avidity and this increase must account for the rosetting of C1q-coated particles with leukocytes. Other than sensitivity of its function to trypsin, nothing is known of the physical nature of the receptor that binds C1q.

C3 RECEPTORS   Four different proteolytic cleavage products of C3—C3b, iC3b, C3d,g, and C3d—bound to the target of complement activation interact with three types of cellular receptors. A generally accepted terminology for these receptors is evolving with increasing knowledge of the biochemistry of their ligands. In this review, a receptor is named for its preferred ligand, if known. Thus, the immune adherence receptor is designated the C3b receptor rather than complement receptor type 1 (CR1). The less descriptive terms, CR2 and CR3, are used for the other C3 receptors because there is some uncertainty in the identification of their biologically relevant ligands.

C3b receptor   The C3b receptor (CR1) was discovered almost 30 years ago when bacteria which had been incubated in serum that contained specific antibody and active complement were shown to adhere to human erythrocytes (83). The immune adherence receptor of the erythrocyte was shown subsequently to be specific for C3b on the immune complex (37) and to reside on other cell types, including neutrophils, eosinophils, monocytes, macrophages, B and some T lymphocytes, mast cells, and glomerular podocytes (7, 30, 43, 59, 69, 95, 118, 126). A population of large granular lymphocytes with natural killer activity and identified by the monoclonal antibody, anti-HNK-1, lacks C3b receptors (114). Although the presence of C3b receptors on these various cells is usually detected in rosette assays with sheep erythrocytes that bear C3b, monospecific polyclonal and monoclonal antibodies to the human C3b receptor have now become available (19, 22, 53).

The C3b receptor was first isolated from human erythrocytes and was shown to be a glycoprotein of M_r 205,000 to 250,000 when assessed by polyacrylamide gel electrophoresis in the presence of sodium dodecyl sulfate (21). The higher estimate of M_r is based on the more recent observation that the purified C3b receptor migrated slightly more slowly than erythrocyte band 1, which has a M_r of 240,000. Monospecific rabbit antibody to this glycoprotein blocked C3b receptor function on erythrocytes, neutrophils, monocytes, and B and T lymphocytes and immunoprecipitated from each of these cell types a single membrane protein identical to the glycoprotein initially isolated from erythrocytes (19, 22, 126). This polyclonal anti-C3b receptor also bound exclusively to glomerular podocytes in renal biopsy specimens (59). Thus, in man the C3b receptor of each of these cells represents a common molecular entity. This conclusion was supported by other studies with monoclonal antibodies to the C3b receptor (53).

Two forms of the C3b receptor differing in their mobility on polyacrylamide gel electrophoresis in sodium dodecyl sulfate have recently been identified. This structural polymorphism represents a stable phenotypic characteristic inherited as an autosomal codominant trait and expressed by receptors on erythrocytes, neutrophils, and mononuclear leukocytes. In the normal population, 65% have only receptors of M_r 250,000, 3% have only receptors of a higher molecular weight, and 32% have both forms of receptors (W. Wong and D. Fearon, unpublished observation).

The number of C3b receptors on cells has been estimated by measuring the uptake at saturating concentrations of ^{125}I-labeled dimeric C3b and anti-C3b receptor (3, 127). Erythrocytes have an average of only 500–600 receptors per cell. The number of receptors expressed by erythrocytes is genetically regulated by two autosomal codominant alleles determining high and low receptor numbers, respectively. Individuals homozygous for the high allele have erythrocytes with an average of 800 receptors per cell, those homozygous for the low allele have cells with 250 receptors, and heterozygotes' erythrocytes have 500 receptors per cell. Of normal individuals, 34 and 12% are homozygous for the high and low alleles, respectively (127). A markedly different distribution of these phenotypes was found among patients with systemic lupus erythematosus, of whom only 5 and 53% were homozygous for the high and low alleles, respectively. Thus, the low numbers of C3b receptors on erythrocytes of lupus patients was found to be inherited, as had been suggested in another study (76), rather than acquired (53), and may represent a predisposing factor in the development of this disease. An additional abnormality of the C3b receptor in lupus is its absence from glomerular podocytes in proliferative glomerulonephritis (59).

Two studies of purified neutrophils have estimated the presence of 20,000–40,000 C3b receptors per cell (3, 22), whereas a third found 140,000 receptors per cell (53). Purified monocytes and B lymphocytes have approximately the same number of receptors as neutrophils, whereas the 5–25% of peripheral blood T cells that express this glycoprotein have fewer receptors (3, 126). Unstimulated neutrophils and monocytes in peripheral blood express only 10–20% of their total number of receptors, and the process of purifying these cells causes temperature-dependent increases of up to eightfold in the number of receptors present on the plasma membrane. For example, neutrophils purified at 4°C expressed only 5000 C3b receptors per cell; incubation of the purified cells at 37°C for 15 min led to the expression of 38,000 receptors per cell. Addition to whole blood of C5a desArg in nanomolar concentrations also caused neutrophils to increase their plasma membrane expression of receptors rapidly in a dose-response manner (24). Thus, a minority of C3b receptors of unstimulated neutrophils and monocytes resides on the plasma membrane while the majority is present in latent, intracellular compartments that can be recruited to the plasma membrane. This observation explains the capacity of chemotactic factors to enhance the formation of rosettes between these cells and sheep erythrocytes (58). In contrast, peripheral blood B lymphocytes express 30,000–40,000 receptors per cell in the absence of stimulation by purification procedures or chemotactic factors (24).

The ontogeny of C3b receptors on human B lymphocytes and neutrophils has been studied. The receptor is present on only 15% of large pre-B cells and 35–50% of small pre-B cells in the bone marrow. Although 60–80% of immature B cells in bone marrow and fetal liver and all peripheral blood mature B cells express C3b receptors, maturation to plasma cells is associated with the loss of these receptors (114). The gradual appearance of C3b receptors during B cell development is consistent with its absence in several lymphoblastoid cell lines and on most B cells of patients with chronic lymphatic leukemia arrested early in differentiation. C3b receptors on neutrophils appear later in development than do CR3 receptors (96).

Equilibrium binding studies, with dimeric C3b either found in purified preparations of the protein or produced by chemical cross-linking, demonstrated that the C3b receptor on erythrocytes, neutrophils, monocytes, and B lymphocytes had an apparent association constant for these ligands of $2–6 \times 10^7$ M^{-1} (3). These experiments probably underestimated the affinity of the receptor for the bivalent ligand since they did not determine the percentage of the ligand capable of specific binding. When the active bindable fraction of dimeric C3b was purified by specific elution from C3b receptors and was reassessed for uptake by receptors on erythrocytes, an association constant of 10^9 M^{-1} was obtained (D. Fearon, unpublished observation). The affinity of the receptor for monomeric C3b was estimated to be only 2×10^6 M^{-1} (3).

The form of C3 recognized by the C3b receptor also can be generated by hydrolysis of the thioester without cleavage of the C3a peptide from the α-polypeptide chain, indicating that the conformational changes occurring in C3 following thioester hydrolysis are similar to those induced by cleavage of the α-polypeptide chain (6). Although particles bearing iC3b do not form C3b receptor-dependent rosettes under conditions of physiological ionicity, rosette formation has been observed under conditions of low ionic strength, which suggests the receptor has weak affinity for this form of C3 (97). This finding may relate to the possible cofactor role of the C3b receptor in the proteolytic processing by factor I of iC3b to C3c and C3d,g (73, 74). High concentrations of C3c, but not C3d, inhibit formation of rosettes between human erythrocytes and particles bearing C3b, a finding that maps the receptor binding site in C3b to its C3c region (99).

Although the preferred ligand for C3b receptors is C3b, the same receptors can also mediate uptake of particles bearing C4b (16). Receptor affinity for C4b is considered to be lower than for C3b, since the purified receptor is only one tenth as efficient in dissociating C2a from C4b as it is in dissociating Bb from C3b (86).

Relatively less is known about C3b receptors in species other than man. Generally, they reside on the same cell types, with possible exceptions of erythrocytes and glomerular podocytes, which appear to lack receptors in non-primate species, and platelets, which appear to have this function, in contrast to man (82). A C3b-binding protein of M_r 64,000 has been isolated from rabbit alveolar macrophages, and a membrane protein of M_r 60,000–63,000 on murine leukocytes cross-reacts with anti-human C3b receptor, but these proteins have not been shown to mediate C3b receptor function (105, 128).

C3 receptor type 2 (CR2)   Indicator particles bearing C3d form rosettes with essentially all Raji and Daudi lymphoblastoid cells, approximately one third of sheep erythrocyte-negative peripheral blood lymphocytes, and most of the tonsil B lymphocytes (19). Thus, in peripheral blood, approximately 50% of the B cells carry only C3b receptors, and the rest express both C3b receptors and CR2; tissue B lymphocytes apparently express both receptors. T lymphocytes, monocytes, and neutrophils are generally considered not to have CR2, based on absence of staining by rabbit anti-CR2 and absence of formation of rosettes with particles coated with C3d, although one report found C3d-dependent rosette formation with monocytes (11, 65). In addition, the capacity of target-bound C3d to augment antibody-dependent cellular cytotoxicity suggests that null cells also have CR2 (92).

The CR2 isolated from culture supernatants of Raji cells was a glycoprotein of M_r 72,000, and rabbit antibody to the purified receptor immunoprecipitated from Raji and BF lymphoblastoid cells a single protein of the same molecular weight as the immunogen. This antibody blocked all C3d-dependent and some iC3b-dependent rosette formation, indicating that the C3d region is sufficiently exposed in iC3b for interaction with CR2 (19, 64). Further studies on ligand specificity have indicated that CR2 is capable of binding all three degradation fragments of C3b, which include iC3b, C3d,g, and C3d (98). Thus, CR2 shares some ligands with CR3, but it is the only receptor capable of binding C3d. The number of receptors on cells and their affinities for these ligands is not known.

C3 receptor type 3 (CR3) The existence of a third type of receptor for a C3 fragment covalently bound to the target of complement activation was suggested by the capacity of sheep erythrocytes bearing iC3b to form rosettes with monocytes, neutrophils, some lymphocytes, and glomerular podocytes (11, 19, 22). The inability of antibodies specific for the C3b receptor and for CR2 to block iC3b-dependent rosette formation of neutrophils, monocytes, and all of the lymphocytes must be considered strong evidence for the presence of this additional receptor (19, 22).

Reports conflict concerning the forms of C3 capable of binding to CR3. In one study with ^{125}I-labeled C3, sheep erythrocytes bearing only iC3b formed rosettes with monocytes and glomerular podocytes (11). Other investigators have prepared indicator particles by treatment of C3b on sheep erythrocytes with purified factors H and I under conditions generally considered to produce iC3b and little, if any, C3d,g. These particles have been shown to adhere to neutrophils and monocytes, implying again that the receptor is capable of binding iC3b (5, 22, 107). However, a recent report comparing the products of treatment of fluid phase and particle-bound C3b with factors H and I found that the former reaction ended at the iC3b stage, whereas the products of the latter reaction were iC3b and an uncharacterized polypeptide of M_r 41,000, which was slightly smaller than the M_r 43,000 fragment of the α-polypeptide disulfide bonded to the β-chain in iC3b. Although this M_r 41,000 fragment resembles the C3d,g fragment by size, this study did not attempt to demonstrate their common identity (98). When iC3b and a mixture containing iC3b and the M_r 41,000 fragment were each assessed for binding to CR3, only the mixture had this function, which suggests the fragment resembling C3d,g was the ligand for this receptor (97, 98).

Direct evidence for the capacity of C3d,g to bind to receptors on neutrophils has been obtained with the purified, soluble form of the protein. C3d,g isolated from aged human serum was reversibly bound in a saturable fashion by human neutrophils, and binding was not inhibited by anti-C3b receptor antibody. The number of receptors for C3d,g ranged from 40,000–60,000 per cell, and the association constant of the receptor for this ligand was $0.5-1 \times 10^7$ M^{-1} (D. Vik and D. Fearon, unpublished observation). The receptor responsible for this interaction is probably CR3, but these studies did not address the question of whether this receptor also binds iC3b. The CR3 could be designated the C3d,g receptor if it can only bind C3d,g.

The lymphocyte subpopulation expressing CR3 includes B cells, although some CR3-positive cells were reactive with the monoclonal antibodies anti-Leu-1 and OKM-1 (97). The former recognizes a T-cell antigen, and the latter recognizes determinants on neutrophils, monocytes, and null cells that have natural killer activity. Less than one third of peripheral blood B lymphocytes and less than half of tonsil B lymphocytes express CR3, indicating CR3 resides on only a subpopulation of B cells. Variable proportions of Raji, Daudi, and BF lymphoblastoid cells express CR3 (19).

The membrane protein named CR3 has recently been suggested to be identical to or spatially associated with the Mac-1 antigen. Rat monoclonal anti-Mac-1 inhibited formation of rosettes between sheep erythrocytes bearing human iC3b and human neutrophils and between antibody-sensitized sheep erythrocytes treated with whole mouse serum complement and mouse peritoneal macrophages (5). There was no inhibition of C3b receptor function. Anti-Mac-1 recognizes an antigen present on murine and human monocytes, macrophages, neutrophils, and a subpopulation of lymphocytes having natural killer activity. However, Mac-1 does not reside on peripheral blood or tissue B lymphocytes (4). This disparity between the different subsets of lymphocytes with functionally detectable CR3 and which express Mac-1 antigen must be resolved before the latter can be considered the CR3.

FACTOR H RECEPTOR   The presence of receptors for factor H on leukocytes has been demonstrated by the adherence to these cells of particles coated with factor H and by the binding of radiolabeled factor H. Two groups of investigators measuring the binding of labeled factor H to Raji cells reported that 4.2×10^6 (63) and 4×10^4 (104) molecules of H were bound per cell, respectively, at equivalent inputs of the ligand. In the former study, saturation was achieved with an approximate input of 3×10^{-7} M of H (63). Subsequently, 157,000 H receptors were estimated to be present on Raji cells by measuring the uptake of a rabbit anti-idiotypic anti-human factor H antibody that had specificity for factor H receptors (66). The reasons for these disparate results are not apparent, although explanations may include the varying degrees of aggregation of the ligand leading to an overestimate of receptors, or the varying receptor number expressed on Raji cells depending on culture conditions and phase of growth cycle. The anti-idiotypic antibody immunoprecipitated from biosynthetically labeled Raji cells a cross-reactive protein comprised of polypeptides of M_r 100,000 and 50,000 in their unreduced states and M_r 50,000 when disulfide-reduced (66). A similar protein in the spent culture media of these cells was adsorbed by agarose to which factor H had been coupled.

The B lymphoblastoid lines Raji and BF, and peripheral blood B cells have been shown to express factor H receptors, whereas a T cell line, HSB, did not (63). Another study found that 94% of tonsil lymphocytes formed rosettes with particles bearing factor H (104), indicating that T cells may also express these receptors. However, lower percentages of peripheral blood lymphocytes, "nearly all" of which had surface immunoglobulin (63, 104), were capable of binding factor H, indicating a possible difference between circulating and tissue T cells. Factor H-coated particles have also been found to bind to granulocytes and monocytes (104), but since the particles used in these experiments also carried C3b, rosette formation may have been mediated by C3b receptors on these phagocytic cells. Moreover, C3b noncovalently modified by factor H has been suggested to bind to novel C3 receptors on Raji cells that are distinct from C3b receptors CR2 and CR3 (87), a finding that further complicates analyses of factor H receptors by use of particles bearing both C3b and H.

Receptors for Diffusable Complement Ligands

C3a AND C4a RECEPTOR   Despite the early recognition that C3a can induce contraction of the guinea pig ileum, increase vascular permeability, and cause degranulation of mast cells (15, 18), and despite the relative ease with which C3a can be purified (51), studies of the binding of C3a to various cell types are incomplete. Therefore, the conclusion that specific receptors for C3a reside on mast cells is still based on indirect experiments. Purified C3a induces secretion of histamine from purified mast cells (57). The guinea pig ileal contraction caused by this peptide is primarily histamine-dependent, and this tissue exhibits tachyphylaxis, or desensitization, to C3a (15, 18). Therefore, the contractile response of the guinea pig ileum is considered secondary to release of histamine caused by interaction of C3a with specific receptors on mast cells in this tissue. No equilibrium-binding studies of the interaction of C3a with mast cells of any species have been performed, but an estimate of the affinity of this putative cellular receptor for C3a can be made based on the capacity of 10^{-8} M human C3a to induce contraction of the guinea pig ileum (15). This concentration of C3a would be achieved by cleavage of less than 1% of C3 in plasma. Secretion of histamine by basophils in preparations of human peripheral blood leukocytes required concentrations of human C3a in the range of 10^{-7}–10^{-6} M (38); the basis for the 10- to 100-fold higher concentration of C3a necessary for the human basophil reaction than for the guinea pig ileal response is not known. The biologic activities of C3a are abolished when the C-terminal arginine is removed, and this is presumed to indicate that C3a desArg cannot bind to the C3a receptors. The primary structural features of C3a essential for its biological activity have been examined by use of synthetic oligopeptides (52).

Autoradiographic studies that involved incubation of ^{125}I-labeled C3a with human peripheral blood leukocytes demonstrated binding of the peptide to eosinophils, neutrophils, and monocytes (38). However, dose-response curves describing the quantitative uptake of ^{125}I-labeled C3a by preparations of neutrophils and monocytes were convex to the abscissa rather than concave, as would have been expected for a saturable reaction, and only 2–3% of the ligand was bound. Therefore, additional studies of the binding of C3a by neutrophils, eosinophils, and monocytes are necessary to establish the presence of C3a receptors on these cells.

High concentrations of C4a induce contraction of the guinea pig ileum, and the contractile response can be abolished by prior incubation of the tissue with C4a or C3a (41). C4a desArg has no activity. This low affinity interaction of C4a with C3a receptors on tissue mast cells probably does not reflect a physiological function of C4a since the micromolar concentrations of C4a required for the response could be achieved only by cleavage of almost the entire content of C4 in plasma. However, this finding reveals a functional homology between C3a and C4a, which also share similar structural features (41).

C5a RECEPTOR   Cell types that exhibit biological responses to C5a are mast cells, basophils, neutrophils, eosinophils, monocytes, and macrophages. Evidence that specific receptors for C5a mediate the response of mast cells and basophils is indirect and is similar to that for C3a receptors on these cells. Treatment of the guinea pig ileum with C3a or C5a induces tachyphylaxis to the same but not the other peptide, suggesting that the receptors on mast cells for these two peptides are distinct. Concentrations of C5a of 10^{-10}–10^{-9} M cause contraction of the guinea pig ileum, and cleavage of the C-terminal arginine from C5a decreases its activity by at least 1000-fold (32). Since the serum concentration of C5, the parent molecule, is approximately 10^{-6} M, C5a desArg might not be present in vivo in concentrations sufficient for interacting with these receptors. The impaired binding of human C5a desArg is caused in part by its single oligosaccharide, because deglycosylation of the peptide augmented its biologic activities (31).

In contrast to the indirect evidence for C5a receptors on mast cells, the presence of specific receptors on neutrophils (14) and macrophages (13) has been established. Human neutrophils have a saturable binding capacity for ^{125}I-labeled C5a, each cell taking up a maximum of 100,000–300,000 molecules of the glycopeptide with an association constant of 0.3–1×10^9 M^{-1}. Similar numbers of receptors on murine peritoneal macrophages bound human C5a with a comparable affinity. C5a desArg also binds to C5a receptors on neutrophils and macrophages as indicated by its capacity to inhibit competitively the uptake of ^{125}I-labeled C5a when present in 100-fold higher concentrations (13, 14, 31). Direct analysis of the binding of ^{125}I-labeled C5a desArg to neutrophils demonstrated the presence of only 40,000–50,000 receptors per cell (24). The difference in the numbers of receptors binding C5a and C5a desArg probably was caused by different experimental conditions: Although binding of C5a desArg was performed at 4°C, C5a was incubated with neutrophils at 24°C, which may have permitted internalization of the ligand or expression of additional receptors cycling between the plasma membrane and intracellular compartments. Saturation of 50% of the neutrophil receptors occurred with 2.5×10^{-8} M C5a desArg, a concentration that would be achieved by cleavage of less than 5% of plasma C5 (24). It is not known whether the C5a receptors on mast cells are similar or identical to those present on neutrophils and macrophages.

C3e RECEPTOR   Incubation of radioiodinated human C3e with rabbit blood for 1 hr at 37°C led to the association of 640,000 molecules of the peptide per polymorphonuclear leukocyte and 73,000 molecules per mononuclear leukocyte (35). However, the specificity, saturability, and reversibility of binding was not examined. More recently, saturable uptake of human C3e by human neutrophils was reported and the presence of 60,000 receptors per cell was estimated (33). The affinity of these cellular receptors for the peptide was not stated.

BIOLOGICAL FUNCTIONS OF COMPLEMENT RECEPTORS

Receptors for Complement Ligands that are Attached to the Target of Complement Activation

C1q RECEPTOR   Interaction of human neutrophils with latex beads coated with purified C1q stimulated a respiratory burst as measured by chemiluminescence and NADPH-oxidase-dependent hexose monophosphate shunt activity (117). No superoxide anion was detected. The cellular response to the C1q-coated beads was partially inhibited by free monomeric C1q. These findings indicate that either C1q receptors on neutrophils, when cross-linked by particle-bound C1q, mediated a respiratory response of the cells or, since latex beads have some intrinsic stimulatory capacity, the receptor-ligand interaction augmented the response to latex beads by promoting their attachment to the neutrophils.

Chicken erythrocytes bearing C1q were lysed by Raji cells and Daudi cells in a 20-hr cytolytic assay (34). For cytolysis to occur, the target cells did not have to be sensitized with antibody, suggesting that the reaction was mediated by C1q receptors rather than Fc receptors on the lymphoblastoid cells. C1q also augmented the cytolysis of antibody-sensitized chicken erythrocytes by peripheral blood lymphocytes purified by rosette formation with sheep erythrocytes. Although the normal effector cells were not characterized, the presence of C1q receptors on null cells lacking both B and T cell markers (116) supports the possibility that some cells mediating antibody-dependent cellular cytotoxicity may express these receptors. The function of C1q receptors on B cells is not known.

C3 RECEPTORS

C3b receptor The C3b receptor of erythrocytes is a relatively static membrane protein in that it does not internalize bound ligand or become attached to cytoskeletal elements following cross-linking by multivalent ligands as do receptors on neutrophils and monocytes. However, greater than 85% of the C3b receptors in peripheral blood are found on erythrocytes, so that immune complexes bearing C3b present in the intravascular space will most likely bind to these cells rather than to leukocytes. Recently, a role for erythrocytes in the clearance of such complexes has been suggested by the finding that IgG-containing immune complexes administered intra-arterially to primates are taken up by these cells and are then removed as they traverse the liver through the portal system (49). The mechanism by which the complexes are released from erythrocytes and transferred to Fc receptor-bearing cells in the liver, which are presumably Kupfer cells, is not known and may be related to the conversion of complex-associated C3b to iC3b and C3d,g by factor I, with the C3b receptor serving as the cofactor.

Additional biologic functions of erythrocyte C3b receptors are related to their factor H-like activity. This receptor was initially isolated from erythrocyte membranes because it dissociated Bb from the C3b,Bb bimolecular complex and served as a cofactor for conversion of C3b to iC3b by factor I (21). The receptor had analogous functions on C4b,2a and C4b (36, 54). Recent studies have also suggested that the C3b receptor mediates the cleavage by factor I of cell-bound iC3b to C3d,g and C3c. (73, 74). Thus, erythrocyte C3b receptors may protect cells in the circulation from attack by the classical or alternative pathways, and they may facilitate the conversion of C3b on immune complexes to iC3b and C3d,g. The latter capability, which would be shared by all cells with C3b receptors, creates the ligands for CR2 and CR3, thereby recruiting the functional participation of these receptors.

C3b receptors on neutrophils, monocytes, and macrophages have long been recognized as enhancing phagocytosis of IgG-bearing particles (20). The synergy of C3b receptor-mediated binding of the target and Fc receptor-mediated phagocytosis suggests a cooperative interaction between the two receptors. The bidirectional co-capping of cross-linked C3b receptors and Fc receptors on human neutrophils (R. Jack and D. Fearon, unpublished observations) indicates that multivalent interaction of particle-bound C3b with receptors on a localized area of the plasma membrane would recruit Fc receptors to the same topographical site and enhance the likelihood of interaction of Fc receptors with particle-bound IgG. The capacity of C3b receptors on monocytes and macrophages to mediate phagocytosis of sheep erythrocytes coated with C3b is dependent on the state of activation or maturation of these cells. Culture of human peripheral blood monocytes for 7 days leads to the acquisition by these cells of a capacity to internalize C3b-coated particles in the absence of IgG (85). In addition, C3b receptors on both neutrophils and monocytes when cross-linked can rapidly internalize soluble ligands by adsorptive pinocytosis through coated pits and coated vesicles (1, 25). Thus, the C3b receptor on these phagocytic cells may mediate several responses, depending on the physical state of material bearing the ligand and on the maturational stage of the cells.

One study reported that serum-treated Sepharose from which all IgG had been removed stimulated a respiratory burst in neutrophils and secretion of granule contents (39). However, these particles probably had iC3b and possibly C3d,g in addition to C3b, so that the cellular response may not have been initiated by C3b receptors. Indeed, a subsequent study utilizing sheep erythrocytes bearing only C3b found that these cells adhered to but did not stimulate neutrophils (84). Direct cross-linking of C3b receptors on neutrophils with F(ab')$_2$ anti-C3b receptor does not induce secretory or respiratory responses (D. Fearon, unpublished observation). Although a recent study stating that dimeric C3b caused release of histaminase from human neutrophils appears to contradict these findings, the concentration of dimeric C3b required to elicit the secretory reaction was 60- to 300-fold more than that which caused saturation of 50% of the C3b receptors, raising the possibility that the effect was mediated by a minor contaminant (75).

The function of C3b receptors on B lymphocytes is unknown. Addition of C3b to guinea pig and human lymphocytes caused secretion of a lymphokine-mediating chemotaxis of monocytes, but the cell source of this activity was not directly shown to be the B cell (60). An early report of mitogenesis of murine spleen cells induced by C3b is now attributed to factor H contaminating the preparation of C3b (46). C3b receptors have recently been found to be present in 15–25% of peripheral blood T lymphocytes. Whether or not these cells are a specialized subset has not been determined, and they are present in both subpopulations of T cells defined by the OKT4 and OKT8 monoclonal antibodies (126). The extensive literature describing depressed antibody responses and development of memory B cells to T-dependent antigens in mice rendered C3-deficient by administration of cobra venom factor cannot be interpreted at the level of cellular receptors (90). Some role for C3 in the localization of antigen to lymphoid follicles has been established, but the particular C3 receptor involved in this function is not known (89).

C3 receptor type 2 (CR2) F(ab')$_2$ antibody to CR2 has been reported to inhibit the proliferative responses of human lymphocytes induced by pokeweed mitogen, antigen, and mixed lymphocyte reactions, but not by phytohemagglutinin or concanavalin A, thus confirming an earlier finding that soluble C3d inhibited antigen-induced proliferation of human lymphocytes (62, 103). Since these assays reflect primarily T rather than B cell responses, and only B cells have been considered to express CR2, the cell circuits involved in the CR2-mediated suppression may be complex. However, enhanced killing by human peripheral blood lymphocytes was observed when antibody-sensitized target cells were also coated with C3d, suggesting that CR2 also resides on some non-B lymphocytes.

C3 receptor type 3 (CR3) Killing of target cells in lymphocyte-mediated antibody-dependent cellular cytotoxicity is enhanced when the targets carry iC3b (92). This finding does not unambiguously demonstrate that the killer cell carried CR3, since iC3b binds also to CR2. The CR3 on human monocytes cultured for 7 days can mediate the phagocytosis by these cells of iC3b-coated sheep erythrocytes (85). Earlier studies that demonstrated that lymphokine-stimulated mouse peritoneal macrophages ingest IgM- and whole mouse serum-sensitized sheep erythrocytes probably were also examining CR3-dependent phagocytosis (42). Human neutrophils incubated with 10^{-8}–10^{-7} M of soluble C3d,g, which probably binds to CR3, selectively secreted specific granule contents, but not azurophil granule enzymes, in the presence or absence of cytochalasin B. These concentrations of C3d,g are similar to those causing 50% saturation of cellular receptors (D. Vik and D. Fearon, unpublished observation). In addition, sheep erythrocyte ghosts bearing iC3b have recently been found to induce a respiratory burst by neutrophils, as assessed by chemiluminescence (107).

FACTOR H RECEPTOR   Binding of presumably polymeric factor H to B lymphoblastoid cell lines and to peripheral blood lymphocytes and tonsil lymphocytes induced release of factor I, which was detected antigenically by immunodiffusion against goat anti-factor I and functionally by conversion of C3b to iC3b in the presence of factor H (63). Release of factor I from these cells was calcium- and energy-dependent (63). This finding has been considered by some to account for the capacity of factor H to induce binding of sheep erythrocytes bearing C3b to Raji cells, which have few, if any, C3b receptors but which express receptors for iC3b/C3d,g and C3d (63, 66). As others have noted, however, Raji cells may bind particles bearing factor H linked to C3b directly by their factor H receptors (104).

Human factor H has recently been shown to induce proliferation of mouse spleen cells (44). The possibility that this effect was mediated by contaminating lipopolysaccharide was excluded since H had a comparable effect on spleen cells from C3H/He mice. Removal of spleen cells that formed rosettes with erythrocytes treated with antibody and complement diminished the proliferative response, suggesting that at least some of the responding cells also had receptors for fragments of C3. Approximately two thirds of the responding blast-like spleen cells could be stained with anti-IgM. As factor H frequently contaminates preparations of C3 and C3b, these authors considered these findings to be the basis for an earlier report of stimulation of murine B cells by human C3b (46). Comparable experiments with human lymphocytes have not been reported.

In support of the finding of factor H receptors on human monocytes (104), incubation of these cells with factor H induced a rapid chemiluminescent response (106). The concentration of factor H eliciting this response was in the appropriate range of 10^{-7} M since this receptor-ligand interaction has an association constant of 3×10^6 M^{-1} (63). The capacity of factor H to induce the respiratory burst in neutrophils was not examined, although this group of investigators had reported the presence of factor H receptors on this cell type (104).

Receptors for Diffusable Complement Ligands

C3a RECEPTOR   The capacity of H1 blocking agents to prevent the contractile response of the guinea pig ileum to C3a had been taken as evidence that the only receptor-mediated biologic response of C3a was the secretion of histamine-containing granules by mast cells. However, some investigators had observed a histamine-independent effect of complement anaphylatoxins on the guinea pig ileum (8), and recently, the contractile response of guinea pig lung strips induced by 10^{-8}–10^{-7} M C3a was shown to be relatively unaffected by H1 or H2 blocking agents (110). It is not known what cell type in this tissue is responding to C3a, nor has the mediator been identified. However, by analogy to the effects of C5a and of C5a desArg on guinea pig lung tissue, the non-histamine mediator may be one of the leukotrienes (94, 109).

C3a has recently been shown to suppress certain lymphocyte functions. This first report suggesting this function for C3a found that addition of a low-molecular-weight C3-derived fragment to human peripheral blood lymphocytes in serum-free culture media suppressed mitogenesis caused by antigen, pokeweed mitogen, concanavalin A and phytohemagglutinin (81). There was no absolute identification of the fragment as being C3a, although the activity of the material resembled that of C3a in being heat- and acid-stable. The presence of human serum in the culture medium reduced but did not abolish the inhibitory effect of the low-molecular-weight fragment(s). The same group also reported that human or rat C3a could inhibit the secondary in vitro antibody response of rat lymphocytes in the presence of fetal calf serum (50). Proliferative responses were not assessed in this study. A third report from these investigators found that C3a had no inhibitory effect on mitogen-induced human lymphocyte proliferation when 5% or higher concentrations of fetal calf serum, rather than autologous serum, were included in the culture medium (122). These findings were extended by the demonstration that highly purified C3a, in concentrations of 0.1–50 μg/ml, suppressed in a dose-related manner the plaque-forming cellular response of human peripheral blood lymphocytes to Fc fragments (77). No inhibition of proliferation was observed, but fetal calf serum was present in the culture media. Suppression was localized to the level of impaired helper T cell function since normal responses occurred if soluble helper factors were added. C3a desArg had no activity in this assay system. This latter finding contrasts with that of the earlier study in which only partial loss of inhibitory activity of the low-molecular-weight fragment of C3, presumed to be C3a, was observed in the presence of human serum containing carboxypeptidase N (50). This discrepancy may be resolved by a preliminary report suggesting that C3e, which is similar to C3a in size and which is resistant to degradation by enzymes present in human serum, may have accounted for the growth-suppressing activity previously ascribed to C3a (123). The mechanism by which C3a might alter lymphocyte function is obscure since these cells are not considered to possess receptors for this peptide; the possible role of basophils and mast cells that might be present in cultures of human peripheral blood lymphocytes and murine and rat spleen cells has not been addressed.

C5a RECEPTOR   As noted previously in this review, binding of C5a to receptors on mast cells and basophils causes secretion of their granules and release of histamine. Recently, two groups of investigators provided evidence that C5a and C5a desArg also cause the release of slow reacting substance of anaphylaxis, which is comprised of leukotrienes, from guinea pig lung and tracheal strips (94, 109, 110). Although the cell type interacting with these peptides is presumably the pulmonary mast cell, macrophages also possess C5a receptors (13) and these cells have been shown in studies utilizing other stimulants to be capable of synthesizing leukotrienes (101). The diverse effects of these products of arachidonic acid (reviewed in 70) broaden the potential pathobiologic consequences of complement activation in the lung and may indicate some of the mechanisms by which C5a and C5a desArg produce acute pulmonary injury (17, 48, 111).

Binding of C5a or C5a desArg to receptors on neutrophils induces an array of cellular reactions: chemotaxis along a gradient of the C5-derived glycopeptides, specific granule secretion, increased adhesive properties, increased oxygen consumption and superoxide generation, and enhanced expression of C3b receptors (24, 26, 47). These cellular responses are not unique to C5a receptor-ligand interactions, as they occur following binding of the synthetic formylated peptides to different receptors. Since there are several comprehensive reviews of this extensively studied area of neutrophil cell biology (29, 120), only a discussion of the relevance of these reactions to host defense and inflammation is presented. Although C5a desArg interacts with the C5a receptors on neutrophils with 25- to 50-fold lower affinity than does C5a, it is probably the form of the peptide that is acting in vivo. In the time required for establishing a concentration gradient and for diffusing from extravascular sites to the intravascular space in which neutrophils normally reside, carboxypeptidase-N would remove the C-terminal arginine from C5a. The serum concentration of C5 is approximately 10^{-6} M, so that sufficient C5a desArg can easily be generated for binding to receptors with an association constant for the peptide in the range of 2–4 $\times$ 10^{-7} M^{-1}. Studies in vitro have demonstrated that maximal neutrophil chemotaxis occurs with approximately 10^{-8} M C5a desArg in the presence of an anionic serum cochemotaxin (91), and that maximal expression of neutrophil C3b receptors is induced with nanomolar concentrations of the peptide.

The increased stickiness of neutrophils induced by C5a desArg is presumably necessary for the cells' capacity to adhere to endothelial cells preparatory to diapedesis and chemotaxis (56). This cellular response may also account for the pulmonary leukosequestration and transient leukopenia observed in animals to which cobra venom factor is administered (71) and in humans undergoing hemodialysis and cardiopulmonary bypass (12, 56), situations in which activation of C5 occurs via the alternative pathway. The C5a desArg-induced rapid increase in the number of C3b receptors available on the plasma membrane of neutrophils (24, 58) would prepare the cell for uptake of C3b-coated complexes when it completes its migration to the site of complement activation.

Less is known about the biologic responses of monocytes and macrophages elicited by occupancy of their C5a receptors. C5a and C5a desArg induce chemotaxis of these cells and the slow secretion of glycolytic and proteolytic enzymes (72, 119). An uncharacterized factor chemotactic for neutrophils is also released by rabbit alveolar macrophages in response to these peptides. It would be of great interest if this factor was shown to be LTB$_4$, an extremely potent chemoattractant, since this would constitute evidence that C5a and C5a desArg induce synthesis of the leukotrienes by a defined cell type.

Addition of nanomolar concentrations of C5a or C5a desArg to murine spleen cells augmented the primary plaque-forming cell response to antigen (40). This effect was also accomplished by adding to the cultures peritoneal macrophages pulsed with the peptides, a result that indicates that the macrophages mediated the C5a response. Spreading of human monocytes induced by factor Bb also has been suggested as being secondary to the action of C5a cleaved from cell-associated C5 by Bb (112).

C3e RECEPTOR   Systemic complement activation in animals and in man is accompanied by a rapid fall in the number of circulating neutrophils, a secondary effect of C5a on these cells, followed by a return to levels above those prior to complement activation. This leukocytosis is thought to be secondary to the C3e fragment of C3, which has been shown to induce release of leukocytes from the perfused, isolated rat femur and to induce leukocytosis in rabbits (35, 100). The absence of leukocytosis accompanying pyogenic infections in some individuals genetically deficient in C3 reflects the inability of these patients to generate C3e (2).

CONCLUSIONS

The complement system is a mechanism for the generation of ligands that elicit a wide range of biologic functions by interacting with specific cellular receptors. These receptors reside in almost all cells that participate in immune and inflammatory reactions. The proteolytic reactions occurring among the complement proteins that lead to the elaboration of these ligands are relatively well understood, but little is known concerning complement receptors other than their distribution among cell types and some of the cellular responses induced by binding of ligands. Only two complement receptors have been isolated, the C3b receptor and CR2. Essential studies, such as calculation of the number and affinities of receptors, have been performed only for the C1q, C5a, and C3b receptors. Biologic studies of complement receptors on neutrophils, macrophages, and mast cells have established their roles in eliciting cellular responses that contribute to phagocytic, immediate hypersensitivity, and acute inflammatory reactions. In contrast, the role of lymphocyte complement receptors, such as the C3b, CR2, and CR3 receptors, in regulating the immune response must be clarified.

ACKNOWLEDGMENTS

This work was supported in part by grants AI-17917, AI-07722, AI-10356, and RR-05669.

Literature Cited

1. Abrahamson, D. R., Fearon, D. T. 1983. Endocytosis of the C3b receptor of complement within coated pits in human polymorphonuclear leukocytes and monocytes. *Lab. Invest.* In press
2. Alper, C. A., Colten, H. R., Rosen, F. S., Rabson, A. R., MacNab, G. M., et al. 1972. Homozygous deficiency of C3 in a patient with repeated infections. *Lancet* 2:1179–81
3. Arnaout, M. A., Melamed, J., Tack, B. F., Colten, H. R. 1981. Characterization of the human complement (C3b) receptor with a fluid phase C3b dimer. *J. Immunol.* 127:1348–54
4. Ault, K. A., Springer, T. A. 1981. Cross-reaction of a rat-antimouse phagocyte-specific monoclonal antibody (anti-MAC-1) with human monocytes and natural killer cells. *J. Immunol.* 126:359–64
5. Beller, D. I., Springer, T. A., Schreiber, R. D. 1982. Anti-MAC-1 selectively inhibits rosetting mediated by the type three complement receptor (CR3). *J. Exp. Med.* 156:1000–9
6. Berger, M., Gaither, T. A., Hammer, C. H., Frank, M. M. 1981. Lack of binding of human C3, in its native state, to C3b receptors. *J. Immunol.* 127:1329–34
7. Bianco, C., Patrick, R., Nussenzweig, V. 1970. A population of lymphocytes bearing a membrane receptor for antigen-antibody-complement complexes. I. Separation and characterization. *J. Exp. Med.* 132:702–20
8. Bodammer, G., Vogt, W. 1970. Contraction of the guinea pig ileum induced by anaphylatoxin independent of hista-
mine release. *Int. Arch. Allergy Appl. Immunol.* 39:648–57
9. Bokisch, V. A., Müller-Eberhard, H. J. 1970. Anaphylatoxin inactivator of human plasma: its isolation and characterization as a carboxypeptidase. *J. Clin. Invest.* 49:2427–36
10. Carlo, J. R., Ruddy, S., Sauter, S., Yount, W. J. 1979. Deposition of β1H in renal biopsies from patients with immune renal disease. *Arthrit. Rheum.* 22:403–8
11. Carlo, J. R., Ruddy, S., Studer, E. J., Conrad, D. H. 1979. Complement receptor binding of C3b-coated cells treated with C3b inactivator, β1H globulin and trypsin. *J. Immunol.* 123:523–28
12. Chenoweth, D. E., Cooper, S. W., Hugli, T. E., Stewart, R. W., Blackstone, E. H., Kirklin, J. W. 1981. Complement activation during cardiopulmonary bypass. *N. Engl. J. Med.* 304:497–503
13. Chenoweth, D. E., Goodman, M. G., Weigle, W. O. 1982. Demonstration of a specific receptor for human C5a anaphylatoxin on murine macrophages. *J. Exp. Med.* 156:68–78
14. Chenoweth, D. E., Hugli, T. E. 1978. Demonstration of specific C5a receptor on intact human polymorphonuclear leukocytes. *Proc. Natl. Acad. Sci. USA* 75:3943–47
15. Cochrane, C. G., Müller-Eberhard, H. J. 1968. The derivation of two distinct anaphylatoxin activities from the third and fifth components of human complement. *J. Exp. Med.* 127:371–86

16. Cooper, N. R. 1969. Immune adherence by the fourth component of complement. *Science* 165:396–98

17. Desai, W., Kreutzer, D. L., Showell, H., Arroyave, C. V., Ward, P. A. 1979. Acute inflammatory pulmonary reactions induced by chemotactic factors. *Am. J. Pathol.* 96:71–83

18. Dias da Silva, W., Lepow, I. H. 1967. Complement as a mediator of inflammation. II. Biological properties of anaphylatoxin prepared with purified components of human complement. *J. Exp. Med.* 125:921–46

19. Dobson, N. J., Lambris, J. D., Ross, G. D. 1981. Characteristics of isolated erythrocyte complement receptor type one (CR₁, C4b-C3b receptor) and CR₁-specific antibodies. *J. Immunol.* 126:693–98

20. Ehlenberger, A. G., Nussenzweig, V. 1977. The role of membrane receptors for C3b and C3d in phagocytosis. *J. Exp. Med.* 145:357–71

21. Fearon, D. T. 1979. Regulation of the amplification C3 convertase of human complement by an inhibitory protein isolated from human erythrocyte membranes. *Proc. Natl. Acad. Sci. USA* 76:5867–71

22. Fearon, D. T. 1980. Identification of the membrane glycoprotein that is the C3b receptor of the human erythrocyte, polymorphonuclear leukocyte, B lymphocyte and monocyte. *J. Exp. Med.* 152:20–30

23. Fearon, D. T., Austen, K. F. 1980. The alternative pathway of complement—a system for host resistance to microbial infection. *N. Engl. J. Med.* 303:259–63

24. Fearon, D. T., Collins, L. A. 1983. Increased expression of C3b receptors on polymorphonuclear leukocytes induced by chemotactic factors and by purification procedures. *J. Immunol.* 130:370–75

25. Fearon, D. T., Kaneko, I., Thomson, G. G. 1981. Membrane distribution and adsorptive endocytosis by C3b receptors on human polymorphonuclear leukocytes. *J. Exp. Med.* 153:1615–28

26. Fernandez, H. N., Henson, P. M., Otani, A., Hugli, T. E. 1978. Chemotactic response to human C3a and C5a anaphylatoxins. I. Evaluation of C3a and C5a leukotaxis *in vitro* and under simulated *in vivo* conditions. *J. Immunol.* 120:109–16

27. Fujita, T., Gigli, I., Nussenzweig, V. 1978. Human C4 binding protein. II. Role of proteolysis of C4b by C3b inactivator. *J. Exp. Med.* 148:1044–51

28. Gabay, Y., Perlmann, H., Perlmann, P., Sobel, A. T. 1979. A rosette assay for the determination of C1q receptor-bearing cells. *Eur. J. Immunol.* 9:797–801

29. Gallin, J. I., Quie, P. G., ed. 1978. *Leukocyte Chemotaxis: Methods, Physiology and Clinical Implications.* New York: Raven

30. Gelfand, M. C., Frank, M. M., Green, I. 1975. A receptor for the third component of complement in the human renal glomerulus. *J. Exp. Med.* 142:1029–34

31. Gerhard, C., Chenoweth, D. E., Hugli, T. E. 1981. Response of human neutrophils to C5a: a role for the oligosaccharide moiety of human C5a desArg-74 but not of C5a in biologic activity. *J. Immunol.* 127:1978–82

32. Gerhard, C., Hugli, T. E. 1981. Identification of classical anaphylatoxin as the des-Arg form of the C5a molecule: evidence of a modulator role for the oligosaccharide unit in human des-Arg⁷⁴-C5a. *Proc. Natl. Acad. Sci. USA* 78:1833–37

33. Ghebrehiwet, B. 1982. C3e-induced lysosomal enzyme release from polymorphonuclear leukocytes. *Fed. Proc.* 41:966 (Abstr.)

34. Ghebrehiwet, B., Müller-Eberhard, H. J. 1978. Lysis of C1q-coated chicken erythrocytes by human lymphoblastoid cell lines. *J. Immunol.* 120:27–32

35. Ghebrehiwet, B., Müller-Eberhard, H. J. 1979. C3e: An acidic fragment of human C3 with leukocytosis-inducing activity. *J. Immunol.* 123:616–21

36. Gigli, I., Fearon, D. T. 1981. Regulation of the classical pathway convertase of human complement by the C3b receptor of human erythrocytes. *Fed. Proc.* 40:1172 (Abstr.)

37. Gigli, I., Nelson, R. A. Jr. 1968. Complement dependent immune phagocytosis. *Exp. Cell Res.* 51:45–67

38. Glovsky, M. M., Hugli, T. E., Ishizaka, T., Lichtenstein, L. M., Erickson, B. W. 1979. Anaphylatoxin-induced histamine release with human leukocytes. *J. Clin. Invest.* 64:804–11

39. Goldstein, I. R., Kaplan, H. B., Radin, A., Frosch, M. 1976. Independent effects of IgG and complement upon human polymorphonuclear leukocyte function. *J. Immunol.* 117:1282–87

40. Goodman, M. G., Chenoweth, D. E., Weigle, W. O. 1982. Potentiation of the primary humoral immune response in vitro by C5a anaphylatoxin. *J. Immunol.* 129:70–75

41. Gorski, J. P., Hugli, T. E., Müller-Eberhard, H. J. 1979. C4a: The third ana-

phylatoxin of the human complement system. *Proc. Natl. Acad. Sci. USA* 76:5299–302

42. Griffin, J. A., Griffin, F. M. Jr. 1979. Augmentation of macrophage complement receptor function in vitro: I. Characterization of the cellular interactions required for the generation of a T-lymphocyte product that enhances macrophage complement receptor function. *J. Exp. Med.* 150:653–75

43. Gupta, S., Ross, G. D., Good, R. A., Siegal, F. P. 1976. Surface markers of human eosinophils. *Blood* 48:755–63

44. Hammann, K. P., Raile, A., Schmitt, M., Mussel, H. H., Peters, H., et al. 1981. β1H stimulates mouse spleen B lymphocytes as demonstrated by increased thymidine incorporation and formation of B cell blasts. *Immunobiology* 160:289–301

45. Harrison, R. A., Lachmann, P. J. 1980. The physiological breakdown of the third component of human complement. *Mol. Immunol.* 17:9–20

46. Hartman, K. U., Bokisch, V. A. 1975. Stimulation of murine B lymphocytes by isolated C3b. *J. Exp. Med.* 142:600–10

47. Henson, P. M., Zanolari, B., Schwartman, N. A., Hong, S. R. 1978. Intracellular control of human neutrophil secretion. I. C5a induced stimulus-specific desensitization and the effects of cytochalasin B. *J. Immunol.* 121:851–55

48. Henson, P. M., McCarthy, K., Larsen, G. L., Webster, R. O., Giclas, P. C., et al. 1979. Complement fragments, alveolar macrophages and alveolitis. *Am. J. Pathol.* 97:93–110

49. Hebert, L. A., Cornacoff, J. B., VanAman, M. E., Smead, W. L., Birmingham, D. J. 1982. In vivo kinetics of the primate erythrocyte-immune complex clearing mechanism. *Clin. Res.* 30:350A (Abstr.)

50. Hobbs, M. V., Feldbush, T. L., Needleman, B. W., Weiler, J. M. 1981. Inhibition of secondary *in vitro* antibody responses by the third component of complement. *J. Immunol.* 128:1470–75

51. Hugli, T. E. 1975. Human anaphylatoxins (C3a) from the third component of complement: primary structure. *J. Biol. Chem.* 250:8293–301

52. Hugli, T. E., Erickson, B. W. 1977. Synthetic peptides with the biological activities and specificity of human C3a anaphylatoxin. *Proc. Natl. Acad. Sci. USA* 74:1826–30

53. Iida, K., Mornaghi, R., Nussenzweig, V. 1982. Complement receptor (CR₁)

deficiency in erythrocytes from patients with systemic lupus erythematosus. *J. Exp. Med.* 153:1427–38

54. Iida, K., Nussenzweig, V. 1981. Complement receptor is an inhibitor of the complement cascade. *J. Exp. Med.* 153:1138–50

55. Isenman, D. E., Kells, D. I. C., Cooper, N. R., Müller-Eberhard, H. J., Pangburn, M. K. 1981. Nucleophilic modification of human complement protein C3: Correlation of conformational changes with acquisition of C3b-like functional properties. *Biochemistry* 20:4458–67

56. Jacob, H. S., Craddock, P. R., Hammerschmidt, D. E., Moldow, C. F. 1980. Complement-induced granulocyte aggregation. *N. Engl. J. Med.* 302:789–94

57. Johnson, A. R., Hugli, T. E., Müller-Eberhard, H. J. 1975. Release of histamine from rat mast cells by the complement peptides C3a and C5a. *Immunology* 28:1067–80

58. Kay, A. B., Glass, E. J., Salter, D. McG. 1979. Leukoattractants enhance complement receptors on human phagocytic cells. *Clin. Exp. Immunol.* 38:294–99

59. Kazatchkine, M. D., Fearon, D. T., Appay, M. D., Mandet, C., Bariety, J. 1982. Immunohistochemical study of the human glomerular C3b receptor in normal kidney and in seventy-five cases of renal diseases. *J. Clin. Invest.* 69:900–12

60. Koopman, W. J., Sandberg, A. L., Wahl, S. M., Mergenhagen, S. E. 1976. Interaction of soluble C3 fragments with guinea pig lymphocytes. Comparison of effects of C3a, C3b, C3c and C3d on lymphokine production and lymphocyte proliferation. *J. Immunol.* 117:331–36

61. Lachmann, P. J., Pangburn, M. K., Oldroyd, R. G. 1982. Breakdown of C3 after complement activation. *J. Exp. Med.* 156:205–16

62. Lambris, J. D., Cohen, P. L., Dobson, N. J., Wheeler, P. W., Papamichail, M., et al. 1982. Effect of antibodies to CR₂ (C3d receptors) on lymphocyte activation. *Clin Res.* 30:514A (Abstr.)

63. Lambris, J. D., Dobson, N. J., Ross, G. D. 1980. Release of endogenous C3b inactivator from lymphocytes in response to triggering by membrane receptors for β1H globulin. *J. Exp. Med.* 152:1625–44

64. Lambris, J. D., Dobson, N. J., Ross, G. D. 1981. Isolation of lymphocyte mem-

brane complement receptor type two (the C3d receptor) and preparation of receptor-specific antibody. *Proc. Natl. Acad. Sci. USA* 78:1828–32

65. Lambris, J. D., Ross, G. D. 1982. Assay of membrane complement receptors (CR$_1$ and CR$_2$) with C3b- and C3d-coated fluorescent microspheres. *J. Immunol.* 128:186–89

66. Lambris, J. D., Ross, G. D. 1982. Characterization of the lymphocyte membrane receptor for factor H (β1H-globulin) with an antibody to anti-factor H idiotype. *J. Exp. Med.* 155:1400–11

67. Law, S. K., Fearon, D. T., Levine, R. P. 1979. Action of the C3b inactivator on cell-bound C3b. *J. Immunol.* 122:759–65

68. Law, S. K., Levine, R. P. 1981. Binding reaction between the third human complement protein and small molecules. *Biochemistry* 20:7457–63

69. Lay, W. H., Nussenzweig, V. 1968. Receptors for complement on leukocytes. *J. Exp. Med.* 128:991–1007

70. Lewis, R. A., Austen, K. F. 1981. Mediation of local homeostasis and inflammation by leukotrienes and other mast cell-dependent compounds. *Nature* 293:103–8

71. McCall, C. E., De Chatelet, L. R., Brown, D., Lachmann, P. J. 1974. New biological activity following intravascular activation of the complement cascade. *Nature* 249:841–43

72. McCarthy, K., Henson, P. M. 1979. Induction of lysosomal enzyme secretion by alveolar macrophages in response to the purified complement fragments C5a and C5a desArg. *J. Immunol.* 123:2511–17

73. Medicus, R. G., Arnaout, M. A. 1982. Surface restricted control of C3bINA dependent C3c release from bound C3bi molecules. *Fed. Proc.* 41:848 (Abstr.)

74. Medof, M. E., Prince, G. M., Mold, C. 1982. A function for immune adherence receptors. *Fed. Proc.* 41:966 (Abstr.)

75. Melamed, J., Arnaout, M. A., Colten, H. R. 1982. Complement (C3b) interaction with the human granulocyte receptor: correlation of binding of fluid phase radio-labelled ligand with histaminase release. *J. Immunol.* 128:2313–18

76. Miyakawa, Y., Yamada, A., Kosaka, K., Tsuda, F., Mayumi, M. 1981. Defective immune-adherence (C3b) receptor on erythrocytes from patients with systemic lupus erythematosus. *Lancet* 2:493–97

77. Morgan, E. L., Weigle, W. O., Hugli, T. E. 1982. Anaphylatoxin-mediated regulation of the immune response. *J. Exp. Med.* 155:1412–26

78. Müller-Eberhard, H. J., Schreiber, R. D. 1980. Molecular biology and chemistry of the alternative pathway of complement. *Adv. Immunol.* 29:1–53

79. Nagasawa, S., Ichihara, C., Stroud, R. M. 1980. Cleavage of C4b by C3b inactivator: production of a nicked form of C4b, C4b', as an intermediate cleavage product of C4b by C3b inactivator. *J. Immunol.* 125:578–82

80. Nagasawa, S., Stroud, R. M. 1977. Mechanism of the C3b inactivator: requirement for a high molecular weight cofactor (C3b-C4bINA cofactor) and production of a new C3b derivative (C3b'). *Immunochemistry* 14:749–56

81. Needleman, B. W., Weiler, J. M., Feldbush, T. L. 1981. The third component of complement inhibits human lymphocyte blastogenesis. *J. Immunol.* 126:1586–91

82. Nelson, D. S. 1963. Immune adherence. *Adv. Immunol.* 3:131–80

83. Nelson, R. A. Jr. 1953. The immune adherence phenomenon. *Science* 118:733–37

84. Newman, S. L., Johnston, R. B. Jr. 1979. Role of binding through C3b and IgG in polymorphonuclear neutrophil function: studies with trypsin-generated C3b. *J. Immunol.* 123:1839–46

85. Newman, S. L., Musson, R. A., Henson, P. M. 1980. Development of functional complement receptors during in vitro maturation of human monocytes into macrophages. *J. Immunol.* 125:2236–44

86. Nicholson-Weller, A., Burge, J., Fearon, D. T., Weller, P. F., Austen, K. F. 1982. Isolation of a human erythrocyte membrane glycoprotein with decay-accelerating activity for C3 convertases of the complement system. *J. Immunol.* 129:184–89

87. Okuda, T., Tachibana, T. 1980. Complement receptors on Raji cells. The presence of a new type of C3 receptor. *Immunology* 41:159–66

88. Pangburn, M. K., Schreiber, R. D., Müller-Eberhard, H. J. 1977. Human complement C3b inactivator: isolation, characterization and demonstration of an absolute requirement for the serum protein β1H for cleavage of C3b and C4b in solution. *J. Exp. Med.* 146:257–70

89. Papamichail, M., Gutierrez, C., Embling, P., Johnson, P., Halborow, E. J.,

et al. 1975. Complement dependency of localization of aggregated IgG in germinal centres. *Scan. J. Immunol.* 4: 343–49

90. Pepys, M. B. 1974. Role of complement in induction of antibody production in vivo. Effect of cobra venom factor and other C3-reactive agents on thymus-independent antibody responses. *J. Exp. Med.* 140:126–45

91. Perez, H. D., Goldstein, I. M., Webster, R. O., Henson, P. M. 1981. Enhancement of the chemotactic activity of human C5a desArg by an anionic ("cochemotaxin") in normal serum and plasma. *J. Immunol.* 126:800–4

92. Perlmann, H., Perlmann, P., Schreiber, R. D., Müller-Eberhard, H. J. 1981. Interaction of target cell-bound C3bi and C3d with human lymphocyte receptors. Enhancement of antibody-mediated cellular cytotoxicity. *J. Exp. Med.* 153: 1592–603

93. Reid, K. B. M., Porter, R. R. 1981. The proteolytic activation systems of complement. *Ann. Rev. Biochem.* 50:433–64

94. Regal, J. F., Pickering, R. J. 1981. C5a-induced tracheal contraction: effect of an SRS-A antagonist and inhibitors of arachidonate metabolism. *J. Immunol.* 126:313–16

95. Reynolds, H. Y., Atkinson, J. P., Newball, H. H., Frank, M. M. 1975. Receptors for immunoglobulin and complement on human alveolar macrophages. *J. Immunol.* 114:1813–19

96. Ross, G. D., Jarowski, C. I., Rabellino, E. M., Winchester, R. J. 1978. The sequential appearance of Ia-like antigens and two different complement receptors during the maturation of human neutrophils. *J. Exp. Med.* 147:730–44

97. Ross, G. D., Lambris, J. D. 1982. Identification of a C3bi-specific membrane complement receptor that is expressed on lymphocytes, monocytes, neutrophils and erythrocytes. *J. Exp. Med.* 155:96–110

98. Ross, G. D., Lambris, J. D. 1982. Identification of three forms of iC3b that have distinct structures and receptor binding site properties. *Mol. Immunol.* (Abstr.) 19:1399

99. Ross, G. D., Polley, M. J. 1975. Specificity of human lymphocyte complement receptors. *J. Exp. Med.* 141: 1163–80

100. Rother, K. 1972. Leukocyte mobilizing factor: A new biological activity from the third component of complement. *Eur. J. Immunol.* 2:550–58

101. Rouzer, C. A., Scott, W. A., Hamill, A. L., Cohn, Z. A. 1982. Synthesis of leukotriene C and other arachidonic acid metabolites by mouse pulmonary macrophages. *J. Exp. Med.* 155:720–33

102. Ruddy, S., Austen, K. F. 1971. C3b inactivator of man. II. Fragments produced by C3b inactivator cleavage of cell-bound or fluid phase C3b. *J. Immunol.* 107:742–50

103. Schenkein, H. A., Genco, R. J. 1979. Inhibition of lymphocyte blastogenesis by C3c and C3d. *J. Immunol.* 122: 1126–33

104. Schmitt, M., Mussel, H. H., Hammann, K. P., Scheiner, O., Dierich, M. P. 1981. Role of β1H for the binding of C3b-coated particles to human lymphoid and phagocytic cells. *Eur. J. Immunol.* 11:739–45

105. Schneider, R. J., Kulczycki, A. Jr., Law, S. K., Atkinson, J. P. 1981. Isolation of a biologically active macrophage receptor for the third component of complement. *Nature* 290:789–92

106. Schopf, R. E., Hammann, K. P., Scheiner, O., Lemmel, E.-M., Dierich, M. P. 1982. Activation of human monocytes by both human β1H and C3b. *Immunology* 46:307–12

107. Schreiber, R. D., Pangburn, M. K., Bjornson, A. B., Brothers, M. A., Müller-Eberhard, H. J. 1982. The role of C3 fragments in endocytosis and extracellular cytotoxic reactions by polymorphonuclear leukocytes. *Clin. Immunopathol.* 23:335–57

108. Sim, R. B., Arland, G. J., Columb, M. G. 1979. C1 inhibitor-dependent dissociation of human complement component C1 bound to immune complexes. *Biochem. J.* 179:449–57

109. Stimler, N. P., Bach, M. K., Bloor, C. M., Hugli, T. E. 1982. Release of leukotrienes from guinea pig lung stimulated by C5a desArg anaphylatoxin. *J. Immunol.* 128:2247–52

110. Stimler, N. P., Brocklehurst, W. E., Bloor, C. M., Hugli, T. E. 1981. Anaphylatoxin-mediated contraction of guinea pig lung strips: a non-histamine tissue response. *J. Immunol.* 126: 2258–61

111. Stimler, N. P., Hugli, T. E., Bloor, C. M. 1980. Pulmonary injury induced by C3a and C5a anaphylatoxins. *Am. J. Pathol.* 100:327–38

112. Sundsmo, J. S., Götze, O. 1981. Human monocyte spreading induced by factor Bb of the alternative pathway of complement activation. *J. Exp. Med.* 154: 763–77

113. Tack, B. F., Harrison, R. A., Janatova, J., Thomas, M. L., Prahl, J. W. 1980. Evidence for the presence of an internal thioester in the third component of complement. *Proc. Natl. Acad. Sci. USA* 77:5764–68

114. Tedder, T. F., Fearon, D. T., Gartland, G. L., Cooper, M. D. 1983. Expression of complement receptor type one (for C3b) on human natural killer cells, B cell and lymphoblastoid cell lines. *J. Immunol.* In press

115. Tenner, A. J., Cooper, N. R. 1980. Analysis of receptor-mediated C1q binding to human peripheral blood mononuclear cells. *J. Immunol.* 125:1658–64

116. Tenner, A. J., Cooper, N. R. 1981. Identification of types of cells in human peripheral blood that bind C1q. *J. Immunol.* 126:1174–79

117. Tenner, A. J., Cooper, N. R. 1982. Stimulation of a human polymorphonuclear leukocyte oxidative response by the C1q subunit of the first complement component. *J. Immunol.* 128:2547–52

118. Vrainian, G. Jr., Conrad, D. H., Ruddy, S. 1981. Specificity of C3 receptors that mediate phagocytosis by rat peritoneal mast cell. *J. Immunol.* 126:2302–6

119. Ward, P. A. 1968. Chemotaxis of mononuclear cells. *J. Exp. Med.* 128:1201–21

120. Ward, P. A., Hugli, T. E., Chenoweth, D. E. 1979. Complement and chemotaxis. In *Handbook of Inflammation,* ed. L. E. Glynn, J. G. Houck, G. Weissmann, 1:153–78. New York: Springer

121. Wautier, J. L., Souchon, H., Cohen Solal, L., Peltier, A. P., Caen, J. P. 1976.

122. Weiler, J. M., Ballas, Z. K., Feldbush, T. L., Needleman, B. W. 1982. Inhibition of human lymphocyte blastogenesis by C3: the role of serum in the tissue culture medium. *Immunology* 45:247–52

123. Weiler, J. M., Needleman, B. W., Hobbs, M. V., Feldbush, T. L. 1982. Two separate fragments of the third component of complement inhibit immune responses. *Fed. Proc.* 41:849 (Abstr.)

124. West, C. D., Davis, N. C., Forristal, J., Herbst, J., Spitzer, R. 1966. Antigenic determinants of human βIC and βIG globulins. *J. Immunol.* 96:650–58

125. Whaley, K., Ruddy, S. 1976. Modulation of the alternative complement pathway by β1H globulin. *J. Exp. Med.* 445:1147–63

126. Wilson, J. G., Tedder, T. F., Fearon, D. T. 1982. Antigenic and functional detection of C3b receptors on human peripheral blood T lymphocytes. *Fed. Proc.* 41:965 (Abstr.)

127. Wilson, J. G., Wong, W. W., Schur, P. H., Fearon, D. T. 1982. Mode of inheritance of decreased C3b receptors on erythrocytes of patients with systemic lupus erythematosus. *N. Engl. J. Med.* 307:981–86

128. Wong, W. W., Fearon, D. T. 1982. Antigenic cross-reactivity between human C3b receptor and a murine lymphocyte surface protein. *Fed. Proc.* 41:965 (Abstr.)

129. Ziccardi, R. J., Cooper, N. R. 1979. Active disassembly of the first complement component, C1, by C1 inactivator. *J. Immunol.* 123:788–92.

C1 and human platelets. *Immunology* 31:595–99

Ann. Rev. Immunol. 1983. 1:273–306
Copyright © 1983 by Annual Reviews Inc. All rights reserved

CYTOLYTIC T LYMPHOCYTES

M. Nabholz

Genetics Unit, Swiss Institute for Experimental Cancer Research, CH–1066
Epalinges, Switzerland

H. R. MacDonald

Ludwig Institute for Cancer Research, Lausanne Branch, CH–1066
Epalinges, Switzerland

PREFACE

The purpose of this review is twofold. In the first part we trace the history
of cytolytic T lymphocytes (CTL) in the context of the development of
immunology as a whole; i.e. we analyze why the known characteristics of
CTL were discovered at a particular point in time. We think that such an
analysis is not only interesting but also useful, because the context of the
discovery of a phenomenon determines sometimes to a large extent its
interpretation, and the distinction between the experimental observations
and their interpretation becomes blurred with time.

A re-evaluation of the data on CTL may become important at a time
when the emphasis of the questions asked about these cells changes. We
think the discovery of T-cell growth factor and the design of methods that
have allowed the maintenance of CTL clones in permanent culture have
ushered in a period in which the established conceptual and experimental
framework of cellular immunology is rapidly ceding to a much more cell-
biologically oriented approach to CTL. Therefore, in the second part, we
identify some of the questions that arise from the work with CTL clones
and that represent this new outlook.

Our references are quite eclectic, and for some areas— e.g. lytic mechanisms— we refer mainly to some of the comprehensive reviews that have
appeared recently.

0732-0582/83/0410-0273$02.00

THE HISTORY OF CTL

Graft Rejection and Cytotoxic Cells

The observation that tissue grafted from one individual to another is rejected is very old (see 83 for review). That it was mediated by lymphoid cells was apparently first suggested at the beginning of this century, when the phenomenon of accelerated rejection of secondary grafts was also first observed. It seems somewhat surprising, therefore, that it took almost 50 years before it became firmly established, as a consequence of the transfer experiments of Mitchison (103) and the studies of Algire and his collaborators (149) with diffusion chambers, that in most situations cells but not antibodies mediated graft rejection. The main reason for this delay was probably that after the discovery of antibodies and the appreciation of their incredible discriminatory power, the temptation to ascribe all immune phenomena to these molecules was, for most people, irresistible. (Those who insist that T-cell specificity must be controlled by genes of the immunoglobulin gene family may be fighting the last battle of this controversy.)

In 1948 Medawar looked, unsuccessfully, for evidence for the destruction of tissue by cells from appropriately immunized animals. Several other investigators also failed (see 123 for references), but in 1960 Govaerts (54) reported that thoracic duct lymphocytes from dogs that had received a kidney graft were cytotoxic for kidney epithelial cells from the donor, but not for cells from "irrelevant" dogs. This report was rapidly confirmed (see 155 for review), and it was shown that (a) the cytotoxic effects were immunologically specific, (b) that they required the contact between living effector and target cells, (c) that serum from the immunized animal was not required, and (d) that specific cytotoxic lymphocytes could be raised not only against allografts but also against syngeneic tumors (124).

Although the first reports on cytotoxicity of leukocytes for cultured cells clearly suggested that such effects were mediated by lymphocytes that specifically recognized the target cells, the picture subsequently became blurred by reports (see 116, 117 for reviews) that lymphocytes or other leukocytes from unimmunized individuals could lyse different types of target cells nonspecifically, especially when they were stimulated with mitogens, and that normal lymphocytes could lyse cells precoated with antitarget antibodies. In addition, immune lymphocytes could release nonspecifically cytotoxic molecules (lymphotoxins) when incubated with the relevant antigen.

But the evidence for the existence of specifically cytotoxic cells in animals immunized against foreign tissues was sufficiently strong to survive this period of confusion. From 1968 on, the situation gradually became clearer for two reasons: the introduction of a simple, rapid, quantitative assay for specific cytotoxic cells, and the identification of cytotoxic cells as T lymphocytes.

In the early experiments, the cytotoxic activity of the lymphoid populations was measured by direct microscopic observation of the target cells, usually monolayers of adherent cells (see 155 for review). In the search for methods that make reliable quantitation possible, the solution that has imposed itself is the measurement of lytic activity as defined by the release of ^{51}Cr from prelabeled target cells (20). This assay was first used to measure complement-dependent, antibody-mediated lysis of nucleated cells (see 117 for references). When the cytotoxic activity of a lymphocyte population against tumor target cells as measured by this assay is compared with the capacity of the same population to reduce the plating efficiency of the targets, a good correlation is obtained (20, 82).

The ^{51}Cr release has the advantage of allowing precise quantitative comparison of the lytic activity of different cell *populations* against the same target cells, but neither this nor any other method can reliably enumerate the *number* of cytolytic cells in a population.

In 1968 the idea that there were two antigen-specific lymphocyte lineages, B cells and T cells, was finally accepted thanks to the experiments of Miller & Mitchell (101) on T-B cell collaboration. Thus, a clear interpretation of all the earlier experiments pointing to a dichotomy between cellular and humoral immunity imposed itself, and Brunner & Cerottini (19) rapidly demonstrated that specific cytotoxicity depended on T cells, and provided some evidence that other cell types were not required for the induction of target lysis. In addition, they showed that animals primed with allogeneic cells give a faster and stronger cytotoxic response to a boost with the same cells. This immunologic memory is long lasting.

Several questions presented themselves at this point. What was the in vivo role of CTL? Were they indeed the effector cells in graft rejection, and possibly in other immune phenomena such as the graft-versus-host reaction? How were CTL generated? Did the precursors of active CTL (CTL-P) express the same specificity as their progeny? And what were the mechanisms by which they induced target cell lysis?

CTL and the Major Histocompatibility Complex

That tissue graft rejection was due to an immune response to alloantigens on normal or tumor cells was established by the work of Gorer and of Medawar and his colleagues in the 1930s and early 1940s (see 76 for review). The early work on the genetics of histocompatibility (*H*) loci had already shown that there was a very large number of such genes (the estimates turned around 30). In 1956, Snell concluded from his work with strains congenic for different *H* genes that there were strong and weak *H* genes, and subsequently it became clear that *H-2* is the only strong *H* locus in the mouse. The cytotoxic lymphocyte populations that could be obtained from animals immunized with allogeneic tissue were soon shown to be specific for determinants controlled by this locus (18). The serological and histogenetic analysis of *H-2* revealed that the number of alleles at this locus was very large (this was at a time when it was still thought that most individuals of a species were homozygous for the same allele at most loci). This finding and the bewildering serological complexity of *H-2* turned its study into a more and more esoteric field, comprehensible, if at all, only to an exclusive club of specialists. But in the middle of the 1960s a renaissance (76) of *H-2* was in the offing, stimulated by several developments.

Through improvements in surgical techniques, routine kidney transplants became possible, and the clinical importance of graft rejection increased dramatically. Search for human *H* genes rapidly led to the discovery that in man, also, a single complex genetic system, *HLA,* was responsible for strong tissue incompatibility (see 23 for review). This was the first evidence pointing to the phylogenetic convergence of a major histocompatibility gene complex (MHC), which, as we now know, extends all through vertebrate evolution. Then, in 1971, Klein and Shreffler re-interpreted H-2 serology into the current bipartite genetic model with two alloantigen loci, *H-2K* and *H-2D* (77). This model, in general, was compatible with the findings of *HLA* serology. And finally, in 1972, McDevitt and his co-workers (97) reported that a gene controlling the immune response (Ir) to a specific antigen (*Ir* gene) discovered in 1965 was located between *H-2K* and *H-2D,* and subsequent reports showed that the immune response to many antigens was controlled by genes in this *H-2I* region.

From the studies on *H-2* it was already clear that it would be difficult to develop serological methods that allowed one to predict the histocompatibility of a particular kidney donor-recipient combination, and the development of rapid in vitro assays for this purpose seemed urgent. In an outbred species such as man, experiments in which the proliferation of lymphocytes was stimulated with phytohemagglutinin inevitably led to the discovery that proliferation occurred also when leukocytes from different individuals were mixed (10, 63). This reaction, called the mixed leukocyte culture (MLC) response, was immediately interpreted as an in vitro analogue of an allograft reaction. Experiments in the mouse and in man soon established that strong proliferative MLC responses were associated with MHC incompatibility (8, 36) and obeyed the laws of transplantation. [To avoid two-way reactions in mixtures of allogeneic cells, one population, the stimulator, is made unresponsive by blocking its DNA synthesis (see 110 for references).] But would an MLC response lead to the generation of the cytotoxic cells described, in other systems, as possible candidates for the effector cells in graft rejection?

In 1968, Ginsburg (48) had shown that rat lymphocytes cultured on mouse embryo fibroblast monolayers developed H-2-specific cytolytic capacity, and in 1970, using the ^{51}Cr release assay and cultured cell lines as targets, four groups (57, 61, 64, 139) showed that cytolytic cells were generated also in MLC. However, the requirement for fibroblasts or established cell lines as targets effectively precluded an evaluation of the prognostic capacity of CTL generation in MLC for the outcome of organ transplants. Normal lymphocytes were relatively refractory to CTL-induced lysis and showed a high spontaneous ^{51}Cr release, but mitogen-activated lymphocytes proved to be adequate targets (100).

Now the stage was set for a genetic analysis of the genes that controlled the proliferative MLC response, the generation of CTL, and the target determinants of the latter. After the finding of recombination events between the serologically defined *HLA* genes and those coding for MLC incompatibility in man by Yunis & Amos (157), analysis of MLC and CTL responses between mice carrying recombinant *H-2* haplotypes led to the conclusion that strong MHC-controlled MLC responses were elicited by *H-2I* and to a lesser extent by *H-2D*-region incompatibility in the mouse (9, 98), and by *HLA-DR* in man (156). CTL, on the other hand, recognize mainly determinants coded for by the genes controlling the classical, serologically defined MHC antigens (*H-2K, H-2D, HLA-A,B*) or by genes very closely linked to these loci (4, 39, 111, 144).

The demonstration of *I*-region-controlled MLC responses coincided with the discovery of *I*-region-associated (Ia) antigens (33, 56, 59, 129), and after a period of considerable controversy it is now generally accepted, but has not been proved, that the genes coding for these antigens and MLC determinants largely overlap (see 156 for review). Ia antigens are expressed predominantly on B cells and macrophages and early experiments comparing the stimulatory capacity of different leukocytes had indeed suggested that both of these cell types were efficient stimulators. (When Ia antigens were first detected, the idea that *H-2*-linked *Ir* genes coded for the T cell receptors seemed so convincing that the first reports on Ia antigens suggested that they were expressed predominantly on T cells, and much less on B cells. The opposite turned out to be true.) However, more recent studies with more rigorous separation methods show that Ia$^+$ macrophages and/or dendritic cells (141) may be responsible for most of the stimulating activity in spleen cell populations (1, 2).

Further work made it clear, in any case, that the separation between the genes coding for MLC incompatibility and for CTL determinants was not complete (see 110 for review). The clearest evidence for an overlap comes from the study of mutations in the *H-2K* gene, many of which lead both to MLC incompatibility and the appearance of new CTL determinants (45, 151).

In the meantime, sufficient evidence was accumulating to convince almost everybody that the cells responding specifically to antigenic stimulation in MLC by proliferation or generation of cytotoxic cells were T lymphocytes (see 110 for discussion). The genetic complexity of the response was the first hint at a corresponding cellular complexity, a topic to which we return later. One important puzzle raised by these studies concerned the progeny of cells of a given individual responding to a given allogeneic MHC haplotype (see 110 for review). Estimates of this frequency were obtained by determining the proportion of cells that would respond to allogeneic stimulators by DNA synthesis or division. There was no way of measuring the fraction of CTL in this population, but, recently, limiting dilution assays (see below) of CTL-P have indicated that they represent about 50% of all cells stimulated in a MLC (86). This problem was clearly a new aspect of the older question of the biological significance of graft rejection and transplantation antigens. The findings that as many different MHC haplotypes could be defined by MLC responses as by serology, and that more than 1% of all T cells responded to a single allogeneic haplotype whereas only MHC locus differences would elicit a measurable MLC response by cells of an unprimed individual, provoked speculations that alloreactive T cells were not monospecific. (In the mouse there is at least one locus, not linked to *H-2,* that controls the expression of strong MLC determinants, the *Mls* locus. The number of alleles at this locus is small and homologous genes have not been found in other species.) But all attempts to demonstrate poly-specificity failed, and the conclusion that most if not all T cells were specific for MHC alloantigens seemed inescapable but met understandably with a great deal of resistance. (One way of solving this dilemma was obviously through the identification and characterization of the structures that were the basis of antigen recognition by T cells. The enormous amount of effort spent on searching for the T-cell receptors since the early 1970s has resulted, so far, in negative or controversial results. No quantitative antigen-binding assay for T cells has been developed and their specificity can still be determined only through assays of their function.) One hypothesis that provided an explanation for the biological significance of alloreactivity was Jerne's (67; see also 110) postulate that the germ-line genes for lymphocyte antigen receptors coded for molecules that recognized the MHC antigens of the species, and that the repertoire of an adult was generated by somatic mutations in the genes coding for receptors specific for the individual's own MHC determinants. One of the starting points of Jerne's speculation was the observation that in xenogeneic MLC an inverse correlation existed between the magnitude of the response and the phylogenetic distance between the mixed cells (7), the so-called allo-aggression phenomenon, which clearly indicated an interaction between the genes coding for MHC antigens and those controlling T-cell specificity.

The most important advance in our understanding of the biological role of MHC antigens has certainly been the discovery of the phenomenon we call MHC restriction. The first experimental evidence for a role of MHC antigens in cell-cell interactions in the immune response came from studies (69, 75) showing that efficient cooperation between T and B lymphocytes in an antibody response, in which allogeneic effects were prevented, required compatibility of the cooperating cells in the *H-2I* region. Katz et al suggested, quite naturally, that efficient cooperation depended on positive recognition of self by the interacting cells, self being defined by *H-2I* region-coded cell surface molecules that would interact, e.g. in the way enzyme subunits self-assemble to form oligomers (69). In the meantime, the realization that in antiviral immunity cell-mediated phenomena played an important role (see 3 for review) induced Zinkernagel & Doherty (160) to look for virus-specific CTL in infected animals. They could demonstrate virus-specific cytotoxicity against syngeneic-infected but not allogeneic (*H-2* incompatible)-infected target cells. Shearer et al (133), in the same year, reported that the same held true for hapten-specific CTL generated in vitro by stimulating lymphocytes with syngeneic haptenated cells.

Interpretations of these data in terms of positive recognition of self MHC determinants were soon swept away by experiments that demonstrated that a virus-specific CTL from an *H-2* heterozygous F1 animal would recognize infected target cells of only one parent (161) [see also the earlier findings in the guinea pig (125, 135)], and that in irradiated F1 mice reconstituted with parental bone marrow cells these would give rise to virus- or hapten-specific CTL-P able to recognize infected or haptenated targets bearing the allogeneic parental MHC haplotype (146, 158): i.e. during their differentiation into CTL-P, stem cells "learn" to restrict their responses to non-MHC determinants so that they will recognize such antigens only on cells expressing the MHC haplotypes encountered during maturation (16, 147, 159). It was quickly appreciated that the rules of MHC restriction applied not only to virus- and hapten-specific CTL responses, but also to those against minor *H-* (15, 53) and tumor virus-specific antigens (118). Furthermore, the proliferative responses to soluble antigens that could be induced by exposing T cells from immunized animals to antigen-pulsed macrophages were also MHC restricted (see 145 for review). Whereas the determinants restricting CTL were coded for, mostly, by the *H-2K* and *H-2D* genes (or their homologues in other species), antigen-induced proliferation was restricted, in most cases, by genes mapping in the *I* region, most likely coding for the same molecules as those carrying Ia-determinants.

These findings provided an explanation of why the activity of virus- or hapten-specific CTL could not be inhibited by non-cell-bound antigen (an obvious a priori requirement for effector lymphocytes). They also offered, at last, some insight into the biological role of MHC antigens and stimulated the elaboration of new hypotheses on the evolutionary origin of the MHC polymorphism (110, 114). The attempts to elucidate the structural basis of MHC restriction gave rise to two types of models (see 138 for discussion): One postulates that T-cell receptors are specific for "altered self" determinants formed by interactions between MHC and surface-bound virus, hapten, or cell surface molecules; the other claims that T cells carry a "dual recognition" system consisting of receptors specific for self MHC determinants and other non-MHC antigen-specific ones. To us, it seems more simple to explain allo-aggression, the MHC polymorphism, and the large fraction of T cells that respond to MHC alloantigens, as well as the cross-reactivity of (self) MHC-restricted cells specific for non-MHC determinants with MHC alloantigens (see below), in the context of an altered-self model. However, it seems clear, by now, that a definitive answer will come only from the structural characterization of the molecules forming the T cell receptor(s).

The Nature of the Antigenic Stimulus for T Cells

One of the obstacles to progress in the identification of the T-cell receptor(s) is the lack of a quantitative specific binding assay with characterizable antigenic material, even in allogeneic situations where genetic analysis allows precise identification of the genes coding for the determinants recognized. Although it is possible to demonstrate that activated alloreactive T lymphocytes specifically bind membrane fractions from appropriate cells (113), it has not been shown that this works with purified MHC molecules. One report indicates CTL can induce specific ^{51}Cr release from artificial liposomes containing cellular glycoproteins (65), but several other groups have had problems reproducing these results. Attempts to inhibit CTL with antigen-bearing membrane fractions have resulted in failures in most laboratories, whereas inhibition with whole cold-target cells has been repeatedly demonstrated.

These difficulties in defining the molecular nature of the determinants recognized by T cells may be related to the ones encountered in the investigations into the nature of the entities stimulating the activation (proliferation or acquisition of cytolytic activity) of T cells. Although in one of the first descriptions of the proliferative MLC response it was claimed that extracts from allogeneic cells could stimulate this reaction (63), this turned out not to be true: In unprimed cells a proliferative or CTL response can only be induced with living metabolically active stimulators, although these can be irradiated with lethal doses of X rays or treated with mitomycin C. But UV-irradiation abolishes their stimulatory capacity (see 110 for references). In cell populations from immunized animals a CTL response can be induced by subcellular antigenic membrane preparations (42), but only when adherent accessory spleen cells are present (34). Accessory spleen cells, probably Ia^+ macrophages, are also required for the generation of CTL in a primary MLC against allogeneic or haptenated syngeneic cells (2). They may be allogeneic or syngeneic with the responder cells. Analogy with the proliferative response of primed T cells to soluble protein antigens suggests that one function of the accessory cells may be antigen presentation, but a definitive picture of the role of these and the other cells participating in a MLC will almost certainly emerge only when we understand the molecular bases of their interactions.

Evidence for Cell-Cell Interaction in the Generation of CTL

The use of mitogen-induced blast cells as targets made possible the genetic analysis not only of the target determinants of CTL, but also of the incompatibilities required for their generation in a MLC. It soon became apparent that in some experimental systems induction of CTL activity required, as expected, differences at loci that coded for the target determinants but, in addition, incompatibility that induced a strong proliferative MLC response (4, 38). At the same time evidence for the synergistic involvement of two sets of T cells in lethal graft-versus-host reactions was reported, as well as data suggesting the possibility that the cells that respond mainly by proliferation to allogeneic stimulators were different from the CTL-P and that CTL generation was the result of a cooperation between the two different sets (see 22 for references). Experiments in which responder cells were mixed with two different stimulators made it clear that the determinants inducing a strong proliferative response and those recognized by the CTL generated in the MLC did not have to be on the same cell (38, 130): i.e. this system did not follow the then-accepted requirements for hapten-carrier linkage in T-B cell collaboration. It became apparent quickly, however, that the influence of incompatibilities inducing a strong proliferative response on CTL generation was more quantitative than qualitative (41, 111) and could not be shown in some systems. Furthermore, the genes coding for determinants recognized by the proliferating cells and those controlling CTL determinants were found to overlap (112, 151). Convincing evidence for the existence of two subsets of alloreactive T cells was provided by Cantor et al (21, 22). They found that murine CTL and CTL-P expressed the cell surface antigens Lyt 2 and Lyt 3 (their results suggesting that CTL and CTL-P did not express Lyt 1 turned out to be a reflection of only quantitative variation in the expression of this antigen), that the bulk of the proliferating cells in a MLC did not express Lyt 2 or Lyt 3 and were not cytotoxic, and that the two populations would show synergism during CTL generation—the Lyt 2^-3^- MLC-responsive cells would amplify the CTL response. They had the same Lyt phenotype as helper T cells required for many antibody responses.

Did they belong to the same population and, if so, did they exert their influence on B cells and CTL by the same mechanism?

T-Cell Growth Factor

The suggestion that the synergistic effects between different T-cell subsets during CTL generation were mediated by soluble factors without antigenic specificity was taken seriously surprisingly late, in spite of the accumulating evidence for the existence of several biological activities in MLC supernatants (5, 37, 68, 119). To some extent the belated recognition of the importance of such molecules is almost certainly due to the obsession of immunologists with antigenic specificity. Rather than admit the existence of such uninteresting hormone-like agents, some preferred to suggest, for example, that the T-cell growth-stimulating activity in supernatants of mitogen-activated leukocytes was a mixture of antigenically specific molecules. Two papers finally ushered in what one might call the age of factors (or lymphokines or interleukins or whatever name you prefer). One was contributed by a group (105) whose main concern had been the problem of growing normal lymphocytes in long-term culture for cell biological and oncological studies. The impact of their discovery that normal T cells required a factor present in supernatants from T-cell mitogen-stimulated leukocytes for long-term proliferation hit immunologists really only when Gillis & Smith (46) reported that with this T-cell growth factor (TCGF or interleukin-2) they could maintain cytolytically active CTL in culture for several months. Virtually at the same time Ryser et al (126) showed that in vitro-primed MLC cells (i.e. cells stimulated several weeks earlier in a MLC) could be induced to re-express specific CTL activity by exposure to supernatants from short-term secondary MLC cultures. The active principle in this latter system almost certainly includes TCGF.

Further work on the role of TCGF in the proliferative responses to mitogenic lectins has shown that resting mature T cells can only respond to TCGF if they are previously activated, e.g. by a short-term exposure to concanavalin A (Con A) (81, 136), and it has been assumed that antigen plays the same role in a specific response, although some evidence that this may be so has only recently been published (62). Con A may render cells responsive to TCGF by inducing them to express TCGF receptors: several groups showed that Con A-activated but not resting T cells absorb TCGF (e.g. 81), a result recently confirmed by binding assays with purified labeled growth factor (122). TCGF itself is a T cell product (136), and work with murine T-cell clones has shown that most TCGF-producing cells belong to the non-cytolytic, Lyt 2⁻3⁻ class (52, 72). [It is possible that TCGF-producing cells themselves require TCGF for proliferation (137), i.e. the proliferative response of all T cells in a MLC may depend entirely on this growth factor, but this hypothesis is not supported by any direct evidence.] TCGF production, in turn, requires the presence of another lymphokine, lymphocyte activating factor (or interleukin-1), a macrophage product (115, 136). This explains, at least in part, the requirement for macrophage-like accessory cells in the MLC response. It appears, in addition, that lymphocyte-activating factor production may depend on an interaction between antigen-specific T cells and antigenic or antigen-presenting macrophages. (One notes that use of the terms "stimulation" and "responding cells" in their original definition becomes meaningless.)

The experimental support for many aspects of this type of model for T-cell activation as proposed by several groups at the same time (81, 115, 136) is still shaky, and further probing will undoubtedly reveal additional interactions. But at least it has become impossible to ignore the complexity of the T-cell response to antigenic stimulation, and the importance of understanding the molecular bases of the cellular interactions involved is increasingly apparent.

T-Cell Clones

The discovery of TCGF made possible two developments: the estimation of the frequencies of CTL-P of a given specificity in limiting dilution assays; and the establishment, in culture, of cell lines derived from CTL that continue to express the capacity to recognize and lyse appropriate target cells.

Attempts to estimate CTL-P frequencies by stimulating them at limiting dilutions had already been made prior to the discovery of TCGF, but the results of such experiments only became convincing when the distribution of the amount of ^{51}Cr released remained bimodal regardless of the number of responder cells/well, the positive events were distributed according to one-hit Poisson statistics, and the frequency estimates stopped increasing from one publication to the next, when the cultures, in 0.2-ml wells of microtiter plates, contained, besides irradiated spleen cells as stimulators, TCGF-containing supernatants from secondary MLC- or Con A-stimulated spleen cells (127).

The first attempts to maintain CTL in culture for prolonged periods were made by re-exposing the cells obtained from a primary MLC periodically to fresh stimulators (87). Each restimulation induced a wave of proliferation that peaked on day 3 or 4 and then declined rapidly. During the first few restimulations the proliferative response was accompanied by a strong rise in cytolytic activity, which reached its maximum on day 5 and then abated more slowly than DNA synthesis. Later on, the cytolytic but not the proliferative response would rapidly become smaller and would disappear completely after about six to eight restimulations (60). Analysis of the Lyt phenotype of the responding cells suggests that this decline is due to a gradual dilution of the Lyt 2^+3^+ CTL-P during such culture regimes (H. R. MacDonald, unpublished results).

The report by Gillis & Smith (46) that CTL-containing MLC populations could be grown for several months in TCGF-containing media, in the absence of stimulator cells, without loss of specific cytolytic activity was quickly followed by the demonstration that CTL-like clones could be isolated from such long-term cultures and could be established as permanent lines (11, 109, 148). Soon afterward, Glasebrook & Fitch (50) showed that CTL clones obtained by the limiting dilution technique described above could also be passaged for apparently unlimited periods.

We have compared the mouse CTL lines obtained by the two types of protocols elsewhere (106), and we restate here only the most prominent observations plus our conclusions. Both types of lines depend on TCGF, but the ones obtained by immediate cloning in limiting dilution cultures (which we have called CTL-A clones) also require the addition of irradiated spleen cells (filler cells) for continued growth. Although the original reports suggested that in the presence of an excess of TCGF-containing supernatants syngeneic filler cells were sufficient (50, 84), recent papers show that at least some CTL-A clones proliferate only if the filler cells are antigenic (6, 121). CTL-A clones resemble normal CTL more than the lines derived from populations maintained in TCGF-containing media prior to cloning; the latter—we have called them CTL-B lines—may be derived from filler cell-independent variants that arise and are selected for during the first months of culture (55).

The possibility to clone CTL and CTL-P, and to establish cell lines with the functional characteristics of CTL, has been exploited in different ways: (a) it provides a tool for an analysis of the specificity repertoire of CTL. (b) It allows, combined with the use of monoclonal antibodies and fluorescence-activated cell sorting, a more rigorous characterization of their cell surface antigen phenotype. (c) Assays of CTL clones for other functions, in particular production of soluble mediators, have led to the realization that CTL are a much more heterogeneous cell population than was previously suspected. (d) Extension of such analyses to T cell populations from various organs at different stages of development has already become important in the study of the ontogeny of the immune system. (e) The possibility to clone CTL-P and the availability of pure TCGF is giving new impetus to investigations on the requirements for the complex set of events lumped together as T cell activation or CTL generation. Finally, (f) CTL lines are being used for attempts to identify the genetic basis of CTL function by somatic cell genetic or molecular biology techniques. In the following sections we discuss some of these topics.

THE CELLULAR BIOLOGY OF CTL

The Phenotype of CTL Clones and their Precursors

SPECIFICITY REPERTOIRE   The frequency estimates for anti-allogeneic CTL-P among lymphoid cells from unprimed mice obtained when limiting numbers of lymphocytes were cultured in the presence of H-2-incompatible stimulator cells and TCGF-containing supernatants varied from 1 per 170 to 1 per 1000 spleen cells, or 1 per 40 to 1 per 147 lymph node cells (86). Specific CTL-P represent about 50% of all cells that proliferate in this system in response to alloantigen (127); thus these estimates agree very well with the earlier ones based on the fraction of cycling cells in MLCs. Most, if not all, of the CTL-P responding to stimulator cells from completely unrelated strains are directed against H-2 determinants, and the finding that the frequencies of CTL-P responding to determinants coded for by H-$2K$ or H-$2D$ alleles, or by H-$2K$ gene mutations, are all very similar (86, 152) suggests that most CTL-P are specific for H-2K or D determinants. Although CTL populations and clones specific for H-2I determinants have been described, the frequency of H-2I-specific CTL-P has not been determined. The frequencies of CTL-P reactive against minor H or tumor virus antigens could only be accurately determined in cells from immunized animals (86).

The possibility of expanding clones obtained in the limiting dilution cultures was exploited to examine their activity against several targets. The results of these experiments indicate that the cross-reactivity patterns observed in uncloned populations of CTL (see 138 for review) reflect the frequency of cross-reactive clones (143). [The term cross-reactivity is used entirely in an operational sense: none of the experiments we discuss rules out the theory that the cross-reactivity of an individual clone is due to the expression of several receptor populations with different specificities. Furthermore, the cross-reactivity of a clone may depend on its expression of molecules, such as Lyt 2, that do not contribute to the receptor specificity (see below).] When the cross-reactivity patterns of cells responding to an unrelated H-$2K$ allele against a series of mutations of this allele were compared, a high proportion of all possible different cross-reactivity patterns was detected (134). (The conclusions that can be drawn from these experiments are limited because the estimate of the CTL-P frequency was about 10 times lower than that obtained in other laboratories, and the monoclonality of the positive cultures was not ascertained.) The reactivity of anti-MHC clones does not depend on the non-MHC background of the target cells (80), a finding that argues strongly against the hypothesis of Matzinger & Bevan (96), according to which alloreactive T cells are specific for non-MHC antigens restricted by the allogeneic MHC determinants.

Among MHC-restricted CTL, cross-reactivity between non-MHC (influenza virus) antigens (17), as well as cross-reactivity with regard to the restriction element, has been observed (55, 150); and there are several cases of clones specific for non-MHC antigens on syngeneic targets that cross-react with allogeneic targets not carrying the relevant non-MHC antigen (17, 27, 148). These experiments suggest that the distinction of allogeneic MHC antigens, non-MHC antigens, and self-MHC restricting determinants is quite artificial: CTL clones do not seem to distinguish among them.

PRODUCTION OF SOLUBLE MEDIATORS   The role of soluble mediators in cell-mediated immune responses had been recognized very early, and soon after the discovery of the MLC reaction the presence, in MLC supernatants, of biological activities such as blastogenic factor (68) and an activity replacing T cells in B-cell responses was noted (37). But none of these activities was biochemically characterized, and although it was often assumed that they were produced by the responding T cells, no clear evidence to this effect was provided. Most cellular immunologists were not interested in nonspecific soluble mediators and assumed that lymphocyte proliferation was induced directly by their interaction with antigen. (Mitogenic plant lectins were thought to be universal antigens.) This view became untenable with the discovery of TCGF, and it is now clear that the growth of many if not all mature T cells depends on this lymphokine, which itself was soon shown to be a product of T cells (51, 136).

In 1979, Glasebrook & Fitch (50) isolated the first TCGF-producing alloreactive T cell clone, and subsequently several groups started to test whether T cell clones produced other soluble mediators [TCGF, interferon (IFN), granulocyte-macrophage colony-stimulating factor (GM-CSF), macrophage activating factor (MAF), and factors involved in the proliferation and maturation of B cells (B cell helper factors) (see 51 for references)]. Initially, the main purpose of such screenings was to identify the cellular source of various biological activities. The results showed that T cell clones were very heterogeneous when they were classified according to the biological activities they produced (51, 71, 78), and this type of analysis would sometimes allow a functional separation of the molecules mediating two different biological activities by the fact that certain clones produced only one of them.

After the division of T cells into subclasses (21, 22), it had been widely assumed that the biological activities present in MLC supernatants were products of non-cytolytic proliferating Lyt 2^-3^- cells specific for I-region determinants. Comparison of estimates of the frequency of MHC-reactive T cells that proliferate but do not give rise to CTL (127) with those of the frequency of cells that give rise to TCGF-producing clones (85) suggests that the two populations largely overlap. The majority of the TCGF producers are Lyt 2⁻ (72) and, in a response to *H-2*-incompatible stimulators, they respond to *I*-region determinants, but among them are about 10–20% cells specific for *K*- or *D*-region antigens (85). Among *K/D*-region-reactive Lyt 2$^+$ CTL clones isolated by micromanipulation from a 5-day MLC, 13% produced detectable quantities of TCGF (A. Glasebrook, personal communication). They were included in a larger fraction, 25%, which did not secrete detectable amounts of TCGF but still proliferated in response to stimulator cells without addition of growth factor. It seems likely that such clones produce sufficient TCGF to stimulate their own growth, but not enough to be detectable in the culture fluid.

Recent analyses (51, 72) show that most CTL clones produce one or more factors, and that, as far as tested, CTL can produce the same biological activities as non-cytolytic Lyt 2⁻3⁻ MLC responders, although at least for MAF and GM-CSF the frequency of alloreactive (*H-2 + Mls*) splenic T cells giving rise to producer clones is about three times higher among Lyt 2⁻ negative cells (about 1/100) than among Lyt 2$^+$ ones (72). In any case, clones with almost any possible combination of functional activities can be found. The one exception is IFN and MAF, where a strict correlation has been observed so far (71), but there is no convincing biochemical evidence that these two factors are not the same.

The kinetics of appearance of MAF, GM-CSF, and B-cell helper factor in clone supernatants is similar; they reach a plateau about 24 hr after exposure to allogeneic cells or Con A (51, 71). TCGF activity, on the other hand, declines rapidly after a peak at around the same time. It seems likely that this decrease, which does not depend on the Lyt 2 phenotype of the producer clone, is due to absorption and consumption by the producing cells themselves (122). Comparison of the kinetics of factor accumulation with that of proliferation of TCGF-independent clones in response to stimulator cells suggests that factor production ceases one to several days before proliferation stops.

At present we find it difficult to suggest what the biological significance of the bewildering multitude of T-cell phenotypes that emerges from these studies might be. Is it a culture-induced artifact? Does it reflect a further splitting up of the T-cell lineages into sublineages? Or do the different phenotypes represent different maturation stages of the same lineage? So far there is little evidence for changes in the phenotype of a clone, but the data are not yet very extensive. It seems likely that several of the mediators produced by lymphocytes will be purified, and their genes will be cloned in the next few years. Thus, the most promising way to approach the questions raised above may be to use such cloned probes to analyze the mechanisms by which the expression of the corresponding genes is regulated.

CELL SURFACE MARKERS   In spite of very intensive screenings in several laboratories for CTL-specific monoclonal antibodies, no such reagents have been reported to date. Like most other T cells, murine CTL express the so-called lymphocyte functional antigen-1 (LFA-1) (140), and probably all CTL as well as their precursors belong to the population of T cells expressing the Lyt 2 and Lyt 3 determinants (for review, see 104). This includes CTL specific for *I*-region determinants or for non-MHC antigens (102). The only reported cases of Lyt 2⁻-specific CTL concern antigen loss variants (35, 49, 107) or, in one case, a cell population maintained in culture for several years before its Lyt 2 expression was tested (142) (see p. 306). Cerottini et al (28) found, however, that a fraction of alloreactive Lyt 2⁻ clones obtained at limiting dilutions of normal cells was able to induce ^{51}Cr release from targets when Con A or phytohemagglutinin was added to the assay, but it is not clear whether or not the mechanism by which isotope release is induced by these cells is the same as for Lyt 2⁺ CTL (44). Antibodies against Lyt 2, Lyt 3, and LFA-1 can block CTL activity, and we discuss them in a later section on the lytic mechanism.

In 1978, Kimura & Wigzell (73) reported that CTL could be distinguished from CTL-P and other non-CTL activated in a MLC by their expression of a glycoprotein, T145. A lectin from *Vicia villosa* (VV) was shown to have a preferential affinity for this molecule when it was derived from Con A-activated T cell blasts (74), and VV-affinity columns were claimed to retain preferentially the CTL in a T cell population activated by in vitro stimulation. Other groups have reported results that seem contradictory in that they found no difference between the VV-binding capacities of CTL and of non-cytolytic blasts activated in a MLC (90), and that not all types of CTL populations express T145 (70). Comparison of CTL-B with T lymphoma lines (31) showed that only the former bound and were sensitive to growth inhibition by VV, and that they express a glycoprotein with approximately the same molecular weight as T145. Lymphoma cells, on the other hand, expressed a glycoprotein resembling T130 that is found on thymocytes and is probably structurally related to T145. The number of VV-molecules bound per CTL-B cell was about 10^7, and affinity chromatography of cell lysates showed that VV bound not only the T145-like molecules but many other surface glycoproteins from these cells. VV-resistant mutants of a CTL-B line that bound reduced or nondetectable amounts of the lectin were still cytolytically active and TCGF dependent, but they also still expressed a T145-like molecule (31). On the basis of these results, Conzelmann (29) has suggested that in some but not in all situations CTL activation is correlated with a changing pattern of glycosyl-transferase activities that results in an increased density of terminal α-linked D-N-acetyl galactosamine residues on O-linked carbohydrates of many glycoproteins. The most prominent result of this change would be the altered lectin affinity and the changed apparent molecular weight of T130, which becomes T145. Similar alterations affecting other, less abundant molecules may be difficult to detect on Con A blasts, but they are obvious on CTL-B cells.

The fact that the intensive searches for CTL-specific monoclonal antibodies have remained unsuccessful so far, even when the screening was based on inhibition of CTL activity, does not mean that CTL-specific cell surface structures do not exist. The relative frequencies with which hybridomas that secrete monoclonal antibodies against different cell surface antigens are obtained depends on their antigenicity and relative abundance on the immunogen and, probably, on the immunization schedule; therefore, CTL-specific hybridomas may not yet have been isolated because their incidence was too low.

It is possible, on the other hand, that CTL use the same surface molecules as other cells, but with different effects because, for example, they interact with different molecules on the cytoplasmic side of the cell membrane. Thus, maybe we should look there for CTL-specific molecules. Of course, technically this is more difficult.

Not surprisingly, our criteria for distinguishing CTL-P from other T lymphocytes are even more shaky than the ones applicable to CTL. About all we know is that most of them belong to the Lyt 2^+3^+ subpopulation (21, 94). We assume that CTL-P have the same specificity as their mature progeny, but this has still not been proved. These observations at least indicate that CTL-P do exist as cells committed to becoming CTL, and that they do not have the alternative of giving rise to T helper cells whose precursors are mainly Lyt 2^-3^-.

The Cytolytic Mechanism

Following the first demonstration of cell-mediated target cell destruction it gradually emerged that several different types of leukocytes could lyse target cells (for review, see 116): these include macrophages, granulocytes, and three types of lymphoid cells—CTL, natural killer cells, and K cells that lyse antibody-coated target cells. The ontogenetic relationships between the lymphoid cytotoxic cells is not clear, and to some degree these relationships may overlap. It is not known whether or not their effector mechanisms are different, nor has it been ruled out that CTL can induce ^{51}Cr release by several different mechanisms.

The studies on these mechanisms have been covered by several recent reviews (12, 92), so we only summarize the findings concerning CTL. Soon after their discovery it was shown that CTL kill autonomously: i.e. they do not require the assistance of other cells or products to inflict lethal damage.

The lytic event requires contact between the CTL and its target, and innocent bystanders are not affected. This includes the CTL itself, which is able to interact sequentially with several targets: CTL recycle. But they are not immune to lysis by other CTL. In an interaction between two CTL, one of which specifically recognizes the other, only the latter is lysed. These observations seem to rule out killing by a short-lived, locally acting toxin.

Inhibitor studies have allowed the dissection of the CTL-target interaction into three stages: binding or conjugation, lethal hit or programming for lysis, and target cell lysis or zeiosis. But the nature of the lethal hit is still entirely mysterious. It is generally assumed to be a cell surface-mediated event as it does not affect CTL, and it does not necessarily lead to the immediate destruction of the target. The approaches in this field, which includes morphological studies of CTL-target conjugates, use of metabolic inhibitors, and antibodies has not changed greatly during the last 10 years. Although the availability of cloned CTL helps to clarify some of the hanging issues, they do not as yet provide a direct approach to the problem, unless somebody invents a method for the isolation of non-cytolytic variants by selection for this very deficiency.

There is no evidence that the different stages of the lytic interaction involve molecular structures, which would form a lytic machinery at the surface of CTL distinct from the determinant-specific receptors, but the morphological studies suggest that the CTL-target interaction is a very complex process. Recently, Ryser et al (128) have observed that CTL-target conjugate formation is accompanied by the polarization of CTL actin and by rapid movements of the CTL membrane in the contact area. These morphological phenomena have not been shown to be required for ^{51}Cr release or target cell lysis, or to be specific for the interaction of cytolytic cells with their targets. But they do suggest that the recent model of Berke & Clark (13) is a bit too simple. These authors have suggested that "the imposition of what is in effect a two-dimensional cross-linking grid consisting of the CTL surface and its embedded MHC receptors creates instabilities at the target cell MHC protein bilayer lipid interface that lead to increased membrane permeability." This speculation, by itself (i.e. without ad hoc assumptions), does not explain why CTL-P do not lyse targets, or what the differences are between cytolytic and non-cytolytic T-cell clones that recognize determinants on the same target cell molecules.

One approach that would simplify the study of the CTL-target interaction would be the construction of artificial targets, and we have already referred to the reports that CTL-containing lymphocyte populations induced ^{51}Cr release from liposomes containing MHC molecules (65). There are a number of possible reasons for the apparent difficulty of reproducing these results in other laboratories (see 99), and this approach may well eventually be successful.

Inasmuch as target cell lysis by CTL is mediated by cell surface interactions, the use of antibodies for the analysis of CTL activity has been an obvious approach for a long time, and antibodies against the MHC determinants of the target cells have been helpful in establishing the nature of the target structures.

On the other hand, the usefulness of conventional anti-effector antisera, which block killing activity, has been hampered by the impossibility of adequately clarifying the specificity of such reagents. As discussed above, the monoclonal antibodies that block CTL activity have been directed, so far, either against the molecular complex bearing the Lyt 2 and Lyt 3 determinants or against LFA-1.

The characteristics of these antigens have recently been reviewed (88, 140), so that we restrict ourselves to summarizing the reported findings.

1. Both antigens are expressed not only on CTL but also in other cells: Lyt 2/3 expression is restricted to T cells, CTL-P, suppressor T cells, and the majority of thymocytes. LFA-1 is expressed on all lymphocytes, including B cells, and the majority of bone marrow cells, but not on cells outside the hemapoietic system.

2. Antibodies against both antigens inhibit not only CTL activity, but also other functions of antigen reactive cells: Anti-Lyt 2/3 antibodies can both block the proliferation of antigen-responsive CTL clones, and can inhibit the release of lymphokines. Anti-LFA-1 reagents seem to block all T cell-dependent immune responses.

3. Antibodies against either antigen block conjugate formation and do not inhibit the so-called programming-for-lysis step. Their blocking activity can be overcome by addition of Con A or phytohemagglutinins.

4. Expression of Lyt 2/3 is not an absolute requirement for specific CTL activity: certain highly active CTL populations cannot be inhibited, and some CTL lines or CTL hybrids do not, or no longer, express Lyt 2/3, but still are able to lyse targets specifically. Comparing the effects of controlled trypsin treatment on Lyt 2 expression, and of anti-Lyt 2 antibodies on the CTL activity of a number of different CTL clones, MacDonald et al (88) have arrived at the conclusion that the dependence of the lytic interaction on Lyt 2/3, at least in an in vitro situation, probably depends on the affinity of the CTL for its target. The strongest support in favor of this interpretation comes from experiments showing that the activity of several clones against their specific targets does not depend on Lyt 2/3, whereas activity of the same clones against a cross-reacting target does. Similar studies with regard to LFA-1 have not yet been reported.

5. The molecule bearing LFA-1 determinants consists of two noncovalently associated subunits, one of which is apparently shared with another antigen, Mac-1, expressed not on T cells but on macrophages, monocytes, and granulocytes. Anti-Mac-1 antibodies inhibit the binding of erythrocyte-

immunoglobulin M-complement complexes to such cells, probably by blocking their receptors for complement component 3. These findings emphasize the possibility, mentioned above, that different cell types may use the same or similar molecules for different functions.

CTL Induction

Leukocyte populations from normal animals have no demonstrable specific cytolytic activity prior to stimulation. Most of us have interpreted this to mean that there are non-cytolytic CTL-P that have to receive certain stimuli to become cytolytically competent. But most protocols for the generation of CTL activity induce a strong proliferation (i.e., enrichment of the specific cells), and the only experiments that clearly indicate that inactive CTL-P indeed exist are those in which activity is induced while DNA synthesis is prevented (89). We have chosen to call this process induction and not differentiation (or maturation) for the following reasons. We believe that the term differentiation should be applied to irreversible or virtually irreversible events by which a cell becomes committed to a certain lineage or stage. The molecular bases for such a commitment are probably heritable events at the level of the genome itself, e.g. demethylation of specific sites in certain genes (see 154 for review). Although as yet there is very little evidence concerning the mechanisms by which CTL-P become CTL, there is some reason to believe that this process may, to a large extent, be reversible and may resemble more the induction of the synthesis of specific enzymes in differentiated cells of a particular tissue.

More for a priori reasons than because of experimental evidence, it seems probable that the first step in the induction of CTL-P is their interaction with antigen. The hypothesis that antigen induces the expression of TCGF receptors on CTL-P and renders them responsive to the growth factor is based largely on an analogy with the experiments (discussed above) that show that Con A has these effects: short-term exposure (4 hr) to Con A certainly renders a purified T cell population TCGF-responsive (81). But these experiments have not been done with single cells and pure TCGF. Thus, they do not rule out that this effect of exposure to Con A (or antigen) requires yet other cellular interactions or other factors in the medium or the TCGF preparation. Recent experiments from Mescher's laboratory (62) have shown that in vivo-primed spleen cells from which adherent cells had been removed and that were exposed for 12 hr at 37°C with liposomes containing purified H-2 molecules, and then cultured for 4 days, developed strong CTL activity. This was four times higher (in terms of lytic units) when the medium during the second culture period was supplemented with proteins from supernatants of T-cell mitogen-activated leukocytes. The same authors show that the efficiency with which liposomes activate CTL-P is enhanced if they contain detergent-insoluble membrane matrix elements. These results are promising, but they do not prove that antigenic liposomes exert their effect directly by their interaction with CTL-P, and they certainly do not yet shed any light on the mechanism by which antigen or Con A activate cells.

The models concerning the role of TCGF in CTL generation are, to a large part, based on experiments with crude culture supernatants or ammonium sulfate-precipitated material as sources of biological activity, and it is not clear whether or not their effects on CTL-generation are due to a single factor. The purification of TCGF was based on its capacity to maintain proliferation of CTL lines, and it shares many functional properties with other growth factors, such as epidermal growth factor and platelet-derived growth factor, in that it seems to be required for the passage of dependent cells from G1 (or G0) into S-phase (132). Other polypeptide hormones and growth factors possess multiple biological activities, and TCGF may not only act as a growth factor but also, for example, as an inducer of the expression of cytolytic activity in CTL-P. Induced CTL do not seem to require TCGF to remain cytolytic: cloned CTL lines can be maintained for several days in the absence of exogenous TCGF without a decrease of the lytic activity per cell (131, 132). But we cannot yet rule out that these lines produce small amounts of TCGF, insufficient to allow growth but enough to maintain their cytolytic activity. This possibility seems not at all unlikely given the finding that some CTL clones do produce TCGF after antigenic stimulation.

Recent evidence from several laboratories indicates that TCGF and stimulator cells are not sufficient to induce CTL-P to become cytolytic. Raulet & Bevan (120) have shown that when spleen cells rendered TCGF-responsive by incubation with Con A are cultured in supernatant from Con A-stimulated spleen cells (CS), they acquire strong cytolytic activity. But when supernatant from a TCGF-producing cell line was used instead of CS, no lytic activity was induced. Further experiments suggest that the CTL-inducing activity in CS is a protein, different from lymphocyte activating factor or TCGF. Although the authors (120) show that CS from which the growth-promoting activity, i.e. TCGF, has been absorbed are still capable of inducing generation of CTL activity, they cannot rule out that TCGF is also required, because it is probably produced by the Con A-treated cells themselves. Kanagawa (manuscript in preparation) has shown that cells treated with leukoagglutinin, which renders cells TCGF-responsive, but unlike Con A does not induce its production, proliferate in culture media from TCGF-producing tumor cell lines as well as in secondary MLC supernatants, but generate CTL activity only in response to the latter. The proportion of Lyt 2^+ cells is the same in both populations. This suggests that mitogen-treated CTL-P can proliferate in response to TCGF but require additional factors to become cytolytic. It remains to be shown whether or not the same holds for CTL-P that respond to an antigenic stimulus rather than a mitogenic lectin. The CTL-inducing activity is found in supernatants from a fraction of alloreactive T-cell clones after stimulation with spleen cells and thus may well be a T-cell product. CTL lines or induced CTL populations maintain their cytolytic activity when they are grown in a source of TCGF that does not contain detectable inducing activity.

The Fate of Activated CTL

After in vivo immunization with appropriate antigens (allogeneic or tumor cells, or viruses), CTL activity can be detected only during a short period (19), but the primed animals show long-lasting immunologic memory in the sense that subsequent in vivo or in vitro challenges with the same stimulators result in an accelerated and stronger CTL response which reflects, in part, an increased frequency of CTL-P (86). But whether or not these memory CTL-P are derived from the CTL activated during priming, and whether or not they have different biological properties from virgin CTL-P, is not clear.

Activated CTL are not terminally differentiated cells: A CTL that has killed one or more target cells can divide and give rise to a clone (N. Thiernesse, unpublished results). Together with the high plating efficiency of CTL clones, close to 1, this finding renders very improbable any models that view CTL clones as continuously differentiating systems in which always only a fraction of cells becomes cytolytic during every cell generation.

As we described above, in a MLC both the proliferative and the CTL response are transient. In a primary MLC lytic activity per cell reaches a peak after about 5 days and then declines fairly rapidly. Restimulation with the same stimulator cells induces a faster and higher wave of CTL activity, which can probably be accounted for entirely by an enrichment of specific CTL-P (93). Also it is not clear in this case whether or not these primed CTL-P differ in any way from their virgin progenitors.

Cell fractionation studies have unambiguously shown that the CTL-P responding to secondary stimulation are small nondividing cells derived from the blast cell population that contains the CTL at the peak of the response (87). When MLC-responsive clones were screened for their CTL-P content, these were found only in clones that themselves had demonstrable CTL activity (91). These results suggest that active CTL can revert to small inactive lymphocytes that are CTL-P. However, we can not rule out that during stimulation in each clone the cells that have acquired

CTL activity die, possibly because TCGF becomes limiting, while blast cells that give rise to the secondary CTL response do not become cytolytically active.

In CTL clones, whether they belong to the filler cell-dependent (CTL-A) type or to the independent (CTL-B) type, the lytic activity per cell remains constant, even when growth is stopped by removal of TCGF (131, 132): they do not revert to CTL-P, and they do not survive for prolonged periods in the absence of TCGF-containing supernatants, i.e. without proliferation, whereas in vitro-primed CTL-P can survive for several weeks in the absence of any detectable TCGF. If the primed CTL-P generated in a MLC are indeed derived from the CTL activated during the primary response, then we must conclude that this reversion is not spontaneous but is induced by signals produced in a MLC, athough not in the conditions used for culturing CTL clones.

THE ONTOGENY OF CTL-P

In the adult mouse, CTL-P can be detected in the thymus as well as in peripheral lymphoid tissues such as spleen, lymph node, and blood (86). During ontogeny, CTL-P activity can first be detected in the thymus at or shortly after birth (24, 153). CTL-P in the spleen are not detected until 4–5 days of age. Frequency studies of CTL-P in the developing neonatal thymus provide strong (although indirect) evidence that in situ differentiation of these cells is occurring. Thus, thymic CTL-P frequencies increase by 10 to 40-fold in the 24 to 48-hr period following birth in C57BL/6 mice (153). Clearly, a synchronous immigration of these cells from some other organ cannot be excluded; however, more recent studies with organ-cultured thymic rudiments (25) provide direct evidence that CTL-P arise within the thymus itself. The latter system should provide a useful model for detailed analysis of the intrathymic differentiation pathway of CTL-P.

Although there is good experimental evidence that CTL-P arise in the thymus during ontogeny, the absolute necessity of this organ for CTL-P differentiation can be called into question. Recent studies from several laboratories (66, 95) have shown that CTL-P do develop in congenitally athymic (*nu/nu*) mice, albeit at a much slower rate than in their normal (*nu/+*) counterparts. Surface phenotype analysis of CTL-P in *nu/nu* mice has shown further that these cells are not of maternal origin (66) and that they express the Thy-1 and Lyt-2 antigens characteristic of cells of the CTL-P lineage in normal animals (94). However, some differences in antigenic specificity of CTL-P have been observed in *nu/nu* versus *nu/+* animals (66). These data provide direct evidence for the existence of an extrathymic differentiation pathway for CTL-P and hence raise the possibility that the thymus may have a quantitative rather than qualitative role in CTL-P development.

The conversion of immature T cells into CTL-P is a process that almost certainly would fall under our definition of differentiation, i.e. it probably reflects heritable changes in the state of a set of genes, the expression of which is required for the manifestation of the CTL phenotype. Thus genetic analysis of the differences between CTL and cell lines representing immature T cells should allow us to identify such genes, and the results of recent experiments involving crosses between murine thymoma and CTL lines suggest this is indeed the case (30, 32, 108). The relevant observations have been that (a) CTL lines differ from AKR-thymomas with regard to the following characteristics—they are TCGF-dependent, VV-sensitive, glucocorticoid-resistant, and cytolytically active; (b) hybrids between the two cell types behave like either one or the other parent; and (c) circumstantial evidence suggests that thymoma-like hybrids are derived from CTL-like ones by loss of a CTL chromosome, but this hypothesis has not yet been proved. We have suggested that the correlation of the expression of several CTL characteristics in the hybrids reflects the existence of one, or several, linked genes that control these traits and are expressed in CTL but not in the AKR-thymomas, and that heritable activation of this (these) gene(s) may be one of the steps in the differentiation of immature T cells into CTL-P.

One of the assumptions in this speculation is that AKR-thymomas indeed represent an immature stage of T-cell differentiation (and not, for example, another type of mature T cells). If, as is generally believed, AKR-thymomas are the results of a retrovirus-mediated activation or introduction of a transacting oncogene, then our speculation predicts that the immature progenitor cells of such thymomas are not dependent on TCGF, but that the transformation event abolishes some other growth requirement. Unfortunately, as yet no convincing set of criteria exist by which T-cell lines can be classified, but recent evidence suggests that the stage of maturation represented by lymphoid tumor cell lines is correlated with the presence of stage-specific transforming genes in such cells (79). Confirmation of these results may provide us with a very clear-cut system for T cell-line classification. In addition, analogy with other systems suggests the possibility that the transforming genes characteristic for a certain stage may be derived from cellular genes controlling its expression (14).

THE PHYSIOLOGICAL SIGNIFICANCE OF CTL

In recent years, numerous studies have addressed themselves to the potential physiological role of CTL. Although CTL were originally discovered (and are still most frequently studied) in allograft reactions (26), it is clear that this recent experimental invention cannot be the selective mechanism responsible for the maintenance of these cells throughout vertebrate evolution. The discovery that CTL were also generated in the course of immune responses to pathogenic viruses (161) provides a more plausible explanation (in evolutionary terms) for the existence of these potentially destructive cells. As discussed in detail previously, CTL responding to viruses behave in a MHC-restricted fashion in the sense that they preferentially lyse infected target cells bearing compatible (self) MHC antigens (160). Although this degree of specificity would seem to be at odds with the alloreactivity exhibited by CTL, recent studies of CTL clones have demonstrated a considerable degree of cross-reactivity between syngeneic virus-infected target cells and allogeneic uninfected target cells (17, 28). Thus, it is possible that all CTL do potentially recognize virus-modified syngeneic target cells, and that allogeneic cells represent a universe of antigenic determinants that resemble these modified self determinants.

Although the presence of CTL can be demonstrated in viral infections and allograft responses, and even at the site of rejection of certain allogeneic or virus-induced syngeneic tumors (58), it has not been a simple matter to demonstrate that CTL are directly involved in the pathology of these phenomena. At least two major difficulties are encountered in the interpretation of experiments in which some aspect of cellular immune function (skin graft or tumor graft rejection, delayed type hypersensitivity, lethal graft-versus-host disease, etc) is assessed in vivo following transfer of purified subpopulations of T lymphocytes. First, the large number of cells that must ordinarily be transferred, together with the long time period required to observe the effect of the transferred cells, combine to make it difficult to rule out the possibility that activity is mediated by rare contaminants selected for in vivo. This problem can be overcome, at least in theory, by the utilization of cloned populations of T lymphocytes; however, at the present time, there are few successful reports of systemic immunity mediated by cloned T cells (vide infra).

A second difficulty, which is not solved by the use of homogeneous T-cell populations, is the inherent complexity of cellular immune responses in vivo. Thus, graft rejection and other inflammatory-type reactions involve the participation of non-lymphoid cells (such as neutrophils and cells of the monocyte/macrophage series), as well as T lymphocytes. The underlying mechanisms of such reactions remain poorly understood (see 58), despite (or perhaps because of) the impressive array of lymphokines shown to be released from mixtures of sensitized T cells and appropriate antigen-presenting cells. Furthermore, it has proved extremely difficult (and maybe impossible) to eliminate the participation of endogenous (host-derived) cells in the inflammatory process by irradiation or drug treatment. In view of this complexity, the only conclusion that can be firmly derived from a cell transfer experiment would be that a particular subclass (or clone) of T lymphocytes is capable of provoking a particular immune reaction in vivo. The participation of other (host-derived) cells in this reaction cannot be ruled out, neither (more importantly) can it necessarily be excluded that another T-cell subclass (or clone) might be capable of mediating the same phenomenon.

In considering the evidence for a participation of CTL in graft or tumor rejection in vivo, it is clear that the above problems have frequently been ignored. Thus, investigators have concluded that CTL do or do not mediate a particular in vivo function simply on the basis of whether the test population transferred did or did not have detectable CTL activity. Studies of T cell subpopulations and their potential role in graft rejection are too voluminous to review here. Interested readers are advised to consult a recent volume on this topic (43). Insofar as data with cloned CTL populations are concerned, several workers have been successful in eliminating allogeneic or syngeneic (virus-infected) tumor cells by injecting the cloned CTL simultaneously at the site of tumor challenge in a Winn type assay (47). Clearly, such tests do not assess the ability of the CTL to recirculate and home to the site of tumor challenge in vivo. More recently, Engers et al (40) have obtained evidence that certain CTL clones are able to provoke the elimination of allogeneic tumor cells implanted at a distant site. In their studies, C57BL/6 anti-DBA/2 CTL clones were injected intravenously into sublethally irradiated syngeneic animals that received [125]IUdR-labeled P-815 mastocytoma (DBA/2) cells intraperitoneally. Tumor cell elimination was followed kinetically with a whole-body counting procedure. The results indicated that complete elimination of allogeneic tumor cells (and concomitant survival) could be achieved with some CTL clones, whereas no effect was seen with other clones of similar specificity. At present, the basis for this heterogeneity in the protective effects in vivo is not known, although preliminary studies suggest that it may be correlated with the ability of some CTL clones to proliferate specifically in response to antigen in the absence of exogenously added TCGF. Furthermore, as emphasized above, the actual mechanism whereby graft rejection occurs in such systems (i.e. via direct cytotoxicity or via indirect activation of other non-lymphoid cells) may prove difficult to assess, since cloned CTL are capable of releasing a variety of lymphokines upon exposure to appropriate antigens (51). In any event, many more data concerning the recirculation and homing patterns of CTL clones are required before any rational approach to immunotherapy with these cells can be devised.

Literature Cited

1. Ahman, G. B., Naeller, P. I., Binkraut, A., Hodes, R. J. 1979. T cell recognition in the mixed lymphocyte response. I. Non-T, radiation-resistant spleen adherent cells are the predominant stimulators in the murine mixed lymphocyte reaction. *J. Immunol.* 123: 903–9

2. Ahman, G. B., Naeller, P. I., Binkraut, A., Hodes, R. J. 1981. T cell recognition in the mixed lymphocyte response. II. Ia-positive splenic adherent cells are required for non-I region-induced stimulation. *J. Immunol.* 127:2308–13

3. Allison, A. C. 1974. Interactions of antibodies, complement components and various cell types in immunity against viruses and pyrogenic bacteria. *Transplant. Rev.* 19:3–55

4. Alter, B. J., Schendel, D. J., Bach, M. L., Bach, F. H., Klein, J., Stimpfling, J. 1973. Cell-mediated lympholysis. Importance of serologically defined H-2 regions. *J. Exp. Med.* 137:1303–9

5. Altman, A., Cohen, I. R. 1975. Cell-free media of mixed lymphocyte cultures augmenting sensitization in vitro of mouse T lymphocytes against allogeneic fibroblasts. *Eur. J. Immunol.* 5:437–44

6. Andrew, M. E., Braciale, T. J. 1981. Antigen-dependent proliferation of cloned continuous lines of H-2-restricted influenza virus-specific cytotoxic T lymphocytes. *J. Immunol.* 127:1201–4

7. Asantila, T., Vahala, J., Toivanen, P. 1974. Response of human fetal lymphocytes in xenogeneic mixed leukocyte culture: Phylogenetic and ontogenetic aspects. *Immunogenetics* 1:272–90

8. Bach, F. H., Amos, D. B. 1967. HU-1: Major histocompatibility locus in man. *Science* 156:1506–8

9. Bach, F. H., Widmer, M. B., Bach, M. L., Klein, J. 1972. Serologically defined and lymphocyte-defined components of the major histocompatibility complex in the mouse. *J. Exp. Med.* 136:1420–44

10. Bain, B., Vaz, M. R., Lowenstein, L. 1964. The development of large immature mononuclear cells in mixed lymphocyte cultures. *Blood* 23:108–16

11. Baker, P. E., Gillis, S., Smith, K. A. 1979. Monoclonal cytolytic T-cell lines. *J. Exp. Med.* 149:273–78

12. Berke, G. 1980. Interaction of cytolytic T lymphocytes and target cells. *Prog. Allergy* 27:69–133

13. Berke, G., Clark, W. R. 1982. T lymphocyte-mediated cytolysis—a comprehensive theory. I. The mechanism of CTL-mediated cytolysis. *Adv. Exp. Med. Biol.* 146:57–68

14. Beug, H., Palmieri, S., Freudenstein, Zentgraf, H., Graf, T. 1982. Hormone-dependent terminal differentiation in vitro of chicken erythroleukemia cells transformed by ts mutants of avian erythroblastosis virus. *Cell* 28:907–19

15. Bevan, M. J. 1975. Interaction antigens detected by cytotoxic T cells with the major histocompatibility complex as modifier. *Nature* 256:419–21

16. Bevan, M. J. 1977. In a radiation chimera, host H-2 antigens determine immune responsiveness of donor cytotoxic cells. *Nature* 269:417–18

17. Braciale, T. J., Andrew, M. E., Braciale, V. L. 1981. Heterogeneity and specificity of cloned lines of influenza-virus-specific cytotoxic T lymphocytes. *J. Exp. Med.* 153:910–23

18. Brondz, B. D. 1972. Lymphocyte receptors and mechanisms of in vitro cell-mediated immune reactions. *Transplant. Rev.* 10:112–51

19. Brunner, K. T., Cerottini, J.-C. 1971. Cytotoxic Lymphocytes as effector cells of cell mediated immunity. In *Progress in Immunology*, ed. B. Amos, pp. 385–98. New York: Academic

20. Brunner, K. T., Mauel, J., Cerottini, J.-C., Chapuis, B. 1968. Quantitative assay of the lytic action of immune lymphoid cells on ^{51}Cr-labeled allogeneic target cells in vitro; inhibition by isoantibody and by drugs. *Immunology* 14: 181–96

21. Cantor, H., Boyse, E. A. 1975. Functional subclasses of T lymphocytes bearing different Ly antigens. I. The generation of functionally distinct T cell subclasses is a differentiative process independent of antigen. *J. Exp. Med.* 141:1376–89

22. Cantor, H., Boyse, E. A. 1975. Functional subclasses of T lymphocytes bearing different Ly antigens. II. Cooperation between subclasses of Ly+ cells in the generation of killer activity. *J. Exp. Med.* 141:1390–99

23. Ceppelini, R. 1971. Old and new facts and speculations about transplantation antigens of man. In *Progress in Immunology*, ed. B. Amos, pp. 973–1025. New York: Academic

24. Ceredig, R. 1979. Frequency of alloreactive cytotoxic T cell precursors in the mouse thymus and spleen during ontogeny. *Transplantation* 28:377–81

300   NABHOLZ & MACDONALD

25. Ceredig, R., Jenkinson, E. J., Mac-Donald, H. R., Owen, J. J. T. 1982. Development of cytolytic T lymphocyte precursors in organ-cultured mouse embryonic thymus rudiments. *J. Exp. Med.* 155:617–22
26. Cerottini, J.-C., Brunner, K. T. 1974. Cell-mediated cytotoxicity, allograft rejection and tumor immunity. *Adv. Immunol.* 18:67–132
27. Cerottini, J.-C., MacDonald, H. R. 1981. Limiting dilution analysis of alloantigen-reactive T lymphocytes. V. Lyt phenotype of cytolytic T lymphocyte precursors reactive against normal and mutant H-2 antigens. *J. Immunol.* 126:490–96
28. Cerottini, J.-C., MacDonald, H. R., Brunner, K. T. 1982. Alloreactivity of self H-2 restricted cytolytic T lymphocytes. *Behr. Inst. Mitt.* 70:53–58
29. Conzelmann, A. 1982. *Somatic cell genetic analysis of cytolytic T-lymphocytes.* PhD thesis. Univ. Lausanne, Switzerland
30. Conzelmann, A., Corthésy, P., Nabholz, M. 1982. Correlation between cytolytic activity, growth factor dependence and lectin resistance in cytolytic T-cell hybrids. In *Isolation, Characterization and Utilization of T Lymphocyte Clones,* ed. C. G. Fathman, F. W. Fitch, pp. 205–15. New York: Academic
31. Conzelmann, A., Pink, R., Acuto, O., Mach, J. P., Dolivo, S., Nabholz, M. 1980. Presence of T145 on cytolytic T-cell lines and their lectin-resistant mutants. *Eur. J. Immunol.* 10:860–68
32. Conzelmann, A., Silva, A., Cianfriglia, M., Tougne, C., Sekaly, R. P., Nabholz, M. 1982. Correlated expression of TCGF-dependence, sensitivity to Vicia Villosa and cytolytic activity in hybrids between cytolytic T cells and T lymphomas. *J. Exp. Med.* In press
33. David, C. S., Shreffler, D. C., Frelinger, J. A. 1973. New lymphocyte antigen system (Lna) controlled by the Ir region of the mouse H-2 complex. *Proc. Natl. Acad. Sci. USA* 70:2509–14
34. Degiovanni, G., Cerottini, J.-C., Brunner, K. T. 1980. Generation of cytolytic T lymphocytes in vitro. XI. Accessory cell requirement in secondary responses to particulate alloantigen. *Eur. J. Immunol.* 10:40–45
35. Dialynas, D. P., Loken, H. R., Glasebrook, A. L., Fitch, F. W. 1981. Lyt 2⁻/Lyt 3⁻ variants of a cloned cytolytic T cell line lack an antigen receptor functional in cytolysis. *J. Exp. Med.* 153:595–604
36. Dutton, R. W. 1966. Spleen cell proliferation in response to homologous antigens studied in congenic resistant strains of mice. *J. Exp. Med.* 123:665–71
37. Dutton, R. W., Falkoff, R., Hirst, J. A., Hoffman, M., Kappler, J. W., Kettman, J. R., Lesley, J. F., Vann, D. 1971. Is there evidence for a non-antigen specific diffusable chemical mediator from the thymus-derived cell in the initiation of the immune response? In *Progress in Immunology,* ed. B. Amos, pp. 355–68. New York: Academic
38. Eijsvoogel, V. P., DuBois, M. J. G. J., Meinesz, A., Bierhorst-Eijlander, J., Zeylemaker, W. P., Schellekens, P. Th. A. 1973. The specificity and the activation mechanism of cell-mediated lympholysis (CML) in man. *Transplant. Proc.* 5:1675–78
39. Eijsvoogel, V. P., DuBois, M. J. G., Melief, C. J. M., De Groot-Kooy, M. L., Koning, C., Van Rood, J. J., Van Leeuwen, A., Du Toit, E., Schellekens, P. Th. A. 1972. Position of a locus determining mixed lymphocyte reaction (MLR), distinct from the known HL-A loci, and its relation to cell-mediated lympholysis (CML). *Histocompatibility Testing 1972, Munskgaard,* pp. 501–8
40. Engers, H. D., Glasebrook, A. L., Sorenson, G. D. 1982. Allogeneic tumor rejection induced by the intravenous injection of Lyt-2⁺ cytolytic T lymphocyte clones. *J. Exp. Med.* 156:1280–85
41. Engers, H. D., MacDonald, H. R. 1976. Generation of cytolytic T lymphocytes in vitro. *Contemp. Topics Immunobiol.* 5:145–90
42. Engers, H. D., Thomas, K., Cerottini, J.-C., Brunner, K. T. 1975. Generation of cytotoxic T lymphocytes in vitro. V. Response of normal and immune mouse spleen cells to subcellular alloantigen. *J. Immunol.* 115:356–60
43. Fefer, A., Goldstein, A. L., eds. 1982. The potential role of T cells in cancer therapy. *Progr. Cancer Res. Therapy* 22:1–297
44. Fischer Lindahl, K., Nordin, A. A., Schreier, M. H. 1982. Lectin-dependent cytolytic and cytolymic T helper clones and hybridomas. *Curr. Top. Microbiol. Immunol.* 100:1–10
45. Forman, J., Klein, J. 1975. Analysis of H-2 mutants: Evidence for multiple CML target specificities controlled by

the H-2K^b gene. *Immunogenetics* 1:469–81

46. Gillis, S., Smith, K. A. 1977. Long-term culture of tumour-specific cytotoxic T-cells. *Nature* 268:154–56

47. Gillis, S., Watson, J. 1981. Interleukin-2 dependent culture of cytolytic T cell lines. *Immunol. Rev.* 54:81–109

48. Ginsburg, H. 1968. Graft versus host reaction in tissue culture. I. Lysis of monolayers of embryo mouse cells from strains differing in the H-2 histocompatibility locus by rat lymphocytes sensitized in vitro. *Immunology* 14:621–35

49. Giorgi, J., Zawadzki, J. A., Warner, N. L. 1982. Cytotoxic T lymphocyte lines reactive against murine plasmacytoma antigens: dissociation of cytotoxicity and Lyt-2 expression. *Eur. J. Immunol.* In press

50. Glasebrook, A. L., Fitch, F. W. 1979. T cell lines which cooperate in generation of specific cytolytic activity. *Nature* 278:171–74

51. Glasebrook, A. L., Kelso, A., Zubler, R. H., Ely, J. M., Prystowsky, M. B., Fitch, F. W. 1982. Lymphokine production by cytolytic and noncytolytic alloreactive T cell clones. In *Isolation, Characterization and Utilization of T Lymphocyte Clones,* ed. C. G. Fathman, F. W. Fitch, pp. 341–54. New York: Academic

52. Glasebrook, A. L., Sarmiento, M., Loken, M. R., Dialynas, D. P., Quintans, J., Eisenberg, L., Lutz, C. T., Wilde, D., Fitch, F. W. 1981. Murine T-lymphocyte clones with distinct immunological functions. *Immunol. Rev.* 54:225–66

53. Gordon, R. D., Mathieson, B. J., Samelson, L. E., Boyse, E. A., Simpson, E. 1976. The effect of allogeneic presensitization on H-Y graft survival and in vitro cell-mediated responses to H-Y antigen. *J. Exp. Med.* 144:810–20

54. Govaerts, A. 1960. Cellular antibodies in kidney homotransplantation. *J. Immunol.* 85:516–20

55. Haas, W., Mathur-Rochat, J., Pohlit, H., Nabholz, M., v. Boehmer, H. 1980. Cytotoxic T cell responses to haptenated cells. III. Isolation and specificity analysis of continuously growing clones. *Eur. J. Immunol.* 10:828–34

56. Hämmerling, G. J., Deak, B. D., Mauve, G., Hämmerling, U., McDevitt, H. O. 1974. B lymphocyte alloantigens controlled by the I region of the major histocompatibility complex in mice. *Immunogenetics* 1:68–81

57. Hardy, D. A., Ling, N. R., Wallin, J. 1970. Destruction of lymphoid cells by activated human lymphocytes. *Nature* 227:723–25

58. Haskell, J. S., Häyry, P., Radov, L. A. 1978. Systemic and local immunity in allograft and cancer rejection. *Contemp. Topics Immunobiol.* 8:107–70

59. Hauptfeld, V., Klein, D., Klein, J. 1973. Serological identification of an Ir-region product. *Science* 181:167–69

60. Häyry, P. 1976. Anamnestic responses in mixed lymphocyte culture-induced cytolytic (MLC-CMC) reaction. *Immunogenetics* 3:427–53

61. Häyry, P., Defendi, V. 1970. Mixed lymphocyte cultures produce effector cells: Model in vitro for allograft rejection. *Science* 168:133–35

62. Herrman, S. H., Weinberger, O., Burakoff, S. J., Mescher, M. F. 1982. Analysis of the two-signal requirement for precursor cytolytic T lymphocyte activation using H-2K^k in liposomes. *J. Immunol.* 5:1968–74

63. Hirschhorn, K., Bach, F., Kolodny, R. L., Firschein, I. L., Hashem, N. 1963. Immune response and mitosis of human peripheral blood lymphocytes in vitro. *Science* 142:1185–87

64. Hodes, R. J., Svedmyr, E. A. J. 1970. Specific cytotoxicity of H-2-incompatible mouse lymphocytes following mixed culture in vitro. *Transplantation* 9:470–77

65. Hollander, N., Mehdi, S. G., Weissman, I. L., McConnell, H. M., Kriss, J. P. 1979. Allogeneic cytolysis of reconstituted membrane vesicles. *Proc. Natl. Acad. Sci. USA* 76:4042–45

66. Hünig, T., Bevan, M. T. 1980. Specificity of cytotoxic T cells from athymic mice. *J. Exp. Med.* 152:688–702

67. Jerne, N. K. 1971. The somatic generation of immune recognition. *Eur. J. Immunol.* 1:1–9

68. Kasakura, S., Lowenstein, L. 1965. A blastogenic factor in the medium of peripheral leukocyte cultures. *Histocompatibility Testing, Munskgaard,* pp. 203–9

69. Katz, D. H., Hamaoka, T., Dorf, E. D., Maurer, P. H., Benacerraf, B. 1973. Cell interactions between histocompatible T and B lymphocytes. IV. Involvement of the immune response (Ir) gene in the control of lymphocyte interactions in responses controlled by the gene. *J. Exp. Med.* 138:734–39

70. Kaufmann, Y., Berke, G. 1981. Cell surface glycoproteins of cytotoxic T

lymphocyte induced in vivo and in vitro. *J. Immunol.* 126:1443–46

71. Kelso, A., Glasebrook, A. L., Kanagawa, O., Brunner, K. T. 1982. Production of macrophage activating factor by T lymphocyte clones and correlation with other lymphokine activities. *J. Immunol.* 129:550–56

72. Kelso, A., MacDonald, H. R. 1982. Precursor frequency analysis of lymphokine-secreting alloreactive T lymphocytes: Dissociation of subsets producing interleukin-2, macrophage activating factor and granulocyte-macrophage colony stimulating factor on the basis of Lyt-2 phenotype. *J. Exp. Med.* 156:In press

73. Kimura, A., Wigzell, H. 1978. Cell surface glycoproteins of murine cytotoxic T lymphocytes. I. T145, a new cell surface glycoprotein selectively expressed on Lyt 1⁻2⁺ cytotoxic T lymphocytes. *J. Exp. Med.* 147:1418–34

74. Kimura, A., Wigzell, H., Holmquist, G., Ersson, B., Carlsson, P. 1979. Selective affinity fractionation of murine cytotoxic T lymphocytes (CTL): Unique lectin specific binding of CTL associated surface glycoprotein, T145. *J. Exp. Med.* 149:473–84

75. Kindred, B., Shreffler, D. C. 1972. H-2 dependence of cooperation between T and B cells in vivo. *J. Immunol.* 109: 940–43

76. Klein, J. 1976. *Biology of the Mouse Histocompatibility-2 Complex.* New York: Springer

77. Klein, J., Shreffler, D. C. 1971. The *H-2* model for major histocompatibility systems. *Transplant. Rev.* 6:3–29

78. Krammer, P. H., Marcucci, F., Waller, M., Kirchner, H. 1982. Heterogeneity of soluble T cell products. I. Precursor frequency and correlation analysis of cytotoxic and immune interferon (IFN-)-producing spleen cells in the mouse. *Eur. J. Immunol.* 12:200–4

79. Lane, M. A., Sainten, A., Cooper, G. M. 1982. Stage-specific transforming genes of human and mouse B- and T-lymphocyte neoplasms. *Cell* 28:873–80

80. Langhorne, J., Fischer Lindahl, K. 1982. Role of non-H-2 antigens in the cytotoxic T cell response to allogeneic H-2. *Eur. J. Immunol.* 12:101–6

81. Larsson, E.-L., Coutinho, A., Martinez, C. 1980. A suggested mechanism for T lymphocyte activation: Implications on the acquisition of functional activities. *Immunol. Rev.* 51:61–91

82. Lees, R. K., MacDonald, H. R., Sinclair, N. R. 1977. Inhibition of clone formation as an assay for T cell-mediated cytotoxicity: Short-term kinetics and comparison with ⁵¹Cr release. *J. Immunol. Methods* 16:233–44

83. Levey, R. H. 1974. Cellular aspects of transplantation immunity. In *Mechanisms of Cell-Mediated Immunity,* ed. R. T. McCluskey, S. Cohen, pp. 257–87. New York: Wiley

84. Lutz, C. T., Glasebrook, A. L., Fitch, F. W. 1981. Alloreactive Cloned T Cell Lines. IV. Interaction of alloantigen and T cell growth factors (TCGF) to stimulate cloned cytolytic T lymphocytes. *J. Immunol.* 127:391–92

85. Lutz, C. T., Glasebrook, A. L., Fitch, F. W. 1981. Enumeration of alloreactive helper T lymphocytes which cooperate with cytolytic T lymphocytes. *Eur. J. Immunol.* 11:726–34

86. MacDonald, H. R., Cerottini, J.-C., Ryser, J.-E., Maryanski, J. L., Taswell, C., Widmer, M. B., Brunner, K. T. 1980. Quantitation and cloning of cytolytic T lymphocytes and their precursors. *Immunol. Rev.* 51:93–123

87. MacDonald, H. R., Engers, H. D., Cerottini, J.-C., Brunner, K. T. 1974. Generation of cytotoxic T lymphocytes in vitro. II. Effect of repeated exposure to alloantigens on the cytotoxic activity of long-term mixed leukocyte cultures. *J. Exp. Med.* 140:718–30

88. MacDonald, H. R., Glasebrook, A. L., Bron, C., Kelso, A., Cerottini, J.-C. 1982. Clonal heterogeneity in the functional requirement for Lyt-2/3 molecules on cytolytic T lymphocytes (CTL): Possible implications for the affinity of CTL antigen receptors. *Immunol. Rev.* 68:89–115

89. MacDonald, H. R., Lees, R. K. 1980. Dissociation of differentiation and proliferation in the primary induction of cytolytic T lymphocytes by alloantigens. *J. Immunol.* 124:1308–13

90. MacDonald, H. R., Mach, J.-P., Schreyer, M., Zaech, P., Cerottini, J.-C. 1981. Flow cytofluorometric analysis of the binding of Vicia Villosa lectin to T lymphoblasts: Lack of correlation with cytolytic function. *J. Immunol.* 126: 883–86

91. MacDonald, H. R., Ryser, J.-E., Engers, H. D., Brunner, K. T., Cerottini, J.-C. 1978. Differentiation pathway of murine cytolytic T lymphocytes: Analysis by limiting dilution. In *Human Lymphocyte Differentiation: Its Application to Cancer,* ed. B. Serrou, C. Rosenfeld, pp. 23–28. Amsterdam: Elsevier/North Holland Biomed.

92. Martz, E. 1977. Mechanism of specific tumor-cell lysis by alloimmune T lymphocytes: Resolution and characterization of discrete steps in the cellular interaction. *Contemp. Topics Immunobiol.* 7:301–61
93. Maryanski, J. L., MacDonald, H. R., Cerottini, J.-C. 1980. Limiting dilution analysis of alloantigen-reactive T lymphocytes. IV. High frequency of cytolytic T lymphocyte precursor cells in MLC blasts separated by velocity sedimentation. *J. Immunol.* 124:42–47
94. Maryanski, J. L., MacDonald, H. R., Sordat, B., Cerottini, J.-C. 1981. Cell surface phenotype of cytolytic T lymphocyte precursors in aged nude mice. *Eur. J. Immunol.* 11:968–72
95. Maryanski, J. L., MacDonald, H. R., Sordat, B., Cerottini, J.-C. 1981. Cytolytic T lymphocyte precursor cells in congenitally athymic C57BL/6 nu/nu mice: quantitation, enrichment and specificity. *J. Immunol.* 126:871–76
96. Matzinger, P., Bevan, M. J. 1977. Why do so many lymphocytes respond to major histocompatibility antigens? *Cell. Immunol.* 29:1–5
97. McDevitt, H. O., Deak, B. D., Shreffler, D. C., Klein, J., Stimpfling, J. H., Snell, G. D. 1972. Genetic control of the immune response. Mapping of the *Ir-1* locus. *J. Exp. Med.* 135:1259–78
98. Meo, T., Vives, G., Rijnbeck, A. M., Miggiano, V. C., Nabholz, M., Shreffler, D. C. 1973. A bipartite interpretation and tentative mapping of *H-2*-associated MLR determinants in the mouse. *Transplant. Proc.* 5:1339–50
99. Mescher, M. F., Balk, S. P., Burakoff, S. J., Hermann, S. H. 1982. Cytolytic T lymphocyte recognition of subcellular antigen. *Adv. Exp. Med. Biol.* 146:41–55
100. Miggiano, V. C., Bernoco, D., Lightbody, J., Trinchieri, G., Ceppelini, R. 1972. Cell-mediated lympholysis in vitro with normal lymphocytes as target: Specificity and cross-reactivity of the test. *Transplant. Proc.* 4:231–37
101. Miller, J. F. A. P., Mitchell, G. F. 1968. Cell to cell interaction in the immune response, I. Hemolysin forming cells in neonatally thymectomized mice reconstituted with thymus or thoracic duct lymphocytes. *J. Exp. Med.* 128:801–20
102. Miller, R. A., Stutman, O. 1982. Monoclonal antibody to Lyt 2 antigen blocks H-2I- and H-2K-specific mouse cytotoxic T cells. *Nature* 296:76–78
103. Mitchison, N. A. 1954. Passive transfer of heterozygous tumors as material for

the study of cell heredity. *Proc. R. Phys. Soc.* 250:45–48.
104. Möller, G., ed. 1982. Effects of antimembrane antibodies on killer T cells. *Immunol. Rev.* 68:1–218
105. Morgan, D. A., Ruscetti, F. W., Gallo, R. C. 1976. Selective in vitro growth of T-lymphocytes from normal human bone marrows. *Science* 193:1007–8
106. Nabholz, M. 1982. The somatic cell genetic analysis of cytolytic T-lymphocyte functions. In *Isolation, Characterization and Utilization of T Lymphocyte Clones,* ed. C. G. Fathman, F. W. Fitch, pp. 165–81. New York: Academic
107. Nabholz, M., Conzelmann, A., Acuto, O., North, M., Haas, W., Pohlit, H., v. Boehmer, H., Hengartner, H., Mach, J. P., Engers, H., Johnson, J. P. 1980. Established murine cytolytic T-cell lines as tools for a somatic cell genetic analysis of T-cell functions. *Immunol. Rev.* 51:125–56
108. Nabholz, M., Conzelmann, A., Tougne, C., Corthésy, P., Silva, A. 1982. Cytolytic hybrids between murine CTL-lines and mouse or rat thymomas. In *B and T Cell Tumors: Biological and Clinical Aspects, UCLA Symposia on Molecular and Cellular Biology,* ed E. Vitetta, C. F. Fox, 24:115–25. New York: Academic
109. Nabholz, M., Engers, H. D., Collavo, D., North, M. 1978. Cloned T-cell lines with specific cytolytic activity. *Current Topics Microbiol. Immunol.* 81:176–87
110. Nabholz, M., Miggiano, V. C. 1977. The biological significance of the mixed leukocyte reaction. In *B and T cells in Immune Recognition,* ed. F. Loor, G. E. Roelants, pp. 261–89. New York: Wiley
111. Nabholz, M., Vives, J., Young, H. M., Meo, T., Miggiano, V., Rijnbeck, A., Shreffler, D. C. 1974. Cell mediated lysis in vitro: Genetic control of killer cell production and target specificities in the mouse. *Eur. J. Immunol.* 4:378–87
112. Nabholz, M., Young, H., Rijnbeck, A., Boccardo, R., David, C. S., Meo, T., Miggiano, V., Shreffler, D. C. 1975. I-region associated determinants: Expression on mitogen stimulated lymphocytes and detection by cytotoxic T-cells. *Eur. J. Immunol.* 5:594–99
113. Nagy, Z., Elliott, B. E., Nabholz, M. 1976. Specific binding of *K*- and *I*-region products of the H-2 complex to activated thymus-derived (T) cells belonging to different Lyt subclasses. *J. Exp. Med.* 144:1545–53

114. Ohno, S. 1977. The original function of MHC antigens as the general plasma membrane anchorage site of organogenesis-directing proteins. *Immunol. Rev.* 33:56–59

115. Paetkau, V., Shaw, J., Mills, G., Caplan, B. 1980. Cellular origins and targets of costimulator (IL-2). *Immunol. Rev.* 51:157–75

116. Perlman, P., Cerottini, J.-C. 1979. Cytotoxic lymphocytes. In *The Antigens*, ed. M. Sela. 5:173–281. New York: Academic

117. Perlman, P., Holm, G. 1969. Cytotoxic effects of lymphoid cells in vitro. *Adv. Immunol.* 11:117–93

118. Plata, F., Jongeneel, V., Cerottini, J.-C., Brunner, K. T. 1976. Antigenic specificity of the cytolytic T lymphocyte (CTL) response to murine sarcoma virus (MSV)-induced tumors. I. Preferential reactivity of in vitro generated secondary CTL with syngeneic tumor cells. *Eur. J. Immunol.* 6:823–29

119. Plate, J. M. D. 1976. Soluble factors substitute for T-T-cell collaboration in generation of T-killer lymphocytes. *Nature* 260:329–31

120. Raulet, D. H., Bevan, M. J. 1982. A differentiation factor required for the expression of cytotoxic T-cell function. *Nature* 296:754–57

121. Raulet, D. H., Bevan, M. J. 1982. Helper T cells for cytotoxic T lymphocytes need not be *I* region restricted. *J. Exp. Med.* 155:1766–84

122. Robb, R. J., Munck, A., Smith, K. A. 1981. T cell growth factor receptors. Quantitation, specificity, and biological relevance. *J. Exp. Med.* 154:1455–74

123. Rosenau, W., Moon, H. D. 1961. Lysis of homologous cells by sensitized lymphocytes in tissue culture. *J. Nat. Cancer Inst.* 27:471–83

124. Rosenau, W., Morton, D. L. 1966. Tumor-specific inhibition of growth in methyl cholanthrene-induced sarcomas in vivo and in vitro by sensitized isologous lymphoid cells. *J. Nat. Cancer Inst.* 36:825–36

125. Rosenthal, A. S., Shevach, E. M. 1973. Function of macrophages in antigen recognition by guinea pig T-lymphocytes. I. Requirement for histocompatible macrophages and lymphocytes. *J. Exp. Med.* 138:1194–212

126. Ryser, J.-E., Cerottini, J.-C., Brunner, K. T. 1978. Generation of cytolytic T lymphocytes in vitro. IX. Induction of secondary CTL responses in primary long-term MLC by supernatants from secondary MLC. *J. Immunol.* 120:

370–77

127. Ryser, J.-E., MacDonald, H. R. 1979. Limiting dilution analysis of alloantigen-reactive T lymphocytes. I. Comparison of precursor frequencies for proliferative and cytolytic responses. *J. Immunol.* 122:1691–96.

128. Ryser, J.-E., Rungger-Brändle, E., Chaponnier, C., Gabbiani, G., Vassalli, P. 1982. The area of attachment of cytotoxic T lymphocytes to their target cells shows high motility and polarization of actin, but not myosin. *J. Immunol.* 128:1159–62

129. Sachs, D. H., Cone, J. L. 1973. A mouse B-cell alloantigen determined by gene(s) linked to the major histocompatibility complex. *J. Exp. Med.* 138:1289–304

130. Schendel, D. J., Alter, B. J., Bach, F. H. 1973. The involvement of LD- and SD-region differences in MLC and CML: A three-cell experiment. *Transplant. Proc.* 5:1651–55

131. Sekaly, R. P., MacDonald, H. R., Zaech, P., Glasebrook, A. L., Cerottini, J.-C. 1981. Cytolytic T lymphocyte function is independent of growth phase and position in the mitotic cycle. *J. Exp. Med.* 154:575–80

132. Sekaly, R. P., MacDonald, H. R., Zaech, P., Nabholz, M. 1982. Cell cycle regulation of cloned cytolytic T cell by T cell growth factor: Analysis by flow microfluorometry. *J. Immunol.* 129:1407–15

133. Shearer, G. M., Rehn, T. G., Garbarino, C. A. 1975. Cell-mediated lympholysis of trinitrophenyl-modified autologous lymphocytes: Effector cell specificity to modified cell surface components controlled by the H-2K and H-2D serological regions of the murine major histocompatibility complex. *J. Exp. Med.* 141:1348–64

134. Sherman, L. A. 1980. Dissection of the B10·D₂ anti-H-2Kᵇ cytolytic T lymphocyte receptor repertoire. *J. Exp. Med.* 151:1386–97

135. Shevach, E. M., Rosenthal, A. J. 1973. Function of macrophages in antigen recognition by guinea pig T-lymphocytes. II. Role of the macrophage in the regulation of genetic control of the immune response. *J. Exp. Med.* 138:1213–29

136. Smith, K. A. 1980. T-Cell Growth Factor. *Immunol. Rev.* 51:336–57

137. Smith, K. A. 1982. T-cell growth factor and glucocorticoids: Opposing regulatory hormones in neoplastic T-cell growth. *Immunobiology* 161:157–73

138. Snell, G. D. 1978. T Cells, T cell recognition structures and the major histocompatibility complex. *Immunol. Rev.* 38:3–69

139. Solliday, S., Bach, F. H. 1970. Cytotoxicity: Specificity after in vitro sensitization. *Science* 170:1406–9

140. Springer, T. A., Davignon, D., Ho, M. K., Kürzinger, K., Martz, E., Sanchez-Madrid, F. 1982. LFA-1 and Lyt-2,3 molecules associated with T lymphocyte-mediated killing; and Mac-1, an LFA-1 homologue associated with complement receptor function. *Immunol. Rev.* 68:171–95

141. Steinman, R. M., Witmer, M. D. 1978. Lymphoid dendritic cells are potent stimulators of the primary mixed leukocyte reaction in mice. *Proc. Natl. Acad. Sci. USA* 75:5132–36

142. Swain, S. L., Dennert, G., Wormsley, S., Dutton, R. W. 1981. The Lyt phenotype of a long-term allospecific T cell line. Both helper and killer activities to I-A are mediated by Ly-1 cells. *Eur. J. Immunol.* 11:175–80

143. Taswell, C., MacDonald, H. R., Cerottini, J.-C. 1980. Clonal analysis of cytolytic T lymphocyte specificity. I. Phenotypically distinct sets of clones as the cellular basis of cross-reactivity to alloantigens. *J. Exp. Med.* 151:1372–85

144. Trinchieri, G., Bernoco, D., Curtoni, S. E., Miggiano, V. C., Ceppelini, R. 1972. Cell mediated lympholysis in man: Relevance of HL-A antigens and antibodies. *Histocompatibility Testing 1972, Munskgaard,* pp. 509–19

145. Unanue, E. R. 1981. The regulatory role of macrophages in antigenic stimulation. II. Symbiotic relationship between lymphocytes and macrophages. *Adv. Immunol.* 31:1–136

146. von Boehmer, H., Haas, W. 1976. Cytotoxic T lymphocytes recognize allogeneic tolerated TNP-conjugated cells. *Nature* 261:141–42

147. von Boehmer, H., Haas, W., Jerne, N. K. 1978. Major histocompatibility complex linked immune responsiveness is acquired by lymphocytes of low-responder mice differentiating in the thymus of high-responder mice. *Proc. Natl. Acad. Sci. USA* 75:2439–42

148. von Boehmer, H., Hengartner, H., Nabholz, M., Lernhart, W., Schreier, M. H., Haas, W. 1979. Fine specificity of a continuously growing killer cell clone specific for H-Y antigen. *Eur. J. Immunol.* 9:592–97

149. Weaver, J. M., Algire, G. H., Prehn, R. T. 1955. The growth of cells in vivo in diffusion chambers. II. The role of cells in the destruction of homografts in mice. *J. Nat. Cancer Inst.* 15:1737–58

150. Weiss, A., Brunner, K. T., MacDonald, H. R., Cerottini, J.-C. 1980. Antigenic specificity of the cytolytic T lymphocyte (CTL) response to murine sarcoma virus-induced tumors. III. Characterization of CTL clones specific for Moloney leukemia virus associated cell surface antigens. *J. Exp. Med.* 152:1210–25

151. Widmer, M. B., Alter, B. J., Bach, F. H., Bach, M. L. 1973. Lymphocyte reactivity to serologically undetected components of the major histocompatibility complex. *Nature* 242:239–41

152. Widmer, M. B., MacDonald, H. R. 1980. Cytolytic T lymphocyte precursors reactive against mutant K^b alloantigens are as frequent as those reactive against a whole foreign haplotype. *J. Immunol.* 124:48–51

153. Widmer, M. B., MacDonald, H. R., Cerottini, J.-C. 1981. Limiting dilution analysis of alloantigen reactive T lymphocytes. IV. Ontogeny of cytolytic T lymphocyte precursors in the thymus. *Thymus* 2:245–55

154. Wigler, M. 1981. The inheritance of methylation patterns in vertebrates. *Cell* 24:282–85

155. Wilson, D. B., Billingham, R. E. 1967. Lymphocytes and transplantation immunity. *Adv. Immunol.* 7:189–273

156. Winchester, R. J., Kunkel, H. G. 1979. The human Ia-system. *Adv. Immunol.* 28:222–92

157. Yunis, E. J., Amos, D. B. 1971. Three closely linked genetic systems relevant to transplantation. *Proc. Natl. Acad. Sci. USA* 68:3031–35

158. Zinkernagel, R. M. 1976. Virus-specific T cell-mediated cytotoxicity across the H-2 barrier to virus altered alloantigen. *Nature* 261:139–41

159. Zinkernagel, R. M., Callahan, G. N., Althage, A., Cooper, S., Klein, P. A., Klein, J. 1978. On the thymus in the differentiation of "H-2 self-recognition" by T cells: Evidence for dual recognition? *J. Exp. Med.* 147:882–96

160. Zinkernagel, R. M., Doherty, P. C. 1974. Restriction of in vitro T cell mediated cytotoxicity in lymphocytic choriomeningitis within a syngeneic or semiallogeneic system. *Nature* 248:701–2

161. Zinkernagel, R. M., Doherty, P. C. 1974. Immunological surveillance against altered self-components by sensitized T lymphocytes in lymphocytic choriomeningitis. *Nature* 251:547–48

(*continued*)

NOTE ADDED IN PROOF   (see p. 288)

It is now clear that Lyt-2⁻ CTL clones specific for I-region determinants do exist, as well as their human counterparts (ie OKT8⁻ CTL clones specific for HLA-DR determinants). The frequency of such clones has not yet been determined, and we therefore do not know what proportion of I-region-specific CTL-P is of this phenotype.

Ann Rev. Immunol. 1983. 1:307–33

REGULATION OF B-CELL GROWTH AND DIFFERENTIATION BY SOLUBLE FACTORS

Maureen Howard and William E. Paul

Laboratory of Immunology, National Institute of Allergy and Infectious Diseases, National Institutes of Health, Bethesda, Maryland 20205

INTRODUCTION

Interleukins are a family of molecules that transmit growth and differentiation signals between various types of leukocytes and thus presumably are major effectors of immune regulation. Cell biologists and immunologists have directed much attention to the nature of interleukins that operate on T lymphocytes (84, 85) and macrophages (15, 106). Recently, it has become clear that such antigen-nonspecific cofactors also operate on B lymphocytes (5, 27, 49, 51, 52, 56, 63, 73, 97, 99, 104, 110, 129, 135, 141, 142). However, as is common in an emerging field, an overall understanding of the involvement of soluble factors in B-cell responses has been hampered by competing nomenclatures and by the use of differing operational criteria to identify, describe, enumerate, and characterize the various B cell-specific interleukins. Here we review a variety of observations derived from both single and multicellular in vitro assays, using both mouse and human lymphocytes and numerous sources of interleukins, in an effort to derive a unified concept of factor-mediated B-cell development. Such an understanding will require an appreciation of the substantial functional heterogeneity of B cells, e.g. subsets, lineages, differentiation stages, and individual specificity. Thus, it is desirable to preface our review with a brief description of the physiology of this system.

Splenic B lymphocytes of the mouse may be divided into two major subpopulations of approximately equal size that differ from one another in the membrane antigens they express, in the immunogens to which they

307

0732-0582/83/0410-0307$02.00

respond, and in their sensitivity to distinct regulatory mechanisms. The appreciation that such subpopulations exist is based, to a very large extent, on studies of mice that have an immunologic defect determined by the *xid* gene (1, 22, 53, 54, 89, 100, 117, 128). This X chromosome gene, when present in the hemizygous or homozygous state, leads to a series of B-lymphocyte abnormalities, among which are unresponsiveness to soluble polysaccharides and other type II antigens (89), depressed serum IgM and IgG3 concentrations (100), and failure of B lymphocytes to proliferate upon stimulation with anti-immunoglobulin (Ig) antibodies (117). Mice with the *xid* defect lack B lymphocytes that bear the Lyb3 (54), Lyb5 (1), Lyb7 (128), and Ia.W39 (53) antigens. Lymphocytes bearing these differentiation antigens are referred to as Lyb5$^+$ B lymphocytes. Lyb5$^+$ B lymphocytes are found in all normal strains and the defects of *xid* mice can be accounted for by the absence of these cells. The properties assigned to these two major B cell subpopulations (i.e. Lyb5$^+$ and Lyb5$^-$) are based on studies either of the responsiveness of B cells from *xid* mice, or of normal Lyb5$^-$ B cells prepared by treating B cells from normal mice with anti-Lyb5 antiserum and complement.

Of particular interest for this review, a major feature that distinguishes Lyb5$^+$ and Lyb5$^-$ B cells is their requirements for activation and for proliferative responses to antigen. According to the work of Singer, Hodes and colleagues (8, 120), Lyb5$^+$ B cells are readily responsive to the combination of antigen and soluble helper factors. In contrast, Lyb5$^-$ B cells require a contactual major histocompatibility complex (MHC)-restricted (cognate) interaction with helper T cells. Similar conclusions have recently been reached by Takatsu and colleagues (134, 138). Using either normal mice in which the ability to respond to a soluble T cell-derived helper factor had been suppressed via neonatal treatment with an alloantiserum directed against the acceptor site for this factor (134) or, alternatively, utilizing the mutant mouse strain DBA/2Ha, which is unresponsive to this soluble helper factor (138), they clearly demonstrated a distinction between B cells that are responsive to factors alone and those subject to cognate T-cell interactions. Both subsets appear to require soluble differentiation factors [T cell replacing factors (TRFs)] to drive proliferating cells into Ig-secreting cells (ISC). These concepts are summarized schematically in Figure 1. Based on such a compartmentalization of B-cell function, this review focuses on three main areas of factor regulation: (*a*) the nature of soluble helper factors utilized by Lyb5$^+$ B cells for initial activation and proliferative events; (*b*) the nature of differentiation factors that drive proliferating B cells into ISC; and (*c*) a consideration of the role of soluble factors in the B-cell-responses dependent upon cognate interactions with helper T cells.

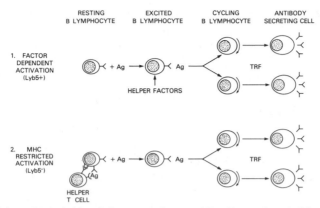

Figure 1 Schematic depiction of the two pathways of B-cell growth and differentiation. MHC-restricted activation may also be designated cognate interaction.

MHC-UNRESTRICTED B-CELL TRIGGERING

Evidence for Unique B-Cell Proliferation Cofactors

According to the basic tenets of clonal selection (16), immune responses involve the activation of a small number of precursor cells committed in specificity to the invading antigen, which results in their proliferation and differentiation into a population of functional effector cells. Early experimental evidence indicating that B-cell responses included a significant proliferative response in addition to obvious maturation was obtained by demonstrating autoradiographically that most antibody-secreting cells (ASC) had incorporated tritiated thymidine into their DNA when exposed to the labeled nucleotide 1 or 2 days preceding the assay (42) and by the demonstration that bromodeoxyuridine and light markedly diminished the number of plaque-forming cells appearing in primary in vitro responses to sheep erythrocytes (28). More recently, sophisticated B-cell cloning assays have been developed that not only demonstrate specific antigen-initiated B cell proliferation, but that actually measure the number of ASC derived from a single precursor, thus providing a minimum estimate of the degree of antigen-driven proliferation of precursors of ASC (2, 80, 94, 95).

With the ultimate goal of regulating B-cell function, considerable attention has been focused on the nature of signals governing B-cell proliferation and differentiation. The discovery that discrete TRFs were involved in the production of ASC (27, 110) seemed adequately to accommodate the well-recognized nonspecific role of T cells in B-cell immunity. This, together with reports that antigen alone induced B-cell proliferation (25, 26) and that TRF functioned when added after B-cell exposure to antigen (110), led to the proposal of a model in which nonspecific TRF induced antibody secretion by acting on B cells already proliferating in response to antigen (26,

111). Data from a number of laboratories would now suggest this model is oversimplified. As is outlined in more detail below, Andersson, Melchers and their colleagues were among the first to propose that the occupancy of membrane Ig by antigen and/or cognate T-cell interactions involving the Ia antigens of resting B cells resulted in functional expression of receptors for soluble factors and that it was these factors that in turn stimulated B cells to replicate and then to differentiate (5, 6). Although the expression of growth-factor receptors on B cells has yet to be demonstrated experimentally, the general notion is supported by numerous subsequent reports. A particularly valuable system for addressing events related to B-cell proliferation has been the use of purified anti-Ig antibody as a polyclonal analogue for antigen, first developed in the rabbit system by Sell & Gell (115) and adapted to mouse lymphocytes by Parker (96) and Weiner et al (143). Anti-Ig combines the advantages that it acts upon the receptors for antigen of virtually all B cells and that the resultant activation is restricted to DNA synthesis without development of ISC (63, 99). Using this system, Parker showed that soluble T cell-derived factors, which had no apparent effect on resting B cells, enhanced anti-Ig-induced B-cell proliferation particularly when relatively low concentrations of anti-Ig were used (97, 98). This was the first clear demonstration of soluble factors that could influence the proliferative component of a B-cell response. More recently, definitive evidence that antigen-induced B-cell proliferation requires soluble cofactors has been provided by Nossal and colleagues (101, 141). Using the elegant single cell cloning assay of Wetzel & Kettman (144), they cultured single affinity-purified hapten-specific B cells and showed that antigen-specific B-cell proliferation was only obtained in the presence of exogenously supplied cofactor-rich supernatants.

A major problem in precisely identifying factors involved in B-cell proliferation has been the lack of simple functional assays in which contaminating accessory cells capable of mediating secondary effects have been excluded. Even some B-cell cloning assays do not escape this dilemma as they often depend on filler cell supplements. In an attempt to overcome this problem, we have developed a short-term B-cell costimulator assay (49) in which highly purified B cells are cultured at low cell density, thus greatly reducing the possibility of accessory cell contamination. Encouraged by the observations of Parker (97, 98), we selected affinity-purified anti-IgM antibodies as a polyclonal analogue for antigen. Under appropriate culture conditions, anti-IgM causes approximately 50% of resting splenic B cells to synthesize DNA (22). As mentioned above, these responsive cells are absent from *xid* mice and they appear to correspond to the Lyb5[+] population of normal B cells (1). Thus, the use of anti-IgM antibodies seemed an appropriate choice for defining the soluble proliferation cofactors shown in Figure 1.

Using this low-cell-density B-cell costimulator assay, and measuring DNA synthesis via [³H]-thymidine uptake after 2.5 days of culture, we found that three separate stimuli were required to cause the Lyb5⁺ subset of small B lymphocytes to enter S phase: a signal delivered by antibodies specific for the Ig receptor expressed on the B-cell membrane; a signal delivered by a T cell-derived factor, which we designated B-cell growth factor (BCGF); and a signal delivered by the macrophage-derived factor interleukin 1 (IL1) (49, 52). BCGF and IL1 acted in an antigen-nonspecific, synergistic manner. The requirement for BCGF and IL1 had not previously been observed when B cells were cultured at high density, presumably reflecting accessory cell contamination and endogenous factor production at such cell densities. Full details of our current progress in characterization of BCGF are provided in the following section. Identification of the macrophage product involved in anti-Ig-induced B-cell proliferation as IL1 is based on correlation of B cell and thymocyte costimulating activities following a series of biochemical purification procedures (52). Both murine IL1 purified to apparent homogeneity by Mizel (83) and human IL1 purified to high specific activity by Lachman (65, 67) showed excellent B-cell costimulating activity in our assay system. Indeed, in the final step in the mouse IL1 purification scheme, B-cell and thymocyte costimulatory activity migrated identically. Although our data provide compelling support for the notion that BCGF and IL1 are the proliferation cofactors illustrated in Figure 1, we wish to emphasize that ultimate proof that these agents act directly on B cells will await reagents capable of demonstrating specific factor-receptor binding.

Our results are consistent with many previous observations documented in the literature and help resolve some areas of conflict. In particular, the involvement of accessory cells in anti-IgM-induced B-cell proliferation has previously been a matter of controversy. It had been proposed by some laboratories that anti-Ig-induced proliferation proceeded independently of T cells and other accessory cells, as vigorous depletion of such cell types failed to prevent activation (96, 117). However, involvement of macrophages had been reported by Mongini et al (86) and Parker's more recent work on T cell-derived factors suggested a need for T cells (97). It would now seem that failure to observe the need for one or both cell types reflects the use of assay conditions in which small numbers of one or both accessory cells are present with consequent endogenous factor production. Our own experiments show a need for both exogenous BCGF and IL1 only at low cell densities (2 X 10⁴ cells/well and below), despite the rigorous B-cell purification procedure we have employed. Thus, it would appear that very small numbers of accessory cells can provide B cell-stimulatory cofactors.

The proposal that IL1 is involved in B-cell function is not a new concept (34, 46–48, 76, 112, 148–150). However, whereas others have previously suggested such a role for IL1, their experiments measured ASC responses in high cell density culture systems and thus failed to distinguish a direct action on B cells versus secondary actions via other cell types. A particular problem in this respect is that IL1 appears responsible for the generation of at least some T cell-derived lymphokines, e.g. interleukin 2 (IL2) (35, 38, 71, 122, 123) and colony-stimulating factor (88). With the recent identification of several distinct B cell-specific lymphokines (5, 27, 49, 51, 52, 56, 63, 73, 97, 99, 104, 110, 129, 135, 141, 142), one could certainly envisage IL1-mediated factor cascades in many biological assays for B-cell function. Although such objections may also be leveled at the experiments from our laboratory outlined above, we believe they are not likely to apply here for the following reasons: (a) We have used very highly purified B cells cultured at low cell densities, thus greatly reducing the possibility of accessory cell contamination; (b) as is discussed in the following section, we show IL1 may be added quite late in the G_1 phase of the B-cell cycle, thus providing little opportunity for a factor cascade to develop; (c) we have shown that several other T cell-derived B cell-specific lymphokines, e.g. BCGF, and two distinct TRFs [EL-TRF and B15-TRF (see 93 for nomenclature)] fail to replace the need for IL1. Again, however, ultimate proof that IL1 acts directly on B cells will await reagents capable of demonstrating specific binding.

Characteristics of BCGF

In cultures of mouse B lymphocytes, the term BCGF is currently used to designate the T cell-derived B cell costimulator that synergizes with affinity-purified anti-IgM antibodies to induce polyclonal B-cell proliferation (49). Although IL1 is not generally added to these cultures, the conditions usually employed for assays of B-cell activation probably allow its endogenous production. BCGF is found in culture supernatants of a subline of the thymoma, EL4, which has been stimulated with phorbol myristate acetate (49) and in antigen-induced supernatants from normal T-cell helper lines (50) propagated in long-term culture, as described by Kimoto & Fathman (60). Surprisingly, it is not detected in supernatants of spleen cells cultured for 24 hr with concanavalin A or phytohemagglutinin. BCGF has little or no apparent action on resting B cells and its effect on anti-IgM-activated B cells is polyclonal, not MHC-restricted, and seemingly is limited to DNA synthesis without the production of ASC (49). It also acts as a proliferation costimulant for some, but probably not all, B cells activated with lipopolysaccharide. When cultures are supplemented with additional differentiation cofactors, BCGF is found to be an essential component of both antigen-specific (49) and polyclonal anti-Ig-induced (93) ASC responses (see MHC-Unrestricted B-Cell Differentiation). Preliminary experiments indicate that

BCGF is required for antigen-specific proliferation of affinity-purified hapten-specific B cells stimulated by hapten conjugated to the semisynthetic polysaccharide Ficoll (52a).

BCGF can be separated from IL2 by several different methods, including gel filtration, phenylsepharose chromatography, sodium dodecyl sulfate (SDS)-polyacrylamide gel electrophoresis (PAGE), tris(hydroxymethyl)aminomethane (Tris)-glycinate-PAGE, and isoelectric focusing, as well as by cellular absorption (31, 33, 49). BCGF can be distinguished both functionally and by molecular weight from other previously described T-cell factors, namely TRF (27, 109, 110, 142), colony-stimulating factor (15), interleukin 3 (44), B cell replication and maturation factor (5), and B cell differentiation factor (BCDFμ) (56). It does not cause proliferation of phytohemagglutinin-stimulated thymocytes, indicating that it is also distinct from IL1 (67). The proportion of B cells responsive to BCGF has not yet been established. However, the levels of proliferation obtained in B-cell costimulator experiments are consistent with involvement of the entire anti-IgM-responsive B-cell pool, a pool recently shown to consist of approximately 50% of normal B cells (22). Splenic B cells from *xid* mice fail to synthesize DNA in response to the combination of anti-IgM and BCGF (50). Independent confirmation of the existence of murine BCGF with physical and biological properties similar to those outlined above has recently been provided by Leanderson et al (72).

Farrar et al (31, 33, 49) have studied the physicochemical properties of BCGF. They have shown that BCGF is a polypeptide by virtue of its trypsin sensitivity. BCGF produced by phorbol myristate acetate-induced EL4 cells in serum-containing media is readily separated from the bulk of proteins in such culture supernatants by ammonium sulfate precipitation. More than 90% of the BCGF activity is precipitated between 80 and 100% saturated ammonium sulfate; only 5–9% of the total protein is found in this fraction. By conventional gel filtration on ACA 54 in neutral solution, BCGF elutes as a broad peak between 13,000 and 20,000 daltons. Analysis of BCGF by SDS-PAGE reveals two components that exhibit biologic activity after extraction from the gels and dialysis. One has a monomeric M_r of 11,000, the other 14,000. A degree of microheterogeneity is also observed when BCGF is examined by electrofusing. Two peaks of BCGF activity are observed migrating with a pI of 6.4-6.7 and 7.4-7.6, respectively. BCGF can be completely separated from IL2 by a combination of ammonium sulfate precipitation, phenylsepharose column chromatography, and Tris-glycine gel electrophoresis. Surprisingly, BCGF displays a lower electrophoretic mobility than IL2 on Tris-glycine gels, despite its lower molecular weight; this is probably due to the alkaline pI's of BCGF. Although BCGF has not been purified to homogeneity, it has been brought to very high specific activity by a sequential purification procedure involving ammonium sulfate precipitation, phenylsepharose chromatography, Tris-glycinate PAGE, and SDS-PAGE (33).

Four groups have isolated human B-cell growth factors (17, 37, 64, 75, 90, 91, 126 152). The experimental systems used in these laboratories vary, and it is not yet clear whether the same factor is being investigated by each of the groups, or whether the human BCGFs are analogous to the mouse factor. Nevertheless, some uniform conclusions regarding the nature of human BCGF(s) have emerged. Human BCGF is currently used to designate either a material that costimulates with anti-Ig or other polyclonal B-cell activators to cause DNA synthesis by normal or leukemic human B cells (64, 90, 91, 152), or alternatively a material that maintains the proliferation of activated purified human B lymphocytes in suspension culture (37, 75, 126). Obviously, these two functions may well be mediated by different moieties. Human BCGF defined by both assays can be found in supernatants of lectin-activated normal T lymphocytes (37, 90, 126, 152) and also in T-cell hybridoma supernatants (17, 64). Human BCGF can be distinguished from human IL2 by its relatively delayed appearance after lectin stimulation of human T cells and by cellular absorption studies with either normal human T-cell blasts or long-term IL2-dependent T-cell lines (37, 64, 91, 152). Such cells remove IL2 but do not diminish BCGF activity in culture supernatants. The existence of T hybridomas that secrete BCGF and not IL2 (17, 64), and of conditioned media that lack one or other activity (107), verify this distinction. Biochemical characterization of human BCGF has commenced in three laboratories; by standard gel filtration procedures, material with BCGF activity in 72-hr lectin-stimulated peripheral blood lymphocyte-conditioned media has been shown to have approximate M_r of 17,000–20,000 (17, 152) and 12,000–13,000 (75). Two of these groups have reported the isoelectric point of human BCGF to be approximately 6.5-6.9 (37, 152). As is the case with IL2, human BCGF gives excellent responses in the mouse BCGF assay, but mouse BCGF is not detected in human BCGF assays (52b). Direct binding of human BCGF to B-cell blasts has been demonstrated via its depletion following absorption on human leukemic B cells activated with anti-Ig or anti-idiotypic antibodies (64, 152).

Recently, Kishimoto and colleagues have reported a second distinct human BCGF produced by an IL2-dependent allospecific helper T-cell clone upon stimulation with cells bearing the alloantigen for which the T cells are specific (152). This factor stimulated proliferation of normal B cells cultured with anti-Ig or of leukemic B cells cultured with anti-idiotype antibody. It had a M_r of 50,000 by gel filtration. In 4 M urea, its M_r fell to 19,000. Both size estimates distinguished it from the human BCGF the same group had described in supernatants of peripheral blood T cells stimulated with phytohemagglutinin. This BCGF showed a M_r of 17,000 by conventional gel filtration and of 12,000–14,000 in 4 M urea. Normal T cells stimulated with the combination of phytohemagglutinin and phorbol myristate acetate produced both the 17,000 and 50,000 BCGFs. Most interestingly, these two BCGFs showed a synergistic effect on the proliferation of anti-Ig-stimulated B cells. This suggests that the larger BCGF is not simply an aggregated form of the smaller.

B-Cell Activation: Two-Signal Model Revisited

The understanding of B-lymphocyte activation has been a tantalizing prospect for both immunologists and cell biologists as these cells have the unique characteristics of possessing a well-characterized receptor for, and displaying a clear inducible response to, a specific stimulant. Despite this apparent simplicity of B-lymphocyte systems, progress in understanding their activation has been relatively slow. Some 10 years ago, three major models to explain antigen-specific B-cell activation were proposed: (a) Activation occurs solely as a result of the binding of antigen to membrane (m)Ig (24, 117); (b) activation occurs via an unrelated nonspecific receptor, with mIg serving specifically to focus molecules capable of binding to this receptor (20); and (c) activation requires both a signal resulting from the binding of antigen to the mIg receptor and a signal delivered by an interaction at another, antigen-nonspecific receptor on the B-cell membrane (13, 14).

The recent explosion of information on B-cell proliferation cofactors outlined in the preceding sections promises to clarify this question, although it must be recognized that the events of DNA synthesis and mitosis may be far removed from those signals operating at the resting (G_0) lymphocyte level. For this reason, some groups have attempted to devise experimental systems to analyze the activation events that precede entry into the S phase of the cell cycle. One useful approach has been the monitoring of size enlargement following lectin or antigen stimulation of B lymphocytes. Andersson et al (4) initially probed early activation events by measuring both size enlargement and DNA synthesis in response to a variety of B-cell mitogens. Of particular interest, they found that purified protein derivative of tuberculin (PPD) was capable of driving blasts that had been prepared by lipopolysaccharide stimulation through subsequent rounds of replication, but that PPD could not induce G_0 B cells to complete the first round of replication. This was the first indication that the activation requirements of G_0 B cells and of "activated" B cells were different.

More recently, this theme has been developed and extended by DeFranco et al (22). They found that culture of purified G_0 B cells with anti-IgM for as brief a time as 1 hr led to a distinct enlargement of these cells. This enlargement was progressive and quite synchronous for the first 24 hr of culture provided the anti-IgM was present continuously. It appeared to reflect entry of resting B cells into and transition through G_1 phase as it correlated well with both increased RNA synthesis and acquisition of the ability to enter S phase upon receipt of the appropriate additional signal. At 20–24 hr, a second signal provided by lipopolysaccharide or high concentrations of anti-IgM antibody was required for the completion of G_1 and for the entry of cells into S phase some 6-10 hr later. The concept that distinct mechanisms control entry of G_0 cells into G_1 and late G_1 cells into S was strongly supported by experiments involving the immune-deficient *xid* mice. Although anti-IgM fails to stimulate DNA synthesis by *xid* B cells, it quite effectively transfers these cells from G_0 to G_1 as monitored by size enlargement.

The studies of DeFranco et al prompted speculation that one or both of the soluble helper factors we had shown to be critical to anti-IgM stimulation of DNA synthesis (i.e. BCGF and IL1) might regulate the second control point, i.e. the transition of late G_1 B cells to S phase. Careful kinetic experiments were designed to test this hypothesis (52). Anti-IgM and one cofactor were added to B cells at the initiation of the culture, with the second cofactor being added at various times over the course of 30 hr. In light of the findings of DeFranco et al (22), tritiated thymidine was added at 30 hr and cultures were terminated at 42 hr so that most of the thymidine uptake would represent B cells that had entered S phase for the first time. The somewhat surprising results are illustrated in Figure 2. The cofactor IL1 could be added 16–20 hr after initiation of culture without any significant decrease in final thymidine uptake; it thus seems likely that this factor corresponds to the second control agent suggested by DeFranco's studies. The surprising result was obtained when the time of addition of BCGF was varied. Its effectiveness as a costimulator declined rapidly and linearly as its addition was delayed. If its addition was delayed beyond 10 hr, it was essentially without effect for thymidine uptake between 30–42 hr. This suggests that BCGF acted either on G_0 B cells or during the early (excitation) portion of the G_1 phase of the cell cycle. The time of action and biological function of BCGF thus makes it a prime candidate for the second signal of the original Bretscher/Cohn hypothesis (24, 87). Whether BCGF acts directly on the resting B cell or on the early G_1 B cell, its precise role in B-cell activation and its possible role in B-cell tolerance are the subjects of current investigation.

Triggering Lyb5+ B Cells: Summary

Based on the various experiments described above, we propose the following model, summarized schematically in Figure 3, for the activation and proliferation of Lyb5+ B cells. Resting B cells that bind anti-IgM through their mIgM enter the G_1 phase of the cell cycle. This process presumably requires cross-linking of mIg; multivalent antigens would be expected to have the same effect on those resting B cells that possess receptors specific for that antigen. BCGF acts either together with anti-IgM to cause the $G_0 \rightarrow G_1$ transition, or acts upon the early G_1 cell. It is known that anti-IgM is required continuously during early G_1 (22); whether BCGF is also needed continuously or acts at a single point in early G_1 (or G_0) is not known. As a result of these two stimulants, the B cell progresses through early G_1 and reaches a point at about 16–20 hr when it requires IL1. The presence of IL1 allows the cell to continue to progress through G_1 and then to enter S phase. The first cells to enter S phase do so at about 30 hr after initial stimulation. From the work of DeFranco et al (22), it appears that anti-IgM is not required for cells to progress through late G_1 and to enter S phase. The need for BCGF in late G_1 has not been established. It seems likely that the time of actions of the BCGF and IL1 correlate with the time that receptors for them are expressed on the B cell surface, just as T cell sensitivity to IL2 appears to correlate with the expression of receptors for IL2 on T cells stimulated with lectins or antigen (105). We emphasize, however, that no direct evidence for receptors for BCGF or IL1 on activated mouse B cells has yet been obtained.

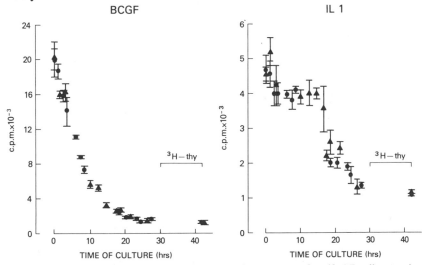

Figure 2 Time of requirements of BCGF and IL1 in responses of purified B cells to anti-μ In the left panel (titled BCGF) B cells were cultured at 5×10^4 cells per 0.2 ml; anti-μ (10 ug/ml) and a source of IL1 were added at the initiation of the culture. BCGF was added at various times, as indicated, and all cultures received 1 μCi of tritiated thymidine 30 hr after initiation of culture. Uptake of tritiated thymidine was measured at 42 hr. The experiment shown in the right panel (titled IL1) was carried out similarly except that cells were cultured at 2×10^4 cells per 0.2 ml and anti-μ and BCGF were added at the beginning of the culture. IL1 was added at various times in the course of the culture.

Thus, the overall concept for activation of Lyb5$^+$ B cells is similar to models proposed for T-cell activation (3, 69, 70). Lafferty, in particular, has emphasized that antigen plus one costimulator acts on the resting T lymphocyte to induce an excited or receptive state, and that this excited lymphocyte proliferates when it receives a second costimulator signal (68, 69).

Long-Term B-Cell Lines

B-cell lymphomas, myelomas, and hybridomas have provided invaluable resources for the study of antibody diversity, idiotypy, and immunoglobulin gene rearrangement and expression (82, 103, 108). The limitation of such tumor lines is that they have relatively little immunocompetence and so do not allow the analysis of many of the cellular and biochemical mechanisms related to B-cell activation, isotype switching, idiotype selection, and immunoregulation. The recent development of long-term cultures of antigen-specific immunocompetent T-cell clones (11, 36, 39, 92, 127) has raised hopes that the same technology might be possible for B lymphocytes. Indeed, much of the impetus for defining B-cell proliferation cofactors as outlined above has been to obtain a growth factor(s) capable of maintaining continuous in vitro growth of normal B cells, just as IL2 maintains continuous in vitro growth of some normal T cells. In many respects, the development of B-cell clones would seem an easier endeavor than that of T-cell clones. The technologies for marked enrichment of antigen-specific B lymphocytes (43) and for isolation of individual immunocompetent antigen-specific clones (102) have been established for some years. The one remaining difficulty in producing long-term cloned lines of antigen-specific immunocompetent B lymphocytes lies in the continuous propagation of these cells. Although it has been possible to propagate individual B-cell clones in vivo for periods of up to 10 months (9, 147), early attempts to culture B lymphocytes for periods longer than 4 weeks were unsuccessful (81). Indeed, the continuous cultivation of B lymphocytes is still some

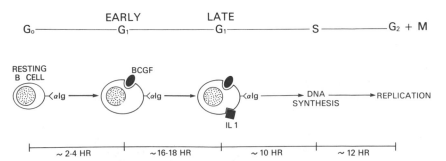

Figure 3   Schematic depiction of the regulation of B-cell activation and replication.

distance from being an established procedure. However, activity in this area has recently flourished, with the emergence of five independent reports of long-term B-cell lines (18, 37, 51, 75, 126, 146). Although these procedures are clearly tedious and difficult to perform on a reproducible and routine basis, it is hoped that they represent a foundation for greater successes ahead.

MHC-UNRESTRICTED B-CELL DIFFERENTIATION

Several Distinct TRFs

It is generally agreed that once B cells are proliferating, their differentiation into ISC requires one or more soluble T cell-derived products, generally designated TRFs. This was first demonstrated in the pioneering studies of Dutton et al (26, 27) and Schimpl & Wecker (110, 111). Both groups observed that the response of T cell-depleted spleen cells to sheep erythrocyte antigens required a soluble factor that could be found in mixed lymphocyte reaction culture supernatants and in supernatants produced by mitogen- or antigen-activated T cells. The picture became more complicated when other antigens were investigated. In general, nonspecific T cell-derived factors are inefficient in supporting primary antibody responses of T cell-depleted populations to soluble proteins and hapten-protein conjugates; the latter generally require either antigen-specific factors and/or direct cognate interactions with antigen-specific MHC-restricted helper T cells. The initial concern that the action of TRF might be unique to the rare erythrocyte antigen system was soon dispelled, however, by two ensuing sets of observations. Firstly, nonspecific soluble factors were found to drive the response of T-depleted hapten-primed B cells into ASC (61, 133, 136). Secondly, with anti-Ig antibodies as a polyclonal analog for antigen, it was shown that the subset of B cells responsive to this reagent could develop into ISC only if T cell-derived soluble factors were additionally provided (63, 98). As the pool of anti-Ig-responsive B cells has recently been quantitated as 50% of normal splenic B cells (22), this latter observation establishes the extensiveness of TRF phenomenology.

To date, little definitive biochemical data on the nature of TRFs have been obtained. Two groups have used standard biochemical procedures to identify in concanavalin A-induced spleen cell supernatants a factor of M_r 30,000–40,000 which supports the ASC response of spleen cells from congenitally athymic donors to sheep erythrocyte antigens (109, 142). However, an increasing number of laboratories are now reporting that ASC responses involve synergism between two or more factors (34, 48, 73, 93, 129). Therefore, the biochemical definition of a single TRF is subject to the concerns that one has identified the chromatographic region where two or more synergizing components overlap. Thus, before accurate molecular characterization is possible, the number of factors required in the assays of TRF function must be elucidated. Once this is achieved, the characterization and purification of TRF should be greatly aided by the recent advent of T-cell clones and hybridomas that secrete B-cell differentiation factors (23, 129, 135). It must be remembered, however, that although such clones represent monoclonal sources of the factors, they are not necessarily sources of single factors. Indeed, evidence from several laboratories suggests that T-cell lines, hybridomas, and tumors are capable of synthesizing numerous lymphokines (23, 31, 41, 56, 113).

One obvious problem to contend with is the recent discovery that the proliferative component of an ASC response requires its own separate cofactors (see Figure 3). Most reports of synergizing factors involved in ASC responses do not distinguish whether this simply reflects a requirement for proliferation and differentiation cofactors or whether in fact there are several differentiation cofactors. In an attempt to clarify this issue, our laboratory has again utilized the anti-IgM polyclonal activation system as an effective means of separating proliferation and differentiation events, and hence the cofactors involved. As outlined above, the proliferation cofactors involved in this system (BCGF and IL1; see Figure 3) support excellent proliferation, but do not lead to production of ISC. Recently, Nakanishi in our laboratory has demonstrated (93) that two additional T cell-derived products must be provided for development of ISC to ensue: these factors are currently designated B151K12-TRF (B15-TRF) and EL-TRF. They are found in supernatants from B151K12 T hybridoma cells (135) and phorbol myristate acetate-induced EL-4 cells (32), respectively. They are further distinguished by kinetic studies that clearly demonstrate their independent roles in ISC development: B15-TRF is required before 48 hr in a 4-day culture, whereas EL-TRF appears to function in the final 24 hr. The actions of B15-TRF and EL-TRF are apparently antigen nonspecific and are dependent on the presence or prior action of BCGF. The cycling status of cells responsive to B15-TRF and EL-TRF, and the biochemical characteristics of these factors, are currently unknown. Figure 4 presents a model for proliferation and differentiation of Lyb5$^+$ B cells based on the concepts developed in our laboratory with the anti-IgM polyclonal activation system.

A major outcome of the studies of Nakanishi et al (93) is the identification of two distinct factors that may be appropriately classified as T cell-replacing (or differentiation) factors. The existence of two TRFs would obviously help resolve current conflicts in the literature concerning the biochemical and biological nature of this factor. Although further information is required to relate B15-TRF and EL-TRF to differentiation factors described in other laboratories, our results are consistent with the following suggestions. On the basis of factor source alone it would appear that the TRF described by Takatsu et al (135) in B151K12 supernatant corresponds to B15-TRF, and that BCDFμ described by Pure et al (104) may correspond to EL-TRF. As the TRF in concanavalin A-induced supernatants described by Schimpl & Wecker (110, 111) is absent from EL-4 supernatant and functions at 24–48 hr of a 5-day culture, it may very well correspond to B15-TRF. The TRFs observed in other concanavalin A-induced supernatants (45, 98, 99, 142), in the supernatant of the C.C3.11.75 cell line (129), and in supernatant from antigen-stimulated T helper cell lines (5) may well include both B15-TRF and EL-TRF.

Numerous reports suggest that T cells are not only responsible for differentiation of activated B cells into ASC, but in fact influence the immunoglobulin isotype expressed in a particular immune response (12, 87, 121, 137, 139). This may be achieved either by expansion of B cells precommitted to a particular isotype, or alternatively through the induction of a "switch" in isotype expression in individual B cells. Although the precise mechanism of this phenomenon remains elusive, several groups have accumulated data that strongly suggest the existence of isotype-specific helper T cells that act in an antigen nonspecific, MHC-unrestricted manner (10, 29, 61, 78, 107). The existence of such cells prompts speculation regarding their secretion of class-specific soluble helper factors, i.e. the existence of a family of TRFs, each of which causes the expression of a particular Ig isotype. There are currently two separate reports in support of this contention. Kishimoto & Ishizaka were first to demonstrate that carrier-specific T helper cells involved in IgE antibody responses were different from those involved in IgG antibody responses to the same antigen (61). Furthermore, the two types of helper cells were found to release different soluble factors that triggered release of the respective Ig class from hapten-primed precursor cells (62, 63). More recently, Vitetta and colleagues have reported a

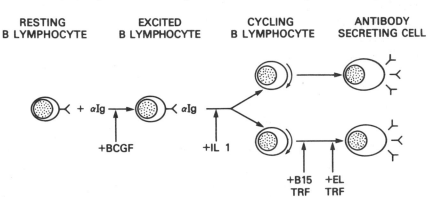

Figure 4   The factor dependent pathway of B-cell activation, proliferation and differentiation.

soluble factor(s) in supernatants produced by certain T-cell hybridomas and T-cell lines that selectively enhances the expression of IgG_1 secretion by lipopolysaccharide-induced blasts (56). This factor can be distinguished from numerous other lymphokines, including both BCGF and the TRFs of Takatsu et al (i.e., B151K12 supernatant) and Swain et al (C.C3.11.75 supernatant), which appear mainly to induce IgM secretion (104). These reports imply a level of factor complexity skillfully camoflaged by the simple term "T-cell replacing factor" (Figure 4).

Are TRFs Also Proliferation Cofactors?

One aspect of factor regulation of B-cell function still requiring significant clarification is whether or not any or all of the T-cell differentiation factors described in the preceding section are additionally capable of stimulating proliferation. There are marginal data both for and against this notion. Against are the observations of Kishimoto et al (64), who in generating factor-secreting human T hybridomas obtained lines that secrete BCGF or TRF, but none that secrete a product(s) that supports both proliferation and differentiation. On the other hand, several laboratories have provided evidence in favor of a role for TRF in stimulating proliferation. Firstly, there is the report of Andersson & Melchers that a 30,000-M_r component released by antigen-activated T-cell lines served as both a replication and maturation factor (5). However, further evidence is required to eliminate the possibility of a separate TRF and BCGF being co-purified in these experiments. Perhaps more persuasive are recent reports by Swain et al (130) and Nakanishi et al (93) that BCGF-free preparations of DL-TRF and/or B15-TRF produce a small but significant level of proliferation in purified B cells (130) or alternatively enhance the proliferation obtained with anti-IgM and BCGF (93). Nakanishi has additionally demonstrated that B15-TRF improves final cell yields in such experiments. These observations prompt speculation that B15-TRF is the mouse analogue of the 50,000-M_r human BCGF recently reported by Kishimoto and colleagues (see Characteristics of BCGF) (see also 152). Although some of the biological characteristics of B15-TRF and 50,000-M_r human BCGF seem disparate, both factors synergize with low-molecular-weight BCGF to cause extensive proliferation of activated B cells.

Finally, in considering the possible function of TRF as a proliferation cofactor, one must consider an elegant study recently reported by Wetzel et al (145). Using a cloning assay in which the responses of single isolated B cells were followed, they showed that DL-TRF increased the frequency of B cells that undergo clonal proliferation. However, it was not clear in this assay whether DL-TRF was acting as a growth factor per se or whether it was activating otherwise resting B cells to respond to the polyclonal B-cell activators lipopolysaccharide and dextran sulfate, which were also present in the culture.

Involvement of IL2 in B-Cell Responses

The preceding discussions on soluble factors that regulate B-cell function have until now avoided a controversial issue in this area, namely whether or not the growth factor for T lymphocytes, IL2, is capable of exerting a direct effect on B lymphocytes. Those proponents of direct action of IL2 on B cells have provided the following lines of evidence to support this concept: depletion of IL2 from cofactor-rich supernatants by absorption on IL2-dependent T-cell lines also removes the B-cell costimulating factor (98) and/or B-cell differentiation factor (73, 76) in these sources; when various cofactor-rich supernatants are analyzed, there is strict correlation between levels of IL2 and of one of two B cell-specific factors required in plaque-forming cell responses (73, 76, 129, 131); this same B cell-specific factor cannot be distinguished from IL2 in terms of molecular weight (73, 76).

Although the possibility that IL2 acts directly on B cells cannot yet be formally excluded, we believe this to be unlikely in view of the following considerations. (a) Excellent reagents for detecting IL2 receptors on cell surfaces are now available (105) and these have failed to demonstrate such receptors on resting or activated B cells. (b) Attempts to absorb IL2 from cofactor-rich supernatants with B cells and B-cell blasts have failed (73). (c) Both mouse (31, 33, 49), and human (17, 37, 64, 75, 90, 91, 126, 152) BCGFs, and the TRFs described by Melchers & Andersson (5), Howard et al (49), Nakanishi et al (93), Schimpl et al (109, 110), Takatsu et al (135, 136), Swain et al (129, 131), Kishimoto et al (63), and Pure, Isakson et al. (56, 104) have been distinguished from IL2 either by biochemical separation, by difference in source, by cellular absorption, or by a combination of these procedures. (d) A recent report by Pike et al (101) demonstrates that clonal expansion and differentiation of individual hapten-specific B cells into ASC requires antigen together with one or more soluble factors distinct from IL2.

Several explanations can be offered to accommodate the observations of those proposing a direct role for IL2 in B-cell function. Firstly, it is now clear that the IL2-containing supernatant generally used in these laboratories (the supernatant of FS6-14.13) also contains BCGF (M. Howard, unpublished data) and possibly one or more of the TRFs that have been described. It would not be surprising if these factors were nonspecifically absorbed by the "sticky" activated target cells used for IL2 depletion. Alternatively, although these laboratories employ vigorous purification procedures to obtain T cell-free B-cell populations for their analyses, the lengthy in vitro assays required for B-cell function provide time for pre-T cells to differentiate into mature lymphocytes, which could then be effectively expanded by IL2 into a significant pool of lymphokine-secreting T cells. Indeed, in support of this view, several laboratories have shown that T-cell lines can be established from nude mouse cells with the combination of antigen or mitogen activation and IL2 (40, 55). Thus although the arguments outlined above challenge the concept that IL2 acts directly on resting or activated B cells, it is quite probable that IL2, added to culture or produced endogenously, may have profound indirect effects on B-cell proliferation and differentiation.

SOLUBLE FACTORS IN MHC-RESTRICTED B-CELL RESPONSES

The previous sections of this review have concentrated on B-cell activation, proliferation, and differentiation stimulated by cross-linkage of membrane Ig together with a series of soluble factors derived from T cells and macrophages. It is now clear that a second principal pathway of B-cell activation exists that depends upon the interaction of a helper T (T_H) cell with a B cell in an antigen-specific, histocompatibility-restricted manner. More precisely, this pathway of B cell stimulation depends upon the capacity of the T_H cell, or its product(s), to recognize both antigen and a class II MHC molecule on the surface of the B cell. Indeed, there had been considerable controversy for several years as to the importance of MHC-restricted T_H-B cell interactions. Several authors had pointed out that much of the available data interpreted as supporting such interactions could actually have reflected MHC-restricted interactions between antigen-presenting cells (APC) and T_H cells, with the ensuing production of soluble factors acting upon B cells activated as a result of receptor cross-linkage that resulted from the binding of antigen (30, 79, 116, 118, 119). This seemed particularly plausible since, in most experiments, APC and B cells were derived from the same donors. However, careful experiments with both in vivo (124, 125) and in vitro (132, 151) systems now make it clear that MHC-restricted T_H-B cell interactions leading to antibody synthesis do occur and are responsible for a significant component of the total antibody response. There is mounting evidence, mainly derived from the work of Singer, Hodes and their colleagues (120), that MHC-restricted cellular interactions are mainly a feature of the subset of B lymphocytes found in mice with the *xid*-determined immune defect. That is, Lyb5$^-$ B cells can respond to "thymus-dependent" antigens only through MHC-restricted T_H-B cell interactions. Moreover, it appears that the very same T_H cells can participate in both the MHC-restricted T_H-B cell interaction and in the factor-mediated activation of Lyb5$^+$ B cells described in previous portions of this review, although the conditions required to activate T_H cells for these two types of interactions appear to be quite different. Low concentrations of antigen appear to lead to MHC-restricted T_H-B-cell interactions whereas high concentrations mainly result in interactions controlled by nonspecific soluble factors (7).

The control of B-cell responses in MHC-restricted T_H-B-cell interactions is still a matter of considerable controversy. Perhaps the most uncertain issue is the need for antigen binding to the mIg receptors of the responding B cells for activation, proliferation, and differentiation to occur. In a study of B-cell proliferation in response to stimulated T_H cells, Tse et al (140) showed that B cells from nonprimed donors could be stimulated to proliferate by T_H cells from donors primed to the terpolymer of $Glu^{60}Ala^{30}Tyr^{10}$ (GAT) in the presence of GAT. This T_H-B-cell response appears to be restricted by products of the I-region of the MHC, since primed T cells from (B10 X B10.S)F_1 donors could cause proliferation by B10 but not B10.S B cells, even if both B cell types were present in the same culture. The response to GAT is controlled by an I-A^b-encoded Ir gene; B10 mice are responders and B10.S mice are nonresponders. Thus, F_1 T cells contain clones that recognize B10 B cells that have bound GAT to their membranes but lack clones that recognize B10.S B cells with bound GAT. Furthermore, the B cells participating in this interaction may be drawn from donors that have not been primed to GAT; and B cells from primed donors do not react to any greater degree than do unprimed B cells. Thus, it seems most likely that the bulk of B cells that respond do not have mIg receptors specific for GAT. Rather, they appear to have simply bound GAT to their membranes and to present it with I-A-encoded class II gene products to T cells in a manner analogous to that used by APC. Thus, B cell activation for proliferation mediated by a MHC-restricted interaction with T_H cells does not appear to require binding of antigen to B cell receptors. These findings have been confirmed with resting B cells and cloned long-term T cell lines both for size enlargement ($G_0 \rightarrow G_1$ transition) (21) and for proliferation (58). Similarly, Coutinho and colleagues have reported that a T_H line specific for a minor histocompatibility antigen expressed on the B cell surface can stimulate the appearance of very substantial numbers of ISC despite the fact that no antigen is provided to cross-link receptors of the vast majority of cells (19, 77).

Groups that have studied antibody responses to specific antigens in MHC-restricted T_H-B cell interactions have, by contrast, emphasized the need for the binding of receptors of precursor cells to obtain measureable responses (6, 59, 153, 154). Indeed, one group has reported that even the polyclonal generation of ISC in MHC-restricted interactions between lines of T_H cells specific for minor histocompatibility antigens, and B cells that express those antigens is dependent upon the presence of anti-Ig antibodies, presumably to provide the analog of the antigen-dependent stimulation mediated by receptor cross-linkage (59).

A possible explanation to resolve these apparent differences is that an initial round of MHC-dependent B-cell proliferation can be initiated by interactions of B cells with T_H cells in the absence of receptor cross-linkage. Subsequent divisions of precursor cells, which are needed to expand the B cell clone, may require receptor cross-linkage.

The need to use T_H cells to obtain MHC-restricted T_H-B cell interactions tends to obscure any need for lymphokines in this response since the T_H cells, in addition to participating directly in T_H-B interactions, may also produce the soluble factors needed for the response. Some have reported that T_H cells can produce soluble antigen-specific, MHC-restricted factors capable of acting in place of the T_H cell itself (5, 57, 74). A detailed discussion of this important but controversial subject is beyond the scope of this review, which is mainly concerned with nonspecific soluble factors.

Two groups have particularly emphasized a role for soluble factors in the MHC-restricted T_H-B cell interactions that lead to the development of ISC (5, 114, 153, 154). In general, it has been proposed that these soluble factors act upon excited B cells (or B cell blasts) and cause their further proliferation and differentiation. Relatively little has been reported on the number of factors required and on the physicochemical properties of these factors. Such factors appear to be present in the supernatant of some long-term T-cell lines, in the supernatant of the thymoma cell line, EL-4, and in supernatants from secondary in vitro mixed lymphocyte cultures. Andersson & Melchers (5) have reported a factor with a M_r of 30,000 (p30) isolated from the supernatant of a T_H cell line specific for horse erythrocytes. This appeared to be the nonspecific factor active in stimulation of B cells activated by an antigen-specific helper factor. This p30 factor has been designated by them as B-cell replication and maturation factor. Whether it is active in all MHC-restricted interactions and whether it acts in a manner analogous to BCGF or to the early acting B15-TRF discussed earlier has not yet been determined. The involvement of other differentiation factors such as EL-TRF or BCDFμ in this process is also a matter of uncertainty. It can be anticipated that rapid progress in the understanding of the role of nonspecific lymphokines on B cells activated as a result of MHC-restricted T_H-B cell interactions will be made in the next few years.

SUMMARY

The B-lymphocyte family of cells presents one of the most remarkable opportunities for the detailed study of regulation of growth and differentiation. Some members of this cell population have the property that they may be stimulated by ligand-receptor interactions, together with the sequential action of a series of lymphokines, to progress from the resting state, through several rounds of proliferation, and then to differentiate to immunoglobulin secretion. Other cells in this group participate in cognate cellular interactions with helper T cells in which the recognition of both antigen and a class II MHC molecule on the B-cell surface is key to activation. The differentiation of these cells is also controlled by soluble products.

We have reviewed our developing knowledge of the biochemistry and mode of action of the lymphokines that act upon B cells. These include distinct growth and differentiation factors. Among these are the BCGFs of mice and humans and the various TRFs, which include molecules often described as differentiation factors.

The next several years should witness major progress in understanding the physicochemical properties of the B cell-specific factors, their time and nature of action, and the nature of their receptors. In addition, we can anticipate a major effort to understand the intracellular events that flow from the action of specific growth and differentiation factors that act upon B cells. Such information should lead to a new physiologically-based pharmacology for manipulation of antibody responses in human disease and in responses to vaccines. In addition, the fuller understanding of the nature and mode of action of the various growth and differentiation factors should make long-term growth of cloned B cells a procedure that can be routinely used in immunological laboratories for the precise study of the biology of responses by homogeneous populations of B lymphocytes.

Literature Cited

1. Ahmed, A., Scher, I., Sharrow, S. O., Smith, A. H., Paul, W. E., et al. 1977. B-lymphocyte heterogeneity: Development and characterization of an alloantiserum which distinguishes B-lymphocyte differentiation alloantigens. *J. Exp. Med.* 145:101–10
2. Andersson, J., Coutinho, A., Lernhardt, W., Melchers, F. 1977. Clonal growth and maturation to immunoglobulin secretion *in vitro* of every growth-inducible B lymphocyte. *Cell* 10:27–34
3. Andersson, J., Gronvik, K., Larsson, E., Coutinho, A. 1979. Studies on T-lymphocyte activation. I. Requirement for the mitogenic-dependent production of T-cell growth factors. *Eur. J. Immunol.* 9:581–87
4. Andersson, J., Lernhardt, W., Melchers, F. 1979. The purified protein derivative of tuberculin. A B-cell mitogen

that distinguishes in its action resting, small B cells from activated B-cell blasts. *J. Exp. Med.* 150:1339–50
5. Andersson, J., Melchers, F. 1981. T cell-dependent activation of resting B cells: Requirement for both nonspecific unrestricted and antigen-specific Ia-restricted soluble factors. *Proc. Natl. Acad. Sci. USA* 78:2497–501
6. Andersson, J., Schreier, M. H., Melchers, F. 1980. T-cell-dependent B-cell stimulation is H-2 restricted and antigen dependent only at the resting B-cell level. *Proc. Natl. Acad. Sci. USA* 77:1612–16
7. Asano, Y., Shigeta, M., Fathman, C. G., Singer, A., Hodes, R. J. 1982. Lyb5+ and Lyb5- B cells differ in their requirements for restricted activation by cloned helper T cells. *Fed. Proc.* 41:721

8. Asano, Y., Singer, A., Hodes, R. J. 1981. Role of the major histocompatibility complex in T cell activation of B cell subpopulations. MHC restricted and unrestricted B cell responses are mediated by distinct B cell subpopulations. *J. Exp. Med.* 154:1100–15

9. Askonas, B. A., Williamson, A. R., Wright, B. E. G. 1970. Selection of a single antibody-forming cell alone and its propagation in syngeneic mice. *Proc. Natl. Acad. Sci. USA* 67:1398–403

10. Augustin, A. A., Coutinho, A. 1980. Specific T Helper cells that activate B cells polyclonally. *In vitro* enrichment and cooperative function. *J. Exp. Med.* 151:587–601

11. Augustin, A. A., Julius, M. H., Cosenza, H. 1979. Antigen-specific stimulation and trans-stimulation of T cells in long-term culture. *Eur. J. Immunol.* 9:665–70

12. Braley-Mullen, H. 1974. Regulatory role of T cells in IgG antibody formation and immune memory to type III pneumococcal polysaccharide. *J. Immunol.* 113:1909–20

13. Bretscher, P. 1975. The two signal model for B cell induction. *Transplant. Rev.* 23:37–48

14. Bretscher, P., Cohn, M. 1970. A theory of self-nonself discrimination. *Science* 169:1042–49

15. Burgess, A., Metcalf, D. 1980. The nature and action of granulocyte-macrophage colony stimulating factors. *Blood* 56:947–58

16. Burnet, F. M. 1959. *The Clonal Selection Theory of Acquired Immunity.* London: Cambridge Univ.

17. Butler, J., Muraguchi, A., Lane, C., Fauci, A. 1983. Development of a human T-T cell hybridoma secreting B cell growth factor. *J. Exp. Med.* 157:60–8

18. Coutinho, A. Personal communication

19. Coutinho, A., Augustin, A. A. 1980. Major histocompatibility complex-restricted and unrestricted T helper cells recognizing minor histocompatibility antigens of B cell surfaces. *Eur. J. Immunol.* 10:535–41

20. Coutinho, A., Moller, G. 1974. Immune activation of B cells: Evidence for one non-specific triggering signal not delivered by the Ig receptors. *Scand. J. Immunol.* 3:133–46

21. DeFranco, A., Ashwell, J., Schwartz, R. H., Paul, W. E. Manuscript in preparation

22. DeFranco, A., Raveche, E., Asofsky, R., Paul, W. E. 1982. Frequency of B lymphocytes responsive to anti-immunoglobulin. *J. Exp. Med.* 155:1523–36

23. Dennert, G., Weiss, S., Warner, J. 1981. T cells may express multiple activities: Specific allohelp, cytolysis and delayed type hypersensitivity are expressed by a clone T cell line. *Proc. Natl. Acad. Sci. USA* 78:4540–3

24. Diener, E., Feldman, M. 1972. Relationship between antigen and antibody-induced suppression of immunity. *Transplant. Rev.* 8:76–94

25. Dutton, R. W. 1974. T cell factors in the regulation of the B cell response. In *The Immune System, Genes, Receptors, Signals,* ed. E. E. Sercarz, A. R. Williamson, C. F. Fox, pp. 485–96. New York: Academic

26. Dutton, R. W. 1975. Separate signals for the initiation of proliferation and differentiation in the B cell response to Antigen. *Transplant. Rev.* 23:66–77

27. Dutton, R. W., Falkoff, R., Hirst, J. A., Hoffman, M., Kappler, J., et al. 1971. Is there evidence for a non-antigen specific diffusable chemical mediator from the thymus-derived cell in the initiation of the immune response? *Progr. Immunol.* 1:355

28. Dutton, R., Mishell, R. 1967. Cellular events in the immune response. The *in vitro* response of normal spleen cells to erythrocyte antigens. *Cold Spring Harbor Symp. Quant. Biol.* 32:407–14

29. Elson, C. O., Heik, J. A., Strober, W. 1979. T cell regulation of murine IgA synthesis. *J. Exp. Med.* 149:632–43

30. Erb, P., Meier, B., Matunaga, T., Feldmann, M. 1979. Nature of T-cell macrophage interaction in helper-cell induction in vitro. II. Two stages of T-helper-cell differentiation analyzed in irradiation and allophenic chimeras. *J. Exp. Med.* 149:686–701

31. Farrar, J., Benjamin, W., Hilfiker, M., Howard, M., Farrar, J., Fuller-Farrar, J. 1982. The biochemistry, biology, and role of IL-2 in the induction of cytotoxic T cells and antibody-forming B cell responses. *Immunol. Rev.* 63:129–66

32. Farrar, J., Fuller-Farrar, J., Simon, P., Hilfiker, M., Stradler, B., Farrar, W. 1980. Thymoma production of T cell growth factor (interleukin 2). *J. Immunol.* 125:2555–58

33. Farrar, J., Howard, M., Fuller-Farrar, J., Paul, W. E. Submitted for publication

34. Farrar, J., Koopman, W., Fuller-Bonar, J. 1977. Identification and partial purification of two synergistically acting helper mediators in human mixed

leukocyte culture supernatants. *J. Immunol.* 119:47–54

35. Farrar, J. J., Mizel, S. B., Fuller-Bonar, J., Hilfiker, M. L., Farrar, W. L. 1980. Lipopolysaccharide-mediated adjuvanticity: Effect of lipopolysaccharide on the production of T cell growth (interleukin 2). In *Microbiology,* ed. D. Schlessinger. Washington, DC: Am. Soc. Microbiol. 36 pp.

36. Fathman, C. G., Hengartner, H. 1978. Clones of alloreactive T cells. *Nature* 272:617–18

37. Ford, R., Mehta, S., Franzini, D., Montagna, R., Lachman, L., et al. 1981. Soluble factor activation of human B lymphocytes. *Nature* 294:261–63

38. Gillis, S., Mizel, S. B. 1981. T cell lymphoma model for the analysis of interleukin 1 mediated T-cell activation. *Proc. Natl. Acad. Sci. USA* 78:1133–37

39. Gillis, S., Smith, K. A. 1977. Long term culture of tumour-specific cytotoxic T cells. *Nature* 268:154–56

40. Gillis, S., Union, N., Baker, P., Smith, K. 1979. The in vitro generation and sustained culture of nude mouse cytolytic T-lymphocytes. *J. Exp. Med.* 149: 1460–76

41. Glasebrook, A., Sarmiento, M., Loken, M., Dialynas, D., Quintans, J., et al. 1981. Murine T lymphocyte clones with distinct immunological functions. *Immunol. Rev.* 54:225–66

42. Gudat, F. G., Harris, T. N., Harris, S., Hummeler, K. 1971. Studies on antibody producing cells. III. Identification of young plaque-forming cells by thymidine-^{3}H labeling. *J. Exp. Med.* 134: 1155–69

43. Haas, W., Layton, J. E. 1975. Separation of antigen-specific lymphocytes. I. Enrichment of antigen-binding cells. *J. Exp. Med.* 141:1004–14

44. Hapel, A., Lee, J., Farrar, W., Ihle, J. 1981. Establishment of continuous cultures of thyl1.2⁺, Lyt1⁺2⁻ T cells with purified interleukin 3. *Cell* 25:179–86

45. Harwell, L., Kappler, J., Marrack, P. 1976. Antigen-specific and nonspecific mediators of T cell/B cell cooperation. III. Characterization of the nonspecific mediator(s) from different sources. *J. Immunol.* 116:1379–84

46. Hoffman, M. 1980. Macrophages and T cells control distinct phases of B cell differentiation in the humoral immune response *in vitro. J. Immunol.* 125: 2076–81

47. Hoffman, M. K., Koenig, S., Mittler, R. S., Oettgen, H. F., Ralph, P., et al. 1979. Macrophage factor controlling differen-

tiation of B cells. *J. Immunol.* 122:497–502

48. Hoffman, M. K., Watson, J. 1979. Helper T cell-replacing factors secreted by thymus-derived cells and macrophages: Cellular requirements for B cell activation and synergistic properties. *J. Immunol.* 122:1371–75

49. Howard, M., Farrar, J., Hilfiker, M., Johnson, B., Takatsu, K., et al. 1982. Identification of a T-cell derived B-cell growth factor distinct from interleukin 2. *J. Exp. Med.* 155:914–23

50. Howard, M., Johnson, B., Yu, A., Ansel, J., Cohen, D., Nakanishi, K., Paul, W. E. Manuscript in preparation

51. Howard, M., Kessler, S., Chused, T., Paul, W. E. 1981. Long-term culture of normal mouse B lymphocytes. *Proc. Natl. Acad. Sci. USA* 78:5788–92

52. Howard, M., Mizel, S. B., Lachman, L., Ansel, J., Johnson, B., Paul, W. E. Submitted for publication

52a. Howard, M., Pillai, S., Scott, D., Paul, W. E. Manuscript in preparation

52b. Howard, M., Sredni, B. Unpublished observations

53. Huber, B. 1979. Antigenic marker on a functional subpopulation of B cells, controlled by the I-A subregions of the H-2 complex. *Proc. Natl. Acad. Sci. USA* 76:3460

54. Huber, B., Gershon, R. K., Cantor, H. 1977. Identification of a B cell surface structure involved in antigen-dependent triggering: Absence of this structure on B cells from CBA/N mutant mice. *J. Exp. Med.* 145:10–20

55. Hunig, T., Bevan, M. J. 1980. Specificity of cytotoxic T cells from athymic mice. *J. Exp. Med.* 152:688–702

56. Isakson, P., Pure, E., Vitetta, E., Krammer, P. 1982. T cell-derived B cell differentiation factor(s). Effect on the isotype switch of murine B cells. *J. Exp. Med.* 155:734–48

57. Jaworski, M. A., Shiozawa, C., Diener, E. 1981. Triggering of affinity-enriched B cells. Analysis of B cell stimulation by antigen-specific helper factor or lipopolysaccharide. I. Dissection into proliferative and differentiative signals. *J. Exp. Med.* 155:248–63

58. Jones, B., Janeway, C. A. Jr. 1981. Cooperative interaction of B lymphocytes with antigen-specific helper T lymphocytes is MHC restricted. *Nature* 292:547–49

59. Julius, M. H., von Boehmer, H., Sidman, C. L. 1982. Dissociation of two signals required for activation of resting

B cells. *Proc. Natl. Acad. Sci. USA* 79:1989–93
60. Kimoto, M., Fathman, C. G. 1980. Antigen-reactive T cell clones. I. Transcomplementing hybrid I-A region gene products function effectively in antigen presentation. *J. Exp. Med.* 152:759–70
61. Kishimoto, T., Ishizaka, K. 1973. Regulation of antibody response in vitro. VI. Carrier-specific helper cells for IgG and IgE antibody response. *J. Immunol.* 111:720–32
62. Kishimoto, T., Ishizaka, K. 1973. Regulation of antibody response *in vitro*. VII. Enhancing soluble factors for IgG and IgE antibody response. *J. Immunol.* 111:1194–205
63. Kishimoto, T., Ishizaka, K. 1975. Immunologic and physiochemical properties of enhancing soluble factors for IgG and IgE antibody responses. *J. Immunol.* 114:1177–84
64. Kishimoto, T., Yoshizaki, K., Okada, M., Miki, Y., Nakagawa, T., et al. 1982. Activation of human monoclonal B cells with anti-Ig and T cell derived helper factor(s) and biochemical analysis of the transmembrane signaling in B cells. *Proc. UCLA Symp. Molecular Cell. Biol.* 24:375–89
65. Lachman, L. B., Hacker, M. P., Handschumacher, R. E. 1977. Partial purification of lymphocyte-activating factors (LAF) by ultrafiltration and electrophoretic techniques. *J. Immunol.* 119: 2019–23
66. Lachman, L. B., Moore, J. O., Metzgar, R. S. 1978. Preparation and characterization of lymphocyte-activating factor (LAF) from monocytic leukemia cells. *Cell. Immunol.* 41:199–206
67. Lachman, L. B., Page, S. O., Metzgar, R. S. 1980. Purification of human interleukin 1. *J. Supramol. Struct.* 13: 457–66
68. Lafferty, K., Andrus, L., Prowse, S. 1980. Role of lymphokine and antigen in the control of specific T cell responses. *Immunol. Rev.* 51:279–314
69. Lafferty, K., Warren, H., Woolnough, J., Talmage, D. 1978. Immunological induction of T lymphocytes: Role of antigen and the lymp hocytes co-stimulator. *Blood Cells* 4:395–404
70. Larsson, E., Coutinho, A. 1979. On the role of mitogenic lectins in T-cell triggering. *Nature* 280:239–41
71. Larsson, E.-L., Iscove, N. N., Coutinho, A. 1980. Two distinct factors are required for induction of T cell growth. *Nature* 283:664–66

72. Leanderson, T., Lundgren, E., Ruuth, E., Borg, H., Persson, H., Coutinho, A. 1982. B cell growth factor (BCGF): Distinction from T cell growth factor and B cell maturation factor. *Proc. Natl. Acad. Sci. USA.* 79:7455–9
73. Leibson, H., Marrack, P., Kappler, J. 1981. B cell helper factors. I. Requirement for both IL-2 and another 40,000 M_r factor. *J. Exp. Med.* 154:1681–93
74. Lonai, P., Puri, J., Hammerling, G. 1981. H-2-restricted antigen binding by a hybridoma clone that produces antigen-specific helper factor. *Proc. Natl. Acad. Sci. USA* 78:549–53
75. Maizel, A., Sahasrabuddhe, C., Mehta, S., Morgan, C., Lachman, L., Ford, R. 1982. Isolation of a human B cell mitogenic factor. *Proc. Natl. Acad. Sci. USA.* In press
76. Marrack, P., Graham, S., Kushnir, E., Leibson, J., Roehm, N., et al. 1982. Nonspecific factors in B cell responses. *Immunol. Rev.* 63:33–49
77. Martinez-Alonso, C., Coutinho, A. 1981. B-cell activation by helper cells is a two-step process. *Nature* 290:60–61
78. Martinez-Alonso, C., Coutinho, A., Augustin, A. A. 1980. Immunoglobulin C-gene expression. I. The commitment to IgG subclass of secretory cells is determined by the quality of the nonspecific stimuli. *Eur. J. Immunol.* 10:698–702
79. McDougal, J. S., Cort, S. P. 1978. Generation of T helper cells in vitro. IV. F_1 T helper cells primed with antigen-pulsed parental macrophages are genetically restricted in their antigen-specific activity. *J. Immunol.* 120:445–51
80. Melchers, F. 1977. B lymphocyte development in fetal liver. II. Frequencies of precursor B cells during gestation. *Eur. J. Immunol.* 7:482–87
81. Melchers, F., Coutinho, A., Heinrich, G., Andersson, J. 1975. *Scand. J. Immunol.* 4:853–58
82. Melchers, F., Potter, M., Warner, N. 1978. *Current Topics in Microbiology and Immunology,* Vol. 81. Berlin: Springer
83. Mizel, S. B., Mizel, D. 1981. Purification to apparent homogeneity of murine interleukin 1. *J. Immunol.* 126:834–37
84. Moller, G., ed. 1980. T cell stimulating growth factors. *Immunol.* 51:1–357
85. Moller, G., ed. 1982. Interleukins and lymphocyte activation. *Immunol. Rev.* 63:1–209
86. Mongini, P., Friedman, S., Wortis, H. 1978. Accessory cell requirement for

anti-IgM induced proliferation of B lymphocytes. *Nature* 276:709–11

87. Mongini, P. K. A., Stein, K. E., Paul, W. E. 1981. T cell regulation of IgG subclass antibody production in response to T-independent antigens. *J. Exp. Med.* 153:1–12

88. Moore, R., Hoffeld, J., Farrar, J., Mergenhagen, S., Oppenheim, J., Shadduck, R. 1981. Role of colony stimulating factors as primary regulators of macrophage functions. *Lymphokines* 3:119–48

89. Mosier, D. E., Zitron, I. M., Mond, J. J., Ahmed, A., Scher, I., Paul, W. E. 1977. Surface immunoglobulin D as a functional receptor for a subclass of B lymphocytes. *Immunol. Rev.* 37:89–104

90. Muraguchi, A., Fauci, A. 1982. Proliferative responses of normal human B lymphocytes. Development of an assay system for human B cell growth factor (BCGF). *J. Immunol.* 129:1104–8

91. Muraguchi, A., Kasahara, T., Oppenheim, J., Fauci, A. 1982. Human BCGF and TCGF are distinct molecules. *J. Immunol.* 129:2486–9

92. Nabholz, M., Engers, H., Collavo, D., North, M. 1978. Cloned T-cell lines with specific cytolytic activity. *Curr. Top. Microbiol. Immunol.* 81:176–87

93. Nakanishi, K., Howard, M., Muraguchi, A., Farrar, J., Takatsu, K., Hamaoka, T., Paul, W. E. 1983. Soluble factors involved in B cell differentiation: identification of two distinct T cell replacing factors. *J. Immunol.* In press

94. Nossal, G. J. V., Pike, B. L. 1975. Evidence for the clonal abortion theory of B-lymphocyte tolerance. *J. Exp. Med.* 141:904–17

95. Nossal, G. J. V., Pike, B. L. 1976. Single cell studies on the antibody-forming potential of fractionated, hapten-specific B lyphocytes. *Immunology* 30:189–202

96. Parker, D. 1980. Induction and suppression of polyclonal antibody responses by anti-Ig reagents and antigennonspecific helper factors. *Immunol. Rev.* 52:115–39

97. Parker, D. C. 1975. Stimulation of mouse lymphocytes by insoluble anti-mouse immunoglobulins. *Nature* 258:361–63

98. Parker, D. C. 1982. Separable helper factors support B cell proliferation and maturation to Ig secretion. *J. Immunol.* 129:469–74

99. Parker, D. C., Fothergill, J. J., Wadsworth, D. C. 1979. B lymphocyte activation by insoluble anti-immuno-globulin: Induction of immunoglobulin secretion by a T cell-dependent soluble factor. *J. Immunol.* 123:931–41

100. Perlmutter, R. M., Nahm, M., Stein, K. E., Slack, J., Zitron, I., et al. 1979. Immunoglobulin subclass-specific immunodeficiency in mice with an X-linked B-lymphocyte defect. *J. Exp. Med.* 149:993–98

101. Pike, B., Vaux, D., Clark-Lewis, I., Schrader, J., Nossal, G. 1982. Proliferation and differentiation of single, hapten-specific B lymphocytes promoted by T cell factor(s) distinct from T cell growth factor. *Proc. Natl. Acad. Sci. USA.* 79:6350–4

102. Pillai, S., Scott, D. 1981. Hapten-specific murine colony-forming B cells: *In vitro* response of colonies to fluoresceinated thymus independent antigens. *J. Immunol.* 126:1883–86

103. Potter, M. 1977. Antigen-binding myeloma proteins of mice. *Adv. Immunol.* 25:141–211

104. Pure, E., Isakson, P. C., Takatsu, K., Hamada, T., Swain, S. L., et al. Induction of B cell differentiation by T cell factors. Stimulation of IgM secretion by products of a T cell hybridomas and a T cell line. *J. Immunol.* 127:1953–58

105. Robb, R., Munck, A., Smith, K. 1981. T cell growth factor receptors. Quantation, specificity and biological relevance. *J. Exp. Med.* 154:1455–74

106. Rocklin, R. E., Bendtzen, K., Greineder, D. 1980. Mediators of immunity: lymphokines and monokines. *Adv. Immunol.* 29:55–136

107. Rosenberg, Y. J., Chiller, J. M. 1979. Ability of antigen-specific helper cells to effect a class-restricted increase in total Ig-secreting cells in spleen after immunization with the antigen. *J. Exp. Med.* 150:517–30

108. Sakano, H., Maki, R., Kurosawa, Y., Reeder, W., Tonegawa, S. 1980. Two types of somatic recombination are necessary for the generation of complete immunoglobulin heavy-chain genes. *Nature* 286:676–83

109. Schimpl, A., Hubner, L., Wong, C., Wecker, E. 1980. Distribution between T helper cell replacing factor (TCGF) and T cell growth factor (TCGF). *Behring Inst. Mitt.* 67:221.

110. Schimpl, A., Wecker, E. 1975. A third signal in B cell activation given by TRF. *Transplant. Rev.* 23:176–88

111. Schimpl, A., Wecker, W. 1972. Replacement of T-cell function by a T-cell product. *Nature New Biol.* 237:15–17

112. Schrader, J. W. 1973. Mechanism of activation of the bone marrow-derived lymphocyte. III. A distinction between a macrophage-produced triggering signal and the amplifying effect on triggering B lymphocytes of alogenic interactions. J. Exp. Med. 138:1466–80

113. Schrader, J., Arnold, B., Clark-Lewis, I. 1980. A Con A stimulated T cell hybridoma releases factors affecting hematopoietic colony-forming cells and B-cell antibody responses. Nature 283: 176–77

114. Schreier, M. H., Andersson, J., Lernhardt, W., Melchers, F. 1980. Antigen-specific T-helper cells stimulate H-2 compatible and H-2 incompatible B-cell blasts polyclonally. J. Exp. Med. 151: 194–203

115. Sell, S., Gell, P. G. H. 1965. Studies on rabbit lymphocytes in vitro. I. Stimulation of blast transformation with an anti-allotype serum. J. Exp. Med. 122: 423–39

116. Shih, W. H., Matzinger, P. C., Swain, S. L., Dutton, R. W. 1980. Analysis of histocompatibility requirements for proliferative and helper T cell activity. T cell populations depleted of alloreactive cells by negative selection. J. Exp. Med. 152:1311–28

117. Sieckmann, D. G., Scher, I., Asofsky, R., Mosier, D. E., Paul, W. E. 1978. Activation of mouse lymphocytes by anti-immunoglobulin. II. A thymus-dependent response by a mature subset of B lymphocytes. J. Exp. Med. 148:1628–43

118. Singer, A., Hathcock, K. S., Hodes, R. J. 1979. Cellular and genetic control of antibody responses. V. Helper T-cell recognition of H-2 determinants on accessory cells but not B cells. J. Exp. Med. 149:1208–26

119. Singer, A., Hathcock, K. S., Hodes, R. J. 1980. Cellular and genetic control of antibody responses. VIII. MHC restricted recognition of accessory cells, but not B cells, by parent-specific subpopulations of normal F_1 T helper cells. J. Immunol. 124:1079–85

120. Singer, A., Morrissey, P. J., Hathcock, K. S., Ahmed, A., Scher, I., et al. 1981. Role of the major histocompatibility complex in T cell activation of B cell subpopulations. Lyb5+ and Lyb5- B cell subpopulations differ in their requirement for major histocompatibility complex-restricted T cell recognition. J. Exp. Med. 154:501–16

121. Slack, J., der-Balian, G. P., Nahm, M., Davie, J. M. 1980. Subclass restrictio of murine antibodies. II. The IgG plaque-forming cell response to thymus-independent Type 1 and Type 2 antigens in normal mice and mice expressing an X-linked immunodeficiency. J. Exp. Med. 151:853–62

122. Smith, K. A., Gilbride, K. J., Favata, M. F. 1980. Lymphocyte activating factors promotes T cell growth factor production by cloned murine lymphoma cells. Nature 287:353–54

123. Smith, K. A., Lachman, L. B., Oppenheim, J. J., Favata, M. F. 1980. The functional relationship of the interleukins. J. Exp. Med. 151:1551–56

124. Sprent, J. 1978. Restricted helper function of F_1 hybrid T cells positively selected to heterologous erythrocytes in irradiated parental strain mice. I. Failure to collaborate with B cells of the opposite strain not associated with active suppression. J. Exp. Med. 147: 1142–58

125. Sprent, J. 1978. Restricted helper function of F_1 hybrid T cells positively selected to heterologous erythrocytes in irradiated parental strain mice. II. Evidence for restrictions affecting helper cell induction and T-B collaboration, both mapping to the K-end of the H-2 complex. J. Exp. Med. 147:1159–74

126. Sredni, B., Sieckmann, D., Kumagai, S., House, S., Green, I., et al. 1981. Long-term culture and cloning of nontransformed human B lymphocytes. J. Exp. Med. 154:1500–16

127. Sredni, B., Tse, H. Y., Schwartz, R. H. 1980. Direct cloning and extended culture of antigen-specific MHC-restricted, proliferating T lymphocytes. Nature 283:581–83

128. Subbarao, B., Ahmed, A., Paul, W. E., Scher, I., Lieberman, R., et al. 1979. Lyb 7, a B cell alloantigen controlled by genes linked to the IgC_H locus. J. Immunol. 122:2279–85

129. Swain, S., Dennert, G., Warner, J., Dutton, R. 1981. Culture supernatant of a stimulated T cell line have helper activity that syngengizes with IL-2 in the response of B cells to antigen. Proc. Natl. Acad. Sci. USA 78:2517–21

130. Swain, S., Dutton, R. 1982. A B cell growth promoting activity DL(BCGF) from a cloned T cell ine and its assay on the BCL_1 B cell tumor. J. Exp. Med. 156:1821–34

131. Swain, S., Wetzel, G., Sovbiran, P., Dutton, R. 1982. T cell replacing factors in the B cell response to antigen. Immunol. Rev. 63:111–28

132. Swierkosz, J. E., Rock, K., Marrack, P., Kappler, J. W. 1978. The role of H-2 linked genes in helper T cell function. II. Isolation of antigen-pulsed macrophages of two separate populations of F₁ helper T cells each specific for antigen and one set of parental H-2 products. *J. Exp. Med.* 147:554–70

133. Takatsu, K., Haba, S., Aoki, T., Kitagawa, M. 1974. Enhancing factor on anti-hapten antibody response released from PPDs-stimulated Tubercle bacilli-sensitized cells. *Immunochemistry* 11:107–9

134. Takatsu, K., Sano, Y., Tomita, S., Hashimoto, N., Hamaoka, T. Submitted for publication

135. Takatsu, K., Tanaka, K., Tominaga, A., Kumahara, Y., Hamaoka, T. 1980. Antigen-induced T cell-replacing factor (TRF). III. Establishment of T cell hybrid clone continuously producing TRF and functional analysis of released TRF. *J. Immunol.* 125:2646–53

136. Takatsu, K., Tominaga, A., Hamaoka, T. 1980. Antigen-induced T cell-replacing factor (TRF). I. Functional characterization of helper T lymphocytes and genetic analysis of TRF production. *J. Immunol.* 124:2414–22

137. Taylor, R. B., Wortis, H. H. 1968. Thymus dependence of antibody: Variation with dose of antigen and class of antibody. *Nature* 220:927–28

138. Tominaga, A., Takatsu, K., Hamaoka, T. 1980. Antigen-induced T cell-replacing factor. II. X-linked gene control for the expression of TRF acceptor site(s) on B lymphocytes and preparation of specific antiserum to that acceptor. *J. Immunol.* 124:2423–29

139. Torrigiani, G. 1972. Quantitative estimation of antibody in the immunoglobulin classes of the mouse. II. Thymus-dependence of the different classes. *J. Immunol.* 108:161–64

140. Tse, H. Y., Mond, J. J., Paul, W. E. 1981. T lymphocyte-dependent B lymphocyte proliferative response to antigen. I. Genetic restriction of the stimulation of B lymphocyte proliferation. *J. Exp. Med.* 153:871–82

141. Vaux, D. L., Pike, B. L., Nossal, G. J. V. 1981. Antibody production by single hapten-specific B lymphocytes: An antigen-driven cloning system free of filler or accessory cells. *Proc. Natl. Acad. Sci. USA* 78:7702–6

142. Watson, J., Gillis, S., Marbrook, J., Mochizuki, D., Smith, K. A. 1979. Biochemical and biological characteriza-tion of lymphocyte regulatory molecules. I. Purification of a class of murine lymphokines. *J. Exp. Med.* 150:849–61

143. Weiner, H. L., Moorehead, J. W., Claman, H. 1976. Anti-immunoglobulin stimulation of murine lymphocytes. I. Age dependency of the proliferative response. *J. Immunol.* 116:1656–61

144. Wetzel, G. D., Kettman, J. R. 1981. Activation of murine B lymphocytes. III. Stimulation of B lymphocyte clonal growth with lipopolysaccharide and dextran sulfate. *J. Immunol.* 126:723–28

145. Wetzel, G. D., Swain, S. L., Dutton, R. W. 1982. A monoclonal T cell replacing factor, (DL)TRF, can act directly on B cells to enhance clonal expansion. *J. Exp. Med.* 156:306–11

146. Whitlock, C., Witte, O. 1982. Long-term culture of B lymphocytes and their precursors from murine bone marrow. *Proc. Natl. Acad. Sci. USA* 79:3608–12

147. Williamson, A., Askonas, B. 1972. Senescence of an antibody-forming cell clone. *Nature* 238:337–339

148. Wood, D. D. 1979. Purification and properties of human B cell activating factor. *J. Immunol.* 123:2395–99

149. Wood, D. D., Cameron, P. M., Poe, M. T., Morris, C. A. 1976. Resolution of a factor that enhances the antibody response of T cell-depleted murine splenocytes from several other monocytes products. *Cell. Immunol.* 21:88–96

150. Wood, D. D., Gaul, S. L. 1974. Enhancement of the humoral response of T cell-depleted murine spleens by a factor derived from human monocytes *in vitro*. *J. Immunol.* 113:925–33

151. Yamashita, U., Shevach, E. M. 1978. The histocompatibility restrictions on macrophage T-helper cell interaction determine the histocompatibility restrictions on T-helper cell B-cell interaction. *J. Exp. Med.* 148:1171–85

152. Yoshizaki, K., Nakagawa, T., Fukunaga, K., Kaieda, T., Maruyama, S., Kishimoto, S., Yamamura, Y., Kishimoto, T. Submitted for publication

153. Zubler, R. H., Glasebrook, A. L. 1982. Requirement for three signals in "T-independent" (lipopolysaccharide-induced) as well as in T-dependent B cell responses. *J. Exp. Med.* 155:666–80

154. Zubler, R. H., Kanagawa, O. 1982. Requirement for three signals in B cell responses. II. Analysis of antigen- and Ia-restricted T helper cell-B cell interaction. *J. Exp. Med.* 156:415–29

Ann. Rev. 1983. Immunol. 1:335–59
Copyright © 1983 by Annual Reviews Inc. All rights reserved

MEDIATORS
OF INFLAMMATION

Gary L. Larsen and Peter M. Henson

Department of Pediatrics, National Jewish Hospital and Research
Center/National Asthma Center, and Departments of Pediatrics, Pathology, and
Medicine, University of Colorado School of Medicine, Denver, Colorado 80206

INTRODUCTION

Inflammation is a term used to describe a series of responses of vascularized tissues of the body to injury. The clinical signs of this phenomenon can now be related to increased flow in local blood vessels (calor and rubor), increased vascular permeability and/or cellular infiltration (tumor), and release of a variety of materials at the site of inflammation that induce pain (dolor). Loss of function was later considered an additional cardinal sign of inflammation. However, mechanisms by which the function of a given vascularized tissue is impaired depend greatly on the nature of the tissue and on the detailed processes that contribute to the inflammatory reaction.

The subject of this review is the mediators that produce these alterations. However, it is important to indicate at the outset that the inflammatory process is both varied and extremely complex. We emphasize the broad outlines of (*a*) alterations in hemodynamics and vessel permeability, and (*b*) infiltration by inflammatory cells as representing a common theme for the process. However, each site and stimulus may result in a different mix of these elements, a different time course, and a different outcome. The inflammatory process is also influenced by the nutritional and hormonal status of the individual, as well as by genetic factors. In addition, the processes themselves represent an extensive network of interacting mechanisms, mediators, and cells (85). This represents a high degree of redundancy in the system, which provides a mechanism both for amplification and for preservation of the response if one component of the system is deficient or inactivated. For example, we present an argument that biologically active fragments of C5 are critical initiators of the acute inflammatory

335

0732-0582/83/0410-0335$02.00

reaction. Nevertheless, genetic C5 deficiency is not a major impairment, suggesting that collateral mechanisms can compensate (58). By contrast, certain elements of the inflammatory process are critical. Thus, deficiencies in neutrophil function result in defective inflammatory responses and insufficient protection against infection and, without aggressive chemotherapy, represent lethal mutations (48).

These observations suggest that inflammation is a beneficial reaction of tissues to injury. In fact, it normally leads to removal of the inciting agent and repair of the injured site. This may involve temporary discomfort and loss of function, but in the end it is protective. Nevertheless, a large number of human diseases represent uncontrolled inflammatory reactions that may continue unchecked (rheumatoid arthritis), may induce permanent tissue destruction (emphysema), or may heal, but only by the inappropriate deposition of collagen (pulmonary or hepatic fibrosis).

The immune system has long been linked with inflammation. Both may be seen as protective mechanisms. However, major elements of the inflammatory reaction precede the immune system in evolution. In fact, a key element in the former, chemotaxis of wandering cells, was presumably a property of the earliest eukaryotes. The immune system confers specificity on the reaction and the inflammatory process may be seen to represent an effector arm of the immune system. Thus, antibody-antigen complexes readily initiate generation and release of many of the mediators of inflammation. In addition, antigen-stimulated lymphocytes provide a wealth of mediator functions themselves, as well as modulate the immune response (98).

A short review of a subject as broad and complex as mediators of the inflammatory process will be incomplete and will contain material biased towards the authors' general views on the subject. In addition, the choice of reference material cannot be comprehensive. This work is no exception. Our goal is, thus, to give an overview, stressing the complexity of this process by pointing out many potential interactions of the mediators described, while providing references of both a general and a specific nature for the readers wanting more information. Concepts will receive more emphasis than details of any one mediator. For reasons of simplicity, space, and amount of available information, the acute inflammatory reaction is emphasized. This is not to belittle the importance of the more chronic reactions, especially those associated with cell-mediated immunity. Rather, it is a reflection of the complexity of the latter subject and the recognition that we are just at the brink of molecular definition of the mediators involved. Thus, the field of study of lymphokines, interleukins, other "growth" factors, and macrophage-derived "activating" factors is just entering a new and exciting growth period, but one that requires a separate review(s) of itself.

THE INFLAMMATORY RESPONSE

As indicated above, inflammation is a nonspecific response of tissues to diverse stimuli or insults. However, despite this complexity and variability, the common themes seen in the process argue for common groups of mediators. In particular, we should look for molecules that enhance blood flow, increase vessel permeability, and induce emigration of inflammatory cells from the blood. Once they have reached the site of injury, the cells are stimulated to phagocytose bacteria and debris, and also to secrete many of their preformed or newly synthesized constituents. Molecules promoting these events may be classified as "mediators." In addition, the reactions are controlled throughout by positive and negative feedback reactions, again involving "mediators" and "inhibitors." Inflammatory reactions generally resolve. Molecules promoting resolution and healing may not be, strictly speaking, mediators, but none the less they are of critical importance to the overall process.

The hemodynamic changes associated with inflammation are often the first to be manifested. They are of critical importance. Cohnheim's dictum "without vessels no inflammation" was later parodied by Metchnikoff as "there is no inflammation without phagocytes" (69). Vasodilation, increased flow, and permeability increases are key elements in inflammation. It may be presumed that their "purpose" is to allow maximum opportunity to recruit inflammatory cells and to bring plasma proteins to the site of injury. These latter could contribute to the initiation or perpetuation of the response (complement fragments) or the resolution of the response (plasma protease inhibitors). How these vascular alterations are achieved, however, is very poorly understood (71, 88, 123). In part, this reflects the dynamic nature of the systems, in part the inability to study readily the processes involved in vitro, and in part the considerable variation from site to site in the body in the involvement of the microcirculation in inflammation. For example, in skin (44), mesentery (65), hamster cheek pouch (72), and ear chamber (28), the events occur predominately in the postcapillary venules. In the pulmonary vasculature, however, the capillary may be the major site of leak and emigration (17, 41, 94). Whether this reflects physical effects, differences in endothelial cells, or different types of mediators is not yet known.

The histologic changes associated with a circumscribed acute inflammatory reaction in the lung are shown in Figure 1. The response was induced by instillation into the airways of a phlogistic fragment from the fifth component of complement, C5a desArg (55). This same type of response may be seen in other tissues after exposure to C5a desArg, or in the same tissue (lung) with other stimuli such as immune complexes (57, 90) or bacterial infection (56, 86). This simple example is thus meant to demon-

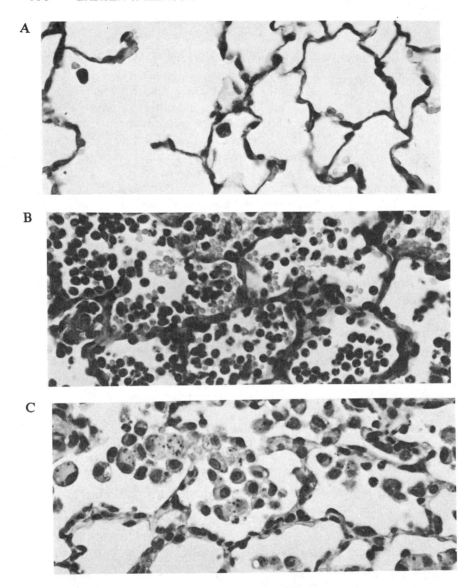

Figure 1 The inflammatory response produced by instillation of C5a desArg into rabbit airways is displayed. In the normal lung (*A*), resident alveolar macrophages are present in some alveoli, but neutrophils are not seen. Six hours after administration of C5a desArg (*B*), granulocytes (predominantly neutrophils) and protein-rich fluid accumulate in the airspaces. By 24 to 48 hr (*C*), the neutrophilic infiltration is replaced by mononuclear cells (macrophages). Over several days, the alveoli clear and the alveolar walls return to a normal thickness and cellularity (hematoxylin and eosin stain; X400).

strate a normal inflammatory response. The sequence of permeability, neutrophil accumulation, and mononuclear phagocyte (macrophage) infiltration that occurs before resolution is apparent. In Figure 2 may be seen a similar inflammatory reaction in one capillary examined ultrastructurally at the time of change from a granulocytic to a monocytic infiltrate. Extensive tissue damage has occurred.

For this site to be repaired, the mononuclear influx of new cells must be halted, injurious oxygen radicals and proteases must be inactivated, fluids and proteins must be removed or reabsorbed, inflammatory cells and debris must be removed, and damaged cells (in this case epithelial cells) must be replaced. This last results from the induction of replication of the epithelial cells and laying down of new basement membrane material. In other circumstances, tissue fibroblasts may be similarly induced to divide and secrete connective tissue elements such as collagen. These subjects, and the mediators that induce them, are poorly studied to date and are just beginning to receive specific emphasis.

MEDIATORS OF INFLAMMATION

Definition of Mediators

In simplest terms, an inflammatory mediator can be thought of as a chemical messenger that will act on blood vessels and/or cells to contribute to an inflammatory response. As is discussed below, a multitude of agents fulfill this definition. The mediator may be distinguished from a neurotransmitter on the one hand, and a hormone on the other, largely by the extent of its sphere of influence (neither at a nerve ending nor alternatively throughout the body). Douglas has introduced the term "autocoids" (20) to refer to many of the substances to be discussed here. As he defined the word, it refers to a variety of substances of intense pharmacologic activity that are normally present in the body and cannot be conveniently classed as neurohumors or hormones. This is very similar to the use of the term mediator indicated above, which is used throughout this review.

Sources of Mediators

The mediators of the inflammatory response can be classified in several ways. One of the most helpful subdivisions is source. First, the chemical messenger may be an exogenous mediator (coming from outside the body) or an endogenous mediator. Although this review concentrates on the latter, it is important to acknowledge the existence and importance of the former, which include bacterial products and toxins (53, 115). When considering endogenous mediators, several other classifications according to source can be made. First, is the origin of the mediator the plasma, blood

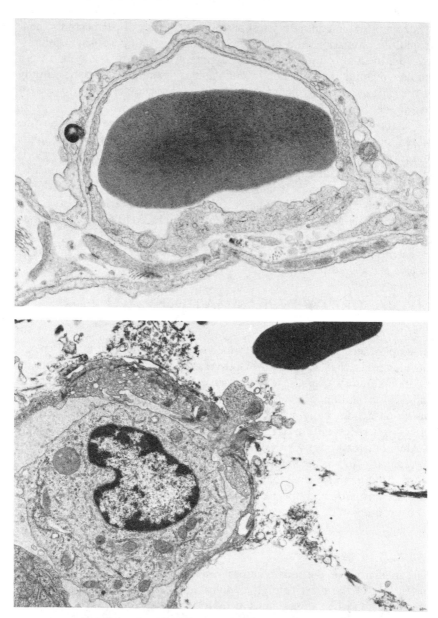

Figure 2 An electron micrograph of a normal alveolar capillary unit (*A*) is shown. By comparison (*B*), an inflammatory reaction in a capillary unit at the time of emigration of a monocyte through the capillary walls exhibits extensive damage, which occurred during the prior neutrophil emigration. Evidence of activation of the coagulation system is manifest by fibrin strands in the extravascular space. Top, X11,500; bottom, X8335)

cells, or tissues? The plasma contains three major mediator-producing systems (kinin, coagulation, complement), which interact in defined manners to generate phlogistic compounds. Other mediators are cell derived and, within the cells of origin, may be preformed and stored in granules (histamine in mast cells, cationic proteins in neutrophils) or may be newly synthesized by the cells (interleukin 1, leukotrienes, platelet-activating factor). The importance of these distinctions lies in part in the rapidity of release of the molecules, but also in therapeutic approaches that may be taken to modify their effects.

Many peptide mediators are generated as a consequence of multiple enzymatic steps involving sequential activation of molecules by limited proteolysis (complement and coagulation systems). The small fragments derived from such steps often exhibit biological activity and may be extremely important mediators (C5a, C3a). Depending on one's point of view, they could be seen as byproducts of complement activation or as key derivatives of the proteolytic sequence.

Lipid mediators are generally synthesized de novo within cells when these cells are activated. Little is known as yet about the mechanisms of secretion, but the biochemical synthetic pathways of a number of important molecules (the eicosanoids and platelet-activating factor) are beginning to be understood. Arachidonic acid is the precursor of a wide variety of molecules collectively termed eicosanoids. It is derived from membrane phospholipids by the direct action of phospholipase A_2 (PLA_2) or indirectly following the effect of phospholipase C. The arachidonic acid may then be converted into prostaglandins and thromboxanes via cyclooxygenase and subsequent enzymes or into hydroperoxy and hydroxyeicosatetraenoic acids (HPETEs and HETEs) by lipoxygenase enzymes. Leukotrienes C_4, D_4, and E_4 (slow reacting substances) as well as B_4 (a potent chemotactic agent) are also derivatives of the lipoxygenase pathways. These metabolic events appear to be of great importance in the inflammatory process. For example, our most potent anti-inflammatory drugs are known to inhibit some or all of these events. For consideration of this fast-moving field, the reader is referred to recent reviews by Goetzl, Parker, and Samuelsson and co-workers (31, 79, 89).

The Mediators

Examples of some of the mediators of inflammation are listed in Table 1. The list is not exhaustive in terms of the mediators listed, and the actions given for any chemical or group of compounds are incomplete. The table does provide, however, an overview of the types (chemical nature), sources, and actions of some central compounds.

Table 1 Representative mediators of inflammation

Mediator	Structure/chemistry	Source[a]	Effects[b]	Reference
Histamine	β-Imidazolylethylamine	Mast cells, basophils	Increase vascular permeability (venules), chemokinesis, mucus production, smooth muscle contraction	3, 15, 64
Serotonin	5-Hydroxytryptamine	Mast cells (rodent), platelets, cells of enterochromaffin system	Increase vascular permeability (venules), smooth muscle contraction	3, 4
Bradykinin	Nonapeptide	Kininogen (by proteolytic cleavage)	Vasodilation, increase in vascular permeability, production of pain, smooth muscle contraction	14, 101, 122
C3a	77-Amino-acid peptide	C3 complement protein	Degranulation of mast cells, smooth muscle contraction	42
C5 fragments C5a C5a desArg	74-Amino-acid peptide 73-Amino-acid peptide	C5 complement protein	Degranulation of mast cells, chemotaxis of inflammatory cells, oxygen radical production, neutrophil secretion, smooth muscle contraction	42, 55, 116
Neutrophil chemotactic factor	$M_r \geqslant 750{,}000$	Mast cells (?)	Chemotaxis of neutrophils	2, 4, 59
Vasoactive intestinal peptide	28-Amino-acid peptide	Mast cells, neutrophils, cutaneous nerves	Vasodilation, potentiate edema produced by bradykinin and C5a des Arg	126

Mediator	Structure	Probable source[a]	Effects[b]	Ref.
Prostaglandin E_2		Arachidonic acid (cyclo-oxygenase pathway)	Vasodilation, potentiate permeability effects of histamine and bradykinin, increase permeability when acting with leukotactic agent, potentiate leukotriene effect	45, 117, 128
Leukotriene B_4		Arachidonic acid (lipoxygenase pathway)	Chemotaxis of neutrophils, increase vascular permeability in the presence of PGE_2	12, 31
Leukotriene D_4		Arachidonic acid (lipoxygenase pathway)	Smooth muscle contraction, increase vascular permeability	21, 31
Platelet-activating factor		Basophils, neutrophils monocytes, macrophages	Release of mediators from platelets, neutrophil aggregation, neutrophil secretion, superoxide production by neutrophils, increase vascular permeability, smooth muscle contraction	83
γ Interferon	Glycoprotein	T lymphocytes	Activation of macrophages, modulation of immune reactions	75, 98
Interleukin 1	Peptide, $M_r = 12{-}16{,}000$	Macrophages	Fever, fibroblast proliferation, induction of collagenase and prostaglandin production	95

[a] Listed are the most probable sources of the mediator in an inflammatory reaction.
[b] Effects listed are limited to those most important to an inflammatory reaction.

EFFECT OF MEDIATORS
ON THE INFLAMMATORY PROCESS

The above paragraphs have dealt with the inflammatory process and mediators of inflammation in general terms. We now address the individual components of this process and look specifically at the actions of the mediators in producing these effects. It is important to point out that several mediators produce the same effect, and the relative importance of one mediator versus another is often difficult to assess. In addition, it is also probable that mediators with similar actions have different levels of importance depending on the tissue involved and the stimulus for injury. To complicate the picture further, certain mediators can be responsible for more than one of the vascular or cellular changes central to the process (see Table 1).

Vasodilation and Hyperemia

Of the features associated with the inflammatory process, vasodilation has been the least studied. Despite this fact, it is generally considered of central importance in the development of acute inflammatory reactions because local blood flow determines in large degree the amount of exudate produced (88). It is known that the flare after histamine injection is at least partially under central nervous system control (1), but the subject of involvement of the neuronal system in inflammatory processes is underdeveloped. A number of mediators, including histamine and various eicosanoids, are also clearly involved in regulation of blood flow, even if the mechanisms are not well understood. Injection of phospholipase A_2 into the skin induces a hyperemia very similar to that seen in inflammation (105). Since PLA_2 may be expected to cause release of arachidonic acid, products of cyclooxygenase or lipoxygenase pathways of arachidonate metabolism are implicated. In fact, prostaglandins are turning out to be important mediators of inflammation and may exert their effects in part by modulating blood flow. It is interesting to note that although prostaglandins were first described approximately one-half century ago by von Euler in 1934 (110) and Goldblatt in 1935 (34), it was not until the last decade that they were proposed as inflammatory mediators (107, 129, 130). Even now, their exact contribution to the process is poorly understood. In large measure this is because the administration of prostaglandins does not induce inflammation directly, and indeed, at some concentrations, some of the compounds, such as PGE_1 (which is not made in mammals) or PGI_2, actually inhibit the function of inflammatory cells (11, 61, 134). However, it has recently become apparent that prostaglandins such as PGE_2 and PGI_2 can have a marked proinflammatory effect as potentiators of the effect of other mediators.

The evidence supporting a role for this group of compounds is threefold.

First, PGE_2 and PGI_2 injections induce vasodilation, presumably by an action on cells of the blood vessel wall (54). Second, prostaglandins, especially PGE_2, have been detected in inflammatory exudates (125, 130). Their accumulation is usually 6 to 24 hr later than histamine or bradykinin. Third, drugs with known anti-inflammatory properties such as aspirin inhibit the production of prostaglandins by acting on the enzyme cyclooxygenase (107, 108). At physiologic concentrations, vasodilator prostaglandins have been reported to enhance the permeability effects of histamine and bradykinin (128).

A similar potentiating effect of vasodilator prostaglandins has been noted with the permeability-inducing effects of LTB_4, C5 fragments, and N-formyl-methionyl-leucyl-phenylalanine (FMLP) (12, 117, 127). These are all chemotactic agents, and the effect has been attributed to an interaction between neutrophils, chemoattractants, and prostaglandin. Local vasodilation is suggested to enhance the effect of neutrophils on the blood vessel wall, and their emigration. In fact, Williams (126) has shown a similar potentiation of permeability by chemoattractants using an unrelated vasodilator, vasoactive intestinal peptide (VIP) (see Table 1). Nevertheless, the prostaglandins might also affect other components of these systems, including the inflammatory cells, the presentation of chemotactic factors, local shear stresses, or the leukocytes adhering to the endothelium or endothelial junctions. This area requires more investigation. It is of critical importance, not only because it is an integral part of the inflammatory process, but also because of the widespread use of nonsteroidal anti-inflammatory drugs that prevent prostaglandin production.

It should be noted that other products of the cyclooxygenase system (thromboxane A_2) have vasoconstricting activity (4). Critical here is to determine the overall effect of such mediators on the blood flow through the different portions of the vascular bed. Moreover, prostaglandins have been reported to inhibit leukocyte and mast cell secretion by increasing intracellular cAMP (11, 61, 134), thus preventing inflammatory reactions. Indeed, stable forms of PGE_1 have been used experimentally to prevent inflammation (23). Although these reports would suggest an anti-inflammatory effect of the prostaglandins, Tauber and co-workers (102) noted low concentrations of prostaglandins (PGE_1 or $PGF_{2\alpha}$) enhanced antigen-induced release of histamine from lung tissue whereas high concentrations had the opposite effect. These results underscore the complexity of these interactions and point out that quantity of mediator present may determine if the effect is pro- or anti-inflammatory. In addition, the actions of the various prostanoids may vary among different vascular beds. Moreover, different cell types, including those lining vessel walls, produce a different mix of prostanoids both intrinsically and after exposure to different stimuli. This complexity argues for a careful examination of each tissue to determine the fine tuning of the regulatory events involved. Despite these conflicting effects, the overall response and the means to achieve it are similar in kind from tissue to tissue.

Vasopermeability

When considering various mediators that may cause increased permeability, it is important to remember that these alterations occur at various times after the injury. In many types of tissue injury, increased permeability has been noted to occur in at least two phases: an early, transient phase, occurring almost immediately after an injury; and a late or second phase, beginning after a variable latent period, but then persisting for hours or days. As discussed by Wilhelm in his review of these and other patterns (123), the types of response in terms of the time course or pattern of permeability noted are functions of the type and intensity of the injury as well as the species of animal being studied. There is good evidence that the early, transient permeability seen with certain types of injury (heat) are due to release of histamine, in that the early permeability, as measured by skin bluing, may be suppressed by antihistamines (99, 124). Mediation of the delayed phase of exudation is more controversial and complex and has been attributed in part to kinins (123), prostaglandins (19), neutrophils (74, 104), and, more recently, lipoxygenase products of arachidonic acid metabolism (54). (See also the discussion of eicosanoid and neutrophil effects noted above.) The mediators to be discussed may act together or sequentially in an inflammatory reaction to alter permeability.

Histamine is the one mediator most often associated with the early increase in permeability after various types of injury. The physiologic and pharmacologic properties (contraction of smooth muscle and vasodilation) of this vasoactive amine were first described by Dale and associates (15, 16). Histamine is widely distributed, but that contained in mast cells and basophil leukocytes (plus platelets in the rabbit) provides the main source in acute inflammation. Tissue mast cells are commonly located around blood vessels, suggesting the possibility of extremely local effects. The stimuli causing release of histamine from mast cells are many, including a variety of low-molecular-weight substances, compound 48/80, physical agents including heat and trauma, various drugs, IgE and other immunologic reactions, and anaphylatoxin (C3a and C5a). The mechanisms of histamine release and regulation have been well studied (60–62). Recently, it has been suggested that arachidonic acid metabolism within the cell is important in both basophil and mast cell secretion. A lipoxygenase product(s) may be required for histamine release from human basophils and may blunt endogenous control mechanisms (66). For example, 5-hydroperoxyeicosa tetraenoic acid (5-HPETE) has been shown to induce both a dose-dependent enhancement of histamine release and a reversal of the inhibition of histamine release caused by agonists acting via adenylate cyclase (82). Peters et al (82) also report that inhibitors of phospholipase A_2 activation and inhibitors of arachidonic acid metabolism through the lipoxygenase pathway act not only on IgE-mediated release of histamine but also on release initiated by anaphylatoxins, f-met peptides, and ionophore A23187. Thus, these data suggest cell membrane phospholipid processes are essential final common pathways leading to secretion of histamine. However, it should be emphasized that although studies of cell functions with eicosanoid metabolism inhibitors are widespread, they must be interpreted with caution. The inhibitors are not very specific, the intracellular biochemical events are not understood, and in most cases one cannot determine whether the arachidonate derivative acts intracellularly as a true messenger or is acting extracellularly on the cell of origin or other cells in the area of the reaction.

It has been suggested that histamine increases vascular permeability by inducing contraction of the endothelial cells of the postcapillary venule, thus allowing the passage of fluid and proteins through the opening of interendothelial junctions (44, 64). However, this view has been challenged (36) and the question of endothelial contractibility is controversial. Although gaps in the endothelium can certainly be seen, the exact site of the barriers to protein transudation in nonglomerular vessels is not yet fully understood and the effect of vasopermeability agents requires more study. Charge effects may be of critical importance in the passage of proteins through blood vessels. When leukocytes are involved, the impact of proteases on the endothelium and the basement membrane must also be considered.

Serotonin, like histamine, is a vasoactive amine that will contract smooth muscle and increase vascular permeability, although its importance may be more as a neurotransmitter. It has been implicated, by the use of inhibitors, as an inflammatory mediator in the early phase of increased vascular permeability in certain inflammatory responses in mice and rats (70, 73). Its role as an inflammatory agent in man is unclear.

The actions of bradykinin include its ability to increase vascular permeability (122). The permeability increase of kinins has been attributed to endothelial cell separation with gap formation in postcapillary venules (101). The generation of this mediator is complex and involves several steps and pathways. First, Hageman Factor (factor XII of the clotting system) is activated by contact with negatively charged surfaces (glass) or contact with a variety of biological material (14). This enzyme then activates (and is activated by) plasma kallikrein. The kinin is cleaved from kininogen by this kallikrein (133) or kallikrein from tissues (122), as well as possibly by other proteases such as plasmin (35).

Kinins, like most potent inflammatory mediators, are rapidly broken down in the plasma and tissues by kininases and within the circulation undergo almost complete inactivation during one passage through the pulmonary circulation (27). However, several pathophysiologic changes may alter these normal control mechanisms. For example, hypoxia may decrease pulmonary angiotensin-converting enzyme activity, leading to impaired clearance of bradykinin (which is cleaved by this enzyme) in the lung (100). Alternatively, hypercapnic acidosis can lead to activation of the kallikrein-kinin system, with subsequent bradykinin production (78). When the animals (sheep) used for these studies were both acidotic and hypoxic, both pulmonary artery and aortic bradykinin concentrations were elevated, and pulmonary vascular permeability was increased. This example serves to illustrate some of the general control mechanisms for inflammatory mediators and how disease states may readily alter these—leading to enhanced pathologic events.

Platelet-activating factor (PAF) or acetyl glyceryl ether phosphorylcholine (AGEPC) is another recently described mediator that can cause increases in vascular permeability. First described in terms of its biological effects on platelets (6, 38) following earlier studies of leukocyte-platelet interactions, this molecule has now been shown to have multiple effects and to be derived from neutrophils and mononuclear phagocytes, as well as from rabbit basophils (83). It now must be considered as a broad spectrum mediator of inflammatory reactions. Definition of its chemical structure (7, 10, 18, 37) has allowed a more detailed investigation of its phlogistic properties (83). PAF produces a wheal and flare reaction in man that, on a molar basis, is 100 to 1000 times more potent than histamine. It also induces non-histamine-dependent increased vascular permeability in the skin of various species. The question of whether or not the mediator acts directly on some other molecule is as yet unanswered. However, it is noteworthy that the permeability induced by PAF in isolated perfused rat lungs may be dependent on release of leukotriene C4 and D4 (109).

Many other molecules have been shown to induce increased vascular permeability. They include fibrinopeptides (88), fibrin degradation products (5), lymphokines, (22, 63), and anaphylatoxins (C3a and C5a) (42). These last probably act indirectly via histamine release and/or leukocyte attraction. This point emphasizes the need for (a) careful studies of the direct or indirect effects of these mediators, (b) determination of the relative importance of a given mediator, as well as (c) defined structural identification of the molecules involved. Finally, a group of molecules derived from leukocytes should be mentioned. Four cationic vasoactive peptides from rabbit neutrophils received considerable attention in the 1960s (84, 93) but have largely been ignored for the last 10 years. As interest in neutrophil-mediated vascular permeability is revived, most emphasis has been on leukocyte-derived toxic oxygen radicals (which certainly can alter endothelial permeability) or leukotrienes and PAF. However, it would seem important to reinvestigate the role of these cationic peptides and determine whether or not analogues are present in the leukocytes of other species. Three of these proteins were suggested to act directly (this would have to be re-evaluated in the light of today's knowledge of lipid mediators) and one via mast cell degranulation.

Leukocyte Emigration and Chemotaxis

As displayed in Figures 1 and 2, one of the most marked histologic changes noted in an inflammatory response is the accumulation of cells within the tissues. Early in the reaction the cellular infiltrate is predominately neutrophils, later followed by monocytes. The exact molecular mechanisms by which leukocytes migrate out of blood vessels are poorly understood. Much emphasis has been placed on the phenomenon of chemotaxis, which probably represents only a small part of the whole process of emigration in vivo. Nevertheless, molecules chemoattractive in vitro can induce leukocyte accumulation in tissues. Some of the most potent of these for neutrophils are C5 fragments. These fragments are low molecular weight factors produced through cleavage of C5 by a variety of endopeptidases. C5 convertases derived from classical or alternative complement pathways cleave a 74 amino acid terminal fragment termed C5a. Other proteases including plasmin, trypsin, kallikrein, and bacterial proteases may cleave at the same or different sites on C5, (112, 113, 121), but generate fragments with similar biological activities. In addition, enzymes that cleave C5 have been found in lysosomes of neutrophils (114) and platelets (119). The C5a, like C3a, will cause release of histamine from mast cells, can alter vascular permeability, and is capable of constricting smooth muscle. In addition to these activities, however, the molecule is also chemotactic for neutrophils and will induce oxygen radical generation as well as neutrophil granule exocytosis (42). Human C5a desArg, the fragment derived from C5a after the serum enzyme carboxypeptidase N removes the C-terminal arginine, is poorly active in terms of increasing vascular permeability and as a spasmogen, or as a stimulus for neutrophils in vitro, but will cause secretion from mononuclear phagocytes (67, 116). By contrast, studies in vivo have shown human C5a desArg to be as or more active at inducing inflammation than C5a (55). The explanation for this observation may relate to a serum helper cofactor for C5a desArg (26, 81, 132) which would be present in the in vivo experiments. Alternatively, differential binding of C5a and C5a desArg to tissue structures may be important. It is interesting in this regard that C5a and C5a desArg from pig and guinea pig are more equivalent in activity and also lack a carbohydrate moiety (42).

The relative importance of C5 and its fragments as mediators of neutrophil influx have been examined in various types of experiments. Snyderman and co-workers (96) found serum from C5-deficient mice gave no chemotactic response when treated with immune complexes or endotoxin, but found the activity was restored by adding purified C5. Intraperitoneal injection of endotoxin into C5-deficient mice resulted in little chemotactic activity and a marked depression of the early neutrophil accumulation. The C5 deficient mice also exhibit a delay or abrogation of the neutrophil accumulation into the lung following intratracheal challenge with immune complexes (57), *Pseudomonas aeroginosa* (56), or exposure to hyperoxia (80). Injection of C5-sufficient plasma restores the response. It is important to point out, however, that even though one can argue that the presence of C5 is important for early neutrophil influx in these systems, neutrophil influx does occur in its absence, thus pointing to the fact that other factors are present that are also neutrophil chemotoxins. Within the lung, chemotactic factors produced by alveolar macrophages after various challenges almost certainly act in concert with the complement system to mount a full inflammatory response (43, 52). This may also be true elsewhere in the body but is more difficult to prove. The alveolar macrophages synthesize and secrete a small molecular weight protein chemoattractant (68) and a low molecular weight lipid (106). With an organism like *Pseudomonas aeroginosa,* another factor that can cause neutrophil emigration must be considered: chemotactic factors derived from the bacterium (53, 115). Some of these may represent products of normal prokaryotic protein synthesis: n-formylated methionyl peptides (91). Such peptides have received enormous attention recently as definite probes of neutrophil function. Receptors have been well characterized on neutrophils for these peptides and they are proving valuable in basic studies of chemotaxis, stimulus-secretion coupling, receptor down regulation, and cellular oxidative metabolism.

C5 fragments are also chemotactic in vitro for monocytes, (97) eosinophils (49, 51), and basophils (50). This raises two important points. First, the demonstration that a molecule is attractive to a cell in the in vitro test system is far from demonstrating its efficacy in vivo. In fact, few molecules have been definitively shown to have significant chemotactic effects in vivo. Even with the C5 fragments it is not yet clear whether this is a direct effect or an action via some other, as yet undefined, mediator.

Second, we must consider the specificity of chemotactic factors. Instillation of C5a desArg into the lung results first in an accumulation of neutrophils and then an accumulation of monocytes. Is the latter due to a slower response of the monocyte to C5a desArg itself or due to some other factor possibly derived from neutrophils (114)? Due to relative potencies in vivo, molecules such as C5a may exhibit increased cell specificity.

Few highly specific chemoattractants for inflammatory cells have been described. Among the most interesting are lymphokines that are chemoattractants in vitro for neutrophils, monocytes, eosinophils, and basophils, and also to lymphocytes themselves (87). Recent studies with T cell hybridomas suggest segregation of different cell-specific chemotactic lymphokines may be possible; that is, each may attract a separate cell type. Other sources of interest are connective tissue breakdown products. Thus, fibronectin fragments may specifically attract monocytes (77). A variety of mast-cell-derived eosinophilotactic molecules are known (4) and, recently, a mast-cell-associated neutrophil chemotactic factor has been described (see Table 1) (2, 4, 59).

Lipid mediators may also be chemotactic. Much the most potent lipid chemotactic for human polymorphonuclear leukocytes so far studied has been leukotriene B4 (LTB$_4$) (29, 32). Mono-HETEs are less active (33). Injection of PAF (AGEPC) into the skin has been noted to cause a neutrophil infiltrate, and the molecule also activates neutrophils in vitro, although rather high concentrations are required (83).

Tissue Damage

Damage to tissue in an inflammatory reaction can be by several mechanisms. Two of the more prominent reasons for tissue injury are the release of enzymes from the granules and lysosomes of infiltrating cells as well as the production of oxygen radicals by these cells. As reviewed above, the central feature of inflammatory processes is the infiltration of cells into the injured tissues. In a simplified acute inflammatory lesion, the most prominent cell types include polymorphonuclear leukocytes at early time points followed by mononuclear phagocytes as the lesion evolves. Both of these cell types carry within them lysosomes with many potential inflammatory mediators. The evidence for the participation of the lysosomal constituents in inflammatory processes include the observation of increased lysosomal activity in inflamed tissues. For example, degranulation of polymorphonuclear leukocytes has been demonstrated in injured tissues along with release of degradative proteases (40). In addition, as first pointed out by Thomas (103), extracts of lysosomes are capable of producing an inflammatory response. Lysosomal enzymes and proteases are secreted during phagocytosis (118) or when such cells are exposed to stimuli on surfaces too large to ingest (39). The mechanism of secretion of lysosomal acid hydrolases and neutral proteases is a subject of much investigation, but is beyond the scope of this review. It is of interest, however, that in addition to phagocytosable stimuli, the same molecules that induce chemotaxis can also induce secretion of enzymes and oxygen radicals.

The contribution of lysosomal components to the inflammatory process has been discussed in several recent reviews (30, 88, 131). Many of these enzymes are only optimally active at acid pH. Whether the pH in inflammatory lesions gets low enough for a significant contribution of the acid hydrolases is as yet unclear. This has led to the suggestion that neutral proteases, enzymes active at neutral and alkaline pHs, are of most importance in terms of modulating the inflammatory reaction and possibly contributing to tissue destruction, whereas acid proteases may function primarily within lysosomal vacuoles to digest phagocytosed material (46). These neutral proteases include collagenases, a serine elastase (neutrophils and monocytes) and a metallo-elastase (macrophages) (120), and cathepsin G. We and others have suggested that not only are they important in the tissue damage of inflammation, but they also allow leukocytes to emigrate through the basement membrane into the lesions in the first place. A degree of "injury" may therefore be a requirement for initiation of inflammation. In addition, these proteases can generate a host of inflammatory mediators by limited proteolysis of precursor proteins, as noted above. The efficient inhibition of the enzymes, therefore, becomes a critical control mechanism to prevent unbridled digestion of the tissues. It may be argued that a key "purpose" for the vasopermeability in inflammation is to flood the lesion with such inhibitors.

Oxygen radicals are derived from reduction of molecular oxygen by inflammatory cells. They are highly bacteriocidal and are also extremely toxic for cells, and in addition they may modify (inactivate) a key protease inhibitor, alpha 1 antitrypsin (13). Much of the early damage in inflammation as well as some of the vascular permeability may be attributed to this group of compounds, and there is great current interest in this field (25). However, these molecules cannot readily be classed as mediators. For this review, it is perhaps more pertinent to emphasize once again that many molecules discussed under "Chemotaxis" can cause release of oxygen radicals from phagocytes.

Resolution

The factors that orchestrate the resolution of an inflammatory response are poorly understood and are just now beginning to receive attention. The delineation of normal repair mechanisms may help define basic abnormalities in disease states where persistent inflammation leads to tissue destruction. The paragraphs below cite examples of recent experimental work dealing with resolution of inflammation.

As mentioned earlier, for repair to begin at an inflammatory site, several events have to occur. First, influx of cells must be halted. This probably occurs in various ways, depending on the insult leading to the inflammation and the tissues involved. For example, chemotactic factor inactivator (8, 24) activity found in human serum irreversibly inhibits C5 fragment and other chemotactic activity and can inhibit cell infiltration in immune complex lung disease (47).

Once this influx of cells has abated, and antiproteases and oxygen scavengers from the serum have arrived at the site, inflammatory cells and other debris must be removed. Newman and co-workers (76) recently showed that mature inflammatory macrophages, but not freshly isolated monocytes or resident alveolar macrophages, are able to phagocytose aged neutrophils. The presence of serum was not required for neutrophil aging or this ingestion, and was demonstrated in autologous as well as heterologous systems employing rabbit and human cells. It is suggested that in a normally evolving inflammatory reaction, incoming monocytes develop into inflammatory macrophages capable of removing not only bacteria and cellular debris, but also effete neutrophils. The stimuli (mediators) that induce this monocyte maturation are yet to be defined.

During the resolution process, the connective tissue response to inflammatory stimuli may determine if the injured tissue will return to a normal functional state, or whether fibrosis may result because of an expanded or hypersecreting population of fibroblasts. Recently, Schmidt and co-workers (92) reported that interleukin 1, a macrophage-derived growth factor that augments replication of certain lymphocytes, may be one factor regulating fibroblast proliferation, and speculated that local production of this mediator by infiltrating mononuclear cells may contribute to the fibrosis observed in chronic inflammatory disease states. Bitterman and co-workers (9) have also recently described another growth factor for fibroblasts that is produced by human alveolar macrophages when those macrophages are stimulated by particulates and immune complexes. The factor had an apparent molecular weight of 18,000 and appeared distinct from other characterized growth factors including interleukin 1. Overproduction of this factor could also be important in the pathogenesis of fibrotic lung disease. Both of these observations point out that cells that produce mediators that help initiate an inflammatory response also produce substances that can either help or hinder in the repair of the tissue. Delineation of the factors that determine this outcome will shed important light on the inflammatory process.

CONCLUSIONS

The study of inflammatory mediators is in a state of transition. Many of the molecules involved have only just been structurally characterized and many others can still be identified only by their activity. Before we can really piece together the complex interactions between these mediators and the cells upon which they act, such molecular characterization is essential. Also needed is detailed study of the interaction of the mediators with their target cells. In addition, it is no longer enough to demonstrate in vitro that one more compound is chemotactic, or induces leukocyte secretion, or contracts smooth muscle. It is time to determine which of these mediators actually participate in inflammatory reactions in vivo. These studies are immensely difficult and require not only precise assays for the mediators or their breakdown products in biologic fluids, but also assessment of their effect on cells or tissues. Even their presence does not prove significant function, so highly specific inhibitors are also required. Most useful in this regard are molecular analogues of the molecule in question, or monospecific antibodies. Genetic deletions of the material or its precursor (as for C5) are also useful. Thus, an interplay between in vitro and in vivo experiments, and between biochemistry, cell biology, physiology, and pathology will be necessary for the further pursuit of this critically important and fascinating field.

ACKNOWLEDGMENTS

We wish to thank Georgia Wheeler, Lydia Titus, and Jaunita Graves for their assistance in the preparation of this manuscript. This work was supported by Grants No. HL-21565, HL-27063, HL-28510, and GM-24834 from the National Institutes of Health.

Literature Cited

1. Appenzeller, O., McAndrews, E. J. 1966. The influence of the central nervous system on the triple response of Lewis. *J. Nerv. Ment. Dis.* 143:190–94
2. Atkins, P. C., Norman, M., Weiner, H., Zweiman, B. 1977. Release of neutrophil chemotactic activity during immediate hypersensitivity reactions in humans. *Ann. Intern. Med.* 86:415–18
3. Austen, K. F. 1982. Biological implications of the structural and functional characteristics of the chemical mediators of immediate-type hypersensitivity. *Harvey Lect.* 73:93–161
4. Bach, M. K. 1982. Mediators of anaphylaxis and hypersensitivity. *Ann. Rev. Microbiol.* 36:371–413
5. Belew, M., Gerdin, B., Porath, J., Saldeen, T. 1978. Isolation of vasoactive peptides from human fibrin and fibrinogen degraded by plasmin. *Thromb. Res.* 13:983–94
6. Benveniste, J., Henson, P. M., Cochrane, C. G. 1972. Leukocyte-dependent histamine release from rabbit platelets. The role of IgE, basophils, and a platelet-activating factor. *J. Exp. Med.* 136:1356–77
7. Benveniste, J., Tence, M., Varenne, P., Bidault, J., Boullet, C., Polonsky, J. 1979. Semisynthese et structure proposee du facteur activant les plaquettes (P.A.F.): PAF-acether, un alkyl ether analogue de la lysophosphatidylcholine. *C.R. Acad. Sci. (Paris)* 289:1037–40
8. Berenberg, J. L., Ward, P. A. 1973. Chemotactic factor inactivator in normal human serum. *J. Clin. Invest.* 52:1200–6
9. Bitterman, P. B., Rennard, S. I., Hunninghake, G. W., Crystal, R. G. 1982. Human alveolar macrophage growth factor for fibroblasts. *J. Clin. Invest.* 70:806–22
10. Blank, M. L., Snyder, F., Byers, L. W., Brooks, B., Muirhead, E. E. 1979. Antihypertensive activity of an alkyl ether analog of phosphatidylcholine. *Biochem. Biophys. Res. Commun.* 90:1194–200
11. Bourne, H. R., Lehrer, R. I., Cline, M. J., Melmon, K. L. 1971. Cyclic 3', 5'-adenosine monophosphate in the human leukocyte: synthesis, degradation, and effects on neutrophil candidacidal activity. *J. Clin. Invest.* 50:920–29
12. Bray, M. A., Cunningham, F. M., Ford-Hutchinson, A. W., Smith, M. J. H. 1981. Leukotriene B₄: a mediator of vascular permeability. *Br. J. Pharmacol.* 72:483–86
13. Carp, H., Janoff, A. 1979. In vitro suppression of serum-elastase-inhibitory capacity by reactive oxygen species generated by phagocytosing polymorphonuclear leukocytes. *J. Clin. Invest.* 63:793–97
14. Cochrane, C. G., Revak, S. D., Wuepper, K. D., Johnston, A., Morrison, D. C., Ulevitch, R. 1974. Soluble mediators of injury of the microvasculature: Hageman factor and the kinin forming, intrinsic clotting and the fibrinolytic systems. *Microvasc. Res.* 8:112–21

15. Dale, H. H., Laidlaw, P. P. 1910. The physiological action of B-iminazolylethylamine. *J. Physiol.* 41:318–44

16. Dale, H. H., Richards, A. N. 1918. The vasodilator action of histamine and of some other substances. *J. Physiol.* 52:110–65

17. Damiano, V. V., Cohen, A., Tsang, A. L., Batra, G., Petersen, R. 1980. A morphologic study of the influx of neutrophils into dog lung alveoli after lavage with sterile saline. *Am. J. Pathol.* 100:349–64

18. Demopoulos, C. A., Pinckard, R. N., Hanahan, D. J. 1979. Platelet-activating factor. Evidence for 1-0-alkyl-2-acetyl-sn-glyceryl-3-phosphorylcholine as the active component (a new class of lipid chemical mediators). *J. Biol. Chem.* 254:9355–58

19. Di Rosa, M., Giroud, J. P., Willoughby, D. A. 1971. Studies of the mediators of the acute inflammatory response induced in rats in different sites by carrageenan and turpentine. *J. Pathol.* 104:15–29

20. Douglas, W. W. 1980. Autacoids. In *The Pharmacological Basis of Therapeutics,* ed. A. G. Gilman, L. S. Goodman, A. Gilman, pp. 608–9. New York: MacMillan. 1843 pp.

21. Drazen, J. M., Austen, K. F., Lewis, R. A., Clark, D. A., Goto, G., et al. 1980. Comparative airway and vascular activities of leukotrienes C-1 and D in vivo and in vitro. *Proc. Natl. Acad. Sci. USA* 77:4354–58

22. Ewan, V., Yoshida, T. 1979. Lymphokines and cytokines. In *Chemical Messengers of the Inflammatory Process,* ed. J. C. Houck, pp. 197–227. Oxford: Elsevier/North-Holland. 421 pp.

23. Fantone, J. C., Kunkel, S. L., Ward, P. A., Zurier, R. B. 1980. Suppression by prostaglandin E_1 of vascular permeability induced by vasoactive inflammatory mediators. *J. Immunol.* 125:2591–96

24. Fantone, J., Senior, R. M., Kreutzer, D. L., Jones, M., Ward, P. A. 1979. Biochemical quantitation of the chemotactic factor inactivator activity in human serum. *J. Lab. Clin. Med.* 93:17–24

25. Fantone, J. C., Ward, P. A. 1982. Role of oxygen-derived free radicals and metabolites in leukocyte-dependent inflammatory reactions. *Am. J. Pathol.* 107:397–418

26. Fernandez, H. N., Henson, P. M., Otani, A., Hugli, T. E. 1978. Chemotatic response to human C3a and C5a anaphylatoxins. I. Evaluation of C3a and C5a leukotaxis in vitro and under simulated in vivo conditions. *J. Immunol.* 120:109–15

27. Fishman, A. P., Pietra, G. G. 1974. Handling of bioactive materials by the lung. *N. Engl. J. Med.* 291:884–90; 953–59

28. Florey, H. W., Grant, L. H. 1961. Leucocyte migration from small blood vessels stimulated with ultraviolet light: an electron-microscope study. *J. Pathol. Bacteriol.* 82:13–18

29. Ford-Hutchinson, A. W., Bray, M. A., Doig, M. V., Shipley, M. E., Smith, M. J. H. 1980. Leukotriene B, a potent chemokinetic and aggregating substance released from polymorphonuclear leukocytes. *Nature* 286:264–65

30. Gleisner, J. M. 1979. Lysosomal factors in inflammation. In *Chemical Messengers of the Inflammatory Process,* ed. J. C. Houck, pp. 229–60. Oxford: Elsevier/North Holland. 421 pp.

31. Goetzl, E. J. 1981. Oxygenation products of arachidonic acid as mediators of hypersensitivity and inflammation. *Med. Clin. North Am.* 65:809–28

32. Goetzl, E. J., Pickett, W. C. 1980. The human PMN leukocyte chemotactic activity of complex hydroxy-eicosatetraenoic acids (HETEs). *J. Immunol.* 125:1789–91

33. Goetzl, E. J., Woods, J. M., Gorman, R. R. 1977. Stimulation of human eosinophil and neutrophil polymorphonuclear leukocyte chemotaxis and random migration by 12-L-hydroxy-5, 8, 10, 14-eicosatetraenoic acid (HETE). *J. Clin. Invest.* 59:179–83

34. Goldblatt, M. W. 1935. Properties of human seminal plasma. *J. Physiol.* 84:208–18

35. Habal, F. M., Burrowes, C. E., Movat, H. Z. 1976. Generation of kinin by plasma kallikrein and plasmin and the effect of alpha$_1$antitrypsin and antithrombin III on the kininogenases. In *Kinins, Pharmacodynamics and Biological Role,* ed. F. Sicuteri, N. Back, G. L. Haberland, pp. 23–36. New York: Plenum. 398 pp.

36. Hammersen, F. 1980. Endothelial contractility—does it exist? In *Advances in Microcirculation: Vascular Endothelium and Basement Membranes,* ed. B. M. Altura, pp. 95–134. Basel: Karger. 345 pp.

37. Hanahan, D. J., Demopoulos, C. A., Liehr, J., Pinckard, R. N. 1980. Identification of platelet-activating factor isolated from rabbit basophils as acetyl glyceryl ether phosphorylcholine. *J. Biol. Chem.* 255:5514–16

38. Henson, P. M. 1970. Release of vasoactive amines from rabbit platelets induced by sensitized mononuclear leukocytes and antigen. *J. Exp. Med.* 131: 287–306

39. Henson, P. M. 1971. The immunologic release of constituents from neutrophil leukocytes. I. The role of antibody and complement on nonphagocytosable surfaces or phagocytosable particles. *J. Immunol.* 107:1535–46

40. Henson, P. M. 1972. Pathologic mechanisms in neutrophil-mediated injury. *Am. J. Pathol.* 68:593–612

41. Henson, P. M., Larsen, G. L., Webster, R. O., Mitchell, B. C., Goins, A. J., Henson, J. E. 1982. Pulmonary microvascular alterations and injury induced by complement fragments: synergistic effect of complement activation, neutrophil sequestration, and prostaglandins. *Ann. NY Acad. Sci.* 384:287–300

42. Hugli, T. E., Muller-Eberhard, H. J. 1978. Anaphylatoxins: C3a and C5a. *Adv. Immunol.* 26:1–53

43. Hunninghake, G. W., Gallin, J. I., Fauci, A. S. 1978. Immunologic reactivity of the lung: the in vivo and in vitro generation of a neutrophilic chemotactic factor by alveolar macrophages. *Am. Rev. Respir. Dis.* 117:15–23

44. Hurley, J. V. 1963. An electron microscopic study of leukocyte emigration and vascular permeability in rat skin. *Aust. J. Exp. Biol. Med. Sci.* 41:171–86

45. Issekutz, A. C. 1981. Effect of vasoactive agents on polymorphonuclear leukocyte emigration in vivo. *Lab. Invest.* 45:234–40

46. Janoff, A. 1972. Neutrophil proteases in inflammation. *Ann. Rev. Med.* 23: 177–90

47. Johnson, K. J., Anderson, T. P., Ward, P. A. 1977. Suppression of immune complex-induced inflammation by the chemotactic factor inactivator. *J. Clin. Invest.* 59:951–58

48. Johnston, R. B. Jr., Newman, S. L. 1977. Chronic granulomatous disease. *Pediatr. Clin. North Am.* 24:365–76

49. Kay, A. B. 1970. Studies on eosinophil leukocyte migration. II. Factors specifically chemotactic for eosinophils and neutrophils generated from guinea-pig serum by antigen-antibody complexes. *Clin. Exp. Immunol.* 7:723–37

50. Kay, A. B., Austen, K. F. 1972. Chemotaxis of human basophil leukocytes. *Clin. Exp. Immunol.* 11:557–63

51. Kay, A. B., Stechschulte, D. J., Austen, K. F. 1971. An eosinophil leukocyte chemotactic factor of anaphylaxis. *J. Exp. Med.* 133:602–19

52. Kazmierowski, J. A., Gallin, J. I., Reynolds, H. Y. 1977. Mechanism for the inflammatory response in primate lungs: demonstration and partial characterization of an alveolar macrophage-derived chemotactic factor with preferential activity for polymorphonuclear leukocytes. *J. Clin. Invest.* 59:273–81

53. Keller, H. U., Sorkin, E. 1967. On the chemotactic effect of bacteria. *Int. Arch. Allergy Appl. Immunol.* 31:505–17

54. Kuehl, F. A., Egan, R. W. 1980. Prostaglandins, arachidonic acid, and inflammation. *Science* 210:978–84

55. Larsen, G. L., McCarthy, K., Webster, R. O., Henson, J., Henson, P. M. 1980. A differential effect of C5a and C5a des Arg in the induction of pulmonary inflammation. *Am. J. Pathol.* 100:179–92

56. Larsen, G. L., Mitchell, B. C., Harper, T. B., Henson, P. M. 1982. The pulmonary response of C5 sufficient and deficient mice to Pseudomonas aeruginosa. *Am. Rev. Respir. Dis.* 126:306–11

57. Larsen, G. L., Mitchell, B. C., Henson, P. M. 1981. The pulmonary response of C5 sufficient and deficient mice to immune complexes. *Am. Rev. Respir. Dis.* 123:434–39

58. Larsen, G. L., Parrish, D. A., Henson, P. M. 1983. Lung defense: the paradox of inflammation. *Chest.* In press

59. Lee, T. H., Nagy, L., Nagakura, T., Walport, M. J., Kay, A. B. 1982. Identification and partial characterization of an exercise-induced neutrophil chemotactic factor in bronchial asthma. *J. Clin. Invest.* 69:889–99

60. Lichtenstein, L. M. 1969. Characteristics of leukocytic histamine release by antigen and by anti-immunoglobulin and anti-cellular antibodies. In *Cellular and Humoral Mechanisms in Anaphylaxis and Allergy*, ed. H. Z. Movat, pp. 176–86. Basel: Karger. 288 pp.

61. Lichtenstein, L. M., DeBernardo, R. 1971. The immediate allergic response: in vitro action of cyclic AMP-active and other drugs on the two stages of histamine release. *J. Immunol.* 107:1131–36

62. Lichtenstein, L. M., Margolis, S. 1968. Histamine release in vitro: inhibition by catecholamines and methylxanthines. *Science* 161:902–3

63. Maillard, J. L., Pick, E., Turk, J. L. 1972. Interaction between "sensitized lymphocytes" and antigen in vitro. V. Vascular permeability induced by skin-reactive factor. *Int. Arch. Allergy* 42:50–68

64. Majno, G., Shea, S. M., Leventhal, M. 1969. Endothelial contraction induced by histamine-type mediators: an electron microscopic study. *J. Cell Biol.* 42:647–72

65. Marchesi, V. T. 1961. The site of leucocyte emigration during inflammation. *Q. J. Exp. Physiol.* 46:115–18

66. Marone, G., MacGlashan, D. W., Kagey-Sobotka, A., Lichtenstein, L. M. 1980. Desensitization and phospholipid metabolism in human basophils. In *Advances in Allergology and Applied Immunology*, ed. A. Oehling, I. Glazer, E. Mathov, C. Arbesman, pp. 147–154. Oxford: Pergamon. 795 pp.

67. McCarthy, K., Henson, P. M. 1979. Induction of lysosomal enzyme secretion by alveolar macrophages in response to the purified complement fragments C5a and C5a des Arg. *J. Immunol.* 123: 2511–17

68. Merrill, W. W., Naegel, G. P., Matthay, R. A., Reynolds, H. Y. 1980. Alveolar macrophage-derived chemotactic factor. Kinetics of in vitro production and partial characterization. *J. Clin. Invest.* 65:268–76

69. Metchnikoff, E. 1891. Lecture XII. Transl. F. A. Starling, E. H. Starling, 1968, in *Lectures on the Comparative Pathology of Inflammation*, pp. 180–97. New York: Dover. 224 pp. (From French)

70. Mota, I. 1964. The mechanism of anaphylaxis. I. Production and biological properties of 'mast cell sensitizing' antibody. *Immunology* 7:681–99

71. Movat, H. Z. 1979. The acute inflammatory reaction. In *Inflammation, Immunity and Hypersensitivity*, ed. H. Z. Movat, pp. 1–161. Hagerstown, Md: Harper and Row. 689 pp.

72. Movat, H. Z., Fernando, N. V. P. 1963. Acute inflammation. The earliest fine structural changes at the blood-tissue barrier. *Lab. Invest.* 12:895–910

73. Movat, H. Z., Macmorine, D. R. L., Takeuchi, Y. 1971. The role of PMN-leukocyte lysosomes in tissue injury, inflammation and hypersensitivity. VIII. Mode of action and properties of vascular permeability factors released by PMN-leukocytes during 'in vitro' phagocytosis. *Int. Arch. Allergy Appl. Immunol.* 40:218–35

74. Movat, H. Z., Udaka, K., Takeuchi, Y. 1970. Polymorphonuclear leukocyte lysosomes and vascular injury. *Thromb. Diath. Haemorrh.* 40:211–24

75. Neta, R., Salvin, S. B. 1982. Lymphokines and interferon: similarities and differences. In *Lymphokines 7*, ed. E. Pick, M. Landy, pp. 137–63. New York: Academic. 277 pp.

76. Newman, S. L., Henson, J. E., Henson, P. M. 1982. Phagocytosis of senescent neutrophils by human monocyte-derived macrophages and rabbit inflammatory macrophages. *J. Exp. Med.* 156:430–42

77. Norris, D. A., Clark, R. A. F., Swigart, L. M., Huff, J. C., Weston, W. L., Howell, S. E. 1982. Fibronectin fragment(s) are chemotactic for human peripheral blood monocytes. *J. Immunol.* 129: 1612–18

78. O'Brodovich, H. M., Stalcup, S. A., Pang, L. M., Lipset, J. S., Mellins, R. B. 1981. Bradykinin production and increased pulmonary endothelial permeability during acute respiratory failure in unanesthetized sheep. *J. Clin. Invest.* 67:514–22

79. Parker, C. W. 1982. The chemical nature of slow-reacting substances. In *Advances in Inflammation Research*, ed. G. Weissmann, 4:1–24. New York: Raven. 200 pp.

80. Parrish, D. A., Mitchell, B. C., Henson, P. M., Larsen, G. L. 1982. The pulmonary response of C5 sufficient and deficient mice to hyperoxia. *Am. Rev. Respir. Dis.* 125(2):52 (Abstr.)

81. Perez, H. D., Goldstein, I. M., Chernoff, D., Webster, R. O., Henson, P. M. 1980. Chemotactic activity of C5a des Arg: evidence of a requirement for an anionic peptide "helper factor" and inhibition by a cationic protein in serum from patients with systemic lupus erythematosus. *Mol. Immunol.* 17:163–69

82. Peters, S. P., Siegel, M. I., Kagey-Sobotka, A., Lichtenstein, L. M. 1981. Lipoxygenase products modulate histamine release in human basophils. *Nature* 292:455–57

83. Pinckard, R. N., McManus, L. M., Hanahan, D. J. 1982. Chemistry and biology of acetyl glyceryl ether phosphorylcholine (platelet-activating factor). In *Advances in Inflammation Research*, ed. G. Weissmann, 4:147–80. New York: Raven. 200 pp.

84. Ranadive, N. S., Cochrane, C. G. 1968. Isolation and characterization of permeability factors from neutrophils. *J. Exp. Med.* 128:605–22

85. Ratnoff, O. D. 1971. A tangled web. The interdependence of mechanisms of blood clotting, fibrinolysis, immunity and inflammation. *Thromb. Diath. Haemorrh.* 45:109–18

86. Rehm, S. R., Gross, G. N., Pierce, A. K. 1980. Early bacterial clearance from murine lungs. Species-dependent phagocyte response. *J. Clin. Invest.* 66:194–99

87. Rocklin, R. E., Bendtzen, K., Greineder, D. 1980. Mediators of immunity: lymphokines and monokines. *Adv. Immunol.* 29:55–136

88. Ryan, G. B., Majno, G. 1977. Acute inflammation, a review. *Am. J. Pathol.* 86:185–276

89. Samuelsson, B., Hammarstrom, S., Borgeat, P. 1979. Pathways of arachidonic acid metabolism. In *Advances in Inflammation Research,* ed. G. Weissman, B. Samuelsson, R. Paoletti, 1:405–12. New York: Raven. 640 pp.

90. Scherzer, H., Ward, P. A. 1978. Lung and dermal vascular injury produced by preformed immune complexes. *Am. Rev. Respir. Dis.* 117:551–57

91. Schiffmann, E., Aswanikumar, S., Venkatasubramanian, K., Corcoran, B. A., Pert, C. B., et al. 1980. Some characteristics of the neutrophil receptor for chemotactic peptides. *FEBS Lett.* 117:1–7

92. Schmidt, J. A., Mizel, S. B., Cohen, D., Green, I. 1982. Interleukin 1, a potential regulator of fibroblast proliferation. *J. Immunol.* 128:2177–82

93. Seegers, W., Janoff, A. 1966. Mediators of inflammation in leukocyte lysosomes. VI. Partial purification and characterization of a mast cell rupturing component. *J. Exp. Med.* 124:833–49

94. Shaw, J. O. 1980. Leukocytes in chemotactic-fragment-induced lung inflammation. Vascular emigration and alveolar surface migration. *Am. J. Pathol.* 101:283–302

95. Simon, P. L., Willoughby, W. F. 1982. Biochemical and biological characterization of rabbit interleukin-1 (IL-1). In *Lymphokines 6, Lymphokines in Antibody and Cytotoxic Responses,* ed. S. B. Mizel, pp. 47–63. New York: Academic. 318 pp.

96. Snyderman, R., Phillips, J. L., Mergenhagen, S. E. 1971. Biological activity of complement in vivo. Role of C5 in the accumulation of polymorphonuclear leukocytes in inflammatory exudates. *J. Exp. Med.* 134:1131–43

97. Snyderman, R., Shin, H. S., Hausman, M. H. 1971. A chemotactic factor for mononuclear leukocytes. *Proc. Soc. Exp. Biol. Med.* 138:387–90

98. Sonnenfeld, G. 1980. Modulation of immunity by interferon. In *Lymphokine Reports 1,* ed. E. Pick, M. Landy, pp.

99. Spector, W. G., Willoughby, D. A. 1959. Experimental suppression of the acute inflammatory changes of thermal injury. *J. Pathol. Bacteriol.* 78:121–32

100. Stalcup, S. A., Lipset, J. S., Legant, P. M., Leuenberger, P. J., Mellins, R. B. 1979. Inhibition of converting enzyme activity by acute hypoxia in dogs. *J. Appl. Physiol. Resp. Environ. Exercise Physiol.* 46:227–34

101. Svensjo, E., Arfors, K.-E., Raymond, R. M., Grega, G. J. 1979. Morphological and physiological correlation of bradykinin-induced macromolecular efflux. *Am. J. Physiol.* 236:H600–6

102. Tauber, A. I., Kaliner, M., Stechschulte, D. J., Austen, K. F. 1973. Immunologic release of histamine and slow reacting substance of anaphylaxis from human lung. *J. Immunol.* 111:27–32

103. Thomas, L. 1964. Possible role of leukocyte granules in the Schwartzman and Arthus reactions. *Proc. Soc. Exp. Biol. Med.* 115:235–40

104. Uriuhara, T., Movat, H. Z. 1967. Role of PMN-leukocyte lysosomes in tissue injury, inflammation and hypersensitivity. V. Partial suppression in leukopenic rabbits of vascular hyperpermeability due to thermal injury. *Proc. Soc. Exp. Biol. Med.* 124:279–84

105. Vadas, P., Wasi, S., Movat, H. Z., Hay, J. B. 1981. Extracellular phospholipase A2 mediates inflammatory hyperaemia. *Nature* 293:583–85

106. Valone, F. H., Franklin, M., Sun, F. F., Goetzl, E. J. 1980. Alveolar macrophage lipoxygenase products of arachidonic acid: isolation and recognition as the predominant constituents of the neutrophil chemotactic activity elaborated by alveolar macrophages. *Cell Immunol.* 54:390–401

107. Vane, J. R. 1971. Inhibition of prostaglandin synthesis as a mechanism of action for aspirin-like drugs. *Nature New Biol.* 231:232–35

108. Vane, J. R. 1976. The mode of action of aspirin and similar compounds. *J. Allergy Clin. Immunol.* 58:691–712

109. Voelkel, N. F., Worthen, S., Reeves, J. T., Henson, P. M., Murphy, R. C. 1982. Nonimmunological production of leukotrienes induced by platelet-activating factor. *Science* 218:286–88

110. von Euler, U.S. 1934. Zur kenntnis der pharmakologischen wirkungen von nativsekreten und extracten mannlicher accessorischer geschlechtsdrusen. *Arch. Exp. Pathol. Pharmakol.* 175:78–84

113–31. New York: Academic. 259 pp.

111. Ward, P. A. 1968. Chemotaxis of mononuclear cells. *J. Exp. Med.* 128: 1201–21
112. Ward, P. A. 1971. Leukotactic factors in health and disease. *Am. J. Pathol.* 64:521–30
113. Ward, P. A., Chapitis, J., Conroy, M. C., Lepow, I. H. 1973. Generation by bacterial proteinases of leukotactic factors from human serum and C3 and C5. *J. Immunol.* 110:1003–9
114. Ward, P. A., Hill, J. H. 1970. C5 chemotactic fragments produced by an enzyme in lysosomal granules of neutrophils. *J. Immunol.* 104:535–43
115. Ward, P. A., Lepow, I. H., Newman, L. J. 1968. Bacterial factors chemotactic for polymorphonuclear leukocytes. *Am. J. Pathol.* 52:725–36
116. Webster, R. O., Hong, S. R., Johnston, R. B., Henson, P. M. 1980. Biological effects of the human complement fragments C5a and C5a des Arg on neutrophil function. *Immunopharmacology* 2:201–19
117. Wedmore, C. V., Williams, T. J. 1981. Control of vascular permeability by polymorphonuclear leukocytes in inflammation. *Nature* 289:646–50
118. Weissmann, G., Zurier, R. B., Hoffstein, S. 1972. Leukocytic proteases and the immunologic release of lysosomal enzymes. *Am. J. Pathol.* 68:539–64
119. Weksler, B. B., Coupal, C. E. 1973. Platelet-dependent generation of chemotactic activity in serum. *J. Exp. Med.* 137:1419–30
120. Werb, Z., Banda, M. J., McKerrow, J. H., Sandhaus, R. A. 1982. Elastases and elastin degradation. *J. Invest. Dermatol.* 79:154s–59s
121. Wiggins, R. C., Giclas, P. C., Henson, P. M. 1981. Chemotactic activity generated from the fifth component of complement by plasma kallikrein of the rabbit. *J. Exp. Med.* 153:1391–1404
122. Wilhelm, D. L. 1971. Kinins in human disease. *Ann. Rev. Med.* 22:63–84
123. Wilhelm, D. L. 1973. Mechanisms responsible for increased vascular permeability in acute inflammation. *Agents Actions* 3:297–306
124. Wilhelm, D. L., Mason, B. 1960. Vascular permeability changes in inflammation: the role of endogenous permeabil-

ity factors in mild thermal injury. *Br. J. Exp. Pathol.* 41:487–506
125. Williams, T. J. 1977. Oedema and vasodilatation in inflammation. The relevance of prostaglandins. *Postgrad. Med. J.* 53:660–62
126. Williams, T. J. 1982. Vasoactive intestinal polypeptide is more potent than prostaglandin E_2 as a vasodilator and oedema potentiator in rabbit skin. *Br. J. Pharmacol.* 77:505–9
127. Williams, T. J., Jose, P. J. 1981. Mediation of increased vascular permeability after complement activation. Histamine-independent action of rabbit C5a. *J. Exp. Med.* 153:136–53
128. Williams, T. J., Peck, M. J. 1977. Role of prostaglandin-mediated vasodilation in inflammation. *Nature* 270:530–32
129. Willis, A. L. 1969. Release of histamine, kinin and prostaglandin during carrageenin-induced inflammation in the rat. In *Prostaglandins, Peptides and Amines*, ed. P. Mantegazza, E. W. Horton, pp. 31–38. New York: Academic. 191 pp.
130. Willis, A. L. 1970. Identification of prostaglandin E_2 in rat inflammatory exudate. *Pharmac. Res. Comm.* 2:297–304
131. Wintroub, B. U. 1982. Neutrophil-dependent generation of biologically active peptides. In *Advances in Inflammation Research*, ed. G. Weissmann, 4:131–45. New York: Raven. 200 pp.
132. Wissler, J. H., Stecher, V. J., Sorkin, E. 1972. Chemistry and biology of the anaphylatoxin related serum peptide system. III. Evaluation of leucotactic activity as a property of a new peptide system with classical anaphylatoxin and cocytotaxin as components. *Eur. J. Immunol.* 2:90–96
133. Wuepper, K. D., Cochrane, C. G. 1972. Plasma prekallikrein: isolation, characterization, and mechanism of activation. *J. Exp. Med.* 135:1–20
134. Zurier, R. B., Weissmann, G., Hoffstein, S., Kammerman, S., Tai, H. H. 1974. Mechanism of lysosomal enzyme release from human leukocytes. II. Effect of cAMP and cGMP, autonomic agonists, and agents which affect microtubule function. *J. Clin. Invest.* 53:297–309

Ann. Rev. Immunol. 1983. 1:361–92

THE ROLE OF CELL-MEDIATED IMMUNE RESPONSES IN RESISTANCE TO MALARIA, WITH SPECIAL REFERENCE TO OXIDANT STRESS

Anthony C. Allison and Elsie M. Eugui

Institute of Biological Sciences, Syntex Research, Palo Alto, California 94304

SUMMARY

Asexual blood forms of malaria parasites are microaerophilic and sensitive to oxidant stress. *Plasmodium falciparum* and some other species of malaria parasites undergo schizogony attached to endothelial cells of postcapillary venules, where oxygen tensions are low.

Acquired immune responses to all forms of malaria parasites so far investigated are thymus dependent. Animals deprived of T lymphocytes do not recover from the infections and cannot be immunized against malaria parasites. In contrast, animals unable to make antibodies recover normally from some primary infections, e.g. with *Plasmodium chabaudi*, and when rescued by chemotherapy from other species of malaria parasite develop lasting, nonsterile immunity. Immunity to malaria can be transferred in mice by T lymphocytes of the Ly1+ phenotype, but transfer of B lymphocytes together with this T-cell subset increases the effectiveness of immunity to *Plasmodium yoelii*. Thus, antibodies facilitate recovery from some primary malaria infections and increase the effectiveness of cell-mediated immune responses in these infections. Mice of the A strain are highly susceptible to malaria and are unable to increase the number of mononuclear cells in the spleen during the course of the infections.

It is postulated that T lymphocytes responding to parasite antigens release factors that stimulate the proliferation of effector cell precursors and

361

0732-0582/83/0410-0361$02.00

their recruitment into the red pulp of the spleen. In this site, the liver and probably the peripheral circulation, effector cells bind to the surface of parasitized erythrocytes and are activated to release superoxide (O_2^-). The consequent exposure to oxidant stress can lead to degeneration of parasites in erythrocytes. This effect on the parasites can be prevented by agents chelating metals, which suggests that iron-catalyzed lipid peroxidation and consequent K^+ loss, or inactivation of metal-containing enzymes, may be the mechanism by which oxidant stress kills the intracellular parasites. Antibodies on the surface of schizont-infected cells could facilitate binding of effector cells and trigger O_2^- release, thereby acting synergistically with cell-mediated immunity.

Inherited traits, such as abnormal hemoglobins and G-6-PD deficiency, and acquired cell-mediated immunity both subject malaria parasites to oxidant stress and may reinforce one another, increasing the chances of survival of children bearing these traits during the dangerous years of first exposure to malaria in areas where the disease is endemic.

INTRODUCTION

The relationships between plasmodium parasites, anopheline vectors, and mammalian hosts are complex and varied (6). Considering human malarias, the situation differs in various countries and in geographical regions of the same country. Where *Plasmodium falciparum* is holoendemic, the greatest effects of malaria are manifested in children between the ages of 6 months and 5 years, after which the disease becomes progressively less severe. Resistance to malaria develops in stages. First, the severity of clinical symptoms and later levels of parasitemia fall and splenomegaly decreases (91, 92). By the time they are adults, most persons living in endemic areas show little parasitemia, although they are frequently bitten by infected mosquitoes; but after as little as 6 months abroad, or after elimination of parasites by chemotherapy, they may again become susceptible to severe malaria (94). These observations suggest gradually acquired, adaptive immunity dependent on frequent antigenic stimulation and persistence of parasites. Such nonsterilizing immunity to parasites has been termed premunition (135). As discussed below, nonsterilizing immunity is observed in experimental malarias when the immune response is purely cell mediated, which is one reason that type of response is of interest.

The number of children dying directly or indirectly from malaria is not accurately known. Certainly the majority survive to school age, by which time parasite rates and counts have fallen as a result of acquired immunity. Untreated infections with the same parasites in European visitors are often lethal. Several factors probably contribute to the relative resistance of African children to falciparum malaria. Some of these, such as HbS and G-6-PD deficiency, have been investigated in detail and provide information of general interest from the point of view of host-parasite relationships (6). This review is confined to mechanisms of acquired immunity to malaria parasites, which are most easily studied in experimental animals, especially mice, in which it is possible to suppress either T-lymphocyte or B-lymphocyte responses and to characterize the isotypes of antibodies and phenotypes of cells transferring immunity. Furthermore, the availability of mice with other selective defects of immune function, such as T cell-dependent recruitment of mononuclear cells, is useful.

Mouse malarias are also convenient experimentally. They are divided into two groups: one includes *Plasmodium berghei* and *Plasmodium yoelii*, and the other, *Plasmodium vinckei* and *Plasmodium chabaudi*. The similarities within and distinctions between the two groups are manifest in the structure and behavior of blood-stage parasites, serology, patterns of cross protection, and isoenzyme types (21). Within each group it is possible to obtain infections from which most strains of mice recover and are fully immune, so that after homologous challenge no parasitemia is detectable (e.g. *P.y. yoelii* 17XNL, *P.c. chabaudi*, and *P. vinckei petteri*), or strains that are lethal in unprotected mice (e.g. *P.y. yoelii* 17XL or *P.v. vinckei*). Immunization against the asexual blood stages of the latter can be achieved by infection and treatment (36), by infection with a nonlethal parasite in the same group (e.g. *P.y. yoelii* 17XNL protecting against *P.y. yoelii* 17XL, or *P.c. chabaudi* protecting against *P.v. vinckei*), or by immunization with homologous crude vaccines (114) or purified parasite antigens (71). The availability of cross-reacting virulent and avirulent strains of parasites allows experimental analysis of mechanisms of immunity. As described below, protective immunity to *P. chabaudi* appears to be antibody independent, whereas recovery from primary infections with *P. yoelii* is antibody dependent. Thus, different mechanisms of immunity can be illustrated by mouse malarias, although care must be taken when attempting to generalize from these results to human malarias.

Acquired immunity is stage specific. Mice infected with X-irradiated sporozoites of *P. berghei* are resistant to sporozoite infections but not to challenge with asexual blood stages (109). Selective immunity can also be achieved by immunization with gametocytes (22) and can develop without any effect on asexual blood stages, being manifested by lack of infectivity of parasitized blood for mosquitoes.

In *Plasmodium knowlesi* infections of *Macaca mulatta*, classical antigenic variation, demonstrable by the sequential appearance of antibodies agglutinating schizont-infected cells, has been demonstrated (16). The repeated *P. falciparum* infections observed in African children over several years have been attributed to the presence of many antigenic variants, the existence of which can be shown by precipitation in gels (157). However, the relationship between such soluble antigens and those on the surface of infected schizonts, which are presumed targets of protective immunity (74), remains to be established.

ACQUIRED IMMUNITY TO MALARIA

Role of Antibodies

It has long been believed that antibodies opsonize malaria-parasitized erythrocytes for phagocytosis by macrophages (140). More direct evidence for a role of antibodies in protection came from observations by Cohen et al (33) that administration of gamma-globulin fractions from pools of adult African sera accelerated the recovery of African children from falciparum malaria. IgG fractions of immune serum were likewise found to protect rats from *P. berghei* infections (40, 41). More recently, monoclonal antibodies have been used to define antigens specific for developmental stages of the parasite (sporozoites, asexual blood forms, and gametocytes) and to analyze mechanisms of humoral immunity (56, 115, 120). Only a few examples can be mentioned in this paper. A monoclonal antibody against a protein of molecular weight 42,000 in the membrane of *P. berghei* sporozoites has been shown to protect mice against mosquito-transmitted infection (115) and prevent the entry of sporozoites into cultured cells (72). This antibody does not protect against blood-transmitted infection. However, monoclonal antibodies that recognize antigens of the asexual blood forms of *P.y. yoelii* can protect mice against blood-transmitted infection (56). Holder & Freeman (71) have used monoclonal antibodies to identify and purify antigens of *P.y. yoelii* membranes. They found that mice immunized with either a 235,000-molecular-weight merozoite-specific protein or a 230,000-molecular-weight schizont protein and its derivatives were protected against infection with asexual blood forms of *P.y. yoelii*. These observations show that antibodies against stage-specific antigens of the parasite can have a protective role, although this does not appear to be exclusive. For example, mice with humoral responses suppressed by neonatal injection of anti-μ serum can be immunized with irradiated sporozoites against mosquito challenge (23).

Thymus Dependence of Immunity to Malaria Parasites

The first observations showing the importance of thymus-dependent (T) lymphocytes in recovery from malaria infections were made by Brown et al (14). Rats inoculated with *P. berghei* when 13 weeks of age recovered from the infections, but in neonatally thymectomized rats we found higher and more prolonged parasitemias and an appreciable mortality (14). These observations were independantly confirmed (138).

With the availability of mice congenitally deprived of mature T-lymphocytes (nude or *nu/nu* mice), and a strain of *P. yoelii* producing infections from which intact mice recover, the problem was taken up again (27). In nude mice infected with *P. yoelii,* escalating parasitemias were observed. This parasite can invade reticulocytes, and in infected mice there was a progressive increase in susceptible cells, until with a parasitemia of 70–80% the animals all succumbed.

The extreme susceptibility of *nu/nu* mice to *P. yoelii* infections was confirmed (122, 152); moreover, it was shown that parasitemia recrudesced in mice after termination of the acute disease by treatment with clindamycin (122). Recrudescence was not observed in *nu/nu* mice that had been grafted with thymic tissue. Intact CBA mice recovered from *P. chabaudi* infections, whereas *nu/nu* mice on a CBA background developed lethal infections (46) (Figure 1).

The consistent aggravation of nonlethal malaria infections by deprivation of T cells should be contrasted with the delayed mortality observed in T cell-deprived mice infected with one strain of *P. berghei* (44), which is explicable by thymus-dependent immunopathology.

Antibody-Independent Resistance of Experimental Animals to Malaria

Rank & Weidanz (119) made chickens agammaglobulinemic by combined chemical bursectomy (treatment with testosterone and cyclophosphamide).

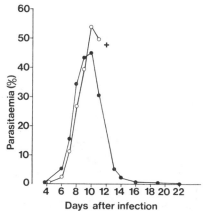

Figure 1   Parasitemias in CBA *nu/nu* (open circles) and CBA *nu/+* (closed circles) mice at different intervals after intraperitoneal inoculation of 10^5 erythrocytes infected with *P. chabaudi* AS. In each group, five female mice aged 6 to 10 weeks were used. All *nu/nu* mice had died by day 12–13 after infection. (From 46.)

When such chickens were infected with *Plasmodium gallinaceum* they died with fulminating parasitemias. However, when rescued from otherwise fatal infections by chloroquine therapy, the infection was controlled and the B cell-deficient chickens resisted challenge infection with the same parasite. Thus, the B cell-deficient chickens could produce a protective immune response, although their parasites were not completely cleared.

In mice made B cell deficient by neonatal injections of anti-μ serum, infections with the normally avirulent *P. yoelli* were found to produce lethal disease (123). However, antimalarial chemotherapy of these mice led to a nonsterilizing protective immunity (124). With another murine parasite, *P. chabaudi adami,* infection of B cell-deficient mice was found in the absence of chemotherapy to activate a T cell-dependent immune mechanism that terminated the malaria in a manner similar to that seen in immunologically intact mice (65) (Figure 2). The immunized B cell-deficient mice were resistant to homologous challenge, as well as to infections initiated with *P. vinckei,* but not to challenge with *P. yoelii* or *P. berghei.* These observations emphasize the importance of thymus-dependent, antibody-independent mechanisms of immunity to rechallenge with all species of malaria parasites so far tested, and the ability of this mechanism to bring about recovery from *P. chabaudi* without chemotherapy. It may be relevant that *P. chabaudi* elicits a powerful, thymus-dependent increase in spleen cell cellularity and nonspecific cytotoxic activity (46).

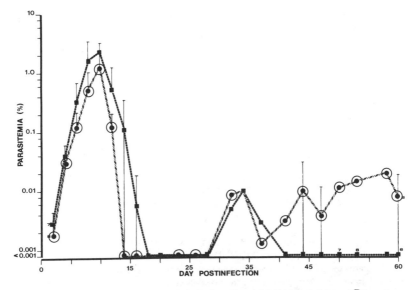

Figure 2   The course of *P. chabaudi* infections in eight B cell-deficient ((●)) and seven immunologically intact (■) mice. (From 65.)

Further evidence for a role of T lymphocytes in immunity to blood stages of murine malaria parasites came from observations of Finerty & Krehl (50) that infections with a lethal strain of *P. yoelii* (17XL) could be converted into nonlethal infections by pretreatment with cyclophosphamide. The cyclophosphamide, administered 2 days before infection, increased delayed-type hypersensitivity to parasite antigens (presumably from the elimination of suppressor cells), but antibodies against blood-stage parasites were not detectable during the recovery period.

It has long been recognized that immunization of rhesus monkeys against *P. knowlesi* requires complete Freund's adjuvant (15, 18), which elicits cell-mediated immunity, and resistance to challenge is not well correlated with the presence of antibodies inhibiting parasite reinvasion and replication in vitro (83).

T-Lymphocyte Responses in Malaria

Observations in several laboratories showed by other methods that malaria infections elicit T-lymphocyte responses. In mice carrying chromosome markers for T lymphocytes and infected with *P. yoelii,* proliferation of these cells in the spleen was observed (81). T lymphocytes (sensitive to anti-Thy1 serum) from spleens of mice immune to *P. yoelii* showed early proliferative responses when cultured with infected erythrocytes or soluble parasite antigens (151). Likewise, T lymphocytes from humans that had been infected with malaria showed proliferative responses to *P. falciparum* antigens (166).

Mechanisms by Which T Lymphocytes Participate in Recovery from Malaria Infections

The first possibility that must be considered is a helper effect on antibody formation, as postulated by Brown (15). He made the ingenious suggestion that T-lymphocyte responses to common antigenic determinants accelerate the formation of protective antibodies to sequentially appearing variant-specific antigens. Mice deprived of T lymphocytes showed diminished germinal center responses and IgG1 antibody formation when infected with *P. yoelii* (81). The strict thymus dependence of the formation of antibodies against *P. yoelii* antigens in mice and in cultured murine cells has been well documented (142).

As already discussed, B cell-deficient mice are unusually susceptible to infection with this parasite (152). T cell-deficient nude mice showed recrudescence of *P. yoelii* infections after treatment, but this was not observed in mice given hyperimmune serum (122). A synergistic effect of T cells and antibody-forming B lymphocytes in immunity to *P. yoelii* 17X has been postulated by Jayawardena et al (80). T cells (Ly1+ phenotype) from CBA mice that had recovered from *P. yoelii* infections transferred immunity to nonimmune recipients, but the addition of immune B lymphocytes to the Ly1 cells increased their ability to transfer immunity and enhance antibody production. B cells from CBA/N mice (which are defective in producing high-affinity antibodies as well as those of the IgG3 isotype) had no such effect.

In contrast to the observations with *P. yoelii,* B cell-deprived mice recover from primary infections with *P. chabaudi* and are immune to challenge, so that antibody-independent mechanisms of immunity appear to be of major importance with this parasite (65).

The thymus dependence of recovery must therefore be manifested by some function of T cells other than helper effects in antibody formation. One possibility is that the cytotoxic subset of T lymphocytes can lyse parasites or parasitized cells. Our attempts to demonstrate such a mechanism failed and are in any case inconsistent with the observation of parasites degenerating within circulating erythrocytes, as discussed below. The finding that the Ly-23 subset is not required for the effector phase of immunity to *P. yoelii* further supports the interpretation that cytotoxic T cells are not important defensive elements in malarial infections (80).

Another possibility is that specifically sensitized T lymphocytes, stimulated by parasite antigens, release a mediator that inhibits parasite growth; for example, γ-interferon or lymphotoxin are possible candidates. Finally, lymphocytes stimulated by parasite antigens may produce a factor that activates a distinct population of cells, such as macrophages or NK (natural killer) cells, which in turn release another effector molecule acting on parasites or parasitized cells.

The third hypothesis seems most likely, because the effects of nonspecific immunity, which can be induced by *Corynebacterium parvum* even in thymus-deprived nude mice, appear to be equivalent to those of thymus-dependent specific immunity. This is true of both the sensitivity of different species of *Plasmodium* to nonspecific immunity and its principal manifestation, death of parasites within intact, nonphagocytized erythrocytes.

Nonspecific Immunity

Powerful effects of nonspecific immunity on asexual blood forms of malaria parasites of mice were demonstrated in our laboratory (28). Mice previously inoculated intravenously with live BCG were found to develop low, transient parasitemias when infected with *P. vinckei,* which produces lethal infections in unprotected mice, and were subsequently shown to have life-long immunity to challenge with this parasite. It was then found that injections of killed *C. parvum* produced similar protection against hemoprotozoa (29).

Nude mice can be protected against *P. vinckei* and *P. chabaudi* infections by prior injection of *C. parvum* but not of *Myobacterium bovis* BCG (47). This suggests that the effector phase of immunity activated by *C. parvum* bypasses a requirement for T lymphocytes. Immunity to tumors can also be elicited by *C. parvum* in nude mice (160). *C. parvum* exerts some protection against *P. berghei* sporozoites, but there is little effect on the blood-stream forms of this parasite (108). In general, blood-stream forms of parasite strains not multiplying in reticulocytes are more sensitive to the suppressive effects of nonspecific immunity than those that can multiply in reticulocytes.

Susceptibility of Different Mouse Strains to Malaria

Genetic analysis of susceptibility to infections is a potent way of revealing the underlying mechanisms. For example, a single autosomal gene confers susceptibility of mice to infections with *Leishmania donovani* (13) and another confers susceptibility to *Leishmania tropica* (73). Greenberg & Kendrick (63) showed differences in the time to death of several mouse strains infected with *P. berghei.* We observed more striking differences in the susceptibility of different mouse strains to *P. chabaudi;* in the A/He strain, the infections were always lethal, whereas in other strains the infections were self-limiting (46). Investigations of congenic strains suggested that susceptibility does not correlate with H-2 haplotype. The F_1 offspring of susceptible A and resistant B10.A mice proved to be resistant, and some offspring of backcrosses to the susceptible parental strain were susceptible. Our observations have been confirmed and extended (139). Segregation analysis of backcrosses and F_2 progeny derived from A/J mice susceptible to *P. chabaudi* and resistant B10.A parental mice suggested that host resistance is controlled by a single, dominant, non-H-2-linked gene.

Strain Differences in Mononuclear Cell Recruitment

Since macrophages recently immigrated into organs have much greater capacity than resident cells to produce O_2^-, which is likely to be important in immunity to malaria parasites, factors affecting recruitment of these cells are relevant. During the course of malaria infections in most strains of mice, there is a rapid increase in the number of mononuclear cells in the spleen, rising to a peak (some fivefold) about the time of maximum parasitemia (46); thereafter, there is a marked increase in the number of erythropoietic precursors (55). In CBA *nu/nu* mice infected with *P. chabaudi,* there is no increase in mononuclear cells in the spleen and their capacity to kill tumor cells (46), showing the thymus dependence of the phenomenon. In mice of the A/He strain infected with *P. chabaudi,* there was no increase in the number of leukocytes in the spleen (46) (Table 1). Hence, A mice appear

Table 1 Changes in the number of nucleated cells in the spleen, and in NK activity, of different strains of mice infected with *P. chabaudi*

Mouse strain	Parasitemia[a] (%)	Spleen cells ($\times 10^7$)	NK activity[b] (% specific lysis)
C57B1/6	0	5.9 (5.8–6.0)	10.9 (10.7–11.1)
	13–28	22.4 (12.6–32.0)	30.2 (26.7–33.0)
A/He	0	12.3 (11.3–13.4)	8.2 (7.2–9.4)
	56–77	11.0 (9.4–14.0)	12.9 (9.6–19.4)

[a] Samples were taken at peak parasitemia, 8 or 9 days after infection.
[b] NK activity was measured using 100 spleen cells to one YAC–1 target cell, in a ^{51}Cr release assay. Arithmetic means and ranges (in brackets) are taken from Eugui & Allison (46).

to be defective in the production by T lymphocytes of factors stimulating the proliferation and/or recruitment of mononuclear cells or in the response to such factors.

Consistent with this interpretation is the observation that strain A/J mice inoculated with BCG are not protected against *P. berghei,* whereas other strains are (103). In contrast, strain A/J mice can be protected against *Babesia microti* infections by *C. parvum* (158); recruitment of mononuclear cells by this agent, but not by BCG, is thymus independent. The possible relationship of the defective response of mononuclear cells in A strain mice to their low NK activity (68) and low mononuclear cell-mediated cytotoxicity of tumor cells (2) is discussed later.

Immune spleen cells implanted into the peritoneal cavity in mill'pore chambers together with parasitized erythrocytes were found to elicit a large macrophage infiltrate around the chambers (7). Presumably, this was due to the release from antigen-activated lymphocytes of factors stimulating the proliferation of mononuclear cells and their recruitment. Possibly, the monocyte chemotactic factor and macrophage migration inhibition factor described in the spleens of infected rodents (34, 162) were involved.

The finding of splenomegaly during acute malaria infections in humans (94) suggests that a similar recruitment occurs. Splenomegaly is marked in children under school age exposed to repeated malaria infections, and thereafter the proportion of persons with palpable spleens declines (92).

Changes in Lymphocyte Subsets During Malaria Infections

Changes in proportions and absolute numbers of lymphocyte subsets in human and experimental malarias have been described. The number of circulating T cells in human peripheral blood decreases, and the number of null cells (lacking B or T markers) increases, during malaria infections (64, 161); B cells may decrease (64) or remain unchanged (161). In rhesus monkeys infected with lethal *P. knowlesi,* a marked decrease in the number of T cells, and some decrease in B cells, was observed in peripheral blood (143).

Complex changes reported in rodents infected with lethal and nonlethal parasites include increases in T and B cells, granulocytes, and especially mononuclear phagocytes in spleens of mice; these are more marked in nonlethal *P. yoelii* than in lethal *P. berghei* infections (55, 113). Increased numbers and activity of macrophages in the bone marrow have been described (54), and protective immunity is correlated with peripheral blood monocytosis (84). These observations are consistent with the interpretation that during malaria infections T cells, and to a lesser extent B cells, migrate from the peripheral blood into the spleen and liver, where they proliferate as a result of antigenic stimulation. Macrophage percursors proliferate in the bone marrow, appear as monocytes in peripheral blood, and migrate into the spleen and liver; there may also be local proliferation of these cells.

Death of Parasites Within Circulating Erythrocytes

It is well known that degenerate forms of malaria parasites are observed inside red cells after chemotherapy. In 1944, Taliaffero & Taliaffero (141), studying *Plasmodium brasilianum* in *Cebus capucinus* monkeys, noted the appearance of "crisis forms" of parasites inside circulating red cells about the time that host immunity was becoming established. *B. microti* produces a high parasitemia in mice, followed by natural recovery during which we observed many condensed forms of the parasite inside circulating erythrocytes (31). Systematic electron microscope observations confirmed that these were degenerating intra-erythrocytic parasites. Similar changes were subsequently observed in mice infected with *P. chabaudi* and *P. vinckei.* Degenerate forms are rare when the parasitemia is rising but predominate when it is falling during natural recovery or protection by BCG or *C. parvum.*

In the absence of evidence for killing of these parasites elsewhere in the body, it appears that the infections of mice are normally terminated by their death within red cells. This implies that inhibition of merozoite attachment to erythrocytes, and opsonization of parasitized erythrocytes for phagocytosis, cannot be major mechanisms of immunity to these parasites in mice. Similar degenerating intra-erythrocytic parasites are seen during recovery from *P. berghei* infections in the rat (163), and the observations on *P. brasilianum* quoted above show that the same is true of at least some primate malarias. Since the sensitive stages of *P. falciparum* occur in erythrocytes attached to postcapillary venules, major effects of oxidant stress would be expected locally rather than being manifested in circulating erythrocytes.

Role of Phagocytosis

For a long time it was believed that phagocytosis of infected cells or free parasites was the principal mechanism by which immunity was effected (140). Macrophages of mice with *P. berghei* infections showed enhanced phagocytic activity toward infected and uninfected erythrocytes in culture (136) and in millipore chambers implanted into the peritoneal cavity (37). Immune serum increases phagocytosis by macrophages from animals with malaria (37) and normal animals (24, 145). However, recovery from nonlethal *P. berghei* and *P. chabaudi* does not correlate with increased phagocytosis (19, 87) but is coincident with the presence of degenerate parasites in intact erythrocytes. Thus, increased phagocytosis is certainly not the only mechanism of immunity in malaria, and often it appears to be secondary to intra-erythrocytic death of parasites or extracellular lysis of infected cells.

The Spleen as an Effector Organ in Immunity to Malaria

A role of the spleen in resistance to malaria parasites has long been recognized. The narrow host range of most species of *Plasmodium* is partly due to the requirement for erythrocyte receptors, an appropriate Hb type, and nutritional factors (99), but can in many cases be extended by splenectomy (164). Examples are the transfer of *P. gallinaceum* from chickens into the Norwegian rat, *P. vinckei* and *P. chabaudi* from the African thicket rat into the Norwegian rat, *P. berghei* from the African thicket rat into the rhesus monkey, and *P. falciparum, P. vivax,* and *P. ovale* from man into the chimpanzee (164). In other cases, infection with a parasite highly virulent to the natural host can result in brief, low parasitemia in unnatural hosts (e.g. *P. falciparum* in the owl monkey, *Aotus trivirgatus*). Splenectomy in the unnatural host may convert the infection to a malignant course like that in the natural host (164).

These observations suggest that the spleen efficiently filters out parasitized cells as soon as they can be recognized as foreign. Rat erythrocytes parasitized by *P. berghei* are more rapidly removed from the circulation into the spleen than are uninfected cells (116, 165). A unique structural feature of the spleen not found in other lymphoid organs, the red pulp, is likely to be the site where this filtration occurs (155). As shown in Figure 3, arterioles open into cords that are connected with sinuses. The macrophage is the dominant leukocyte in the cords; blood mononuclear cells entering the cords through arteriolar terminations appear to be selectively held in the interstices of the reticular meshwork. In other words, immigrant mononuclear cells with high capacity for a respiratory burst are concentrated in the post-arteriolar region, and parasitized erythrocytes entering

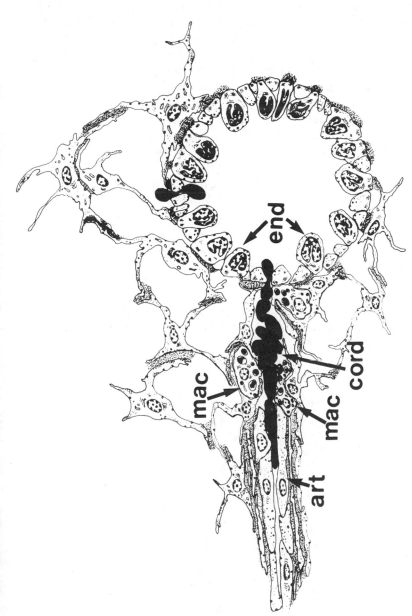

Figure 3   Diagram showing a common type of intermediate circulation in the red pulp of the spleen (155). An arteriole (art) opens into part of the cord. Macrophages (mac) are concentrated around the termination of the arteriole. Erythrocytes (black) leave the arteriole in an area where macrophages are concentrated and pass from the cord between endothelial cells (end) to enter the sinus.

through arterioles must pass through this region in close apposition to these effector cells, which can bombard them with O_2^- (Figure 4). If the parasitized erythrocytes are retained in the post-arteriolar region (for example, by binding of IgG antibodies on their surface to Fc receptors on the macrophages), the dose of O_2^- they receive will be increased. High O_2 tensions in the post-arteriolar region facilitate O_2^- formation.

Erythrocytes or bacteria bearing small numbers of IgG molecules are removed from the circulation into the spleen but not into the liver or other organs (53). Hence, cells with high-affinity receptors for IgG are concentrated in the spleen. Furthermore, as discussed below, immune complexes can activate O_2^- production by effector cells. Thus, synergistic effects of antibodies and cell-mediated immunity might occur. Products of activated T-lymphocytes could stimulate the proliferation and recruitment of a subset of mononuclear cells with high capacity to produce O_2^-. Antibodies of high affinity and appropriate isotype could facilitate binding of these effectors to target cells and activate O_2^- formation by the effectors. Thus, the initial effector mechanism would be of antibody-dependent cellular cytotoxicity (ADCC) rather than NK type, although the same effector cell might be involved.

A role of antibodies in recovery from primary infections with *P. yoelii* in the mouse and *P. gallinaceum* in the chicken (119, 124) has been discussed above. In other infections, such as with *P. chabaudi* in the mouse, cell-

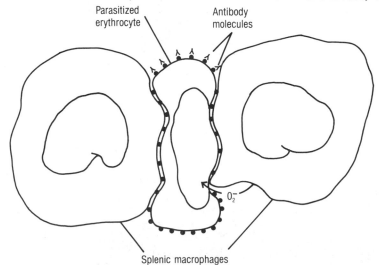

Figure 4   The splenic effector system. Parasitized erythrocytes passing between macrophages in the post-arteriolar region of the red pulp are detained and bombarded with O_2^-. Antibody molecules on the surface of parasitized cells bind to Fc receptors on the macrophages and facilitate this interaction.

mediated immunity can perform this function efficiently in the absence of specific antibodies, perhaps because the parasitized cells bind to the surface of the effectors and activate O_2^- production by an antibody-independent mechanism.

The second filter system of the red pulp must also be mentioned. Blood leaving the cord enters the lumen of the vascular sinus by passing between endothelial cells (155). This constitutes a slit-like space, and erythrocytes passing through it must be pliant. If the deformability of the erythrocytes is reduced, as when they harbor malaria parasites, their passage is delayed and a portion of the cell containing a rigid body, such as a degenerate parasite, may break off and remain in the cords, just outside the sinus. "Crisis forms" of parasites persist in the circulation of splenectomized animals longer than in animals with intact spleens (7).

Mononuclear Phagocytes as Splenic Effectors

"Mononuclear phagocyte" is used broadly to cover phagocytic cells other than polymorphonuclear leukocytes. Van Furth and his colleagues (148) applied the designation to a single lineage of cells originating in the bone marrow from common stem cells (CFU-GM, precursors of granulocytes and monocytes) and differentiating into promonocytes and monocytes; the latter enter the circulating blood and then tissues, becoming Kupffer cells as well as peritoneal, alveolar, and other resident macrophages.

Not all authorities accept this widely held view. From ^{3}H-labeling studies it has been concluded that resident macrophages have the capacity to proliferate locally, although they can be replenished from a special subset of bone marrow precursors (39, 149). Daems (38) drew attention to the different pattern of peroxidase activity in monocytes and resident macrophages (Kupffer cells and peritoneal macrophages). In monocytes, the peroxidase reaction product is observed in cytoplasmic granules (peroxisomes), whereas in the resident cells it is observed in the nuclear membrane and endoplasmic reticulum. The substrate specificity and thermal stability of peroxidase in the two cell types are different, and enzyme activity is inhibited by aminotriazole only in the resident cells. No cells with transitional characteristics, which would suggest that monocytes can become resident macrophages, were observed.

Mononuclear cells in mice or guinea pigs elicited into the peritoneal cavity or lungs by *C. parvum* or BCG, when appropriately stimulated, show much greater chemiluminescence and H_2O_2 production than do resident cells (17, 82, 106). Activation by lymphokines of mononuclear cells, so that they can kill tumor cells, red blood cells, and intracellular parasites such as *Leishmania* is well documented (105, 118). Mononuclear cells recently recruited into the peritoneal cavity by lymphokines are much more cytolytic for tumor cells than are resident cells (98, 106). In the killing by mononuclear phagocytes of tumor cells (1, 125) and *Leishmania* promastigotes and amastigotes (67, 104), H_2O_2 appears to be a more potent mediator than O_2^- since cytotoxicity can be inhibited by catalase but not superoxide dismutase (SOD). A soluble cytolytic factor, believed to be a serine esterase, can act synergistically with H_2O_2 in mononuclear cell-mediated killing of tumor cells (1).

If mononuclear cells are able to function as NK or K cells in malaria, it would be expected that they would bind to infected cells and that such binding would activate a respiratory burst. Preliminary indications that this may be the case have been published (95). Mouse peritoneal cells exposed to *P. berghei*-infected erythrocytes or lysates sometimes responded by luminol-enhanced chemiluminescence, and in the presence of immune serum the reaction was consistently observed and of increased magnitude.

The situation in vivo is complex (68), and complementary information can be obtained with cells cloned in culture. Human mononuclear cell lines HL60 (153) and U937 (118), treated with phorbol myristate acetate (PMA) or lymphokines, differentiated into macrophage-type cells (phagocytic, with Fc receptors) that were cytotoxic for tumor cells. The lymphokine triggering differentiation in these cells appeared to be distinct from colony-stimulating activity (which stimulates proliferation of CFU-GM), and cell lines with myelomonocytic markers did not acquire NK or ADCC activity when cultured with the same lymphokine. One interpretation of these observations, consistent with that of Volkman (149) and Daems & De Bakker (39) mentioned above, is that a lineage of mononuclear cells, distinct from immediate precursors of monocytes and granulocytes, can be triggered by products of activated T-lymphocytes to differentiate into macrophage-like cells with NK and ADCC activity. If these cells become cytotoxic in the circulating blood before they become phagocytic, they might be a major subset of cells with NK and ADCC activity.

Sequestration of Parasitized Erythrocytes

Erythrocytes bearing malaria parasites are often differentially sequestered in vascular beds. This depends on preferential adhesion to different types of endothelial cells of parasitized erythrocytes at particular stages of development. Thus, erythrocytes infected with *P. falciparum, P. fragile,* or *P. coatneyi* adhere preferentially to vessels in heart and adipose tissue, and erythrocytes bearing *P. falciparum* adhere to cultured human endothelial cells (147). Young forms of *P. berghei* are sequestered in the deep vasculature of the heart and kidney in the rat, whereas schizonts of *P. berghei* are concentrated in the bone marrow in rats and in spleen and in the bone marrow in mice, but show no preferential localization in hamsters. In venous sinuses of the bone marrow of mice, Weiss (154) observed reticulo-cytes parasitized with *P. berghei* NK-65 adherent to endothelial cells and to nonparasitized reticulocytes. He suggests that merozoites may enter adherent reticulocytes without significant extracellular exposure. The re-lease of factors from parasitized cells or lymphocytes might explain the early reticulocytosis observed in this disease before the hematocrit falls greatly; thus, new reticulocytes would be released in a site favorable for parasitization. By this anatomical arrangement the spleen is largely avoided, and the parasitized cells are resistant to oxidant stress, so that cell-mediated immunity is less effective than in *P. chabaudi,* which shows far less predilection for the bone marrow and multiplies in mature ery-throcytes.

A Peripheral Effector System

The spleen is not the only site in which immunity to hemoprotozoa is manifested. Splenectomized animals can often recover from these infections (7). A second organ that may be involved is the liver, since migration of mononuclear cells into the liver occurs during infection (42), and partial hepatectomy can aggravate experimental malaria (20). However, an addi-tional possibility is that effector cells can recognize malaria-parasitized erythrocytes in the peripheral circulation (Figure 5). This mechanism might be especially effective when schizonts develop attached to endothelial cells in postcapillary venules, as in the case of *P. falciparum.* Parasite antigens appear on the surface of infected erythrocytes only about the time of schizogony, so that infected erythrocytes become recognizable as foreign only at the stage of development when they are secluded. Small amounts of IgG antibodies on the surface of infected erythrocytes might provide a recognition mechanism through Fc-IgG receptors on effector cells, but the possible existence of recognition systems not dependent on antibodies must also be considered in the case of some infections. The effector cells in such a system should be present in the peripheral blood and should have Fc-IgG receptors or bind malaria-parasitized cells in the absence of antibodies, and such binding should trigger O_2^- production. NK cells are candidates, and their possible role in immunity to malaria is considered in the next section.

NK and ADCC Activity in Immunity to Malaria

In 1980, we suggested that NK cells might play a role in immunity to malaria (46). Mice of the A/He strain were found to be more susceptible to *P. chabaudi* infections than other strains, and we found two defects in them. First, the usual thymus-dependent increase in mononuclear cell num-ber in the spleen of malaria-infected mice did not occur in A mice, as already mentioned. Second, mice of the A strain have a well-known defect

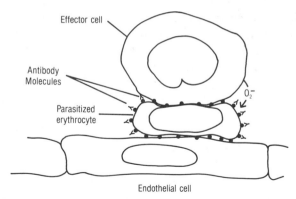

Figure 5 The peripheral effector system. An erythrocyte parasitized with *P. falciparum* adheres to a postcapillary venous endothelial cell through surface knobs. Antibody molecules binding to the knobs interact with Fc receptors on an effector cell, triggering O_2^- formation at the site of interaction.

of NK activity, as measured with several target cells (68), and we found no increase in NK activity during the course of malaria infections in A mice (46).

Wood & Clark (159) have concluded that NK cells are irrelevant to resolution of infections with *P. vinckei petteri* and *B. microti.* Increased NK activity during the infections preceded the fall in parasitemia, and infections followed a normal course in ^{89}Sr-treated mice and in *bg/bg* mice (see also 127). However, these criticisms are irrelevant if the subset of NK cells not depleted in *bg/bg* mice and sensitive to [89]Sr (88) is that which functions as as an effector against blood-stage forms of hemoprotozoa.

BCG can augment NK activity in *bg/bg* mice (11) and can protect mice other than the A strain against *P. berghei* infections (103). Thus, a subset of NK cells unaffected by the *bg/bg* mutation and responsive to BCG, but defective in mice of the A strain, could participate in immunity to malaria. Genetic analysis of basal NK activity against YAC-1 target cells in backcross and congenic mice will likewise be uninformative. What is required is information about capacity of animals to increase NK activity against relevant targets during the course of infection.

Roder et al (125) have characterized cells from human peripheral blood with capacity to kill K562 and provided evidence that release of O_2^- is involved in the killing mechanism. The leukocytes were depleted of monocytes and were separated on a discontinuous Percoll gradient. More than 50% of the cells in the fraction with NK activity expressed OKMI and HNK-1 markers (the latter believed to be NK specific) and Fc-IgG. In contrast, < 0.5% of the cells expressed the Mo2 marker believed to be specific for human monocytes, were phagocytic, or stained for acid esterase.

In the presence of K562, but not insensitive targets, the NK cells released considerable amounts of O_2^-, as shown by chemiluminescence and cytochrome c reduction, and SOD (but not heated SOD or catalase) inhibited their capacity to kill K562 cells. Additional evidence that human peripheral blood leukocytes with receptors for Fc-IgG and reacting with the monoclonal antibody OKM1 mediate NK activity has been presented, and a cloned NK cell line, derived from the "null cell" fraction of human peripheral blood, lacking T3, T8, B1, and Mo2 markers and able to lyse K562, has been obtained (69).

Increased NK activity of peripheral blood cells, assayed against K562 target cells, has been observed in children infected with *P. falciparum* (111). The sera of most infected children were found to have antiviral activity, predominantly α-interferon. Addition of exogenous interferon increased NK activity of peripheral blood cells in normal children whereas it had no significant effect on cells from malaria-infected children, suggesting prior in vivo activation of the latter.

When NK cells have Fc receptors they may also display ADCC activity, and an increase in this activity of human peripheral blood cells during malaria infections has been described (64). Increased ADCC activity has likewise been observed in the spleens of mice during the second week of *P. chabaudi* infections (90).

The properties of human peripheral blood NK or ADCC cells, as defined above, would make them efficient peripheral effectors in malaria, and this possibility should be tested directly by culturing such cells with synchronized *P. falciparum* schizont-infected human erythrocytes in the presence and absence of immune IgG and interferon. Surface markers defined by monoclonal antibodies, and tests for capacity to generate O_2^- after binding to infected cells, should establish the lineage of the cells and their capacity to produce oxidant stress. Likewise, the isolation of effector cells from spleens of infected mice, and similar tests with *P. chabaudi*-bearing mouse erythrocytes, should establish their lineage and capacity to generate O_2^-, thereby imposing oxidant stress on the intra-erythrocytic parasites.

Claësson & Olsson (25) have reported NK activity in colonies of murine bone-marrow cells growing in semisolid agar. The colonies were distinct morphologically from typical macrophage and mixed macrophage-granulocyte colonies. These results suggest that certain bone-marrow cells, possibly distinct from monocyte-granulocyte precursors, are cytotoxic for susceptible target cells such as syngeneic fibroblasts. A working hypothesis is that this distinct cell type is the precursor of NK cells lacking the NK 1.2 marker and not depleted by ^{89}Sr, because the precursors either are relatively radioresistant or are present also in extramedullary sites. Characterization of this lineage of murine NK cells, and ascertaining whether they are able, like human NK cells bearing OKM and HNK-1 markers, to release O_2^- following binding to susceptible targets, is a high priority task for the future.

Soluble Factors

We have recently reviewed the role of interferon (IFN) in immunity to malaria (47), and only the principal findings are summarized here. IFN is observed in the serum soon after *P. berghei* blood-stage infections, reaching a maximum at 35 hr (75). Injections of virus-induced IFN slightly delayed the rise of parasitemia and death in this infection (133), whereas injection of antibodies against this type of IFN slightly accelerated the rise of parasitemia and death (129). These observations suggest a modest role for IFN (α or β) in retarding the proliferation of lethal *P. berghei*. IFN induced by Sendai virus had no effect on blood-stage infection with *P. vinckei* (31, 35), and potent antibodies against α- and β-IFN had no effect on *P. vinckei petteri* or *P. chabaudi* infections (31, 46). Hence, it is unlikely that α- or β-IFN plays a significant part in these infections, which does not exclude a possible role of lymphocyte-derived γ-IFN. Several IFN inducers were found to protect mice significantly and sometimes completely against sporozoite-induced infections (77, 78), the protection being greatest against the late pre-erythrocytic (liver) stage. These effects may have been due to IFN-mediated activation of NK cells or macrophages.

Emphasis has been placed recently on the role of lipopolysaccharide endotoxins (LPS) in malaria infections and the release of tumor-necrosis factor (TNF) from macrophages (3, 26, 93, 113). It is highly improbable that so chemically specialized a product as LPS (known only in certain gram-negative bacteria) should be present in plasmodia, and LPS could not be demonstrated in sonicates of large numbers of *P. berghei* blood-stage parasites (49). Conceivably, LPS of bacterial origin might play a role in the immunopathology of some malarial infections, but such a chance event makes it unlikely that LPS contributes significantly to the regular recovery of mice from infections with nonlethal malaria parasites and of humans living in endemic areas from *P. falciparum*. TNF has been found to inhibit the replication of *P. falciparum* in cultures of human erythrocytes (C. Haidaris, D. Haines, M. S. Meltzer, A. C. Allison, unpublished results). Polyamine oxidase in the presence of spermine has the same effect, with the appearance of degenerate parasites in the cells (A. Ferrante, C. Rzepczyk, A. C. Allison, unpublished results). Polyamine oxidase is found in activated macrophages (101). TNF activity requires O_2 and can be inhibited by metal chelation (128), which suggests it may be a dioxygenase. A possible common factor in all of these mediators is the imposition of oxidant stress on parasitized erythrocytes.

PRODUCTION AND REACTIONS OF O_2^- AND H_2O_2

When neutrophils and macrophages are stimulated by phagocytosis, there is a burst of increased metabolism by the hexose monophosphate shunt pathway, with the production of O_2^- and H_2O_2 (8). At the same time, chemiluminescence is observed and can be amplified by luminol and other compounds (3, 82). Production of O_2^- can also be increased by immune complexes (156). In chronic granulomatous disease, production of O_2^- is impaired (8, 134), and patients are susceptible to infections with organisms possessing catalase (96).

O_2^- can interact with many biological molecules (70, 130). Of special import is the interaction with redox-active metal ions, for example (70):

$$Fe(II) + O_2^- \rightarrow Fe(O_2)^+.$$

The oxidative addition of O_2^- to Fe(II)EDTA in aqueous solution produces a ferric peroxo complex, Fe(III)EDTA (O_2^{2-}) (130). The formation of such powerfully oxidant complexes is relevant to iron-catalyzed lipid peroxidation and to inactivation of metal-containing or metal-activated enzymes considered below. O_2^- does not oxidize unsaturated lipids such as linoleic acid; however, this reaction can be catalyzed not only by iron, but also by lipid hydroperoxide (144). Thus, chain reactions, initiated by an iron oxidant system and perpetuated by hydroperoxide, can occur. Autoxidation of polyunsaturated fatty acids may be one of the principal reactions by which free radicals produce damage in biological systems. Some lipid hydroperoxides, e.g. those produced by the lipoxygenase and cyclo-oxygenase systems, have potent pharmacological activity and may regulate biochemical pathways in parasites, as they do in other cell types. Exogenous hydroperoxides, such as t-butylhydroperoxide, can impose oxidant stress on cells by consuming reduced glutathione through glutathione peroxidase (52) and thereby oxidizing NAD(P)H. An example is the effect of t-butylhydroperoxide on Ca^{2+} retention by mitochondria (86).

The importance of the conjugate acid of O_2^-, the perhydroxyl radical $(HO_2^{\cdot})$, is now being appreciated. In aqueous solutions, $HO_2^{\cdot}$ is in equilibrium with O_2^- (12):

$$O_2^- \ pKa = 4.8 \ HO_2^{\cdot}.$$

As a consequence of this equilibrium, nearly 1% of any O_2^- formed under physiological conditions (pH near 7.0) will be present as $HO_2^{\cdot}$. In acidic microenvironments (such as phagocytic vacuoles or where negatively charged effector and target cell membranes are closely apposed), the proportion of $HO_2^•$ will be much greater. In many cases, $HO_2^•$ reacts orders of magnitude faster than O_2^- or is the sole reactant with a biological compound (e.g. vitamin E) (12). Since $HO_2^•$ is uncharged, it could readily diffuse into hydrophobic regions of lipid bilayers or through membranes, whereas passage of O_2^- through membranes is confined to anion channels (89).

Most metal ions and complexes are not oxidized by O_2^- in aqueous solution at rates competitive with its disproportion, even when such oxidations are thermodynamically favored (130). The probable explanation is that O_2^- must first be coordinated and reduced in the first coordination sphere of the metal. Oxidations by $HO_2^•$ are not subject to the same restriction and readily occur: for example, of aqueous Fe(II) (130). When Fe(II) is already coordinated, as in heme-containing enzymes and Hb, $HO_2^•$ will be much more likely than O_2^- to oxidize Fe.

Fe can also catalyze the formation of highly reactive hydroxyl radicals (OH·) from O_2^- and H_2O_2, and lipid peroxidation by this mechanism can be inhibited by the iron chelator deferoxamine (66).

EFFECTS OF OXIDANT STRESS

Erythrocytes

The principal manifestations of oxidant stress on erythrocytes are reduction in their content of reduced glutathione, lipid peroxidation, Hb denaturation, and the formation of inclusions known as Heinz bodies. K^+ is lost from the cells (51, 150), which eventually lyse. Erythrocytes damaged by oxidants but not lysed are rapidly removed from the circulation by the spleen and liver (9). Erythrocytes deficient in glucose-6-phosphate dehydrogenase (G-6-PD$^-$) are more sensitive than those with a normal content of this enzyme (G-6-PD$^+$). Since activity of G-6-PD, SOD and other protective enzymes (57) declines as erythrocytes age (10, 97), their sensitivity to oxidant stress increases with age.

Agents known to produce oxidant damage in erythrocytes include phenylhydrazine (79) and alloxan, which forms a redox system with dialuric acid (48, 126). In both cases production of O_2^- (57) and its interaction with a hydrogen atom transferred from the organic molecule could generate powerfully oxidant radicals (130). Damage by alloxan of pancreatic islet β-cells can be prevented by SOD, and is thought to be mediated through O_2^- production (62).

Baehner et al (9) incubated G-6-PD$^-$ erythrocytes with normal human leukocytes in which the respiratory burst had been triggered by phagocytosis. The erythrocytes were lysed, or when labeled and injected into human subjects were rapidly removed in the spleen and liver. When chronic granulomatous disease leukocytes were incubated with $G-6-PD^-$ erythrocytes, or normal leukocytes were incubated with $G-6-PD^+$ erythrocytes, damage was not seen. Hence, activation of a respiratory burst in leukocytes, with concomitant production of O_2^- and H_2O_2, can impose oxidant stress on erythrocytes. The relevance of this mechanism to immunity against malaria parasites will be evident.

Asexual Forms of Malaria Parasites in Erythrocytes

P. falciparum is microaerophilic: in cultures of human erythrocytes, it grows best at 3% O_2 and it is inhibited by higher concentrations of O_2 (132). However, there is a critical level ($< 0.5\%$) below which the parasite will not survive, in contrast to true anaerobic forms of life. From these observations and the lack of a Pasteur effect, Scheibel & Adler (131) concluded that in the malaria parasite O_2 may not play the same role in electron transport as it does in many mammalian cells, and they suggested that O_2 may act through a distinct type of metalloprotein oxidase enzyme system. They found that the metal chelators diethyldithiocarbamate, 8-hydroxyquinoline, and 2-mercapto-pyridine-N-oxide inhibit *P. falciparum* replication in culture (131). This effect was attributed to favorable lipid-water partition coefficients and binding constants allowing inhibition of metalloprotein oxidases.

The importance of metalloenzymes in protozoan parasites is illustrated by the iron-sulfur ferredoxin system of anaerobe *Tritrichomonas,* which confers sensitivity to oxidant damage as well as to drugs such as metronidazole (102). The existence of such an enzyme system in *Leishmania* has been postulated on the basis of inhibitory effects of redox-active dyes (117).

Although suggestions that hemoprotozoa and leishmania have metalloprotein oxidases are stimulating, their existence is at present conjectural. Metal chelators could have other effects: for example, heme is bound to at least three soluble proteins in *P. falciparum* (51). In addition to functioning as the active site of enzymes and O_2-combining sites of Hb and myoglobin, heme is known to have major regulatory effects, including activation of a factor (eIF2α) required for the initiation of protein synthesis (110) and of guanylate cyclase (76). Thus, heme could participate in the regulation of metabolism in *Plasmodium* parasites. Chloroquine is thought to function as a heme chelator (51), and other metal chelators may also bind heme.

Etkin & Eaton (45) presented evidence that in the erythrocytes of mice infected with *P. berghei,* H_2O_2 is generated and that the ability of infected cells to prevent or repair oxidant damage is decreased. They observed in *P. berghei*-infected mice an increase in steady-state metHb concentration directly proportional to parasitemia. Catalase in infected erythrocytes was inhibited by 3-amino-1,2,4-triazole (an inhibition requiring the presence of H_2O_2), and infected erythrocytes accumulated more metHb than uninfected cells when exposed to oxidants generated by ascorbic acid. In vitamin E-deficient mice, *P. berghei* infections were found to develop more slowly than in normal mice (43). In the vitamin E-deficient mice, parasites were found only in reticulocytes and anemia was marked, which suggests protection was mediated by premature lysis of oxidant-sensitive erythrocytes.

Inhibition of catalase by aminotriazole has also been used by Friedman (61) as a measure of H_2O_2 production by *P. falciparum* cultured in human erythrocytes. The catalase inhibition was found to be proportional to the partial pressure of O_2 in the culture medium over the range 10–30%.

The sensitivity of malaria parasites to oxidant stress is emphasized by the damaging effect on intra-erythrocytic parasites of agents that exert such stress on erythrocytes. In 1950, Rigdon et al (121) reported that administration of phenylhydrazine to rhesus monkeys infected with *P. knowlesi* resulted in the appearance of degenerate forms of the parasite within circulating erythrocytes. His finding attracted little attention and remained without explanation for 30 years.

Our interest in this subject arose from the observation that incubation of mouse erythrocytes infected with *P. chabaudi* in the presence of a dioxygenase enzyme system generating H_2O_2 and other products (bovine serum polyamine oxidase and spermine) abolished their infectivity (100). Clark & Hunt (30) injected alloxan intravenously into mice infected with *P. vinckei* and observed a fall in parasitemia and the appearance of degenerate parasites within circulating erythrocytes; this effect was prevented by prior injection of the selective iron chelator deferoxamine, or the less selective metal chelator diethyldithiocarbamate (DDC).

Oxidant effects of t-butylhydroperoxide on erythrocytes have been described (35, 137, 146). We have found that injection of alloxan or t-butylhydroperoxide produces a dramatic fall in parasitemia in mice infected with *P. chabaudi* or a lethal strain of *P. yoelii* (17XL), with the appearance of crisis forms of parasites in circulating erythrocytes (Figure 6). The treatment did not kill all parasites, and transient parasitemia recurred afterwards; however, most animals recovered and were resistant to challenge, showing that exposure to t-butylhydroperoxide did not interfere with the induction of protective immunity against the parasite. Again, prior administration of deferoxamine or DDC prevented the effect of t-butylhydroperoxide on the parasite.

Why asexual forms of malaria parasites in erythrocytes are sensitive to oxidant stress is at present unknown. Conceivably, the primary metabolic effect of oxidants is on the host erythrocyte, with secondary effects on the

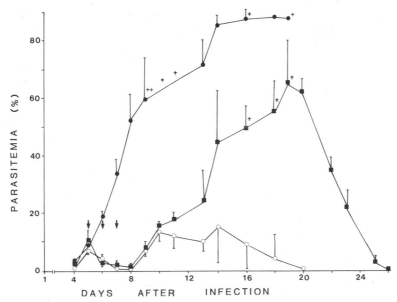

Figure 6 Effect of t-butylhydroperoxide on the course of *P. yoelii* 17XL infections in male C57B1/6 mice. In control animals (●), parasitemia increased progressively and all died. Following injections of t-butylhydroperoxide (0.2 ml of a 1% solution), parasitemia decreased. In treated animals with low initial parasitemia (◯), a moderate parasitemia recurred, but all six recovered. In animals with higher initial parasitemia, severe parasitemia recurred after treatment, and three out of five mice died (■).

parasite. For example, oxidant drugs produce loss of K^+ from erythrocytes (48, 150), and this could reduce their capacity to support parasite multiplication (60). However, the inhibition by DDC as well as deferoxamine of oxidant damage to plasmodia raises the interesting possibility that a metalloenzyme (e.g. Fe-Cu or Fe-Mo) is irreversibly damaged by strong oxidants; reversible binding of deferoxamine or DDC during the period of exposure to a strong oxidant could protect the enzyme from permanent damage. Whatever the explanation, oxidant effects are likely to be greater on parasites multiplying in mature erythrocytes than in reticulocytes.

The degeneration of parasites in circulating erythrocytes after exposure to oxidants is closely analogous to the crisis forms observed during recovery from malaria infections as a result of cell-mediated immunity. A reasonable conclusion is that a major effector mechanism of cell-mediated immunity is the imposition of oxidant stress on parasitized cells.

WORKING HYPOTHESIS

From available evidence, it is possible to formulate a working hypothesis about immunity to asexual blood forms of malaria parasites, bearing in mind that this is a framework for future research rather than a definitive statement (Figure 7). Parasite antigens efficiently stimulate a subset of T lymphocytes (Ly1+ in the mouse). Presumably, the antigens are presented to T cells by Ia-bearing cells such as splenic dendritic cells (107) and Langerhans cells. Clones of T cells with receptors for parasite antigens proliferate in the spleen and other lymphoid organs, and circulating T cells are diverted to the spleen and liver.

T cells contribute to immunity against malaria parasites in two ways. The formation of antibodies against parasite antigens requires T cell help, and such antibodies are necessary for recovery from primary infections with some parasites, e.g. *P. yoelii* in the mouse and *P. gallinaceum* in the chicken. Second, activated T lymphocytes release factors stimulating the proliferation of mononuclear cells with high capacity to synthesize O_2^-. Proliferation of the same subset of cells can be stimulated by *C. parvum* independently of T-lymphocytes. When these cells are released into the peripheral blood they have not yet acquired phagocytic capacity, but have Fc-IgG receptors and can be triggered to produce O_2^- by binding to target cells, thereby exerting NK and ADCC activity. One such target is erythrocytes bearing *P. falciparum* schizonts attached to postcapillary endothelial cells; antibodies combining with parasite antigens on the surface of infected erythrocytes facilitate such binding and activation of effector cells. The same subset of T-lymphocytes activated by parasite antigens releases a factor chemotactic for the O_2^--producing subset of mononuclear cells. The latter migrate selectively into sites such as the post-arteriolar region of the red pulp of the spleen, where they come into close contact with malaria-parasitized erythrocytes, bind to their surfaces, and release O_2^-.

Malaria parasites in erythrocytes are sensitive to oxidant stress. They are relatively resistant to H_2O_2, which can actually be produced in parasitized cells. They are more sensitive to O_2^-, perhaps in the form of $HO_2^{\cdot}$, and are highly sensitive to radicals produced by interaction of O_2^- and organic compounds such as phenylhydrazine. Oxidants produce K^+ leakage from cells, which is unfavorable for parasite growth, and the parasites may have a metalloprotein electron transport system that is inactivated by strong oxidants.

It follows that cells that exert cytotoxic effects through release of O_2^-, such as NK cells, may be more efficient as effector cells against malaria-parasitized erythrocytes than are monocytes, which release H_2O_2. Subsets of NK cells resistant to the bone-seeking isotope ^{89}Sr and unaffected by the beige mutation in mice, and bearing the HNK-1 and OMK-1 markers in humans, could be important in immunity to malaria. Mice of the A strain are highly susceptible to malaria and are unable to recruit the mononuclear cells required as effectors of cell-mediated immunity. A speculative interpretation is that a distinct lineage of macrophage precursors in the bone marrow

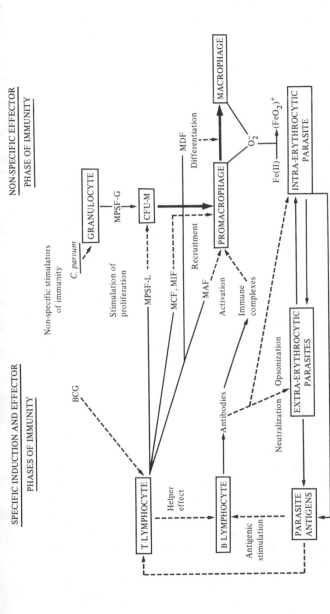

SPECIFIC INDUCTION AND EFFECTOR
PHASES OF IMMUNITY

NON-SPECIFIC EFFECTOR
PHASE OF IMMUNITY

Figure 7 **Working hypothesis for cellular interactions in immunity to hemoprotozoa.** T lymphocytes help antibody formation against parasite antigens, which may inhibit parasite replication and opsonize parasites or parasitized cells for phagocytosis. T lymphocytes produce a macrophage precursor-stimulating factor (MPSF-L), which stimulates the proliferation of a distinct lineage of macrophage precursors (CFU-M) in peripheral tissues as well as the bone marrow. These first differentiate into cells with Fc-IgG receptors and capacity to liberate O_2^-, which have NK and macro-ADCC activity. Promacrophages are recruited into sites of infection by lymphokines known as macrophage chemotactic factor (MCF) and macrophage migration-inhibition factor (MIF). Under the influence of a lymphokine [macrophage differentiation factor (MDF)], promacrophages differentiate further into phagocytic cells that are resident macrophages. These also have high capacity to form O_2^- after binding to target cells, which can be facilitated by antibody. During this process immune complexes are formed, and they can trigger O_2^- formation. O_2^- exerts oxidant stress or parasitized erythrocytes, producing intra-erythrocytic parasite degeneration. Some nonspecific stimulators of immunity, such as BCG, act through T lymphocytes, whereas others, e.g. *C. parvum*, can act independently, most efficiently following ingestion by granulocytes. This suggests that another factor (MPSF-G), released by appropriately activated granulocytes, can stimulate the proliferation of macrophage precursors. Mice of the A strain are unable to form MPSF-L or respond to it. (Thick arrows represent cell lineage, thin continuous arrows represent cell products, and discontinuous arrows represent influences on cell or parasite function.)

can be seeded out to the spleen, peritoneal cavity, and other sites. In these locations they are still able to respond to lymphokines and other stimuli by proliferation, so that they are not bone marrow dependent (and therefore insensitive to ^{89}Sr). Promacrophages of this lineage in the circulation are termed NK cells. When they settle down in the spleen and other sites they become phagocytic, this differentiation being promoted by lymphokines (as in the U937 cell line). This lineage of cells, characterized by peroxidase in the endoplasmic reticulum and high O_2^- production, is an important mediator in cell-mediated immunity to malaria and other infections, whereas cells of the promonocyte-monocyte lineage, with peroxidase in peroxisomes, are not.

Malaria-parasitized erythrocytes adhere to endothelial cells in characteristic patterns. Schizonts of *P. berghei* in the mouse adhere to sinus endothelial cells in bone marrow and merozoites enter reticulocytes adherent to schizonts, which makes them relatively resistant to immune attack. *P. falciparum* schizonts complete their replication attached to postcapillary endothelial cells, where oxygen tensions are low.

Under these conditions, *P. falciparum* does not multiply well in human erythrocytes bearing HbS (58, 112), which may explain the relative resistance of carriers of this trait to this parasite (4, 5). *P. falciparum* grows poorly in G-6-PD$^-$ and β-thalassemic erythrocytes when exposed to oxidant stress (59). Such stress would be minimal when parasite replication is completed in postcapillary venules, but would be increased if adherent effector cells released O_2^-. Thus, resistance due to inherited erythrocyte defects could be reinforced by cell-mediated immunity, increasing the chances of survival of children bearing these traits during the critical years of first exposure to malaria.

The experiments reviewed suggest that immunity based solely on T cell and macrophage activation may not be sterile. Presumably, some parasites escape from oxidant damage. As the parasitemia falls, the T cell and macrophage activation decline. If the parasitemia rises (from persistent parasites or reinfection), memory T-lymphocytes are able to respond quickly and immunity is reactivated. This is the basis of premunition (135), which is observed in infections where cell-mediated immunity plays a predominant role. Human malaria infections may fall into this category, although interactions of cell-mediated and humoral immunity are likely to promote survival of hosts under natural conditions. Parasite survival depends on the suppression of immune responses against antigens essential for its replication, such as receptors on merozoites for erythrocytes. That is the principal reason why immunity to malaria is not more effective, requiring many infections over a period of several years, and rather easily lost following chemotherapy or residence in a non-malarious region.

ACKNOWLEDGMENTS

We are indebted to J. Grun and W. Weidanz, and to L. Weiss, for permission to reproduce Figures 2 and 3, respectively.

Literature Cited

1. Adams, D. O., Johnson, W. J., Fiorito, E., Nathan, C. F. 1981. *J. Immunol.* 127:1973–77
2. Adams, D. O., Marino, P. A., Meltzer, M. S. 1981. *J. Immunol.* 126:1843–47
3. Allen, R. C., Stjernholm, R. I., Steele, R. H. 1972. *Biochem. Biophys. Res. Commun.* 47:679–84
4. Allison, A. C. 1954. *Brit. Med. J.* i:290–93
5. Allison, A. C. 1964. *Cold Spring Harbor Symp. Quant. Biol.* 29:137–49
6. Allison, A. C. 1982. In *Quantitative Aspects of Host-Parasite Relations*, ed. S. Berhnard, In press. Berlin: Springer
7. Allison, A. C., Christensen, J., Clark, I. A., Elford, B. C., Eugui, E. M. 1979. In *The Role of the Spleen in the Immunology of Parasitic Diseases*, ed. G. Torrigiani, pp. 151–82. Basel: Schwabe
8. Badwey, J. A., Karnovsky, M. L. 1980. *Ann. Rev. Biochem.* 46:695–726
9. Baehner, R. L., Nathan, D. G., Castle, W. B. 1971. *J. Clin. Invest.* 50:2466–73
10. Bartosz, G., Tannist, C., Fried, R., Leyko, W. 1978. *Experientia* 34:1464
11. Beck, B. N., Henney, C. S. 1981. *Cell. Immunol.* 61:343–52
12. Bielsky, N. 1983. In *Superoxide and Superoxide Dismutase*, ed. G. Cohen, In press. New York: Raven
13. Bradley, D. J., Taylor, B. A., Blackwell, J., Evans, E. P., Freeman, J. 1979. *Clin. Exp. Immunol.* 37:7–14
14. Brown, I. N., Allison, A. C., Taylor, R. B. 1968. *Nature* 219:292–93
15. Brown, K. N. 1971. *Nature* 30:163–65
16. Brown, K. N. 1977. *Immun. Parasit. Dis. Colloq. INSERM Paris* 72:46–58
17. Bryant, S. M., Lynch, R. E., Hill, H. R. 1982. *Cell. Immunol.* 69:46–58
18. Butcher, G. A., Mitchell, G. H., Cohen, S. 1978. *Immunology* 34:77–86
19. Cantrell, W., Elko, E. E., Hopff, B. M. 1970. *Exp. Parasitol.* 28:291–97
20. Cantrell, W., Moss, W. G. 1963. *J. Infect. Dis.* 112:67–71
21. Carter, R., Diggs, C. L. 1977. In *Parasitic Protozoa*, ed. J. P. Kreier, 111:359–466. New York: Academic
22. Carter, R., Gwadz, R. W., Green, I. 1979. *Exp. Parasitol.* 47:194–208
23. Chen, D. E., Tigelar, R. E., Weinbaum, F. I. 1977. *J. Immunol.* 118:1322–27
24. Chow, J. S., Kreier, J. P. 1972. *Exp. Parasitol.* 31:13–18
25. Claësson, M. H., Olssen, L. 1980. *Nature* 283:578–80
26. Clark, I. A. 1978. *Lancet* ii:75–76
27. Clark, I. A., Allison, A. C. 1974. *Nature* 252:328–29
28. Clark, I. A., Allison, A. C., Cox, F. E. G. 1976. *Nature* 259:309–11
29. Clark, I. A., Cox, F. E. G., Allison, A. C. 1977. *Parasitology* 74:9–17
30. Clark, I. A., Hunt, N. H. 1983. *Infect. Immun.* 39:1–6
31. Clark, I. A., Richmond, J. E., Willis, E. J., Allison, A. C. 1977. *Parasitology* 75:189–96
32. Clark, I. A., Virelizier, J. L., Carswell, E. A., Wood, P. A. 1981. *Infect. Immun.* 32:1058–66
33. Cohen, S., McGregor, I. A., Carrigton, S. 1961. *Nature* 192:733–37
34. Coleman, R. M., Bruce, A., Rencricca, N. J. 1976. *J. Parasitol.* 62:137–46
35. Corry, W. D., Meiselman, H. J., Hochstein, P. 1980. *Biochim. Biophys. Acta* 597:224–34
36. Cox, H. W. 1964. *J. Parasitol.* 50:23–29
37. Criswell, B. S., Butler, W. T., Rossen, R. D., Knight, V. 1971. *J. Immunol.* 107:212–21
38. Daems, W. T. 1980. In *The Reticuloendothelial System, A Comprehensive Treatise*, ed. I. Carr, W. T. Daems, 1:57–75. New York: Plenum
39. Daems, W. T., De Bakker, J. M. 1982. *Immunobiology* 161:204–11
40. Diggs, C. L., Osler, A. G. 1969. *J. Immunol.* 102:298–305
41. Diggs, C. L., Osler, A. G. 1975. *J. Immunol.* 114:1243–47
42. Dockrell, H. M., De Souza, J. B., Playfair, J. H. L. 1980. *Immunology* 41:421–30
43. Eaton, J. W., Eckman, J. R., Berger, E., Jacob, H. S. 1976. *Nature* 264:758–60
44. Eling, W. M. C. 1982. In *Immune Reactions to Parasites*, ed. W. Frank, pp. 141–56. Stuttgart: Gustav Fischer
45. Etkin, R. L., Eaton, J. W. 1975. In *Erythrocyte Structure and Function*, ed. G. J. Brewer, pp. 219–34. New York: Liss
46. Eugui, E. M., Allison, A. C. 1980. *Parasite Immunol.* 2:277–92

47. Eugui, E. M., Allison, A. C. 1982. In *NK Cells and Other Natural Effector Cells*, ed. R. Herberman, pp. 1491–502. New York: Academic
48. Fee, J. A., Bergamini, R., Briggs, R. G. 1975. *Arch. Biochem. Biophys.* 169:160–167
49. Felton, S. C., Prior, R. B., Spagna, V. A., Kreier, J. P. 1980. *J. Parasitol.* 66:846–47
50. Finerty, J. F., Krehl, E. P. 1976. *Infect. Immun.* 14:1103–5
51. Fitch, C. D. 1983. In *Malaria and the Red Cell, Ciba Found. Symp.*, 94:222–32. London: Pitman.
52. Flohé, L. 1979. In *Oxygen Free Radicals and Tissue Damage, Ciba Found. Symp.* 65:95–115. Amsterdam: Excerpta Medica
53. Frank, M. M. 1975. Complement. In *Current Contents Series*. Kalamazoo, MI: Upjohn
54. Frankenberg, S., Londner, M. V., Greenblatt, C. V. 1980. *Cell. Immunol.* 55:185–90
55. Freeman, R. R., Parish, C. R. 1978. *Immunology* 35:479–84
56. Freeman, R. R., Trejdosiewicz, A. J., Cross, G. A. M. 1980. *Nature* 284:366–68
57. Fridovitch, I. 1976. In *Free Radicals in Biology*, ed. W. A. Prior, 1:239–77. New York: Academic
58. Friedman, M. J. 1978. *Proc. Natl. Acad. Sci. USA* 75:1944–97
59. Friedman, M. J. 1979. *Nature* 280:245–47
60. Friedman, M. J. 1981. In *Biochemistry and Physiology of Protozoa*, ed. M. Levandowsey, S. H. Hutner, 4:463–93. New York: Academic
61. Friedman, M. J. 1983. *Malaria and the Red Cell, Ciba Found. Symp.*, 94. London: Pitman. In press
62. Grankvist, K., Marklund, S., Sehlin, J., Täljedal, I-B. 1979. *Biochem. J.* 182:17–25
63. Greenberg, J., Kendrick, L. P. 1958. *J. Parasitol.* 44:592–98
64. Greenwood, B. M., Oduloju, A. J., Stratton, D. 1977. *Trans. R. Soc. Trop. Med. Hyg.* 71:408–10
65. Grun, J. L., Weidanz, W. P. 1981. *Nature* 290:143–45
66. Gutteridge, J. M. C., Richmond, R., Halliwell, B. 1979. *Biochem. J.* 184:469–72
67. Haidaris, C. G., Bonventre, P. F. 1982. *J. Immunol.* 129:850–55
68. Herberman, R. B., ed. 1982. *NK Cells and Other Natural Effector Cells*, p. 1566. New York: Academic

69. Hercend, T., Meuer, S., Reinherz, E. L., Schlossman, S. L., Ritz, J. 1982. *J. Immunol.* 129:1299–303
70. Hill, H. A. O. 1979. In *Oxygen Free Radicals and Tissue Damage, Ciba Found. Symp.*, 65:5–11. Amsterdam: Excerpta Medica
71. Holder, A. A., Freeman, R. R. 1981. *Nature* 294:361–64
72. Hollingdale, M. R., Zavala, F., Nussenzweig, R. S., Nussenzweig, V. 1982. *J. Immunol.* 128:1929–30
73. Howard, J. G., Hale, C., Liew, F. Y. 1980. *Parasite Immunol.* 3:303–14
74. Howard, R. J. 1983. *Contemp. Topics Immunol.* 12:In press
75. Huang, K. Y., Schultz, W. W., Gordon, F. B. 1968. *Science* 162:123–24
76. Ignarro, L. J., Degnan, J. N., Baricos, W. H., Radowitz, R. J., Wolin, M. S. 1982. *Biochim. Biophys. Acta* 718:49–59
77. Jahiel, R. I., Nussenzweig, R. S., Vanderberg, J., Vilcek, J. 1981. *Nature* 220:710–11
78. Jahiel, R. I., Vilcek, J., Nussenzweig, R. S. 1970. *Nature* 227:1350–51
79. Jain, S. K., Hochstein, P. 1979. *Biochim. Biophys. Acta* 586:128–36
80. Jayawardena, A. N., Murphy, D. B., Janeway, C. A., Gershon, R. K. 1982. *J. Immunol.* 129:377–81
81. Jayawardena, A. N., Targett, G. A. T., Carter, R. L., Leuchars, E., Davies, A. J. S. 1977. *Immunology* 32:849–59
82. Johnston, R. B. Jr. 1982. In *Lymphokines*, ed. E. Pick, 3:33–56. New York: Academic
83. Langhorne, J., Butcher, G. A., Mitchell, G. H., Cohen, S. 1979. In *Role of the Spleen in the Immunology of Parasitic Diseases*, pp. 205–25. Basel: Schwabe
84. Lelchuk, R., Taverne, J., Agomo, P. U., Playfair, J. H. L. 1979. *Parasite Immunol.* 1:61–78
85. Livingston, F. B. 1967. *Hemoglobins in Human Populations*. Chicago: Aldine
86. Lötscher, H. R., Winterhalter, K. H., Carafoli, E., Richter, C. 1979. *Proc. Natl. Acad. Sci. USA* 76:4340–44
87. Lucia, H. L., Nussenzweig, R. S. 1969. *Exp. Parasitol.* 25:319–23
88. Lust, J. A., Kumar, V., Burton, R. C., Bartlett, S. P., Bennett, M. 1981. *J. Exp. Med.* 154:306–17
89. Lynch, R. E., Fridovitch, I. 1978. *J. Biol. Chem.* 253:4697–99
90. McDonald, V., Phillips, R. S. 1978. *Clin. Exp. Immunol.* 34:159–63
91. McGregor, I. A. 1972. *Brit. Med. Bull.* 28:22–27
92. McGregor, I. A., Gilles, H. M., Wal-

ters, J. H., Davies, A. H., Pearson, F. A. 1956. *Brit. Med. J.* 2:686–92
93. MacGregor, R. R., Sheagren, J. N., Wolff, S. M. 1969. *J. Immunol.* 102:131–38
94. Maegraith, B. G. 1974. In *Medicine in the Tropics,* ed. A. W. Woodruff, pp. 27–73. Edinburgh: Churchill Livingstone
95. Makimura, S., Brunkmann, V., Mossmann, H., Fischer, H. 1982. *Infect. Immun.* 37:800–4
96. Mandell, G. L., Hook, G. W. 1969. *J. Bacteriol.* 100:531–32
97. Marks, P. A., Johnson, A. B., Hirschberg, E. 1958. *Proc. Natl. Acad. Sci. USA* 44:529–32
98. Meltzer, M. S., Ruco, L. P., Boraschi, D., Nacy, C. A. 1979. *J. Reticuloendothel. Soc.* 26:403–15
99. Miller, L. H., Carter, R. 1976. *Exp. Parasitol.* 40:132–46
100. Morgan, D. M. L., Christensen, J. R., Allison, A. C. 1981. *Biochem. Soc. Trans.* 9:563–64
101. Morgan, D. M. L., Ferluga, J., Allison, A. C. 1980. In *Polyamines in Biomedical Research,* ed. J. Gaugas, pp. 303–8. New York: John Wiley
102. Müller, M. 1981. *Scand. J. Infect. Dis.* 26 (Suppl.):31–41
103. Murphy, J. R. 1981. *Infect. Immun.* 33:199–211
104. Murray, H. W. 1982. *J. Immunol.* 129:351–54
105. Murray, H. W., Masur, H., Keithly, J. S. 1982. *J. Immunol.* 129:344–50
106. Nathan, C. F., Root, R. K. 1977. *J. Exp. Med.* 146:1648–62
107. Nussenzweig, M. D., Steinman, R. M., Gutchinkov, B., Cohn, E. A. 1980. *J. Exp. Med.* 152:1070–84
108. Nussenzweig, R. S. 1967. *Exp. Parasitol.* 21:224–31
109. Nussenzweig, R. S., Vanderberg, J., Most, H., Orton, C. 1969. *Nature* 222:488–89
110. Ochoa, S. 1981. *Eur. J. Cell. Biol.* 26:212–16
111. Ojo-Amaize, E. A., Salimonu, L. S., Williams, A. I. O., Akinwolere, O. A. O., Shabo, R., Alm, G. V., Wigzell, H. 1981. *J. Immunol.* 127:2296–300
112. Pasvol, G., Wetherall, D. J., Wilson, R. J. M. 1978. *Nature* 274:701–3
113. Playfair, J. H. L. 1982. *Brit. Med. Bull.* 38:153–59
114. Playfair, J. H. L., De Souza, J. B., Cottrell, B. J. 1977. *Immunology* 32:681–87
115. Potocnjak, P., Yoshida, N., Nussen-
zweig, R. S., Nussenzweig, V. 1980. *J. Exp. Med.* 151:1504–13
116. Quinn, T. C., Wyler, D. J. 1979. *J. Clin. Invest.* 63:1187–94
117. Rabinovitch, M., Dedet, J. P., Ryter, A., Robineaux, R., Topper, G., Brunet, E. 1982. *J. Exp. Med.* 155:415–31
118. Ralph, P., Williams, N., Moore, M. A. S., Litcofsky, P. B. 1982. *Cell. Immunol.* 71:215–23
119. Rank, R. G., Weidanz, W. P. 1976. *Proc. Soc. Exp. Biol. Med.* 151:257–59
120. Rener, J., Carter, R., Rosenberg, Y., Miller, L. H. 1980. *Proc. Natl. Acad. Sci. USA* 77:6797–99
121. Rigdon, R. H., Micks, D. W., Breslin, D. 1950. *Am. J. Hyg.* 52:308–22
122. Roberts, D. W., Rank, R. G., Weidanz, W. P., Finerty, J. F. 1977. *Infect. Immun.* 16:821–26
123. Roberts, D. W., Weidanz, W. P. 1978. *Infect. Immun.* 20:728–31
124. Roberts, D. W., Weidanz, W. P. 1979. *Am. J. Trop. Med. Hyg.* 28:1–3
125. Roder, J. C., Helfand, S. L., Werkmeister, J., McGarry, R., Beaumont, T. J., Duwe, A. 1982. *Nature* 298:569–72
126. Rose, C. S., György, P. 1952. *Am. J. Physiol.* 168:414–20
127. Ruebush, M. J., Burgess, D. E. 1982. In *NK Cells and Other Natural Effector Cells,* ed. R. Herberman, pp. 1483–89. New York: Academic
128. Ruff, M. R., Gifford, G. E. 1981. In *Lymphokines,* ed. E. Pick, 2:235–72. New York: Academic
129. Sauvagier, F., Fauconnier, B. 1978. *Biomedicine* 29:184–87
130. Sawyer, D. T., Valentine, J. S. 1981. *Acc. Chem. Res.* 14:393–400
131. Scheibel, L. W., Adler, A. 1980. *Mol. Pharmacol.* 18:320–25
132. Scheibel, L. W., Ashton, S. H., Trager, W. 1979. *Exp. Parasitol.* 47:410–18
133. Schultz, W. W., Huang, K. Y., Gordon, F. B. 1968. *Nature* 220:709–10
134. Segal, A. W., Allison, A. C. 1979. In *Oxygen Free Radicals and Tissue Damage, Ciba Found. Symp.,* 65:205–19. Amsterdam: Exerpta Medica
135. Sergent, E. 1963. In *Immunity to Protozoa,* ed. P. C. C. Garnham, A. E. Pierce, I. M. Roitt, pp. 39–47. Oxford: Blackwell Sci.
136. Shear, H. L., Nussenzweig, R. S., Bianco, C. 1979. *J. Exp. Med.* 149:1288–98
137. Srivastava, S. K., Awasthi, Y. C., Beutler, E. 1974. *Biochem. J.* 139:289–95
138. Stechschulte, D. J. 1969. *Proc. Soc. Exp. Biol. Med.* 131:748–52

392 ALLISON & EUGUI

139. Stevenson, M. M., Lyang, J. J., Ska-mene, E. 1983. *J. Immunol.* In press
140. Taliaferro, W. H. 1929. *The Immunology of Parasite Infections.* New York: Century
141. Taliaferro, W. H., Taliaferro, L. G. 1944. *J. Infect. Dis.* 75:1–17
142. Taylor, D. W., Bever, C. T., Rollwagen, F. M., Evans, C. B., Asovsky, R. 1982. *J. Immunol.* 128:1854–59
143. Taylor, D. W., Crum, S. R., Kramer, K. J., Siddiqui, W. A. 1980. *Infect. Immun.* 28:502–7
144. Thomas, M. J., Mehl, K. S., Pryor, W. A. 1982. *J. Biol. Chem.* 257:8343–47
145. Tosta, C. E., Wedderburn, N. 1980. *Clin. Exp. Immunol.* 42:114–20
146. Trotta, R. J., Sullivan, S. G., Stern, A. 1982. *Biochem. J.* 204:405–15
147. Udeinya, I. J., Schmidt, J. A., Aikawa, M., Miller, L. H., Green, I. 1981. *Science* 213:555–57
148. Van Furth, A., Cohn, Z. A., Hirsch, J. G., Humphrey, J. H., Spector, W. G., Langevoont, H. L. 1973. *Bull. Wld. Hlth. Org.* 46:845–50
149. Volkman, A. 1976. *J. Reticuloendoth. Soc.* 19:249–68
150. Weed, R. I. 1961. *J. Clin. Invest.* 40:140–43
151. Weinbaum, F. I., Evans, C. B., Tigelaar, R. E. 1976. *J. Immunol.* 116:1280–83
152. Weinbaum, F. I., Evans, C. B., Tige-laar, R. E. 1976. *J. Immunol.* 117:1999–2005
153. Weinberg, J. B. 1981. *Science* 213:655–57
154. Weiss, L. 1983. *J. Parasitol.* In press
155. Weiss, L. P. 1979. In *Role of the Spleen in the Immunology of Parasitic Diseases,* ed. G. Torrigiani, pp. 6–20. Basel: Schwabe
156. Weiss, S. J., Ward, P. A. 1982. *J. Immunol.* 129:309–13
157. Wilson, R. J. M. 1977. In *Immunity in Parasitic Diseases, Colloq. INSERM Paris* 72:87–101
158. Wood, P. R., Clark, I. A. 1982. *Infect. Immun.* 35:52–57
159. Wood, P. R., Clark, I. A. 1982. *Parasite Immunol.* 4:319–27
160. Woodruff, M. F. A., Warner, N. L. 1977. *J. Natl. Cancer Inst.* 58:111–16
161. Wyler, D. J. 1976. *Clin. Exp. Immunol.* 23:471–76
162. Wyler, D. J., Gallin, J. I. 1977. *J. Immunol.* 118:478–84
163. Wyler, D. J., Miller, L. H., Schmidt, L. H. 1977. *J. Infect. Dis.* 135:86–93
164. Wyler, D. J., Oster, C. N., Quinn, T. C. 1979. In *Role of the Spleen in the Immunology of Parasitic Diseases,* ed. G. Torrigiani, pp. 183–204. Basel: Schwabe
165. Wyler, D. J., Quinn, T. C., Chen, L. T. 1981. *J. Clin. Invest.* 67:1400–4
166. Wyler, D. J., Weinbaum, F. I., Herrod, H. R. 1979. *J. Infect. Dis.* 140:215–21

Ann. Rev. Immunol. 1983. 1:393–422
Copyright © 1983 by Annual Reviews Inc. All rights reserved

BIOSYNTHESIS AND REGULATION OF IMMUNOGLOBULINS

Randolph Wall

Molecular Biology Institute and Department of Microbiology and Immunology, University of California School of Medicine, Los Angeles, California 90024

Michael Kuehl

National Cancer Institute-Naval Medical Oncology Branch, National Naval Medical Center, Tower 1, Room 415, Bethesda, Maryland 20814

INTRODUCTION

Immunoglobulin gene expression exhibits a number of intriguing regulatory features. Immunoglobulin genes undergo dynamic DNA rearrangements and somatic mutations. The expression of immunoglobulin genes exhibits allelic exclusion and isotypic exclusion. The developmental regulation of immunoglobulin gene expression during B cell differentiation is characterized by (*a*) the differential onset of heavy and light chain production, (*b*) great changes in the level of expression, and (*c*) transition from the insertion of immunoglobulin as antigen receptors in the lymphocyte membrane to its active secretion.

Darnell (40, 41) and Brown (27) have recently reviewed the variety of transcriptional, post-transcriptional (e.g. RNA processing), and translational mechanisms employed in the regulation of eukaryotic gene expression. This review summarizes recent progress towards understanding the molecular mechanisms in immunoglobulin gene expression and regulation in the context of the developments in other eukaryotic genes. The complex array of lymphoid cell and lymphokine factor interactions that affect immunoglobulin gene expression are not considered here.

393

0732-0582/83/0410-0393$02.00

IMMUNOGLOBULIN STRUCTURE

Most immunoglobulin molecules consist of two identical light (L) chains and two identical heavy (H) chains (74). Immunoglobulin L and H chains are encoded by three unlinked gene families—one each for κ and λ L chains, and one for H chains—which are present on mouse chromosomes 6, 16, and 12, respectively (reviewed in 43). Each chain is composed of a series of homologous domains, the NH_2-terminal domain is the variable (V) region and the remainder form the constant (C) region. The combination of H and L-chain V regions are responsible for antigen binding and may include thousands of different sequences. In contrast to V regions, the C regions have only a few alternative sequences. The various classes (IgM, IgD, IgG, IgE, IgA) and subclasses (e.g. IgG1, IgG2, IgG3, IgG4) of immunoglobulins with different biological functions are distinguished by different heavy chains (μ, δ, γ, ϵ, α) defined by their C regions (C_μ, C_δ, C_γ, C_ϵ, C_α). All classes of immunoglobulin contain light chains of either κ or λ isotype (containing C_κ or C_λ regions).

The immunoglobulin (Ig) molecule can exist in two very different environments: in the lymphocyte cell membrane as a surface antigen receptor, and in the circulation as a secreted antibody. For most immunoglobulin classes, the Ig molecules are secreted as four-chain disulfide-linked monomers (H_2L_2). However for IgM and IgA, the heavy chains carry a short COOH-terminal sequence, which permits polymerization of the monomers. These polymeric immunoglobulin molecules are covalently linked by disulfide bonds from their H-chain COOH-terminal segments to a single J (or joining) chain molecule per IgM or IgA polymer (85, 100). Thus, IgM is secreted as a pentamer $(\mu_2L_2)_5J$ and IgA can be secreted as a monomer (α_2L_2), dimer $(\alpha_2L_2)_2J$, or trimer $(\alpha_2L_2)_3J$.

Immunologists have long known that the proliferation and maturation of B lymphocytes in response to antigen (and perhaps other external signals such as anti-idiotypic antibody) is triggered by binding of antigen to membrane Ig displayed on the surfaces of B cells. It is likely that all classes and subclasses of Ig can exist in a membrane form as well as a secreted form. In contrast to secreted Ig molecules, all membrane Ig molecules, including IgM and IgA, are present as monomeric structures (H_2L_2) (110). Studies from a number of laboratories have demonstrated that membrane μ (μ_m) and secreted μ (μ_s) chains have different molecular properties. As would be expected for integral membrane proteins, μ_m chains exhibit hydrophobic behavior (123, 176). Protein and nucleic acid structural studies demonstrate identical sequences up to their COOH-terminal segments, but unique COOH-terminal sequences for both μ_s and μ_m H chains (2, 48, 77, 136, 196). More recent studies have shown different membrane and secreted species for murine γ_3, γ_1, γ_{2a}, γ_{2b}, δ, and α H chains (58, 82, 88, 97, 102, 112, 116, 121, 159, 188). The COOH-terminal sequences of the membrane forms of γ_3, γ_1, γ_{2a}, γ_{2b}, δ, and α H chains have been inferred from nucleic acid sequences of cDNA or genomic DNA clones (33, 125, 140, 171, 189, 193). The COOH-terminal sequence of the membrane forms of H chains includes (a) an acidic segment of 12–25 amino acids, which has a net charge of -6; (b) a highly conserved sequence of 26 amino acids, which is designated the hydrophobic segment and is thought to anchor the H chain into the plasma membrane; and (c) an intracellular segment of variable length, which begins with the residues Lys-Val-Lys. The transmembrane peptide sequences of all H chain M exons show an unexpectedly high level of sequence conservation. This sequence conservation is the basis of a two-chain model for transmembrane insertion of surface immunoglobulins (136, 140). The short, positively charged intracellular sequence constitutes the complete intracellular segment of μ and δ H chains, whereas γ_1, γ_{2b}, and γ_{2a} (and presumably γ_3) H chains have conserved intracellular segments of 28 residues, which include the Lys-Val-Lys sequence. It is not known if the different intracellular segments exert different activation functions in cells expressing membrane IgG or IgA as compared to cells expressing membrane IgM or IgD.

DIFFERENTIAL IMMUNOGLOBULIN EXPRESSION DURING B-LYMPHOCYTE DEVELOPMENT

The developmental stages in B-lymphocyte development into Ig-secreting plasma cells are distinguished by differences in the organization of H and L-chain genes, the kinds of Ig chains expressed, and the levels of Ig gene expression (see Figure 1). It appears that a pluripotential hematopoietic stem cell generates lymphoid stem cells that subsequently generate pre-B cells, i.e. the first stage of B cell development in which cells express immunoglobulin chains (29, 95, 154). As a lymphoid stem cell differentiates to a pre-B cell, there is μ H chain gene rearrangement, transcription, translation, and expression of cytoplasmic μ H chains, but no expression of L chains. Early pre-B cells have not rearranged their L-chain genes, but are proliferating cells presumed eventually to rearrange and express different L-chain genes together with the H chain to which the pre-B cell is already committed (29, 95, 134b). A possible late pre-B cell developmental stage is suggested by several recent examples of pre-B-cell lines that have functionally rearranged κ L chain genes, but do not transcribe L chain mRNA until the cells are stimulated by bacterial lipopolysaccharide (101a, 128, 142a). This presumed late pre-B-cell developmental stage (i.e., with L chain DNA rearrangement but no L chain expression) has not been clearly identified in normal cell populations and may represent a transient stage of normal B cell development.

As the pre-B cell develops to an immature B cell, L chains are expressed and IgM monomers (H_2L_2) are displayed on the cell surface as antigen receptors (177). The immature B cell matures to a B cell that co-expresses IgM and IgD on the B lymphocyte surface (57). Co-expressed μ and δ chains in IgM and IgD, respectively, have the same V region, and their expression exhibits allelic exclusion (36, 64, 133). Up to this stage, B-lymphocyte development is antigen independent, although immature B cells that express surface IgM are presumably capable of being affected by contact with antigen or anti-idiotypic antibody. Maturation beyond the B cell stage is thought to require activation by mitogens or an appropriate combination of antigen, T cells or T cell factors, and macrophages (100a, 109a). The activated B cell that secretes significant amounts of IgM matures into either a memory B cell that continues to express surface Ia antigens but expresses a new isotype of surface immunoglobulin (IgG, IgE, or IgA), or a terminally differentiated plasma cell that no longer expresses surface Ia antigens but expresses large quantities of secreted IgM, IgG, IgE, or IgA with little if any surface Ig (190, 122). Antigenic stimulation of memory B cells generates plasma cells secreting IgG, IgE, or IgA.

Either memory B cells or plasma cells can express immunoglobulin isotypes other than IgM or IgD. The expression of different H chain C regions (C_γ, C_ϵ, or C_α) occurs with the same V region associated with C_μ in IgM on the surface of immature B cells and is a result of a process called H chain "class switching." Heavy chain class switching is thought to be an antigen- and T cell-dependent event mediated by DNA rearrangements that delete the C_μ region and bring the H-chain V region next to a C_γ, C_ϵ, or C_α region gene segment (42, 71, 191, 192).

In addition to differential expression of H and L chains, there is differential expression of J chain during B-cell development (100). There is little or

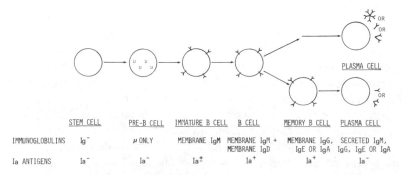

	STEM CELL	PRE-B CELL	IMMATURE B CELL	B CELL	MEMORY B CELL	PLASMA CELL
IMMUNOGLOBULINS	Ig⁻	μ ONLY	MEMBRANE IgM	MEMBRANE IgM + MEMBRANE IgD	MEMBRANE IgG, IgE OR IgA	SECRETED IgM, IgG, IgE OR IgA
Ia ANTIGENS	Ia⁻	Ia⁻	Ia±	Ia⁺	Ia⁺	Ia⁻

Figure 1 A simplified scheme of B-cell differentiation showing stages in immunoglobulin and Ia antigen gene expression (100a).

no J chain expression in the pre-B cell, immature B cell, or B cell stages of development, but J chain expression is prominent in all plasma cells and apparently occurs in all B cells that secrete immunoglobulins (100, 105a, 134). Similar to the situation for H chain and L chain expression, J chain expression is elevated substantially (ca 10- to 100-fold) if one compares cell lines that represent early and late stages of B cell development.

Tumors of the B cell lineage corresponding to various developmental stages have proven to be of great value in the analysis of immunoglobulin gene expression. The pre-B cell is represented by the carcinogen-induced 70Z cell line (122a), as well as by Abelson virus-transformed lymphoid cell lines (6, 142a). Immature B cells, B cells, and memory B cells are represented by B cell lymphomas (57, 92, 100a, 159, 180, 188). Plasma cells are represented by plasmacytomas (myeloma) tumors (132). Needless to say, it is critical to study normal cells from different B lymphocyte developmental stages to verify that cell lines and tumors represent valid models for B lymphocyte development. Although some studies of normal cells have been possible, isolation of pure populations of cells representing a particular B lymphocyte developmental stage previously has been difficult (155, 164). The recent development of continuous cloned lines of normal mouse lymphoid cells and their precursors is an extremely important breakthrough for future studies on B cell development (72, 183).

STRUCTURAL FEATURES IN IMMUNOGLOBULIN GENE CONTROL

Regulation of Immunoglobulin Gene Formation: Allelic and Isotypic Exclusion

Functional H chain and L chain expression is subject to allelic exclusion, in both normal B cells and B-cell tumors or tumor cell lines. Allelic exclusion is reflected in the random expression within a single cell of either the paternal or the maternal allele for each H or L chain. Although rare exceptions that may violate allelic exclusion have been reported for B-cell tumors, it is likely that allelic exclusion applies to greater than 99% of B cells. Allelic exclusion has not been described for other eukaryotic autosomal genes. In addition to allelic exclusion, functional L chain expression in normal B cells and B-cell tumors or cell lines is subject to isotypic exclusion, so that a single cell generally expresses either κ or λ L chains. Although some cloned murine B cell lines simultaneously express κ and λ L chains, it remains to be determined if both kinds of L chain are functional (3, 46, 92). As a result of allelic exclusion and L chain isotypic exclusion, a single B cell usually expresses functional immunoglobulin molecules (comprised of associated H chains and L chains) with only one combining site specificity.

As a consequence of the Dreyer & Bennett (45) hypothesis that immuno-globulin genes are created somatically by joining V and C gene segments, it was proposed that the creation of a joined V + C gene may generate a signal (perhaps the immunoglobulin chain itself) that inhibits subsequent joining of V and C genes in the same family (45, 141). The finding that cloned B cells can simultaneously express both normal and abnormal L chains, apparently encoded by two rearranged genes, required modification of this idea so that the signal that prevents further joining events is not activated by an abnormal joining event but only by a normal joining event. A large number of studies from many laboratories have resulted in substantial support for this modified hypothesis to explain allelic exclusion for both H and L chain as well as L-chain isotypic exclusion.

Analysis of L-chain genes in normal B cells, hybridomas, and B-cell tumors or cell lines representing different stages of B cell development can be briefly summarized: (a) In approximately two thirds of normal or transformed B cells producing κ L chains, one κ allele is functionally rearranged and one allele remains in a germline configuration; (b) in approximately one third of normal or transformed B cells, one κ allele is functionally rearranged and one allele is aberrantly (see below) rearranged or deleted; (c) with one exception, all κ alleles are deleted or aberrantly rearranged in normal or transformed B cells that express functional λ L chains but no functional κ L chain; (d) λ alleles remain in a germline configuration in B cells that express a functional κ L chain; and (e) λ alleles can be aberrantly rearranged in B cells that express a functional λ L chain (3, 37, 46, 84, 118). Aberrant rearrangements of κ (or λ) alleles can be of several types: (a) joining of a normal V gene segment to a non-J or pseudo-J segment: (b) joining of a pseudo-V or non-V segment to a normal J segment: and (c) joining of a normal V segment to a normal J segment so that codons are deleted from the VJ junction or an out-of-phase reading frame is established beyond the VJ junction (3, 7, 17, 22, 34, 46, 87, 90, 108, 129, 148–150). In addition, several possible examples (grouped with aberrant rearrangements above) of B cells express a normal κ (or λ) L chain and also express an abnormal second L chain despite an apparently correct VJ joining event (22, 148). Aberrantly rearranged L chain genes may be transcribed and translated, but they do not encode functional L chains (i.e. L chains that efficiently assemble to H chains for expression as membrane or secreted immunoglobulin).

These data on L-chain gene expression have led to two current hypotheses regarding L chain isotypic and allelic exclusion. First, it has been proposed that there is a hierarchy of L-chain gene formation in which κ alleles are rearranged before λ alleles. However, more complicated possibilities cannot be ruled out with certainty (e.g., κ genes are specifically deleted or can undergo aberrant rearrangements in B cells that express a functional λ L chain). Second, Alt et al (3) have proposed that further L-chain gene rearrangements are actively prevented when a functional L chain (see above) is expressed. However, the possibility that a functional VJ junctional DNA or RNA sequence somehow provides the signal to prevent further L chain rearrangements cannot be rigorously disproved at present.

Although less is known about allelic exclusion of H-chain expression, some differences and similarities to L chain are apparent: (a) In 90% or more of normal or transformed B cells representing a variety of B cell developmental stages expressing H chains, both H-chain alleles are rearranged (presumably one functionally and the other aberrantly); (b) in 90% or more of Abelson virus-transformed null lymphoid cell lines, all detectable H chain alleles are rearranged (presumably aberrantly) or deleted, even though no functional H chain is expressed; (c) an Abelson virus-transformed lymphoid cell line that has undergone a potentially functional D-J_H joining event generates subclones with different function V_H-D-J_H combinations; (d) L-chain genes are never present in a rearranged form unless at least one H chain gene is rearranged; and (e) as noted for L-chain genes, transcription and translation of some aberrantly rearranged H-chain genes can occur (1, 5, 6, 29, 131, 140, 179, 193).

These results have led to the following proposals for regulation of H chain gene formation. First, H-chain genes are rearranged at an earlier stage of B-cell development than are L-chain genes. Second, formation of functional H-chain genes is much less efficient than is formation of functional L chain genes, possibly as a consequence of the more complex events involved (i.e. $D \rightarrow J_H$ plus $V_H \rightarrow D$ joining, with the possibility of $D \rightarrow D$ joining and somatic mutation as well). A significant population of null B lymphoid cells seems to result from the inefficient formation of functional H-chain genes (6). Third, functional H-chain gene formation seems to be necessary before further B-cell development and L-chain rearrangement can occur. Fourth, allelic exclusion of H chain may be an active process in which expression of a functional H chain prevents additional H-chain gene rearrangements. The possiblity that functional V_H-D-J_H junctional DNA or RNA sequences provide a signal for allelic exclusion seems unlikely on the basis of a recent report (5).

In summary, although the nature and hierarchy of H and L chain gene rearrangements are relatively well established, more direct proof of the hiearchy of κ and λ L-chain gene rearrangements and of allelic exclusion of immunoglobulin gene expression is needed. In addition, much remains to be learned about the molecular mechanisms and events that control these processes.

Transcriptional Control Signals in Immunoglobulin Genes

The initiation sites for both light and heavy chain transcription units exhibit "promoter" elements similar to these in other eukaryotic genes (reviewed in 25) located 5' to the leader coding sequence in V-region gene segments (Figure 2). These sequences function in phasing correct initiation of transcription (25). These sequences in immunoglobulin genes are not altered by somatic mutation, even though nearby V regions show up to 5% base changes (35). Thus, somatic mutations of V regions is apparently not a mechanism used in controlling levels of immunoglobulin gene transcription.

Enhancer sequences are a newly discovered class of control elements, which appear to affect the efficiency or level of transcription and which are located several hundred nucleotides upstream (i.e. to the 5' side) of the "TATA box" and cap site in the few viral and cellular genes where they have been identified (10, 11, 44, 61, 62, 94, 111, 165, 166). These so-called activator or enhancer sequences may be involved in RNA polymerase II recognition of eukaryotic transcription units in chromosomes, and appear to affect the efficiency of transcription (10, 11, 25, 44, 61, 62, 94, 111, 165, 166). Two features also suggest that these regions may be important for specific gene regulation. The first is the finding that these enhancer sequences exhibit no significant sequence homology, as would be expected for specific rather than universal gene control regions such as the TATA box. Second, in virus systems these enhancer sequences are functionally interchangeable (10, 44, 94) and their activity is correlated with host range restriction and the specificity of viral gene expression. Polyoma and SV40 mutants with altered host range from the wild-type virus have minor sequence alterations exclusively restricted to these specific regions (54, 63, 76, 152).

The enhancer sequences in viral genomes and in cellular histone genes are located some distance 5' of the TATA box and cap site (61, 62). Several lines of evidence raise the intriguing possibility that enhancer sequences in immunoglobulin genes might not be conventionally located in the flanking DNA 5' to the TATA box but instead are associated with the C region. First, immunoglobulin germline V regions contain 5' flanking sequences with all of the general signals (TATA box, cap site) needed for correct initiation of transcription (13, 18, 21, 75, 79, 108, 167), because specific initiation of transcription is observed in vitro and in *Xenopus* oocytes (13). Nonetheless, unrearranged V regions are not transcribed in vivo (106), whereas rearranged V + C genes are actively transcribed in myeloma cells (56, 104). Second, unrearranged C_κ regions are transcribed in myeloma cells at levels comparable to active rearranged light chain genes (173). Tran-

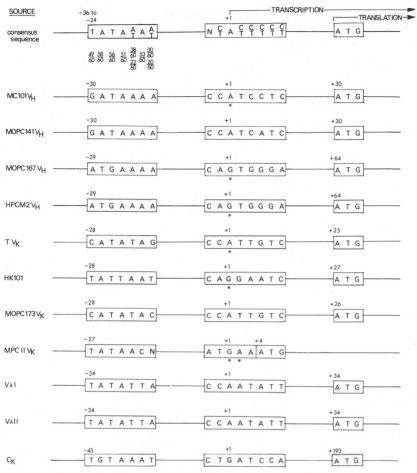

Figure 2   Comparison of the 5' flanking sequences of representative immunoglobulin V region gene segments. Selected immunoglobulin light and heavy chain sequences are aligned to indicate the conserved placement of presumed promoter sequences in relation to the consensus sequence of Breathnach & Chambon (25). Experimentally determined transcription initiation sites (i.e. cap sites) are denoted by asterisks. The other immunoglobulin sequences shown are aligned on the basis of homology with experimentally established transcription initiation sites. References for V region 5' flanking sequences are $MC101V_H$ and $MOPC141V_H$ (75), $MOPC167V_H$ and $HPCM2V_H$ (35), T V_κ (21), HK101 (13), $MOPC173V_\kappa$ (108), $MPC11V_\kappa$ (79), $V_{\lambda I}$ (18), $V_{\lambda II}$ (167), C_κ (173).

scripts of the unrearranged C_κ-gene segment begin near a sequence strongly resembling a TATA box. This strongly suggests that the C_κ (and possibly other immunoglobulin C regions) may contain an enhancer-like sequence. Finally, sequence comparisons of mouse and human J + C_κ germline gene segments have revealed a region of strikingly conserved sequence homology approximately 0.7 kb 5' to the C_κ region (68). This is striking because the only other areas of sequence conserved between the mouse and human genes are the J and C_κ coding regions. Interestingly, we have noted that this highly conserved intron sequence 5' to the C_κ region contains short sequences related to the SV40 enhancer sequences and to the specific polyoma enhancer sequences from polyoma host range variants (R. Deans, R. Wall, personal communication). This conserved intron sequence is present in an aberrantly rearranged κ light chain gene (from MPC 11), which is the smallest active immunoglobulin transcription unit known. This truncated κ gene is as actively transcribed as the normally rearranged V + C_κ gene in MPC 11 cells (34, 150). Heavy chains also appear to contain control sequences in the $J_H \rightarrow C_H$ intron sequence (5, 35, 80, 81). Interestingly, large deletions in this intron region in 18–81 pre-B cell lines are correlated with the loss of μ expression, which is restored by LPS stimulation (4). If this novel prediction is confirmed, then the V region (with the general promoter elements essential for correct transcript initiation) and the C region (with enhancer function) both contribute important functions for activating immunoglobulin genes.

Methylation and Gene Expression

IMMUNOGLOBULIN GENES

The demethylation of methylated cytosines in chromosomal DNA is closely correlated with gene expression as in many eukaryotic gene systems (reviewed in 134a, 139, 185). Most of the methylation in mammalian DNA is 5-methylation of cytosines in the symmetrical dinucleotide C-G, and most of the C-G sequences contain 5-methylcytosine (C^m-G). The sequence C-C-G-G is cut by restriction enzymes Hpa II and Msp I, whereas C-C^m-G-G is cut only by Msp I. Therefore, parallel Southern blots (161) of chromosomal DNA cut with HpaII or Msp I can be used to show whether these restriction sites are methylated or not.

   Methylation patterns have been analyzed in mouse heavy chain genes (C_μ, C_δ, and $C_{\gamma 1}$, $C_{\gamma 2b}$) that are activated by different mechanisms. The C_μ gene is expressed as a result of the DNA rearrangements that join the V_H, D_H, and J_H gene segments (V region formation) (reviewed in 46). The C_δ gene is expressed by transcription through from the rearranged C_μ gene, in a VDJ_H-C_μ-C_δ complex transcription unit (33, 102, 112). The $C_{\gamma 1}$ and $C_{\gamma 2b}$ genes are expressed by means of DNA rearrangements, which bring the VDJ_H close to either the $C_{\gamma 1}$ or $C_{\gamma 2b}$ gene segment (class switching) (reviewed in 46, 103).

The HpaII/MspI restriction mapping technique has been used to determine the methylation of immunoglobulin heavy chain genes. The C_μ, C_δ, $C_{\gamma1}$, $C_{\gamma2b}$, and C_α regions all are methylated in cells that do not express them but that are demethylated when they are expressed (39, 139, 190). In particular, the δ gene remains methylated, and thus presumably untranscribed, in a lymphoma cell line that probably represents an early stage of B cell differentiation and produces only μ heavy chains (139). Because μ and δ RNAs are co-transcribed from a single complex transcription unit at a later stage of B-cell differentiation, this finding suggests that the μ-plus-δ-complex transcription unit is of variable length regulated by C_δ methylation.

These results show complete correlation between demethylation and expression of immunoglobulin C_H genes, in agreement with results in numerous other eukaryotic gene systems (reviewed in 134a, 139, 185, 190).

Selective demethylation of one heavy chain C-region allele represents one possible molecular mechanism for allelic exclusion. However, in the B cell lymphoma W279, which makes only a single allotype of μ chain (153a, 180), both alleles of the C_μ gene are demethylated. Furthermore, both $C_{\gamma1}$ regions are also fully demethylated in IgG-secreting P3K myeloma cells (139). If these findings in established tumor cell lines are relevant to regulation in normal B cells, it would appear that selective demethylation of one chromosome is not the mechanism of allelic exclusion.

J-CHANGING

Secreted pentameric IgM (and multimeric IgA) is assembled by covalent linkage to a small protein called the J chain. J-chain synthesis is induced in B cells after antigenic stimulation (85, 100). Its production is initiated by the onset of J-chain gene transcription (105a). J-chain genes are not rearranged when germline DNA is compared to DNA from lymphoid cells expressing J chain. The J-chain gene is methylated in tumor cell lines representative of pre-B cells; immature and mature B-lymphocytes do not express J chain. However, the J-chain gene is demethylated in immunoglobulin-secreting cell lines representative of plasma cells (190). J-chain expression is thus closely correlated with demethylation, just as are immunoglobulin genes.

It is not clear whether or not demethylation precedes immunoglobulin gene and J-chain gene transcription and is a mechanism that initiates expression. However, recent studies showing that estrogen induction causes demethylation of the 5'-flanking regions of hen vitellogenin (185) and that in vitro methylation of a single HpaII site in SV40 inhibits late but not early

SV40 expression (53) strongly support the likelihood that demethylation (at least in 5' control regions) precedes the onset of transcription.

Chromatin Structure in Immunoglobulin Gene Expression

As first demonstrated by Weintraub & Groudine (181), genes in chromosomes that are expressed (i.e. active chromatin) are preferentially digested by treatment of nuclei with the sequence nonspecific nuclease DNase I. Subsequent studies in a number of eukaryotic gene systems have confirmed that transcribed regions, as well as considerable flanking sequences of expressed genes, are more sensitive to nuclease digestion than are unexpressed genes or bulk chromatin (reviewed in 99, 107). Because immunoglobulin genes undergo both productive and aberrant DNA rearrangements, it is of interest to compare Ig gene expression and chromatin configuration in germline and various rearranged states in different lymphoid cells. Storb et al (163) have examined DNase I sensitivity of light-chain nuclear V_κ and C_κ gene segments in nuclei from myeloma cells (κ^+), B-lymphoma cells (κ^+), and liver cells (κ^-). They found that C_κ genes are sensitive to DNase I in both rearranged and germline configurations in myeloma tumors and a B lymphoma. Unrearranged V_κ regions are insensitive to DNase I, but become DNase I-sensitive when rearranged. No difference in DNase I sensitivity could be detected between the two rearranged C_κ regions of MOPC-21. In this cell line, one C_κ gene is productively rearranged and makes functional κ light chains; the other C_κ gene is aberrantly rearranged with a misalignment of V-J, which leads to a shift in translation reading frame (177a). This nonproductive C_κ gene is transcribed into a nonfunctional κ mRNA.

Restriction enzyme digestion of nuclei has shown that rearranged expressed C_κ regions are relatively more accessible to enzyme than are germline C_κ regions (130).

Many expressed eukaryotic genes in nuclei also contain sites (often in the flanking sequences 5' to the TATA box) that inhibit nuclease sensitivity 10-fold greater than transcribed gene sequences, which in turn are 10-fold more sensitive than is bulk chromatin (reviewed in 49, 99, 107). These so-called hypersensitive sites are apparently nucleosome-free in viral minichromosomes, and include the enhancer control regions important in regulating transcription (63, 73, 145, 174, 175). Furthermore, nuclease-hypersensitive sites in certain eukaryotic chromosomal genes contain DNA sites whose methylation pattern is correlated with tissue-specific gene expression (70, 99).

Chromatin from T lymphocytes from thymus exhibits a DNase I hypersensitivity in the sequences between C_μ and J_H (162). This may be related to the widespread occurrence of aberrant μ RNA species in T cells. These μ RNAs do not contain V_H regions and appear to initiate 5' to the J_H

region, where Clarke et al (35) have evidence for a transcription origin in unrearranged μ genes from certain myeloma cells.

Parslow & Granner (124) have recently determined the DNase I hypersensitivity patterns of the C_κ regions in the pre-B-cell line, 70Z (124). The 70Z cells only synthesize detectable κ mRNA with LPS (128), even though they contain a productively rearranged V + C_κ gene (in addition to an unrearranged C_κ region) (101a). LPS activation of κ mRNA transcription was associated with the appearance of a DNase I hypersensitive site approximately 0.7 kb 5' to the C_κ region exon (124). This sequence exhibits the hypersensitivity of known enhancer regions and coincides with the highly conserved sequence in the J → C_κ introns of mouse and human C_κ regions (68). This study found no evidence of hypersensitive sites 5' to the V region in the LPS-activated light chain gene. This interesting result further supports the proposal that enhancer sequences in immunoglobulin genes might be adjacent to the C regions. Weischet et al (182) have reported that DNase I hypersensitive sites are present in the 5' sequences upstream of both transcribed and untranscribed rearranged light chain genes in a myeloma cell.

CONTROL OF IMMUNOGLOBULIN GENE EXPRESSION BY RNA PROCESSING

Membrane and Secreted Heavy Chains

Membrane and secreted immunoglobulin heavy chains are coded by heavy chain mRNA species with different 3' ends generated by RNA processing mechanisms (reviewed in 140, 177). This was first shown for μ mRNA. Various B lymphoma cells and myeloma cells were found to contain two prominent μ mRNA's at 2.7 and 2.4 kb, which coded for membrane (μ_m) and secreted (μ_s) chains (2, 80, 128, 136, 158). Analysis of μ cDNA clones of both μ mRNA species revealed that the μ_s and μ_m mRNA's were identical through the end of the $C_\mu 4$ constant region coding sequence, but thereafter contained very different COOH-terminal coding segments and 3'-untranslated regions (3'-UT) (2, 136). The μ_s COOH-terminal sequence encoded a 20-residue hydrophilic segment identical to the μ_s COOH-terminal amino acid sequence determined by Kehry et al (1978). The μ_m COOH-terminal sequence encodes 41 residues (designated the M or membrane region) with the properties of a transmembrane protein (48, 136).

The locations of the COOH-terminal segments of μ_m or μ_s mRNA's were established by R-loop electron microscopy and nucleotide sequencing of the cloned μ gene (48, 136). The secreted COOH-terminus coding sequence and the 3' untranslated region of μ_s mRNA is encoded contiguous with 3' end of the $C_\mu 4$ domain (Figure 2). The μ_m COOH-terminus and 3'-UT region is encoded in two exons (the M exons) located 3' to the

$C_\mu 4$ domain. These two M exons are joined to the $C_\mu 4$ domain by two RNA splices that replace the μ_s COOH-terminus and 3'-UT region sequences to generate μ_m mRNA. The chromosomal gene codons at the 3' end of the $C_\mu 4$ domain where the M exons are spliced have the sequences G/GTAAA encoding Gly-Lys. This sequence in the μ chromosomal gene (30, 48) is identical to the consensus sequence for an "upstream" RNA splicing site, G/GTAAG (93, 138, 151). The underlined GT has been universally found at the exon/intron juncture of "upstream" splicing sites in eukaryotic genes (reviewed in 25).

Because all immunoglobulin heavy chains contain the sequence Gly-Lys at the end of the last domain encoded by a potential site for RNA splicing (G/GTAAA or G/GTAAG), Rogers et al (136) predicted that all other immunoglobulin heavy chain genes would also contain M gene segments and would generate membrane mRNA species by RNA splicing. It has now been confirmed that all B-cell lines that produced either γ_1, γ_{2a}, γ_{2b}, or γ_3 secreted heavy chains contained both γ_s mRNA and a minor γ mRNA species (γ_m) with a spliced structure analogous to that of the μ_m mRNA (135, 137, 140, 171). Mapping and sequencing of the γ_1, γ_{2a}, γ_{2b}, and γ_3 M gene segments (136, 171, 193) have located the M exons in these heavy chain genes. Membrane and secreted forms of α heavy chain mRNA have recently been reported (82, 189). The M exon of the α heavy chain gene segment has now been determined (189).

The C_δ gene is organized differently from other heavy chain genes in that it contains an extended hinge and lacks an internal $C_H 2$ domain (98, 169). Unlike all other heavy chain genes where the secreted C-terminal sequences are continuous with the final C region domain, the δ gene has both its secreted and membrane exons in separate noncontiguous gene segments 3' to the $C_{\delta 3}$ domain (33, 98, 102, 112, 169, 170). The M exon sequences (33) of δ_m chains also exhibit a relatively high degree of homology to the M exon sequences of other heavy chains (140).

Mouse B lymphoma cells co-expressing membrane μ and δ chains contain both μ_m and μ_s mRNA, as well as two δ_m mRNA species (102, 112, 116). Both these δ_m mRNA species apparently contain identical coding sequences but terminate at different poly(A) sites and therefore have different 3'-UT sequences (33, 102). An analogous situation has been reported for two H-chain α_m mRNA species for α heavy chains (189) and dihydrofolate reductase (153). There is no evidence for functional differences between these two forms of δ and α mRNA's.

Heavy Chain Genes are Complex Transcription Units

Most eukaryotic genes now studied (including immunoglobulin light chain genes) (127, 178) are simple transcription units that code for a single mRNA (reviewed in 40, 41). Simple transcription units contain a single poly(A) addition site, and the pattern of RNA splicing is invariant except in unusual cases where alterations in splicing sites generate aberrant splicing patterns. Complex transcription units contain multiple poly(A) addition sites and/or alternative RNA-splicing sites, which are used to generate multiple mRNA species from a single gene (40, 41, 197). Complex transcription units have been found in adenovirus, SV40, polyoma virus (40, 41, 197), and retroviruses (165, 166). The mapping of membrane and secreted mRNAs established immunoglobulin heavy chain genes as the first-known nonviral complex transcription units. The genes for yeast invertase (31) and rat calcitonin (8) have recently been shown to be complex transcription units. Immunoglobulin heavy chain complex transcription units contain at least two poly(A) addition sites. When nuclear RNA precursors have poly(A) added at the end of the secreted COOH-terminal sequence, RNA splicing then acts on all exons to yield a secreted heavy chain mRNA (Figure 3). Nuclear RNA precursors that have poly(A) added at the end of the M exons undergo two additional RNA splices that delete the secreted C-terminal sequence and connect the final C region domain to the M exons to generate membrane heavy chain mRNA's.

Even though membrane heavy chain mRNAs are produced by two RNA splicing events more than are required for secreted heavy chain mRNAs, it is believed that the processing pathways leading to secreted or membrane heavy chain mRNA's are determined by the choice of poly(A) addition sites (48, 136, 140, 178). This proposal is based on the general finding that polyadenylation precedes splicing under normal conditions (reviewed in 40, 41).

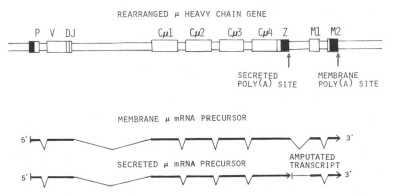

Figure 3   RNA species and splicing events in the generation of membrane and secreted heavy chain mRNA. The upper line depicts the structure established for the μ gene (30, 48). Boxes are exons and untranslated regions are shaded. The lower two lines depict the two polyadenylated alternative primary transcripts made from a single gene, with their splicing patterns. The amputated transcript is a novel polyadenylated RNA species generated in the addition of poly(A) to the secreted mRNA precursor. Although shown here for the μ gene, all heavy chain genes have a similar organization encoding membrane and secreted mRNA species.

Several studies have reported large nuclear RNA precursors of sizes expected for primary transcripts and nuclear RNA processing intermediates to secreted and membrane heavy chain mRNA's (35, 104, 128, 140, 147). However, because of the low levels of the presumptive primary transcripts and difficulties in reproducibly detecting splicing intermediates, the differential processing pathways leading to membrane and secreted heavy chain mRNA's have not been resolved. However, Rogers & Wall (137, 140) have obtained another line of evidence in support of the proposal that membrane and secreted heavy chain mRNA's are processed from the transcripts of a complex transcription unit. As a general rule, it appears that poly(A) addition occurs at points of cleavage in nuclear RNA molecules rather than through termination of transcription (40, 69, 117). Indeed, transcription apparently continues some distance past poly(A) addition sites (several kilobases) before cleavage of the nascent transcript occurs. Assuming that poly(A) addition occurs at points of cleavage in immunoglobulin heavy chain gene transcripts, it was reasoned that transcription past the poly(A) addition site for secreted mRNA precursors would produce a novel polyadenylated species called the "amputated transcript." This predicted polyadenylated RNA species should contain intron sequences between the secreted poly(A) site and the membrane exons (Figure 2). Since the M1 and M2 exons are separated by complete splicing signals in the amputated transcript, it is likely that this intron would be spliced out. Finally, the amputated transcript should have a 5' terminus beginning in the sequence following the secreted poly(A) addition site and should not contain any heavy chain V or C_H region exons. Discrete nuclear RNA species with precisely these predicted properties have now been detected and characterized in both μ- and γ_{2b}-producing cells (137, 140). The confirmation of the amputated transcript as predicted establishes that heavy chain genes are complex transcription units, and reaffirms that poly(A) is added by a mechanism involving RNA cleavage.

Co-Expression of Different Heavy Chain Classes

It was predicted that the co-expressed μ and δ heavy chains in surface IgM and IgD would be generated from a single large complex transcription unit by RNA processing mechanisms (55, 178). This prediction was based on the findings that co-expressed μ and δ chains appeared to have the same variable region (reviewed in 57), exhibited allelic exclusion, and were encoded on the same chromosome (23, 64). The C_μ and C_δ regions are closely linked. The C_δ gene is separated from the μ_m exons only by approximately 2 kb of DNA (33, 98, 102, 112, 169).

If μ and δ are co-expressed from a single gene, then the δ gene should not be rearranged in IgM + IgD-producing lymphocytes. This prediction was confirmed by mapping studies that showed that the C_δ gene was in the germline configuration in $\mu + \delta$-producing cells (83, 102, 112). These data clearly indicated that IgD expression in $\mu + \delta$ B cells did not involve a V_H to C_δ DNA class-switch DNA rearrangement. Accordingly, the simultaneous expression of C_μ and C_δ with a single V_H region appears to be mediated by alternative routes of RNA processing of a very large primary nuclear transcript that contains the V_H, C_μ, and C_δ gene segments (reviewed in 102, 140).

The $\mu + \delta$ complex transcription unit is approximately 25 kb long, with five known alternative polyadenylation sites and more than a dozen potential exons for splicing. Large nuclear RNA species approaching the 25-kb size estimated for the $\mu + \delta$ primary transcript have eluded direct detection. These RNAs are likely to be present in <10 copies/cell, and this low abundance may preclude their direct detection. A complete primary transcript for the major late adenovirus transcription unit (32 kb long) was never detected directly (40, 41, 197). Instead, confirmation of this major transcription unit was obtained by indirect means, including pulse-labeling studies and UV transcript mapping (40, 41). Such indirect means appear likely to be required for confirming the large $\mu + \delta$ complex transcription unit. As in heavy chain transcription units for membrane and secreted mRNA's, the choice of poly(A) sites in the $\mu + \delta$ transcription unit is presumed to determine the pattern of RNA splicing. The RNA splicing events in this complex system are more complicated than in membrane and secreted heavy chain mRNA processing where all complete RNA splicing sites are used. When the V_H region is spliced to the C_δ region to make δ mRNA, RNA splice sites in the C_μ exons are apparently ignored.

How C_μ or C_δ exons are chosen for splicing remains a mystery. A model for RNA splicing has been proposed in which U-1, a ubiquitous small nuclear RNA in eukaryotic cells, base-pairs with both ends of an intron and aligns them precisely in register for cutting and splicing (93, 138). Yang et al (195) have obtained preliminary evidence in support of this model. However, this model does not explain how μ exon splice sites that function in making μ mRNA's are ignored in splicing δ mRNA's. It seems plausible that the secondary structure of the nuclear RNA molecule might direct the course of RNA splicing, but this hypothesis is not yet amenable to experimental testing.

IgD-secreting myeloma cells contain a single prominent δ_s mRNA species (52, 102, 112, 116). Rare instances of IgD secretion by myeloma cells in mice apparently do not occur through RNA splicing, but rather appear to result from a DNA rearrangement that brings the rearranged VDJ_H region into proximity to the C_δ gene with the deletion of the C_μ gene segments (102, 112). Furthermore, the δ_s mRNA is not detectable in B lymphomas, making μ_s, μ_m, and the two δ_m mRNA species. These findings indicate that two different mechanisms are employed in δ gene expression and the production of IgD. RNA processing alterations in B cells produce membrane δ mRNA's along with μ mRNA's from a single large complex transcription unit, whereas a rare DNA rearrangement in mouse plasma cells produces secreted δ mRNA.

Rare lymphoid cells have been reported that apparently contain two classes of surface immunoglobulins other than IgM and IgD. These cells are of considerable interest because they may represent memory B cell transitional stages in the process of class switching, leading to antibody secretion (reviewed in 103). Such cells have recently been isolated and their immunoglobulin gene organization has been analyzed. Perlmutter & Gilbert (125) have shown that splenic B lymphocytes exhibiting surface IgM and IgG1 contain two C_μ genes and $C_{\gamma 1}$ genes that show no evidence of rearrangement. Similarly, $\mu^+\epsilon^+$ splenic B lymphocytes isolated by fluorescence-activated cell sorting show no evidence of DNA rearrangements between J_H and the C_ϵ gene segment (194). These two reports conclude that the co-expression of different H chain mRNA's involves extremely large nuclear RNA precursors. Both these reports are based on the presumption that the isolated B lymphocytes exhibiting two different heavy chains by staining are simultaneously synthesizing both chains. This is partially satisfied for the $\mu^+\epsilon^+$ cells, but this point needs to be rigorously confirmed.

This reservation is not confronted in other studies on cloned B lymphocyte lines derived from an Abelson virus-transformed pre-B cell line called 18–81, which express two H chain classes. As characteristic for a pre-B cell, primary 18–81 subclones only express low levels of μ mRNA but no light chain mRNA. Long-term tissue-culture-adapted subclones of 18–81 (18–81 A-2) apparently undergo a switch from C_μ to $C_{\gamma 2b}$ expression (4, 28). The level of γ_{2b} mRNA appears to be inducible with a B cell mitogen, lipopolysaccharide. Furthermore, μ chain expression is undetectable without LPS but is restored by LPS treatment of these cells. These cells contain C_μ regions. Deletion events occurring within the J_H-C_μ intron are correlated with the loss of constitutive μ gene expression and the $\mu \rightarrow \gamma_{2b}$ switch (4). The $J_H \rightarrow C_\mu$ intron has been shown to contain repetitive DNA sequences that readily undergo deletion events (46, 103). These experiments have not shown that the δ, γ_3, and γ_1 gene segments are present in their germline context. If this is the case, the $\mu \rightarrow \gamma_{2b}$ switch in 18–81 A-2 cells may result from the differential RNA splicing of a large (>100 kb) multi-C_H-gene transcript (4). Given the large sizes of the predicted $\mu + \gamma_1$, $\mu + \epsilon$, or $\mu + \gamma_{2b}$ nuclear RNA precursors (100–200 kb), and the experimental inability to detect even the 25-kb $\mu + \delta$ co-transcript, experimental confirmation of these large precursors will require ingenious approaches. It also remains to be established that such cells represent transitional stages in heavy chain class switching. Immunoglobulin gene introduction into lymphoid cells together with defined conditions that induce class switching (mitogens, T cell factors, etc) should provide further insights into the molecular mechanisms and dynamics of class switching.

QUANTITATIVE CHANGES IN LEVEL OF IMMUNOGLOBULIN EXPRESSION DURING DEVELOPMENT

As noted above, there is differential expression of H, L, and J chains during development. In addition, the expression of all three products is amplified substantially during B cell maturation. In general, when both H and L chains are expressed (i.e. in immature B cells and later stages of development), they are synthesized at roughly equimolar levels even as the levels of expression are increased several orders of magnitude (86, 101, 105a, 113, 128, 134, 153b, 154). In contrast, the level of J chain expression is not tightly coordinated with the level of H and L chain expression, despite the fact that in plasma cells, J chain is expressed at levels roughly comparable to the levels of H and L chain expression (85, 100, 134). Cloned murine pre-B-cell lines and some immature B-cell lines may synthesize as little as 0.01–0.1% of their protein as μ only or as μ plus L chains (101, 153b, 154), whereas murine myeloma cells may synthesize as much as 20–30% of their protein as H and L chains (86, 113). Similar amplification of H and L chain expression in late stages of B lymphocyte development is also seen in studies of normal cells (109, 155,). The levels of H and L chain expression in most immature B-cell, B-cell, and memory B-cell lines appear to be roughly intermediate between the extremes cited above, although there are few critical measurements for the intermediate stages of B-cell development (100, 134, 188).

Various B-cell lines and normal B cells have been shown to amplify both H and L chain impression approximately two to ten times when stimulated in vitro with mitogens (e.g. lipopolysaccharide), anti-immunoglobulins, or appropriate combinations of antigen and lymphocyte supernatants (24, 60, 101, 109, 122b, 154). A number of pre-B cell lines stimulated with LPS begin to express L chain at levels comparable to H chain, with little stimulation of the level of H chain synthesis (101, 122b, 142a). In contrast, after LPS stimulation, several Abelson virus-transformed pre-B-cell lines are stimulated to synthesize H chains at an approximately 10-fold increased level but still do not express L chains (4, 5). It has been noted for these Abelson virus-transformed pre-B-cell lines that LPS stimulation of H-chain synthesis occurs only in sublines or clones that spontaneously have decreased the level of H-chain synthesis substantially (about 10-fold). The decrease in the unstimulated level of H-chain synthesis is correlated with a substantial deletion in the J_H-C_μ intron 5' of the μ switch region. Based on these results, it has been suggested that LPS stimulation somehow compensates for the loss of a DNA sequence that has a positive effect on H-chain production. Myeloma cells with smaller deletions in the J_H-C_μ intron seem to be unaffected in their level of H-chain expression.

Somatic cell hybrids between mouse myeloma cells and mouse or human pre-B cells, immature B cells, B cells, or memory B cells co-dominantly express H and L chains at approximately the same high level of expression observed in the myeloma fusion partner. In all reported studies of this type, the overall phenotype of immunoglobulin expression (i.e. quantity of H and L chains and secreted versus membrane form of H chain) is determined by the more differentiated myeloma cell fusion partner (50, 91, 96, 134, 134b, 188).

A number of studies with murine cell lines have demonstrated that the qualitative and quantitative levels of H, L, and J chain expression at different stages of B cell development are determined largely, if not entirely, by the amount of cytoplasmic H, L, and J chain mRNA's, respectively (38, 56, 104, 128, 134, 147). The few studies on normal B lymphocytes are consistent with this conclusion (168).

The rates of in vivo H and L chain RNA synthesis and processing has been studied extensively in MOPC 21 and MPC 11 murine myeloma cell lines by the laboratories of Wall and Perry, respectively (56, 104, 147). These studies, together with other studies on the turnover of cytoplasmic mRNA in MOPC 21 myeloma cells (38), provide insight into how a single functional H- or L-chain gene can account for as much as 10% of the protein synthesized by a rapidly growing myeloma cell. The following conclusions apply to both myeloma cell lines: (a) The rate of transcription is greater than 20–30 transcripts per minute (which is comparable to the apparent maximal rates of transcription observed for ribosomal RNA); (b) processing of the putative primary transcripts into mRNA and transport of mRNA from the nucleus to cytoplasm is rapid and essentially quantitative; (c) cytoplasmic mRNA is very stable, so that it is minimally metabolized during the 20-hr cell generation time; and (d) the fraction of cellular mRNA encoding H and L chains is comparable to the fraction of cellular protein synthesis comprised by H and L chains. Thus, it appears that the level of H- and L-chain gene expression in myeloma cells may well approach the maximum level possible for a higher eukaryotic cell, with an approximate 20-hr generation time.

We know much less about regulatory steps that result in a 10- to 1000-fold lower expression of H and L chain mRNA (and protein) in B cells representing earlier stages of B lymphocyte development. However, some preliminary information is available from a study comparing steady-state nuclear and cytoplasmic H chain mRNA contents of MPC 11 murine myeloma cells and the 70Z murine pre-B cell line (128). MPC 11 cells contain about 30,000 copies of H chain mRNA or about 300 times as much cytoplasmic H-chain mRNA as 70Z cells, consistent with the fact that H chain comprises about 10–12% and 0.02–0.10% of protein synthesis in MPC 11 and 70Z cells, respectively (86, 101). In contrast to the large difference in cytoplasmic H-chain mRNA content, only six times as much nuclear H-chain mRNA occurs in MPC 11 myeloma cells as in 70Z cells. The 50-fold higher ratio of nuclear to cytoplasmic H-chain mRNA in 70Z cells compared to myeloma cells is probably a consequence of differences in one or more of the following post-transcriptional events: (a) rate of H-chain nuclear RNA precursor processing; (b) extent of intranuclear H-chain mRNA degradation; (c) rate of transport of mature H-chain mRNA from the nucleus to the cytoplasm; and (d) rate of cytoplasmic H chain mRNA turnover. It is of interest that the ratio of nuclear to cytoplasmic nonimmunoglobulin poly(A)-containing RNA in either 70Z or MPC 11 myeloma cells is essentially the same as the ratio of nuclear to cytoplasmic H-chain mRNA in the 70Z cells. In view of the apparent differential post-transcriptional events specifically affecting H-chain mRNA metabolism, it seems unlikely that differences in transcription rates are fully responsible for the vastly different levels of cytoplasmic H-chain mRNA in these two cell lines. Obviously more studies of the transcription rates and metabolism of H- and L-chain mRNA's are required to understand how the levels of H- and L-chain mRNA (and thus H and L chains) are regulated during B lymphocyte development.

TRANSLATIONAL AND POST-TRANSLATIONAL REGULATION OF IMMUNOGLOBULIN EXPRESSION

Initiation of translation of H- and L-chain mRNA's appears to be much more efficient than the average cell mRNA in that H and L chains represent a larger fraction of the newly synthesized protein when there is a nonspecific decrease in initiation of translation (e.g. mitosis, starvation, viral infection) (119, 120, 160). However, there is no convincing evidence for regulation of the efficiency of mRNA translation in B cells of different developmental stages. Microsomal localization of H, L, and J chain mRNA's to the rough endoplasmic reticulum and translocation of H, L, and J chains into the cisterna of the rough endoplasmic reticulum seems to be determined by the amino terminal signal sequence, which is proteolytically removed from the primary translation product prior to chain termination and release from the ribosome (reviewed in 86). The signal sequence functions normally even in an unusual situation where the signal sequence is contiguous with the C_κ sequence rather than a V_κ sequence, as found for normal κ L-chain mRNA's (51, 141). There are no known examples of either cell variants or normal regulatory events that alter binding of mRNA to the rough endoplasmic reticulum or the vectorial transport of immunoglobulin chains into the cisterna.

Glycosylation of all normal H chains as well as L chains that contain the Asn-X-Thr (or Ser) tripeptide N-glycosylation acceptor site in the amino terminal V region domain usually occurs by transfer of the core oligosaccharide (glucose$_3$-N acetyl-glucosamine$_2$-mannose$_{11}$) from a lipid carrier to a nascent polypeptide (15, 16). The transfer of the core oligosaccharide to a growing nascent H chain apparently slows polypeptide elongation sufficiently so that the net rate of H chain synthesis is decreased 10–25%, a phenomenon that may contribute to the slight imbalance of H and L chain synthesis observed in both normal and malignant B lymphocytes (9, 16, 86). Parameters that affect the efficiency of N-glycosylation or potential tripeptide acceptor sites of immunoglobulin chains are reviewed elsewhere (16). There is no evidence indicating that core glycosylation is regulated differentially in B cells of different developmental stages or in different physiological states. Most processing of the core oligosaccharide, as well as addition of terminal sugars, is thought to occur during the brief period the glycosylated immunoglobulin chain traverses the Golgi apparatus (16, 164). The decision process for generating simple versus complex asparagine-linked oligosaccharides is poorly understood. The best example of possible regulation of glycosylation involves a block in converting a core oligosaccharide to a complex oligosaccharide on the secreted H chains in less differentiated B cells (see below for a more complete discussion of this interesting regulatory event).

Intramolecular folding, including covalent disulfide bond formation, occurs to a significant extent on nascent L chains, and presumably on nascent H chains as well (14). Intermolecular assembly of immunoglobulin chains also begins on nascent H chains, with some classes of immunoglobulin forming H-H disulfide bonds and other classes of immunoglobulin forming H-L disulfide bonds on nascent H chains (15, 146). However, most assembly occurs between completed chains, with the pathway of assembly dependent on the class of Ig synthesized by the cell (19, 86, 146). Assembly of H chains is restricted to the chains of the same class, so that cells expressing μ and γ H chains, for example, do not form covalent μ-γ heterodimers (105).

The co-expression of secretory and membrane forms of H chain (e.g. μ_m and μ_s) in various ratios in the same cell at different B lymphocyte developmental stages raises the question as to whether or not the secretory and membrane forms of H chains form heterodimers. The best studies to address this issue involve B lymphomas and myeloma cells, which simultaneously express γ_m and γ_s. In two different murine B lymphomas that express γ_{2am} and γ_{2as} in roughly equivalent amounts, no γ_{2am}-γ_{2as} heterodimers were seen (121, 188). Analysis of labeled cell surface Ig of MPC 11 myeloma cells revealed no detectable γ_{2bs} chains expressed on the cell surface (88). These results then suggest that membrane and secreted H-chain heterodimer covalent assembly is rare or non-existent in B cells where the membrane and secreted chains are co-expressed. Lack of significant heterodimer assembly of these chains could be due to one or more of the following: (*a*) preferential assembly of H chains synthesized by the ribosomes on the same mRNA molecule; (*b*) preferential assembly of H chains selectively partitioned into the membrane (H_m) instead of into the cisterna (H_s) of the rough endoplasmic reticulum; and (*c*) preferential assembly of homodimers due to intrinsic properties of the H chains. In contrast to the results cited above, recent work suggests that covalent heterodimer formation and membrane insertion of γ_{1s} and γ_{1m} H chains occurs in MOPC 21 myeloma cells, a cell line in which γ_{1s} is synthesized in a vast molar excess compared to γ_{1m} (58). This result must be questioned in view of the results in B lymphoma and other myeloma cells cited above. The putative γ_{1s} chains in labeled surface Ig of MOPC 21 cells could, for example, represent partially degraded γ_{1m} chains.

Differential intracellular degradation of mutant and normal H and L chains represents a well-documented example of post-translational regulation. For example, MOPC 173 and MOPC 21 myeloma lines synthesize a molar excess of L chains over H chains, but degrade the unassembled L chains (9). A MOPC 21 variant that synthesizes L chains but not H chains does not secrete the L chains, but quantitatively degrades them within the cell (184). Fusion of this MOPC 21 L chain-producing variant to a different myeloma cell permits assembly of the MOPC 21 L chain to a heterologous H chain. This assembly prevents the MOPC 21 L chain from being degraded intracellularly (184). However, in most lymphoid cells, unassembled L chains are not degraded intracellularly (9). Variant or normal cells that express H chain in the absence of L chain expression may or may not rapidly degrade the H chains within the cell (113, 154). Variant myeloma cells that express L chains and J chains, but not H chains, degrade the J chains (113–115). Finally, there is very little information regarding regulation of H, L, or J chain degradation in cells representing different B-lymphocyte developmental stages.

The final step in expression of Ig by B lymphocytes is cell surface insertion of membrane Ig or secretion of secreted Ig. In general, free L chains are secreted, whereas normal H chains are not secreted unless linked to L chains (86, 113). Thus, pre-B cells generally do not secrete their H chains (101, 122b, 142, 154). A single report of human pre-B cells actively secreting H chain in the absence of L chain (95) needs to be rigorously documented by further experiments. In contrast, there are a number of examples of lymphoid cells that secrete mutant H chains in the absence of L chain expression (113). As noted above, unassembled MOPC 21 L chains are not secreted, but are degraded intracellularly. However, translation of MOPC 21 L chain mRNA in *Xenopus* oocytes results in an L chain that is neither secreted nor degraded (172). However, the MOPC 21 L chain is secreted if MOPC 21 H chain mRNA is simultaneously translated in oocytes (172). If another L chain MRNA is co-injected into oocytes with MOPC 21 mRNA, only the non-MOPC 21 L chain is secreted (172). Thus, it appears that the lack of secretion of free MOPC 21 L chain from MOPC 21 myeloma cells may not result from the rapid intracellular degradation of this L chain. Several variant myeloma cell lines synthesize mutant L chains or H chains that are not secreted even when assembled normally (113, 115). The mutations are not localized to a unique region. This suggests that these mutations may cause conformational changes in immunoglobulin chains that are incompatible with secretion.

Expression of membrane Ig seems to require simultaneous expression and co-assembly of H and L chains. Pre-B cells generally express cytoplasmic μ but no surface or secreted μ chains (95, 101, 122b, 153c, 154). There are several recent reports of murine and human pre-B lymphomas that express μ_m on the external cell surface in the absence of L chain expression (58a, 122a). The murine cell line (70Z) usually expresses cytoplasmic μ, but can express cell surface μ spontaneously or after dextran sulfate stimulation. If this result can be validated for normal B cells, it has important implications for potential regulation of pre-B cells by anti-idiotypic antibody or antigen.

A plethora of studies has attempted to determine whether glycosylation of normal or variant H chains is necessary for cell surface expression and/or secretion of Ig. In general, glycosylation of H chains is not necessary for cell surface expression (67, 153b). The results are more complicated for secretion of nonglycosylated Ig: (*a*) Secretion is blocked or markedly inhibited for tunicamycin-treated myelomas that synthesize IgM, IgE, and IgA, but a tunicamycin-treated hybridoma secreted a different IgA molecule at apparently normal rates; (*b*) secretion from tunicamycin-treated myelomas is little affected for IgG or IgD; (*c*) a B lymphoma (WEHI 279) secretes low levels of IgM at only a marginally slower rate after tunicamycin treatment (20, 65, 66, 82, 113, 153b, 155, 156, 157, 187). Some of the results obtained above do not depend on the cells per se, since the same hybrid cell is shown to secrete nonglycosylated IgG but not to secrete nonglycosylated IgM. In sum, whether or not nonglycosylated Ig is secreted depends on the class of Ig, the amount of Ig synthesized, and possibly the V_H gene segment in the H chain. As noted above, there is no evidence for regulation of core glycosylation in B lymphocytes.

The most intriguing example of differential post-translational regulation of Ig at different stages of B cell development involves expression of H chains in secreted and surface Ig. As noted previously, B lymphocytes that synthesize H chain mRNA contain both membrane and secreted H-chain mRNA's. The ratio of secreted/membrane mRNA increases during B-cell development (140). From limited data, it appears that the ratio of intracellular membrane and secreted polypeptide chains reflects the ratio of the two H-chain mRNA's. Approximately 20–50% of the H chain is μ_s in pre-B or immature B cell lines and small resting B lymphocytes from animals (2, 4, 5, 48, 101, 136, 122b, 153b). Yet these cells secrete little (or none) of the μ_s polypeptide translation product (101, 122b, 155). In contrast, the μ_s translation product is quantitatively secreted from myeloma cells, B-cell lines and normal resting B cells stimulated with mitogens or T cell factors (86, 113, 122b). The lack of secretion of μ_s at early B cell developmental stages is not due to a lack of core glycosylation, since the unsecreted μ_s chains contain core oligosaccharide (but lack terminal sugars). Neither is it due to a block in H and L chain assembly, since IgM with μ_s chains is formed in these cells. Similar results have been found for a B cell line that expresses both α_s and α_m (82, 159), although the lack of secretion here is less well documented than the examples cited above.

The mechanism of this specific post-translational regulation of secretion is unclear. It might be due to any or a combination of the following: (a) differential oligosaccharide processing and terminal glycosylation of the core oligosaccharide; (b) failure of transport into the Golgi apparatus, which would explain the glycosylation deficiencies; (c) lack of J chain (100). The apparent lack of quantitative secretion of α_s from a B lymphoma argues against a role for J chain since IgA can be secreted as a monomer (82, 159). Similarly, the normal secretion of nonglycosylated IgM from a B lymphoma cell line (153b) argues that glycosylation does not explain this situation.

Studies on certain lines of 70 Z and 18–81 cells suggest that the expression of surface IgM may require more than the expression of μ_m and L chains (26, 122a, 142). The molecular mechanisms that control IgM secretion in early B cells represent interesting questions yet to be resolved.

FUTURE DIRECTIONS

The studies reviewed here clearly provide considerable insights into the events in immunoglobulin gene expression and regulation. Nonetheless, many of the molecular mechanisms affecting the changing patterns of immunoglobulin gene expression in B-cell development remain to be resolved.

The powerful resolution of recombinant DNA cloning and sequencing will continue to provide insights into immunoglobulin gene structure. Several exciting recent developments provide molecular immunologists even more penetrating approaches for dissecting immunoglobulin gene control. These include (*a*) the establishment of cloned normal pre-B cell and B cell lines that undergo developmental changes in culture (183a), (*b*) the successful introduction and expression of cloned immunoglobulin genes transfected into lymphoid cells (42a), and (*c*) the increasing refinement of culture conditions and factors that stimulate B cell growth and affect changes in immunoglobulin gene expression (55a, 93a, 124a).

ACKNOWLEDGMENTS

Michael Kuehl has been supported by USPHS grants AI 12525, AI 17748, and AI 00293 at the University of Virginia Medical School, Department of Microbiology. Randolph Wall has been supported by USPHS grants AI 13410 and CA 12800, and by NSF grant PCM 79-24876.

Literature Cited

1. Alt, F., Baltimore, D. 1982. *Proc. Natl. Acad. Sci. USA* 79:4118–22
2. Alt, F. W., Bothwell, A. L. M., Knapp, M., Siden, E., Mather, E., et al. 1980. *Cell* 20:293–301
3. Alt, F., Enea, V., Bothwell, A. L. M., Baltimore, D. 1980. *Cell* 21:1–12
4. Alt, F. W., Rosenberg, N., Casanova, R. J., Thomas, E., Baltimore, D. 1982. *Nature* 296:325–31
5. Alt, F. W., Rosenberg, N., Enea, V., Siden, E., Baltimore, D. 1982. *Mol. Cell. Biol.* 2:386–400
6. Alt, F. W., Rosenberg, N., Lewis, S., Thomas, E., Baltimore, D. 1981. *Cell* 27:391–400
7. Altenburger, W., Steinmetz, M., Zachau, H. G. 1980. *Nature* 287:603–7
8. Amara, S., Jonas, V., Rosenfeld, M. G., Ong, E. S., Evans, R. M. 1982. *Nature* 298:240–44
9. Baumal, R., Scharff, M. D. 1973. *J. Immunol.* 111:448–56
10. Banerji, J., Rusconi, S., Schaffner, W. 1981. *Cell* 27:299–308
11. Benoist, C., Chambon, P. 1981. *Nature* 290:304–10
12. Bentley, D. L., Farrell, P. J., Rabbitts, T. H. 1982. *Nucl. Acids Res.* 10:1841–56
13. Bentley, D. L., Rabbitts, T. H. 1980. *Nature* 288:730–33
14. Bergman, L. W., Kuehl, W. M. 1979. *J. Biol. Chem.* 254:8869–76

15. Bergman, L. W., Kuehl, W. M. 1979. *J. Supramol. Struct.* 11:9–24
16. Bergman, L. W., Kuehl, W. M. 1982. In *The Glycoconjugates,* 3:81–98. New York: Academic
17. Bernard, O., Gough, N. M., Adams, J. M. 1981. *Proc. Natl. Acad. Sci. USA* 78:5812–16
18. Bernard, O., Hozumi, N., Tonegawa, S. 1978. *Cell* 15:1133–44
19. Bevan, M., Parkhouse, R. M. E., Williamson, A. R., Askonas, B. A. 1972. *Prog. Biophys. Mol. Biol.* 25:131–62
20. Blatt, C., Haimovich, J. 1981. *Eur. J. Immunol.* 11:65–66
21. Bodary, S., Mach, B. 1983. *EMBO Journal* 6:In press
22. Bothwell, A. L. M., Paskind, M., Schwartz, R. C., Sonenshein, G. E., Gefter, M. L., Baltimore, D. 1981. *Nature* 290:65–67
23. Bourgois, A., Kitajima, K., Hunter, I. R., Askonas, B. A. 1977. *Eur. J. Immunol.* 7:151–56
24. Boyd, A. W., Goding, J. W., Shrader, J. W. 1981. *J. Immunol.* 126:2451–65
25. Breathnach, R., Chambon, P. 1981. *Ann. Rev. Biochem.* 50:349–83
26. Brock, E. J., Kuehl, W. M. Unpublished observation
27. Brown, D. 1981. *Science* 211:667–74
28. Burrows, P. D., Beck, G. B., Wabl, M. R. 1981. *Proc. Natl. Acad. Sci. USA* 78:564–68

29. Burrows, P., Le Jeune, M., Kearney, J. F. 1979. *Nature* 180:838–41
30. Calame, K., Rogers, J., Early, P., Davis, M., Livant, P., et al. 1980. *Nature* 284:452–55
31. Carlson, M., Botstein, D. 1982. *Cell* 28:145–54
32. Cebra, J., Colberg, J., Dray, S. 1966. *J. Exp. Med.* 123:547–58
33. Cheng, H.-L., Blattner, F. R., Fitzmaurice, L., Mushinski, J. F., Tucker, P. W. 1982. *Nature* 296:410–15
34. Choi, E., Kuehl, M., Wall, R. 1980. *Nature* 286:776–79
35. Clarke, C., Berenson, J., Goverman, J., Boyer, P. D., Crews, S., et al. 1982. *Nucl. Acids Res.* 10:7731–49
36. Coffman, R. L., Cohn, M. J. 1977. *J. Immunol.* 118:1806–15
37. Coleclough, C., Perry, R., Karjalainen, K., Weigert, M. 1981. *Nature* 290: 372–78
38. Cowan, N. J., Milstein, C. 1974. *J. Mol. Biol.* 82:469–81
39. Dackowski, W., Morrison, S. 1981. *Proc. Natl. Acad. Sci. USA* 78:7091–95
40. Darnell, J. E. 1979. *Prog. Nucl. Acids Res. Mol. Biol.* 22:327–53
41. Darnell, J. E. 1982. *Nature* 297:365–71
42. Davis, M., Kim, S., Hood, L. 1980. *Cell* 22:1–2
42a. Deans, R., Denis, K., Hermanson, G., Lernhardt, W., Carter, C., et al. Submitted for publication
43. D'Eustachio, P., Bothwell, A., Takaro, A., Baltimore, D., Ruddle, F. 1981. *J. Exp. Med.* 153:793–800
44. de Villiers, J., Schaffner, W. 1981. *Nucl. Acids Res.* 9:6251–64
45. Dreyer, W. J., Bennett, J. C. 1965. *Proc. Natl. Acad. Sci. USA* 54:864–69
46. Early, P., Hood, L. 1981. *Cell* 24:1–3
47. Early, P., Huang, H., Davis, M., Calame, K., Hood, L. 1980. *Cell* 19:981–92
48. Early, P., Rogers, J., Davis, M., Calame, K., Bond, M., et al. 1980. *Cell* 20:313–19
49. Elgin, S. C. R. 1981. *Cell* 27:413–15
50. Eshhar, Z., Blatt, C., Bergman, Y., Haimovich, J. 1979. *J. Immunol.* 122:2430–34
51. Fetherston, J., Boime, I. 1982. *Biochem. Biophys. Res. Commun.* 104:1630–37
52. Fitzmaurice, L., Owens, J., Blattner, F. R., Cheng, H., Tucker, P. W., Mushinski, J. F. 1982. *Nature* 296:459–62
53. Fradin, A., Manley, J. L., Prives, C. L. 1982. *Proc. Natl. Acad. Sci. USA* 79: 5142–46
54. Fujimura, F., Linney, E. 1982. *Proc. Natl. Acad. Sci. USA* 79:1479–83
55. Gilbert, W. 1978. *Nature* 271:501
55a. Gillis, S., Scheid, M., Watson, J. 1980. *J. Immunol.* 125:2570–78
56. Gilmore-Hebert, M., Wall, R. 1979. *J. Mol. Biol.* 135:879–91
57. Goding, J. W. 1978. *Cont. Topics Immunobiol.* 8:203–43
58. Goding, J. W. 1982. *J. Immunol.* 128:2416–21
58a. Gordon, J., Hamblin, T., Smith, J., Stevenson, F., Stevenson, G. 1981. *Blood* 58:552–56
59. Gough, N., Bernard, O. 1981. *Proc. Natl. Acad. Sci. USA* 78:509–13
60. Gronowicz, E. S., Doss, C. A., Howard, F. D., Morrison, D. C., Strober, S. 1980. *J. Immunol.* 125:976–80
61. Grosschedl, R., Birnstiel, M. L. 1980. *Proc. Natl. Acad. Sci. USA* 77:7102–6
62. Gruss, P., Dhar, R., Khoury, G. 1981. *Proc. Natl. Acad. Sci. USA* 78:943–47
63. Herbomel, P., Saragosti, S., Blangy, D., Yaniv, M. 1981. *Cell* 25:651–58
64. Herzenberg, L. A., Herzenberg, L. A., Black, S. J., Loken, M. R., Okumura, K., et al. 1977. *Cold Spring Harbor Symp. Quant. Biol.* 41:33–45
65. Hickman, S., Kornfeld, S. 1978. *J. Immunol.* 121:990–96
66. Hickman, S., Kulczycki, A. Jr., Lynch, R. G., Kornfeld, S. 1977. *J. Biol. Chem.* 252:4402–8
67. Hickman, S., Wong-Yip, Y. P. 1979. *J. Immunol.* 123:389–95
68. Hieter, P. A., Max, E. E., Seidman, J. G., Maizel, J. V., Leder, P. 1980. *Cell* 22:197–207
69. Hofer, E., Darnell, J. E. 1981. *Cell* 23:585–93
70. Hofer, E., Hofer-Warbinek, R., Darnell, J. E. 1982. *Cell* 29:887–93
71. Honjo, T., Nakai, S., Nishida, Y., Kataoka, T., Yamawaki-Kataoka, Y., et al. 1981. *Immunol. Rev.* 59:33–67
72. Howard, M., Kessler, S., Chused, T., Paul, W. E. 1981. *Proc. Natl. Acad. Sci. USA* 78:5788–92
73. Jakobovits, E. B., Bratosin, S., Aloni, Y. 1980. *Nature* 285:263–65
74. Kabat, E. A. 1976. *Structural Concepts in Immunology and Immunochemistry*. New York: Holt Rinehart. 2nd ed.
75. Kataoka, T., Nikaido, T., Miyata, T., Moriwaki, K., Honjo, T. 1982. *J. Biol. Chem.* 257:277–85
76. Katinka, M., Vasseur, M., Montreau, N., Yaniv, M., Blangy, D. 1981. *Nature* 290:720–22
77. Kehry, M., Ewald, S., Douglas, R., Sibley, C., Raschke, W., et al. 1980. *Cell* 21:393–406

78. Kehry, M., Sibley, C., Fuhrman, J., Schilling, J., Hood, L. E. 1979. *Proc. Natl. Acad. Sci. USA* 76:2932–36
79. Kelley, D. E., Coleclough, C., Perry, R. P. 1982. *Cell* 29:681–89
80. Kemp, D. J., Harris, A. W., Adams, J. M. 1980. *Proc. Natl. Acad. Sci. USA* 77:7400–4
81. Kemp, D. J., Harris, A. W., Cory, S., Adams, J. M. 1980. *Proc. Natl. Acad. Sci. USA* 77:2876–80
82. Kikutani, H., Sitia, R., Good, R. A., Stavnezer, J. 1981. *Proc. Natl. Acad. Sci. USA* 78:6436–40
83. Knapp, M. R., Liu, C., Newell, N., Ward, R. B., Tucker, P. W., et al. 1982. *Proc. Natl. Acad. Sci. USA* 79:2996–3000
84. Korsmeyer, S., Hieter, P., Ravetch, J., Poplack, D., Waldmann, T., Leder, P. 1981. *Proc. Natl. Acad. Sci. USA* 78:7096–100
85. Koshland, M. E. 1975. *Adv. Immunol.* 20:41–69
86. Kuehl, W. M. 1977. *Curr. Topics Microbiol. Immunol.* 76:1–47
87. Kuehl, M. 1981. *Trends Biochem. Sci.* 6:206–8
88. Kuehl, M. W., Word, C. J. Unpublished results
89. Kurosawa, Y., Tonegawa, W. 1982. *J. Exp. Med.* 155:201–18
90. Kwan, S.-P., Max, E. E., Seidman, J. G., Leder, P., Scharff, M. D. 1981. *Cell* 26:57–66
91. Laskov, R., Kim, J. K., Asofsky, R. 1979. *Proc. Natl. Acad. Sci. USA* 76:915–19
92. Laskov, R., Kim, J. K., Woods, V. L., McKeever, P. E., Asofsky, R. 1981. *Eur. J. Immunol.* 11:462–68
93. Lerner, M. R., Boyle, J. A., Mount, S. M., Wolin, S. L., Steitz, J. A. 1980. *Nature* 283:220–24
93a. Lernhardt, W., Corbel, C., Wall, R., Melchers, F. 1982. *Nature.* 300:355–7
94. Levinson, B., Khoury, G., Vande Woude, G., Gruss, P. 1982. *Nature* 295:568–72
95. Levitt, D., Cooper, M. D. 1980. *Cell* 19:617–25
96. Levy, R., Dilley, J. 1978. *Proc. Natl. Acad. Sci. USA* 75:2411–15
97. Lifter, J., Higgins, G., Choi, Y. S. 1980. *Mol. Immunol.* 17:483–90
98. Liu, C., Tucker, P. W., Mushinski, J. F., Blattner, F. R. 1980. *Science* 209:1348–52
99. McGee, J. D., Wood, W. I., Dolan, M., Engel, J. D., Felsenfeld, G. 1981. *Cell* 27:45–55

100. McHugh, Y., Yagi, M., Koshland, M. E. 1981. In *B Lymphocytes in the Immune Response: Functional, Developmental, and Interactive Properties,* ed. N. Klinman, D. Mosier, I. Sher, E. Vitetta, pp. 467–74. New York: Elsevier
100a. McKenzie, I., Potter, T. 1979. *Adv. Immunol.* 27:179–338
101. Mains, P. E., Sibley, C. H. 1982. *J. Immunol.* 128:1664–70
101a. Maki, R., Kearney, J., Paige, C., Tonegawa, S. 1980. *Science* 209:1366–69
102. Maki, R., Roeder, W., Traunecker, A., Sidman, C., Wabi, M., et al. 1981. *Cell* 24:353–65
103. Marcu, K. B. 1982. *Cell* 29:719–21
104. Marcu, K. B., Schibler, U., Perry, R. P. 1978. *J. Mol. Biol.* 129:381–400
105. Margulies, D., Kuehl, W. M., Scharff, M. D. 1976. *Cell* 6:405–415
105a. Mather, E. L., Alt, F. W., Bothwell, A. L. M., Baltimore, D., Koshland, M. E. 1981. *Cell* 23:369–78
106. Mather, E., Perry, R. 1981. *Nucl. Acids Res.* 9:6855–67
107. Mathis, D., Oudet, P., Chambon, P. 1980. *Prog. Nucl. Acids Res. Mol. Biol.* 24:1–55
108. Max, E. E., Seidman, J. G., Miller, H., Leder, P. 1980. *Cell* 21:793–99
109. Melchers, F., Andersson, J. 1974. *Eur. J. Immunol.* 4:181–88
109a. Melchers, F., Andersson, J., Lernhardt, W., Schreier, M. H. 1980. *Immunol. Rev.* 52:89–114
110. Melchers, U., Eidels, L., Uhr, J. W. 1975. *Nature* 258:434–35
111. Mellon, P., Parker, V., Gluzman, Y., Maniatis, T. 1981. *Cell* 27:279–88
112. Moore, K. W., Rogers, J., Hunkpiller, T., Early, P., Nottenburg, C., et al. 1981. *Proc. Natl. Acad. Sci. USA* 78:1800–4
113. Morrison, S. L., Scharff, M. D. 1981. *CRC Crit. Rev. Immunol.* 3:1–22
114. Mosmann, T. R., Gravel, Y., Williamson, A. R., Baumal, R. 1978. *Eur. J. Immunol.* 8:94–101
115. Mosmann, T. R., Williamson, A. R. 1980. *Cell* 20:283–92
116. Mushinski, J. F., Blattner, F. R., Owens, J. D., Finkelman, F. D., Kessler, S. W., et al. 1980. *Proc. Natl. Acad. Sci. USA* 77:7405–9
117. Nevins, J. R., Darnell, J. E. 1978. *Cell* 15:1477–93
118. Nottenberg, C., Weissman, I. L. 1981. *Proc. Natl. Acad. Sci. USA* 78:484–88
119. Nuss, D. L., Koch, G. 1976. *J. Virol.* 19:572–78

120. Nuss, D. L., Oppermann, H., Koch, G. 1975. *J. Mol. Biol.* 102:601–12
121. Oi, V. T., Bryan, V. M., Herzenberg, L. A., Herzenberg, L. A. 1980. *J. Exp. Med.* 151:1260–74
122. Okumura, K., Julius, M. H., Tsu, T., Herzenberg, L. A., Herzenberg, L. A. 1976. *Eur. J. Immunol.* 6:467–72
122a. Paige, C., Kincade, P., Ralph, P. 1981. *Nature* 292:631–33
122b. Paige, C. J., Schreier, M. H., Sidman, C. L. 1982. *Proc. Natl. Acad. Sci. USA* 79:4756–60
123. Parkhouse, R. M. E., Lifter, J., Choi, Y. S. 1980. *Nature* 284:280–81
124. Parslow, T. G., Granner, D. K. 1982. *Nature* 299:449–51
124a. Paul, W. E., Sredni, B., Schwartz, R. H. 1981. *Nature* 294:697–99
125. Perlmutter, H., Gilbert, W. 1983. *Proc. Natl. Acad. Sci. USA* 80:In press
126. Pernis, B. G., Chiappino, G., Kelus, A. S., Gell, P. G. H. 1965. *J. Exp. Med.* 122:853–76
127. Perry, R. P. 1981. *J. Cell Biol.* 91:28s–38s
128. Perry, R. P., Kelley, D. E. 1979. *Cell* 18:1333–39
129. Perry, R. P., Kelley, D. E., Coleclough, C., Seidman, J. G., Leder, P., et al. 1980. *Proc. Natl. Acad. Sci. USA* 77:1937–41
130. Pfeiffer, W., Zachau, H. G. 1980. *Nucl. Acids Res.* 8:4621–38
131. Ponte, P. A., Siekevitz, M., Schwartz, R. C., Gefter, M. L., Sonenshein, G. E. 1981. *Nature* 291:594–96
132. Potter, M. 1972. *Physiol. Rev.* 52:632–719
133. Raschke, W. C. 1978. *Curr. Topics Microbiol. Immunol.* 81:70–76
134. Raschke, W. C., Mather, E. L., Koshland, M. E. 1979. *Proc. Natl. Acad. Sci. USA* 76:3469–73
134a. Razin, A., Riggs, A. D. 1980. *Science* 210:604–10
134b. Riley, S. C., Brock, E. J., Kuehl, W. M. 1981. *Nature* 289:804–6
135. Rogers, J., Choi, E., Souza, L., Carter, C., Word, C., et al. 1981. *Cell* 26:19–27
136. Rogers, J., Early, P., Carter, C., Calame, K., Bond, M., et al. 1980. *Cell* 20:303–12
137. Rogers, J., Komaromy, M., Wall, R. Manuscript in preparation
138. Rogers, J., Wall, R. 1980. *Proc. Natl. Acad. Sci. USA* 77:1877–79
139. Rogers, J., Wall, R. 1981. *Proc. Natl. Acad. Sci. USA* 78:7497–501
140. Rogers, J., Wall, R. 1983. *Contemp. Topics Mol. Immunol.* 10: In press
141. Rose, S. M., Kuehl, W. M., Smith, G. P. 1977. *Cell* 12:453–62
142. Rosenberg, N. Personal communication
142a. Rosenberg, N., Siden, E., Baltimore, D. 1979. In *B Lymphocytes in the Immune Response*, ed. M. Cooper, D. Mosier, A. Scher, E. S. Vitetta, pp. 379–86. New York: Elsevier
143. Sakano, H., Huppi, K., Heinrich, G., Tonegawa, S. 1979. *Nature* 280:288–94
144. Sakano, H., Maki, R., Furasaw, Y., Roeder, W., Tonegawa, S. 1980. *Nature* 286:676–83
145. Saragosti, S., Moyne, G., Yaniv, M. 1980. *Cell* 20:65–73
146. Scharff, M. D. 1974. *Harvey Lecture* 69:125–42
147. Schibler, U., Marcu, K. B., Perry, R. P. 1978. *Cell* 15:1495–509
148. Schwartz, R. C., Sonenshein, G. E., Bothwell, A., Gefter, M. L. 1981. *J. Immunol.* 126:2104–8
149. Seidman, J. G., Leder, P. 1978. *Nature* 276:790–95
150. Seidman, J. G., Leder, P. 1980. *Nature* 286:779–83
151. Seif, I., Khoury, G., Dhar, R. 1979. *Nucl. Acids. Res.* 6:3387–98
152. Sekikawa, K., Levine, A. J. 1981. *Proc. Natl. Acad. Sci. USA* 78:1100–4
153. Setzer, D., McGrogan, M., Nunberg, J. H., Schimke, R. T. 1980. *Cell* 22:361–70
153a. Sibley, C. H., Ewald, S. J., Kehry, M. R., Douglas, R. H., Raschke, W. C., Hood, L. E. 1980. *J. Immunol.* 125:2097–105
153b. Sibley, C. H., Wagner, R. A. 1981. *J. Immunol.* 126:1868–73
153c. Siden, E., Alt, F. W., Shinefeld, L., Sato, V., Baltimore, D. 1981. *Proc. Natl. Acad. Sci. USA* 78:1823–27
154. Siden, E. J., Baltimore, D., Clark, D., Rosenberg, N. E. 1979. *Cell* 16:389–96
155. Sidman, C. 1981. *Cell* 23:379–89
156. Sidman, C. 1981. *J. Biol. Chem.* 256:9374–76
157. Sidman, C., Potash, M. J., Köhler, G. 1981. *J. Biol. Chem.* 256:13180–87
158. Singer, P. A., Singer, H. H., Williamson, A. R. 1980. *Nature* 285:294–300
159. Sitia, R., Kikutani, H., Rubartelli, A., Bushkin, Y., Stavnezer, J., Hammerling, U. 1982. *J. Immunol.* 128:712–16
160. Sonenshein, G. E., Brawerman, G. 1976. Biochemistry. 15:5497–506
161. Southern, E. 1975. *J. Mol. Biol.* 98:503–17
162. Storb, U., Arp, B., Wilson, R. 1981. *Nature* 294:90–92

163. Storb, U., Wilson, R., Selsing, E., Walfield, A. 1981. *Biochemistry* 20:990–96
164. Tartakoff, A., Vassalli, P. 1979. *J. Cell Biol.* 83:284–99
165. Temin, H. M. 1981. *Cell* 27:1–3
166. Temin, H. M. 1982. *Cell* 28:3–5
167. Tonegawa, S., Maxam, A. M., Tizard, R., Bernard, O., Gilbert, W. 1978. *Proc. Natl. Acad. Sci. USA* 75:1485–89
168. Tsuda, M., Honjo, T., Shimizu, A., Mizuno, D., Natori, S. 1978. *J. Biochem.* 84:1285–90
169. Tucker, P. W., Liu, C., Mushinski, J. F., Blattner, F. R. 1980. *Science* 209:1353–60
170. Tucker, P. W., Slighton, J. L., Blattner, F. R. 1981. *Proc. Natl. Acad. Sci. USA* 78:7684–88
171. Tyler, B. M., Cowman, A. F., Gerondakis, S. D., Adams, J. M., Bernard, O. 1982. *Proc. Natl. Acad. Sci. USA* 79:2008–12
172. Valle, G., Besley, J., Colman, A. 1981. *Nature* 291:338–40
173. Van Ness, B. G., Weigert, M., Coleclough, C., Mather, E., Kelley, D. E., Perry, R. P. 1981. *Cell* 27:593–602
174. Varshavsky, A. J., Sundin, O. H., Bohn, M. J. 1978. *Nucl. Acids Res.* 5:3469–77
175. Varshavsky, A. J., Sundin, O. H., Bohn, M. J. 1979. *Cell* 16:453–66
176. Vassalli, P., Tedghi, K., Lisowski-Bernstein, B., Tartakoff, A., Jaton, J. C. 1979. *Proc. Natl. Acad. Sci. USA* 76: 5515–19
177. Vitetta, E. S., Uhr, J. W. 1975. *Science* 189:964–68
177a. Walfield, A., Selsing, E., Arp, B., Storb, U. Unpublished results
178. Wall, R., Choi, E., Carter, C., Kuehl, M., Rogers, J. 1980. *Cold Spring Harbor Symp. Quant. Biol.* 45:879–85
179. Wallach, M., Ishay-Michaeli, R., Givol, D., Laskov, R. 1982. *J. Immunol.* 128:684–90
180. Warner, N. L., Leary, J. F., McLaughlin, S. 1979. In *B Lymphocytes in the Immune Response,* ed. M. Cooper, D. Mosier, I. Scher, E. Vitetta, pp. 371–78. Amsterdam: Elsevier/North Holland: 371–378
181. Weintraub, H., Groudine, M. 1976. *Science* 193:848–56
182. Weischet, W. O., Glotoz, B., Schnell, H., Zachau, H. 1982. *Nucl. Acids Res.* 10:3627–42
183. Whitlock, C., Witte, O. N. 1982. *Proc. Natl. Acad. Sci. USA* 79:3608–12
183a. Whitlock, C. A., Ziegler, S., Treiman, L., Stafford, J., Witte, O. Submitted for publication
184. Wilde, C. D., Milstein, C. 1980. *Eur. J. Immunol.* 10:462–67
185. Wilks, A. F., Cozens, P. J., Mattaj, I. W., Jost, J. 1982. *Proc. Natl. Acad. Sci. USA* 79:4252–55
186. Williams, P. B., Kubo, R. T., Grey, H. M. 1978. *J. Immunol.* 121:2435–39
187. Williamson, A. R., Singer, H. H., Singer, P. A., Mosmann, T. R. 1980. *Biochem. Soc. Trans.* 8:168–70
188. Word, C. J., Kuehl, M. W. 1981. *Mol. Immunol.* 18:311–22
189. Word, C., Mushinski, J. F., Tucker, P. W. 1983. *Cell* 32: In press
190. Yagi, M., Koshland, M. E. 1981. *Proc. Natl. Acad. Sci. USA* 78:4907–11
191. Yamawaki-Kataoka, Y., Kataoka, T., Takahashi, N., Obata, M., Honjo, T. 1980. *Nature* 283:786–89
192. Yamawaki-Kataoka, Y., Miyata, T., Honjo, T. 1981. *Nucl. Acids. Res.* 9:1365–81
193. Yamawaki-Kataoka, Y., Nakai, S., Miyata, T., Honjo, T. 1982. *Proc. Natl. Acad. Sci. USA* 79:2623–27
194. Yaoita, Y., Kumagai, Y., Okumura, K., Honjo, T. 1982. *Nature* 297:697–99
195. Yang, V. W., Lerner, M. R., Steitz, J. A., Flint, S. J. 1981. *Proc. Natl. Acad. Sci. USA* 78:1371–75
196. Yaun, D., Uhr, J. W., Vitetta, E. S. 1980. *J. Immunol.* 125:40–46
197. Ziff, E. B. 1980. *Nature* 287:491–99

Ann. Rev. Immunol. 1983. 1:423–38

THE BIOCHEMISTRY OF ANTIGEN-SPECIFIC T-CELL FACTORS

D. R. Webb

Roche Institute of Molecular Biology, Nutley, New Jersey 07110

Judith A. Kapp and Carl W. Pierce

Department of Pathology and Laboratory of Medicine, The Jewish Hospital of St. Louis, and the Departments of Microbiology and Pathology, Washington University School of Medicine, St. Louis, Missouri 63110

INTRODUCTION

Soluble factors produced by immunocompetent cells in response to antigenic stimulation represent one communications link by which the immune system regulates its activities. These factors may be specific for the inducing antigen or may be non-antigen-specific. Other chapters in this volume discuss in detail the cellular aspects of immune regulation. This communication focuses on the biochemical and serological nature of antigen-specific regulatory molecules. These molecules are exclusively the products of T-cells (immunoglobulins are obviously antigen-specific and may also serve a regulatory function, but they are discussed elsewhere). They may be divided into two categories: helper factors (T_HF), which help T effector cells or B cells to respond (e.g. become cytotoxic or make antibody); and suppressor factors (T_SF), which suppress T and/or B cell responses. These factors are extremely potent in terms of their specific activity, and are present in picogram quantities in lymphoid tissues. For these reasons, studies on the biochemical nature of these substances have progressed very slowly. A limited number of these factors have been purified to any significant degree and only a few laboratories have reported purification to homogeneity (11, 21). This state of affairs has changed recently with application of somatic

423

0732-0582/83/0410-0423$02.00

cell hybridization methodology to T cells (3) and the use of microanalytical protein purification and analysis to characterize the hybridoma T-cell product (16, 21, 49, 50). Somatic T-cell hybrids provide reasonably stable, rapidly growing cell lines, which constituitively produce T-cell factors. Primary T-cell clones have also been produced, but so far have been less useful for the extensive biochemical characterization of T-cell factors, due to culture limitations and stability (12). Most of the studies discussed here have not been carried out on fully purified factors. Thus, conclusions regarding the precise biochemical properties of most of these substances should be regarded as tentative. The reason for such caution becomes apparent with the serological evidence implicating IgV_H region genes as the source of a portion of antigen-specific T-cell factors. What follows, then, is in the nature of a progress report on what is known about antigen-specific factors, both their structure and their genetic origin. We close with a consideration of what questions need to be addressed in the immediate future.

ANTIGEN-SPECIFIC HELPER FACTORS (T_HF)

In 1980, several very excellent reviews appeared, which discussed the progress that had been made on understanding the biology, genetics, and biochemistry of antigen-specific factors (1, 37, 45). In the ensuing 2 years it is safe to say the progress has been uneven. This is particularly true of the antigen-specific helper factors. Despite the development of T-cell somatic hybrid technology, it is still difficult to make antigen-specific T_HF-producing T-cell hybrids. Because of this, and the fact that those hybridomas that have been made still produce relatively small amounts (micrograms/liter) of T_HF, biochemical characterization has proceeded slowly. The T_HF most extensively characterized are presented in Table 1. We have not attempted to list all of the T_HF that have been reported; rather, we have tried to include those factors for which there are more data than were available in 1980 (1, 37, 45). As may be seen, the list is short. However, some generalities about structure and function are beginning to emerge, and some conclusions reached from earlier studies seem to be holding up.

First, those factors that have been tested react with an antisera specific for V_H-framework determinants; many T_HF also react with anti-idiotype-specific antisera (Id+). This implies that T cells use V_H genes or a portion of V_H genes to specify T_HF. Recent results from several laboratories raise some doubt about the validity of this conclusion. Following the observations of Tonegawa and his colleagues (32), as well as others (33), that mature B cells contain rearranged V_L and V_H genes, several groups analyzed T-cell DNA (7, 19). The results have been quite consistent. T cells

Table 1 Antigen-specific helper factors[a]

Antigen	Specificity	Source	Type of response	Ig deter.	H-2 deter.	MW	Structure	pI	Specific activity	CHO	Ref.
(T,G)-A--L	(T,G)-A--L (phe,G)-A--L (T-T-G-G)-A--L	E-9M(+) (C$_3$H/SW)	IgG	V_H^+ Id$^+$ IgC$^-$	H-2^{b+}	14–17K 43–67K	1–2 chains	ND	ND	ND	2, 3, 8
Streptococcal antigen		Monkey	IgG	IgM(F$_c$)$^+$ RAMHF$^+$ $\alpha\beta_2$microglob$^+$	ND	70K	ND	4.9–5.2	ND	+	22, 53
Streptococcal antigen		Mouse	IgG	$\alpha\beta_2$microglob$^+$ RAMHF$^+$	I-A$^+$ K-J$(-)$	70K	ND	6.4–6.7	ND	+	2, 53
CGG	CGG	B10.BR T-cell clone T85–109–45	IgG	V_H^+ V_L^-	I-A$^+$ (A$_\beta$) I-E$^+$ (E$_\alpha$)	60–70K 25–32K	ND	ND	ND	ND	23–25

[a] (T,G)-A--L, (Tyrosine, glutamic acid)-alanine-lysine; (phe,G)-A--L, (phenylalanine, glutamic acid)-alanine-lysine; (T,G)-A--L, (tyrosine-tyrosine-glutamic acid-glutamic acid)-alanine-lysine; V$_H$, variable region, heavy chain; V$_L$, variable region, light chain; IgC, immunoglobulin constant region; Id, idiotype; RAMHF, rabbit antimouse helper factor antisera; $\alpha\beta_2$microglob, anti-β_2-microglobulin antisera prepared from rabbits, guinea pig, rat, goat, chicken, or human sources; ND, not determined; CHO, carbohydrate.

may rearrange D-region and J-region sequences, but no T cell has yet been found that has a productively rearranged (e.g. functional) V_H or V_L gene. This leads to the conclusion that whatever T cells use to make antigen-specific proteins, they do not need rearranged V_H or V_L genes. Furthermore, molecular probes that represent the gene sequences specifying certain idiotypes have been utilized in studies of RNA and DNA from serologically idiotype-positive (Id^+) T cells. Using these probes, the two groups conducting these studies have failed to detect functional rearrangements of the V_H genes in Id^+ T cells; neither have they been able to show the existence of mRNA coding for Id^+ proteins in the serologically Id^+ T-cell lines studied so far (27; E. Kraig, personal communication). This would suggest that the Id^+ T cells may bear a determinant that cross-reacts with anti-idiotype antisera but is not encoded by the gene specifying the original idiotype.

A fascinating result is the observation by Lamb et al (22) that the streptococcal antigen (SA)-specific T_HF from monkey and mouse react with xenoantisera against human β_2 microglobulin. Since β_2 microglobulin is homologous with constant region domains, once again there is a suggestion of common genetic origin for these T-cell factors and Ig (not to mention Thy1, H-2, etc). It is also possible that in the case of the SA-specific T_HF, β_2 microglobulin constitutes part of the molecule. In the case of mouse SA-specific T_HF, it reacts with chicken, rat, and guinea pig anti-β_2 microglobulin but not monoclonal anti-human β_2 microglobulin; monkey SA-specific T_HF react with chicken, rabbit, and rat anti-β_2 microglobulin, as well as with monoclonal anti-human β_2 microglobulin (22, 53). These data cannot discriminate between the possibility that the SA-specific T_HF are associated with β_2 microglobulin or that they have sequence homology with β_2 microglobulin, although Lamb et al favor the latter possibility. All of the factors tested react with antisera specific for H-2 loci. This raises the issue as to whether or not the mixing of putative gene products of chromosome 12 with those of chromosome 17 can occur. Since anti-Id antisera may not be detecting a chromosome 12 gene product, there may be no problem at all. Also, if some factors contain two chains, then one can easily concoct a two-gene two-polypeptide chain hypothesis to explain the results. This may be the case for one class of two-chain polypeptide suppressor factors (see below). One-chain T_HF, containing apparently both V_H and H-2 determinants, would present the biggest problem, but no T_HF of this type have yet been reported. The solution is simply to wait until more data become available concerning the nature of the genes that specify these molecules. It is notable that none of the studies summarized in Table 1 has assigned a specific activity to the factor being examined. This is unfortunate. The measurement of biological activity by dilution analysis, coupled with careful estimates of protein concentration, give one of the best measures of relative purity when one calculates activity/unit weight (e.g. milligrams or micrograms). Purity, or its lack, represents the biggest obstacle to clear-cut interpretations of data. This is particularly important when drawing conclusions about molecular weight or structural composition. For this reason, the data summarized in Table 1 must be regarded as tentative, pending more complete purification.

ANTIGEN-SPECIFIC SUPPRESSOR FACTORS

In contrast to the antigen-specific helper factors, more progress has occurred in the biochemical characterization of antigen-specific suppressor factors. Reports are beginning to appear on purification and concommitant biochemical analyses of the isolated factors (11, 21). In addition, at least two groups have begun studies aimed at the cloning of the C-DNA derived from the mRNA that codes for T_SF, which would for the first time allow us to look directly at the genes responsible for these factors (44, 51).

A large number of T_SF have been described, and new reports continue to appear (1, 37, 45). The progress has been made possible both by the continued improvement in our capacity to grow cloned T cells that maintain their function and by the use of hybridoma technology (6). Most reports of factor-producing hybridomas that have appeared since 1980 are concerned with suppressor factors rather than antigen-specific helper factors (6). Whether this is related to the almost exclusive reliance on BW5147 as a fusion partner or to other reasons is not clear. In addition to these techniques, the successful production of radiation leukemia virus-transformed T-cell lines with suppressor function has been presented recently (30, 31). The fact that such lines produce large amounts of antigen-specific T_SF, are easy to grow, and have apparently a normal chromosome complement may make them increasingly attractive to workers in this field.

Since other authors in this volume cover the regulatory aspects of suppressor and helper factors, we confine ourselves, as with T_HF, to a consideration of the biochemistry of T_SF. It is known that antigen-specific suppressor factors could be elaborated to many kinds of antigens. For example, several different synthetic polypeptides, the immune response to which is known to be under Ir gene control, induce suppressor cells. In many cases these cells have been shown to release suppressor factors that block the appearance of antibody-forming cells (18, 20, 34, 48). Suppressor factors that block the humoral immune response to natural proteins such as KLH or HGG have also been reported (39, 40). Cellular immune responses to alloantigens or tumor antigens may be shown to be regulated by suppressor cells and T_SF (15, 38). Also, haptens, either on proteins or on autologous cells, may induce specific T_SF (14, 54). It may be concluded that virtually every type of immune response results in the generation of suppressor cells that may release antigen- or determinant-specific T_SF. It was also shown that such factors could be idiotype specific rather than antigen specific and still be able to induce antigen-specific suppression (36). Many of the factors reacted with both anti-V region antisera (under which heading for convenience we include anti-idiotype) and anti-I-region-specific antisera; and none had Ig constant-region markers, as they are defined serologically (1, 37, 45).

Rather than present a complete listing of all of the suppressor factors described since 1980, once again we focus on those factors for which we now have substantial biochemical information. A compilation of these factors and their characteristics is listed in Table 2. There are four groups of investigators who have been carrying out the most extensive biochemical analyses. They include (a) our group, which is studying factors specific for the synthetic copolymers GAT, GA, and GT (16, 21, 49); (b) Cantor and his associates, studying a SRBC-specific factor (10–12); (c) Taniguchi and Tada and their co-workers, studying KLH-specific suppressor factors (41–43, 46, 47); and (d) Dorf, Greene, Benacerraf, and their associates, who have done extensive serological studies and perhaps have the most complete set of suppressor factors specific for NP (26, 28, 29). Many other investigators have performed studies on suppressor factors, but they have emphasized the more biological aspects of these agents and for that reason these other factors are not listed here (1, 37, 45).

Fresno et al have studied T_SF obtained from a cloned T-cell line using radiolabeled T_SF. The T_SF in Table 2 have been shown to suppress primary humoral immune responses, secondary humoral immune responses, or cell-mediated responses. All of the factors react with antisera specific for one or another determinant on antibody V_H chains. All but one factor, the glycophorin T_SF, react with H-2I region-specific antisera. Where study has been possible, many of the factors show MHC restriction or Igh restriction, or both. In terms of molecular weights, these molecules range between 19,000–80,000 and may be composed of one or two chains. One of the more interesting features of the table is the listing of specific activities. Of the four research groups, only two have made estimates of specific activity. In our opinion, such estimates are crucial to any biochemical study, since such estimates represent the only way of assessing the degree of purity.

Very little is known concerning structure-function relationships of the majority of the T_SF. This is no doubt largely because biochemical studies are quite recent. Fresno et al (10) have reported that the glycophorin T_SF may be dissociated into an antigen-binding peptide of 25,000-MW and a 45,000-MW protein that retains the capacity to be suppressive; however, it now lacks antigenic specificity. The studies of MHC and immunoglobulin

Table 2 Antigen-specific suppressor factors[a]

Antigen	Specificity	Source	Type of response	Ig deter.	H-2 deter.	MHC/Igh restric.	MW	Structure	pI	Specific act. S_{50} U/µg	CHO	Reference
GAT	GAT/GA	hybridoma (258C4.4)	1° IgG	Id+ IgC-	H-2I-J^q	none/none	29,000	1 chain, hydrophobic	ND	7.8×10^7	+	16, 21 49, 50
GAT	GAT/GT	hybridoma (342B1.11)	1° IgG	Id+ IgC-	H-2I-J^s	none/none	29,000	1 chain, hydrophobic	ND	2.6×10^8	+	16, 49, 50
GAT	GAT/GT	hybridoma (372B3.5)	1° IgG	Id+ IgC-	H-2I-J^b	none/none	29,000 (19,000)	1 chain, hydrophobic	6.2–6.4	14.2×10^7	+	16, 49, 50
GT	GT/GAT	hybridoma (395A4.4)	1° IgG	Id+ IgC-	H-2I-J^s	none/none	29,000 (19,000)	1 chain, hydrophobic	ND	9.0×10^7	ND	16, 49, 50
SRBC	Glycophorin (sheep)	T-cell line C1Ly23/4	1° IgM	IgC- IgV$_H^+$	Ia$^-$	ND/ND	70,000	2 subunits 45K, 25K 1 chain	4.9–5.1	1.0×10^5*	+	10–12
KLH	KLH	hybridoma	2° IgG	Igh+	K-J^{b+}	I-J^b/Igh^b	42–68,000	2 chains 1 Ag binding Igh^{b+}; 1 I-J^{b+}	ND	ND	ND	41–43 46, 47
NP	NP	hybridoma (TS$_1$)	DTH	IgC- Id+ IghC- IghV+	I-J$^+$	none/IghV	ND	ND	ND	ND	ND	26, 28, 29
NP	NP^b idiotype	hybridoma (TS$_2$)	DTH	IgC-	I-J$^+$	I-J/Igh	ND	ND	ND	ND	ND	26, 28, 29
NP	NP	hybridoma (TS$_3$)	DTH	IgC- Id+	K-J$^+$	I-J/Igh	ND	ND	ND	ND	ND	26, 28, 29

[a] GAT, Glutamic acid^{60}-alanine^{30}-tyrosine^{10}; GT, glutamic acid-tyrosine; GA, glutamic acid-alanine; 1°, primary immune response; 2°, secondary immune response; DTH, delayed-type hypersensitivity; Id, idiotype; IgC, Ig-constant region; IgV$_H$, Ig heavy chain variable region; Igh, immuno-globulin heavy chain allotype; ND, not determined; CHO, carbohydrate.

*Specific activity was calculated from data presented in reference 11.

allotype restriction suggest that both H-2 determinants and Igh determinants present on T_SF are required for function. In this regard, the recent results of Sorensen et al are of interest (17). They have obtained two-chain GAT-T_SF from a hybridoma (372D6.5) in which they have confirmed the observations of Taniguchi et al (41–43, 46, 47). That is, one chain carries the antigenic specificity and one chain bears an I-J determinant (I-J^b). When the two chains are dissociated by reduction and alkylation, they may still be shown to have suppressive activity when added together. What is most surprising, the antigen-binding chain may be added to spleen cell cultures given antigen; 24–48 hrs later, the I-J-bearing chain may be added and suppression of the anti-GAT response is still observed (17). This result suggests that in tissue culture, at least, antigen-binding chains may function in the absence of I-J chain and the I-J chain may perform its function possibly without the antigen-binding chain. That such complex structure-function relationships are the norm rather than the exception is also suggested by recent data reported by Gershon and his associates (9, 52), who propose that suppressor-inducer factors may form molecular associations with "schlepper" or carrier moieties which help to direct them to the proper cellular receptor.

Another aspect of structure-function is specificity. In the synthetic polypeptide system (GAT, GA, or GT), it may be seen that hybridomas derived from fusions with GAT as the immunizing antigen may be either GAT/-GA- or GAT/GT- specific. Of interest are the two hybridomas derived from fusions with H-2^S splenic T cells. Whether the immunizing antigen was GT or GAT, the binding specificity to GAT-, GA-, or GT-Sepharose is the same: both 342.B1.11 and 395A4.4 T_SF bind GAT or GT. In fact, 395A4.4 may be analyzed using GAT-MBSA-primed cells with identical results to GT-MBSA-primed cells. The amino terminal sequences currently available through residue 30 show few differences between GAT- or GT-specific T_SF (D. Webb, manuscript in preparation). Where the specificity-determining regions are in these molecules remains to be determined.

The SRBC-induced T_SF-producing clone C1Ly23/4 was shown by Fresno et al (11) to have specificity for glycophorin. Interestingly, these T_SF can discriminate (apparently) glycophorins from different species of red blood cells (e.g. sheep versus burro). To date, no studies have been performed to determine the precise specificity of the KLH-T_SF.

In addition to the results of studies shown in Table 2, two groups have also characterized to a similar extent the mRNA that codes for their respective suppressors. A summary of these results is presented in Table 3. Wieder, in our laboratory (51), partially purified polyA$^+$ RNA from hybridoma 258C4.4 by subjecting it to sucrose density gradient centrifugation. Fractions were collected and were translated in a rabbit reticulocyte lysate

Table 3 Characterization of mRNA specific for T_SF

Antigen	Specificity	Source	mRNA size	In vitro translation	MW	Cell-free translation product[a]				Reference
						Ig deter.	MHC deter.	Ag binding	Structure	
GAT	GAT/GA	hybridoma 258C4.4	14–16S	rabbit reticulocyte lysate	19,000[b] 28,000[c]	ND	I-J^q	+	1 chain hydrophobic	51
GAT	GAT/GT	hybridoma 342B1.11	16S	rabbit reticulocyte lysate	21,500[c] 84,000[c]	ND	ND	ND	ND	50
GAT	GAT/GT	hybridoma 372B3.5	19S	rabbit reticulocyte lysate	34,000 80,000	ND	I-J^{b+}	ND	ND	35
KLH	KLH	hybridoma	11S 13S 18S	frog oocyte	25–27,000[d] 29–34,000[d] 45–62,000[d]	Ig$^-$ Igh$^+$ Igh$^+$	I-J$^+$	– + +	2 chains I-J$^+$ and Ag$^+$Igh$^+$	44

[a] ND, Not determined.
[b] Based on bioactivity as determined by SDS-PAGE following affinity chromatography and HPLC.
[c] Based on bioactivity recovered from HPLC gel permeation columns.
[d] Based on SDS-PAGE analysis and immune precipitation.

translation system, and the translated products were assayed for GAT-specific suppression. RNA that had a sedimentation coefficient of 16S was able to code for GAT-T_SF. The molecular properties of this translated product indicate that it is similar if not identical to supernatant GAT-T_SF from 258C4.4. Other investigators in our group (35; K. Krupen, unpublished data; E. Gerassi, unpublished data; C. Healy, submitted for publication; C. Sorensen, manuscript in preparation) have translated polyA$^+$ RNA from T-cell hybridomas and have obtained similar results. An intriguing feature in some hybridomas is the appearance in the cell-free translation products of a larger molecular weight form of T_SF (80,000). What relationship this has to the lower molecular weight form remains to be determined. In another study, Taniguchi and co-workers (44) isolated polyA$^+$ RNA from a KLH-T_SF-producing hybridoma and subjected it to sucrose density centrifugation. They were able to isolate three species of RNA. Two of these RNAs, 13S and 18S, when translated in a frog oocyte system, yielded the antigen-binding chain, and the I-J-bearing chain was derived from an mRNA with a sedimentation coefficient of 11S. The properties of the translated products are similar if not identical to those of T_SF obtained from culture supernatants of the hybridomas. Both our group and that of Taniguchi are involved in the molecular cloning of their respective mRNAs. This suggests that in the not-too-distant future we should be able to study directly the genes involved in the synthesis of regulatory T-cell products. Such molecular probes will undoubtedly be useful in deciphering the nature of the antigen receptor on T cells.

Having outlined the biochemical data currently available on antigen-specific T_SF, it is necessary to address the significance of these results. Some notable differences shown in Table 2 are the different molecular weights that have been determined for the different factors and the differing structures (e.g. one chain or two). A plausible reason for these differences may be derived from studies in the GT system as well as the data obtained with the NP hapten, which suggest that several types of suppressor factors exist. Indeed, Germain & Benacerraf, in a recent review (13), have concluded that a suppressor cell "cascade" exists in which T_SF serve as the links between different suppressor cell subclasses, as well as acting on nonsuppressor targets, T-helper cells, or B cells. Although it may be that the details of the Germain & Benacerraf "consensus pathway" are incorrect, the general idea seems to fit the available biochemical data. In the GT system, it is possible to show that single-chain, non-MHC-restricted, antigen-specific factors (T_SF_1) induce the appearance of MHC-restricted (presumably two chain) antigen-specific suppressor factors (T_SF_2). In the NP system, a cascade has been established involving an antigen-specific idiotype-positive suppressor inducer T_SF_1, which induces an anti-idiotype T_SF_2 which induces antigen-specific T_SF_3. It is the T_SF_3 that directly interacts with T_H or B cells. If one accepts this kind of scheme, it is possible to assign the KLH-T_SF to the T_SF_2 class, based on homology to GAT-T_SF_2 currently being studied by C. Sorensen, D. Webb, and C. Pierce (manuscript in preparation). This T_SF_2 is composed of two chains, one antigen-specific and one I-J$^+$, and is induced by T_SF_1. Fresno et al (12) have preliminary data that suggest that the glycophorin T_SF can directly affect B cells, which would place them in the T_SF_3 class. Note that the T_SF_2 may be either antigen-specific or idiotype-specific. Although these explanations remain hypothetical to a large extent, they serve to illustrate the point that some of the different factors presented in Table 2 almost certainly represent different classes of T_SF.

An issue related to molecular and structural heterogeneity is the question of serological or immunochemical analysis. Virtually all the serological evidence for Igh, V_H, idiotype determinants, or H-2 determinants must be regarded as tentative. The principal reason for making so sweeping a statement is that for each case mentioned above, the antisera were initially raised with another antigen. Thus, reactions of T_SF with these sera are cross-reactive. In most instances it is not known whether the antisera recognize sequence-specific determinants or conformationally derived determinants.

As mentioned earlier in the discussion of T_HF, it is possible to demonstrate expression of idiotype-positive material by T cells serologically in the absence of any detectable mRNA that can hybridize to the IgV$_H$ gene probes that encode idiotype-positive Ig. In the case of the GAT-T_SF, for example, E. Kraig (personal communication) has shown that RNA extracted from several GAT-T_SF-producing hybridomas will not hybridize a molecular probe that codes for idiotype-bearing anti-GAT antibody (IgG). Nonetheless, the GAT-T_SF will react with xenoantisera specific for this idiotype (GAT). Furthermore, preliminary sequence analysis of three GAT-T_SF from three different hybridomas (258C4.4, 372B3.5, 395A4.4; Table 2) shows no homology so far, with either anti-GAT monoclonal antibody or any known V_L or V_H region sequences. Such data suggest one cannot be too cautious about drawing conclusions from serological data in the absence of corroborating information.

In some of the cases shown in Table 2, corroboration does exist. In the NP system, the existence of I-J reactivity on the T_SF and an I-J restriction at the cell target level supports the notion of a true I-J determinant on NP-T_SF. Likewise, the Igh determinants and Igh restriction imply a strong likelihood that an Igh-linked gene or genes plays a role in the structure and function of T_SF. The fact that both Tada (42) and Taniguchi (47) can show by differential absorption the presence of unique I-region gene products as well as Igh gene products on T cells strongly suggests that both sets of genes are important in determining the structure and function of T_HF and T_SF.

The possible existence of two determinants on a single polypeptide (GAT-T_SF), each specified by genes that lie on separate chromosomes, is a difficult problem. One possible solution, particularly where anti-idiotypic antisera are used, is that the apparent presence of the idiotype on the factor is misleading. In the case of the GAT-T_SF, the anti-idiotypic antisera, anti-cGAT and anti-GA-1, have not been adsorbed with idiotype first and then tested for binding to GAT-T_SF; thus, the idiotype-positive GAT-T_SF may turn out either to be cross-reactive or to reflect the limited heterogeneity of amino acids allowed in the combining site for any GAT-specific protein. The case for the I-J determinant reflecting a real rather than a cross-reacting epitope is stronger in those cases where I-J restriction has been noted (e.g. NP and KLH). It is also more convincing that the epitope is at least "I-J like" in those instances where monoclonal antibodies are available. This is the case for two of the GAT-T_SF, the KLH-T_SF, and the NP-T_SF molecules. The experiments with the cell-free translation product of GAT-T_SF mRNA are also of interest in this regard. Wieder et al (51) showed that GAT-T_SF generated from mRNA translated in a rabbit reticulocyte lysate system were bound by anti-I^a-Sepharose. Since little if any glycosylation takes place in the microsome-free, rabbit reticulocyte lysate system, these results suggest that the I-region determinant recognized is not carbohydrate. These results were confirmed by Sorensen et al (35) using GAT-T_SF (H-2^b) translated in a rabbit reticulocyte lysate system, which was bound by monoclonal anti-I-J^b-Sepharose (provided by C. Waltenbaugh).

THE T-CELL ANTIGEN RECEPTOR AND ANTIGEN-SPECIFIC FACTORS

Part of the reason for studying T_SF or T_HF is the possibility that such factors will be related to the elusive T-cell receptor for antigen. Indeed, there are many serological similarities between TsF and T-cell membrane moieties. For example, both T cells and factors may bear idiotype, I-J, or Igh determinants (1, 37, 45), and Binz and Wigzell and their colleagues (3–5) have implicated idiotype-positive molecules as being part of the T-cell antigen receptor. Several factors (e.g. all of the GAT-T_SF and the KLH-T_SF) can be isolated from membrane extracts of T-cell hybridomas. Indeed, Matsuzawa & Tada (37) showed that I-J$^+$ T cells could bind soluble antigen and that anti-I-J antisera could block antigen-binding to I-J-positive cells. The authors interpreted these results as implicating I-J-bearing molecules in involvement at antigen-binding sites. A major difficulty that arises, however, is the structural heterogeneity of T-cell factors in comparison to the Ig on B cells. Setting aside for the moment the fact that Ig may form dimers or pentamers, the basic structure of Ig is the same from isotype to isotype, e.g. two light chains and two heavy chains linked to one another by disulfide bridges. By contrast, among the antigen-specific T_SF, there are single polypeptide chain molecules of 29,000, an apparently single chain of 70,000, and two-chain T_SF, which may have disulfide bonds or not depending on their cellular location. Do these molecules reflect the antigen receptor on the respective T cells? The question must remain unanswered. It should be noted that C. Healy et al (submitted for publication) have found in a GAT-T_SF-producing hybridoma a membrane form of GAT-T_SF that exists primarily as a dimer of 60,000–70,000 molecular weight (with minor species of higher molecular weight polymers). Since its amino acid content is identical (in moles percent) to the 29,000-molecular-weight form, the 29,000- and 60,000-MW GAT-T_SF appear to be identical. C. Sorensen (manuscript in preparation), in a related study of GAT-T_SF from a responder strain (H-2^b) finds that GAT-T_SF in their pure state have a strong propensity to aggregate. Thus, multimers of the basic 29,000-MW protein begin to appear and ultimately cause the protein to precipitate from solution even though the solution contains a hydrophobic solute (propanol).

All of the characteristics outlined above should serve as a caution to the unwary that considerable work remains to be done to unravel the nature of the T-cell antigen receptor. The best prediction at this time suggests that such a receptor will reflect the nature of the cell from which it is derived, as well as reflecting the nature of that cell's products. It is unlikely that different T-cell subsets use the same antigen receptor, if we use the antigen-specific T-cell factors as a guide.

Additional questions concerning T-cell product and receptor diversity, and the extent of T-cell repertoires, the influence of haplotype on these, and their relationship to the T-cell antigen receptor must await the biochemical analysis that is just beginning. It is likely in the next 5 years that many of the remaining questions will have found at least partial answers.

Literature Cited

1. Altman, A., Katz, D. H. 1980. Production and isolation of helper and suppressor factors. *J. Immunol. Meth.* 38:9–41
2. Apte, R. N., Löwy, I., DeBaetscher, P., Mozes, E. 1981. Establishment and characterization of continuous helper T cell lines specific to poly (L tyr, L glu)-poly(DL ala)—poly(L lys). *J. Immunol.* 127:25–30
3. Binz, H., Soots, A., Nemlander, A., Wight, E., Fenner, M., et al. 1982. Induction of specific transplantation tolerance via immunization with donor-specific idiotypes. *Ann. NY Acad. Sci.* 392:360–74
4. Binz, H., Wigzell, H. 1976. Specific transplantation tolerance induced by autoimmunization against the individual's own naturally occurring idiotypic antigen-binding receptors. *J. Exp. Med.* 144:1438–57
5. Binz, H., Wigzell, H. 1981. T-cell receptors with allo-major histocompatability complex specificity from rat and mouse. *J. Exp. Med.* 154:1261–78
6. Boehmer, H. V., Haas, W., Köhler, G., Melchers, F., Zeuthen, T., eds. 1982. T-cell hybridomas. *Curr. Top. Microbiol. Immunol.* 100
7. Cory, S., Adams, J. M., Kemp, D. J. 1980. Somatic rearrangements forming

active immunoglobulin μ genes in B and T lymphoid lines. *Proc. Natl. Acad. Sci. USA* 77:4943–47

8. Eshhar, Z., Woks, T., Zinger, H., Mozes, E. 1982. T-cell hybridomas producing antigen-specific factors express heavy chain-variable-region determinants. *Curr. Topics Microbiol. Immunol.* 100:103–9

9. Flood, P., Yamauchi, K., Gershon, R. 1982. Analysis of the interactions between two molecules that are required for the expression of Ly-2 suppressor cell activity. Three different types of focusing events may be needed to deliver the suppressive signal. *J. Exp. Med.* 156:361–71

10. Fresno, M., McVay-Boudreau, L., Cantor, H. 1982. Antigen-specific T-lymphocyte clones III. Papain splits purified T suppressor molecules into two functional domains. *J. Exp. Med.* 155:981–93

11. Fresno, M., McVay-Boudreau, L., Nabel, G., Cantor, H. 1981. Antigen-specific T lymphocyte clones. II. Purification and biological characterization of an antigen-specific suppressive protein synthesized by cloned T-cells. *J. Exp. Med.* 153:1246–59

12. Fresno, M., Nabel, G., McVay-Boudreau, L., Furthmayer, H., Cantor, H. 1981. Antigen-specific T lymphocyte clones. I. Characterization of a T lymphocyte clone expressing antigen-specific suppressive activity. *J. Exp. Med.* 153:1260–74

13. Germain, R. N., Benacerraf, B. 1981. A single major pathway of T lymphocyte interactions in antigen-specific immune suppression. *Scand. J. Immunol.* 13:1–10

14. Greene, M. I., Bach, B. A., Benacerraf, B. 1979. Mechanisms of regulation of cell mediated immunity. III. The characterization of Azobenzenearsonate-specific suppressor T-cell derived suppressor factors. *J. Exp. Med.* 149:1069–83

15. Greene, M. I., Fujimoto, S., Sehon, A. H. 1977. Regulation of the immune response to tumor antigens. III. Characterization of thymic suppressor factor(s) produced by tumor bearing hosts. *J. Immunol.* 119:757–63

16. Healy, C. T., Kapp, J. A., Stein, S., Brink, L., Webb, D. R. 1983. Comparative analysis of a monoclonal antigen-specific T suppressor factor obtained from supernatant, membrane or cytosol of a T-cell hybridoma. In *Ir Genes: Past, Present and Future,* ed. C. W. Pierce, S.

E. Cullen, J. A. Kapp, B. D. Schwartz, and D. C. Schreffler. Clifton, NJ: Humana. In press

17. Kapp, J. A., Araneo, B. A., Sorensen, C. M., Pierce, C. W. 1982. Identification of H-2 restricted suppressor T-cell factor specific for L-glutamic acid^{50}-L-tyrosine^{30} (GT) and L-glutamic acid^{60}-L-alanine^{30}-L-tyrosine^{10} (GAT). In *Ir Genes: Past, Present and Future,* ed. C. W. Pierce, S. E. Cullen, J. A. Kapp, B. D. Schwartz, and D. C. Shreffler. Clifton, NJ: Humana. In press

18. Kapp, J. A., Pierce, C. W., DeLa Croix, F., Benacerraf, B. 1976. Immunosuppressive factor(s) extracted from lymphoid cells of nonresponder mice primed with L-glutamic acid^{60}-L-alanine^{30}-L-tyrosine^{10} (GAT). I. Activity and antigenic specificity. *J. Immunol.* 16:306–15

19. Kemp, D. J., Wilson, A., Harris, A. W., Shortman, K. 1980. The immunoglobulin β constant region gene is expressed in mouse thymocytes. *Nature* 286:169–70

20. Kontiainen, S., Howie, S., Maurer, P. H., Feldmann, M. 1979. Suppressor cell induction in vitro. VI. Production of suppressor factors to synthetic poly peptide GAT and (T,G)-A-L from cells of responder and non-responder mice. *J. Immunol.* 122:253–59

21. Krupen, K., Araneo, B., Brink, L., Kapp, J. A., Stein, S., et al. 1982. Purification and characterization of a monoclonal T cell suppressor factor specific for L-glutamic acid^{60}-L-alanine^{30}-L-tyrosine^{10} (GAT). *Proc. Natl. Acad. Sci. USA* 79:1254–59

22. Lamb, J. R., Zanders, E. D., Sanderson, A. R., Ward, P. J., Feldmann, M., et al. 1981. Antigen-specific helper factors reacts with antibodies to human β_2 microglobulin. *J. Immunol.* 127:231–34

23. Lonai, P., Arman, E., Bitton-Grossfield, J., Grooten, J., Hammerling, G. 1982. H-2 restricted helper hybridomas: One locus or two control dual specificity? *Curr. Top. Microbiol.* 100:97–102

24. Lonai, P., Puri, J., Bitton, S., Ben-Meriah, Y., Givol, D., Hammerling, G. 1981. H-2 restricted helper factor secreted by cloned hybridoma cells. *J. Exp. Med.* 154:942–49

25. Lonai, P., Puri, J., Hammerling, G. 1981. H-2 restricted antigen-binding by a hybridoma clone that produces antigen-specific helper factor. *Proc. Natl. Acad. Sci. USA* 78:549–53

26. Minami, M., Okuda, K., Furusawa, S., Benacerraf, B., Dorf, M. E. 1981. Ana-

lysis of T-cell hybridomas. I. Characterization of H-2 and Igh-restricted monoclonal suppressor factors. *J. Exp. Med.* 154:1390–401

27. Nakanishi, K., Sugimura, K., Yaoita, Y., Maeda, K., KashiWamura, S., Honjo, T., Kishimoto T., 1982. A T15 idiotype positive T suppressor hybridoma does not use the T15 V_H gene segment. *Proc. Natl. Acad. Sci. USA* 79:6984–8

28. Okuda, K., Minami, M., Furusaura, S., Dorf, M. E. 1981. Analysis of T cell hybridomas. II. Comparisons among three distinct types of monoclonal suppressor factors. *J. Exp. Med.* 154:1838–51

29. Okuda, K., Minami, M., Sherr, D. H., Dorf, M. E. 1981. Hapten-specific T cell responses to 4-hydroxy-3-nitrophenyl acetyl. XI. Pseudogenetic restrictions of hybridoma suppressor factors. *J. Exp. Med.* 154:468–79

30. Ricciardi-Castagnoli, P., Doria, G., Adorini, L. 1981. Production of antigen-specific suppressive T cell factor by radiation leukemia virus-transformed suppressor T cell. *Proc. Natl. Acad. Sci. USA* 78:3804–8

31. Ricciardi-Castagnoli, P., Robliati, F., Barbanti, E., Doria, G., Adorini, L. 1982. Establishment of functional, antigen-specific T cell lines by RadLV-induced transformation of murine T lymphocytes. *Curr. Top. Micro. Immunol.* 100:89–96

32. Sakano, H., Maki, R., Kurosama, Y., Roeder, W., Tonegawa, S. 1980. Two types of somatic recombination are necessary for the generation of complete immunoglobulin heavy chain genes. *Nature* 286:676–83

33. Seidman, J. G., Leder, A., Nan, M., Norman, B., Leder, P. 1978. Antibody diversity. *Science* 202:11–17

34. Sorensen, C. M., Pierce, C. W. 1982. Antigen-specific suppression in genetic responder mice to L-glutamic acid^{60}-L-alanine^{30}-L-tyrosine^{10} (GAT). Characterization of conventional and hybridoma derived factors produced by suppressor T cells from mice injected as neonates with syngeneic GAT-macrophages. *J. Exp. Med.* 156:1691–710

35. Sorensen, C. M., Webb, D. R., Pierce, C. W. 1982. Characterization of an antigen-specific suppressor factor generated by cell-free translation. In *Ir Genes: Past, Present and Future*, ed. C. W. Pierce, S. E. Cullen, J. A. Kapp, B. D. Schwartz, and D. C. Shreffler. Clifton, NJ: Humana. In press

36. Sy, M-S., Dietz, M. H., Germain, R. N., Benacerraf, B., Greene, M. I. 1980. Antigen and receptor driven regulatory mechanisms. IV. Idiotype bearing I-J$^+$ suppressor T-cell factors induce second order suppressor cells which express anti-idiotype receptors. *J. Exp. Med.* 151:1183–95

37. Tada, T., Okumura, K. 1980. The role of antigen-specific T cell factors in the immune response. *Adv. Immunol.* 28:1–87

38. Takei, F., Levy, J. G., Killburn, D. G. 1978. Characterization of a soluble factor that specifically suppresses the in vitro generation of cells cytotoxic for syngeneic tumor cells in mice. *J. Immunol.* 120:1218–24

39. Takemori, T., Tada, T. 1975. Properties of antigen-specific suppressive T-cell factor in the regulation of antibody response in the mouse. I. In vivo activity and immunochemical characterizations. *J. Exp. Med.* 142:1241–53

40. Taniguchi, M., Miller, J. F. A. P. 1978. Specific suppression of the immune response by a factor obtained from spleen cells of mice tolerant to human γ-globulin. *J. Immunol.* 120:21–26

41. Taniguchi, M., Saito, T., Takei, I., Tokuhisa, T. 1981. Presence of interchain disulfide bonds between two gene products that compose the secreted form of an antigen-specific suppressor factor. *J. Exp. Med.* 153:1672–77

42. Taniguchi, M., Saito, T., Tada, T. 1979. Antigen-specific suppressive factor produced by a transplantable I-J bearing T-cell hybridoma. *Nature* 278:555–58

43. Taniguchi, M., Takei, I., Tada, T. 1980. Functional and molecular organization of an antigen-specific suppressor factor from a T-cell hybridoma. *Nature* 283:227–28

44. Taniguchi, M., Tokuhisa, T., Tanno, M., Yaoita, Y., Shimizu, A., Honjo, T. 1982. Reconstitution of antigen-specific suppressor activity with translation products of mRNA. *Nature* 298:172–74

45. Taussig, M. 1980. Antigen-specific T-cell factors. *Immunology* 41:759–85

46. Tokuhisa, T., Taniguchi, M. 1982a. Two distinct allotypic determinants on the antigen-specific suppressor and enhancing T-cell factors that are encoded by genes linked to the immunoglobulin heavy chain locus. *J. Exp. Med.* 155:126–39

47. Tokuhisa, T., Taniguchi, M. 1982b. Constant region determinants on the

antigen-binding chain of the suppressor T-cell factor. *Nature* 298:174–76

48. Waltenbough, C., Debre, P., Theze, J., Benacerraf, B. 1977. Immunosuppressive factor(s) specific for L-glutamic acid^{50}-L-tyrosine^{50} (GT). I. Production, characterization and lack of H-2 restriction for activity in recipient strain. *J. Immunol.* 118:2073–77

49. Webb, D. R., Araneo, B. A., Healy, C., Kapp, J. A., Krupen, K., et al. 1982. Purification and biochemical analysis of antigen-specific suppressor factors isolated from T-cell hybridomas. *Curr. Top. Micro. and Immunol.* 100:53–59

50. Webb, D. R., Araneo, B. A., Gerassi, E., Healy, C., Kapp, J. A., et al 1983. Purification and analysis of antigen-specific suppressor proteins derived from T-cell hybridomas. *Second Int. Conf. on Immunopharmacol.* In press

51. Wieder, K. J., Araneo, B. A., Kapp, J. A., Webb, D. R. 1982. Cell-free transla-

tion of a biologically active, antigen-specific suppressor T-cell factor. *Proc. Natl. Acad. Sci. USA* 79:3599–603

52. Yamauchi, K., Chao, N., Murphy, D. B., Gershon, R. K. 1982. Molecular composition of an antigen-specific, Ly-1 T suppressor inducer factor: One molecule binds antigen and is I-J$^-$; another is I-J$^+$ does not bind antigen, and imparts an Igh-variable region-linked restriction. *J. Exp. Med.* 155:655–65

53. Zanders, E. D., Lamb, J. R., Kontiainen, S., Lehner, T. 1980. Partial characterization of murine and monkey helper factor to a Streptococcal antigen. *Immunology* 41:587–96

54. Zimbala, M., Asherson, G. I. 1974. T cell suppression of contact sensitivity in the mouse. II. The role of soluble suppressor factor and its interaction with macrophages. *Euro. J. Immunol.* 4:794–804

Ann. Rev. Immunol. 1983. 1:439–63

IMMUNOREGULATORY T-CELL PATHWAYS

Douglas R. Green, Patrick M. Flood, and Richard K. Gershon

Department of Pathology, The Howard Hughes Medical Institute for Cellular Immunology, Yale University School of Medicine, New Haven, Connecticut 06510

INTRODUCTION

Towards the end of the 18th and the beginning of the 19th century an alteration occurred in Western thinking that allowed radical changes in many fields of study, including the life sciences.[1] Prior to this time, the study of nature was predominantly hermeneutic, concerned with classification and interpretation of "signs" that would show one the relationship of a plant or animal to its place in the universe (for example, its role in a medical cure). This perception of the order of reality persisted through the 18th century, until a new view emerged, in which underlying causes and processes were examined. Thus, nature and life moved into the abstraction; the word biology was coined in 1800 by Burdach and extended in 1802 to the study of living bodies by Treviranus and again by Lamark. It was also at this time (1803) that the germ theory of disease first appeared (Med. J. IX, 484). In 1826, Fidele Bretonneau concluded that diseases are specific, developing through specific reproducing agents. Agostino Bassi (1836) isolated a parasitic fungus as the causative agent in a silkworm disease, and John Snow (1849) postulated similar agents operative in the communication of cholera. Pasteur began his studies in fermentation in 1854, which led to those studies that, together with those of Koch, resulted in the acceptance of the germ theory. Thus, disease came to be understood as an invasion of a healthy body by an infectious agent.

Our understanding of immunology is firmly based upon this view. Although Pasteur generalized the vaccination procedure of Jenner to include

---

[1]The views presented in this introduction were largely influenced by Foucault (1). Other references to these historical perspectives are found elsewhere (2,3).

0732-0582/83/0410-0439$02.00

many diseases, it was Kitasato and von Behring who proposed the humoral theory of immunity, which was given a molecular basis by Paul Ehrlich.

Thus, the view of immunity as recognition and response to an invasion by a foreign substance has dominated the study of the immune system. The mechanism by which self and non-self are discriminated has only recently become a pressing one in spite of Ehrlich's conception of "horror autotoxicus," the body's (usual) failure to react against itself.

Most immunologists today would agree with this view of immunology. Yet, science in the 20th century is undergoing another alteration in view. Not only can nature be unified by virtue of underlying processes, but processes themselves can be understood in abstract terms (4). Information theory and cybernetics present ways of looking at complex systems in a novel way.

The immune system can be viewed as being in a complex homeostasis, in which recognition of self (rather than non-self) is a key feature. This concept represents a paradigm shift, with resulting shifts in emphasis. Thus, the immune response can be viewed as a heuristic process in which foreign antigen in the context of particular self antigens perturbs equilibrium, producing a regulatory shift leading ultimately to positive responses (immunity) or negative ones (tolerance). Immune regulation emerges, then, as a central area in immunology. It is the complex set of regulatory interactions that allows the immune system to "think" about its primary decision to generate a response that produces maximum destruction of a potential pathogen with minimum damage to self tissues.

Immune regulation is a young field of study, and as usually happens in new areas, terminology is not yet standardized. Different groups have adopted different terms for the systems under study, and attempts to reconcile all of these terms and systems (5) may be premature until they are better understood.

In this review we consider only one of several outlooks on immune regulation, that of our laboratory and perhaps a few others. It is our hope that this will aid the commendable attempts to unify immunoregulatory theory.

IMMUNOREGULATORY CIRCUITS AND THEIR DISSECTION

The regulation of the immune response occurs by the interactions of different subsets of lymphocytes, especially T cells. Regulation may be negative (suppressive) or positive, as is discussed. Each regulatory T-cell function stems from interactions of cells that form circuits with a consistent structure. Activation of a circuit involves an induction event, usually a signal from an inducer-cell subset. This signal is accepted by another subset, which transduces that signal (thus "transducer" subset) into an effect, via effector cells. Functionally, then, inducer cells act indirectly and thus are defined by the requirement for transducer and/or effector cells for their activity to become manifest. Transducer cells, in turn, are defined as those cells that accept the inducer cell signals. This admittedly circular definition is none-theless useful. Effector cells are defined by their production of active mole-cules that contain the final information signal of the circuit.

In each case, functions can be ascribed to T cells that may correlate with discrete subpopulations of cells, or which may represent multiple capabili-ties of a single T-cell set. To distinguish between these possibilities, it is necessary to develop means to separate T-cell populations and match each to its function.

In multicellular organisms, cells differentiate to a point where their activ-ity extends beyond cellular survival to that of the overall individual. A cell achieves this by the activation of discrete families of genes designed for its specific function ("luxury" genes, because they are not vital to the cell's immediate livelihood and therefore they tend to be shut down during cellu-lar crisis). In the course of this activation, a profile of molecular entities encoded by these genes may appear on the cell surface. Thus, the surface phenotype of a cell (such as a particular T lymphocyte) correlates with its specialized function in the whole organism. With the use of specific reagents directed against such surface labels, unique profiles of cellular antigens can be associated with distinct functional populations. Note that there is no requirement in such an approach for knowledge of what each surface label is actually doing there, only that it is there on one type of cell and not on another (of course such knowledge may be extremely valuable in under-standing the mechanisms of a cellular activity).

Examples of surface markers that are important to the study of regula-tory circuits in the murine immune system are the Ly antigens (6), particu-larly Ly-1 and Ly-2; the Qa antigens (7), particularly Qa-1; antigen associated with the I region of of the H-2, such as I-A (8) and I-J (9); and specific lectin receptors (10).

Throughout this review a dominant theme is the correlation of a unique cell surface profile with a unique function to define cell subsets in regulatory circuits.

ORGANIZATION OF IMMUNOREGULATORY CIRCUITS

In this review, regulatory T-cell circuits have been organized into levels of activity. This has been done acording to a simple scheme. When antigen perturbs homeostasis, there are cellular interactions that are necessary and sufficient for a response to occur (e.g. helper T-cell activation of a B cell to produce specific antibody). In the absence of one of these necessary cell populations because of, say, a genetic deficiency, no response will occur. This, then, would be a failure to respond at the ground state (level 0). Many theories of Ir gene function postulate genetic control at this level (e.g. failure in ability to present antigen) (11).

Activation by antigen perturbation raises the system to level 1. At this level, failure to respond may be due to concomitant activation of suppressor T cells (via the level 1 suppressor circuit). This, again, agrees with many models of Ir gene function (in which unresponsiveness is mediated by suppressor T cells) (12).

Responsiveness can proceed in the face of active suppressor cell function via the activation of contrasuppressor T cells, which protect the targets of suppressor cell signals (13). Such contrasuppressor cell activation raises the system to level 2. At this level, the suppressor cells mentioned above (level 1 suppressors) are relatively ineffective (14, 15). Unresponsiveness at this level is postulated to be mediated principally by an amplification of suppression that we refer to as the level 2 suppressor circuit. Such an activity has been identified and seems to correlate with a previously described suppressor cell circuit. This is discussed in some detail in a section devoted to level 2 regulation.

These immunologic eigen states (anteigen states?) form the basis of a broad view of immune regulation. This quantum view suggests that energy is required to go to higher levels; information theory dictates that energy is similarly required to increase order. The higher levels are those required to tune the immune response exquisitely to protect maximally with minimal autoimmune dysfunction. The greater the threat, the more metabolic energy will be expended in the process of this tuning. In the final section we discuss some evidence that is compatible with these conclusions.

LEVEL 1 SUPPRESSION: CELLULAR ASPECTS

The Level 1 Suppressor Cell Circuit (Feedback Suppression)

Both the afferent and efferent components of the "feedback suppressor circuit" have been described extensively (16, 17). Basically, a signal produced by an antigen-stimulated inducer cell is transduced and amplified by a second population of non-immune regulatory cells, resulting in the generation of effector cells of level 1 suppression (Figure 1). The new designation "level 1 suppression" for the feedback suppression circuit is based upon a more unified concept of immunoregulation, discussed in the previous section.

In the in vitro response to sheep red blood cells (SRBC), addition of feedback inducer cells results in elimination of plaque-forming cell genera-

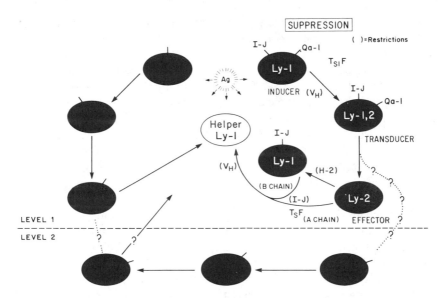

Figure 1 Level 1 suppression. The interaction of an I-J$^+$, Qa-1$^+$, Ly-1 T inducer cell with an I-J$^+$, Qa-1$^+$, Ly-1,2 T transducer cell results in an Ly-2 T effector cell (I-J$^-$). The latter, together with a naive I-J$^+$, Ly-1 T cell acts to inactivate target helper T cells.

tion (16). The activity of this cell population requires a transducer population that transfers the suppressor signal from the inducer cell to the effector population. In cultures that lack transducer cells, the decrease in the amplitude of the response by the inducer cells is very inefficient.

The functional activity of suppressor induction at this level is correlated with an Ly-1 cell (Ly-1$^+$,2$^-$) with an I-J$^+$, Qa-1$^+$ surface profile. The Ly-1 marker is also found on other inducer T cells, such as those that activate B cells to make antibody (18), activate macrophages in inflammation (17), and activate cytotoxic T cells (19). In addition to the Ly-1$^+$,2$^-$ phenotype, the above suppressor inducer cells also bear products of the I-J subregion and the Qa-1 locus, a profile that distinguishes them from the helper cell in the in vitro antibody response to SRBC (20).

The population that transduces the induction signal into one of direct suppression is an Ly-1,2 cell (Ly-1$^+$,2$^+$) with the I-J$^+$, Qa-1$^+$ phenotype (20). Although the mechanism by which these cells convert the inducing signal into suppression is not well understood, at least one mode of action may be the differentiation of these cells into mature suppressor effector cells in the presence of the inducing signal (21).

The interaction between the Ly-1 inducer cell and the Ly-1,2 transducer cell is antigen specific and is restricted by genes linked to the V region of the immunoglobulin heavy chain locus. That is, these populations must be matched at this locus as a minimum requirement for the induction of suppression. How the interaction between these cells relates to idiotype-anti-idiotype regulation (22) is discussed in some detail below and in the next section.

The effector cells of the level 1 suppressor circuit are functionally defined by their ability to suppress Ly-1 T cells directly. In addition to exerting their effects upon helper cells, these cells can also act upon the feedback inducer cells themselves and thus cause the inactivation of both new antibody synthesis (which requires T-cell help) and the recruitment of additional suppressor cell function (23). The latter fact illustrates the appropriateness of the term feedback as applied to this circuit. The level 1 suppressor effector cells bear the Ly-2 surface marker, but unlike the inducer and transducer subsets do not carry the same I-J gene product, at least in the case of the suppressor of the primary response to SRBC in vitro (14).

The interaction between the suppressor effector cell and its target is restricted by genes mapping to the major histocompatibility locus, especially the I-J subregion. The molecular basis for this restriction is understood and is best explained in a subsequent section.

Recent studies have indicated that an I-J$^+$, Ly-1 T cell is required for the action of the suppressor effector cell on its targets (24). This requirement is also best understood in molecular terms to be discussed in a subsequent section.

The Nature of V_H Restriction

Methylcholanthrene A (MCA) has been used to induce a large number of sarcomas, all of which possess unique transplantation antigens (TSTA). Old et al (25) have managed to generate an antiserum to a unique antigen on the surface of the MCA sarcoma and subsequently mapped the antigen to the short arm of chromosome 12 in the mouse (26). This region also houses the immunoglobulin heavy chain complex, a fact that prompted Flood et al (27) to investigate the activity of anti-MCA sera in cellular interactions between T cells restricted to V_H-linked genes. Antisera raised in homozygous syngeneic recipients all possessed the ability to block the interaction between the level 1 inducer and transducer cell. This ability was only dependent on the V_H locus of the interacting cells: if the cells came from mice that shared only Igh-linked genes and no other polymorphisms with Balb/c mice (the strain in which the antiserum was produced), the anti-MCA serum blocked the reaction. In addition, the activity of the serum was independent of the antigen used to induce the suppressor cells.

The MCA antigens were found to be present on all methylcholanthrene tumors tested. Taken together, these observations argue strongly that shared surface antigens on methylcholanthrene-induced sarcomas are identical to or highly cross-reactive with determinants found on molecules involved in V_H-restricted T-cell interactions. Since it is very unlikely that MCA-induced tumors carry the entire array of "idiotypes" expressed in the mouse, one must postulate that these cross-reactive determinants represent constant regions that are independent of the V_H-encoded hypervariable regions used by antibody and (probably) T-cell derived molecules to recognize and bind antigen. This introduces the possibility that V_H restriction is not always dependent on individual recognition of a wide variety of idiotypes, but may be accomplished by recognition of a much smaller number of molecular determinants.

V_H restriction and the activity of the anti-MCA antisera is considered further in somewhat more molecular terms in the following section. The question of why a cellular interaction molecule for immune function should appear on a fibrosarcoma is open for speculation (27), but certainly a closer examination of this genetic region should yield fascinating information.

LEVEL 1 SUPPRESSION: MOLECULAR ASPECTS

Suppressor Factors

In this section we discuss the components and action of two factors believed to mediate the functions of the inducer and effector cells of the level 1 suppressor circuit. Splenic T cells from animals hyperimmunized with xenogeneic red blood cells served as a source of these factors. The immunized T cells were treated with antisera plus complement to produce active, secreting populations. The production of Ly-1 TsiF (suppressor inducer factor) involved treatment of the T cells with anti-Ly-2. Alternatively, Ly-2 TsF (suppressor factor) required treating the T cells with anti-Ly-1. In both cases, the enriched subsets were then cultured without antigen for 48 hr, after which time the supernatant (containing the molecular activity) was harvested (28, 29).

In addition to this procedure, a molecule with the chemical and biological properties of the Ly-2 TsF was isolated from serum of hyperimmunized animals (30). This material was found to possess a unique idiotypic determinant (ShId) that allowed its purification. Since the preparation of the Ly-2 TsF contains, in addition to the $ShId^+$ material, other regulatory factors (to be discussed), the serum material has been designated $ShId^+$ TsF. It is likely, however, that the Ly-2 TsF (level 1 suppressor factor) and the $ShId^+$ TsF are identical.

Ly-1 TsiF

The Ly-1 TsiF is dependent upon I-J^+ Ly-1 T cells for its generation and (like the suppressor inducer cell) requires I-J^+, Ly-1,2 transducer cells for suppression to occur. This interaction is restricted by genes mapping to the

V region, and the effects are specific for the priming antigen (28). The Ly-1 TsiF must be added to cultures prior to day 2 if suppression is to be seen. Based on such observations, it was concluded that this material mediates the activity of the suppressor inducer cell.

The Ly-1 TsiF was found to be composed of two separate molecules (Figure 2). One can be removed by absorption with the appropriate antigen (hence dubbed antigen-binding chain). The other binds to and can be eluted from a column coupled with a genetically appropriate anti I-J reagent (I-J$^+$ chain). The I-J$^+$ chain has no antigen specificity and cannot be absorbed with antigen. Both chains of the Ly-1 TsiF appear to be 68-72 kilodaltons on SDS-PAGE, have a low pI, and are hydrophobic. Recent evidence suggests that each chain is produced by a different cell (or a cell in two stages of differentiation), one I-J$^+$ (producing the I-J$^+$ chain), the other I-J$^-$ (producing the antigen-binding chain) (31).

In experiments with I-J$^+$ chains and antigen-binding chains from Ly-1 TsiF's of different specificities and produced by different strains of mice, it was found that although the antigen-binding chain determined the antigen specificity of the factor, it was the I-J$^+$ chain that determined the V_H restriction of the inducer activity (32). This observation helps to confirm those described in the previous section in that the restriction has no antigen specificity (antigen specificity might be expected for an idiotype-anti-idiotype interaction). In support of these ideas, the anti-MCA antiserum described above was found to bind the I-J$^+$ chain but not the antigen-binding chain (unpublished observation).

Thus, apparently one molecule of Ly-1 TsiF (I-J$^+$ chain) seems to possess a determinant encoded by a locus on the 17th chromosome (I-J) and one on the 12th chromosome (V_H). These cannot be separated by SDS-PAGE or reduction and alkylation (unpublished observations), which raises the possibility that an excitingly novel genetic mechanism may be at work. Either the structural gene for I-J is actually on the 12th chromosome (under control by the 17th), or else gene products from two separate chromosomes

Figure 2   Biologically active molecules mediating the induction of level 1 suppression (Ly-1 TsiF). See text for details.

can be linked by translocation, RNA processing, or a specific covalent protein interaction [such as can be observed in the complement system (33)]. The possibility that the MCA structural gene is on chromosome 17 has been ruled out by somatic cell genetics (26).

These speculations are further complicated by observations of Ly-1 TsF made in F_1 animals. (Balb/c X B6)F_1 animals (CB6F$_1$) make an Ly-1 TsF that suppresses either Balb/c (I-J^d, Igh-V^a), or C57B1/6 (I-J^b,Igh-V^b). Surprisingly, however, removal of the I-J^b chain from the CB6F$_1$-only factor removed activity for C57B1/6 cells, whereas removal of the I-J^d chain resulted in loss of activity on Balb/c cells (31). This indicates that no (or fewer) I-J^b chains were constructed with Igh-V^a activity, neither were detectable I-J^d chains produced with the Igh-V^b restriction (recall that the factors are V_H restricted, showing no H-2 restrictions). This paradox is illustrated in Figure 3.

Ly-2 TsF

The Ly-2 TsF is produced by I-J$^-$, Ly-2 T cells and is capable of directly suppressing helper T cells in the absence of Ly-1,2 T cells. The suppression is antigen specific. These observations support the idea that this factor mediates the function of level one suppressor effector cells.

The suppressor effector molecule appears to be a single chain of 68 kilodaltons. In SDS-PAGE, it has a low pI (4.9), is hydrophobic, and possesses a blocked N-terminus. It binds antigen and, in the case of SRBC-specific factor, bears a recognizable idiotype (ShId). This material had been identified in Ly-2 TsF, ShId$^+$ material from hyperimmune serum, and in

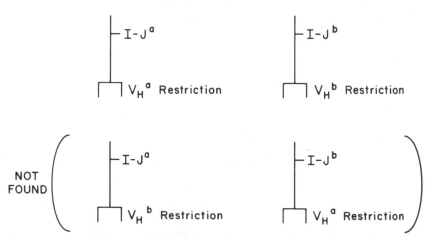

Figure 3   *cis* association of parental gene products from different chromosomes in Ly-1 TsiF from an F_1 animal.

supernatants of an Ly-2 T suppressor cell clone (30; unpublished observation). In hyperimmune serum, ShId$^+$ material can be found up to concentrations of approximately 0.1 mg/ml, and to be precipitable in high salt (28% NH_3SO_4). This raises the possibility that some of the suppressive activity attributed to antibody in hyperimmune serum (34) may actually be due to products of suppressor T cells.

As observed for level 1 suppressor effector cells, the Ly-2 TsF requires I-J$^+$, Ly-1 T cells to effect suppression on I-J$^-$, Ly-1 T cells (24). The Ly-2 TsF, in the presence of antigen, induces the production of a molecule from I-J$^+$, Ly-1 cells, which represents the second functional chain of the suppressor factor. The induction of this second chain requires H-2 homology between the I-J$^+$ Ly-1 T cells and the Ly-2 TsF (manuscript in preparation). This induced molecule is antigen-nonspecific (and does not bind antigen) and bears an I-J subregion-encoded determinant. To function, it must genetically match the target helper T (I-J$^-$ cell) at V_H. The functional interaction of the induced I-J$^+$ chain and the Ly-2 TsF is restricted by the I-J subregion of the H-2. A minor number of anti-MCA antisera discussed above will block the suppressive activity of the Ly-2 TsF (27). A number of anti-I-J monoclonal antibodies have also been found to block Ly-2 TsF activity (35) (Figure 4).

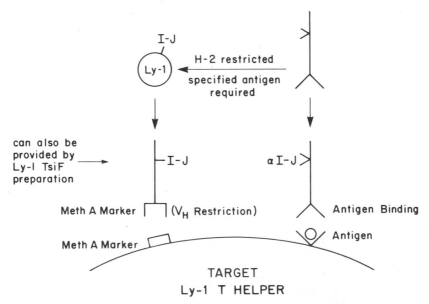

Figure 4 Biologically active effector molecules mediating level 1 suppression (Ly-2 TsF). See text for details.

An I-J$^+$ chain that will function with the Ly-2 TsF can also be found in Ly-1 TsiF (24). This suggests either that the I-J$^+$ molecule for both factors is interchangeable or that there is more than one type of I-J$^+$ molecule in the Ly-1 TsiF. The latter case is supported by three lines of evidence: (a) Certain anti-MCA serum preparations are capable of blocking only Ly-1 TsiF activity and not Ly-2 TsF, and vice versa; (b) the target cell of the two factors is different, an Ly-1,2 cell for Ly-1 TsiF and an Ly-1 cell for the Ly-2 TsF; and (c) an I-J$^+$ induced from Ly-1 cells by the Ly-2 TsF was found to function with the Ly-2 TsF to effect suppression but not with the antigen-binding chain of the Ly-1 TsiF to induce suppression (unpublished observation). The latter observation may be due to the quantity of I-J$^+$ chain required by each factor rather than the existence of two different I-J$^+$ chains. Further work is required to characterize the differences in the two I-J$^+$ molecules.

Speculation on the subsequent action of the Ly-2 TsF, following binding of both chains to the helper cell target, points toward the possibility that the active molecule is internalized. Similar internalization has been observed for certain bacterial toxins (36), ribosome inhibitory proteins from plants (37), and several peptide hormones, including insulin (36), and suggests a similarity of action between these and the Ly-2 TsF.

This similarity with toxin is enticing. Many sophisticated hormones that function in man have been identified as communication molecules in prokaryotes, an observation that suggests an evolutionary continuity in the development of hormones for different compartments of the physiology of higher organisms. Toxins may serve as defense mechanisms in some prokaryotes (38) and as a form of transplantation rejection against interspecies grafts in certain plants (37). With the appearance of the vertebrate immune system, the role of toxin may have inverted, becoming directed at elements of the immune system itself. If so, then the mechanism of action of Ly-2 TsF may well be the same as that of many bacterial toxins and plant ribosome-inactivating proteins (RIPs), i.e. inhibition of protein translation.

TsF "toxicity" depends, like other toxins, upon targeting the active fragment (antigen-binding chain) with a binding chain (I-J$^+$ chain) directed at the target (via V_H recognition). It may be that specificity depends solely upon this targeting. This would explain why the cells that produce the active chain (Ly-2 TsF) seem to be resistant to its toxic effects. It does not appear, however, that target cells, in the case of Ly-2 TsF, are actually killed. Cantor et al (39) present evidence that suppressor factor inhibits the secretion of certain proteins from helper-cell clones, without affecting others. One possibility is that translation of certain luxury gene products depends upon specialized or specially compartmentalized ribosomes, and that it is only these ribosomes that are suppressed. Specialized translational mechanisms for small numbers of induced proteins have been demonstrated in eukaryote heat shock (40). Plant RIPs are known to be increasingly toxic as one moves away from the species of origin (the ribosomes of which are completely resistant), to closely related species (partially susceptible), to distant species (totally susceptible) (37). This model predicts, among many other things, that whereas murine TsF is known to suppress murine cells without killing them (unpublished observations), human TsF (suppressor effector factor), properly targeted, might be found to be toxic to the appropriate mouse cells. Human TsF has been shown to suppress murine responses (41). It should be a simple matter to test it upon a mouse helper cell clone if the appropriate "schlepper" (I-J$^+$, V$_H$ restricted) chain can interact with it.

Genetically Constructed Competitive Inhibitors of Ly-2 TsF

One test of the model of two-chain interaction presented above involves the construction of competitive inhibitors for the Ly-2 TsF. According to the model, the I-J$^-$, antigen-binding Ly-2 TsF induces the production of an I-J$^+$ molecule with which it interacts. This interaction is restricted by polymorphic I-J subregion gene products. The I-J$^+$ chain, in turn, interacts with its helper cell target via a V$_H$-controlled interaction (Figure 4). If one were to add another I-J$^+$ chain, bearing only one appropriate and one inappropriate restriction element, then this should act as a competitive inhibitor of Ly-2 TsF-mediated suppression (Figure 5).

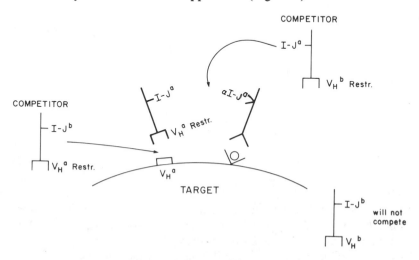

Figure 5   Genetically constructed competitive inhibitors of Ly-2 TsF.

This is exactly the case (42). With a C57Bl/6 Ly-2 TsF (SRBC) acting upon C57Bl/6 Ly-1 T cells, suppression was blocked by addition of I-J$^+$ chain (antigen-absorbed Ly-1 TsiF) from CB20 (matched at V_H but mismatched at H-2) or from BC9 (matched at H-2 but mismatched at V_H). No effect of a Balb/c factor was seen, as this is mismatched at both loci and therefore could not complete.

The model presented above was therefore employed to construct effects that otherwise would be unexplainable (i.e., the inhibition of an Ly-2 TsF by a genetically altered Ly-1 TsF). Although this demonstration does not prove the model of Ly-2 TsF action, it does illustrate its predictive power.

Adaptive Differentiation of Genetic Restrictions

In the previous sections we described two genetically restricted immunoregulatory interactions between T-cell subsets in the generation of level 1 suppression. Ly-1 TsiF, under normal circumstances, cannot induce suppressive activity in the transducer cells if these cells express Igh-V-linked polymorphisms that are different from those of the Ly-1 T cell used to produce the factor. Ly-2 TsF is also genetically restricted, requiring H-2 homology between the Ly-2 cells that produced the factor and the Ly-1 T cells upon which the factor will act (to both induce and interact with the cooperating I-J$^+$ chain).

The genotypes of the factor-producing cells do not completely govern these restrictions. When homozygous ("A") T cells mature in an (A X B)F$_1$ radiation chimera or in an (A X B)F$_1$ thymus grafted to a homozygous A nude mouse, these cells will produce factors restricted to both A and B (43, 44). This applies not only to the H-2 restriction of the Ly-2 TsF, but also to the Igh-V restriction of the Ly-1 TsiF. Additionally, (A X B)F$_1$ T cells differentiating in an A-type thymus produce factors restricted to A only and are unable to interact with B-type cells.

An extensive literature exists on the role of the thymus in the generation of H-2 restrictions (45), but the adaptive differentiation of Igh-V-linked recognition in the thymus is novel. The F$_1$-into-parent chimera experiments make it unlikely that the thymus is passively "armed" by circulating molecules (such as antibody) that might act as the selection elements. Given these striking similarities in the development of recognition of Igh-V and MHC gene products as self, one should consider the possibility that Igh-V-linked gene products, expressed by cells in the thymus, function to select for those Ly-1 TsiF producer cells that can recognize these Igh-V-linked cell interaction structures as self. This might then be added to the list of similarities between genetic products of the MHC and Igh regions, which include roles in graft rejection and cellular interaction, and genetic homology (46).

Although the above observations are certainly remarkable, they are not particularly difficult to envision. An even more surprising observation was made in examining the interaction of Ly-2 TsF and its target cells in the types of chimeras mentioned above. It was found that A cells that had matured in an A X B thymus had also gained the ability to be suppressed by both Ly-1 TsiF and Ly-2 TsF made by B animals. Thus, not only did the factor-producing cells learn a new self recognition in the F_1 thymus, but these cells "taught" their transducer cells this new self specificity. This phenomenon has been called tandem adaptive differentiation. The implications have been discussed in detail (44). One particularly extraordinary implication derives from the analysis that has shown that the Ly-2 TsF recognizes private regions of I-J on the I-J$^+$ molecule. Since anti-I-J antibodies directed against the B-type I-J determinants do not detect any alteration in I-J, the work suggests that a variable (adaptable) portion of the I-J molecule is recognized by the Ly-2 TsF but not by antibodies (44). Attempts are now being made to produce antibodies to this adaptable region of I-J-encoded products.

CONTRASUPPRESSION: CELLULAR AND MOLECULAR ASPECTS

Contrasuppression and the regulatory T-cell interactions that produce it have been reviewed in depth (47, 48). Nevertheless, this activity serves as a distinguishing feature between level 1 and level 2 suppression, and we believe it to be a crucial feature of immunoregulation, which should be discussed herein.

Contrasuppression is defined as an activity that interferes with level 1 suppression to allow dominant immunity. Technically, then, the competitive inhibition experiments described in the last section could be considered as a form of contrasuppression (one that may, perhaps, give us insights into the mechanism of physiologic contrasuppression). In this section, however, we are concerned with an immunoregulatory T cell circuit that possesses this activity (with syngeneic action). The circuit is illustrated in Figure 6.

The first cell identified as a member of this circuit is an immune I-J$^+$, Ly-2 T cell, the contrasuppressor inducer (14). It is likely that other regulatory cells might be involved in activating this cell, but this is currently only speculation. The contrasuppressor inducer cell produces a factor (TcsiF) that bears an I-J$^+$ subregion-encoded determinant and binds antigen (49). Whether the TcsiF is composed of one or two chains is unknown. The antigen-binding capability is more cross-reactive than that of the Ly-2 TsF, such that a SRBC-induced factor can be absorbed on horse red blood cells (leaving the SRBC-specific TsF behind).

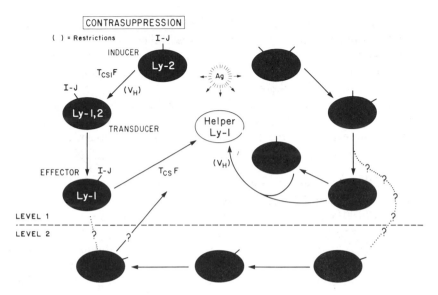

Figure 6 Contrasuppression. The interaction of an I-J$^+$, Ly-2 T inducer cell and an I-J$^+$, Ly-1,2 T transducer cell results in an I-J$^+$, Ly-1 effector cell. The latter acts to render the targets of level 1 suppression (e.g. helper T cells) resistant to suppressor cell signals.

Both the contrasuppressor inducer cell and factor interact with an I-J$^+$, Ly-1,2 T contrasuppressor transducer cell to activate contrasuppression (14, 49). Recent experiments indicate that this interaction is restricted by genes mapping to the Ig-V$_H$ region, and some anti-MCA antisera (discussed previously) are capable of blocking this induction (50). These antisera, unlike those discussed that block induction of suppression, are more commonly produced in F$_1$ animals, which are semi-syngeneic for the fibrosarcoma. MCA fibrosarcomas grown in F$_1$ animals have been found to express much higher levels of surface antigens, which absorb this contrasuppressor blocking activity, than those grown in the parental strain. Although this elevation might account for the appearance of the antibody predominantly in F$_1$ animals, it leaves the puzzle of the elevation itself. This might be a phenomenon related to that previously described by Barret & Derringer (51) in which tumors appeared to acquire antigens when passaged in F$_1$ animals.

The effector cell of contrasuppression is an I-J$^+$, Ly-1 T cell that adheres to the *Vicia villosa* lectin. This cell is present in the spleens of hyperimmune animals (15) and can be generated in cultures of spleen cells from neonatal animals (13), old animals (52), adult animals treated so as to remove certain inhibitory populations (discussed later), and animals manipulated to become autoimmune (53).

The contrasuppressor effector cell acts directly upon helper T cells to render them resistant to subsequent suppressor T-cell signals (13, 15). A cell-free product of the effector cells with this activity has been identified and has been found to bear an I-J subregion-encoded determinant (52). Its mechanism of action is unknown.

The I-J-encoded determinant found on all of the cells and factors of the contrasuppressor circuit can be distinguished from that on the cells and factors of the level 1 suppressor circuit by the use of monoclonal anti-I-J antibodies (54).

The Physiologic Role of Contrasuppression

We have suggested that contrasuppression exists as a sort of bridge between level 1 and level 2 immunoregulation, interferring with the former and being regulated, in turn, by the latter. At this point, therefore, we propose to examine the role of contrasuppression in the regulation of immunity, which will also serve to underline the potential importance of level 2 suppression (in balancing contrasuppression), for which relatively little information exists.

Basically, contrasuppression is viewed as functioning in at least three areas of immunoregulation. These are (a) as a sequel to general suppression (leading to responsiveness), (b) in the localization of immune responses to discrete regions, and (c) in the establishment of suppression-resistant states.

A state of general unresponsiveness associated with suppressor-cell activity (probably level 1 suppression) accompanies the earliest stages of neonatal life (55), parasitic infections (56), many tumors (57), and severe trauma (58). In each case, recovery of immune function and establishment of dominant immunity often requires activity of contrasuppressor cells (reviewed in 48).

A similar contrasuppressive activity might serve to produce a local resistance to general suppression, thus focusing an immune response into one discrete area. The Peyer's patches of the gut-associated lymphoid tissue (GALT) contain cells of both the suppressor and the contrasuppressor circuits (59). Contrasuppression might permit GALT responses to proceed against ingested antigen, whereas suppression prevents systemic immunity to any antigen that reaches the circulation. Such GALT immunity in the face of systemic tolerance has been observed (60). The skin (61–63) and bone marrow (64) are two other environments in which local activation of contrasuppression may have an important role of maintaining immunity in the presence of suppression. Specialized antigen-presenting cells, such as Langerhans cells and dendritic cells, have been implicated in this local activation (63, 65). Other local mediators that might serve to activate contrasuppression are interferon (66) and histamine (67).

Several immune conditions lead to what can be observed as a suppression-resistant state. Although hyperimmune animals show no sign of suppression or unresponsiveness, it is well established that serum from such animals contains a potent suppressive substance (34) that is antigen specific and T-cell derived (30). A similar situation is seen in some forms of autoimmunity, i.e. autoimmunity may proceed in the face of systemic nonspecific suppression. Examples are the MRL murine model of systemic lupus erythematosis (68), old age (69), and autoimmunity associated with malaria (70). In each case of the suppression-resistant state contrasuppression has been implicated in some way (48,52), and it remains to be seen whether or not a cause and effect relationship can be established.

LEVEL 2 SUPPRESSION

Evidence for a Second "Level" of Suppression

THE REGULATION OF CONTRASUPPRESSOR CELL GENERATION
Many investigators have observed the generation of potent suppressor activity upon culture of normal, adult spleen cells (71, 72). When cells from very young animals are cultured, however, dominant contrasuppression, rather than suppression, is produced (13). A mixture of adult and neonatal cells was found to produce no contrasuppressor cells, an effect requiring a Thy-1$^+$, I-J$^+$, Ly-1,2 cell in the adult population (73). The presence in the adult population of a cell that prevents the generation of contrasuppression raised the possibility that specific removal of this population would allow contrasuppressor cells to appear in cultures of adult spleen. A novel monoclonal anti-Thy-1 (F7D5) was employed, which at low concentrations reacted with a small population of adult spleen cells, including the cells responsible for the above inhibitory effects on contrasuppressor cell generation. Treated adult spleen cells were found to produce I-J$^+$, Ly-1$^+$ contrasuppressor cells upon culture (73).

Thus, the presence of one or more cell populations (including an I-J$^+$, Ly-1,2 population, but perhaps other cells as well) prevents the spontaneous generation of contrasuppression by adult cells in culture (and allows dominant suppressor activity). This activity appears ontogenetically with the loss of ability to produce contrasuppressor cells in untreated cell cultures (increasing from around day 8 to day 15 of life).

PRODUCTION OF DOMINANT CONTRASUPPRESSOR CELL EFFECTS
When Ly-2 TsF is produced and tested on cells of most strains of mice, suppression tends to be the dominant activity observed, even when the factor contains Ly-2 TcsiF and the cultures contain contrasuppressor trans-

ducer cells. Exceptions to this are the B10 congenic mouse strains [notable because these were used to characterize the Ly-2 TcsiF (49)] and the A-strain congenics. Any genetically matched Ly-2 factor will produce dominant contrasuppression on B10 cells (such that suppression may only be seen upon removal of the TcsiF or the contrasuppressor transducer cell), whereas factors produced in A strain mice seem to be contrasuppressive on otherwise suppressible recipients (for example, an F_1 recipient) (74).

These effects were mimicked by treatment with F7D5. Cells from C57BL/6 animals were treated with F7D5, then were used to produce an Ly-2 factor. This factor suppressed cultures in the absence of Ly-2$^+$ cells in the recipients, or in the presence of monoclonal anti-I-J reagents known to block contrasuppression, but otherwise it produced no effects in cultures of unfractionated cells (73). Thus, removal of an F7D5$^+$ cell from the Ly-2 population resulted in the production of an Ly-2 factor resembling the one made by cells from untreated A strains.

Similarly, treatment of C57BL/6 spleen cells with a low concentration of F7D5 caused these cells to be resistant to suppression unless the contrasuppressor circuit was quenched. Thus, F7D5 caused B6 strain cells to behave like B10 strain cells (unpublished observations).

Therefore, F7D5$^+$ cells exist that function to override the contrasuppressor circuit, in the absence of which the induction of contrasuppression by the Ly-2 TcsiF is dominant.

Contrasuppression Versus the Amplification of Suppression

An interesting symmetry has appeared between systems that under some conditions are deficient in certain cellular members of the suppressor cell circuit described by Tada et al (75) and those with a dominant contrasuppressive activity. The suppressor cell circuit involves an Ly-2 inducer cell that is I-J$^+$. This acts via an I-J$^+$, Ly-1,2 transducer population, resulting in active I-J$^+$, Ly-2 suppressor effector cells. This interaction produces an amplification of suppression (Figure 7).

Mice of the A strain background fail to produce this KLH-specific suppressor-including factor (76). As previously discussed, A strain mice do produce a suppressor factor for primary anti-SRBC response, but one in which contrasuppression dominates. Similarly, B10 strains fail to accept this Ly-2 suppressor inducer factor and also show dominant contrasuppression in the SRBC system. Experiments on absorption of the KLH suppressor inducer factor demonstrated that B10 mice are indeed deficient in this acceptor (transducer) cell (76).

The examination of the cell that accepts the inducer factor in Tada's system revealed an I-J$^+$, Ly-1,2 cell, which bore a high density of Thy-1 antigen, as detected with a heteroantiserum (75). It is possible, therefore,

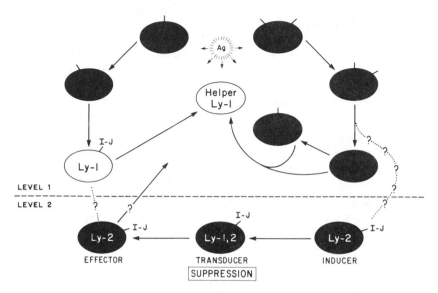

Figure 7 Level 2 suppression. This suppressor-amplification circuit involves an I-J$^+$, Ly-2 inducer cell, an I-J$^+$, Ly-1,2 transducer (acceptor) cell, and an I-J$^+$, Ly-2 effector cell.

that this cell (at least) will prove to be sensitive to a low concentration of F7D5.

Observations similar to the above were made in another system. Peyer's patch T cells from mice bearing the d allele at the H-2D locus will suppress the primary SRBC response of normal spleen cells in vitro. This suppression required the interaction of an Ly-2 cell in the Peyer's patch population and an Ly-2$^+$ cell (presumably Ly-1,2 cell) in the splenic population (77). Peyer's patch T cells from non-H-2D^d animals induced a dominant contrasuppression (induced by an I-J$^+$, Ly-2 cell), but also suppressed if the contrasuppressor circuit was severed (59). The resulting suppression was effected by an I-J$^-$, Ly-2 cell (level 1 suppressor). Thus, again, in cases where suppressor amplification did not seem to operate, contrasuppression was dominant. When this contrasuppression was removed, level 1 suppression was found to produce dominant unresponsiveness.

The resulting symmetry, i.e. contrasuppression dominates where the "Tada" suppressor circuit is absent, resembles the situation discussed in the last section. This leads to the possibility that the regulation of contrasuppression and the amplification of suppression are mediated by the same cellular circuit, i.e. the level 2 suppressor circuit (Figure 7).

Regulatory Aspects of Level 2 Suppression

We have preliminary evidence that the level 2 suppressor circuit starts to appear at about day 8 post-birth in the mouse. Its appearance may represent a control for the contrasuppression that appears earlier (around day 2) and may be required for the maturation of normal responsiveness (suppressed in neonates by level 1 suppressor cells). Premature introduction of adult spleen cells (which contain the cells of the level 2 circuit) at or before day 8 blocked normal immune development (52, 78). Treatment of these adult cells with a low concentration of F7D5 completely removed this blocking activity (52). Thus, it seems likely that the sequential appearance of level 1 suppression, then contrasuppression, then level 2 suppression must proceed on this schedule for normal immune function to mature.

Polyclonal B-cell activators generally induce suppression in vivo. When animals were thymectomized on day 3 of life, however, they were found to produce normal T cell-dependent immune responses, but became autoimmune in response to polyclonal B cell activators (53). The thymectomized animals were capable of producing level 1 suppressor cells in vitro, but the autoimmune animals also possessed a potent contrasuppressive activity, sensitive to treatment with anti-I-J (79). Thus, removal of the thymus at a time after the maturation of cells of the level 1 suppressor circuit and the contrasuppressor circuit, but before the appearance of level 2 regulatory cells, leads in maturity to a state prone to autoimmunity. The potential importance of the level 2 suppressor circuit cannot be overemphasized.

CONCLUSION: LEVELS UPON LEVELS

In the introduction we proposed that the regulatory levels we discussed become activated at the expense of increased metabolic energy. Thus, the greater the threat, the greater the activation of a higher level. This might explain the well-known (but poorly understood) dose response of the immune system (Figure 8, *top*).

Generally, immune responses have an optimal antigen dose range. Sub- and supra-optimal doses of antigen produce suppression and unresponsiveness. Recent studies of antigen-induced regulatory activities in mouse (80) and man (81; T. Lehner, personal communication) have indicated that dominant contrasuppression is activated only in the optimal dose range. This contrasuppressive activity can overcome low-dose suppression, but not high dose (80). We postulate that low-dose suppression is associated with level 1, whereas high-dose suppression is due to activation of the level 2 circuit (Figure 8, *bottom*). Experiments are in progress to test this notion.

If higher antigen dose can be considered an increased threat, then this is compatible with our notion of higher level activation. High-affinity antibody is more cross-reactive than low-affinity antibody (82, 83), thus maximal, high-affinity responses entail an increased risk of self destruction. Activation of the immune response in the absence of level 2 regulatory cells

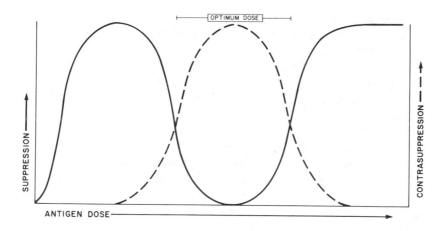

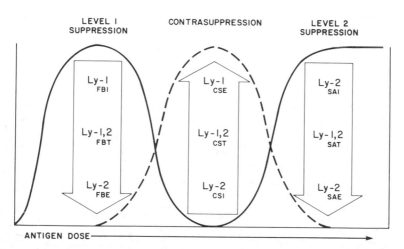

Figure 8 (*Top*) Dose response and immunoregulation. (*Bottom*) Dose response and immunoregulation. I, inducer; T, transducer; E, effector; FB, feedback (level 1) suppression; CS, contrasuppression, SA, (level 2) suppression amplification.

can lead to a potent autoimmune response (53, 79). Further studies on the relationship between level 2 regulation and autoimmunity are needed, as are studies on the role of contrasuppression and level 2 suppression in maturation of high-affinity responses without autoimmunity.

The view that autoimmunity may be produced by a failure to control high-affinity responses to foreign antigens properly is not a new one (84). The scheme put forward here seeks to ascertain those regulatory sites at which control may go awry. In addition, our view places a greater emphasis on homeostasis, so that although antigens may perturb a defective system toward autoimmunity, disruptions other than those produced by antigen may also produce autoimmunity.

Are there more levels? Perhaps so, but studies on relatively simple antigen challenges may fail to reveal these additional complexities. As we increase our understanding of immune responses to tumors and parasitic infections, however, we might well find that such levels exist. Alternatively, it may be that the immune system's failure to protect successfully against such intrusions may be linked to the absence of higher levels. Thus, the immune system may not be capable of producing an effectively strong response in some cases, precisely because it lacks sufficient heuristic power in its regulatory circuitry. If so, then by gaining insight into the levels of regulation that do exist, we may be able artificially to add our mental processes to those of the immune system in producing therapeutic manipulations. The prospects of such immunoengineering are not far off.

ACKNOWLEDGMENT

D. R. Green and P. M. Flood are supported by National Institutes of Health Training Grant number AI 07019.

Literature Cited

1. Foucault, M. 1973. *The Order of Things.* New York: Vintage Books
2. *Oxford English Dictionary.* 1971. Oxford: Oxford Univ.
3. Bynum, W. F., Browne, E. J., Porter, R., eds. 1981. *Dictionary of the History of Science.* Princeton: Princeton Univ.
4. Weiner, N. 1948. *Cybernetics.* Cambridge: Technology Press
5. Germaine, R. N., Benacerraf, B. 1981. A single major pathway of T-lymphocyte interactions in antigen-specific immune suppression. *Scand. J. Immunol.* 13:1
6. Cantor, H., Boyse, E. A. 1975. Functional subclasses of T lymphocytes bearing different Ly antigens. *J. Exp. Med.* 141:1376
7. Stanton, T. H., Boyse, E. A. 1976. A new serologically defined locus, Qa-1, in the Tla region of the mouse. *Immunogenetics* 3:525
8. Shreffler, D. C., David, C. S. 1975. The H-2 major histocompatibility complex and the I immune response region: Genetic variation, function and organization. *Adv. Immunol.* 20:125
9. Murphy, D. B. 1978. The I-J Subregion of the Murine H-2 Gene Complex. *Springer Sem. Immunopath.* 1:111
10. Kimura, A. K. 1982. The fine specificity of lectins applied to the study of lymphocyte membrane structure and immunological reactivity. *Riv. Immunol. Immunofarm.* 3:117
11. Shevach, E. M., Rosenthal, A. S. 1973. Function of macrophages in antigen recognition by guinea pig T-lymphocytes. II. Role of the macrophage in the regulation of genetic control of the immune response. *J. Exp. Med.* 138:1213
12. Debre, P., Kapp, J. A., Dorf, M. E., Benacerraf, B. 1975. Genetic control of immune suppression. II. H-2 linked dominant genetic control of immune suppression by the random copolymer GT. *J. Exp. Med.* 142:1447
13. Green, D. R., Eardley, D. D. Kimura, A., Murphy, D. B., Yamauchi, K., Gershon, R. K. 1981. Immunoregulatory circuits which modulate responsiveness to suppressor cell signals: Characterization of an effector cell in the contrasuppressor circuit. *Eur. J. Immunol.* 11:973
14. Gershon, R. K., Eardley, D. D., Durum, S., Green, D. R., Shen, F. W., Yamauchi, K., Cantor, H., Murphy, D. B. 1981. Contrasuppression: A novel

immunoregulatory activity. *J. Exp. Med.* 153:1533

15. Green, D. R., Gershon, R. K. 1982. Hyperimmunity and the decision to be intolerant. *Ann. N. Y. Acad. Sci.* 392: 318

16. Eardley, D. D., Hugenberger, J., McVay-Boudreau, L., Shen, F. W., Gershon R. K., Cantor, H. 1978. Immunoregulatory circuits among T-cell sets. I. T-helper cells induce other T-cell sets to exert feedback inhibition. *J. Exp. Med.* 147:1106

17. Cantor, H., Gershon, R. K. 1979. Immunological circuits: Cellular composition. *Fed. Proc.* 38:2058

18. Davies, A. J. S. 1969. The thymus and the cellular basis of immunity. *Transplant. Rev.* 1:43

19. Cantor, H., Boyse, E. A. 1977. Lymphocytes as models for the study of mammalian cellular differentiation. *Immunol. Rev.* 33:105

20. Eardley, D. D., Murphy, D. B., Kemp, J. D., Shen, F. W., Cantor, H., Gershon, R. K. 1980. Ly-1 inducer and Ly-1,2 acceptor cells in the feedback suppression circuit bear an I-J subregion controlled determinant. *Immunogenetics* 11:549

21. McDougal, J. S., Shen, F. W., Cort, S. P., Bard, J. 1980. Feedback suppression: Phenotypes of T cell subsets involved in the Ly-1 T-cell induced immunoregulatory circuit. *J. Immunol.* 125:1157

22. Jerne, N. 1974. Towards a network theory of the immune system. *Ann. Immunol.* 125:373

23. Green, D. R., Gershon, R. K., Eardley, D. D. 1981. Functional deletion of different Ly-1 T cell inducer subset activities by Ly-2 suppressor T lymphocytes. *Proc. Natl. Acad. Sci. USA* 78:3819

24. Flood, P., Yamauchi, K., Gershon, R. K. 1982. Analysis of the interactions between 2 molecules that are required for the expression of Ly-2 suppressor cell activity; 3 different types of focusing events may be needed to deliver the suppressive signals. *J. Exp. Med.* 156:361

25. DeLeo, A. B., Shiku, H., Takakarki, T., John, M., Old, L. J. 1977. Cell surface antigens of chemically induced sarcomas of the mouse. I. Murine leukemia virus-related antigens and alloantigens on cultured fibroblasts and sarcoma cells: Description of a unique antigen on Balb/c Meth A sarcoma. *J. Exp. Med.* 146:720

26. Pravtcheva, D. D., DeLeo, A. B., Ruddle, F. H., Old, L. J. 1981. Chromosome assignment of the tumor specific antigen of a 3-methylcholanthrene-induced mouse sarcoma. *J. Exp. Med.* 154:964

27. Flood, P. M., DeLeo, A. B., Old, L. J., Gershon, R. K. 1983. The relation of cell surface antigens on methylcholanthrene-induced fibrosarcomas to cell Igh-V linked cell interaction molecules. *Proc. Nat. Acad. Sci. USA.* In press

28. Yamauchi, K., Murphy, D. B., Cantor, H., Gershon, R. K. 1981. Analysis of an antigen specific H-2 restricted, cell free product(s) made by "I-J" Ly-2 cells (Ly-2 TsF) that suppresses Ly-2 cell depleted spleen cell activity. *Eur. J. Immunol.* 11:913

29. Yamauchi, K., Murphy, D. B., Cantor, H., Gershon, R. K. 1981. Analysis of antigen specific, Ig restricted cell free material made by I-J⁺ Ly-1 cells (Ly-1 TsiF) that induces Ly-2⁺ cells to express suppressive activity. *Eur. J. Immunol.* 11:905

30. Iverson, G. M., Eardley, D. D., Janeway, C. A., Gershon, R. K. 1983. The use of anti-idiotype immunosorbents to isolate circulating antigen specific T-cell derived molecules from hyperimmune Sera. *Proc. Nat. Acad. Sci. USA.* In press

31. Chao, N. 1981. *Analysis of an antigen specific, Ig restricted cell free material made by I-J⁺, F₁ Ly-1 cells.* thesis. Yale Univ., New Haven, CT

32. Yamauchi, K., Chao, N., Murphy, D. B., Gershon, R. K. 1982. Molecular composition of an antigen-specific, Ly-1 T suppressor inducer factor: One molecule binds antigen and is I-J⁻; another is I-J⁺, does not bind antigen, and imparts an Igh-variable region linked restriction. *J. Exp. Med.* 155:655

33. Hostetter, M. K., Thomas, M. L., Rosen, F. S., Tack, B. F. 1982. Binding of C3b proceeds by a transesterification reaction at the thiolester site. *Nature* 298:72

34. Haughton, G., Nash, D. R. 1969. Specific immunosuppression by minute doses of passive antibody. *Transplant Proc.* 1:616

35. Golde, W. T., Green, D. R., Waltenbaugh, C. R., Gershon, R. K. 1982. Different I-J specificities define different T-cell regulatory pathways. *Fed. Proc.* 41:190 (ABSTR.)

36. Middlebrook, J. L., Kohn, L. D., eds. 1981. *Receptor-Mediated Binding and*

Internalization of Toxins and Hormones. New York: Academic

37. Stirpe, F. 1982. On the action of ribosome inactivating proteins: Are plant ribosomes species specific? *Biochem. J.* 202:279

38. Janzen, D. 1977. Why fruits rot, seeds mold, and meat spoils. *Am. Naturalist* 111:691

39. Clayberger, C., Cantor, H. 1982. Generation, specificity and function of 4-hydroxy-3-nitrophenyl (acetyl) specific clones. *Fed. Proc.* 41:1213 (ABSTR.)

40. Storti, R. V., Scott, M. P., Rich, A., Pardue, M. L. 1980. Translational control of protein synthesis in response to heat shock in D. melanogaster cells. *Cell* 22:825

41. Kontiainen, S., Feldmann, M. 1979. Structural characteristics of antigen-specific suppressor factors: Definition of "constant" region and "variable" region determinant. *Thymus* 1:59

42. Green, D. R., Flood, P. M., Gershon, R. K. Genetically constructed competitive inhibitors of suppressor activity. Manuscript in preparation

43. Yamauchi, K., Flood, P., Singer, A., Gershon, R. K. 1983. Homologies between cell-interaction molecules controlled by MHC and Igh-V linked genes that T cells use for communication: both molecules undergo adaptive differentiation in the thymus. *Eur. J. Immunol.* In press

44. Flood, P., Yamauchi, K., Singer, A., Gershon, R. K. 1982. Homologies between cell interaction molecules controlled by MHC and Igh-V linked genes that T cells use for communication: Tandem "adaptive differentiation" of producer and acceptor cells. *J. Exp. Med.* 156:1390

45. Moller, G., ed. 1979. Acquisition of the T cell repertoire. *Immunol. Rev.* Vol. 42

46. Steinmetz, M., Frelinger, J. G., Fisher, D., et al. 1981. Three cDNA clones encoding mouse transplantation antigens: Homology to immunoglobulin genes. *Cell* 24:125

47. Green, D. R. 1982. Contrasuppression: Its role in immunoregulation. In *The Potential Role of T-Cells in Cancer Therapy,* ed. A. Fefer, A. Goldstein. New York: Raven

48. Green, D. R., Gershon, R. K. Contrasuppression. Submitted for publication

49. Yamauchi, K., Green, D. R., Eardley, D. D., Murphy, D. B., Gershon, R. K. 1981. Immunoregulatory circuits that modulate responsiveness to suppressor cell signals: The failure of B10 mice to respond to suppressor factors can be overcome by quenching the contrasuppressor circuit. *J. Exp. Med.* 153:1547

50. Flood, P. M., DeLeo, A. B., Old, L. J., Gershon, R. K. 1983. Antisera against tumor-associated surface antigens on methylcholanthrene-induced sarcomas inhibit the induction of contrasuppression. *Mod. Trends Hum. Leukemia* 5: In press

51. Barret, M. K., Deringer, M. K. 1950. An induced adaptation in a transplantable tumor of mice. *J. Natl. Cancer Inst.* 11:51

52. Green, D. R. 1981. *Contrasuppression: an immunoregulatory T cell activity.* PhD thesis. Yale Univ., New Haven, CT

53. Smith, H. R., Green, D. R., Raveche, E. S., Smathers, P. A., Gershon, R. K., Steinberg, A. D. 1983. Induction of autoimmunity in normal mice by thymectomy and administration of polyclonal B cell activators: Association with contrasuppressor function. *Clin. Exp. Immunol.* In press

54. Yamauchi, K., Taniguchi, M., Green, D., Gershon, R. K. 1982. The use of a monoclonal I-J specific antibody to distinguish cells in the feedback suppression circuit from those in the contrasuppressor circuit. *Immunogenetics* 16:551

55. Mosier, D. E., Johnson, B. M. 1975. Ontogeny of mouse lymphocyte function. II. Development of the ability to produce antibody is modulated by T lymphocytes. *J. Exp. Med.* 141:216

56. Jayawardena, A. N. 1981. Immune responses in malaria. *Adv. Parasitic Dis.* 1:85

57. Bluestone, J. A., et al. 1979. Suppression of the immune response in tumor bearing mice. *J. Natl. Cancer Inst.* 63:1215

58. McLoughlin, G. A., Wu, A. V., et al. 1979. Correlation between anergy and a circulating immunosuppressive factor following major trauma. *Ann. Surgery* 190:297

59. Green, D. R., Gold, J., St. Martin, S., Gershon, R., Gershon, R. K. 1982. Microenvironmental Immunoregulation: The possible role of contrasuppressor cells in maintaining immune responses in gut associated lymphoid tissues (GALT) (immunosuppression/peyers patches). *Proc. Natl. Acad. Sci. USA* 79:889

60. Challacombe, S. J., Tomasi, T. B. 1980. Systemic tolerance and secretory immu-

nity after oral immunization. *J. Exp. Med.* 152:1459

61. Iverson, G. M., Ptak, W., Green, D. R., Gershon, R. K. The role of contrasuppression in the adoptive transfer of immunity. Submittted for publication

62. Ptak, W., Green, D. R., Durum, S. K., Kimura, A., Murphy, D. B., Gershon, R. K. 1981. Immunoregulatory circuits that modulate responsiveness to suppressor cell signals: Contrasuppressor cells can convert an *in vitro* toleragenic signal into an immunogenic one. *Eur. J. Immunol.* 11:980

63. Ptak, W., Rozycka, D., Askenase, P. W., Gershon, R. K. 1980. Role of antigen-presenting cells in the development and persistence of contact hypersensitivity. *J. Exp. Med.* 151:362

64. Michaelson, J. D. 1982. *The characterization and functional analysis of immunoregulatory cells in murine bone marrow*. Medical school thesis. Yale Univ. New Haven, CT.

65. Britz, J. S., Askenase, P. W., Ptak, W., Steinman, R. M., Gershon, R. K. 1982. Specialized antigen-presenting cells: Splenic dendritic cells, and peritoneal exudate cells induced by mycobacteria activate effector T cells that are resistant to suppression. *J. Exp. Med.* 155:1344

66. Knop, J., Stremmer, R., Neumann, C., DeMaeyer, E., Macher, E. 1982. Interferon inhibits the suppressor T cell response of delayed type hypersensitivity. *Nature* 296:757

67. Seigel, J. N., Schwartz, A., Askenase, P. W., Gershon, R. K. 1983. Suppression and contrasuppression induced respectively by histamine-2 and histamine-1 receptor agonists. *Proc. Natl. Acad. Sci. USA.* In press

68. Gershon, R. K., Horowitz, M., Kemp, J. D., Murphy, D. B., Murphy, D. 1978. The Cellular site of immunoregulatory breakdown in the lpr mouse. In *Genetic Control of Autoimmune Disease*, ed. N. R. Rose, R. E. Bigazzi, N. L. Warner, pp. 223–27. New York: Elsevier

69. Makinodan, T., Kay, M. B. 1980. Age influence on the immune system. *Adv. Immunol.* 29:287

70. Poels, L. G., van Niekerk, C. C. 1977. Plasmodium berghei: Immunosuppression and hyperimmunoglobulinemia. *Exp. Parasitol.* 42:235

71. Janeway, C. A. et al. 1975. T cell population with different functions. *Nature* 253:544

72. Hodes, R. J., Hathcock, K. S. 1976. *In vitro* generation of suppressor cell activity: Suppression of *in vitro* induction of cell mediated cytotoxicity. *J. Immunol.* 116:167

73. Green, D. R., Chue, B., Gershon, R. K. 1983. Two types of T cell suppression discriminated by action on the contrasuppression circuit. *J. Mol. Cell. Immunol.* In press

74. Chue, B., Gershon, R. K., Green, D. R. Apparent failure of A strain mice to produce T suppressor effector factor (Ly-2 TsF) is due to dominant contrasuppressor T cell activity. Manuscript in preparation

75. Tada, T., Okumura, K. 1979. The role of antigen-specific T cell factors in the immune response. *Adv. Immunol.* 28:1

76. Taniguchi, M., Tada, T., Tokuhisa, T. 1976. Properties of the antigen-specific T-cell factor in the regulation of antibody response of the mouse. III. Dual gene control to the T-cell mediated suppression of the antibody response. *J. Exp. Med.* 144:20

77. Green, D. R., St. Martin, S., 1983. Suppression and contrasuppression in the regulation of gut associated immune response. *Ann. NY Acad. Sci.* In press

78. Raybourne, R., Arnold, L. W., Haughton, G. 1981. Perturbation of early development of the immune system by normal adult lymphocytes. *J. Immunol.* 127:1142

79. Smith, H. R., Green, D. R., Raveche, E. S., Smathers, P. A., Gershon, R. K., Steinberg, A. D. 1982. Studies of the induction of anti-DNA in normal mice. *J. Immunol.* 129:2332

80. Schiff, C. 1980. *Regulation of the murine immune response to sheep red blood cells: The effects of educating lymphocytes in vitro with glycoproteins isolated from the sheep erythrocytes membrane*. Thesis. Yale Univ., New Haven, CT.

81. Lehner, T. 1982. The relationship between human helper and suppressor factors to a streptococcal protein antigen. *J. Immunol.* 129:1936

82. Little, J. R., Eisen, H. N. 1969. Specificity of the immune response to the 2,4-dinitrophenyl and 2,4,6-trinitrophenyl groups. Ligand binding and fluorescence properties of cross-reacting antibodies. *J. Exp. Med.* 129:247

83. Underdown, B. J., Eisen, H. N. 1974. Cross-reactions between 2,4-dinitrophenyl and 5-acetourscil groups. *J. Immunol.* 106:1431

84. Asherson, G. L. 1968. The role of micro-organisms in autoimmune responses. *Prog. Allergy* 12:192

Ann. Rev. Immunol. 1983. 1:465-498

THE COMPLEXITY OF STRUCTURES INVOLVED IN T-CELL ACTIVATION

Joel W. Goodman

Department of Microbiology and Immunology, University of California Medical School, San Francisco, California 94143

Eli E. Sercarz

Department of Microbiology, University of California, Los Angeles, California 90024

FOREWORD

Whereas the discussion of epitope-antigen structure relating to B cells would not be very different from a similar consideration a decade ago, the nature of T-cell recognition and activation by "antigen" encompasses a completely altered area of concern. Within the past 10 years, concepts of restriction to major histocompatibility-complex MHC molecules and involvement of T cells in idiotypic interactions have revolutionized the view of the "antigen" that activates the T cells.

In this article we use a special definition for the term antigen structure, which includes all elements shown to be recognized by the diversity of T-cell subpopulations: epitopes (from the nominal antigen), idiotopes (on secreted Ig or on other cell surfaces), and at least two types of MHC determinants, which will be defined later.

We do not undertake a review of the basis and evidence for MHC restriction, or T-cell idiotypic interactions: earlier excellent reviews on the former (57, 139, 148, 186) and latter (47, 82, 83, 175) exist, together with several in the present volume. Our focus is confined to current information describing the complexity of structures involved in T-cell activation.

0732-0582/83/0410-0465$02.00

INTRODUCTION

The antigen specificity of T lymphocytes has been intensively investigated since their recognition as a distinct class of cells, but a clear understanding of the subject has yet to be attained. Early in the game, it was natural enough to assume that T cells recognized antigen using conventional immunoglobulins of perhaps a distinctive subclass, but that supposition has not stood the test of time. There are a number of reasons for the obduracy of the T-cell antigen specificity problem. Until recently, homogeneous T-cell lines analogous to the myeloma cell, which was so crucial to the solution of antibody structure, were not available. Even now, with the advent of antigen-specific normal and neoplastic T-cell lines, it is still difficult to obtain antigen-specific soluble products (secreted proteins) in quantities sufficient to do the essential experiments. This has forced a dependency on assays based on cellular response to antigen rather than on simple ligand-receptor interaction. It is paramount, therefore, at the outset to acknowledge the distinctions between the specificity of ligand binding and the consummate requirements for a cellular response. Consider the revolutionary contribution that hapten binding has made to our understanding of antibody specificity. Consider further that haptens, although they bind readily to the surface receptors of B cells of the appropriate specificity, fail to activate the cells. With some notable exceptions to be dealt with later, haptens also fail to activate T cells, so it has generally been necessary to work with antigens of epitypic complexity, complicating the issue further. Another graphic illustration of the non-identity between binding and activation is the ostensibly equivalent binding of the mitogens concanavalin A (Con A) and phytohem-agglutinin (PHA) by B and T cells (63), although activation of each cell type occurs under distinctly different circumstances (7). The bottom line is that ligand-receptor interaction at the cell surface cannot be equated with cell triggering, the latter involving a different order of complexity.

Another term in the obduracy equation is the subdivision of T cells into functionally distinct subsets with differing activation requirements and, possibly, different antigen receptors as well. A major unresolved question is the mechanism by which most cells, particularly those with cytotoxic or helper function activity, manifest a dual specificity for both nominal antigen and a product of the MHC such that activation takes place only when the T cell is presented with nominal antigen in the context of the appropriate MHC product (18, 90, 138, 153, 157, 188). Suppressor T cells can bind naked nominal antigen under some circumstances and have been enriched by adherence to antigen-coated polystyrene surfaces (109, 166), but the requirements for activation of this class of T cells have not been elucidated (but see below).

The major experimental evidence that T-cell antigen receptors are related to immunoglobulins rests on shared idiotypy. A body of indirect data supports the sharing of V_H regions between antibody molecules and surface receptors or soluble products from T cells of the same antigen specificity. Shared idiotypes have been found on receptors for alloantigens (19, 135), carbohydrates (20), proteins (32) and haptens (42, 131, 133, 166), as well as on antigen-specific helper and suppressor factors (10, 52, 59, 124, 163). There are two critical limitations to the interpretation of these experiments. The first is that serological cross-reactivity cannot be extrapolated to suggest V-region identity. The second is that heterogeneous anti-idiotypic antisera have been employed and it is possible that immunoglobulin and T-cell products are reacting with different components of the reagents. A more convincing case could be made by demonstrating idiotypic sharing using monoclonal anti-idiotypic reagents, although even this would not constitute definitive proof of a common gene pool. The nature of the T-cell antigen receptor, however obscure and fascinating, is not the central theme of this review. Our mission is to summarize the body of evidence identifying structural features of molecules implicated in T-lymphocyte activation and to describe what is known about the nature of the "activating moiety," which, in at least most instances, is not simply nominal antigen itself.

MHC RESTRICTION AND T-CELL RECOGNITION OF ANTIGEN

The discovery that T cells respond to antigen in a genetically restricted fashion changed forever our concept of T-cell antigen recognition. The cytotoxic T cell attacks only targets that express allogeneic MHC markers coded by the K, D, or other class I genes of the MHC or targets that combine nominal antigen with self class I products (18, 153, 189). A variety of helper (or inducer) T cells (Lyt-1$^+$ 23$^-$ cells) respond to antigen only when it is presented in a modified, presumably "processed" form associated with products of the I-region of the MHC (92, 138, 157). Presentation requires the participation of Ia-bearing antigen-presenting cells (APC), which in most, but not all, cases possess the properties of macrophages. The I-region was quickly recognized as the common denominator between antigen recognition by Lyt-1$^+$ T cells and specific immune response (Ir) genes (16). The observations that Ir genes apparently function only in responses to thymus-dependent antigens (16) and manifest remarkable specificity that readily distinguishes between antigens with similar primary structures (71, 136, 168, 169) prompted the hypothesis that the Ir gene product might be the T-cell receptor itself (16). There are compelling arguments opposing this proposal, many of which have been summarized by Benacerraf & Germain (15). The most pervasive of these would appear to be the functional expression of many Ir genes in macrophages and B cells (see 15) and evidence implicating Ia antigens, which are found in greater abundance on macrophage and B cells than on T cells, as the products of Ir genes (108, 155). Especially illuminating in this regard have been studies with the B6.C-H-2^{bm12} strain of mutant mice, which presumptively differ from the parental strain only in the A_α^b chain of the I-A molecule (120, 104). The immune responses of mutant mice to some, but not all, antigens are selectively impaired relative to the parental H-2^b strain (112, 113, 122, 100), furnishing vivid evidence for a causal association between Ir genes, Ia antigens, and T-cell responses.

Recognition by Single or Dual Receptors

A paramount question, then, is the manner in which Ia molecules on macrophages and B cells, or K/D molecules on target cells together with exogenous (nominal) antigen, produce an activating signal for T cells. Various models have been proposed to address this question. They segregate according to whether they postulate independent recognition of antigen and MHC by two distinct receptors on the T cell (dual receptor hypothesis) (21, 38, 84, 96, 105, 177, 187) as opposed to recognition of an interaction product between antigen and MHC by a single receptor (modified self hypothesis) (127, 136, 189). It has been difficult to design experiments that can unambiguously distinguish between the models. Perhaps the closest approach thus far has been the generation of T-cell hybridomas between cells that had different antigen/MHC specificities (92). Thus, hybrids specific for KLH plus H-2^f were fused with T-cell blasts specific for ovalbumin plus H-2^a. "Double" hybrids were obtained that recognized both parental combinations, but none were detected with either of the cross-specificities. In other words, none responded to KLH plus H-2^a or to ovalbumin plus H-2^f. Because the cells expressed receptors for each of the two antigens and the two MHC products, the findings clearly argue against orthodox dual recognition models of MHC restriction. However, the authors carefully point out that dual recognition of an interaction product between antigen and MHC cannot be excluded inasmuch as a given antigen might form different interaction products with MHC molecules of differing haplotype. A definitive decision will probably require direct biochemical characterization of T-cell receptors.

STRUCTURES RECOGNIZED BY T-HELPER AND T-PROLIFERATIVE CELLS

Macrophage-Derived Antigen-Ia Complexes

If the T-cell receptor has resisted all assaults on its character, the molecular entity recognized by the receptor has been of comparable obstinacy. After the basic tenet of genetic restriction in macrophage: T-cell interactions had been extended from the guinea pig to the mouse, it was reported that cell-free supernatants from histocompatible, and only histocompatible, macrophages pulsed with antigen were able to replace intact cells in the generation of helper T cells (50, 51). The macrophage-replacing factor was shown to be removable by either anti-I-A, or antibodies to the antigen, suggesting that it contains elements of both. Its molecular weight is surprisingly uniform (approximately 50–60,000 daltons) considering that antigens of larger and smaller size have been used in its preparation. Lonai and his collaborators have described a similar Ia-antigen complex (IAC) released by antigen-fed macrophages (114, 132). IAC was bound by Lyt-1⁺2⁻ T cells in an H-2 restricted fashion, could suicide antigen-specific helper cells in vitro, and displayed much greater immunogenicity than free antigen in histocompatible recipients (132). Competitive antigen binding by enriched T and B cells with (T,G)-A-L in both free and Ia-coupled forms demonstrated much higher affinity by T cells, but not by B cells, for complexed antigen (114). Furthermore, C3H.SW B cells cross-reacted with a variety of other branched polypeptides, whereas T cells bound only (T,G)-A-L and (Phe,G)-A-L, to both of which they are Ir gene-controlled high responders. The authors concluded that IAC is recognized by T cells with high affinity and may play a major role in defining the specificity, H-2 restriction, and Ir-gene control of T cells. More detailed analysis of the provocative IAC complex or the genetically related factor (GRF) (50, 51) is awaited. Right now, the T-cell recognition problem, remains plagued by the complexity systems used and the paucity of analyzable material.

Antigen Presentation by B Cells

It has recently been found that Ia-positive B cells may also present antigen in a genetically restricted fashion to sensitized T cells. Chesnut & Grey (35) used normal mouse splenic B cells coated with rabbit anti-mouse Ig to stimulate rabbit IgG-primed T cells in a proliferation assay. A variety of controls minimized the likelihood that sources other than B cells might have been responsible for the observed stimulation. This represented a special case inasmuch as anti-mouse Ig will bind to the surface Ig of all B cells. Indeed, protein antigens that lack this property (e.g. ovalbumin) were not effectively presented by normal B cells. However, a series of H-2^d B-cell tumor lines were capable of presenting ovalbumin, KLH, or HGG to I-region-restricted, antigen-specific cloned T-cell hybridomas (177). For the most part, antigen presentation correlated with the expression of Ia antigens on the B cell lines, although a few positive lines failed to present antigen. Functional activity could not be correlated with the stage of differentiation of the lines. All the available tumor lines were of H-2^d haplotype, but the principle of B-cell antigen presentation was extended by Kappler et al (93) by fusing one of the Ia$^+$, H-2^d tumor lines to T-cell-depleted spleen cells from mouse strains of other H-2 haplotypes. Some of the resultant hybridomas presented various antigens to a group of antigen-specific, I-region-restricted T-cell hybridomas with the phenotype of the normal spleen cell partner. In a recent fascinating study, antigen-presenting cell variants with lesions in particular Ia antigenic determinants have been induced by mutation and selected from hybridomas created by fusion of a B-lymphoma line which can present antigen, with normal B cells. Such hybridoma APC mutants now fail to present certain antigens but remain able to present others (61). The availability of homogeneous rapidly growing APC lines (including monocyte lines) (177) should greatly expedite the exploration of I-region-restricted antigen presentation to T cells.

The above findings, which provide a plausible scenario whereby T cells might directly collaborate with B cells that have captured antigen, do raise questions about requirements for antigen processing pursuant to T-cell triggering. It might not be anticipated, a priori, that macrophages and B cells would handle antigens identically in terms of degradation and subsequent association with surface Ia molecules. However, it has been suggested that B cells that bind an exogenous antigen might interiorize, degrade, and process it in the manner of macrophages (174). Final judgment will depend on direct characterization of the structures responsible for T-cell activation.

Cross-Reactivity for Protein Antigens by T Cells and Antibodies

One approach to understanding the antigen specificity of T lymphocytes has been to compare the relative cross-reactivities for protein antigens in helper or delayed hypersensitivity responses with those exhibited by antibodies. A general consensus has not emerged from these studies, although most of the findings are potentially explicable on the basis of T cells recognizing small linear fragments of processed nominal antigen presented in association with MHC gene products on the surfaces of APC, as proposed by Benacerraf (14). An extensive analysis of the specificity of helper T cells for a series of cross-reacting serum albumins led to the conclusion that the specificities of virgin and primed helpers were similar, if not identical, to each other and to that of humoral antibody (134). On the other hand, several studies of the cross-reactivities of erythrocyte antigens suggested a greater discriminatory power for antibodies than for helper T cells, both in vivo (43, 130) and in vitro (53, 74, 75), although the particular antigens involved in the assays were not identified. However, there is a difficulty inherent in this type of comparison: since T cells respond to a more restricted set of epitopes, discrimination on the part of any one T cell may be as acute.

Similar findings were gleaned from a comparison of native and extensively denatured (urea plus reduction and alkylation) forms of the globular proteins ovalbumin and bovine serum albumin, using proliferation as the indicator of T-cell responsiveness (34). This study went a step further by formally establishing that the same T cells responded to both native and denatured forms, an important adjunct in that it eliminated the possibility that a subset of cells recognized the native protein exclusively and, hence, might be specific for conformational features. It is also apparent from studies with T-cell clones specific for lysozyme that certain clones can be propagated on either lysozyme or reduced, carboxymethylated lysozyme (F. Manca, N. Shastri, A. Miller, and E. Sercarz, unpublished data). However, more interesting are the clones that will only grow on one of these forms of the molecule.

Protein Antigens are Recognized as Fragments by T Cells

The comparative specificities of T cells and antibodies for native and denatured forms of the same protein antigen have seemingly pointed to a broader specificity for T cells, in agreement with the results obtained with red cells mentioned above. In retrospect, the evidence does not support the notion that T-cell receptors are actually less precise than their B-cell counterparts. Thus, early experiments demonstrated that guinea pig delayed hypersensitivity reactivity, in contrast to antibody, failed to distinguish between native and extensively denatured protein antigens (56). In another investigation, chemically modified BSA cross-reacted extensively with the unmodified protein at the helper T-cell level, but scarcely at all at the humoral antibody level (146). The list of cases of this kind is long. Other noteworthy examples included antigen E, which in denatured form induces helper T-cell responses to the native antigen without serological cross-reactivity (81); antibodies to staphylococcal nuclease (142) and lysozyme (145) are highly conformation-dependent whereas T-cell responses are not; and helper T cells primed with the unfolded VH and VL domains of MOPC-315 cross-reacted with the native myeloma protein (87). All the findings indicate that the portion of the nominal antigen recognized by helper T cells is sequential and not conformational in character. (It is, however, possible that two nonsequential fragmented peptides derived from the native protein might remain bound to each other and be stabilized by disulfide or other local forces.) These observations, in concert with the MHC-restricted character of T-cell responses, inspired the notion that T cells recognize fragments of antigen following processing and association of antigen to MHC products by the APCs (14). This proposal rests on the likelihood that proteins will lose their native conformation in the acidic environment of the lysosome, resulting in generation of the same peptide fragments from native and denatured antigens. Indeed, even the very few studies that imply a conformational dependence of antigen-specific Th/Tp cell activation can be interpreted along these lines. Consider the proliferative responses of T-cell blasts from BDF1 mice primed to beef apocytochrome c and tested with the homologous apoproteins, the native protein, and cyanogen bromide cleavage products from each protein (28). Cells maintained in long term culture with apocytochrome c responded more vigorously in the proliferative assay to the apoantigen and its cleavage products than to the native protein, suggesting at first glance the possibility of conformational dependence for T-cell recognition. However, a comparison of responses using three different sets of APCs (the syngeneic and two parental types) led to the conclusion that the observed differences were associated with different capacities of the APC to process and present the two forms of antigen. As the authors pointed out, absence of T-cell cross-reactivity between native and denatured forms of a protein might be expected for proteins that are relatively stable in the acidic environment of lysosomes and unlikely to unfold prior to proteolytic degradation. In such instances, the proteolytic fragments obtained will depend on the three-dimensional structure of the substrate. Thus, the differences seen are likely to reside in the APC rather than in the antigen-recognizing properties of the T cell.

Reference was made earlier to the exquisite T-cell specificity manifested toward antigens under Ir gene control. This has been particularly evident with small immunogenic peptides of defined structure and sets of closely related analogues of the immunogen. Thus, guinea pig T lymphocytes sensitized to the octapeptide angiotensin II could readily distinguish single amino acid substitutions, which had no effect on antibody binding (169), when the assay used was the ability to stimulate T-cell proliferation. Similar results were obtained in the same laboratory with human fibrinopeptide B and a series of structural analogues (168). These findings are consistent with the idea that the epitopes recognized by proliferating T cells are dictated primarily by amino acid sequence rather than by conformation, of which there might be expected to be little, in any case, in peptides of this size range. These investigators reasoned that if such be the case, inverting the sequence of amino acid residues in a small peptide antigen should make little difference to the T cell, because the same amino acids are presented in the same spatial order (166). Two assumptions are implicit in this reasoning. One is that there is no polarity or directionality in the presentation of an immunogenic epitope to the T cell. The other is that peptides of the dimension of eight amino acids are not degraded during the course of processing by APC. Should additional degradation occur, then it can be readily foreseen that altering the sequence of a peptide might alter the course of degradation, leading to the formation of different neoantigens. The prediction was tested with a fibrinopeptide fragment of eight amino acids and its inverted sequence peptide in a guinea pig T-cell proliferative assay (166). Strain 2 guinea pig T cells sensitized to the native peptide were completely non-cross-reactive to the inverted peptide; in fact, the inverted peptide was not immunogenic in strain 2 animals. The authors interpreted this to mean that the ordering of the amino acids within the antigen is critical, perhaps because the antigen becomes associated with MHC products in a non-random fashion. The possibility remains that degradation of even small peptide antigens is an intrinsic part of processing and will certainly be affected by amino acid sequence.

Studies with T-Cell Clones

T-cell clones reactive with soluble protein and synthetic polypeptide antigens have recently been propagated in long-term culture (73, 78, 97, 159). Some of these cloned lines have been shown to function as antigen-specific helpers and to obey the rules of MHC restriction. For example, clones reactive to the multichain synthetic polypeptide (T,G)-A-L provided specific help to TNP-primed B cells when challenged in vitro with TNP-(T,G)-A-L and responded in proliferation assays in an I-region-restricted fashion (73). An aspect of the work with cloned, self MHC-restricted, antigen-specific T cells that has an important bearing on the question of the nature of T-cell specificity has been their cross-reactivity with allogeneic MHC determinants. One interpretation of the high frequency of alloreactive T cells, originally noted by Simonsen (156), has been an overlapping of specificities of the alloreactive and nominal antigen-reactive T-cell populations. This issue has been addressed in the past by searching for cross-stimulation of alloreactive cells by antigen and vice versa with heterogeneous T-cell populations (17, 30, 69, 107). Although cross-stimulation was observed in these studies, interpretation was difficult because the possibility of non-specific recruitment could not be excluded. This ambiguity can be circumvented, or at least minimized, by using extensively recloned antigen-reactive T cells. Thus, a DNP-ovalbumin-specific clone from a B10.A mouse proliferated in response to added antigen in conjunction with syngeneic irradiated antigen-presenting cells *as well as* in response to allogeneic spleen cells from B10.S mice in the absence of DNP-ovalbumin or B10.A accessory cells (158). These findings do provide compelling evidence for dually specific T cells responsive to both nominal antigen and alloantigens, but they do not yet permit distinction between hypotheses concerning the receptors involved in antigen recognition. Thus, the findings are compatible with completely independent sets of receptors for alloantigens and for nominal antigens randomly expressed on all T cells (181), with alloantigen recognition through high affinity cross-reactivity with the anti-self MHC receptor (84) or the receptor for nominal antigen (38, 105) in dual receptor models, or by high affinity cross-reactivity with a single receptor for altered self (17, 30). Additional studies with antigen-specific T-cell clones may eventually distinguish between the various models for antigen recognition by T cells.

Cloned T cells will also be useful for settling other unresolved issues concerning antigen recognition and MHC restriction. For example, the specificity of individual T cells for defined epitopes of antigen molecules has been obscured by using heterogeneous populations of cells and relatively complex antigens. The responsiveness of T cells to the compound L-tyrosine-p azobenzenearsenate(ABA-T) furnishes a system for investigating the nature of T-cell antigen recognition in which the antigen is reduced to its simplest possible form. Different functional types of T cells have been induced that are specific for ABA including helpers (5, 6, 62), suppressors (109, 161), killers (154), and mediators of delayed hypersensitivity reactions (11, 169). Very recently, long-term cultures of T cells specifically reactive to ABA-T were established from the lymph node cells of A/J mice, which had been immunized with ABA-T, and the cells were cloned by limiting dilution (70). The MHC restriction of the antigen-induced proliferation response of the clones was assayed with accessory cells from appropriate recombinant strains of mice and monoclonal anti-Ia antibodies as blocking reagents. Fourteen of 15 clones tested proved to be I-A restricted in their response, whereas the other clone was I-E restricted. These findings indicate that the same epitope (ABA-T) can be presented to T cells in the context of more than one MHC product. Although the bulk population of lymph node cells responded when accessory cells compatible at either the I-A or I-E loci were present, individual clones were restricted in their recognition to a single locus. The fine antigen specificity of the clones was analyzed with structural analogues of ABA-T to induce proliferation in conjunction with syngeneic accessory cells. The clones could be divided into three types with respect to responsiveness to ABA-histidine (ABA-H). One type responded about equally strongly to ABA-H and ABA-T. A second group responded to both, but significantly better to ABA-T, whereas the third set did not respond at all to ABA-H. In all cases, the ABA-H-responsive clones discriminated exquisitely between the 2-azo and 4-azo isomers of histidine, responding only to the 4-azo derivative. Thus, at the level of individual clones, T cells display a remarkable specificity for defined epitopes of the nominal antigen in situations where differential processing of antigens is unlikely to be a factor, as it may be when comparing complex antigens differing in primary structure.

The T Proliferative Response is not Identical to the T Helper Cell Response

It is often assumed that cell populations that proliferate in response to a particular epitope in the context of the appropriate MHC-APC are always helper cells. This is surely not the case. For example, in recent work in the beta-galactosidase (GZ) system, T proliferation could be induced in vitro by GZ in most of the populations primed by individual cyanogen-bromide peptides derived from GZ (from 23–95 amino acid residues in length). However, only two of a dozen peptides tested could induce a T-helper effect for the anti-hapten (FITC) response to GZ-FITC (101). The ability to be induced to proliferate by antigen may be shared by many Lyt-1$^+$2$^-$ cells and conceivably by Lyt-1$^-$2$^+$ cells [although in heterogeneous cell populations the suppressor T-cell proliferation might be missed in that the expressed activity of the Ts would preclude the detection of its target population(s).] Surely at the clonal level, many cellular members of the enormous Ts "regulatory army," as well as the Ts itself, might proliferate in conjunction with antigen and/or IL-2. The secretion of IL-2 by a T inducer cell in response to its activation stimulus may represent a major pathway in Ts-cell and other T-effector-cell recruitment.

It has been shown that many of the cells recruited by the proliferating cell are B cells (172). It is possible that upon recruitment and proliferation, a subsequent display of antigen by these B cells will back-recruit a few more antigen-specific T cells. Surely, under the usual Corradin & Chiller (40) conditions with whole, primed lymph nodes, a myriad of recruiting and proliferative events must be occurring.

Determinant Selection on Multideterminant Antigens

The particular choice of determinants on a potentially multideterminant antigen molecule can be attributed to selectivity exercised by a component of the MHC. This concept of determinant selection was elaborated by Rosenthal and his colleagues using the model of insulin recognition by the guinea pig and the mouse (136). Strain 2 and strain 13 guinea pig T cells each responded to a distinct region of beef insulin (137), and this discrimination could also be observed in the mouse (140). In the case of insulin, and others subsequently studied, in which mammals are challenged to respond to related mammalian proteins, it could be argued that, in fact, there is very little "selection": that is, whatever mechanisms purge the repertoire of response against self antigens leave only a few regions that are different between the two proteins, against which a potential response might be directed. This is one reason why results with the bird lysozymes are noteworthy, in that as a group they differ by about 50 residues out of 129 from the mammalian congeners. In fact, almost all of the surface amino acid residues are different. Nevertheless, it is striking that with all this potential "foreignness" only one to three determinant areas are selected by mice of differing haplotypes in the triggering of T-cell helper/proliferative subpopulations (115, 116). In a sense, this reminds us of the concept of immunodominance of certain epitopes in the induction of antibody (88).

At the T-cell level, it has been postulated (14) that what makes a particular determinant area dominant or, rather, responded to at all is the existence of a stretch of amino acids on the nominal antigen that has an affinity for a structure on the MHC restriction element. A useful nomenclature for considering each of the entities involved in the recognition has recently been proposed (144). Thus, the nominal antigenic entity that contacts the MHC has two distinct moieties, at least conceptually: an epitope in contact with subsites on a T-cell receptor, and an agretope (*antigen recognition*), which acts as a site of attachment to the MHC desetope (*determinant selection*). (We might suggest from the limited data available that roughly one agretope exists for every 50 amino acid residues on a protein for Th/Tp cells.) Finally, the additional element on Class I or Class II MHC molecules that is recognized in addition to the epitope is designated a *histotope*.

In work on pigeon cytochrome (64–66) by Schwartz and his colleagues, a potential agretope has been defined as giving the quality of immunogenicity to the C terminal peptide 81–104. Using synthetic peptides, it was shown that residues 100 and 103 were important for antigenic strength, which could be related to interactions of the antigen and MHC structures on the APC. Granting the closeness of the agretope, residue 99 could be considered part of the T-cell contacting epitope related to T-cell memory, since a set of amidinated C-terminal peptides derivatized at position 99 were completely non-cross-reactive with the native peptides but nevertheless showed the same immunogenicity/non-immunogenicity discrimination among strains, which is presumably dependent on the nature of the agretope. In the lysozyme model, a completely analogous result has been reported, in which only one dominant, circumscribed region of the molecule near aa 113–114 is utilized in the B10.D2 haplotype for presentation of the epitope at 113–114 to T cells responding to parallel sets of non-cross reactive bird lysozymes (96).

A case in which an agretope may have been synthetically manufactured appeared in the recent literature. As mentioned earlier, an octafibrinopeptide is immunogenic for strain 2 guinea pig T cells (166). However, strain 13 animals are completely unresponsive to the peptide. When a glycine residue was added to the carboxyl end of the octapeptide, the new nonapeptide induced strain 13 T-cell responses. Although the mechanism of the immunogenic conversion is uncertain, a plausible interpretation is that the glycine residue formed part of an agretope which could productively associate with a strain 13 desetope.

Another candidate for a critical component of an agretope is the side chain of tyrosine in ABA-T. Increasing the hydrophobicity of the side chain decreased immunogenicity without palpably influencing specificity. Thus, the tyrosyl end of the molecule can be visualized as the agretope and the ABA end the epitope, although the precise structures of these entities have not been defined.

If agretope scarcity were responsible for the very limited response to certain antigens, the following consequences might be expected: (a) when an agretope is defined, a diversity of T cells should be available for combination with epitopes associated with that agretope; (b) the distribution of agretopes rather than the repertoire diversity of ambient cells should be the prime factor in determining whether or not a particular antigen will be responded to by the haplotype; and (c) the paucity of appropriate agretopes should lead to some null peptides on a typical immunogen which are not used for addressing any T-cell subpopulation. Some of these corollaries have been observed in the lysozyme system (117). Thus, only a single lysozyme agretope is presumably available for H-2^b strains near aa residue 91, but nevertheless, a heterogeneous array of T proliferative clones of differing fine specificity for epitope can be detected, all of which seemingly utilize the unique agretope. The C-terminal peptide of lysozyme, aa106–129, makes no impact at all on H-2^b mice, neither inducing helper nor suppressor T cells; likewise, no other tryptic peptides than T11 (aa74–96) are immunogenic or reactive in H-2^b strains.

Agretope Limitations and Hierarchies

Hints are available that the scarcity of agretopes may sometimes be apparent rather than real. Accordingly, there may be two or three immunogenic regions on a protein molecule, but a hierarchy of choice can obscure the reactivity to all but a dominant agretope-desetope pair. An example in the lysozyme system (96) involves the subdominant C-terminal peptide in the B10.A strain, reactivity to which cannot be demonstrated after immunization with native lysozyme and challenge with L3. However, initial in vivo immunization with L3 reveals strong responsiveness in this haplotype to either L3 or HEL subsequently used for in vitro challenge. Apparently, a competitive situation exists in which an N-terminal agretope near residues 13–15 takes precedence over two other agretopes that are more C-terminal, or alternatively, T cells possess greater affinities, in general for the dominant epitopes.

SUPPRESSOR T CELLS AND THE RECOGNITION OF ANTIGEN

Does suppressor T-cell activation depend on the same concepts of desetope/agretope interaction we have just discussed for Lyt1$^+$ Th/Tp cells? Do S-agretopes exist that are distinct from H-agretopes? The considerations relevant to this section hinge on the relationship of the MHC to suppressor cell induction and activity, an area that still has many unknowns.

T Suppressor Cell-Inducing Determinants

Suppressor T cells (Ts) play an important role in the expression of the Tp/Th repertoire, in some cases obscuring a latent capacity to produce a response to the native molecule. In Ir gene-controlled responses, it has been reported in a variety of systems that nonresponder mice have the capacity to respond to the antigen in the absence of Ts, or to present the antigen to syngeneic or to Fl(R x NR) T-helper cell populations (9, 90, 98). Disclosure of this latent potential might require the dissociation of the suppressor T cells from the system. In the case of lysozyme, this was achieved by amputation of the suppressor-inducing epitope at the N-terminus of the protein (185). In fact, removal of as little as the terminal tripeptide lys-val-phe by aminopeptidase converts the non-immunogenic lysozyme into a derivative that no longer induces suppression, but rather a helper effect, and T-cell proliferation (A. Miller, L. Wicker, M. E. Katz, and E. Sercarz, unpublished results). By this simple transformation, actually two new determinants were revealed from what would have simply been regarded as a non-immunogenic molecule for H-2^b mice: a suppressor determinant and the previously suppressed, latent responsiveness to the helper epitope on tryptic peptide 11. Of course, the same general problem of paired suppressor and helper epitopes may exist on any "null peptide" such as the C-terminal peptide of lysozyme, and additional experiments can be performed to examine this possibility. Latent responsiveness has also been demonstrated in the cytochrome system (41), and to one of the cyanogen-bromide peptides from beta-galactosidase (U. Krzych, A. Fowler, E. Sercarz, unpublished data).

The fact that a single suppressor-inducing determinant could affect the entire response to a molecule had been recognized with lysozyme (HEL) earlier (1,152). More recently, it has been possible to induce Ts with the CNBr-derived, reduced and carboxymethylated N-terminal dodecapeptide from HEL (N. Shastri, A. Oki, M. Kaplan, D. Kawahara, A. Miller, and E. Sercarz, unpublished results). Earlier evidence of suppressor determinants with beta-galactosidase (150, 173) and poly-glu-ala-tyr (147) also carried the implication that the suppressor- and helper-inducing determinants might be different.

Suppressor and Helper Epitopes are Non-Overlapping

In all protein systems examined, epitopes addressed to the Lyt-1$^-$2$^+$ Tsp population and those activating the Lyt-1$^+$2$^-$Th/Tp subpopulation are different! This was true in the case of the small globular lysozyme molecule

(HEL=129aa) and likewise for the large tetrameric β-galactosidase (GZ=1021 aa/monomer). When the second cyanogen bromide peptide from the N terminus (aa3-92) of GZ was assessed for induction of Th or Ts, both subpopulations were activated. Further analysis showed that a tryptic nonapeptide, amino acids 44 to 52, induced Th but not Ts, and the reverse was true for an 11-amino-acid peptide 27-37. Thus the principal of non-overlap remained intact (102). Similarly, in the response to myelin basic protein in the guinea pig (160) it appears that the suppressor-inducing peptide is different from the encephalitogenic peptide in this species. However, the possibility that the determinants are very close could be concluded from experiments in which specific residue alteration caused a reversal of functional effects induced by the peptides (68, 94). In a similar vein, Th and Ts that recognize the same or very similar epitopes can be induced by ABA-T (109).

Although the picture is not complete, many results can be accommodated by a general model cited below that suggests that functionally and chemically unique antigen-presenting entities (desetopes) are employed in the activation of different T-cell subpopulations. Consequently, it may be the desetope-agretope interaction that controls which T cells will be triggered.

Interactions Within the Suppressive Cell Circuitry

The intereractions among the varied members of the regulatory T-suppressor family have illustrated some new principles in T-cell activation and interaction. In considering the production of antibody, it is evident that members of a single AgTh clone can collaborate with the appropriately specific B cell, leading to B-cell proliferation, differentiation and antibody formation. This is the tip of a cellular iceberg, which hides a multitude of competing regulatory cells, each of which must know with which partner cell it must interact. These rules of intercellular communication, or as Tada has termed it, "immunosemiotic interaction" (162), must eventually lead to cell collaboration that is purposeful rather than haphazard and chaotic.

Evidence indicates that the T-T interactions that accomplish the regulation of suppression are restricted by three types of elements—antigen, MHC and immunoglobulin. More explicitly, these are the suppressor epitopes on the antigen, MHC-presentation elements, and regulatory idiotopes, but in addition there may be MHC-interaction elements and T-cell receptor constant-region determinants that must be recognized. Each member of the system is induced by a specific cell or a factor derived from it: the factors have components that recognize epitopes or idiotopes, in addition to a constant region, coded for by genes on the 12th chromosome in the mouse (170, 171); on two-chain factors a separate peptide bears the MHC element involved, which in the suppressor circuit is a type of I-J.

As suggested in the last paragraph, one of the critical questions that is still undergoing exploration is whether suppressor T cells are MHC restricted at either one of two levels: (a) at the point of presentation of antigen or (b) in the interaction between regulatory and target cells in the collaborative milieu.

Effector-Cell Induction Pathways

T cells which will eventually develop into effectors (Th, Ts, and Tc), seem to be activated by a particular inducer cell (augmenting cells, suppressor inducers and CTL helpers, respectively). It can be predicted that the rules of each partnership are immunosemiotically defined to prevent anomalous induction. The Ts pathways have been most extensively studied and will be discussed in a general, summary way, as other reviews have described their cellular constituents in detail (31, 57).

Lyt-1^+2^-, I-J$^+$suppressor inducer cells (Tsi) were demonstrated (45) to be necessary in interacting with Ts precursor cells, usually Lyt-1^+23^+, to promote them to functional effector status (9, 119). More recently, significant progress has been made in assigning antigenic and idiotypic specificities to these cells or to factors derived from them (46, 164, 184). In general, the restriction in interactions between Tsi and Ts cells appears to be owing to Igh rather than to MHC structures, whereas Ts effector cell interaction with Th targets is MHC-restricted.

It has been suggested that in all suppressor systems, a cascade of three sequential Ts cells (Ts1, Ts2, and Ts3) are induced, originally by the nominal antigen (58). In fact, in the systems which reveal each of the three cell types, the next cell in the sequence is usually induced by a factor derived from the previous cell. However, antigen can also induce Ts3 directly, which then requires a Ts2-derived signal to achieve effector status (123). It appears that the system will employ the effector-cell specificity it needs to control the regulee: if it must down-regulate an idiotype-bearing cell, it will somehow raise idiotype-recognizing T-suppressor cells. On such occasions, if a particular suppressor cell is in short supply, complex circuits involving amplifying cells, whose relationship to the effector cells is based upon idiotype, MHC, or antigenic recognition (31, 162) may be resorted to.

The Ease of Triggering Ts in Non-Responder Strains

Certain well-characterized cases of genetic nonresponsiveness, those concerning the terpolymer glu$_{60}$-ala$_{30}$-tyr$_{10}$ (GAT) and lysozyme, owe their existence to readily activated T-cell suppression (60, 151). In the lysozyme case, a Ts cell is induced in H-2^b mice by chicken lysozyme (HEL) but not by any variant lysozyme that substitutes tyrosine for the phenylalanine at residue 3, such as REL (4). Likewise, H-2^b strains respond to REL but not to HEL, at the antibody formation level. It can be asked whether or not some type of suppressive element exists in regulating the response to REL in the H-2^b mouse? If so, are these Ts simply less readily activated by REL in the non-responder strain, or activated by a parallel but distinct mechanism? Although suppressive circuitry may occur universally, the routes taken to achieve the suppression may be quite different in the nonresponder and responder strains. Possibly, Ts induction to certain epitopes is rapid and effective in the non-responder because of an MHC-restricted activation process that can proceed at a low level in the absence of a T-suppressor inducer cell. In responder strains, stimulation of the idiotype-bearing Ts might occur at a later point, through the parallel mechanism involving a Tsi-delivered induction signal. Such a circuit, for example, might depend on the appearance of the predominant idiotype during B cell maturation, which would arouse idiotype-recognizing Tsi to stimulate suppression (103).

Differences in Th and Ts Recognition of Antigen

At the outset, antigen recognition by Ts cells appears to be quite different from that of Th or Tp cells. Evidence that Ts can bind to native antigen has been used to argue that nominal antigen recognition is independent of the MHC. Isolation of Ts on plates coated with antigen (109, 125, 165) indicates that Ts can recognize native antigen. However, there are several published instances (3, 39, 72, 118) in which Th have likewise been demonstrated to bind to antigen or antigenic peptides.

Endres and Grey (49) found that Ts induced by native or extensively denatured ovalbumin did not functionally cross-react with the alternative form of antigen, in contrast to helper T cells (34). Furthermore, the suppressor cells could be depleted on plastic dishes coated with the homologous but not with the heterologous antigen. These findings suggest a closer kinship of Ts cells to B cells than to Th cells in terms of antigen recognition. However, in the lysozyme system, the native enzyme or its reduced, carboxy-methylated derivative or its fragments (including the N-terminal dodecapeptide) can each induce Ts, although the derivatives surely are not in the native conformation, nor are they cross-reactive at the B-cell level.

Certain authors (34) postulate that Ts cells recognize conformational determinants and that the differences in antigen recognition between Ts and Th are related to the requirements for antigen processing and presentation by accessory cells to Th; this argument neglects the presentation of antigen to Ts. The apparent lack of dependence on adherent cells in the induction of suppression (54, 80, 129) seems to support the idea that antigen processing is not required for the activation of suppressor T cells. However, the existence of special types of APC for different T-cell subpopulations, with different adherence properties, has been documented (29) and would seem a relevant consideration.

The evidence is still concordant with the notion that Ts recognize external epitopes on antigens, owing to the superficial placement of S-agretopes (also see below). This would permit binding to antigen alone, if the affinity of interaction were high enough, but possibly Ts activation would require an additional signal derived via associative recognition.

MHC Restrictedness of Ts Cells

Although the evidence we have sketched suggests that effector Ts may be MHC restricted in the interactions with their inducers and target cells, we will now examine experiments that converge to suggest MHC restriction at the point of presentation of antigen and the induction of Ts precursors.

In the lysozyme system, the ability to induce Ts in the nonresponder H-2^b mouse is limited to an epitope at the N-terminus of several gallinaceous lysozymes, those of hen, bobwhite quail, guinea hen and any other lysozyme having phenylalanine at amino acid residue 3. Interestingly, human lysozyme (HUL) which is almost completely non-cross-reactive at the B cell level, apparently can induce the same Ts, bearing the predominant idiotype, IdXL, which then is equally active on the anti-HEL and anti-HUL in vitro responses. This has a rationale in that not only phe-3 but also eight of the nine N-terminal residues are identical in HEL and HUL. On the other hand, although the H-2^q mouse possesses the same cross-reactive IdXL-bearing Ts, it nevertheless *can* respond to HEL while still being nonresponsive to HUL. This failure to induce Ts by HEL can be attributed to an inability of HEL to interact with the appropriate MHC element in antigen presentation to MHC-restricted Ts (2). Other more direct evidence for MHC restriction in this system was obtained in positive selection of the IdXL$^+$ Ts on HEL-pulsed, presenting-cell monolayers. Only syngeneic monolayers were capable of antigen presentation to Ts, and this positive selection could be inhibited by anti-IdXL serum (8).

In recent work by Jan Klein and his colleagues (12, 13), evidence has been presented for restriction of Ts induction by lactate dehydrogenase B in the context of an I-E^k restriction element. Using nonresponder B10.A(2R) T cells, antigen-presentation by B10.A(4R) lacking the I-E^k element, gives rise to proliferation, but not suppression. Furthermore, in vivo treatment with monoclonal anti-I-E antibody converts the B10.A(2R) into an LDH responder.

Taken together, the evidence points to MHC restriction at the time of initial antigen presentation, and suggests that further study of such early events might reveal additional examples.

DESETOPES AND HISTOTOPES IN CYTOTOXIC T-CELL RECOGNITION

The problems raised by recognition of class I and class II MHC proteins are really quite parallel. Several class I proteins (including K, D, L, M and R) have been implicated in associative recognition with nominal antigens. Just as in the case of Ia-restriction, it is apparent that certain haplotypes utilize particular molecule(s) of the class I set for restricting their CTL responses. One interesting example is provided by the recent work of Ciavarra and Forman (36), studying the restriction elements used for reactivity to vesicular stomatitis virus (VSV) in the $H-2^d$ haplotype. VSV was only recognized in the context of the $H-2L^d$ molecule! This was shown by cold-target inhibition experiments as well as inhibition studies with $H-2L^d$-directed monoclonal antibodies. However, H-2L-restricted responses are not found against trinitrophenol (111) or minor histocompatibility antigens (22). We can discuss these results by employing the same conventions in describing the relationships between MHC class I molecules and nominal antigen as were used earlier in conjunction with Ia restriction elements (see p. 476). The requirements for association between an agretope on the viral antigen and any particular class I desetope may be unusually demanding. Therefore, even though a dozen or more different class I "restriction sites" may be available within a haplotype (if each is unique), a particular viral capsid antigen may not be able to associate productively with most class I molecules. Likewise, hierarchical preferences also seem to be evident in class I T-cell activation.

A related question of some importance is whether a distinct region (a histotope) on the class I or II molecules is necessarily recognized by the T cell *in addition to* that desetope making contact with the nominal epitope at an agretopic site. One experimental system that seems relevant to this question is the recognition of hapten in conjunction with self-class I molecules by Tc cells. Earlier work showed that hapten coupling to class I molecules themselves was unnecessary in achieving a restricted response (143). Molecules of TNP-BSA could insert into the cell membrane and TNP-self recognizing Tc would still respond. This lends some credence to the view that the histotope recognized on a class I molecule may be some distance from the hapten. It has been recently pointed out that the hapten or nominal antigen could also be coupled to a self non-MHC protein (110).

Studies of the sites of allogeneic recognition are also connected to this question. Allogeneic recognition by CTL may be limited to very few areas on the target class I molecules. In a recent report (77), cross-reactivity to mutant mouse class I proteins of known sequence was the device utilized to approach this problem. The 70% cross-reactivity of *b* anti-*a* CTL with *bml* targets could be attributed to a tyr_{155}-tyr_{156} alteration of K^{bml}. Since the identical alteration was uniquely found on the L^d molecule, it apparently accounts for the prominent cross-reactivity. Cold target inhibition studies verified the point. Whether the site functioning as an allogeneic target is related to a desetope or a histotope is an important area of investigation. In this context, evidence provided by Weyand et al (178, 179) shows that 2 distinct polymorphic clusters, A and B, are recognized, both by CTL and monoclonal antibodies. It is interesting but perhaps circumstantial that both alloreactive $H-2^k$-as well as $H-2^d$-restricted self CTL's strongly prefer cluster B.

It appears that a crystal ball is still needed to sort out how many separate elements are recognized by the T cell. No experiments exist limiting the recognition decisively to a fixed number of elements. Within the various cell subpopulations examined, this could range from (a) epitope alone (b) epitope plus some mixture of desetope/agretope to (c) histotope alone or (d) either (a) or (b) plus a distant and distinct histotope.

THE RECOGNITION OF IDIOTYPIC DETERMINANTS BY T CELLS

During the past decade, voluminous evidence has implicated the T cell in intercellular idiotypic interactions (reviewed in 47, 82, 83, 175). We shall consider the nature of the T-cell recognition and in what form the idiotopes are recognized (MHC-restricted, together with antigen, etc.). Although we have discussed some of the Id-anti-Id interactions in the suppressive cell circuitry, we will focus here on the predominant idiotypic determinants displayed on the B-cell surface that seem to play a critical role in establishing the quality of the antibody response.

Predominant Idiotypes

An idiotope on an immunoglobulin molecule can be recognized by an antigen-specific Th cell (AgTh) just as any other immunogenic epitope, in the context of the MHC. Some strains will be high responders to such idiotopes, and others will be low responders, as in any other immune response gene-control system (86). Jorgensen and Hannestad used different portions of MOPC-315 as T-cell priming agents, challenging subsequently with 315 coupled to DNP.

However, we would like to explore the nature of predominant idiotypes on B cells. Why are there certain idiotopes that gain such prominence as the T15 idiotype which appears on more than 90% of the anti-phosphocholine antibodies in the BALB/c (37) strain, and also in several others? In many cases this may be a consequence of the germ-line gene composition (55). The predominance of certain idiotypes can be contrasted to the diversity of private idiotypes that appear on individual hybridomas in many such systems. In several cases (3, 25, 182) it has been demonstrated that T cells that recognize idiotype and can be adsorbed on Id-coated petri dishes are implicated in causing the predominance. One possible scenario would involve a sequential activation of the B cells specific for the antigen by an AgTh that is MHC-restricted. Such AgTh clones (73) have been shown to activate B cells regardless of their idiotype (S. Cammisuli, personal communication), or isotype (149). In the absence of an additional IdX-recognizing cell, this would lead to a nonselective distribution of idiotypes in the expressed repertoire. Subsequently, due to the display of the special predominant idiotope (IdX) on the activated B cell surface, a set of IdX-specific Th cells (IdXTh) would be activated which then would positively select those B cells for expansion that display IdX.

The kinetics of appearance of IdX among the B cells in a variety of systems suggests that in cases where the antigen molecule is ubiquitous or at least available in the self-environment, a prior equilibrium is established between IdX-recognizing Th and Id-bearing Ts which down regulate them.

The Concept of a Regulatory Idiotope

Evidence seems overwhelmingly to favor the proposal by Jerne (85) that the lymphocytes comprising the immune system are linked by an "intricate web of V domains." A balance is established by the historical accidents of extrinsic exposure to immunogens and the receipt of passive maternal Ig transplacentally (180), both playing on the available genetic repertoire of germ-line idiotypes.

The original suggestion by Jerne portrayed an ever-expanding network of idiotype-1 (Ab-1) inducing anti-idiotype (Ab-2), which then influenced the expansion of both Ab-1 and Ab-3 anti-(anti-idiotype)-bearing cells. Ab-3 resembled Ab-1 in idiotypy but not necessarily antigen specificity. An alternative notion of regulatory idiotopes was set forth by Paul and Bona (23, 128) which directly implicates a receptor on an idiotype-recognizing Th (IdXTh) in interactions which determine the nature of the maturing B cells. According to this concept, the network essentially oscillates between idiotype-recognizing and idiotype-bearing lymphocytes.

It is probable that initial triggering of the B cell by antigen and AgTh is necessary before it will respond suitably to ambient IdXTh: otherwise, virgin B cells would be perpetually confronted and distracted by idiotype-recognizing T cells. Since the IdX idiotope must be clearly available for binding by the IdXTh or its specific effector factor, even in the presence of antigen, it has been postulated that such idiotopes must reside outside of the antibody active site (128). These regulatory idiotopes might represent a set of determinants within the framework that are quasi-isotypic, appearing on antibodies with a variety of paratopes.

Starting with the seminal observations by Oudin and Cazenave (33, 126) that antibodies of different specificity can still share idiotypy, this theme has reappeared and has been refined with the study of monoclonal antibodies (98, 106, 121). In these systems it has been shown that monoclonal antibodies with differing peptide specificity for the *same* antigen can each bear the predominant idiotype. This represents a slightly different aspect of Jerne's concept of a non-specific parallel set of antibodies with shared idiotypy but specificity for distinct antigens, evidence for which has accumulated recently in several different anti-hapten systems (44, 46, 48, 75, 76, 183). What is more relevant in our context is whether the perpetuation of antibodies bearing a predominant or partially predominant idiotope is a direct consequence of activation of an IdXTh which then selects from the entire population, those B cells bearing paratopes of various distinct specificities for antigen.

It is probable that the waves of Ab-1, Ab-2, Ab-3, etc. have a role to play in regulating the response, but it remains to be learned how T-cell control and serum antibody control contribute to the final outcome.

Most recently, the question has been addressed of whether any idiotope, even a private one, could subsume the function of being "regulatory." Neonatal animals were treated with a particular silent Id, A48 (141) and this idiotype subsequently became a significant portion of the anti-bacterial levan response. The capacity to promote a silent Id into a predominant one belonged to an idiotype-recognizing Lyt-1$^+$ cell.

The Idiotope as a Restriction Element

On the basis of evidence indicating a concurrent requirement for antigen in the activation of IdXTh (27), Bottomly has suggested that a receptor for antigen as well as for predominant idiotope exists on this cell (26). In such circumstances, the idiotope may serve as a restriction element substituting for the histotope on an MHC molecule. Does the idiotope resemble a histotope? Indeed, it is possible that a receptor for nominal antigen must exist on all T cells in the system, although Janeway reminds us (79) that the MHC-recognizing Th cell can substitute an idiotope for the epitope. Self-idiotopes, as internal images of epitopes, can thereby play a role in repertoire development.

The existence of so many predominant idiotypic systems suggests an evolutionary reason to focus on a limited number of idiotopes, such as the need to allow coordinated and simplified regulation of suppressor T cells (which often bear the predominant IdX of the B-cell pathway (67)). Thus, the involvement of B cells in the triggering of suppressor T cells might serve to initiate the feedback regulation of a well-established response by stimulating an idiotype-recognizing T-suppressor inducer cell (103).

CONCLUDING REMARKS

Perhaps the most intriguing unresolved issues concerning the antigen specificity of T cells center on: (a) the mechanism(s) underlying determinant selection; (b) the chemistry and genetics of T-cell receptors, which should provide the structural basis of genetic restriction; and (c) a structural comparison of receptors from different T-cell subsets with each other and with immunoglobulins. For a complete understanding of the latter two, the receptors themselves (or the genes coding for them) must be delineated. As for determinant selection, the agretope-desetope concept, in which the agretope can be perceived as the "carrier" moiety of T-cell immunogens, provides an attractive framework within which experimental tests can be designed.

The enormous change in our view of T-cell activation in the past decade is almost entirely due to the concepts of MHC restriction and the idiotypic network. It is no longer possible to consider the problem of antigen activation of T cells in a context separate from the recognition of some other element, coded for by the responder's own genome. The difficulty in achieving a complete understanding of T-cell activation is partly due to the tremendous diversity of T regulatory cells that has just recently been unearthed. In the consideration of antibody specificity, and B-cell activation, although it appears that there may be increasing specialization manifested by a large variety of B-cell subpopulations, there is no doubt that each member of the diverse array of B cells bears recognition specificity for a determinant region on the nominal antigen.

This statement cannot be made about the T cell for two separate reasons: (a) the epitope which is recognized may be intricately associated with an MHC molecule and both moieties are simultaneously recognized; (b) some interactions may be restricted by idiotypic elements in conjunction with MHC or antigenic moieties.

In fact, we have mentioned that the feature which makes T recognition so varied is the large number of regulatory T cells each of which must circumscribe its interactions with surrounding cells. To determine its partner unambiguously, principles of simplicity can be invoked to assure coordinated control. One such principle might be that recognition by T cells and specific targeting by T cells can be accomplished by having each cell bind to a combination of elements: an epitope plus idiotope, an epitope plus histotope, or an idiotope plus histotope. A second motif which seems to permeate the relationships among T cells is that antigen-specific T cells throughout the system, which thus bear idiotypic elements in their receptors, will be activated by other T cells which recognize these idiotypic elements. A consequence of this scheme is an alternation of antigen-specific, MHC-restricted cells with idiotype-specific T cells in sequential pathways.

The rules of activation at the molecular level and details about "second messenger" function are bound to be the same for many types of cells, and we have not considered them in this account. We have restricted ourselves to attempting to assemble the information about the state in which epitopes, idiotopes, and histotopes confront the T-cell machinery. If there is a rule about this confrontation which can be discerned, it may be that the choice of epitope which will be available for a T-cell subpopulation will depend on an "antigen-recognition epitope," an agretope, which a presenting cell (or entity such as a soluble factor) can recognize on the antigen. Thus, H-agretopes on the Ag will be recognized by Class II Ia molecules on APC presenting to Th, C-agretopes will interact with an appropriate group or receptor on the Class I molecule found on the surface of cell types recognized by Tc, and S-agretopes will associate with desetopes on MHC molecules which exist on APC of the type which presents antigen to Ts. The list could be extended.

Therefore, the basis for determinant selection would reside in still unknown chemical principles uniting desetope with agretope: desetopes on the MHC molecules used for triggering each important T-cell subpopulation would seek out their complementary agretopes based on a unique but non-clonal chemistry. An example of such a principle might be that the desetopes for H-agretopes recognize a particular grouping of hydrophobic amino acids only, whereas the desetopes for S-agretopes combine with hydrophilic amino acids. This alignment would have utility in explaining two vexing problems. First, how can the system maintain self-non-self discrimination in the face of the enormous excess of self H-agretopes? Probably these H-agretopes remain unexposed until antigenic fragmentation and "processing." Prior T-cell recognition may be required to trigger processing: such may be the role of augmenting or amplifier T cells which would then have to be able to recognize superficial residues on the antigen. Second, since S-agretopes are hydrophilic and external on the antigen, the nearby epitope will be comparably situated: this would explain why Ts can often combine with free antigen.

One of the critical questions that can be approached experimentally is the definition of agretopes for particular haplotypes. Will any chemical structure associated with the agretope in the correct spatial relationship act as an epitope for T cells? This question is being experimentally approached by synthetic chemistry, using as a starting point, for example, the putative agretope at the C-terminus of pigeon cytochrome c, for the B10.A mouse strain. Other H-agretopes on L-tyr-ABA or lysozyme could also be used. It is important to realize that although we may be at the threshold of defining agretopes and synthesizing complete immunogens with a particular agretope as a universal carrier for a series of epitopes, difficulties may abound. Not all epitopes will be permissive for agretope function, as older observations seem to suggest for the small molecule, L-tyr-ABA (6).

Finally, it would seem unlikely that a particular determinant area on any antigen could be designated as immunodominant for T cells in general. However, we may soon arrive at an understanding of the common features of attachment sites (agretopes) chosen by MHC elements for presentation to different T-cell subpopulations. Understanding the nature of specific epitope immunodominance for different T-cell subpopulations that might result from study of the features and constraints of desetope-agretope interaction should permit the logical manipulation of T-cell responsiveness.

ACKNOWLEDGMENTS

This work was supported by NIH grants CA-24442, AI-05664, and AI-11183, and by grants IM-263 and IM-323 from the American Cancer Society.

Literature Cited

1. Adorini, L., M. A. Harvey, A. Miller, and E. E. Sercarz. 1979. The fine specificity of regulatory T cells. II. Suppressor and helper T cells are induced by different regions of hen egg-white lysozyme (HEL) in a genetically nonresponder mouse strain. *J. Exp. Med.* 150: 293–306
2. Adorini, L., M. Harvey, D. Rozycka-Jackson, A. Miller, and E. E. Sercarz. 1980. Differential major histocompatbility complex related activation of idiotypic suppressor T cells. *J. Exp. Med.* 152:521–531
3. Adorini, L., M. Harvey, and E. E. Sercarz. 1979. The fine specificity of regulatory T cells. IV. Idiotypic complementarity and antigen-bridging interactions in the anti-lysozyme response. *Eur. J. Immunol.* 9:906–909
4. Adorini, L., A. Miller, and E. E. Sercarz. 1979. The fine specificity of regulatory T cells. I. Hen-egg-white lysozyme-induced suppressor T cells in a genetically non-responder mouse strain do not recognize a closely related immunogeneic lysozyme. *J. Immunol.* 122:871–877
5. Alkan, S. S., D. E. Nitecki, and J. W. Goodman. 1971. Antigen recognition and the immune response: the capacity of L-tyrosine-azobenzenearsonate to serve as a carrier for a macromolecular hapten. *J. Immunol.* 107:353–8
6. Alkan, S. S., E. B. Williams, D. E. Nitecki, and J. W. Goodman. 1972. Anti-gen recognition and the immune response: humoral and cellular immune responses to small mono-and bifunctional antigen molecules. *J. Exp. Med.* 135:1228–1246
7. Andersson, J., G. M. Edelman, G. Möller, and O. Sjöberg. 1972. Activation of B lymphocytes by locally concentrated concanavalin A. *Eur. J. Immunol.* 2:233–235
8. Araneo, B. A., D. W. Metzger, R. L. Yowell, and E. E. Sercarz. 1981. Positive selection of major histocompatibility restricted suppressor T cells bearing the predominant idiotype in the immune response to lysozyme. *Proc. Natl. Acad. Sci USA.* 78:499–503
9. Araneo, B. A., R. L. Yowell, E. E. Sercarz. 1979. Ir gene defects may reflect a regulatory imbalance. I. Helper T cell activity revealed in a strain whose lack of response is controlled by suppression. *J. Immunol.* 123:961–7
10. Bach, B. A., M. I. Greene, B. Benacerraf, and A. Nisonoff. 1979. Mechanisms of regulation of cell mediated immunity. IV. Azobenzenearsonate (ABA) specific suppressor factor(s) bear cross-reactive idiotype (CRI) determinants the expression of which is linked to the heavy chain allotype linkage group of genes. *J. Exp. Med.* 149:1084–1098
11. Bach, B. A., L. Sherman, B. Benacerraf, and M. I. Greene. 1978. Mechanisms of regulation of cell-mediated immunity. II. Induction and suppression of de-

layed type hypersensitivity to azoben-zenearsonate coupled spleen cells. *J. Immunol.* 121:1460–1468

12. Baxevanis, C. N., N. Ishii, Z. A. Nagy, and J. Klein. 1982. Role of E^k molecule in the generation of suppressor T cells in response to LDH_B. *Scand. J. Immunol.* 16:25–31

13. Baxevanis, C. N., N. Ishii, Z. A. Nagy, and J. Klein. 1982. H-2 controlled suppression of T-cell response to lactate dehydrogenase B. *J. Exp. Med.* 156: 822–833

14. Benacerraf, B. 1978. A hypothesis to relate the specificity of T lymphocytes and the activity of I region-specific Ir genes in macrophages against B lymphocytes. *J. Immunol.* 120:1809–1812

15. Benacerraf, B., and R. N. Germain. 1978. The immune response genes of the major histocompatibility complex. *Immunol. Rev.* 38:70–119

16. Benacerraf, B. and H. O. McDevitt. 1972. The histocompatibility linked immune response genes. *Science.* 175:273–279

17. Bevan, M. J. 1977. Killer cells reactive to altered-self antigens can also be alloreactive. *Proc. Natl. Acad. Sci. USA* 74:2094–2098

18. Bevan, M. J. 1975. The major histocompatibility complex determines susceptibility to cytotoxic T cells directed against minor histocompatibility antigens. *J. Exp. Med.* 142:1349–1364

19. Binz, H., and H. Wigzell. 1977. Antigen-binding, idiotypic T-lymphocyte receptors. *Contemp. Top. Immunobiol.* 7:113–177

20. Black, S. J., G. J. Hammerling, C. Berek, K. Rajewsky, and K. Eichmann. 1976. Idiotypic analysis of lymphocytes in vitro. I. Specificity and heterogeneity of B and T lymphocytes reactive with anti-idiotypic antibody. *J. Exp. Med.* 143:846–859

21. Blanden, R. V., and G. L. Ada. 1978. A dual recognition model of cytotoxic T cells based on thymic selection of precursors with low affinity for self-H-2 antigens. *Scand. J. Immunol.* 7:181–190

22. Blanden, R. V., and U. Kees. 1978. Cytotoxic T-cell responses show more restricted specificity for self than for non-self H-2D coded antigens. *J. Exp. Med.* 147:1661–70

23. Bona, C. A., E. Heber-Katz, and W. E. Paul. 1981. Idiotype-anti-idiotype regulation. I. Immunization with a levan-binding myeloma protein leads to the appearance of auto-anti-(anti-idiotype)

antibodies and to the activation of silent clones. *J. Exp. Med.* 153:951–967

24. Bona, C., P. K. A. Mongini, K. E. Stein, and W. E. Paul. 1980. Anti-immunoglobulin antibodies. I. Expression of cross-reactive idiotypes and *Ir* gene control of the response to IgG2a of the b. allotype. *J. Exp. Med.* 151:1334–1348

25. Bona, C., and W. E. Paul. 1979. Cellular basis of regulation of expression of idiotype. I. T-suppressor cells specific for MOPC 460 idiotype regulate the expression of cells secreting anti-TNP antibodies bearing 460 idiotype. *J. Exp. Med.* 149:592–600

26. Bottomly, K. 1981. Activation of the idiotypic network: environmental and regulatory influences. In *Immunoglobulin Idiotypes,* ed. C. Janeway, E. E. Sercarz, and H. Wigzell, p. 517–532 New York: Academic

27. Bottomly, K., and D. E. Mosier. 1981. Antigen specific helper T cells required for dominant idiotype expression are not H-2 restricted. *J. Exp. Med.* 154: 411–421

28. Britz, J. S., P. W. Askenase, W. Ptak, R. M. Steinman, and R. K. Gershon. 1982. Specialized antigen-presenting cells. Splenic dendritic cells and peritoneal-exudate cells induced by mycobacteria activate effector T cells that are resistant to suppression. *J. Exp. Med.* 155: 1344–1356

29. Buchmüller, T., and G. Corradin. 1982. Lymphocyte specificity to protein antigens. V. Conformational dependence of activation of cytochrome C-specific T cells. *Eur. J. Immunol.* 12:412–416

30. Burakoff, S. J., R. Finberg, L. Glimcher, F. Lemonnier, B. Benacerraf, and H. Cantor. 1978. The biologic significance of alloreactivity. The ontogeny of T-cell sets specific for alloantigens or modified self antigens. *J. Exp. Med.* 148:1414–1422

31. Cantor, H. and R. K. Gershon. 1979. Regulatory T cell circuitry. *Fed. Proc.* 38:2058–2062

32. Cazenave, P.-A., J. M. Cavaillon, and C. Bona. 1977. Idiotypic determinants on rabbit B- and T-derived lymphocytes. *Immunol. Rev.* 34:34–49

33. Cazenave, P.-A. 1978. L'idiotypie comparee des anticorps qui, dans le serum d'un lapin immunise contre la serumalbumine humaine, sont diriges contre des regions differentes de cet antigen. *FEBS Lett.* 31:348–354

34. Chesnut, R. W., R. O. Endres, and H. M. Grey. 1980. Antigen recognition by

T cells and B cells: recognition of cross-reactivity between native and denatured forms of globular antigens. *Clin. Immunol. Immunopathol.* 15:397–408

35. Chesnut, R. W., and H. M. Grey. 1981. Studies on the capacity of B cells to serve as antigen-presenting cells. *J. Immunol.* 126:1075–1079

36. Ciavarra, R., and J. Forman. 1982 H-2L-restricted recognition of viral antigens in the H-2^d haplotype, anti-vesicular stomatitis virus cytotoxic T cells are restricted solely by H-2L. *J. Exp. Med.* 156:778–790

37. Claflin, J. L., and J. M. Davie. 1974. Clonal nature of the immune response to phosphorylcholine. IV. Idioitypic uniformity of binding site associated antigenic determinants among mouse anti-phosphorylcholine antibodies. *J. Exp. Med.* 140:673–86

38. Cohn, M., and R. E. Epstein. 1978. T-cell inhibition of humoral responsiveness. Theory on the role of restrictive recognition in immune regulation. *Cell. Immunol.* 39:125–153

39. Corradin, G. and S. Carel. 1983. In *Protein Conformation as an Immunological Signal,* ed. by F. Celada, V. Schumaker and E. Sercarz. London Plenum: In press

40. Corradin, G., and J. M. Chiller. 1979. Lymphocyte specificity to protein antigens. II. Fine specificity of T-cell activation with cytochrome *c* and derived peptides as antigenic probes. *J. Exp. Med.* 149:436–447

41. Corradin, G., and J. M. Chiller. 1981. Lymphocyte specificity to protein antigens. III. Capacity of low responder mice to beef cytochrome c to respond to a peptide fragment of the molecule. *Eur. J. Immunol.* 11:115–9

42. Cosenza, H., M. H. Julius, and A. A. Augustin. 1977. Idiotypes as variable region markers: analogies between receptors on phosphorylcholine-specific T and B lymphocytes. *Immunol. Rev.* 34:3–33

43. Cunningham, A. J. and E. E. Sercarz. 1971. The asynchronous development of immunological memory in helper (T) and precursor (B) cell lines. *Eur. J. Immunol.* 1:413–418

44. Dzierzak, E. A. and C. A. Janeway, Jr. 1981. Expression of an idiotype (Id-460) during in vivo anti-dinitrophenyl antibody responses III. Detection of Id-460 in normal serum that does not bind dinitrophenyl. *J. Exp. Med.* 154:1442–1454

45. Eardley, D. D., J. Hugenberger, L. McVay-Boudreau, F. W. Shen, R. K. Gershon, and H. Cantor. 1978. Immunoregulatory circuits among T-cell sets. I. T-helper cells induce other T-cell sets to exert feedback inhibition. *J. Exp. Med.* 147:1106–1115

46. Eardley, D. D., F. W. Shen, H. Cantor and R. K. Gershon. 1979. Genetic control of immunoregulatory circuits. Genes linked to the Ig locus govern communication between regulatory T-cell sets. *J. Exp. Med.* 150:44–51

47. Eichmann, K. 1978. Expression and function of idiotypes on lymphocytes. *Adv. Immunol.* 26:196

48. Eichmann, K., A. Coutinho, and F. Melchers. 1977. Absolute frequencies of lipopolysaccharide-reactive B cell producing A5A idiotype in unprimed, streptococcal A carbohydrate-primed, anti-A5A idiotype-sensitized and anti-A5A idiotype suppressed A/J mice. *J. Exp. Med.* 146:1436–1449

49. Endres, R. O., and H. M. Grey. 1980. Antigen recognition by T cells. I. Suppressor T cells fail to recognize cross-reactivity between native and denatured ovalbumin. *J. Immunol.* 125:1515–1520

50. Erb, P., M. Feldmann, and N. Hogg. 1975. The role of macrophages in the generation of T helper cells. III. Influence of macrophage-derived factors in helper cell induction. *Eur. J. Immunol.* 5:759–766

51. Erb, P., M. Feldmann, and N. Hogg. 1976. Role of macrophages in the generation of T-helper cells. IV. Nature of genetically-related factors derived from macrophages incubated with soluble antigens. *Eur. J. Immunol.* 6:365–372

52. Eshhar, Z., R. N. Apte, I. Löwy, Y. Ben-Nerah, D. Givol, and E. Mozes. 1980. T-cell hybridomas bearing heavy chain variable region determinants producing (T,G)-A-L-specific helper factor. *Nature* 286:270–272

53. Falkoff, R., and J. Kettman. 1972. Differential stimulation of precursor cells and carrier-specific thymus-derived cell activity in the in vivo response to heterologous erythrocytes in mice. *J. Immunol.* 108:54–58

54. Feldmann, M., and S. Kontiainen. 1976. Suppressor cell induction in vitro. II. Cellular requirements of suppressor cell induction. *Eur. J. Immunol.* 6:302–305

55. Gearhart, P. J., N. D. Johnson, R. Douglas, and L. Hood. 1981. IgG antibodies to phosphorylcholine exhibit more diversity than their IgM counterparts. *Nature* 291:29–34

56. Gell, P. G. H., and B. Benacerraf. 1959. Studies on hypersensitivity. II. Delayed hypersensitivity to denatured proteins in guinea pigs. *Immunology.* 2:64

57. Germain, R. N., and B. Benacerraf. 1981. Helper and suppressor T cell factors. *Springer Sem. Immunopathol.* 3:93–127

58. Germain, R. N., and B. Benacerraf. 1981. A single major pathway of T-lymphocyte interactions in antigen-specific immune suppression. *Scand. J. Immunol.* 13:1–10

59. Germain, R. N., S. T. Ju, T. J. Kipps, B. Benacerraf, and M. E. Dorf. 1979. Shared idiotypic determinants on antibodies and T-cell-derived suppressor factor specific for the random terpolymer L-glutamic acid[60] -L-tyrosine[40]. *J. Exp. Med.* 149:613–622

60. Gershon, R. K., P. H. Maurer, and C. Merryman. 1973. A cellular basis for genetically controlled immunologic unresponsiveness in mice. Tolerance induction in T cells. *Proc. Natl. Acad. Sci. USA* 70:250–4

61. Glimcher, L. H., T. Hamano, R. Asofsky, D. H. Sachs, L. E. Samelson, S. O. Sharrow and W. E. Paul. 1983. Ia mutant functional antigen presenting cell lines. *J. Exp. Med.,* In press

62. Goodman, J. W., S. Fong, G. K. Lewis, R. Kamin, D. E. Nitecki, and G. Der Balian. 1978. Antigen structure and lymphocyte activation. *Immunol. Rev.* 39:36–59

63. Greaves, M. F., S. Bauminger, and G. Janossy. 1972. Lymphocyte activation. III. Binding sites for phytomitogens on lymphocyte subpopulations. *Clin. Exp. Immunol.* 10:537–554

64. Hansburg, D., T. Fairwell, M. Pincus, R. H. Schwartz, and E. Appella. 1982. The T lymphocyte response to cytochrome c. IV. Distinguishable sites on a peptide antigen which affect antigenic strength and memory. Submitted

65. Hansburg, D., C. Hannum, J. K. Inman, E. Appella, E. Margoliash, and R. H. Schwartz. 1981. Parallel cross-reactivity patterns of 2 sets of antigenically distinct cytochrome c peptides: possible evidence for a presentational model of Ir gene function. *J. Immunol.* 127:1844–1851

66. Hansburg, D., E. Heber-Katz, T. Fairwell, and E. Appella. 1982. MHC-controlled, APC-expressed specificity of T-cell antigen recognition: Identification of a site of interaction and its relationship to *Ir* genes. Submitted

67. Harvey, M. A., L. Adorini, A. Miller, and E. E. Sercarz. 1979. Lysozyme-induced T-suppressor cells and antibodies have a predominant idiotype. *Nature* 281:594–596

68. Hashim, G. A. 1981. Experimental allergic encephalomyelitis: activation of suppressor T lymphocytes by a modified sequence of the T effector determinant. *J. Immunol.* 126:419–423

69. Heber-Katz, E., and D. B. Wilson. 1976. Sheep red blood cell-specific helper activity in rat thoracic duct lymphocyte populations positively selected for reactivity to specific strong histocompatibility alloantigens. *J. Exp. Med.* 143:701–706

70. Hertel-Wulff, B., J. W. Goodman, C. G. Fathman, and G. K. Lewis. 1983. Arsonate-specific murine T cell clones. I. Genetic control and antigen specificity. *J. Exp. Med.* In press

71. Hill, S. W., and E. E. Sercarz. 1975. Fine specificity of the H-2 linked immune response gene for the gallinaceous lysozymes. *Eur. J. Immunol.* 5:317–324

72. Hirayama, A., Y. Dohi, Y. Takagaki, H. Fujio, and T. Amano. 1982. Structural relationships between carrier epitopes and antigenic epitopes on hen egg-white lysozyme. *Immunology.* 46:145–154

73. Hodes, R. J., M. Kimoto, K. S. Hathcock, C. G. Fathman, and A. Singer. 1981. Functional helper activity of monoclonal T cell populations: Antigen specific and H-2 restricted cloned T cells provide help for *in vitro* antibody responses to trinitrophenyl-poly(l-tyr,-glu)-poly (DL-Ag)-poly(L-lys). *Proc. Natl. Acad. Sci. USA* 78:6431–6435

74. Hoffmann, M., and J. W. Kappler. 1972. The antigen specificity of thymus-derived helper cells. *J. Immunol.* 108:261–267

75. Hoffmann, M., and J. W. Kappler. 1973. Regulation of the immune response. III. Kinetic differences between thymus- and bone marrow-derived lymphocytes in the proliferative response to heterologous erythrocytes. *J. Exp. Med.* 137:1325–1337

76. Hornbeck, P. V. and G. K. Lewis. 1983. Idiotype connectance in the immune system I. Expression of a cross-reactive idiotope on induced anti-p-azophenylarsonate antibodies and on endogenous antibodies not specific for arsonate. *J. Exp. Med.* In press

77. Hunt, P., and D. Sears. 1983. CTL crossreactivities reveal shared immunodominant determinants created

by structural homologous regions of MHC class I antigens. *J. Immunol.* In press

78. Infante, A. J., M. Z. Atassi, and C. G. Fathman. 1981. T cell clones reactive with sperm whale myoglobin. Isolation of clones with specificity for individual determinants on myoglobin. *J. Exp. Med.* 154:1342–1356

79. Janeway, C. Jr. 1982. The selection of self-MHC recognizing T lymphocytes: A role for idiotypes. *Immunology Today.* 3:261–265

80. Ishizaka, K. and T. Adachi. 1976. Generation of specific helper cells and suppressor cells *in vitro* for the IgE and IgG antibody responses. *J. Immunol.* 117:40–47

81. Ishizaka, K., T. Kishimoto, G. Delespesse, and T. P. King. 1974. Immunogenic properties of modified antigen E. I. Response of specific determinants for T cells in denatured antigen and polypeptide chains. *J. Immunol.* 113:70–77

82. Janeway, C. A. 1980. Idiotypic control: the expression of idiotypes and its regulation. In *Strategies of Immune Regulation,* ed. by E. E. Sercarz and A. J. Cunningham, New York: Academic 179–98.

83. Janeway, C. A., E. E. Sercarz, and H. Wigzell, 1981. *Immunoglobulin Idiotypes,* New York: Academic

84. Janeway, C. A. Jr., H. Wigzell, and H. Binz. 1976. Two different V_H gene products make up the T-cell receptors. *Scand. J. Immunol.* 5:993–1001

85. Jerne, N. K. 1974. Towards a network theory of the immune system. *Ann. Immunol.* 125C:373–89

86. Jorgensen, T., and K. Hannestad. 1980. H-2-linked genes control immune response to V-domains of myeloma protein 315. *Nature.* 288:396–398

87. Jorgensen, T., and K. Hannestad. 1982. Helper T cell recognition of the variable domains of a mouse myeloma protein (315). Effect of the major histocompatibility complex and domain conformation. *J. Exp. Med.* 155:1587–96

88. Kabat, E. A. 1976. *Structural Concepts in Immunology and Immunochemistry,* New York: Holt, Rinehart and Winston. 2nd ed.

89. Kapp, J. A., B. A. Araneo, and B. L. Clevinger. 1980. Suppression of antibody and T-cell proliferative responses to L-glutamic acid^{60}-L-alanine^{30}-L-tyrosine^{10} by a specific monoclonal T cell factor. *J. Exp. Med.* 152:235–40

90. Kapp, J. A., C. W. Pierce, and B. Benacerraf. 1973. Genetic control of immune responses *in vitro*. II. Cellular requirements for the development of primary plaque forming cell responses in the random terpolymer L-glutamic acid^{60}-L-alanine^{30}-L-tyrosine^{10} (GAT) by mouse spleen cells *in vitro*. *J. Exp. Med.* 138:1121–1132

91. Kappler, J. W., and P. C. Marrack. 1976. Helper T cells require antigen and macrophage surface components simultaneously. *Nature* 262:797–799

92. Kappler, J. W., B. Skidmore, J. White, and P. Marrack. 1981. Antigen-inducible, H-2 restricted, IL-2- producing T cell hybridomas. Lack of independent antigen and H-2 recognition. *J. Exp. Med.* 153:1198–1214

93. Kappler, J., J. White, D. Wegmann, E. Mustain, and P. Marrack. 1982. Antigen-presentation by Ia$^+$ B cell hybridomas to H-2-restricted T cell hybridomas. *Proc. Natl. Acad. Sci. USA* 79:3604–3607

94. Kardys, E. and G. A. Hashim. 1981. Experimental allergic encephalomyelitis in Lewis rats: immunoregulation of disease by a single amino acid substitution in the disease-inducing determinant. *J. Immunol.* 127:862–866

95. Katz, D. H., and B. Benacerraf. 1975. The function and interrelationships of the T cell receptors. Ir genes and other histocompatibility gene products. *Transplant. Rev.* 22:175–95

96. Katz, M. E., R. Maizels, L. Wicker, A. Miller, and E. E. Sercarz. 1982. Immunological focussing by the mouse MHC: Mouse strains confronted with distantly related lysozymes confine their attention to a very few epitopes. *Eur. J. Immunol.* 12:535–540

97. Kimoto, M., and C. G. Fathman. 1980. Antigen-reactive T cell clones. I. Transcomplementing hybrid I-A-region gene products function effectively in antigen presentation. *J. Exp. Med.* 152:759–770

98. Kimoto, M., T. J. Krenz, and C. J. Fathman. 1981. Antigen-reactive T cell clones. III. Low responder antigen-presenting cells function effectively to present antigen to selected T cell clones derived from (high responder X low responder)F1 mice. *J. Exp. Med.* 154:883–891

99. Kohno, Y., I. Berkower, J. Minna and J. A. Berzofsky. 1982. Idiotypes of antimyoglobin antibodies: shared idiotypes among monoclonal antibodies to distinct determinants of sperm whale myoglobin. *J. Immunol.* 128:1742–8

100. Krco, C. J., A. L. Kazim, M. Z. Atassi, R. W. Melvold, and C. S. David. 1981.

Genetic control of immune response to hemoglobin. III. Variant A_β (bm 12) but not Ae (Ea) Ia polypeptide alters immune reactivity towards the a-subunit of human hemoglobin. *J. Immunogenetics* 8:471–476

101. Krzych, U., A. V. Fowler, A. Miller, and E. E. Sercarz. 1982. Repertoires of T cells directed against a large protein antigen, beta-galactosidase. I. Helper cells have a more restricted specificity repertoire than proliferative cells. *J. Immunol.* 128:1529–1534

102. Krzych, U., A. V. Fowler, and E. E. Sercarz. 1983. Antigen structures used by regulatory T cells in the interaction among T suppressor, T helper and B cells. In *Protein Conformation as an Immunological Signal.*, ed. by F. Celada, V. Schumaker and E. E. Sercarz., London: Plenum. In press

103. L'age-Stehr, J., H. Teichmann, R. K. Gershon, and H. Cantor. Stimulation of regulatory T cell circuits by immunoglobulin-dependent structures on activated B cells. 1980. *Eur. J. Immunol.* 10:21–26

104. LaFuse, W. P., J. F. McCormick, R. W. Melvold, and C. S. David. 1981. Serological and biochemical analysis of Ia molecules in the I-A mutant B6.C-H-2^{bm12}. *Transplantation.* 31:434

105. Langman, R. E. 1978. Cell-mediated immunity and the major histocompatibility complex. *Rev. Physio. Biochem. Pharmacol.* 81:1–37

106. Leclerq, L., J.-C. Mazie, G. Commée and J. Theze. 1982. Monoclonal anti-GAT antibodies with different fine specificities express the same public idiotype. *Molecular Immunol.* 19:1001–1010

107. Lemonnier, F., S. J. Burakoff, R. N. Germain, and B. Benacerraf. 1977. Cytolytic thymus-derived lymphocytes specific for allogeneic stimulator cells crossreact with chemically modified syngeneic cells. *Proc. Natl. Acad. Sci. U.S.A.* 74:1229–1233

108. Lerner, E. A., L. Matis, C. A. Janeway Jr., P. P. Jones, R. H. Schwartz, and D. B. Murphy. 1980. Monoclonal antibody against an Ir gene product. *J. Exp. Med.* 152:85–98

109. Lewis, G. K., and J. W. Goodman. 1978. Purification of functional, determinant-specific, idiotype-bearing murine T cells. *J. Exp. Med.* 148:915–924

110. Levy, R. B., and G. M. Shearer. 1982. Can cytotoxic T cells recognize self determinants on molecules lacking polymorphic MHC self determinants? *Immunology Today.* 3:204–205

111. Levy, R. B., G. M. Shearer, and T. H. Hansen. 1978. Properties of H-2L locus products in allogeneic and H-2 restricted, trinitrophenyl specific cytotoxic responses. *J. Immunol.* 121:2263

112. Lin, C.-C. S., A. S. Rosenthal, H. C. Passmore, and T. H. Hansen. 1981. Selective loss of antigen-specific Ir gene function in I-A mutant B6.C-H-2^{bm12} is an antigen presenting cell defect. *Proc. Natl. Acad. Sci. USA* 78:6406–6410

113. Lin, C. S., A. S. Rosenthal, and T. H. Hansen. 1981. I-A mutation resulted in a selected loss of an antigen-specific Ir gene. *J. Supramol. Struct.* 16:115

114. Lonai, P., L. Steinman, V. Friedman, G. Drizlikh, and J. Puri. 1981. Specificity of antigen binding by T cells; competition between soluble and Ia-associated antigen. *Eur. J. Immunol.* 11:382–387

115. Maizels, R. M., J. A. Clarke, M. A. Harvey, A. Miller, and E. E. Sercarz. 1980. Epitope specificity of the T cell proliferative response to lysozyme: Proliferative T cells react predominantly to diferent determinants from those recognized by B cells. *Eur. J. Immunol.* 10: 509–515

116. Maizels, R. M., J. A. Clarke, M. A. Harvey, A. Miller, and E. E. Sercarz. 1980. Ir-gene control of T cell proliferative responses: two distinct expressions of the genetically unresponsive state. *Eur. J. Immunol.* 10:516–520

117. Manca, J., J. A. Clarke, E. E. Sercarz, and A. Miller. 1983. T cells with differing specificity exist for a single determinant on lysozyme. In *Ir Genes: Past, Present and Future,* ed. C. W. Pierce, S. E. Cullen, J. A. Kapp, B. D. Schwartz, and D. C. Shreffler. Clifton, NJ: Humana. In press

118. Maoz, A., M. Feldmann, and S. Kontiainen. 1976. Enrichment of antigen-specific helper and suppressor T cells. *Nature.* 260:324–326

119. McDougal, J. S., F. W. Shen, S. P. Cort and J. Bard. 1980. Feedback suppression: phenotypes of T cell subsets involved in the Lyl T cell induced immunoregulatory circuit. *J. Immunol.* 125:1157–1163

120. McKean, D. J., R. W. Melvold, and C. S. David. 1981. Tryptic peptide comparison of Ia antigen alpha and beta polypeptides from the I-A mutant B6.C-H-2^{bm}12 and its parental strain B6. *Immunogenetics.* 14:41–51

121. Metzger, D. W., A. Miller, E. Sercarz. 1980. Sharing of an idiotypic marker by monoclonal antibodies specific for distinct regions of hen lysozyme. *Nature.* 287:541–542

122. Michaelides, M. M., M. S. Sandrin, G. M. Morgan, I.F.C. McKenzie, R. Ashman, and R. W. Melvold. 1981. Ir gene function in an I-A mutant B6.C-H-2^{bm12}. *J. Exp. Med.* 153:464–469

123. Minami, M., N. Honji, and M. E. Dorf. 1982. Mechanism responsible for the induction of I-J restrictions on Ts_3 suppressor cells. *J. Exp. Med.* 156:1502–1515

124. Mozes, E., and J. Haimovich. 1979. Antigen-specific T-cell helper factor cross reacts idiotypically with antibodies of the same specificity. *Nature* 278:56–57

125. Okumura, K., T. Takemori, T. Tokuhisa, and T. Tada. 1977. Specific enrichment of the suppressor T cell bearing I-J determinants. Parallel functional and serological characterizations. *J. Exp. Med.* 146:1234–1245

126. Oudin, J., P. A. Cazenave. 1971. Similar idiotypic specificities in immunoglobulin fractions with different antibody functions or even without detectable antibody function. *Proc. Natl. Acad. Sci. USA* 68:2616–2620

127. Paul, W. E., and B. Benacerraf. 1977. Functional specificity of thymus-dependent lymphocytes. *Science.* 195:1293–1300

128. Paul, W. E., and C. A. Bona. 1982. Regulatory idiotopes and immune networks: a hypothesis. *Immunol. Today.* 3:230–234

129. Pierres, M., and R. N. Germain. 1978. Antigen-specific T cell-mediated suppression. IV. Role of macrophages in generation of L-glutamic acid -L-alanine -L-tyrosine (GAT)-specific suppressor T cells in responder mouse strains. *J. Immunol.* 121:1306–1314

130. Playfair, J. H. L. 1972. Response of mouse T and B lymphocytes to sheep erythrocytes. *Nature,* 235:115–18

131. Prange, C. A., J. Fiedler, D. E. Nitecki, and C. J. Bellone. 1977. Inhibition of T antigen-binding cells by idiotypic antisera. *J. Exp. Med.* 146:766–778

132. Puri, J. and P. Lonai. 1980. Mechanism of antigen binding by T cells. H-2 (I-A) -restricted binding of antigen plus Ia by helper cells. *Eur. J. Immunol.* 10:273–281

133. Rajewsky, K., and K. Eichmann. 1977. Antigen receptors of T helper cells. *Contemp. Top. Immunobiol.* 7:69–112

134. Rajewsky, K., and R. Mohr. 1974. Specificity and heterogeneity of helper T cells in the response to serum albumins in mice. *Eur. J. Immunol.* 4:111–119

135. Ramseier, H., M. Aguet, and J. Lindenmann. 1977. Similarity of idiotypic determinants of T- and B-lymphocyte receptors for alloantigens. *Immunol. Rev.* 34:50–88

136. Rosenthal, A. S., 1978. Determinant selection and macrophage function in genetic control of the immune response. *Immunol. Rev.* 40:136–152

137. Rosenthal, A. S., M. A. Barcinski, and J. T. Blake. 1977. Determinant selection: A macrophage dependent immune response gene function. *Nature.* 267:156–158

138. Rosenthal, A. S., and E. M. Shevach. 1973. Function of macrophages in antigen recognition by guinea pig T lymphocytes. *J. Exp. Med.* 138:1194–1212

139. Rosenthal, A. S., J. W. Thomas, J. Schroer, and J. T. Blake. 1980. The role of macrophages in genetic control of the immune response. In *Fourth International Congress of Immunol. Progress in Immunology IV.* ed. M. Fougereau and J. Dausset. New York: Academic p. 458–477

140. Rosenwasser, L. J., M. A. Barcinski, R. H. Schwartz, and A. S. Rosenthal. 1979. Immune response gene control of determinant selection. II. Genetic control of the murine T lymphocyte proliferative response to insulin. *J. Immunol.* 123:471–476

141. Rubinstein, L. J., M. Yeh, and C. A. Bona. 1982. Idiotype-anti-idiotype network. II. Activation of silent clones by treatment at birth with idiotypes is associated with the expansion of idiotype-specific helper T cells. *J. Exp. Med.* 156:506–521

142. Sachs, D. H., J. A. Berzofsky, C. G. Fathman, D. S. Pisetsky, A. N. Schechter, and R. H. Schwartz. 1976. The immune response to Staphylococcal nuclease: A probe of cellular and humoral antigen-specific receptors. *Cold Spring Harbor Symp. Quant. Biol.* 41:295–306

143. Schmitt-Verhulst, A.-M., C. B. Pettinelli, P. A. Henkart, J. K. Lunney and G. M. Shearer. 1978. H-2 restricted cytotoxic effectors generated in vitro by the addition of trinitrophenyl-conjugated soluble proteins. *J. Exp. Med.* 147:352–368

144. Schwartz, R. H. 1983. Distinct structures on nominal and MHC antigens

involved in antigen presentation and T cell recognition. In *Ir Genes: Past, Present and Future,* ed. C. W. Pierce, J. A. Kapp, B. D. Schwartz, and D. C. Shreffler, S. E. Cullen. Clifton NJ: Humana In press

145. Scibienski, R. J., V. Klingmann, C. Leung, K. Thompson and E. Benjamini. 1977. In *Recognition of lysozyme by lymphocyte subsets.* ed. M. Z. Atassi and A. B. Stavitsky. *Adv. in Exp. Med. and Biol.* 98:305–317

146. Schirrmacher, V., and H. Wigzell. 1974. Immune responses against native and chemically modified albumins in mice. II. Effect of alteration of electric charge and conformation on the humoral antibody response and on helper T cell responses. *J. Immunol.* 113:1635–1643

147. Schwartz, M., W. C. Waltenbaugh, M. Dorf, R. Cesla, M. Sela, B. Benacerraf. 1976. Determinants of antigen molecules responsible for genetically controlled regulation of immune responses. *Proc. Natl. Acad. Sci. USA* 73:2862–2866

148. Schwartz, R. H. 1982. Functional properties of I region gene products and theories of immune response (Ir) gene function. In *Ia Antigens and their Analogues in Man and Other Animals.* ed. S. Ferrone and C. David, Boca Raton, Florida: CRC In press

149. Seman, M., and J. Theze. 1981. H-2 restricted carrier specific T cell clones do not select immunoglobulin isotypes. In *Immunoglobulin Idiotypes,* ed. C. A. Janeway Jr., E. Sercarz, and H. Wigzell., p. 623–632. New York: Academic

150. Sercarz, E. E., D. T. Corenzwit, D. E. Eardley, and K. M. Morris. 1975. Suppressive versus helper effects in carrier priming. In *Suppressor Cells in Immunity,* ed. S. K. Singhal. Univ. Western Ontario p. 19–25

151. Sercarz, E. E., R. L. Yowell, and L. Adorini. 1977. Immune response genes control the helper suppressor balance. In *The Immune System: Genetics and Regulation.* ed. E. E. Sercarz, L. Herzenberg, and C. F. Fox, 6:497–506. New York: Academic

152. Sercarz, E. E., R. L. Yowell, D. Turkin, A. Miller, M. Araneo, and L. Adorini. 1978. Different functional specificity repertoires for suppressor and helper T cells. *Immunological Rev.* 39:109–137

153. Shearer, G. M., T. G. Rehn, and C. A. Garbarino. 1975. Cell-mediated lympholysis of trinitrophenylated-modified autologous lymphocytes. Effect cell

specificity to modified cell surface components controlled by the H-2K and H-2D serological regions of the major histocompatibility complex. *J. Exp. Med.* 141:1348–1364

154. Sherman, L. A., S. J. Burakoff, and B. Benacerraf. 1978. The induction of cytolytic T-lymphocytes with specificity for p-azophenylarsonate coupled syngeneic cells. *J. Immunol.* 121:1432–1436

155. Shevach, E. M. 1976. The function of macrophages in antigen recognition by guinea pig T lymphocytes. III. Genetic analysis of the antigens mediating macrophage-T lymphocyte interaction. *J. Immunol.* 116:1482–1489

156. Simonsen, M. 1967. The clonal selection hypothesis evaluated by grafted cells reacting against their hosts. *Cold Spring Harb. Symp. Quant. Biol.* 32:517–523

157. Sprent, J. 1978. Restricted helper function of F_1 hybrid T cells positively selected to heterologous erythrocytes in irradiated parental strain mice. II. Evidence for restrictions affecting helper cell induction and T-B collaboration, both mapping to the K-end of the H-2 complex. *J. Exp. Med.* 147:1159–1174

158. Sredni, B., and R. H. Schwartz. 1980. Alloreactivity of an antigen-specific T cell clone. *Nature* 287:855–857

159. Sredni, B., H. Y. Tse, C. Chen, and R. H. Schwartz. 1980. Antigen-specific clones of proliferating T lymphocytes. I. Methodology, specificity and MHC restriction. *J. Immunol.* 126:341–347

160. Swanborg, R. H. 1972. Antigen-induced inhibition of experimental allergic encephalomyelitis. I. Inhibition in guinea pigs injected with non-encephalitogenic modified myelin basic protein. *J. Immunol.* 109:540–546

161. Sy, M.-S., M. H. Dietz, R. N. Germain, B. Benacerraf, and M. I. Greene. 1980. Antigen and receptor-driven regulatory mechanisms. IV. Idiotype-bearing I-J+ suppressor T cell factor induce second-order suppressor T cells which express anti-idiotype receptors. *J. Exp. Med.* 151:1183–1195

162. Tada, T. 1983. I region determinants expressed on different subsets of T-cells: Their role in immune circuits. In *Immunogenetics,* ed. B. Benacerraf, Paris: Masson. In press

163. Tada, T., K. Hayakawa, K. Okumura, and M. Taniguchi. 1980. Coexistence of variable region of immunoglobulin heavy chain and I region gene products on antigen-specific suppressor T cells

and suppressor T cell factor. A minimal model of functional antigen receptor of T cells. *Mol. Immunol.* 17:867–875

164. Takaoki, M., M. S. Sy, B. Whitaker, J. Nepom, R. Finberg, R. N. Germain, A. Nisonoff, B. Benacerraf, and M. I. Greene. 1982. Biologic activity of an idiotype-bearing suppressor T cell factor produced by a long-term T cell hybridoma. *J. Immunol.* 128:49–53

165. Taniguchi, M., and J. F. A. P. Miller. 1977. Enrichment of specific suppressor T cells and characterization of their surface markers. *J. Exp. Med.* 146:1450–1454

166. Thomas, D. W., M. D. Hoffman, and G. D. Wilner. 1982. T lymphocyte recognition of peptide antigens: evidence favoring the formation of neoantigenic determinants. *J. Exp. Med.* 156:289–293

167. Thomas, D. W., K.-H. Hsieh, J. L. Schauster, M. S. Mudd, and G. D. Wilner. 1980. Nature of T lymphocyte recognition of macrophage-associated antigens. V. Contribution of individual peptide residues of human fibrinopeptide B to T lymphocyte responses. *J. Exp. Med.* 152:620–632

168. Thomas, D. W., K.-H. Hsieh, J. L. Schauster, and G. D. Wilner. 1981. Fine specificity of genetic regulation of guinea pig T lymphocyte responses to angiotensin II and related peptides. *J. Exp. Med.* 153:583–594

169. Thomas, W. R., F. I. Smith, I. D. Walker and J. F. A. P. Miller. 1981. Contact sensitivity to azobenzenarsonate and its inhibition after interaction of sensitized cells with antigen-conjugated cells. *J. Exp. Med.* 153:1124–1137

170. Tokuhisa, T., and M. Taniguchi. 1982. Two distinct allotypic determinants on the antigen specific suppressor and enhancing T cell factors that are encoded by genes linked to the immunoglobulin heavy chain locus. *J. Exp. Med.* 155:126–139

171. Tokuhisa, T., Y. Komatsu, Y. Uchida, and M. Taniguchi. 1982. Monoclonal alloantibodies specific for the constant region of T cell antigen receptors. *J. Exp. Med.* 156:888–897

172. Tse, H. Y., J. J. Mond, and W. E. Paul. 1981. T lymphocyte-dependent B-lymphocyte proliferative response to antigen. I. Genetic restriction of the stimulation of B lymphocyte proliferation. *J. Exp. Med.* 153:871–82

173. Turkin, D., and E. E. Sercarz. 1977. Key antigenic determinants in the regulation of the immune response. *Proc. Nat. Sci. USA* 74:3984–3987

174. Unanue, E. R. 1978. The regulation of lymphocyte function by the macrophage. *Immunol. Rev.* 40:227–255

175. Urbain, J., C. Wuilmart, and P.-A. Cazenave. 1981. Idiotypic regulation in immune networks. *Contemp. Top. Molec. Immunol.* 8:113–48

176. von Boehmer, H., W. Haas, and N. K. Jerne. 1978. Major histocompatibility complex-linked immune-responsiveness is acquired by lymphoctyes of low-responder mice differentiating in thymus of high responder mice. *Proc. Natl. Acad. Sci. USA* 75:2439–2442

177. Walker, E., N. C. Warner, R. Chesnut, J. Kappler, and P. Marrack. 1982. Antigen-specific, I region-restricted interactions *in vitro* betweeen tumor cell lines and T cell hybridomas. *J. Immunol.* 128:2164–2169

178. Weyand, C., J. Goronzy, and G. J. Hammerling. 1981. Recognition of polymorphic H-2 domains by T lymphocytes. I. Functional role of different H-2 domains for the generation of alloreactive cytotoxic T lymphocytes and determination of precursor frequencies. *J. Exp. Med.* 54:1717–1731

179. Weyand, C., G. J. Hammerling, and J. Goronzy. 1981. Recognition of H-2 domains by cytotoxic T lymphocytes. *Nature* 292:627–629

180. Wikler, M., C. Demeur, G. Dewasme, and J. Urbain. 1980. Immunoregulatory role of maternal idiotypes: Ontogeny of immune networks. *J. Exp. Med.* 152:1024–1035

181. Woodland, R., and H. Cantor. 1978. Idiotype specific T helper cells are required to induce idiotype positive B memory cells to secrete antibody. *Eur. J. Immunol.* 8:600–606

182. Wilson, D. B. 1974. Immunologic reactivity to major histocompatibility alloantigens: HARC, effector cells and the problem of memory. In *Progress in Immunology*, Vol. 2, ed. L. Brent and J. Holborow, p. 145–156. Amsterdam: Elsevier

183. Wysocki, L. J., and V. L. Sato. 1981. The strain A anti-p-azophenylarsonate major cross-reactive idiotype family includes members with no reactivity toward p-azophenylarsonate. *Eur. J. Immunol.* 11:832–839

184. Yamauchi, K., N. Chao, D. B. Murphy, and R. K. Gershon. 1981. Molecular composition of an antigen specific, Ly-1 suppressor inducer factor. One molecule binds antigen and is I-J⁻, another is I-J⁺, does not bind antigen, and imparts

an IgH-variable region linked restriction. *J. Exp. Med.* 155:655–665
185. Yowell, R. L., B. A. Araneo, A. Miller, and E. E. Sercarz. 1979. Amputation of a suppressor determinant on lysozyme reveals underlying T cell reactivity to other determinants. *Nature* 279:70–71
186. Zinkernagel, R. M., and P. C. Doherty. 1979. MHC-restricted cytotoxic T cells: Studies on the biological role of polymorphic major transplantation antigens determining T-cell restriction-specificity, function, and responsiveness. *Adv. Immunol.* 27:51–177
187. Zinkernagel, R. M., G. N. Callahan, A. Althage, S. Cooper, P. A. Klein, and J. Klein. 1978. On the thymus in the

differentiation of "H-2 self-recognition" by T cells: evidence for dual recognition? *J. Exp. Med.* 147:882–896
188. Zinkernagel, R. M., and P. C. Doherty. 1975. H-2 compatibility requirement for T-cell mediated lysis of target cells infected with lymphocytic choriomeningitis virus: different cytotoxic T-cell specificities are associated with structures coded from H-2K or H-2D. *J. Exp. Med.* 141:1427–1436
189. Zinkernagel, R. M., and P. C. Doherty. 1977. Major transplantation antigens, virus and specificity of surveillance T cells. The "altered self" hypothesis. *Comtemp. Top. Immunbiol.* 7:179

Ann. Rev. Immunol. 1983. 1:499–528

IMMUNOGLOBULIN GENES

Tasuku Honjo

Department of Genetics, Osaka University Medical School, Kita-ku Osaka, 530 Japan

INTRODUCTION

The immunoglobulin gene system is comprised of three separate gene loci of L_κ, L_λ, and H (light and heavy) chain genes, each containing variable (V) and constant (C) genes. The L_κ (157), L_λ (36), and H (35) chain genes are located on mouse chromosomes 6, 16, and 12, respectively [on human chromosomes 2 (95), 22 (104), and 14 (48, 84), respectively)].

Molecular cloning and structural characterization of the immunoglobulin gene have shown that DNA rearrangement plays essential roles in the somatic amplification of the immunoglobulin diversity that manifests in two aspects: the ability to bind an enormous number of antigens, and the ability to trigger a variety of immunologic reactions. Such studies made long-debating immunologists more or less unanimous on how the immunoglobulin diversity is generated. Structural analyses of the immunoglobulin gene have also given new insights into the evolutionary process that contributes to the creation of the antibody diversity.

It is now clear that organization and structure of the immunoglobulin gene per se incorporate the molecular drive that increases the immunoglobulin diversity through the somatic and evolutionary processes. The basic principles that underlie the genetic mechanism to create the antibody diversity are summarized as flexible, random, somatic DNA rearrangements accompanied by deletion and evolutionary mechanisms depending on the frequent recombination and rapid drift. The antibodies thus created contain not only those that are specific to current antigens, but also those that appear useless yet might be reactive to antigens to be encountered in future. Thus, the immunoglobulin gene system has Promethean foresight (116), preparing for unknown antigens by the flexible and random DNA rearrangement. This system has strong advantages over the fixed rigid genetic system that determines all the specifities and physiological functions a priori in the genome, because the latter makes it difficult for the organism to adapt to the drastic change of the environment.

499

0732-0582/83/0410-0499$02.00

In this review I focus on the organizational and structural features of the cloned immunoglobulin genes and on the implications derived from these studies. Transcription of the immunoglobulin gene is not discussed in this review. The encyclopedic citation of references is not the object of this review in view of the limited space for such a broad subject. Unless specified, described information was derived from studies on mouse genes. Several aspects of this gene were reviewed previously (33, 40, 52, 64, 97).

VARIABLE REGION GENES

Structure and Organization

THE VARIABLE REGION IS ENCODED BY TWO OR THREE SEPARATE DNA SEGMENTS   The V region was defined originally by the comparison of amino acid sequences of immunoglobulins. The length of the L-chain V region is rather constant (usually amino terminal 107 residues) whereas that of the H-chain V region is variable, ranging from 107 to 132 residues (74). Cloning and characterization of the $V_{\lambda 2}$ region gene first demonstrated that the V_λ region is encoded by two separate DNA segments, i.e. the V_λ and J_λ segments, as shown in Figure 1 (11, 160). The V_λ segment has two exons separated by a 93-bp intron; the 5' exon codes for a major portion of the leader sequence (residues –19 to –5), which is eventually cleaved off, and the other exon codes for the remaining portion of the leader sequence and the major portion of the V region (residues –4 to 97). The rest of the V region (residues 98–110) is encoded by the J_λ segment, which is located

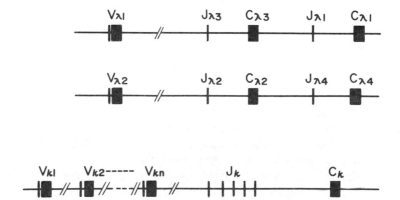

Figure 1   Organization of immunoglobulin L-chain genes. Each exon is shown by a wide rectangle.

close to the C_λ gene. As described below, the expressed V gene is created by the fusion of the germline V and J segments. To distinguish the germline and the complete rearranged forms of the V gene, I refer to the germline form as the V segment throughout this review. The V_κ region is also encoded by the V_κ and J_κ segments with lengths similar to the V_λ and J_λ segments, respectively (102, 137).

By contrast, the V_H region is encoded by the three separate DNA segments, namely, the V_H, D, and J_H segments (38). Consequently, the complete active V_H gene is generated by two somatic recombinations, V_H-D and D-J_H joinings. The D segment encodes primarily the CDR3 and varies considerably in sequence as well as in length (88, 89).

STRUCTURE AND ORGANIZATION OF THE V SEGMENT   There are only two V_λ segments ($V_{\lambda1}$ and $V_{\lambda2}$) in the mouse genome, whereas a large number of V_λ segments are expected to be in the human genome. The $V_{\lambda1}$ segment is genetically linked to the $C_{\lambda3}$ and $C_{\lambda1}$ genes, and the $V_{\lambda2}$ segment is linked to the $C_{\lambda2}$ and $C_{\lambda4}$ genes in the mouse genome (14, 105). The V_λ segments have not been physically linked to any of the C_λ genes.

The germline V_κ segment has the structure essentially similar to the V_λ segment. Multiple V_κ segments are cloned from mouse and man (9, 113, 142). One of them is a pseudogene containing substitutions, deletions, and insertions (9). About 31- to 43-kb DNA fragments carrying two V_κ segments were cloned in a cosmid vector, but the distance between two adjacent V_κ segments is not known precisely (21). Two human genomic fragments carrying two V_κ segments with 12.5- to 15-kb spacers were cloned (126).

Zeelon et al (78) measured the proportion of a distinct V_κ sequence in the total V_κ sequences expressed in mouse spleen and estimated the number of the V_κ group as 280, assuming that all the groups are equally expressed. The total number of the V_κ segment is simply calculated to be about 2000 on the assumption that each group has seven members. Cory et al (30) measured numbers of restriction fragments hybridized with 10 probes and concluded that the total number of the V_κ segments is between 90 and 320. Bentley & Rabbitts (9) estimated that there are only 15–20 V_κ segments in the human genome on the basis of the number of hybridized bands with V_κ probes of three different subgroups.

In view of the presence of pseudogenes and the absence of clear knowledge about the extent of cross-hybridization, these values can not be taken seriously. In addition, the ambiguity about the definition of the V group has to be taken into account. Although the subgroup was originally defined by the amino acid sequence similarity of the V region framework, analysis at the DNA level is consistent with the view that each V subgroup is determined by a single germline V segment. The definition of the group was somewhat arbitrary. For example, V_κ regions varying at three or more of the amino-terminal 23 amino acids were classified into different V_κ groups. The definition of the group at the DNA level is also arbitrary. Several investigators consider that all the V segments cross-hybridizing to a probe belong to the same group. However, cross-hybridization depends on the conditions of hybridization and washing, especially temperature and salt concentration. Evolutionally closely related V segments shoud form a group of V segments. In some cases, however, there would be no discrete boundary between different groups because extensive recombinatorial homogenization of V segments has taken place, as is discussed later.

The V_H segment is also split into two exons by an intron at the identical position to the V_L segment. The lengths of two exons are also similar to those of the V_L segments. Several V_H segments are physically linked (31, 76, 80). The distance between two adjacent V_H segments is 12–15 kb. The average V_H-V_H distance appears to be longer (at least 25 kb) than the above value, because L. Hood and his associates (personal communication) found 21 V_H segments in cloned germline DNA fragments that total 492 kb.

The total number of the V_H segments is not known either. Probably it would fall somewhere between 100 and 1000. Kemp et al (81) estimated that each group contains 15 members by Southern blot hybridization using three probes of different groups. Assuming that the total number of V_H groups is about 10, they calculated that the total number of the V_H segments is 160. Three V_H segments were mapped in the order V_{H76} (inulin binding), V_{T15} (phosphorylcholine binding), and $V_{H1}/A8$ segments by deletion in myeloma cells expressing different V_H segments (81). The members of the same V_H group tend to cluster (31, 76, 80, 131). However, Hood and his associates (personal communication) were unable to link the four V_H members of the V_{T15} group even by cloning the germline DNA segments of 500 kb. It is possible that some members of the group were translocated elsewhere in the V_H cluster.

It is important to stress that members of the same V_H group do not always encode the same antigen specificity. For example, only one of the four members of the V_{T15} group is used for anti-phosphorylcholine antibody, the second is used for anti-hemagglutinin of influenza virus, the third is used for unknown specificity, and the fourth is a pseudogene (31; L. Hood, personal communication).

STRUCTURE AND ORGANIZATION OF THE J SEGMENT   The J_λ segment is located 1.2–1.3 kb 5' to each C_λ gene of mouse (Figure 1). From the genetic linkage between the V_λ and C_λ genes, the $V_{\lambda 1}$ segment is able to associate with either $J_{\lambda 3}$ or $J_{\lambda 1}$ segment, whereas the $V_{\lambda 2}$ segment is to be joined with either the $J_{\lambda 2}$ or the $J_{\lambda 4}$ segment. However, the $V_{\lambda 2}$ was shown to be associated with $J_{\lambda 3}$ and $C_{\lambda 3}$ in a hybridoma, suggesting that the $V_{\lambda 2} J_{\lambda 2} C_{\lambda 2} J_{\lambda 4} C_{\lambda 4}$ cluster is 5' to the $V_{\lambda 1} J_{\lambda 3} C_{\lambda 3} J_{\lambda 1} C_{\lambda 1}$ cluster (43). The $J_{\lambda 4}$ segment is probably inactive because of a 2-bp deletion in the splicing signal. Another pseudo-J segment was identified between the $J_{\lambda 3}$ segment and the $C_{\lambda 3}$ gene (14, 105).

There are four J_κ segments and a pseudo-J_κ segment in the mouse genome. The J_κ segments are clustered 2.6 kb 5' to the C_κ gene (102, 137). The four J_κ segments are presumed to be capable of joining with any of the V_κ segments from amino acid sequence analyses. The human J_κ cluster contains active five J segments about 2.5 kb 5' to the C_κ gene (61). The absence of the mouse $J_{\kappa 3}$ equivalent sequence was explained by a duplication event in the human J_κ cluster by unequal crossing over.

In the H chain gene family there is only one set of the J_H cluster for eight or nine C_H genes. There are four J_H and one pseudo-J_H segments in mouse and six J_H and three pseudo-J_H segments in man (130, 139). Although the length of the mouse J_L segment is constant (13 residues), the J_H segments of mouse and man have varied lengths ranging from 16 to 21 residues.

STRUCTURE AND ORGANIZATION OF THE D SEGMENT   The D segment is characterized by its variability in sequence as well as in length (88, 89). The shortest germline D segment identified so far encodes six amino acid residues, and the longest one encodes 17 residues in mouse. At least some D segments are arranged in a relatively small region of the genome. Clusters of 11 and 4 D segments were found in mouse and man, respectively (88, 150). But not all the amino acid sequences of the H chains are explained by known D segments. Worse than that, the four human D segments reported are not found in known complementarity-determining region (CDR)3 sequences. It is possible that more unknown germline D segments are present in the mouse and human genomes. Kurosawa & Tonegawa (88) proposed the D-D joining as a mechanism to create unexplained expressed D region sequences. However, there is not a single case in which the D-D joining can explain the known D region without the point mutation, insertion, or deletion.

The location of the D segment relative to the C_H and V_H is not known except for one D segment located immediately 5' to the J_H segment (138). Evolutionary consideration led to a proposal that the D segment clusters may be dispersed among V_H clusters (65).

V-(D)-J Recombination

Structural comparison of the germline and expressed V genes clearly demonstrated that a complete V gene is created by recombinational joining of V, (D), and J segments. This recombinational reaction has three important characteristics. First, the combination between different DNA segments seems to be random because the same V_H segments are able to associate with all the J_H segments in the anti-phosphorylcholine antibody H-chain of mouse (53). Secondly, the recombination reaction has flexibility at the joining nucleotide in spite of the strong requirement for the accurate in-frame joining. Thirdly, the recombination is accompanied by deletion.

RECOMBINATIONAL MECHANISM   Nucleotide sequence determination of the germline V and J segments and their flanking regions has revealed that conserved inverted repeat sequences are located immediately 3' and 5' to the V and J segments, respectively (38, 102, 137, 139). The D segment is flanked by the similar sequences at both sides (89, 138). The conserved sequence is comprised of a nonomer and a heptamer separated by 12±1- or 23±1-bp essentially random nucleotides. Sequences of the nonomers and heptamers are not identical but are conserved enough to form stable complementary base pairs between the combining pair of segments, as shown in Figure 2. The nonomer and heptamer sequences are, therefore, considered to be at least a part of the recognition signal for putative V-(D)-J recombinase(s).

The recognition sequence of the V_κ and J_κ segments have 23±1-bp and 12±1-bp spacers, respectively. By contrast, the nonomer and heptamer of the V_H and J_H segments are separated by 23±1-bp spacers. The D segment has a 12±1-bp spacer on each side. Thus, the paired recognition signals have different lengths of the spacer sequences, which is called 12/23-bp spacer rule as originally proposed by Early et al (38). Since one turn of the DNA helix requires about 10.4 bp (168), the conserved spacing of 12 or 23 bp corresponds to either one or two turns of the helix. The two blocks of conserved nucleotides (nonomer and heptamer) are thus located on the

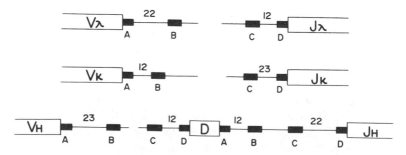

Figure 2   Putative recombinational signal for V-(D)-J recombination. Solid rectangles show nonomer and heptamer recognition signals. The pair A and B form complementary base pairs with C and D. Consensus sequences are A, CACAGTG; B, ACAAAAACC; C, GGTTTTTGT; D, CACTGTG. Numbers indicate average spacer nucleotides.

same side of the helix as if they formed a continuous stretch of 16 conserved nucleotides (38, 139).

The intervening DNA segment between the V and J segments was shown to be deleted in myelomas (27, 137, 145) in the same manner as in the H chain C gene rearrangement (66). The deletion is consistent with a stem-loop structural model based on the inverted repeat structure (102, 137). An alternative model to explain deletion is a sister chromatid exchange model (165) as originally proposed for the H-chain C-gene deletion (115). This model was proposed to account for the fact that DNA segments 5' to the J_κ segment were retained in some myelomas, both of whose J_κ segments were rearranged (156, 165). Most (76%) of the region upstream of J_κ is retained in the population of the C_κ-expressing lymphocytes, even though 68% of the J_κ loci are rearranged. An inversion model was recently proposed to explain the unusual presence of the upstream region of the J segment (see below).

The V-(D)-J joining, in general, is expected to be precise at the nucleotide level. Otherwise, the recombination would often result in the frame-shift mutation. However, the amino acids at position 96 of some myeloma kappa chains differ from that encoded in the corresponding germline DNA segment (102, 137). The V-J joining seems to have flexibility in the exact nucleotide position of the recombination as long as it is in phase with the coding sequence. The four possible alternative sites of the in-frame recombination are shown in Table 1. This flexibility provides further diversity at junctions of germline DNA segments, which are located in the CDR.

In case of V_H genes there are many insertions or deletions at junctions with D segments. These inserted portions, termed N regions, were proposed to be ascribed to the activity of terminal nucleotide transferase during V_H-D-J_H recombination (1).

ABERRANT REARRANGEMENTS   The flexibility at the V-J joining costs the generation of missense genes to the organism. The aberrant V-J joining was reported in the kappa as well as lambda locus (5, 10, 15, 26, 69, 90, 103, 167). Either deletion or insertion of a few bases resulted in out-of-phase J sequences. In many κ-producing myelomas two alleles of kappa genes are rearranged, one of which is usually aberrantly rearranged.

A cell line of MPC11 myeloma produces an unusual fragment of the kappa chain in addition to a normal kappa chain. Amino acid sequence determinations on an in vitro synthesized fragment precursor showed that it consisted of a hydrophobic leader sequence joined directly to the C_κ region and that the sequence corresponding to the V region, was completely lacking (19, 135). The gene encoding this fragment was analyzed, which

Table 1 Flexibility at the recombination site[a]

Germline genes		Codon generated
$V_{\kappa 41}$	$\cdots$ AGT-TCT-CCT-CCCACAGTG $\cdots\cdots$	TGG Trp
	‖ ⫿	CGG Arg
J1	$\cdots\cdots$ CACTGTGG-TGG-ACG $\cdots\cdots\cdots$	CCG Pro
		CCC Pro
$V_{\kappa 41}$	$\cdots$ AGT-TCT-CCT-CCCACAGTG $\cdots\cdots$	TAC Tyr
	‖ ⫿	CAC His
J2	$\cdots\cdots$ CAGTGTGTG-TAC-ACG $\cdots\cdots\cdots$	CCC Pro
$V_{\kappa 41}$	$\cdots$ AGT-TCT-CCT-CCCACAGTG $\cdots\cdots$	TTC Phe
	‖ ⫿	CTC Leu
J4	$\cdots\cdots$ CACTG·GA-TTC-ACG $\cdots\cdots\cdots$	CCC Pro
$V_{\kappa 41}$	$\cdots$ AGT-TCT-CCT-CCCACAGTG $\cdots\cdots$	CTC Leu
	‖‖	CCC Pro
J5	$\cdots\cdots$ CACTGTGG-CTC-ACG $\cdots\cdots\cdots$	

[a] Taken from Max et al (102). Codons are shown by hyphens. Recombination signal underlined; vertical lines indicate possible recombination sites.

showed that it was generated by aberrant recombination of a V_κ segment with the middle of the intervening sequence between the J_κ segments and the C_κ gene (140, 143). The resulting deletion of the J_κ segment, including the splicing signals, allowed an abnormal RNA splicing to create mRNA containing the leader and C_κ region sequences. This abnormal rearrangement seems to be due to homologous recombination between the putative recombinational signal 3' to the V segment and the homologous sequence in the intervening sequence (IVS) between the J_κ segment and the C_κ gene.

Alt & Baltimore (1) found that the inverted D_1-J_{H1} and its 3' flanking sequence is linked to the 5' side of the J_{H3} segment in a spontaneously rearranged clone of Abelson murine leukemia virus-transformed cells. In addition, the inverted segment extends through the inverted J_{H2} nonomer recognition sequence and ends abruptly at the terminus of the inverted J_{H2} heptamer recognition sequence. The inverted J_{H2} sequence is replaced by the noninverted 5' flanking sequence of a D segment at this point (Figure 3). Such an unusual structure has to be explained by intrachromosomal recombination because inverted joinings between sister chromatids would have resulted in one dicentric chromosome and another lacking a centromere. The authors favor the looping-out model for deletion, and argue that the retention of the reciprocal product described above (the region upstream of J) can be explained by inversion. Lewis et al (92) provided further evidence

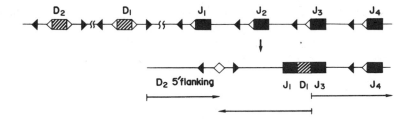

Figure 3   Inverted joining of D and J segments. Top and bottom lines indicate germline and rearranged DNAs, respectively. ◄ and ►, 5' and 3' recognition nanomer, respectively. < and >, 5' and 3' recognition heptamer, respectively. Arrows indicate direction of DNA in germline configuration. Rearranged from Alt & Baltimore (1).

to support the inversion model using a spontaneously rearranging Abelson virus-transformed cell line, which seems to be useful in analyzing differentiation steps.

Some T lymphomas, T hybridomas, and T-cell lines were shown to have rearrangement at the J_H segment (28, 50). Most of them seem to be D-J_H rearrangement (89), indicating that D-J_H recombination initiates at about the same time as the differentiation of T and B lymphocytes.

Somatic Mutation

Studies on the murine $L_{\lambda 1}$ chains produced by myeloma cells provided the first evidence for somatic point mutation of the V-region gene. Amino acid sequences of 19 $V_{\lambda 1}$ chains were determined and were divided into seven groups that differed from a common sequence ($V_{\lambda 0}$) by one to three residues (24, 169). Subsequent DNA sequence determination of a rearranged $V_{\lambda 1}$ gene from H2020 myeloma and the germline $V_{\lambda 1}$ gene demonstrated that the two sequences differ at two positions in CDRs and at one in the flanking region (11). More extensive studies on V_{λ}, V_{κ}, and V_H genes, however, indicate that the substitution is not restricted to either the CDR or the coding region and seems to spread into IVS and flanking regions (16, 31, 54, 69, 76, 83, 123, 139, 146). But no mutations were found in C genes linked to V genes. The C genes of both $\lambda 1$ and $\gamma 2a$ chains of the S43 hybridomas are germline DNA sequences, although both V genes contain mutations (16, 141).

These results are unable to settle the long argument of whether there exists a special enzymic mechanism that preferentially introduces mutations to the V gene or whether the selection at the cellular level is responsible. However, there seems to be some mechanism that accumulates mutations only around the V-coding sequence and not in the C gene. The $V_{\lambda 1}$ gene of MOPC315 myeloma also shows somatic mutations, although it is an aberrantly rearranged gene (69). Similarly, both of the rearranged V_{κ} genes, whether expressed or unexpressed, showed similar levels of somatic mutation (123). There are several expressed V genes that have no somatic mutations (144). It is not clear whether inactive germline V segments in an antibody-producing cell have substitutions or not, although everybody assumes that DNA rearrangement is obligatory for the introduction of somatic mutation into the V segment.

Many V_L and V_H genes in IgM-producing lymphocytes are germline sequence, whereas most of V_L and V_H genes in IgG- or IgA-producing lymphocytes have substitutions (16, 53). These authors claim that class switching may trigger somatic mutation. Since the B lymphocyte is stimulated and divides extensively during or after class switching, the chance of somatic mutation is expected to be higher for non-IgM-producing lymphocytes. It is not known whether or not class switching is coupled with somatic mutation.

Genetic Basis for Variable-Region Diversity

After many years of controversy, DNA cloning succeeded in convincing immunologists unanimously on how V-region diversity is generated in a broad view. Nature seems to be generous to everybody who has proposed any mechanism in such a way that everybody is right in one way or the other (132).

Four types of mechanisms generate the V-region diversity. First of all, the number of the germ-line V segment is reasonably great, probably somewhere between 100 and 1000. This is due to the evolutionary process. Secondly, the recombinational joining of the V, (D), and J segments plays a very important role in amplifying the V-region diversity. The combination of each segment is assumed to be random. Thirdly, the flexibility at the recombination sites augments diversity significantly. Finally, the somatic mutation also contributes to increase the V-region diversity. It is difficult, however, to assess the extent of contribution of each mechanism. The minimal possible diversity due to the three mechanisms other than somatic mutation can be calculated as follows: V_κ diversity of mouse = number of V_κ (100) × number of J_κ (4) × flexibility of joining (2) = 800; and V_H diversity of mouse = number of V_H (200) × number of D (20) × number of J_H (4) × flexibility of joining (2 × 2) = 64,000. Provided any V_κ and V_H can associate to form a different antigen binding activity, the possible diversity would be 5.1 × 10^7 (= 800 × 64,000).

The basic principle for all the genetic mechanisms to amplify the antibody diversity is the random expression of all the possible genetic information by the enormous flexibility inherent in the immunoglobulin gene system. Obviously, the selection of the randomly expressed information at the cellular level is a key to constructing the useful defense system for the organism [clonal selection theory (18)]. The number and sequence of V segments are out of control at the somatic level but certainly are controlled by the balance between evolutionary selection and changes (drift, gene conversion, unequal crossing over, etc).

Allelic Exclusion

In general, only one species of the immunoglobulin is synthesized and secreted by a lymphocyte. The failure of expression of the immunoglobulin genes from more than one chromosome is known as allelic exclusion, the mechanism of which has been the subject of some debate (2, 40, 124). Either a κ or λ light chain is produced, but not both, in a single lymphocyte (isotype exclusion).

It is clear that the exclusion operates at the functional protein but not at transcription of the immunoglobulin genes, as MOPC104E produces both κ and λ chain mRNA's, of which only λ chain mRNA is translated into a stable protein (2, 75, 121). Since many lymphocytes, hybridomas, and myelomas contain at least one unrearranged J_κ allele (72, 124), there must be some mechanism that terminates further V_κ-J_κ rearrangement on the other chromosome. In case the J_κ segments of both chromosomes are rearranged, one of them is usually defective (10, 15, 26). These observations are consistent with a stochastic model: that the V-(D)-J recombination, which has a certain chance of failure, continues until the complete V genes are formed. The formation of the complete V gene, in turn, would shut off the activity of a putative recombinase. In the case of myeloma S107, however, both of the V_κ alleles are rearranged and transcribed, and two different kappa chains are produced (90). Only one of them is able to form a complex with the alpha chain and is secreted into the medium. The result is also explained by the stochastic model, assuming the shut-off signal is not the appearance of the L chain in the cytoplasm but the appearance of a complete immunoglobulin on the membrane or in the cytoplasm.

The H-chain genes have a higher proportion of nonproductive rearrangements in B lymphocytes as well as in hybridomas and myelomas (25, 26, 70, 114). It is not known whether the lower chance of successful joining of the V_H locus is simply due to one additional recombinational event (V-D and D-J), or to the difference in the recombinatorial machinery as compared with the V_L gene.

Recently, Wabl & Steinberg (166) found an H-chain-binding protein that binds to free immunoglobulin H chains in pre-B cells, and proposed that the displacement of the H-chain ' binding protein by the L chain terminates the enzyme system that catalyzes DNA rearrangement at the L-chain loci. To explain exclusion of the H-chain locus they also proposed that the H chains are so toxic that those cells producing H chains from both alleles of the H-chain locus are killed.

Order of the V Gene Rearrangement During Differentiation

During ontogeny, the V_H locus is the first to be rearranged, resulting in the expression of cytoplasmic μ chains in the pre-B cell. Subsequently, the V_L locus undergoes the rearrangement, and IgM molecules are formed to serve as surface receptors. The V_κ gene seems to be rearranged before the V_λ gene, as the J_κ genes were either deleted or rearranged in all 10 of human leukemia cell lines producing λ chains (60). By contrast, the J_λ genes remained unrearranged in eight human B-cell leukemia lines producing κ chains. In mice, the hierarchy is less evident, as a nonproductive V_λ rearrangement in a κ-producing cell line was seen (26). The rearrangement order of the V_κ and V_λ genes may be regulated developmentally (60). Alternatively, the efficiency of recombination may simply favor the V_κ gene locus.

Evolution of V-Region Genes

ORGANIZATION   The β_2 microglobulin L chain of the histocompatibility antigen was shown to have homologous structure with immunoglobulin L chains and was considered to be evolved from a common ancestor. The β_2 microglobulin gene was recently cloned and was shown to be interrupted by IVSs at the amino acid positions +3 and +95 (122). The first intron of the β_2 microglobulin gene corresponds to that between the exons of the signal peptide and the V segment. The location of the second IVS in the β_2 microglobulin gene is similar to the junction between a V_L and a J_L segment or between a V_H and a D segment. The results not only suggest that the prototype of the immunoglobulin gene is a split gene like V-J, but they also imply that the third exon of the β_2 microglobulin gene may be the evolutionary ancestor to both D and J segments. The V segment may have evolved from the first and second exons. If so, one can assume that a prototype V-J gene repeated duplication either as a set or as a separate exon. Prototype J_H segments located within the V_H cluster might have diverged into D segments, whereas prototype J segments closest to the C_μ gene changed into the present J_H segments. From this evolutionary consideration, the homologous V segments are expected to cluster, as shown in several systems (31, 76, 81, 131). Unknown D segments may be found dispersed in the V_H cluster.

EVOLUTIONARY ORIGIN OF V-REGION DIVERSITY   The numbers of the V, D, and J segments and their variability contribute to the V-region diversity enormously. These factors depend entirely on the evolutionary process. What is the mechanism that allows the accumulation of a large number of variables and yet maintains the basic framework structure among the hundreds of related genes? This problem has been discussed theoretically as concerted evolution (51, 154).

Ohta (117) showed that the rate of the amino acid substitution in CDR is as high as that of the fibrinopeptide, the most rapidly diverging protein so far known. The substitution rate in the other portion of the V region is one third of that in CDR, in agreement with the DNA sequence analysis of closely related V_H segments (55). This result suggests that the variability in the V segment is mostly due to the accumulation of random and neutral base changes. In other words, the V segment is under weakest selection pressure at the protein level, as the nucleotide divergence rate at the synonymous position of the V segment is the same as many other genes (107). The homology in the framework region is maintained by domain (or segment) transfer between different loci by either gene conversion or double unequal crossing over, as indicated by computer simulation studies ' ' (13, 153). Assuming that the V segment diverges by the balance of drift and recombinational homogenization, Ohta (117) has evaluated divergence rates of the V segment using a population genetics approach, and her calculated values fit well with observed values. Ohta (118) indicated that a high degree of polymorphism and non-allelic diversity in the major histocompatibility complex can be explained by a similar domain transfer mechanism.

The comparison of the nucleotide sequences of the closely related V_H segments indicates that the rate of the replacement substitutions in CDR1 and CDR2 are greater than in framework regions, indicating that the diversity of the hypervariable region is generated, at least in part, through an evolutionary process (55). The V segment seems to move quickly during evolution since restriction fragments hybridizing to a V_H or V_κ probe are very different among laboratory mouse strains and wild mice (30, 77).

Wu & Kabat (171) found that CDR2 of a human V segment contains 14 nucleotides identical to a human D segment. They proposed that a minigene-like D segment may be involved in the generation of CDR diversity by a gene conversion mechanism. The minigene theory was originally proposed as a somatic mechanism to explain V-region diversity (73), and it predicted the presence of the D segment. Seidman et al (142) proposed that recombination and mismatch repair might be a likely mechanism to explain extensive variability in CDR. They have shown there is extensive homology in the flanking regions of related V segments, which would be consistent with frequent recombination among V segments (23).

In summary, two factors are involved in the evolutionary drive to amplify the V-segment diversity: (a) the frequent recombination (gene conversion and/or double unequal crossing over) facilitated by the organization of the

V-segment family; and (*b*) very low selection pressure at the protein level. These two factors work in opposite directions and reach the equilibrium.

V-REGION GENES IN OTHER SPECIES   The rabbit V_H sequence of a cDNA clone was determined and was shown to contain cDNA sequence homologous to a human D1 in CDR2 (12). The implication would be that the similar D segment is present in the rabbit genome and is involved in gene conversion. Rat J_κ segments were cloned, and their structural analyses showed that six J_κ segments seemed to be evolved by recent two-sequential crossing-over events as compared with the mouse J_κ cluster (20, 147).

CONSTANT-REGION GENES

Structure and Organization

ORGANIZATION OF CONSTANT-REGION GENES   Only one copy of the C_κ gene is represented in the mouse and human genomes. The C_κ genes of mouse and man are located 2.1–2.3 kb 3' to the J segment (61, 102, 137). The distance between the V_κ segments and C_κ gene is unknown. The C_λ locus is much more complex than the C_κ locus because there are four C_λ genes in the mouse genome. The pair of the $C_{\lambda 3}$ and $C_{\lambda 1}$ genes and that of the $C_{\lambda 2}$ and $C_{\lambda 4}$ genes are separately linked (14, 105). Each C_λ gene has its own J_λ segment 1.2–1.3 kb upstream. The $C_{\lambda 4}$ gene seems to be a pseudo-gene because the $J_{\lambda 4}$ has 2-bp deletion in the heptamer of the recombinational signal. Multiple C_λ genes are also found in the human genome (59). Six human C_λ genes were physically linked within a 30-kb region. Each C_λ gene is supposed to have its own J_λ segment. At least five other clones hybridizing with the C_λ probe were isolated, although none of them are linked to the major C_λ cluster. One of them was identified as a processed gene not located on chromosome 22 (63).

The mouse C_H gene locus consists of eight genes (μ, δ, $\gamma 3$, $\gamma 1$, $\gamma 2b$, $\gamma 2a$, ϵ, and α) that cluster in a 200-kb region as shown in Figure 4 (149). The order of the mouse C_H genes 5'-C_μ-C_δ-$C_{\gamma 3}$-$C_{\gamma 1}$-$C_{\gamma 2b}$-$C_{\gamma 2a}$-C_ϵ-C_α-3' is consistent with the order proposed by the deletion profile of the C_H gene myelomas producing different classes of the immunogloblin (66). Unlike the C_L gene, the C_H genes share one set of J_H segments (148), which provides a genetic basis for H-chain class switching, as discussed below. The mouse C_H gene locus does not contain any well-conserved pseudogene.

The human C_H gene family consists of at least nine genes, (μ, δ, $\gamma 1$, $\gamma 2$, $\gamma 3$, $\gamma 4$, ϵ, $\alpha 1$, and $\alpha 2$). The general organization of the human C_H gene cluster is similar to mouse in that the C_δ gene is located several kilobases 3' to the C_μ gene and that at least some C_γ genes are linked (45, 87,

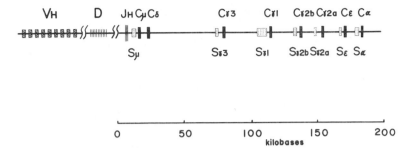

Figure 4 C$_H$ gene organization.

127, 159). However, the C$_\epsilon$-C$_\alpha$ region is duplicated in the human genome (100, 111). As it is not known whether or not all the C$_\gamma$ genes form one cluster, we should consider the possibility that the C$_\gamma$-C$_\epsilon$-C$_\alpha$ region might be duplicated as a set.

In contrast to the mouse C$_H$ gene cluster, the human C$_H$ gene family contains at least three pseudogenes, one C$_\gamma$ (159) and two C$_\epsilon$ (7, 49, 100, 112, 164). The S region of the C$_\gamma$ pseudogene is replaced by completely different reiterated sequences, although the defect in the coding region is not known. The C$_{\epsilon 2}$ pseudogene has deleted the first two exons and the C$_{\epsilon 3}$ is a processed gene.

NUCLEOTIDE SEQUENCES OF CONSTANT-REGION GENES   All the mouse constant-region genes except for the C$_{\lambda 3}$ and C$_{\lambda 4}$ genes were sequenced. The C domain of the L chain is encoded by a single exon (4, 11, 15, 101, 170). As shown in Figure 5, the C$_H$ genes are composed of three (δ, α) or four ($\gamma 1$, $\gamma 2a$, $\gamma 2b$, $\gamma 3$, ϵ, and μ) exons, each encoding a functional and structural unit of the H chain, namely a domain or a hinge region (22, 68, 71, 78, 120, 161, 162, 172, 173; F. Blattner, unpublished data). As an exception, the hinge region of the C$_\alpha$ chain is coded for by the C$_{H2}$ exon (162). Nucleotide sequences of human C$_\kappa$ (62) and C$_\lambda$ (59, 63) genes were determined.

The presence of the membrane-bound form of the immunoglobulin was suggested by analyses of the surface immunoglobulin of the B lymphocyte (79, 119, 151). These results indicate that the membrane-bound form of the immunoglobulin is larger than the secreted form. Exons encoding the hydrophobic transmembrane segment and the short intracytoplasmic segment were found about 2 kb 3' to the exons encoding the last secreted domains of most mouse C$_H$ genes: C$_\mu$ (39, 134), C$_\gamma$ (133, 163, 174), C$_\epsilon$ (71), and C$_\delta$ (22). All the membrane domains are encoded by two exons. The hydrophobic segment of the 26 amino acid residues is highly conserved among all the H chains, suggesting that the membrane

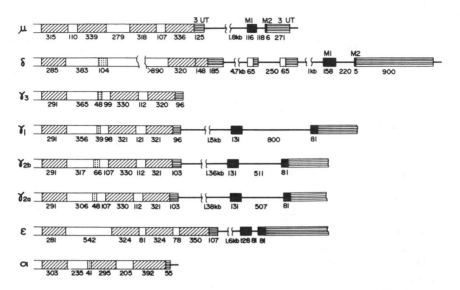

Figure 5 Structures of C_H genes. Oblique-lined, solid, dotted, and lined rectangles indicate domain exons, membrane exons, hinge exons, and 3' untranslated regions, respectively. Numbers indicate nucleotides.

form of the immunoglobulin is anchored by a common membrane protein (163, 174). The intracytoplasmic segments fall into two groups: one group ($\gamma1$, $\gamma2a$, $\gamma2b$, $\gamma3$, and ϵ) has a long segment of 27 residues; the other (μ and δ) has a short two-residue segment. If this portion has anything to do with transmission of the triggering signal of the antigen-antibody interaction, the quality of the signal could be different between two groups. Some of the human C_H genes were completely sequenced: $\gamma1$ (46), $\gamma2$ (45), $\gamma4$ (44), and ϵ (49, 100, 164).

S-S Recombination

DELETION MODEL   During differentiation of a single B lymphocyte, a given V_H gene is first expressed in combination with the C_μ gene of the same allelic chromosome, and later in the lineage of the B lymphocyte, the same V_H gene is expressed in combination with a different C_H gene. This phenomenon is called H-chain class switch.

Hybridization kinetic analyses in which cDNA is used have shown that specific C_H genes are deleted in mouse myelomas, depending on the C_H genes expressed (66). The order of the C_H genes, 5'-V_H gene family-spacer-C_μ-$C_{\gamma3}$-$C_{\gamma1}$-$C_{\gamma2b}$-$C_{\gamma2a}$-C_α-3' was consistent with the assumption that the DNA segment between a V_H and the C_H gene to be expressed is deleted, bringing these genes close to each other. Deletion of the intervening DNA segment during H-chain class switch was confirmed by Southern blot hybridization analyses of myeloma and hybridoma DNAs, in which cloned immunoglobulin genes were used (25, 29, 70, 125, 175, 176). Such studies also support the proposed order of C_H genes.

Comparison of germline and rearranged H-chain genes led to the proposal that a complete H-chain gene is formed by at least two types of recombinational events, as shown in Figure 6 (32, 75, 94, 158). The first type of recombination takes place between a given V_H segment, a D segment, and a J_H segment, completing a V-region sequence. After such recombination, referred to as V-D-J recombination, the V-region sequence is expressed as a μ chain because the C_μ is located closest to the J_H segment. The second type of recombination is called S-S recombination, which takes place between two S regions located at the 5' side of each C_H gene. The initial S-S recombination always involves the S_μ region as one of the pair, and the S region of another class is involved as the other partner, resulting in the switch of the expressed C_H gene from C_μ to another C_H gene without alteration of the V region sequence. The subsequent switch, such as one from C_γ to C_ϵ or C_α, will require another S-S recombination. Both types of rearrangements are accompanied by deletion of the intervening DNA segment from the chromosome. This model postulates several important structural features in the H-chain gene organization: (a) the order of C_H genes 5'-C_μ-C_δ-$C_{\gamma 3}$-$C_{\gamma 1}$-$C_{\gamma 2b}$-$C_{\gamma 2a}$-C_ϵ-C_α-3'; (b) the presence of only one set of the J_H region segment at the 5' side of the C_μ gene; and (c) the presence of the S region before each C_H gene. Studies on the organization of the C_H gene cluster provided direct evidence for the predictions (a) and (b), as described above.

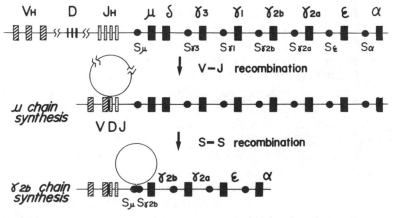

Figure 6   Two types of DNA rearrangements in H-chain gene locus. Taken from Honjo et al (65).

STRUCTURE OF S REGIONS   The S region was originally defined as functionally responsible for the class switch (S-S) recombination (75). Nucleotide sequence determination of the region surrounding the S-S recombination site showed that the structural basis of the S region is the tandem repetition of related nucleotide sequences. The nucleotide sequence of the S_μ region was shown to be represented by $[(GAGCT)_nGGGGT]_m$, where n varies between 1 and 7 (most frequently 3) and m is approximately 150 (109). The other S regions have also tandem repetition of unit sequences that are different from each other but that have in common with the S_μ region the short sequences GAGCT and GGGGT (37, 77, 110). Tandem repetitive sequences of the S region are often deleted partially in different strains of mice (65, 67, 96, 149) and during cloning (109). The S region was found 1–4 kb upstream of each coding region except for the C_δ gene. The length of S region, which varies from about 10 kb ($S_{\gamma1}$) to about 1 kb (S_ϵ), rather than the sequence homology with the S_μ region, is related to the amount of the immunoglobulin class in mouse sera (110). The S regions are generally conserved between human and mouse (112, 129, 158). The S_μ, S_ϵ, and S_α regions, especially, are highly homologous between the two species.

MECHANISM OF S-S RECOMBINATION   There are two different interpretations of the S-region structure. One group (33, 34) considers that the S-region sequences of different C_H genes, say S_μ and S_α, are so different that the homologous recombination is an unlikely mechanism for the class switch. Instead, they propose that class-specific "switching" proteins recognize the S_μ and S_α regions and bind to one another, thus juxtaposing the S_μ and S_α regions, and thereby allowing a recombinational event to join them. The other (77, 109) puts more weight on the short common sequences shared among all the S regions and considers that the recognition of the homologous sequences is the basic mechanism responsible for the class switch, although they do not exclude the presence of specific recombinases for class switching.

The nucleotide sequences surrounding the S-S recombination sites of hybridomas and myelomas have revealed the presence of TGGG or TGAG adjacent to the recombination site (110). Another relatively conserved sequence, YAGGTTG, was found in the vicinity of the recombination site and was dispersed in S regions (98). The S-region sequence contains many Chi sequences, promoters of the generalized recombination in λ phage (82), but the functional significance of such a sequence in the immunoglobulin gene is yet to be seen. Recent studies with the cloned S_μ and S_α regions have shown that preferential recombination can take place between the S_μ and S_α regions in *Escherichia coli* extracts (T. Kataoka et al., submitted for publication). Some cultured cell lines switch classes in the presence or absence of mutagens (42, 128, 155). The mechanism seems to involve recombination between homologous chromosomes, at least in one case (136).

Two alternative mechanisms could be proposed to explain the deletion of the intervening DNA segment associated with the class switch. The first model postulates that the DNA segment is looped out from a chromosome (looping-out model). In this case, the S-S recombination takes place within a single chromosome, which can occur at any stage of the cell cycle. The resultant daughter cells are identical, unless the looping-out occurs between the S and M phases of the cell cycle. The other model explains the deletion of the DNA segment by an unequal crossing-over event between sister chromatids [sister chromatid exchange model (115)]. This mechanism brings about asymmetric segregation, giving rise to one daughter cell with the deletion of C_H genes and to another with the duplication. Most compelling evidence for the sister chromatid exchange model is the presence of a short S_α sequence between the S_μ and $S_{\gamma 1}$ sequences upstream from the expressed γ_1 gene in MC101 myeloma (67, 115).

ABERRANT RECOMBINATION IN OR AROUND S REGIONS   Since the S region is located in the IVS, the S-S recombination does not have to be highly specific to the nucleotides that are joined together. Instead, the S region-mediated recombination is expected to be efficient because the class switch takes place within a short period of time after encountering an antigen. Therefore, not a single example of the aberrant S-S recombination results in crippling the function of the gene. However, a number of abortive S-S recombinations that take place on the unexpressed chromosome are found in myelomas and hybridomas (29, 125, 176). The C_H-gene deletion on the unexpressed chromosome appears to be a secondary reaction because an example in which only an expressed chromosome underwent S-S recombination was clearly demonstrated in BKC#15 myeloma derived from a BALB/c-C57BL F1 animal (175). The myeloma produces the $\gamma 2b$ chain of BALB/c allotype. Using restriction site polymorphism between two parental chromosomes, we have shown that the expressed BALB/c-derived chromosome has deleted the C_μ and $C_{\gamma 1}$ genes, whereas the C_H genes of the unexpressed C57BL-derived chromosome are retained.

IgG2b-producing MPC11 myeloma carries two rearranged $C_{\gamma 2b}$ genes, one of which is the expressed gene. The unexpressed chromosome of MPC11 myeloma underwent an abortive S-S recombination between the $S_{\gamma 3}$ and $S_{\gamma 2b}$ regions, resulting in the deletion of the $C_{\gamma 3}$ and $C_{\gamma 1}$ genes (91).

IgA-producing S107 myeloma also has two rearranged C_α genes, one of which corresponds to the α chain produced by the S107 cell, whereas the other is an aberrant recombinant in which a fragment of DNA, not obviously related to the immunoglobulin C_H gene, has been joined to the S_α

region (85). A similar DNA fragment, presumably unrelated to the immunoglobulin C_H locus, was found to be linked to the S_α regions of an unexpressed C_α gene of an IgA-producing myeloma J558 (57). Harris et al also found that the rearrangement of the DNA segment occurs in 15 out of 20 plasmacytomas tested, including $\gamma 3$, $\gamma 1$, $\gamma 2b$, $\gamma 2a$, and α producers. Most of them, however, do not involve S regions. Harris et al (57) suggested that this non-immunoglobulin-associated rearranging DNA may be related to the frequent translocation of chromosome 15 to chromosome 12 in plasmacytomas (86). Reciprocal translocation between chromosomes 8 and 14 in the Burkitt lymphoma cell line (Daudi) was shown to take place near the V_H cluster (47).

Role of RNA Splicing in Class Switching

It is known that many B lymphocytes carry more than one isotype on their surface. The most frequent double-bearers carry both μ and δ chains with the same V region on their surface. The C_μ and C_δ genes are not rearranged in $\mu^+\delta^+$ lymphoma cells, suggesting that the RNA-splicing mechanism rather than the S-S recombination may be responsible for the simultaneous synthesis of the μ and δ mRNA's (93, 108). The completed V_H gene may be transcribed as a large mRNA precursor containing both C_μ and C_δ sequences. Subsequent processing of the RNA would give rise to mRNA's coding for either μ or δ chain, both having the identical V_H sequence. A δ-producing lymphoma has deleted the C_μ gene (108), although there is no typical S region sequence in the 5' flanking region of the C_δ gene (110).

Yaoita et al (177) analyzed the C_H gene organization of the $\mu^+\epsilon^+$ B cells that were sorted from spleens of *Nippostrogylus*-stimulated SJA/9 mice and found that none of the C_H genes or their flanking regions had undergone DNA rearrangements. The results also suggest that the simultaneous expression of the C_μ and C_ϵ genes is mediated by an RNA-splicing mechanism. Yaoita et al proposed that class switching would proceed by two differentiation steps: the first step involves the activation of the differential splicing, and the second involves the S-S recombination, as shown in Figure 7. The second step is a likely place for T cells to affect B-cell switching, probably by selection. A γ_{2b} chain-producing variant of an Abelson virus-transformed cell line was also shown to retain the C_μ gene (3). These observations are consistent with the above model. None of the above studies, however, have provided the direct evidence for the primary transcript longer than 90 kb, which is presumably in scarce quantity.

Evolution of C_H Genes

IVS-MEDIATED DOMAIN TRANSFER   Extensive nucleotide sequence data allowed the tracing of the evolutionary history of immunoglobulin C_H genes. Miyata et al (106) compared nucleotide sequences of the

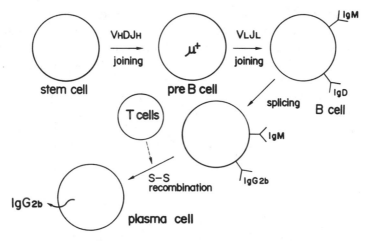

Figure 7 Order of DNA rearrangements during B-cell differentiation.

mouse $C_{\gamma 1}$ and $C_{\gamma 2b}$ genes and found an intriguing homology region containing the C_{H1} domain and the 5' half of the IVS between the C_{H1} domain and hinge region exons. Since they compared only the divergence at the synonymous position, this conservation of the nucleotide sequences is free from selective constraint at the protein level. They proposed that this homology region is not due to the selection pressure at the nucleotide sequence, but to the correction mechanism through recombination such as double unequal crossing over. They have predicted from the comparison of the amino acid sequences of the $\gamma 2b$ and $\gamma 2a$ chains that a similar mechanism must have operated between these two genes. Subsequently, the comparison of the nucleotide sequences of the $C_{\gamma 1}$, $C_{\gamma 2b}$, and $C_{\gamma 2a}$ genes revealed, as expected, the different homology regions depending on the compared pair (120, 141, 173). The homology regions consist of domains and the adjacent IVS. Similarly, the M1 exon and the adjacent intron seem to have undergone IVS-mediated domain transfer between the $C_{\gamma 1}$ and $C_{\gamma 2b}$ genes (174). There are two possible mechanisms for the recombinational correction: unequal crossing-over events, and gene conversion (6, 41, 152). Unfortunately, it is almost impossible to distinguish the two mechanisms experimentally in mammals.

Another example of dynamic exon shuffling has been obtained from structural analyses of human C_{γ} genes (159). The human $C_{\gamma 3}$ gene has four hinge exons, one of which is most homologous to that of the human C_{γ} pseudogene. The remaining three hinge exons are most homologous to the $C_{\gamma 1}$ gene. The results suggest that the human $C_{\gamma 3}$ gene was created by unequal crossing over between the $C_{\gamma 1}$ gene and the C_{γ} pseudogene, followed by the successive duplication of the $C_{\gamma 1}$-type hinge exon.

PSEUDOGENE   Unlike the mouse C_L and C_H genes, the human C gene families contain several pseudogenes. One C_λ pseudogene (63) and one C_ϵ pseudogene (7, 164) are processed genes that lack entire introns and moved to different chromosomes. The evolutionary origin of such genes is an intriguing problem. In view of the fact that the introns are removed precisely at the splicing site, the RNA intermediate seems to be a likely process. A retro-viral, long-terminal, repeat-like structure was found adjacent to the processed C_ϵ pseudogene, suggesting that retro-virus might have been involved in the creation of the processed gene (164).

DIVERGENCE OF EPSILON GENE   Comparison of nucleotide sequences of the human and mouse C_ϵ genes has shown that the epsilon gene is under the weakest selection pressure at the protein level among the immunoglobulin and related genes, although the divergence at the synonymous position is similar (71). The results may suggest that only a small portion of the epsilon chain is important, or that the epsilon gene is dispensable in accordance with the fact that, while IgE has only obscure roles in self-defense, it has an undesirable role as a mediator of hypersensitivity. Another explanation proposed is that the human and murine epsilon genes were derived from different ancestors duplicated a long time ago (paralogous).

CONSTANT-REGION GENES OF OTHER SPECIES   A rabbit γ chain cDNA clone was isolated from a cDNA library of hyperimmunized rabbit spleen mRNA (58, 99). There are at least two C_γ genes in the rabbit genome. A *Xenopus* $C\mu$ chain cDNA clone was isolated from a cDNA library of *Xenopus* spleen mRNA (17). The portion of the H-chain sequence contained in the plasmid (22-amino-acid residues) is 35% homologous to mammalian μ and γ sequences. The mRNA hybridized with this plasmid is 2.5 kb, in agreement with the size of mouse μ mRNA.

ACKNOWLEDGMENTS

I am grateful to Y. Nishida, T. Kataoka, and A. Shimizu for critical reading of the manuscript. I thank F. Oguni, S. Takeda, and T. Noma for their assistance in preparing this manuscript. This investigation was supported by a special grant-in-aid from the Ministry of Education, Science and Culture of Japan.

Literature Cited

1. Alt, F. W., Baltimore, D. 1982. Joining of immunoglobulin heavy chain gene segments: Implications from a chromosome with evidence of three D-J_H fusions. *Proc. Natl. Acad. Sci. USA* 79: 4118–22
2. Alt, F. W., Enea, V., Bothwell, A. L. M., Baltimore, D. 1980. Activity of multiple light chain genes in murine myeloma cells producing a single, functional light chain. *Cell* 21:1–12
3. Alt, F. W., Rosenberg, N., Casanova, R. J., Thomas, E., Baltimore, D. 1982. Immunoglobulin heavy-chain expression and class switching in a murine leukaemia cell line. *Nature* 296:325–30
4. Altenburger, W., Neumaier, P. S., Steinmetz, M., Zachau, H. G. 1981. DNA sequence of the constant gene region of the mouse immunoglobulin kappa chain. *Nucleic Acids Res.* 9: 971–81
5. Altenburger, W., Steinmetz, M., Zachau, H. G. 1980. Functional and nonfunctional joining in immunoglobulin light chain genes of a mouse myeloma. *Nature* 287:603–7
6. Baltimore, D. 1981. Gene conversion: Some implications for immunoglobulin genes. *Cell* 24:592–94
7. Battey, J., Max, E. E., McBride, W. O., Swan, D., Leder, P. 1982. A processed human immunoglobulin ε gene has moved to chromosome 9. *Proc. Natl. Acad. Sci. USA* 79:5956–60
8. Bentley, D. L., Rabbitts, T. H. 1981. Human V_κ immunoglobulin gene number: Implication for the origin of antibody diversity. *Cell* 24:613–23
9. Bentley, D. L., Rabbitts, T. H. 1980. Human immunoglobulin variable region genes-DNA sequences of two V_κ genes and a pseudogene. *Nature* 288: 730–33
10. Bernard, O., Gough, N. M., Adams, J. M. 1981. Plasmacytomas with more than one immunoglobulin κ mRNA: Implications for allelic exclusion. *Proc. Natl. Acad. Sci. USA* 78:5812–16
11. Bernard, O., Hozumi, N., Tonegawa, S. 1978. Sequences of mouse immunoglobulin light chain genes before and after somatic changes. *Cell* 15:1133–44
12. Bernstein, K. E., Reddy, E. P., Alexander, C. N., Mage, R. G. 1982. A cDNA sequence encoding a rabbit heavy chain variable region of the $V_H\alpha2$ allotype showing homologies with human heavy chain sequences. *Nature* 300:74–76
13. Black, J. A., Gibson, D. 1974. Neutral

evolution and immunoglobulin diversity. *Nature* 250:327–28
14. Blomberg, B., Tonegawa, S. 1982. DNA sequences of the joining regions of mouse λ light chain immunoglobulin genes. *Proc. Natl. Acad. Sci. USA* 79: 530–33
15. Bothwell, A. L. M., Paskind, M., Reth, M., Imanishi-Kari, T., Rajewsky, K., Baltimore, D. 1982. Somatic variants of murine immunoglobulin λ light chains. *Nature* 298:380–82
16. Bothwell, A. L. M., Paskind, M., Reth, M., Imanishi-Kari, T., Rajewsky, K., Baltimore, D. 1981. Heavy chain variable region contribution to the NP^b family of antibodies: Somatic mutation evident in a γ_{2a} variable region. *Cell* 24:625–37
17. Brown, R. D., Armentrout, R. W., Cochran, M. D., Cappello, C. J., Langemeier, S. O. 1981. Construction of recombinant plasmids containing *Xenopus* immunoglobulin heavy chain DNA sequences. *Proc. Natl. Acad. Sci. USA* 78:1755–59
18. Burnet, M. 1959. *The Clonal Selection Theory of Acquired Immunity.* England: Cambridge Univ.
19. Burstein, Y., Schechter, I. 1977. Amino acid sequence of the NH$_2$-terminal extra piece segments of the precursors of mouse immunoglobulin λ_1-type and type light chains. *Proc. Natl. Acad. Sci. USA* 74:716–20
20. Burstein, Y., Breiner, A. V., Brandt, C. R., Milcarek, C., Sweet, R. W., Warszawski, D., Ziv, E., Schechter, I. 1982. Recent duplication and germ-line diversification of rat immunoglobulin κ chain gene joining segments. *Proc. Natl. Acad. Sci. USA* 79:5993–97
21. Cattaneo, R., Gorski, J., Mach, B. 1981. Cloning of multiple copies of immunoglobulin variable kappa genes in cosmid vectors. *Nucleic Acids Res.* 9:2777–90
22. Cheng, H-L., Blattner, F. R., Fitzmaurice, L., Mushinski, J. F., Tucker, P. W. 1982. Structure of genes for membrane and secreted murine IgD heavy chains. *Nature* 296:410–15
23. Cohen, J. B., Effron, K., Rechavi, G., Ben-Neriah, Y., Zakut, R., Givol, D. 1982. Simple DNA sequences in homologous flanking regions near immunoglobulin V_H genes: a role in gene interaction? *Nucleic Acids Res.* 10: 3353–70
24. Cohn, M., Blomberg, B., Geckeler, W., Raschke, W., Riblet, R., Weigert, M. 1974. In *First Order Considerations in*

Analyzing the Generator of Diversity in the Immune System, Genes, Receptors, Signals, ed. E. E. Serzarz, A. R. Williamson, C. F. Fox, pp. 89–117. New York: Academic

25. Coleclough, C., Cooper, D., Perry, R. P. 1980. Rearrangement of immunoglobulin heavy chain genes during B-lymphocyte development as revealed by studies of mouse plasmacytoma cells. *Proc. Natl. Acad. Sci. USA* 77:1422–26

26. Coleclough, C., Perry, R. P., Karjalainen, K., Weigert, M. 1981. Aberrant rearrangements contribute significantly to the allelic exclusion of immunoglobulin gene expression. *Nature* 290:372–78

27. Cory, S., Adams, J. M. 1980. Deletions are associated with somatic rearrangement of immunoglobulin heavy chain genes. *Cell* 19:37–51

28. Cory, S., Adams, J. M., Kemp, D. J. 1980. Somatic rearrangements forming active immunoglobulin μ genes in B and T lymphoid cell lines. *Proc. Natl. Acad. Sci. USA* 77:4943–47

29. Cory, S., Jackson, J., Adams, J. M. 1980. Deletions in the constant region locus can account for switches in immunoglobulin heavy chain expression. *Nature* 285:450–56

30. Cory, S., Tyler, B. M., Adams, J. M. 1981. Sets of immunoglobulin V_κ genes homologous to ten cloned V_κ sequences: Implications for the number of germline V_κ genes. *J. Mol. Applied Genet.* 1: 103–16

31. Crews, S., Griffin, J., Huang, H., Calame, K., Hood, L. 1981. A single V_H gene segment encodes the immune response to phosphorylcholine: Somatic mutation is correlated with the class of the antibody. *Cell* 25:59–66

32. Davis, M. M., Calame, K., Early, P. W., Livant, D. L., Joho, R., Weissman, I. L., Hood, L. 1980. An immunoglobulin heavy-chain gene is formed by at least two recombinational events. *Nature* 283:733–39

33. Davis, M. M., Kim, S. K., Hood, L. 1980. Immunoglobulin class switching: Developmentally regulated DNA rearrangements during differentiation. *Cell* 22:1–2

34. Davis, M. M., Kim, S. K., Hood, L. E. 1980. DNA sequences mediating class switching in α-immunoglobulins. *Science* 209:1360–65

35. D'eustachio, P., Pravtcheva, D., Marcu, K., Ruddle, F. H. 1980. Chromosomal location of the structural gene cluster encoding murine immuno-

globulin heavy chains. *J. Exp. Med.* 151:1545–50

36. D'eustachio, P., Bothwell, A. L. M., Takaro, T. K., Baltimore, D., Ruddle, F. H. 1981. Chromosomal location of structural genes encoding murine immunoglobulin λ light chains. *J. Exp. Med.* 153:793–800

37. Dunnick, W., Rabbitts, T. H., Milstein, C. 1980. An immunoglobulin deletion mutant with implications for the heavy-chain switch and RNA splicing. *Nature* 286:669–75

38. Early, P., Huang, H., Davis, M., Calame, K., Hood, L. 1980. An immunoglobulin heavy chain variable region gene is generated from three segments of DNA: C_H, D and J_H. *Cell* 19:981–92

39. Early, P., Rogers, J., Davis, M., Calame, K., Bond, M., Wall, R., Hood, L. 1980. Two mRNAs can be produced from a single immunoglobulin μ gene by alternative RNA processing pathways. *Cell* 20:313–19

40. Early, P., Hood, L. 1981. Allelic exclusion and nonproductive immunoglobulin gene rearrangements. *Cell* 24:1–3

41. Egel, R. 1981. Intergenic conversion and reiterated genes. *Nature* 290: 191–92

42. Eckhardt, L. A., Tilley, S. A., Lang, R. B., Marcu, K. B., Birshtein, B. K. 1982. DNA rearrangements in MPC-11 immunoglobulin heavy chain class-switch variants. *Proc. Natl. Acad. Sci. USA* 79:3006–10

43. Elliott, B. W. Jr., Eisen, H. N., Steiner, L. A. 1982. Unusual association of V, J and C regions in a mouse immunoglobulin λ chain. *Nature* 299:559–61

44. Ellison, J., Buxbaum, J., Hood, L. 1981. Nucleotide sequence of a human immunoglobulin $C_{\gamma 4}$ gene. *DNA* 1:11–18

45. Ellison, J., Hood, L. 1982. Linkage and sequence homology of two human immunoglobulin γ heavy chain constant region genes. *Proc. Natl. Acad. Sci. USA* 79:1984–88

46. Ellison, J., Berson, B. J., Hood, L. E. 1982. The nucleotide sequence of a human immunoglobulin $C_{\gamma 1}$ gene. *Nucleic Acids Res.* 10:4071–79

47. Erikson, J., Finan, J., Nowell, P. C., Croce, C. M. 1982. Translocation of immunoglobulin V_H genes in Burkitt lymphoma. *Proc. Natl. Acad. Sci. USA* 79: 5611–15

48. Erikson, J., Martinis, J., Croce, C. M. 1981. Assignment of the genes for human λ immunoglobulin chains to chromosome 22. *Nature* 294:173–75

49. Flanagan, J. G., Rabbitts, T. H. 1982. The sequence of a human immunoglobulin epsilon heavy chain constant region gene, and evidence for three nonallelic genes. *EMBO J.* 1:655–60

50. Forster, A., Hobart, M., Hengartner, H., Rabbitts, T. H. 1980. An immunoglobulin heavy-chain gene is altered in two T-cell clones. *Nature* 286:897–99

51. Gally, J. A., Edelman, G. M. 1972. The genetic control of immunoglobulin synthesis. *Ann. Rev. Genetics* 6:1–45

52. Gearhart, P. J. 1982. Generation of immunoglobulin variable gene diversity. *Immunol. Today* 3:107–12

53. Gearhart, P. J., Johnson, N. D., Douglas, R., Hood, L. 1981. IgG antibodies to phosphorylcholine exhibit more diversity than their IgM counterparts. *Nature* 291:29–34

54. Gershenfeld, H. K., Tsukamoto, A., Weissman, I. L., Joho, R. 1981. Somatic diversification is required to generate the V_κ genes of MOPC 511 and MOPC 167 myeloma proteins. *Proc. Natl. Acad. Sci. USA* 78:7674–78

55. Givol, D., Zakut, R., Effron, K., Rechavi, G., Ram, D., Cohen, J. B. 1981. Diversity of germ-line immunoglobulin V_H genes. *Nature* 292:426–30

56. Gottlibe, P. D. 1980. Immunoglobulin genes. *Mol. Immunol.* 17:1423–35

57. Harris, L. J., Lang, R. B., Marcu, K. B. 1982. Non-immunoglobulin-associated DNA rearrangements in mouse plasmacytomas. *Proc. Natl. Acad. Sci. USA* 79:4175–79

58. Heidmann, O., Rougeon, F. 1982. Molecular cloning of rabbit γ heavy chain mRNA. *Nucleic Acids Res.* 10:1535–45

59. Hieter, P. A., Hollis, G. F., Korsmeyer, S. J., Waldmann, T. A., Leder, P. 1981. Clustered arrangement of immunoglobulin λ constant region genes in man. *Nature* 294:536–40

60. Hieter, P. A., Korsmeyer, S. J., Waldmann, T. A., Leder, P. 1981. Human immunoglobulin κ light-chain genes are deleted or rearranged in λ-producing B cells. *Nature* 290:368–72

61. Hieter, P. A., Maizel, J. V., Leder, P. 1982. Evolution of human immunoglobulin κ J region genes. *J. Biol. Chem.* 257:1516–22

62. Hieter, P. A., Max, E. E., Seidman, J. G., Maizel, J. V. Jr., Leder, P. 1980. Cloned human and mouse kappa immunoglobulin constant and J region genes conserve homology in functional segments. *Cell* 22:197–207

63. Hollis, G. F., Hieter, P. A., Mcbride, O. W., Swan, D., Leder, P. 1982. Processed genes: a dispersed human immunoglobulin gene bearing evidence of RNA-type processing. *Nature* 296: 321–25

64. Honjo, T. 1982. The molecular mechanisms of the immunoglobulin class switch. *Immunol. Today* 3:214–17

65. Honjo, T., Nakai, S., Nishida, Y., Kataoka, T., Yamawaki-Kataoka, Y., et al. 1981. Rearrangements of immunoglobulin genes during differentiation and evolution. *Immunol. Rev.* 59: 33–68

66. Honjo, T., Kataoka, T. 1978. Organization of immunoglobulin heavy chain genes and allelic deletion model. *Proc. Natl. Acad. Sci. USA* 75:2140–44

67. Honjo, T., Kataoka, T., Yoita, Y., Shimizu, A., Takahashi, N., et al. 1981. Organization and reorganization of immunoglobulin heavy chain genes. *Cold Spring Harbor Symp. Quant. Biol.* 45: 913–23

68. Honjo, T., Obata, M., Yamawaki-Kataoka, Y., Kataoka, T., Takahashi, N., Mano, Y. 1979. Cloning and complete nucleotide sequence of mouse immunoglobulin $\gamma 1$ chain gene. *Cell* 18:559–68

69. Hozumi, N., Wu, G. E., Murialdo, H., Roberts, L., Vetter, D., et al. 1981. RNA splicing mutation in an aberrantly rearranged immunoglobulin λI gene. *Proc. Natl. Acad. Sci. USA* 78:7019–23

70. Hurwitz, J., Coleclough, C., Cebra, J. J. 1980. C_H gene rearrangements in IgM-bearing B cells and in the normal splenic DNA component of hybridomas making different isotypes of antibody. *Cell* 22:349–59

71. Ishida, N., Ueda, S., Hayashida, H., Miyata, T., Honjo, T. 1982. The nucleotide sequence of the mouse immunoglobulin epsilon gene: comparison with the human epsilon gene sequence. *EMBO J.* 1:1117–23

72. Joho, R., Weissman, I. L. 1980. V-J joining of immunoglobulin κ genes only occurs on one homologous chromosome. *Nature* 284:179–81

73. Kabat, E. A., Wu, T. T., Bilofsky, H. 1978. Variable region genes for the immunoglobulin framework are assembled from small segments of DNA-A hypothesis. *Proc. Natl. Acad. Sci. USA* 75:2429–33

74. Kabat, E. A., Wu, T. T., Bilofsky 1979. *Sequences of Immunoglobulin Chains.* U.S. Department of Health, Education

524   HONJO

and Welfare, NIH Publication No. 80:2008

75. Kataoka, T., Kawakami, T., Takahashi, N., Honjo, T. 1980. Rearrangement of immunoglobulin γ1-chain gene and mechanism for heavy-chain class switch. *Proc. Natl. Acad. Sci. USA* 77: 919–23

76. Kataoka, T., Nikaido, T., Miyata, T., Moriwaki, K., Honjo, T. 1982. The nucleotide sequences of rearranged and germline immunoglobulin V_H genes of a mouse myeloma MC101 and evolution of V_H genes in mouse. *J. Biol. Chem.* 10:277–85

77. Kataoka, T., Miyata, T., Honjo, T. 1981. Repetitive sequences in class-switch recombination regions of immunoglobulin heavy chain genes. *Cell* 23:357–68

78. Kawakami, T., Takahashi, N., Honjo, T. 1980. Complete nucleotide sequence of mouse immunoglobulin μ gene and comparison with other immunoglobulin heavy chain genes. *Nucleic Acids Res.* 8:3933–45

79. Kehry, M., Ewald, S., Douglas, R., Sibley, C., Raschke, W., Fambrough, D., Hood, L. 1980. The immunoglobulin μ chains of membrane-bound and secreted IgM molecules differ in their C-terminal segments. *Cell* 21:393–406

80. Kemp, D. J., Cory, S., Adams, J. M. 1979. Cloned pairs of variable region genes for immunoglobulin heavy chains isolated from a clone library of the entire mouse genome. *Proc. Natl. Acad. Sci. USA* 76:4627–31

81. Kemp, D. J., Tyler, B., Bernard, O., Gough, N., Gerondakis, S., et al. 1981. *J. Mol. Applied Genet.* 1:245–61

82. Kenter, A. L., Birshtein, B. K. 1980. Chi, a promotor of generalized recombination in λ phage, is present in immunoglobulin genes. *Nature* 293:402–4

83. Kim, S., Davis, M., Sinn, E., Patten, P., Hood, L. 1981. Antibody diversity: somatic hypermutation of rearranged V_H genes. *Cell* 27:573–81

84. Kirsch, I. R., Morton, C. C., Nakahara, K., Leder, P. 1982. Human immunoglobulin heavy chain genes map to a region of translocations in malignant B lymphocytes. *Science* 216:301–3

85. Kirsch, I. R., Ravetch, J. V., Kwan, S.-P., Max, E. E., Ney, R. L., Leder, P. 1981. Multiple immunoglobulin switch region homologies outside the heavy chain constant region locus. *Nature* 293:585–87

86. Klein, G. 1981. The role of gene dosage and genetic transpositions in carcinogenesis. *Nature* 294:313–38

87. Krawinkel, U., Rabbitts, T. H. 1982. Comparison of the hinge-coding segments in human immunoglobulin gamma heavy chain genes and the linkage of the gamma 2 and gamma 4 subclass genes. *EMBO J.* 1:403–7

88. Kurosawa, Y., Tonegawa, S. 1982. Organization, structure, and assembly of immunoglobulin heavy chain diversity DNA segments. *J. Exp. Med.* 155: 201–18

89. Kurosawa, Y., von Boehmer, H., Haas, W., Sakano, H., Trauneker, A., Tonegawa, S. 1981. Identification of D segments of immunoglobulin heavy-chain genes and their rearrangement in T lymphocytes. *Nature* 290:565–70

90. Kwan, S-P., Max, E. E., Seidman, J. G., Leder, P., Scharff, M. D. 1981. Two kappa immunoglobulin genes are expressed in the myeloma S107. *Cell* 26: 57–66

91. Lang, R. B., Stanton, L. W., Marcu, K. B. 1981. On immunoglobulin heavy chain gene switching: Two γ2b genes are rearranged via switch sequences in MPC-11 cells but only one is expressed. *Nucleic Acids Res.* 10:611–30

92. Lewis, S., Rosenberg, N., Alt, F., Baltimore, D. 1982. Continuing kappa-gene rearrangement in a cell line transformed by abelsoin murine leukemia virus. *Cell* 30:807–16

93. Maki, R., Roeder, W., Traunecker, A., Sidman, C., Wabl, M., et al. 1981. The role of DNA rearrangement and alternative RNA processing in the expression of immunoglobulin delta genes. *Cell* 24:353–65

94. Maki, R., Traunecker, A., Sakano, H., Roeder, W., Tonegawa, S. 1980. Exon shuffling generates an immunoglobulin heavy chain gene. *Proc. Natl. Acad. Sci. USA* 77:2138–42

95. Malcolm, S., Barton, P., Murphy, C., Ferguson-Smith, M. A., Bentley, D. L., Rabbitts, T. H. 1982. Localization of human immunoglobulin κ light chain variable region genes to the short arm of chromosome 2 by *in situ* hybridization. *Proc. Natl. Acad. Sci. USA* 79:4957–61

96. Marcu, K. B., Banerji, J., Penncavage, N. A., Lang, R., Arnheim, N. 1980. 5' Flanking region of immunoglobulin heavy chain constant region genes displays length heterogeneity in germlines of inbred mouse strains. *Cell* 22:187–96

97. Marcu, K. B. 1982. Immunoglobulin heavy-chain constant-region genes. *Cell* 29:719–21

98. Marcu, K. B., Lang, R. B., Stanton, W. L., Harris, L. J. 1982. A model for the molecular requirements of immunoglobulin heavy chain class switching. *Nature* 298:87–89

99. Martens, C. L., Moore, K. W., Steinmetz, M., Hood, L., Knight, K. L. 1982. Heavy chain genes of rabbit IgG: Isolation of a cDNA encoding γ heavy chain and identification of two genomic C_γ genes. *Proc. Natl. Acad. Sci. USA* 79:6018–22

100. Max, E. E., Battey, J., Ney, R., Kirsch, I. R., Leder, P. 1982. Duplication and deletion in the human immunoglobulin ε genes. *Cell* 29:691–99

101. Max, E. E., Maizel, J. V. Jr., Leder, P. 1981. The nucleotide sequence of a 5.5-kilobase DNA segment containing the mouse κ immunoglobulin J and C region genes. *J. Biol. Chem.* 256:5116–20

102. Max, E. E., Seidman, J. G., Leder, P. 1979. Sequences of five potential recombination sites encoded close to an immunoglobulin κ constant region gene. *Proc. Natl. Acad. Sci. USA* 76:3450–54

103. Max, E. E., Seidman, J. G., Miller, H., Leder, P. 1980. Variation in the crossover point of kappa immunoglobulin gene V-J recombination: Evidence from a cryptic gene. *Cell* 21:793–99

104. Mcbride, O. W., Hieter, P. A., Hollis, G. F., Swan, D., Otey, M. C., Leder, P. 1982. Chromosomal location of human kappa and lambda immunoglobulin light chain constant region genes. *J. Exp. Med.* 155:1480–90

105. Miller, J., Selsing, E., Storb, U. 1982. Structural alterations in J regions of mouse immunoglobulin λ genes are associated with differential gene expression. *Nature* 295:428–30

106. Miyata, T., Yasunaga, T., Yamawaki-Kataoka, Y., Obata, M., Honjo, T. 1980. Nucleotide sequence divergence of mouse immunoglobulin γ1 and γ2b chain genes and the hypothesis of intervening sequence-mediated domain transfer. *Proc. Natl. Acad. Sci. USA* 77:2143–47

107. Miyata, T., Yasunaga, T., Nishida, T. 1980. Nucleotide sequence divergence and functional constraint in mRNA evolution. *Proc. Natl. Acad. Sci. USA* 77:7328–32

108. Moore, K. W., Rogers, J., Hunkapiller, T., Early, P., Nottenburg, C., et al. 1981. Expression of IgD may use both DNA rearrangement and RNA splicing mechanisms. *Proc. Natl. Acad. Sci. USA* 78:1800–4

109. Nikaido, T., Nakai, S., Honjo, T. 1981. The switch (S) region of the immunoglobulin C_μ gene is composed of simple tandem repetitive sequences. *Nature* 292:845–48

110. Nikaido, T., Yamawaki-Kataoka, Y., Honjo, T. 1982. Nucleotide sequences of switch regions of immunoglobulin C_ϵ and C_γ genes and their comparison. *J. Biol. Chem.* 257:7322–29

111. Nishida, Y., Hisajima, H., Ueda, S., Takahashi, N., Honjo, T. 1982. In *Organization and Structure of Human immunoglobulin Genes: Proc. Takeda Sci. Found. Symp.*, ed. Y. Yamamura. In press. New York: Academic

112. Nishida, Y., Miki, T., Hisajima, H., Honjo, T. 1982. Cloning of human immunoglobulin ε chain genes: Evidence for multiple C_ϵ genes. *Proc. Natl. Acad. Sci. USA* 79:3833–37

113. Nishioka, Y., Leder, P. 1980. Organization and complete sequence of identical embryonic and plasmacytoma κ V-region genes. *J. Biol. Chem.* 255:3691–94

114. Nottenburg, C., Weissman, I. L. 1981. C_μ gene rearrangement of mouse immunoglobulin genes in normal B cells occurs on both the expressed and nonexpressed chromosomes. *Proc. Natl. Acad. Sci. USA* 78:484–88

115. Obata, M., Kataoka, T., Nakai, S., Yamagishi, H., Takahashi, N., et al. 1981. Structure of a rearranged γ1 chain gene and its implication to immunoglobulin class-switch mechanism. *Proc. Natl. Acad. Sci. USA* 78:2437–41

116. Ohno, S., Epplen, J. T., Matsunaga, T., Hozumi, T. 1981. The curse of Prometheus is laid upon the immune system. *Prog. Allergy* 28:8–39

117. Ohta, T. 1981. In *Evolution and Variation of Multigene Families,* ed S. Lewin. Berlin: Springer

118. Ohta, T. 1982. Allelic and nonallelic homology of a supergene family. *Proc. Natl. Acad. Sci. USA* 79:3251–54

119. Oi, V. T., Bryan, V. M., Herzenberg, L. A. 1980. Lymphocyte membrane IgG and secreted IgG are structurally and allotypically distinct. *J. Exp. Med.* 151:1260–74

120. Ollo, R., Auffray, C., Morchamps, C., Rougeon, F. 1981. Comparison of mouse immunoglobulin γ2a and γ2b chain genes suggests that exons can be exchanged between genes in a multigenic family. *Proc. Natl. Acad. Sci. USA* 78:2442–46

121. Ono, M., Kawakami, M., Kataoka, T., Honjo, T. 1977. Existence of both

kappa and lambda light chain messenger RNA sequences in mouse myeloma, MOPC-104E, known as a lambda chain producer. *Biochem. Biophys. Res. Commun.* 74:796–802

122. Parnes, J. R., Seidman, J. G. 1982. Structure of wild-type and mutant mouse β_2-microglobulin genes. *Cell* 29: 661–69

123. Pech, M., Hochtl, J., Schnell, H., Zachau, H. G. 1981. Differences between germline and rearranged immunoglobulin V_κ coding sequences suggest a localized mutation mechanism. *Nature* 291:668–70

124. Perry, R., Kelley, D., Coleclough, C., Seidman, J., Leder, P., et al. 1980. Transcription of mouse κ chain genes: implications for allelic exclusion. *Proc. Natl. Acad. Sci. USA* 77:1937–41

125. Rabbitts, J. H., Forster, A., Dunnick, W., Bentley, D. L. 1980. The role of gene delection in the immunoglobulin heavy chain switch. *Nature* 283:351–56

126. Rabbitts, T. H., Bentley, D. L., Milstein, C. P. 1981. Human antibody genes: V gene variability and C_H gene switching strategies. *Immunol. Rev.* 59:69–91

127. Rabbitts, T. H., Forster, A., Milstein, C. P. 1981. Human immunoglobulin heavy chain genes: evolutionary comparisons of C_μ, C_δ and C_γ genes and associated switch sequences. *Nucleic Acids Res.* 9:4509–24

128. Radbruck, A., Liesegang, B., Rajewsky, K. 1980. Isolation of variants of mouse myeloma X63 that express changed immunoglobulin class. *Proc. Natl. Acad. Sci. USA* 77:2909–13

129. Ravetch, J. V., Kirsch, I. R., Leder, P. 1980. Evolutionary approach to the question of immunoglobulin heavy chain switching: Evidence from cloned human and mouse genes. *Proc. Natl. Acad. Sci. USA* 77:6734–38

130. Ravetch, J. V., Siebenlist, U., Korsmeyer, S., Waidmann, T., Leder, P. 1982. Structure of the human immunoglobulin μ locus: Characterization of embryonic and rearranged J and D genes. *Cell* 27:583–91

131. Rechavi, G., Bienz, B., Ram, D., Ben-Neriah, Y., Cohen, J. B., et al. 1982. Organization and evolution of immunoglobulin V_H gene subgroups. *Proc. Natl. Acad. Sci. USA* 79:4405–9

132. Robertson, M. 1981. Genes of lymphocytes I: Diverse means to antibody diversity. *Nature* 290:625–27

133. Rogers, J., Choi, E., Souza, L., Carter, C., Word, C., et al. 1981. Gene segments encoding transmembrane carboxyl termini of immunoglobulin γ chains. *Cell* 26:19–27

134. Rogers, J., Early, P., Carter, C., Calame, K., Bond, M., et al. 1980. Two mRNAs with different 3' ends encode membrane-bound and secreted forms of immunoglobulin μ chain. *Cell* 20:303–12

135. Rose, S. M., Kuehl, W. M., Smith, G. P. 1977. Cloned MPC 11 myeloma cells express two kappa genes: a gene for a complete light chain and a gene for a constant region polypeptide. *Cell* 12:453–62

136. Sablitzky, F., Radbruch, A., Rajewsky, K. 1982. Spontaneous immunoglobulin class switching in myeloma and hybridoma cell lines differs from physiological class switching. *Immunol. Rev.* 67:59–72

137. Sakano, H., Huppi, K., Heinrich, G., Tonegawa, S. 1979. Sequences at the somatic recombination sites of immunoglobulin light-chain genes. *Nature* 280:288–94

138. Sakano, H., Kurosawa, Y., Weigert, M., Tonegawa, S. 1981. Identification and nucleotide sequence of a diversity DNA segment (D) of immunoglobulin heavy-chain genes. *Nature* 290:562–65

139. Sakano, H., Maki, R., Kurosawa, Y., Roeder, W., Tonegawa, S. 1980. Two types of somatic recombination are necessary for the generation of complete immunoglobulin heavy-chain genes. *Nature* 286:676–83

140. Schnell, H., Steinmetz, M., Zachau, H. G. 1980. An unusual translocation of immunoglobulin gene segments in variants of the mouse myeloma MPC11. *Nature* 286:170–73

141. Schreier, P. H., Bothwell, A. L. M., Mueller-Hill, B., Baltimore, D. 1981. Multiple differences between the nucleic acid sequences of the IgG2a^a and IgG2a^b alleles of the mouse. *Proc. Natl. Acad. Sci. USA* 78:4495–99

142. Seidman, J. G., Leder, A., Nau, M., Norman, B., Leder, P. 1978. Antibody diversity. *Science* 202:11–17

143. Seidman, J. G., Leder, P. 1980. A mutant immunoglobulin light chain is formed by aberrant DNA- and RNA-splicing events. *Nature* 286:779–83

144. Seidman, J. G., Max, E. E., Leder, P. 1979. A κ-immunoglobulin gene is formed by site-specific recombination without further somatic mutation. *Nature* 280:370–75

145. Seidman, J. G., Nau, M. M., Norman, B., Kwan, S. P., Scharff, M. and Leder,

P. 1980. Immunoglobulin V/J recombination is accompanied by deletion of joining site and variable region segments. *Proc. Natl. Acad. Sci. USA* 77: 6022–26

146. Selsing, E., Storb, U. 1981. Somatic mutation of immunoglobulin light-chain variable-region genes. *Cell* 25:47–28

147. Sheppard, J. W., Gutman, G. A. 1982. Rat kappa-chain J-segment genes: Two recent gene duplications separate rat and mouse. *Cell* 29:121–27

148. Shimizu, A., Takahashi, N., Yamawaki-Kataoka, Y., Nishida, Y., Kataoka, T., Honjo, T. 1981. Ordering of mouse immunoglobulin heavy chain genes by molecular cloning. *Nature* 289:49–53

149. Shimizu, A., Takahashi, N., Yaoita, Y., Honjo, T. 1982. Organization of constant region gene family of the mouse immunoglobulin heavy chain. *Cell* 28: 499–506

150. Siebenlist, U., Ravetch, J. V., Korsmeyer, S., Waldmann, T., Leder, P. 1981. Human immunoglobulin D segments encoded in tandem multigenic families. *Nature* 294:632–35

151. Singer, P. A., Singer, H. H., Williamson, A. R. 1980. Different species of messenger RNA encode receptor and secretory IgM μ chains differing at their carboxy termini. *Nature* 285:294–300

152. Slighton, J. L., Blechl, A. E., Smithies, O. 1980. Human fetal ᴳγ- and ᴬγ-globin genes: complete nucleotide sequences suggest that DNA can be exchanged between these duplicated genes. *Cell* 21:627–38

153. Smith, G. P. 1974. Unequal crossover and the evolution of multigene families. *Cold Spring Harbor Symp. Quant Biol.* 38:507–13

154. Smith, G. P., Hood, L., Fitch, W. M. 1971. Antibody diversity. *Ann. Rev. Biochem.* 40:969–1012

155. Stavnezer, J., Marcu, K. B., Sirlin, S., Alhadeff, B., Hammerling, V. 1982. Rearrangements and deletions of immunoglobulin heavy chain genes in the double-producing B cell lymphoma I29. *Mol. Cell. Biol.* 2:1002–13

156. Steinmetz, M., Altenburger, W., Zachau, H. G. 1980. A rearranged DNA sequence possibly related to the translocation of immunoglobulin gene segments. *Nucleic Acids Res.* 8:1709–20

157. Swan, D., D'eustachio, P., Leinwand, L., Seidman, J., Keithley, D., Ruddle, F. H. 1979. Chromosomal assignment of the mouse κ light chain genes. *Proc. Natl. Acad. Sci. USA* 76:2735–39

158. Takahashi, N., Kataoka, T., Honjo, T. 1980. Nucleotide sequences of class-switch recombination region of the mouse immunoglobulin γ2b-chain gene. *Gene* 11:117–27

159. Takahashi, N., Ueda, S., Obata, M., Nikaido, T., Nakai, S., Honjo, T. 1982. Structure of human immunoglobulin gamma genes: Implications for evolution of a gene family. *Cell* 29:671–79

160. Tonegawa, S., Maxam, A. M., Tizard, R., Bernard, O., Gilbert, W. 1978. Sequence of a mouse germ-line gene for a variable region of an immunoglobulin light chain. *Proc. Natl. Acad. Sci. USA* 75:1485–89

161. Tucker, P. W., Marcu, K., Newell, N., Richards, J., Blattner, F. R. 1979. Sequence of the cloned gene for the constant region of murine γ₂ᵦ immunoglobulin heavy chain. *Science* 206: 1303–6

162. Tucker, P. W., Slighton, J. L., Blattner, F. R. 1981. Mouse IgA heavy chain gene sequence: Implications for evolution of immunolobulin hinge exons. *Proc. Natl. Acad. Sci. USA* 78:7684–88

163. Tyler, B. M., Cowman, A. F., Gerondakis, S. D., Adams, J. M., Bernard, O. 1982. mRNA for surface immunoglobulin γ chains encodes a highly conserved transmembrane sequence and a 28-residue intracellular domain. *Proc. Natl. Acad. Sci. USA* 79:2008–12

164. Ueda, S., Nakai, S., Nishida, Y., Hisajima, H., Honjo, T. 1982. Long terminal repeat-like elements flank a human immunoglobulin epsilon pseudogene that lacks introns. *EMBO J.* 1:1539–44

165. Van Ness, B. G., Coleclough, C., Perry, R. P., Weigert, M. 1982. DNA between variable and joining gene segments of immunoglobulin κ light chain is frequently retained in cells that rearrange the κ locus. *Proc. Natl. Acad. Sci. USA* 79:262–66

166. Wabl, M., Steinberg, C., 1982. A theory of allelic and isotypic exclusion for immunoglobulin genes. *Proc. Natl. Acad. Sci. USA.* 79:6976–78

167. Walfield, A., Selsing, E., Arp, B., Storb, U. 1981. Misalignment of V and J gene segments resulting in a nonfunctional immunoglobulin gene. *Nucleic Acids Res.* 9:1101–9

168. Wang, J. C. 1979. Helical repeat of DNA in solution. *Proc. Natl. Acad. Sci. USA* 76:200–3

169. Weigert, M., Riblet, R. 1976. Genetic control of antibody variable regions.

528   HONJO

Cold Spring Harbor Symp. Quant Biol. 41:837–46

170. Wu, G. E., Govindji, N., Hozumi, N., Murialdo, H. 1982. Nucleotide sequence of a chromosomal rearranged λ_2 immunoglobulin gene of mouse. *Nucleic Acids Res.* 19:3831–43

171. Wu, T. T., Kabat, E. A. 1982. Fourteen nucleotides in the second complementarity-determining region of a human heavy-chain variable region gene are identical with a sequence in a human D minigene. *Proc. Natl. Acad. Sci. USA* 79:5031–32

172. Yamawaki-Kataoka, Y., Kataoka, T., Takahashi, N., Obata, M., Honjo, T. 1980. Complete nucleotide sequence of immunoglobulin γ2b chain gene cloned from mouse DNA. *Nature* 283:786–89

173. Yamawaki-Kataoka, Y., Miyata, T., Honjo, T. 1981. The complete nucleotide sequence of mouse immunoglobulin γ2a gene and evolution of heavy chain genes: further evidence for intervening sequence-mediated domain transfer. *Nucleic Acids. Res.* 9:1365–81

174. Yamawaki-Kataoka, Y., Nakai, S., Miyata, T., Honjo, T. 1982. Nucleotide sequences of gene segments encoding membrane domains of immunoglobulin γ chains. *Proc. Natl. Acad. Sci. USA* 79:2623–27

175. Yaoita, Y., Honjo, T. 1980. Deletion of immunoglobulin heavy chain genes from expressed allelic chromosome. *Nature* 286:850–53

176. Yaoita, Y., Honjo, T. 1980. Deletion of immunoglobulin heavy chain genes accompanies the class switch rearrangement. *Biomed. Res.* 1:164–75

177. Yaoita, Y., Kumagai, Y., Okumura, K., Honjo, T. 1982. Expression of lymphocyte surface IgE does not require switch recombination. *Nature* 297:697–99

178. Zeelon, E. P., Bothwell, A. L. M., Kantor, F., Schechter, I. 1981. An experimental approach to enumerate the genes coding for immunoglobulin variable-regions. *Nucleic Acids Res.* 9:3809–20

Ann. Rev. Immunol. 1983. 1:529–68

GENES OF THE MAJOR HISTOCOMPATIBILITY COMPLEX OF THE MOUSE

Leroy Hood, Michael Steinmetz, and Bernard Malissen

Division of Biology, California Institute of Technology, Pasadena, California 91125

INTRODUCTION

In the past 2 years our understanding of the major histocompatibility complex has advanced dramatically because of the rapid progress made by molecular genetics. Biologists can now begin to dissect the molecular nature of one of the most fundamental problems in eukaryotic biology—the ability of organisms to discriminate between self and nonself. Even the most primitive of metazoa, the sponges, exhibit cell-surface recognition systems capable of identifying and destroying nonself, presumably to preserve the integrity of individuals growing in densely populated environments (34). For example, when two genetically identical sponges are apposed, the individuals fuse to form a single organism. However, when two genetically dissimilar sponges are joined, there is a reaction leading to tissue destruction at the boundary between the two individuals (35, 36). Presumably, cell-surface structures recognize nonself and trigger effector mechanisms that lead to the destruction of the foreign tissue. Perhaps the most striking feature of the molecules involved in these cell-surface recognition phenomena is their enormous diversity or polymorphism. More than 900 genetically distinct sponges have the capacity to reject one another after apposition (35). Self/nonself recognition systems in other invertebrates and vertebrates appear to display similar characteristics. Thus, three features are fundamental to self/nonself recognition systems—cell-surface recognition structures, effector mechanisms that lead to the destruction of nonself, and a high degree of polymorphism in the recognition structures.

Mammals have self/nonself recognition systems encoded by a chromosomal region termed the major histocompatibility complex (MHC) with pre-

529

0732-0582/83/0410-0529$02.00

cisely these features. These systems regulate various aspects of the mammalian immune response. The same recognition systems were first identified in mice because of the existence of inbred (genetically identical) and congenic (genetically identical but for a single chromosomal region) strains of mice. By the grafting of tumors or skin among such mice and following rejection or acceptance of the graft, it was possible to map the rejection of nonself to a region on chromosome 17, which was then denoted the MHC (30, 31). The initial characterization of the cell-surface structures responsible for graft rejection utilized alloantibodies produced by cross-immunization of mice with spleen cells of mice differing only at the MHC. Later, recombinant congenic mice, which had undergone recombination within the MHC, were produced. With these recombinant congenic mice, specific alloantisera to even smaller portions of the MHC could be produced. Thus, inbred, congenic, and recombinant congenic mice with their attendant potential for recombinational analyses and the generation of highly specific alloantisera permitted immunogeneticists (45, 106), later protein chemists (81, 90, 101), and very recently molecular biologists to characterize the MHC of the mouse in considerable detail. In this review, we discuss the results obtained primarily from molecular analyses of the MHC of the mouse. From the data available, the MHC of the mouse appears to be a general model for the MHCs of other mammals and perhaps other vertebrates.

THE MURINE MHC

Three Classes of Molecules

Alloantisera specific for gene products of the murine MHC have permitted the identification of three classes or families of molecules denoted I, II, and III (Figure 1). There are two categories of class I genes. Class I genes located in the left-hand side of the MHC or the H-2 region encode cell-surface molecules termed transplantation antigens, denoted K, D, and L, which mediate the graft rejection assay initially used to define the MHC (Figure 1). The class I genes located in the right-hand portion of the MHC, denoted the Qa-2,3 and Tla regions, encode cell-surface antigens that are structurally related to the transplantation antigens, but that differ in tissue distribution and presumably function (24, 107) (Figure 1). Class II genes, designated A_α, A_β, E_α, and E_β, encode cell-surface Ia antigens that later were found to be identical to the immune response, or Ir genes that control the magnitude of the murine immune responses to different antigens (49). Class III genes encode several components of the activation stages of the complement cascade (1). The availability of recombinant congenic strains and specific alloantisera to various gene products of the MHC of the mouse has permitted the construction of a detailed genetic map (Figure 1). The MHC of man on chromosome 6 is remarkably similar to that of its mouse counterpart, except for a presumed translocation in mice, which led to the separation of the K from other class I loci and the existence of additional class II loci in man whose homologues have not yet been found in mice (Figure 1). Class I and class II molecules and genes isolated from both murine and human cells are closely related.

Polymorphism

The most striking feature of the MHC in mouse and man is the extensive genetic polymorphism exhibited by certain class I and class II genes. For example, the K and D class I genes appear to have 50 or more alleles in both wild and inbred populations of mice (46). The class I genes of the Qa-2,3 and Tla regions are far less polymorphic (24). Certain of the class II genes, A_β and E_β, also appear quite polymorphic, whereas another, E_α, is much less polymorphic (47). The fact that some genes in the class I and II families exhibit extensive polymorphism, whereas others do not, is an interesting feature that may reflect functional requirements for diversity

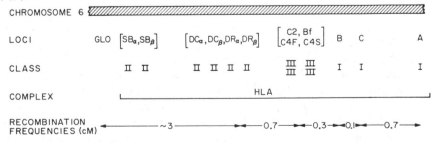

Figure 1 Genetic maps of the MHC in mouse and man. For symbols, see text.

in the polymorphic loci. Individual inbred mice will have distinct combinations or constellations of these alleles. Each individual combination of alleles is termed a haplotype. For example, the BALB/c mouse whose MHC has been most thoroughly studied at the molecular level is of the d haplotype and its gene products are generally denoted by a capital letter with a superscript for the haplotype designation—e.g. K^d, D^d, A_α^d, A_β^d, E_α^d, and E_β^d. The extensive polymorphisms as well as certain other features have led to the hypothesis that interesting gene correction mechanisms are operating within the class I and II gene families (see below) (101, 102).

Evolution

The MHC has been found in all mammals and vertebrates studied to date. In each case with adequate information, there appear to be closely linked class I and class II genes, and in those few cases where the analysis has been carried out, associated class III genes. A controversy has arisen as to whether the class III genes should be regarded as an integral part of the MHC complex or, alternatively, as an inadvertent addition comparable to the MHC-linked genes encoding several apparently unrelated enzymes (48). Since we know little about the structure, organization, or numbers of class III genes in contemporary or primitive vertebrates, it is difficult to argue either side of this issue. The linkage of the class I, class II, and possibly class III genes over the 500 million years of vertebrate divergence poses interesting questions about the selective constraints that have maintained linkage among these gene families.

Size

The MHC spans an enormous chromosomal region, as indicated by the following simplistic calculation. The haploid mouse genome includes about 1600 cM of DNA by genetic analysis and contains 3×10^9 base pairs by biochemical analysis. Therefore, 1 centimorgan equals approximately 2000-kb pairs of DNA. The MHC of the mouse is approximately 1–2 cM in length and, accordingly, should contain about 2000–4000 kb pairs of DNA. The cloning of the MHC has involved a variety of approaches. Various cDNA probes were used to isolate class I and II genes from genomic libraries. Restriction enzyme mapping and chromosomal walking have defined clusters of linked MHC genes. These genes have been mapped into regions of the MHC by gene transfer experiments and restriction enzyme site polymorphisms—as is described subsequently. In the past 2 years, 25–50% of the mouse MHC has been cloned and several new insights concerning the structure and organization of these gene families have emerged.

CLASS I MOLECULES AND GENES

Class I Polypeptides

The most thoroughly characterized class I molecules are the mouse transplantation antigens. These molecules are comprised of a class I polypeptide that is 45,000 daltons and is an integral membrane protein noncovalently associated with β_2-microglobulin, a 12,000-dalton polypeptide encoded by a gene located on chromosome 2 in the mouse (26, 75, 94) (Figure 2). Amino acid sequence analyses have demonstrated that the transplantation antigen is divided into five domains or regions (14, 114). The three external domains, $\alpha 1$, $\alpha 2$, and $\alpha 3$, are each about 90 residues in length. The transmembrane region is about 40 residues and the cytoplasmic domain is about 30 residues in length. The $\alpha 2$ and $\alpha 3$ domains have a centrally placed disulfide bridge spanning about 60 residues and up to three N-linked glycosyl units bound to attachment points in the $\alpha 1$ and $\alpha 2$ (K^b, D^d) and also in the $\alpha 3$ (K^d, L^d, and D^b) domains (68). The transmembrane region contains a hydrophobic and uncharged core of about 25 residues, which presumably transverses the membrane. Binding studies with peptide fragments from class I molecules have shown that the β_2-microglobulin subunit associates with the $\alpha 3$ domain (128). Preliminary X-ray analyses of crystals from a proteolytic fragment of a human transplantation antigen containing the $\alpha 1$, $\alpha 2$, $\alpha 3$, and β_2-microglobulin domains demonstrate that this class I molecule has a twofold axis of symmetry (P. Bjorkman, J. Strominger, and D. Wiley, personal communication) and, accordingly, may be folded so that its four domains are paired in a symmetric manner similar to antibody

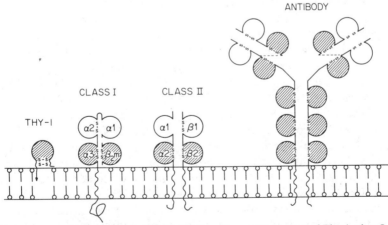

Figure 2 A schematic representation of the membrane orientations of Thy 1, class I, class II, and antibody molecules. Thy 1 is a T-cell differentiation antigen. Shaded domains indicate sequence homologies that suggest a common evolutionary ancestry (see text).

domains (Figure 2). Amino acid sequence analyses suggest that the $\alpha3$ domain (114) and β_2-microglobulin (88) show homology to the constant region domains of immunoglobulins, thus posing the provocative possibility that the genes encoding antibodies and transplantation antigens share a common ancestry.

As mentioned above, the class I genes fall into two categories—those mapping into the H-2 region and encoding transplantation antigens and those mapping into the Qa-2,3 and Tla regions and encoding homologous antigens that appear to be expressed in a tissue-specific manner (Figure 1). Transplantation antigens are found on virtually all nucleated cells of the mouse. Some mice, such as the inbred BALB/c, have three or even more distinct transplantation antigens (K, D, L, and perhaps M or R), whereas other mice appear to express fewer transplantation antigens (21, 33). The cell-surface antigens encoded in the Qa-2,3 and Tla regions can further be distinguished from the classical transplantation antigens because they are significantly less polymorphic (24). Some of the Tla genes encode hematopoietic differentiation antigens, which are found on thymocytes at an early developmental stage as well as on leukemic cells, whereas others appear to be expressed only on leukemic cells. The Qa antigens are found on many different types of lymphoid cells. We denote those class I antigens encoded by the Qa and Tla regions as hematopoietic differentiation antigens, recognizing the provisional nature of this designation.

Class I cDNA Clones

Two procedures were used to isolate mouse and human class I cDNA clones. First, cDNA clones were selected for their ability to bind class I mRNA, which could be identified by in vitro translation and immunoprecipitation of the synthesized polypeptides (54, 89). As an alternative approach, short oligonucleotides were synthesized after reverse translation of the amino acid sequences of class I polypeptides into DNA language and were used as hybridization probes for the identification of class I cDNA clones or as primers to synthesize cDNA's from unpurified mRNA's (91, 108). Furthermore, mouse cDNA clones and human genomic clones also were isolated by cross-species hybridization with cDNA clones (44, 109). Essentially the same approaches were employed for the isolation of mouse and human cDNA clones for β_2-microglobulin (85, 115) and class II polypeptides (3, 32, 51, 60, 64, 113, 118, 124), which is discussed subsequently.

The class I cDNA clones were characterized by DNA sequence analysis to verify their authenticity. Most of the mouse cDNA clones have not been correlated with known class I molecules mainly because of the paucity of amino acid sequence data currently available. However, the cDNA clones for the K^b (91), D^b (93), and K^d (56, 126) molecules have been identified.

The class I cDNA probes were then used to screen genomic libraries to obtain clones that could be used to analyze the structure and organization of class I genes.

Structure of Class I Genes

SEQUENCE ORGANIZATION   Three mouse class I genes have been completely sequenced, and their structural features are depicted in Figure 3A. Two of these genes (27.5 and CH4A-H-2L^d) encode L^d transplantation antigens (23, 76), whereas a third (27.1) has been mapped into the Qa-2,3 region (111). It is not known whether gene 27.1 is expressed. The exon-intron organization of these class I genes is remarkably similar. Each gene is divided into eight exons, which correlate precisely with the domains of the class I polypeptide (111) . The first exon encodes the leader sequence, whereas the second, third, and fourth exons encode the α1, α2, and α3 domains. The fifth exon encodes the transmembrane domain and the sixth, seventh, and eighth exons encode the cytoplasmic domain and 3' untranslated region of these genes (Figure 3A.) The exon-intron organization of a human class I gene is similar to the mouse genes, with the only difference being that the human class I gene contains two rather than three cytoplasmic exons (67). Additional sequence data on human class I genes will be necessary to determine whether or not this feature is characteristic of all human class I genes.

The general features of the class I genes resemble those of other eukaryotic genes. The upstream and downstream intron boundaries typically have the classical GT/AG nucleotides, which appear to be important RNA splicing signals (99). The 5' flanking sequence of gene 27.1 appears to have sequence elements characteristic of other eukaryotic promoter regions— CAAT and TATA boxes at positions 81 and 53 upstream of the AUG initiation codon (111). The consensus recognition sequence, which precedes the site of polyadenylation, AATAAA, appears twice, adjacent to one another approximately 390 nucleotides downstream from the stop codon in exon 8 of the 27.1 gene (111).

According to the nucleotide sequence, the two reported L^d genes are functional in that there are no stop codons at inappropriate positions, the RNA splicing junctions are all characterized by the appropriate signal nucleotides, and there are no deletions or insertions that change the reading frames of the exons (23, 76). In contrast, gene 27.1 appears to be a pseudogene by several criteria (111). First, there is a charged aspartic acid codon in the middle of the highly hydrophobic transmembrane exon. Second, there are termination codons near the end of the transmembrane and seventh exons. Third, the upstream RNA splice signal for the sixth exon does

A.

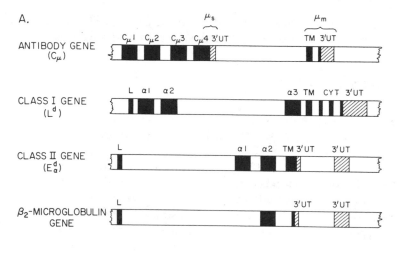

B.

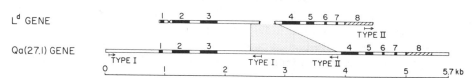

Figure 3 (*A*) A schematic diagram of the exon-intron organization of antibody, class I, class II, and β_2-microglobulin genes. L denotes the exons encoding the signal or leader peptides; $\alpha1$, $\alpha2$, and $\alpha3$ represent exons encoding the external domains; TM designates the transmembrane exons; CYT symbolizes the exons encoding the cytoplasmic domains; and 3' UT denotes the 3' untranslated region. Dark boxes represent coding sequences and hatched boxes 3' UT regions. (*B*) Schematic diagram of the exon-intron organization of two mouse class I genes. Nonhomologous sequences between the L^d and Qa (27.1) genes are indicated by stippling. Exons are numbered; 3' untranslated regions are hatched. Arrows show the localization of type I and type II Alu-like repeat sequences.

not agree with the consensus sequence. Thus, it is clear that gene 27.1 cannot encode a full-length class I polypeptide. However, with the exception of the aspartic acid codon in the transmembrane exon, the aberrations in gene 27.1 are all in the 3' portion downstream from the stop codon at the end of the transmembrane exon. Thus, gene 27.1 may encode a membrane-bound class I polypeptide that is missing the cytoplasmic domain, or alternatively, it may be expressed as a soluble transplantation antigen (39) (see below).

PATTERNS OF VARIABILITY FOR CLASS I GENES AND POLYPEP-TIDES   A detailed comparison of 12 class I amino acid and translated nucleotide sequences has appeared recently (68). The overall homology between six polypeptides for which extensive sequence information exists [K^b, D^b, L^d, K^d, D^d, Qa (27.1)] is 76–94% (Table 1). From these comparisons it is obvious that the alleles K^b and K^d or D^b and D^d are not more related to each other than to other class I polypeptides. This has been denoted a general lack of "D-ness" and "K-ness" and is discussed below. However, close inspection of the sequences reveals the existence of short allele-specific sequences located mostly in exons 4 to 8 (68).

Particularly interesting is the close relationship between the L^d and D^b amino acid sequences (94%), whereas the D^b and D^d polypeptides are only 84% homologous. This observation raises the possibility that the L^d and D^b genes may be alleles (68). Alternatively, the gene correction mechanisms that operate to generate polymorphisms (see below) may be capable of correcting nonallelic genes extensively so that they resemble each other.

The variability in the extracellular part of the analyzed class I molecules appears to be clustered in three regions in the first two external domains (residues 62-83, 95-121, and 135-177)(68). In contrast, the third external domain which interacts with β_2-microglobulin is more highly conserved. Exon 6, encoding part of the cytoplasmic domain, is completely identical in seven functional class I sequences that can be compared (gene 27.1 has a different exon 6 sequence, but also contains a premature stop codon upstream of exon 6). The significance of this finding is unknown.

THE Qa (27.1) GENE SHOWS STRIKING HOMOLOGY TO CLASSICAL TRANSPLANTATION ANTIGENS The overall homology between gene 27.1, which maps into the Qa-2,3 region, and the classical transplantation antigens (K, D, L) is 78–84% (Table 1). Therefore, the Qa-2,3 gene is as closely related to the genes encoding transplantation antigens as they are to one another. These homologies suggest that the Qa and transplantation antigens descended from a common ancestor. It will be interesting to find out whether or not the structural homology is reflected in similar functions as well. Peptide map analyses of Qa-2,3 antigens also have revealed struc-

Table 1 Percent homology between mouse class I molecules[a]

Molecules	H–2K^b	H–2D^b	H–2L^d	H–2K^d	H–2D^d	Qa (27.1)
H–2K^b		83	83	77	89	79
H–2D^b			94	76	84	80
H–2L^d				76	87	79
H–2K^d					76	78
H–2D^d						84
Qa (27.1)						

[a] Amino acid sequences compared ranged from 113 to 342 residues. Table adapted from Maloy & Coligan (68).

tural homology to transplantation antigens (107), whereas TL polypeptides appear to be very different (127). On the other hand, the similar size and binding to β_2-microglobulin of the TL molecules, as well as the strong cross-hybridization of these genes to a probe for the exon encoding the β_2-microglobulin-binding domain again suggest homology and a shared ancestry (112). DNA sequence analyses of the Tla genes are now under way.

ALU REPEATS AND CLASS I GENES   The human genome has an element repeated 300,000 times that is termed the Alu repeat because the restriction enzyme Alu I cleaves within the repeat and thus permits the visualization of this DNA on Southern blots. The mouse genome has an homologous element termed an Alu-like repeat. The 3' untranslated regions of class I genes can be distinguished by the presence or absence of Alu-like repeats (11, 109). Alu-like sequences have the general characteristic of transposons with terminal direct repeats encompassing a central sequence composed of the Alu-type and an A-rich sequence (43). They can be classified into two groups called type I (described above) and type II sequences. Type II Alu sequences contain an additional non-Alu repetitive sequence. Alu-like repeat sequences of type II have been found at the 3' end of the 3' untranslated region of three class I cDNA clones [pH-2II(109), pH-2^d-1(56), and pH-203(93)] and include the AATAAA polyadenylation signal as a part of the A-rich sequence. One of these cDNA clones (pH-203) has been shown to encode the D^b polypeptide (68, 93). The pH-2II clone encodes a polypeptide very similar (except for five positions) to the L^d polypeptide (76, 109) and clone pH-2^d-1 appears to encode the D^d polypeptide (10, 68). A type II Alu-like repeat also has been found at the 3' end of the 3' untranslated region of the L^d gene (76). Thus, class I genes encoded in the D region appear to contain Alu-like repeat sequences in their 3' untranslated regions. The cDNA clones encoding the K^b (92) and the K^d (126) polypeptides apparently lack Alu-like repeat sequences. Their 3' untranslated regions are highly homologous (80–90%) to the D region cDNA clones for the first $\sim$300 bp, whereas the remaining portion of $\sim$120-160 nucleotides encompassing the Alu-like repeat in the D group of genes is completely different for the K genes (93, 126). Indeed, this region of the cDNA clone homologous to the K^d polypeptide has been used as a specific probe for the identification of the K^d gene (126). Gene 27.1 also does not contain the Alu-like repeat at its 3' end (111) but a sequence that is distantly related to the corresponding portion of the K gene, possibly reflecting a closer evolutionary relationship between Qa and K region genes than between K and D region genes. Cross-hybridization of 5' flanking sequences between K and Qa-2,3 genes but not between K and D genes of the b haplotype support this notion (27).

A comparison of the two genomic sequences of the genes 27.1 and L^d has shown that the introns and exons have diverged about equally from one another apart from the insertion of a sequence of approximately 1000 nucleotides in the intron between exons 3 and 4 of gene 27.1 (76)(Figure 3B). This insertion is bounded at its 3' end by a type II Alu-like repeat sequence and contains a type I Alu-like repeat sequence close to its 5' end (Figure 3B). Thus, transposon-like Alu-like repeat sequences may have mediated this insertion of 1000 nucleotides.

ARE THERE TWO Ld GENES? The translated sequences of the two L^d genes correspond to the L^d protein sequence at 77 of 77 positions that can be compared (23, 76) and encode a class I polypeptide that can be detected with anti-L^d monoclonal antibodies after gene transfer into mouse L cells (23, 29) (see below). However, these two genes appear to differ from one another by 20 to 30 nucleotides. These differences might be explained in different ways. First, some of the differences might be due to DNA sequencing errors. However, the differences appear to be too extensive to be explained exclusively in this manner. Second, since these L^d genes were derived from two distinct populations of BALB/c mice, perhaps some genetic polymorphism has already occurred since their separation. Finally, there may be two distinct class I genes that both encode molecules recognized by anti-L^d monoclonal antibodies. This latter hypothesis is being tested by chromosomal walking techniques, which are discussed in a subsequent section.

POSSIBLE SECRETORY CLASS I MOLECULES An unusual category of class I mRNA's with two interesting features has recently been described (16). First, these mRNA's contain several codons for charged amino acids in the transmembrane region, which is usually hydrophobic and uncharged. Second, a termination codon is found at the end of the membrane exon. Gene 27.1 has precisely these same characteristics (111) and thus may be capable of synthesizing a similar mRNA. These unusual class I mRNA's are slightly smaller than the mRNA's encoding full-length class I polypeptides and can be identified by a distinctive probe that again has been derived from the 3' end of the 3' untranslated sequence (17). It appears that these RNA molecules are expressed only in liver cells and not in other mouse tissues that have been tested (17). Although there is to date no formal evidence that these mRNA's are translated into proteins, it is attractive to speculate that they may direct the synthesis of soluble class I gene products.

ALTERNATIVE PATTERNS OF RNA SPLICING AT THE 3' END OF THE CLASS I GENES Antibody heavy chain genes can be transcribed and then differentially processed by RNA splicing at their 3' ends to generate the membrane and secreted forms of immunoglobulins (22) (Figure 3A). Several intriguing observations suggest that class I genes also may undergo alternative patterns of RNA splicing at their 3' ends (Figure 4). First, the C-terminal amino acid sequences of class I polypeptides and translated cDNA clones appear to be heterogeneous in length and they exhibit blocks of amino acids which differ (111) (Figure 4A). These properties would be expected if different coding regions (exons) could be employed for the C termini of these polypeptides. Second, two mouse class I cDNA clones (pH-2I and pH-2II) reveal a notable sequence difference. The 3' ends of these two cDNA clones are homologous to one another, apart from an insertion of 139 bp in the pH-2II clone (111) (Figure 4B). This 139 base pair sequence corresponds precisely to the seventh intron of the class I gene. Thus, it is attractive to suggest that the seventh intron was not removed from the mRNA represented by clone pH-2II, whereas it was for the pH-2I and L^d-like mRNA's (Figure 4C). Third, comparison of the cDNA sequence of pH-202 encoding the K^b molecule with the sequence of gene 27.1 suggests a third pattern of RNA splicing for the seventh intron (92). To give rise to an H-2K^b-like mRNA, exon 7 would be spliced to a sequence 27 nucleotide further upstream of the splicing point used for the generation of pH-2I or L^d-like mRNA's (Figure 4C). These alternative patterns of RNA splicing would lead to C-terminal sequences differing in size and sequence. This particular pattern of RNA splicing also has been described for the gamma gene of fibrinogen (18). It remains to be seen whether these alternative forms of mRNA's result from the transcription of the same class I gene. Nevertheless, the heterogeneity generated by different RNA splicing mechanisms in the cytoplasmic domains of class I molecules might be important for different effector functions through interactions with different components of the cytoskeleton.

Organization of the Class I Gene Family

A MULTIGENE FAMILY   The possible homologies of the class I and antibody genes have raised several questions about the organization of class I genes. First, how many class I genes are encoded in individual mice? Second, do class I genes undergo somatic rearrangements as do their antibody gene counterparts during the differentiation of cells that express class I molecules? Southern blot analyses of the DNAs from various strains of mice with class I cDNA probes display 12–15 bands (13, 69, 86, 100, 104, 109, 111). Therefore, it is presumed that there are 12 or more class I genes in mice. Although most cells including liver express class I molecules, sperm do not express transplantation antigens. There are no gross DNA rearrangements of class I genes in cells actively synthesizing these molecules,

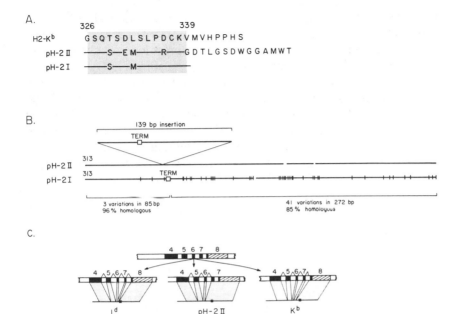

Figure 4 Differential RNA splicing patterns at the 3' end of class I genes. (*A*) Comparison of the C-terminal sequences of the H-2K^b polypeptide and the polypeptides encoded by the class I, pH-2II, and pH-2I cDNA clones. The shaded area indicates the sequences encoded by exon 7. A straight line indicates identity to the H-2K^b sequence. (*B*) Comparison of the DNA sequences at the 3' end of the class I cDNA clones pH-2II and pH-2I, indicating close homology except for a 139-bp insertion in clone pH-2II. Termination codons are boxed. Vertical bars indicate a nucleotide substitution. Gaps were introduced to achieve maximum homology. (*C*) Three possible differential RNA splicing patterns of class I genes consistent with the structures of the L^d polypeptide, the pH-2II-like mRNA's, and the K^b polypeptide. Numbered black boxes indicate exons. The 3' untranslated regions are hatched, and termination codons are shown as black boxes at the mRNA level. (Adapted from 111.)

since liver and sperm DNAs have the same hybridization patterns in Southern blots (109). Furthermore, complete class I genes, uninterrupted by long stretches of noncoding DNA, are present in sperm DNA, as shown by gene cloning and sequencing. Therefore, it is unlikely that major DNA rearrangements of widely separated exons occur in class I genes as they do in antibody genes. Multiple class I genes also have been identified in human and pig DNA (10, 83, 105).

CLUSTERS OF CLASS I GENES   Several laboratories have used the cloning of large DNA fragments (30 to 50 kb in size) in cosmid vectors to study the complexity and linkage relationship of class I genes in mice and man (27, 66, 112). The most extensive study so far has been carried out with

BALB/c DNA (112, 125). Some 56 cosmid clones were identified in a library of BALB/c sperm DNA with class I cDNA clones (112). These clones were mapped with 10 infrequent cutting restriction enzymes to permit the identification of cosmid clones with overlapping DNA inserts. On the basis of these analyses, these clones could be ordered into 13 gene clusters containing 36 distinct class I genes in a total of 840 kb pairs of DNA (Figure 5). Comparison of genomic Southern blots with pooled cosmid DNA analyzed with class I cDNA clones as hybridization probes showed that most of the class I genes of the BALB/c mouse had been cloned.

The largest of these gene clusters, cluster 1, contains seven class I genes spread over 191 kb of DNA (Figure 6) and maps into the Qa-2,3 region

Cluster	Organization	Mapping	Expression	Length(kb)	Overlapping clones
		Location			
	(0 50 100 150 200 kb)				
1	27.1	Qa-2,3		191	17
2	L	D	L	68	2
3		Tla		103	9
4	TLa	Tla	TL	64	9
5	Tla	distal to D	TL	49	4
6	Qa-2,3	between D, Qa-2,3	Qa-2,3	63	3
7		Tla		58	3
8		Tla		47	2
9		Qa-2,3		38	2
10		Tla		42	1
11	K	K	K	43	2
12		distal to D		39	1
13	D	D	D	35	1
	36 class I genes			840	56

Figure 5 A schematic representation of the 13 cosmid clusters and their locations. The locations were determined by genetic mapping through restriction enzyme site polymorphisms and by expression in DNA-mediated gene transfer experiments (see text). The class I genes are indicated as dark boxes. (Adapted from 111.)

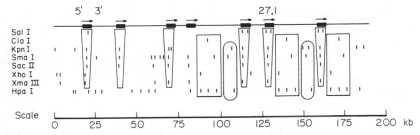

Figure 6   Restriction map of the class I cosmid gene cluster 1 isolated from BALB/c DNA. Class I genes are indicated as black boxes. Homologous regions according to the restriction map are indicated graphically by ovals, rectangles, and wedges (see text). Arrows indicate the 5' to 3' directions of transcription. Sal I, Cla I, etc, are restriction enzymes. The location of gene 27.1 is indicated. (From 111.)

(112) (see below). The spacing between these genes ranges between 7 and 28 kb. The structure of cluster 1 is interesting in that the three class I genes to the 3' side of the cluster as well as their flanking sequences are indistinguishable from one another by restriction map analysis. These genes probably arose by relatively recent duplication events, presumably by homologous unequal crossing-over. The two genes located at the 5' end of cluster 1 also are related to one another, as revealed by cross-hybridization of their 3' flanking sequences. Thus, the genes in cluster 1 suggest that class I genes can be ordered into closely related subgroups of class I genes that are continually undergoing expansion and contraction by homologous unequal crossing-over. In this example, more closely related class I genes seem to be located adjacent to one another.

Seventeen distinct class I genes located on seven cosmid clusters have been isolated and characterized from B10 mouse DNA (27). One of these gene clusters, cluster 3, containing five class I genes maps into the Qa-2,3 region and appears to be the homolog to cluster 1 isolated from BALB/c DNA. Further characterization of the class I genes of B10 mice is necessary to determine whether B10 mice contain only half as many class I genes as BALB/c mice. Such a difference is not apparent from Southern blot analyses (13, 69, 86, 109). If such a difference in class I genes among inbred mice is real, extensive gene expansion and contraction must occur.

Some mouse strains fail to express the L polypeptide at the serological level (e.g. B6, AKR, A.SW). To determine whether or not this failure in expression results from deletion of the L gene, a DNA probe has been isolated from the 5' flanking sequence of the L^d gene (112). According to Southern blot analyses, this sequence is missing in the DNA from mice of the b (B6) and the k (AKR) haplotypes, indicating that nonexpression might arise from deletion of the L gene. Likewise, in mutant BALB/c mice denoted dm1 and dm2, which also fail to express the L^d gene, the same 5' flanking sequence is deleted (H. Sun, personal communication). It will be interesting to determine the precise 5' and 3' end points of these deletions by chromosomal walking experiments, since a hybrid class I molecule comprising the N-terminal part of the D^d molecule and the C-terminal part of the L^d molecule appears to be present in BALB/c dm1 mice (123). These data also support the hypothesis that class I genes are undergoing continual gene expansion and contraction.

GENETIC MAPPING BY RESTRICTION ENZYME SITE POLYMOR-PHISMS

Strategy   The most effective approach for mapping the class I clusters to their respective locations in the MHC has been genetic mapping by restriction enzyme site polymorphisms (27, 125). This approach employs three steps. First, single- or low-copy probes, which will hybridize only to one or a few genomic restriction fragments, are isolated from each of the cosmid clusters. Second, these probes are used to examine the DNAs of congenic and recombinant congenic mice for restriction enzyme site polymorphisms. Theoretically, three types of polymorphisms can be observed—a change in restriction fragment size (mutation), a loss of the restriction fragment (deletion), or an increase in the number of restriction fragments (duplication). Third, these restriction enzyme site polymorphisms are then correlated with the serological polymorphisms of the various congenic and recombinant congenic mouse strains analyzed to allow the mapping of a particular site to one of the four class I regions—K, D, Qa-2,3 and Tla (Figure 7).

Molecular map of class I genes   Single- or low-copy probes were isolated from each of the 13 class I gene clusters from the BALB/c mouse and were used to map the gene clusters with respect to the genetic map of the MHC (125) (Figure 8). Four striking results emerged. First, all 36 class I genes could be mapped to the MHC. No class I genes have been found so far mapping to chromosomes other than 17. In contrast, processed pseudogenes from other multigene families such as the antibody (6, 37), globin (59), and tubulin (121) gene families map to chromosomes other than those containing their functional counterparts. Second, 31 of 36 class I genes map to the Qa-2,3 and Tla regions. Thus, these two regions contain more than 80% of the class I genes. Three gene clusters containing five class I genes map to the K and D regions, which encode the classical transplantation antigens. These data are in agreement with mapping studies of class I gene clusters carried out for B10 mouse DNA (27) and with previous conclusions derived from Southern blot analyses that most of the class I genes are located in the Qa-2,3 and Tla regions (69, 86). However, these DNA blot

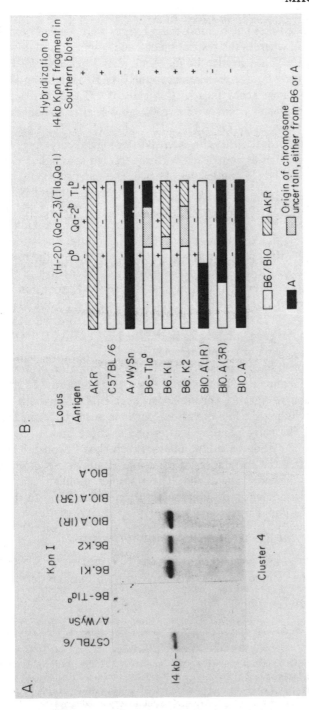

Figure 7 Mapping of the class I gene cluster 4 to the Tla region by restriction enzyme site polymorphisms. (*A*) Kpn I digests of various mouse strains were hybridized with a single-copy probe isolated from the 5' flanking sequence of gene 1 in cluster 4 (125). (*B*) The presence or absence of the 14-kb Kpn I fragment indicative of cluster 4 is compared against the serological polymorphism of MHC antigens in various mouse strains. The presence of the 14-kb Kpn I fragment correlates with the presence of the Tla^b allele, but not with the presence of any other allele of the MHC in these strains. Cluster 4 therefore maps to the Tla region. (Adapted from 125.)

analyses also suggested that one or more class I genes were located outside of the MHC (27, 86), whereas the restriction enzyme mapping studies with cloned BALB/c genes failed to identify any such class I genes (125). Perhaps more than 36 class I genes are present in the BALB/c mouse and one or more of the uncloned genes falls in this category. Third, restriction enzyme site polymorphisms occur with considerably greater frequency in the K and D regions than in the Qa-2,3 and Tla regions (125). This observation is in complete accordance with the extensive serological polymorphism of the K and D molecules as contrasted with the limited serological polymorphism of Qa-2,3 and TL molecules. Hence, the extensive polymorphisms exhibited by the K and D genes also are seen in their flanking sequences, whereas the more limited polymorphisms of the Qa-2,3 and Tla genes also are reflected by a lack of polymorphism in their flanking sequences. Hence, these differences may reflect the differential operation of mechanisms for generating polymorphisms (see below). Finally, the results obtained using 13 single- or low-copy probes to analyze the number of class I bands in four different inbred strains suggest that modest expansion and contraction of the class I gene family occurs, presumably by homologous unequal crossing-over (125). These observations are in accord with those discussed earlier on the gene homologies in the cluster 1 of BALB/c DNA and the loss of the L gene in certain strains.

An elegant approach has been used to study the organization of human class I genes (83). Treatment of human cell lines with gamma rays leads to the random deletion of various chromosomal segments throughout the human genome, including the loss of various portions of the human MHC or HLA complex. The corresponding loss of one or more human class I genes can be followed by Southern blot analyses with class I probes. These studies have permitted the correlation of particular bands detected on the Southern blot with certain HLA alleles. These studies should reveal the location of the as yet unidentified human class I genes equivalent to those of the mouse Qa-2,3 and Tla regions.

Mechanisms for the Generation of Polymorphism

COMPLEX ALLOTYPES, SPECIES-ASSOCIATED RESIDUES, NO "K-NESS" OR "D-NESS"   Three observations made with the initial N-terminal

Figure 8   Locations of the 13 class I gene clusters in the MHC. The order of the clusters in a region is not known. Clusters 5 and 12 map either to the Qa-2,3 or Tla regions. Arrows point to the locations of the probes used for mapping. Open boxes indicate expressed genes by DNA-mediated gene transfer, whereas dark boxes represent nonexpressed genes. (From 125.)

protein sequence data on transplantation antigens suggested that the individual members of the class I multigene family could be rapidly expanded and contracted or corrected against one another (101, 102). First, the alleles of the K (and D) locus differ from one another by multiple residues. These multiply substituted alleles are denoted complex allotypes (102), and they represent a contrast to the typical simple allotypes of the globins, which differ generally by single amino acid substitutions. Thus, a special genetic mechanism appears necessary to account for the extensive allelic differences. Second, the class I polypeptides of human and mouse are distinguished from one another by species-associated residues at various positions (102). Thus, the multiple mouse (and human) class I genes must evolve in parallel by a process termed coincidental evolution (38). Once again, coincidental evolution among multiple genes requires the operation of mechanisms for gene correction (38). Finally, the N-terminal regions of the K alleles and the D alleles exhibited no "K-ness" or "D-ness"; that is, in a pool of sequences for K and D alleles, no features permitted the K alleles to be distinguished from their D counterparts (101, 102) (see also 81). These observations suggest that there is a rapid exchange of information between K and D genes so that they fail to evolve independent characteristics. The mechanism of gene conversion—the correction of one sequence against a second—is also suggested by data on the mutant transplantation antigens.

MUTANT TRANSPLANTATION ANTIGENS   Mutant transplantation antigens were detected by carrying out reciprocal skin grafts among siblings in an inbred strain of mice (80). Occasionally one animal will reject the graft from his siblings. Subsequent characterization of these mutants often reveals that there is an amino acid sequence variation in one transplantation antigen that leads to the subsequent graft rejection. Mutants of the K^b transplantation antigen appear to be remarkable in several contexts (Figure 9 A). First, they are very frequent, occurring at a frequency of approximately 5×10^{-4}/cell/generation. Second, there are generally two or more amino acid substitutions that are relatively closely spaced to one another in each mutant antigen. It appears improbable that one isolated mutation would always be followed by a second closely linked mutation. Third, the same independently derived variants are seen repeatedly. Finally, virtually all of the K^b mutations occur at positions where the mutant residues are observed in other class I molecules. For example, the L^d polypeptide appears to contain most of the substituted amino acids found in the K^b series of mutants (23, 87). Presumably one or more class I genes containing these residues also exist in the b haplotype mouse. These data suggest that the K^b gene can undergo a single gene conversion event with one of the other class I genes leading to variants containing multiple amino acid substitutions (23, 56).

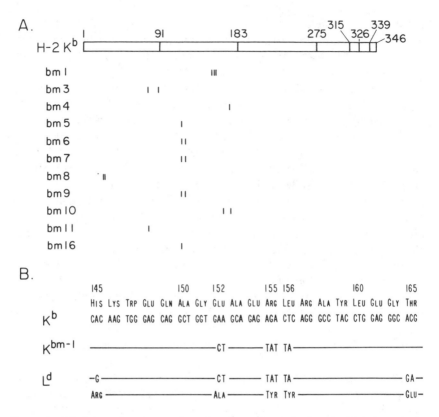

Figure 9 (*A*) Schematic representation of amino acid substitutions in 11 independently isolated K^b mutants. The top illustration shows the presumptive exon organization of the K^b mRNA. Substituted residues in the mutants are indicated by vertical bars. (Adapted from 80.) (*B*) Sequence comparison of the K^b, K^bm1, and L^d genes from codon position 145 to 165. Horizontal lines indicate identity to the K^b protein or DNA sequences. (Adapted from 120.)

The class I gene encoding the mutant K^bm1 transplantation antigen has recently been isolated (97, 119). This mutant class I polypeptide differs at positions 152, 155, and 156 from the wild-type K^b antigen. DNA sequence analyses of the K^b, K^bm1, and L^d genes demonstrate that the K^bm1 mutant has seven nucleotide substitutions arising over a stretch of 13 nucleotides, which lead to the three amino acid substitutions (Figure 9*B*). All seven of these substitutions are seen in the L^d gene, which is identical to the K^bm1 gene, over a stretch of 52 nucleotides encompassing this region. This observation raises the possibility that the mutant K^bm1 gene arose by gene conversion against a b haplotype class I gene identical in sequence to the L^d gene in this region. These data also may be explained by a double recombinational event, which appears less likely. Since this mutant sequence is not located in the single class I gene closely linked to K^b (119), the putative gene conversion of the K^b gene presumably occurs with class I genes located 0.3 or more cM away (Figure 1). Thus, gene conversion apparently may occur over extensive distances (41, 42).

GENE CONVERSION AND GENE EXPANSION AND CONTRACTION CREATE POLYMORPHISMS   Gene correction mechanisms such as gene expansion and contraction or gene conversion are required to explain complex allotypes, species-associated residues, the lack of "D-ness" and "K-ness," and the patterns of substitutions seen in the mutant transplantation antigens (see 38 for a discussion of these issues). Both mechanisms appear to play a fundamental role in generating the complex allotypes of immunoglobulin C_H genes (4, 81a, 96). The analyses of class I gene numbers with low- and single-copy probes clearly indicate that duplication and deletion of class I genes occurs, presumably by homologous unequal crossing over. Furthermore, the mutant bm1 gene appears to have arisen by a gene conversion event. Since many mutant class I polypeptides have properties similar to those of bm1, it is logical to conclude that they also arose by gene conversion and that this process is common throughout the class I gene family. Gene conversion also appears to account for interesting amino acid sequence relationships in human class I molecules (65). Apparently gene conversion cannot occur among all class I genes with the same efficiency. For example, the D^b gene shows a 10-fold lower mutation rate than the K^b gene (45). The important point is that both gene expansion and contraction and gene correction mechanisms play a fundamental role in generating the extensive polymorphism of certain class I genes. An intriguing question is whether one or both of these gene correction mechanisms also give rise to the extensive flanking region polymorphisms of the K and D gene clusters, or alternatively, whether these polymorphisms reflect the operation of yet a third unknown genetic mechanism. One must ask why certain regions and subregions [K, D, I-A (see below)] appear to exhibit extensive polymorphisms, whereas others [Qa, Tla, I-E, S (see below)] do not.

Expression of Class I Genes After Gene Transfer

APPROACH   The unequivocal identification of serologically defined class I genes requires gene transfer experiments in which the cloned gene is expressed in a foreign cell so that its gene product can be identified by serologic techniques. Class I genes have been successfully transferred into mouse L cells lacking the thymidine kinase gene by co-transfer with a herpes virus thymidine kinase gene with the calcium phosphate transfer technique (5, 23, 29, 63, 74, 105). Under appropriate selective conditions, cells expressing the herpes thymidine kinase gene will grow (1 in 10^4–10^5 L cells), and about 90% of the thymidine kinase-positive cells express the newly introduced class I gene product on the cell surface. Mouse L cells are derived from C3H mice of the k haplotype, which appear to express only two transplantation antigens, K^k and D^k. Monoclonal antibodies can readily distinguish each of the BALB/c (d haplotype) serologically defined class I products from their C3H counterparts. The new (BALB/c) class I molecule can be identified by radioimmunoassay, by the fluorescence-activated cell sorter, by two-dimensional gel electrophoresis, and by peptide map analysis. Data obtained by these means can then be correlated with the results obtained by DNA sequence analysis and the mapping of the gene by restriction enzyme site polymorphisms.

IDENTIFICATION OF SEROLOGICALLY DEFINED CLASS I GENES
The gene transfer technique has been used to identify all of the serologically defined class I genes in the BALB/c mouse except for the Qa-1 gene (28). There is no monoclonal antibody to the Qa-1 antigen, and identification of class I gene products in mouse L cells after gene transfer is difficult with heterogeneous alloantisera. Six serologically defined genes have been located in their corresponding cosmid clusters and are identified in Figures 5 and 8. The K^d, D^d, L^d, Qa-2,3, and two Tla genes have been identified. In each case the mapping by restriction enzyme site polymorphisms correlates with the mapping by serological identification after gene transfer (125).

The K^b gene isolated from the C57BL/6 mouse also has been identified by gene transfer experiments (74). The K^b gene resides in a class I gene cluster containing two class I genes, which is similar in organization to the BALB/c K^d gene cluster (27, 112, 125).

Genomic clones for human class I molecules also have been transferred into mouse L cells to study expression of HLA antigens (5, 62, 63). These studies have revealed that human class I genes can be expressed in mouse fibroblasts and that these products associate efficiently with mouse β_2-microglobulin and in some cases induce conformational changes of mouse β_2-microglobulin (62). Four human class I genes encoding HLA-A2, HLA-A3, HLA-B7, and HLA-CW3 antigens were identified with monoclonal antibodies (5; and F. Lemonnier, personal communication).

NOVEL CLASS I GENE PRODUCTS   Six of the 36 class I genes isolated from BALB/c DNA appear to encode serologically defined antigens (28). To determine whether any of the remaining class I genes are expressed, a radioimmunoassay was developed for cell-surface β_2-microglobulin (28). Mouse L cells express relatively constant amounts of the K^k and D^k molecules on their cell surface. Since each known class I molecule is associated with β_2-microglobulin, the amount of β_2-microglobulin associated with endogenous class I gene products on the cell surface does not vary. When foreign class I genes are transferred into mouse L cells, the amount of surface β_2-microglobulin generally increases, because foreign class I gene products also must associate with β_2-microglobulin to be expressed on the cell surface, and excess β_2-microglobulin appears to be synthesized. Thus, each of the 36 class I genes was independently transferred into mouse L cells and 10 novel gene products were identified by the β_2-microglobulin immunoassay (Figure 8). This assay would not identify class I gene products expressed in low levels or not associated with β_2-microglobulin. All of these genes map to the Qa-2,3 and Tla regions. Indeed, the two additional class I genes in the K and L gene clusters do not appear to be expressed by this gene transformation assay.

One of the transformed L cells expressing a novel gene product was used to immunize C3H mice, the strain from which the L cells were derived (28). A specific alloantiserum to the novel gene product was raised. Analysis by two-dimensional gel electrophoresis after immunoprecipitation with this antiserum established that the novel gene product was 45,000 daltons and was associated with β_2-microglobulin. Thus, the novel gene product has the typical features of a class I molecule. Furthermore, this novel gene product appears to be present on spleen cells but not on cells from other tissues in the mouse. Alloantisera to these novel gene products should be useful in defining their tissue distribution, their developmental regulation, and ultimately their functions.

Function of Class I Genes

Cytotoxic or killer T cells carry out an immunosurveillance to eliminate host cells that have been infected with virus (129). The T-cell receptor must recognize both the foreign viral antigen and a self transplantation antigen. The transplantation antigen is termed a restricting element, which permits the T cell to detect foreign antigens in the context of self. Accordingly, the transplantation antigen plays an important role in permitting the cytotoxic T-cell system to distinguish self from nonself. DNA-mediated gene transfer permits the functional analysis of individual class I genes in two ways. First, the restricting elements (K, D or L) that permit different viruses to be eliminated by cytotoxic T cells can be determined. Second, in vitro mutagenesis either by the shuffling of exons between different class I genes or by mutation of individual nucleotides can be used to perturb T-cell recognition and killing functions.

Cellular assays such as the one depicted in Figure 10 have been employed to analyze cytotoxic T cells raised in response to a particular viral infection against mouse L cells expressing various class I genes (74, 82). Cytotoxic T cells raised in a mouse of a particular haplotype can only employ restricting elements from the same haplotype. Thus, cytotoxic T cells raised in BALB/c mice against the LCM virus can kill LCM virus-infected mouse L cells only if they have been transformed with the appropriate BALB/c transplantation antigen. A summary of the results of T-cell killing studies from several laboratories is presented in Table 2. Several striking results emerge. First, a particular virus often uses a particular transplantation antigen as restriction element (e.g. the LCM virus infection in the BALB/c mouse employs the L^d element and the influenza infection in the C57BL/10 mouse employs the K^b element). Second, cytotoxic T cells raised in a mutant d haplotype animal lacking the L gene (BALB/c^{dm2}) do not employ the L^d gene as a restricting element (82). Thus, the dm2 mice appear to be capable of employing another class I molecule as restriction element to

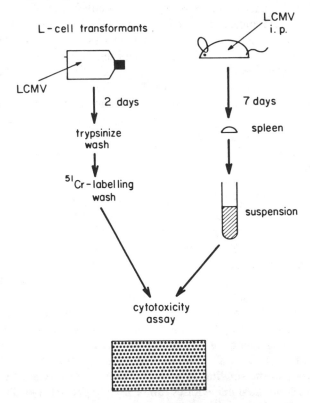

Figure 10 A schematic of the procedures used in the testing of LCM virus-specific H-2 restriction with L-cell transformants. L cells transformed with the L^d gene are infected with LCM virus and then are labeled with ^{51}Cr. Cytotoxic T cells isolated from LCM virus-infected BALB/c mice are then used to show by ^{51}Cr release that the L^d molecule on the surface of the transformed L cells can serve as a restricting element for LCM virus.

Table 2 Products of transfected H–2 class I genes act as restriction element for anti-viral cytotoxic T cells

Transfected gene	Virus	References
H–2K^b	Influenza	74
H–2L^d	Lymphocytic choriomeningitis virus	82
H–2L^d	Vesicular stomatitis virus	25

achieve T-cell recognition and killing of LCM virus-infected cells. Therefore, the system is flexible. Third, cloned T cells behave essentially the same way as their heterogeneous counterparts. Cloned T cells will be essential reagents in the in vitro mutagenesis experiments described below.

Since a transplantation antigen is required to evoke a cytotoxic T-cell response against a particular virus, this element may be mutated to determine which portions of the molecule are important for the recognition and effector functions of T-cell killing. Preliminary exon shuffling experiments between the L^d gene and gene 27.1 have been carried out (I. Stroynowski, personal communication). The L^d element can no longer serve as a restricting element if both the $\alpha1$ and $\alpha2$ exons are exchanged with homologous 27.1 exons. These data suggest that the T-cell receptor recognizes antigenic determinants on the $\alpha1$ and/or $\alpha2$ domains. More extensive in vitro mutagenesis and exon shuffling experiments for class I genes are being carried out now in a number of different laboratories.

Structure and Expression of the β_2-Microglobulin Gene

The exon-intron organization of the β_2-microglobulin gene, which appears to be a single-copy gene in contrast to the class I multigene family, has been determined by DNA sequencing (84). This gene consists of four exons, with most of the coding region present in one exon (amino acids 3–95) (Figure 3A). From the comparison in Figure 3A, it is obvious that the β_2-microglobulin gene has a very similar exon-intron structure to the antibody, class I, and class II genes. This homology also is reflected at the amino acid (88) and DNA sequence level (73, 84) and indicates an evolutionary relationship between these genes (see below). The gene for β_2-microglobulin is located in a region on chromosome 2 encoding several non-MHC-linked transplantation antigens and several immune response genes (75).

Although β_2-microglobulin and class I genes are located on different chromosomes, they are coordinately regulated. Both genes are not expressed in teratocarcinoma stem cells, which are presumably equivalent to early embryonic cells, whereas both are expressed in differentiated cell lines derived from the teratocarcinoma stem cells (19, 77). The regulation of these genes appears to be controlled transcriptionally.

Analyses of t Haplotype Mice with Class I cDNA Probes

Proximal to the MHC on chromosome 17 is a large genetic region termed the t complex (7, 8, 103). Some 25% of wild mice carry a mutation in this region that results in early embryonic death when it is present on both chromosomes. These mutations fall into eight complementation groups (7, 8). A chromosome 17 that carries one of these mutations has been termed a t chromosome and its DNA is said to be of t haplotype origin. All t haplotypes show several unusual properties. First, males transmit the t chromosome to their progeny with a much higher frequency than the normal chromosome 17. Second, genetic recombination between the t and the normal chromosome 17 is suppressed over a region of 20 cM that starts from the centromere and includes the H-2 complex (7, 8, 103). A variety of biochemical, serological, and genetic data have suggested that the t complex may be evolutionarily related to the MHC complex (2). To analyze the t complex, a chromosomal walk from the H-2 complex to the t complex may be feasible (110). As a first approach, class I cDNA clones have been used to analyze class I genes by Southern blot analyses in different t haplotype mice that express antigenically distinct class I molecules (100, 104). Surprisingly, restriction enzymes that revealed a considerable polymorphism between the class I genes of different normal inbred strains of mice failed to show a similar kind of polymorphism between different t haplotypes. This finding has been interpreted to indicate a close evolutionary relationship of all t haplotypes. The significance of these observations, as well as the possibility of an evolutionary relationship between the t complex and the MHC, remains for future studies to elucidate.

CLASS II GENES AND POLYPEPTIDES

Class II Polypeptides

The I region of the MHC encodes the class II molecules and has been divided by recombinational analyses into five distinct subregions denoted I-A, I-B, I-J, I-E, and I-C (78) (Figure 11). Two of these subregions, I-A and I-E, contain genes for the class II molecules, which have been well characterized by serological and biochemical methods. The two class II molecules encoded in the I-A and I-E subregions are both heterodimers composed of alpha and beta chains (Figure 11). The alpha chains range in molecular weight from 30,000–33,000 and the beta chains range in molecular weight from 27,000–29,000. The difference in molecular weight between the α and β chains is primarily due to differences in glycosylation. The α

chain contains two carbohydrate units, whereas only one is bound to β chains. The A_α, A_β, and E_β genes are contained within the I-A subregion, whereas the E_α gene is located within the I-E subregion (Figure 11).

A third subregion, I-J, appears to encode polypeptides that are subunits of the suppressor factors secreted by suppressor T cells. Although the I-J polypeptides have been characterized in functional assays, their biochemical characterization is only beginning (53, 117, 120). The characterization of suppressor factors and their genes has been of vital interest because it represents a unique opportunity to study one form of the elusive T-cell receptor. The I-B and I-C subregions have been defined exclusively on the basis of their ability to modulate certain immune responses in mice, and some immunologists question whether these subregions exist (48).

The structures of the class II polypeptides have been determined primarily from the analysis of human cDNA clones (3, 32, 51, 57, 58, 60, 64, 113, 118, 123) and more recently from the analysis of mouse and human genomic clones (9, 50, 61, 71, 73). Only very limited amino acid sequence information is available for mouse class II molecules (116), whereas more extensive sequence information has been obtained for human class II polypeptides (98). These data suggest that the mouse I-A and I-E molecules are homologous to the human DC and DR molecules, respectively (12) (Figure 1). The counterpart to the third human class II molecule, SB (40), has not been found in the mouse. The three human class II molecules each have α and β chains denoted DC_α, DC_β, DR_α, DR_β, SB_α, and SB_β. Each class II polypeptide is composed of two external domains, each about 90 amino acid residues in length, $\alpha1$ and $\alpha2$, or $\beta1$ and $\beta2$, a transmembrane region of about 30 residues, and a very short cytoplasmic region of about 10–15 residues (Figure 2). Three of the four external domains have centrally placed disulfide bridges ($\alpha2$, $\beta1$, and $\beta2$). Thus, the class I and class II molecules appear to be very similar in overall structure and domain organization (Figure 2).

The class II molecules serve as restricting elements that permit regulatory T cells (helper, suppressor, amplifier) to view antigen in the context of self on the surface of other T cells, macrophages, or B cells (49, 79). Class II genes appear to control the proliferation of regulatory T cells as well as the effector reactions carried out by these T cells, such as the amplification of other T-cell subsets and the promotion of B-cell differentiation.

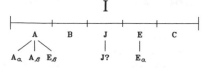

Figure 11   Genetic map of the I region (see text).

Structure of Class II Genes

DNA sequences of the E_α^d (73), E_α^k (71), and A_β^d (M. Malissen, personal communication) genes have been determined and the exon-intron organization of these genes is illustrated in Figure 12. Once again there is a striking correlation between the organization of exons and structural domains of the class II genes and molecules. Both genes contain separate exons for the leader peptide and the $\alpha 1$ and $\alpha 2$ domains. However, the E_α gene employs a single exon to encode the transmembrane and the cytoplasmic domains, whereas the A_β gene appears to contain an intervening sequence in the cytoplasmic portion separating the exons encoding the transmembrane and part of the cytoplasmic domains. The E_α gene has a fourth intron located in the 3' untranslated region (71). Thus, the structures of the class II α and β genes appear to differ in several regards. The gene for β_2-microglobulin shows an exon-intron organization very similar to that of the E_α gene (Figure 3A). For both genes, the exon encoding the leader peptide is separated by a long intervening sequence from the rest of the coding sequence [2.8 kb for β_2-microglobulin (77) and about 2.3 kb for E_α (71)] and an intervening DNA sequence is found in the beginning of the 3' untranslated region. The A_β gene displays a structure more similar to a class I gene in that transmembrane and at least part of the cytoplasmic regions are separated by an intron. These data suggest that the class II α genes are more closely related to β_2-microglobulin, whereas class II β genes are more closely related to class I genes. The human DR_α gene has been sequenced and has the same exon-intron organization as the E_α gene (50, 61).

Comparison of the amino acid sequences encoded by the A_α^k and E_α^k genes has revealed an overall homology of 50% between the two proteins (9). Quite surprisingly, the most conserved region is the transmembrane domain, which exhibits 78% homology, perhaps indicating structurally and functionally important interactions of these parts of the molecules with the β chains or other transmembrane proteins.

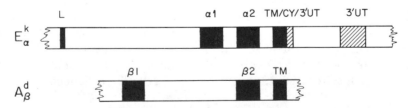

Figure 12   Schematic representation of the exon-intron organization of the E_α^k and A_β^d genes. Black boxes show exons; 3' untranslated regions (3' UT) are hatched; TM denotes transmembrane; and CY indicates cytoplasmic. The exons encoding the leader peptide and part of the cytoplasmic region and the 3' untranslated region have not yet been identified by DNA sequencing for the A_β gene. (Adapted from 70; and M. Malissen, personal communication.)

Homology of Class I, Class II, β_2-Microglobulin, Antibody, and Thy 1 Genes

There are several similarities between class I, class II, β_2-microglobulin, and antibody genes (Figures 2 and 3A). First, all genes exhibit a precise correlation between exons and the protein domains they encode. Second, the sizes of the external domains and the central placement and size of the disulfide bridges are similar. Third, RNA splicing always occurs between the first and second base of the junctional codons according to the GT/AG rule for both gene families. In other genes and gene families, RNA splicing also can occur after the second and third base of the junctional codons (99). Fourth, X-ray data (P. Bjorkman, J. Strominger, and D. Wiley, personal communication) suggest that the class I and antibody molecules have a twofold axis of symmetry and, accordingly, probably exhibit paired domains (Figure 2). Fifth, the DNA sequences of the class I α3 exons, the class II α2 and β2 exons, and β_2-microglobulin are homologous to the exon sequences encoding antibody constant regions (9, 11, 50, 57, 58, 71, 73, 84, 109) (Figure 2). A low level of homology also has been observed between the class I α1 and the class II β1 domains (58). Sixth, the class I α1 and α2 and the class II α1 exons do not show significant sequence homology to one another nor to the antibody genes. However, their sizes are similar to those of the other class I, class II, and antibody exons. Furthermore, the class I α2 exon encodes a centrally placed disulfide bridge. These observations suggest that the complete class I, class II, and antibody genes shared a common ancestor and marked changes occurred after divergence of the genes to fulfill different functions. Finally, homology to the T-cell differentiation antigen, Thy 1, also is observed at the protein level (122). Because of their sequence and structural relationships, it is attractive to postulate that all of these genes —class I, class II, β_2-microglobulin, immunoglobulin, and Thy 1—must have descended from a common ancestor and are therefore members of a supergene family. Two qualifications must be raised. First, it is impossible to distinguish between divergent and convergent evolution. Accordingly, the membrane-proximal domains of these molecules may have arisen from distinct genes that converged toward one another because of common functional constraints. If this hypothesis is correct, one would have to argue that convergent evolution also generated molecules with external domains of similar size and with similar disulfide bridge placement. This possibility appears unlikely, but cannot be formally excluded. Second, perhaps only the membrane proximal domains evolved from a common ancestor, whereas the remaining portions of these genes could have independent origins. Once again, the shared size and centrally placed disulfide bridges of the external domains make this an unattractive but formally possible hypothesis.

Linkage Relationship of Class II Genes

A continuous stretch of about 230 kb of DNA has been isolated from the I region by first screening a BALB/c sperm cosmid library with a human DR_α cDNA probe and then by using single-copy probes for chromosomal walking (110). The cloned region appears to contain genes encoding all four serologically defined class II polypeptides. The class II genes have been identified through the use of synthetic DNA probes whose sequences were generated from protein data, mouse B cell-specific cDNA probes, and human cDNA probes that cross-hybridized to the mouse class II genes. These data are depicted in Figure 13.

This 230-kb region of DNA includes the right-hand boundary of the I region, because the structural gene for the complement component C4 mapping into the S region of the MHC could be identified (M. Steinmetz, L. Fors, A. Örn, unpublished results). The C4 gene lies about 90 kb distal to the E_α gene and was identified with a synthetic oligonucleotide probe specific for the N-terminal portion of the C4 α subunit. The left-hand boundary of the I region cannot be defined at present because the cosmid clones do not extend into the K region.

Five class II genes, extending over a stretch of DNA encompassing approximately 90 kb, have been identified. The E_α (73), A_α (20), and A_β (M. Malissen, personal communication) genes have been identified by direct DNA sequence analysis, the E_β gene has been identified by hybridization with a specific oligonucleotide probe (110), and the $E_{\beta 2}$ gene has been identified by cross-hybridization to a human DC_β chain cDNA clone and the mouse E_β gene (110). Identification of the E_β gene was confirmed by the use of restriction enzyme site polymorphisms to localize a serologically defined E_β recombinant to the middle of the E_β gene (110). Whether the $E_{\beta 2}$ gene is functional or represents a pseudogene is unknown. The A_β and E_β genes have the same 5' to 3' orientation, whereas the E_α gene shows an opposite orientation. The orientation of the A_α gene has not yet been determined.

The location of the A_α gene between the A_β and E_β genes is different from the gene order of A_α-A_β-E_β proposed from peptide map analyses of the A_α polypeptide (95). Strain A.TL, which is an intra-H-2 recombinant derived from a cross between B10.S (A_α^s), and A.AL (A_α^k) was found to bear an A_α chain whose peptide fragments appeared to be a composite of A_α^s and A_α^k peptides. If this interpretation of the data is correct, the A_α gene would map proximal to the A_β gene. However, other investigators could not find a difference between the A_α polypeptides isolated from A.TL and B10.A(1R) mice (15, 72). Until the recombination point in the A.TL mice has been mapped at the DNA level this will remain con-

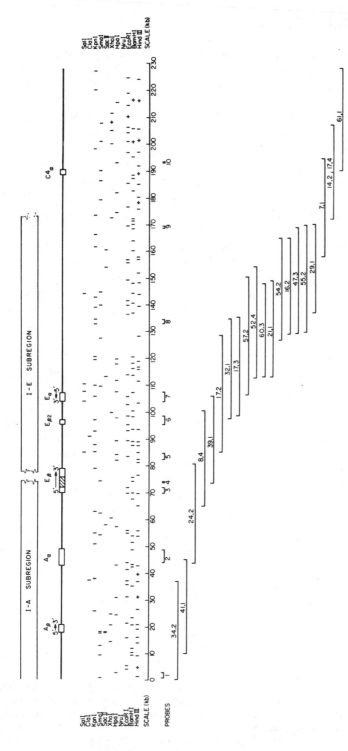

Figure 13 Molecular map of the I region of the BALB/c mouse. Boxes indicate identified class II genes and give the maximum region that hybridizes to the probes used for identification. Overlapping cosmid clones that define the stretch of 230 kb of DNA are listed at the bottom. Ten single-copy probes were used to define the location of the I-A and I-E subregions as indicated. Sal I, Cla I, etc, are restriction enzymes. (Adapted from 110.)

troversial. However, the gene cloning data unequivocally suggest that the class II gene order in BALB/c mice is 5'-A_β-A_α-E_β-E_α-3'.

The region corresponding to that containing the five class II genes in BALB/c DNA (d haplotype) also has been isolated from the DNA of AKR mice (k haplotype) by screening an AKR cosmid library (M. Steinmetz, L. Fors, A. Örn, unpublished results). Two important points arise from a comparison of the I regions from the k and d haplotypes (Figure 14). First, the same five class II genes are present in the k haplotype DNA. Moreover, their organization is virtually identical in both haplotypes without any major insertions, deletions, or rearrangements. Second, a striking boundary of polymorphism proximal to the E_α gene is observed when the restriction maps of d and k haplotype DNA are compared. Although the two DNAs are almost identical to one another distal to this boundary, they show extensive restriction enzyme site polymorphism proximal to this point. Interestingly, this polymorphism breakpoint correlates with a hot spot of recombination in the I region (see below). This difference in restriction enzyme site polymorphisms has been noted previously by Southern blot analyses of various mouse strains (110) and correlates nicely with the serological polymorphisms of the corresponding genes (47). Only two alleles each are known for the E_α, C4, and Slp genes, whereas the A_β and E_β genes appear to be as polymorphic as the class I genes K and D. The important point is that the sequence conservation or restriction enzyme site polymorphisms are not confined to coding sequences but cover large regions of flanking DNA. This observation obviously has important implications for evolutionary mechanisms that attempt to explain the generation of the extensive polymorphism of certain MHC alleles (see earlier discussion).

Southern blot analyses have been carried out on mouse DNA with α and β probes under conditions that favor extensive cross-hybridization in an

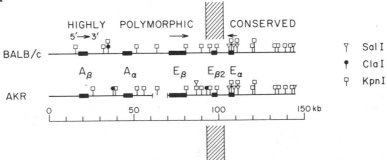

Figure 14   A polymorphic breakpoint in the I region. Cloned regions of the I region from BALB/c (d haplotype) and AKR (k haplotype) mice are compared over a stretch of 150 kb of DNA. The region to the left of the breakpoint exhibits extensive restriction enzyme site polymorphisms, whereas the region to the right does not. The polymorphism breakpoint is located close to the recombinational hot spot at the 3' end of the E_β gene. (From M. Steinmetz, L. Fors and A. Örn, unpublished results.)

attempt to define the total number of α and β genes (110). These data suggest that there are two α genes and four to six β genes in the mouse genome. Thus, the total number of class II genes that can cross-react with the available class II DNA probes will be less than 10, although the existence of more distantly related class II genes in the mouse cannot be excluded. This striking paucity of class II genes stands in sharp contrast to the exquisite specificity of immune reponsiveness regulated by the class II genes. These data suggest that the specificity of immune responsiveness is likely to be governed by the repertoire of T-cell receptors.

Correlation of the Molecular with the Genetic Map of the I Region

Ten single-copy DNA probes were isolated from various portions of the 230 kb of cloned DNA in the I region (110) (see Figure 13). These probes were used to search for restriction enzyme site polymorphisms in various inbred, congenic, and recombinant congenic strains of mice. These polymorphisms were then correlated with the serological polymorphisms, which define the various subregions of the I region. In this manner the molecular map of the I region (Figure 13) could be correlated with the genetic map produced by recombinational analysis of serological polymorphisms (Figure 11). These analyses mapped each of the four serologically defined class II genes to the appropriate subregions (Figure 15). Thus, the A_β, the A_α, and at least the 5' half of the E_β genes are all encoded in the I-A subregion, whereas the E_α gene is encoded in the I-E subregion. The genetic and molecular maps are consistent in the locations of these genes.

An inconsistency exists between the genetic and molecular maps for the I-B and I-J subregions. The analysis of nine recombinant congenic mice defining the right-hand and left-hand boundaries of the I-A and I-E subregions, respectively, have led to the conclusion that these two subregions are separated by no more than 3.4 kb of DNA (110) (Figure 15). To complicate matters further, the 3' half of the E_β gene resides in at least 2.4 kb of this 3.4-kb region. Thus, about 1 kb of DNA remains for the I-B and I-J subregions. Indeed, it is possible that the I-A and I-E subregions join directly to one another. Two observations exclude the possibility that the BALB/c mouse has deleted or rearranged the I-B and I-J subregions in comparison to other mice. First, extensive Southern blotting data indicate that the organization of BALB/c DNA around the 3.4-kb region is no different from that found in three other mouse strains (b, k, s haplotypes), which have been used to define the I-B and I-J subregions (110). Second, the corresponding regions in the cloned k and d haplotype DNAs are co-linear (M. Steinmetz, L. Fors, A. Örn, unpublished results) (Figure 14).

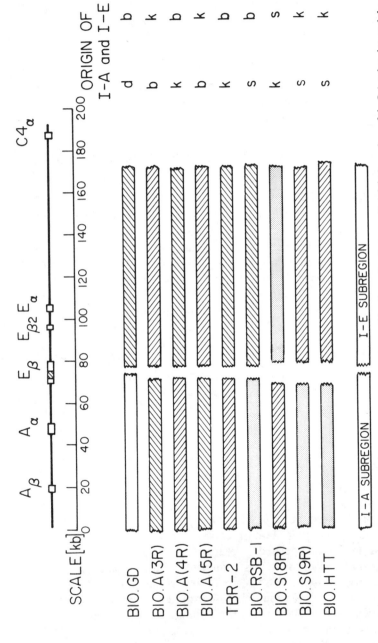

Figure 15   A molecular map of nine inbred mice exhibiting recombination points between the right boundary of the I-A subregion and the left boundary of the I-E subregion. All recombination events have occurred within a stretch of 8 kb of DNA or less.

These observations pose a paradox for explaining the location of the genes of the I-B and I-J subregions. Others have suggested that the I-B subregion can be explained by gene complementation between the I-A and I-E subregions (48). Our data supported the notion that the I-B subregion does not exist. However, it is more difficult to dismiss the I-J subregion because I-J polypeptides have recently been identified by monoclonal antibodies (T. Tada, personal communication).

The Paradox of the I-J Subregion

How can the expression of the I-J polypeptides be encoded by 3.4 kb or less of DNA? There are two possible categories of explanations (110). First, the J gene or genes may be encoded outside that segment of DNA residing between the I-A and I-E subregions. For example, the interpretation of recombinant events in the recombinant congenic mice could be incorrect in that these mice may have undergone multiple recombinational rather than the usually assumed single recombinational events within the MHC. Alternatively, this 3.4-kb segment of DNA may contain a regulatory element that controls the expression of a battery of polymorphic J genes located elsewhere in the mouse genome. A second category of hypotheses suggests that at least part of the J polypeptides are encoded in this 3.4-kb region. The I-J gene might be identical with the E_β gene and appropriate post-translational modification (e.g. glycosylation, etc) could generate the serological determinants of the J polypeptide. Alternatively, the J polypeptides may be encoded by some E_β exons, which are then linked to distinct I-J exons by alternative patterns of RNA splicing much as alternative splicing exists for $\mu_{membrane}$ and $\mu_{secreted}$ mRNA's of antibody heavy chain genes. Finally, the J gene may be transcribed off the DNA strand opposite that which encodes the E_β gene. The possibility that the J gene is encoded entirely in the remaining 1 kb of DNA in this region appears remote since most eukaryotic genes require far more DNA.

Evidence has already been obtained that argues against the second category of explanations (52). RNA isolated from a number of suppressor T-cell lines does not show hybridization to DNA sequences covering this 3.4-kb region. These RNAs also fail to hybridize with extensive regions of DNA upstream and downstream from the 3.4-kb region. These data suggest that the structural gene for I-J is not located between the I-A and I-E subregions.

The location of the J gene is a question of considerable importance because this polypeptide is expressed alone or in conjunction with another polypeptide in two distinct kinds of T-suppressor factors. These T-suppressor factors have been viewed by many as the easiest way to approach the structure and molecular biology of the T-cell receptor. Hence, isolation of the I-J gene could be of considerable significance as an approach to the cloning of the T-cell receptor.

Nonlinearity of Recombination

Nine recombinant haplotypes for the I-A and I-E subregions have been examined by mapping of restriction enzyme site polymorphisms and in all strains the point of crossing-over falls within a region of 8 kb of DNA (110) (Figure 15). Indeed, these recombinations may all have occurred at precisely the same point. These observations suggest that recombination in the eukaryotic genome is not randomly scattered throughout but, at least in this case, is highly localized to a discrete region. The occurrence of hot spots of recombination has obvious implications for any attempts to translate centimorgans of DNA from the genetic or recombinational maps into kilobases of DNA from molecular maps. Such a conversion must assume that recombination is randomly scattered throughout the mouse genome. Since the A_β, E_α, and C4 genes have been linked by molecular cloning, recombination frequencies for this region can be compared to physical distances: A_β and E_α are 0.1 cM apart by recombinational mapping and are separated by 85 kb of DNA. The E_α and C4 genes separated by 0.11 cM are 90 kb from one another. Earlier calculations based on the assumption of random recombination suggested that 1 cM should equal 2000 kb. Thus the actual distances are shorter by about a factor of two to three than the theoretical distances. It will be interesting to examine other chromosomal regions of the eukaryotic genome where multiple recombinations have been mapped to determine whether the occurrence of recombinational hot spots is specific to the I region or whether this is a much more general phenomenon.

Expression of Class II Genes

Four inbred strains of mice (b, f, q, s haplotypes) do not express the I-E molecule on the cell surface. Two of these (b and s) do not express the E_α polypeptide but have cytoplasmic E_β chains, and two (f and q) neither express E_α nor E_β. By Southern blot analyses with specific probes, it appears that the failure to express these genes is not due to a complete deletion of these genes in the respective haplotypes (11). More detailed analyses of the structure and transcription of the E_α gene in these inbred strains has revealed that at least two different mechanisms are responsible (70). No transcription of the E_α gene is observed in mice of the b and s haplotypes, presumably because a small deletion occurred in the promotor region of the gene. Obviously defective RNA transcripts of various sizes are found in mice of the f and q haplotypes.

No one has succeeded unequivocally in expressing class II genes in mouse L cells after DNA-mediated gene transfer. This might be due to the presence of a weak promotor regulating class II gene transcription, inappropriate splicing, or instability of mRNA's. Furthermore, the expression of a third chain, called the invariant (Ii) chain, might be necessary for the transport of class II α and β chains to the cell surface (55). The powerful approaches of molecular biology will certainly lead to class II gene expression within the relatively near future. The long-term goal is to transfer class II genes into biologically relevant cells such as macrophages, B cells, and T cells so that their functions can be assessed and perturbed by in vitro mutagenesis.

THE FUTURE

Molecular genetics has given us striking insights into the structures and organization of genes encoded by the MHC. The tools of molecular biology, when employed in conjunction with the well-developed assays of cellular immunology, provide a unique opportunity to dissect the functions of these multigene families involved in self-nonself recognition phenomena. It also will be interesting to determine whether or not additional multigene families fall into the supergene family, which includes the antibody, class I, and class II genes. Future experiments will show the extent to which strategies of regulation and information expansion are shared by these homologous gene families.

ACKNOWLEDGMENTS

We thank R. Flavell, B. Mach, H. McDevitt, and S. Nathenson for sending preprints of their work prior to publication, B. Larsh for her careful preparation of the manuscript, and E. Kraig and J. Parnes for thoughtful comments. M.S. was supported by a Senior Lievre Fellowship from the California Division of the American Cancer Society and B.M. was supported by a CNRS and DGRST Fellowship from the French Government. The work of L.H. and M.S. was supported by grants from the National Institutes of Health.

Literature Cited

1. Alper, C. A. 1980 *The Role of the Major Histocompatibility Complex in Immunology,* ed. M. E. Dorf, pp. 173–220. New York: Garland STPM
2. Artzt, K., Bennett, D. 1975. *Nature* 256:545–47
3. Auffray, C., Korman, A. J., Roux-Dosseto, M., Bono, R., Strominger, J. L. 1982. *Proc. Natl. Acad. Sci. USA* 79: 6337–41
4. Baltimore, D. 1981. *Cell* 24:592–94
5. Barbosa, J. A., Kamarck, M. E., Biro, P. A., Weissman, S. M., Ruddle, F. H.

1982. *Proc. Natl. Acad. Sci. USA* 79: 6327–31
6. Battey, J., Max, E. E., McBride, O. W., Swan, D., Leder, P. 1982. *Proc. Natl. Acad. Sci. USA* 79:5956–60
7. Bennett, D. 1975. *Cell* 6:441–54
8. Bennett, D. 1980. *Harvey Lectures* 74:1–21
9. Benoist, C. O., Mathis, D. J., Kanter, M. R., Williams, V. E., McDevitt, H. O. 1983. *Proc. Natl. Acad. Sci. USA.* In press
10. Biro, P. A., Pereira, D., Sood, A. K., DeMartinville, B., Francke, N., Weiss-

566   HOOD ET AL

man, S. M. 1981. *Immunoglobulin Idiotypes, ICN-UCLA Symposium on Molecular and Cell Biology*, ed. C. Janeway, E. E. Sercarz, H. Wigzell, 20:315–26. New York: Academic
11. Brégégère, F., Abastado, J. P., Kvist, S., Rask, L., Lalanne, J. L., et al. 1981. *Nature* 292:78–81
12. Bono, M. R., Stominger, J. L. 1982. *Nature* 299:836–38
13. Cami, B., Brégégère, F., Abastado, J. P., Kourilsky, P. 1981. *Nature* 291: 673–75
14. Coligan, J. E., Kindt, T. J., Uehara, H., Martinko, J., Nathenson, S. G. 1981. *Nature* 291:35–39
15. Cook, R. G., Capra, J. D., Bednarczyk, J. L., Uhr, J. W., Vitetta, E. S. 1979. *J. Immunol.* 123:2799–803
16. Cosman, D., Khoury, G., Jay, G. 1982. *Nature* 295:73–76
17. Cosman, D., Kress, M., Khoury, G., Jay, G. 1982. *Proc. Natl. Acad. Sci. USA* 79:4947–51
18. Crabtree, G. R., Kant, J. A. 1982. *Cell* 31:159–66
19. Croce, C. M., Linnenbach, A., Huebner, K., Parnes, J. R., Margulies, D. H., et al. 1981. *Proc. Natl. Acad. Sci. USA* 78:5754–58
20. Davis, M. M., Nielsen, E., Cohen, D., Steinmetz, M., Hood, L. 1983. *Proc. Natl. Acad. Sci. USA*. In press
21. Démant, P., Iványi, D., Oudshoorn-Snoek, M., Calafat, J., Roos, M. H. 1981. *Immunol. Rev.* 60:5–22
22. Early, P., Rogers, J., Davis, M., Calame, K., Bond, M., et al. 1980. *Cell* 20:313–19
23. Evans, G. A., Margulies, D. H., Camerini-Otero, R. D., Ozato, K., Seidman, J. G. 1982. *Proc. Natl. Acad. Sci. USA* 79:1994–98
24. Flaherty, L. 1980. *The Role of the Major Histocompatibility Complex in Immunology*, ed. M. E. Dorf, pp. 33–57. New York: Garland STPM
25. Forman, J., Goodenow, R. S., Hood, L., Ciavarra, R. 1983. *J. Exp. Med.* In press
26. Goding, J. W. 1981. *J. Immunol.* 126:1644–46
27. Golden, L., Mellor, A. L., Weiss, E. H., Bullman, H., Bud, H., et al. Submitted for publication
28. Goodenow, R. S., McMillan, M., Nicolson, M., Sher, B. T., Eakle, K., et al. 1982. *Nature* 300:231–37
29. Goodenow, R. S., McMillan, M., Örn, A., Nicolson, M., Davidson, N., et al. 1982. *Science* 215:677–79

30. Gorer, P. A. 1938. *J. Pathol. Bacteriol.* 47:231–52
31. Gorer, P. A., Lyman, S., Snell, G. D. 1948. *Proc. R. Soc. London Ser. B* 135:499–505
32. Gustafsson, K., Bill, P., Larhammar, D., Wiman, K., Claesson, L., et al. 1982. *Scand. J. Immunol.* 16:303–8
33. Hansen, T. H., Ozato, K., Melino, M. R., Coligan, J. E., Kindt, T. J., et al. 1981. *J. Immunol.* 126:1713–16
34. Hildemann, W. H., Clark, F. A., Raison, R. L. 1981. *Comprehensive Immunogenetics.* New York: Elsevier
35. Hildemann, W. H., Johnston, I. S., Jokiel, P. L. 1979. *Science* 204:420–22
36. Hildemann, W. H., Jokiel, P. L., Bigger, C. H., Johnston, I. S. 1980. *Transplantation* 30:297–301
37. Hollis, G. F., Hieter, P. A., McBride, O. W., Swan, D., Leder, P. 1982. *Nature* 297:83–84
38. Hood, L., Campbell, J. H., Elgin, S. C. R. 1975. *Ann. Rev. Genet.* 9:305–53
39. Hood, L., Steinmetz, M., Goodenow, R. 1982. *Cell* 28:685–87
40. Hurley, C. K., Shaw, S., Nadler, L., Schlossman, S., Capra, J. D. 1982. *J. Exp. Med.* 156:1557–62
41. Hurst, D. D., Fogel, S., Mortimer, R. K. 1972. *Proc. Natl. Acad. Sci. USA* 69:101–5
42. Jackson, J. A., Fink, G. R. 1981. *Nature* 292:306–11
43. Jelinek, W. R., Schmid, C. W. 1982. *Ann. Rev. Biochem.* 51:813–44
44. Jordan, B. R., Brégégère, F., Kourilsky, P. 1981. *Nature* 290:521–23
45. Klein, J. 1975. *Biology of the Mouse Histocompatibility Complex.* Berlin: Springer
46. Klein, J. 1979. *Science* 203:516–21
47. Klein, J., Figueroa, F. 1981. *Immunol. Rev.* 60:23–57
48. Klein, J., Figueroa, F., Nagy, Z. A. 1983. *Ann. Rev. Immunol.* 1:119–42
49. Klein, J., Juretic, A., Baxevanis, C. N., Nagy, Z. A. 1981. *Nature* 291:455–60
50. Korman, A. J., Auffray, C., Schambock, A., Strominger, J. L. 1982. *Proc. Natl. Acad. Sci. USA* 79:6013–17
51. Korman, A. J., Knudsen, P. J., Kaufmann, J. F., Strominger, J. L. 1982. *Proc. Natl. Acad. Sci. USA* 79:1844–48
52. Kronenberg, M., Steinmetz, M., Kobori, J., Kraig, E., Kapp, J., et al. Manuscript in preparation
53. Krupen, K., Araneo, B. A., Brink, L., Kapp, J. A., Stein, S., et al. 1982. *Proc. Natl. Acad. Sci. USA* 79:1254–58
54. Kvist, S., Brégégère, F., Rask, L., Cami,

B., Garoff, H., et al. 1981. *Proc. Natl. Acad. Sci. USA* 78:2772–76

55. Kvist, S., Wiman, K., Claesson, L., Peterson, P. A., Dobberstein, B. 1982. *Cell* 29:61–69
56. Lalanne, J. L., Brégégère, F., Delarbre, C., Abastado, J. P., Gachelin, G., Kourilsky, P. 1982. *Nucleic Acids Res.* 10:1039–49
57. Larhammar, D., Gustafsson, K., Claesson, L., Bill, P., Wiman, K., et al. 1982. *Cell* 40:153–61
58. Larhammar, D., Schenning, L., Gustafsson, K., Wiman, K., Claesson, L., et al. 1982. *Proc. Natl. Acad. Sci. USA* 79: 3687–91
59. Leder, A., Swan, D., Ruddle, F., D'Eustachio, P., Leder, P. 1981. *Nature* 293:196–200
60. Lee, J. S., Trowsdale, J., Bodmer, W. F. 1982. *Proc. Natl. Acad. Sci. USA* 79: 545–49
61. Lee, J. S., Trowsdale, J., Travers, P. J., Carey, J., Grosveld, F., et al. 1982. *Nature* 299:750–52
62. Lemonnier, F. A., Le Bouteiller, P., Malissen, B., Golstein, P., Malissen, M., et al. 1983. *J. Immunol.* In press
63. Lemonnier, F. A., Malissen, M., Golstein, P., LeBouteiller, P., Rebai, N., et al. 1982. *Immunogenetics* 16:355–61
64. Long, E. O., Wake, C. T., Strubin, M., Gross, N., Accolla, R. S., et al. 1982. *Proc. Natl. Acad. Sci. USA.* 79:7465–9
65. Lopéz de Castro, J. A., Strominger, J. L., Strong, D. M., Orr, H. T. 1982. *Proc. Natl. Acad. Sci. USA* 79:3813–17
66. Malissen, M., Damotte, M., Birnbaum, D., Trucy, J., Jordan, B. R. 1982. *Gene.* 20:485–9
67. Malissen, M., Malissen, B., Jordan, B. R. 1982. *Proc. Natl. Acad. Sci. USA* 79:893–97
68. Maloy, W. L., Coligan, J. E. 1982. *Immunogenetics* 16:11–22
69. Margulies, D. H., Evans, G. A., Flaherty, L., Seidman, J. G. 1982. *Nature* 295:168–70
70. Mathis, D. J., Benoist, C. O., Williams, V. E., Kanter, M., McDevitt, H. O. 1983. *Proc. Natl. Acad. Sci. USA.* In press
71. Mathis, D. J., Benoist, C. O., Williams, V. E., Kanter, M., McDevitt, H. O. Submitted for publication
72. McMillan, M., Frelinger, J. A., Jones, P. P., Murphy, D. B., McDevitt, H. O., Hood, L. 1981. *J. Exp. Med.* 153: 936–50
73. McNicholas, J., Steinmetz, M., Hunkapiller, T., Jones, P., Hood, L. 1982. *Science.* 218:1229–32
74. Mellor, A. L., Golden, L., Weiss, E., Bullmann, H., Hurst, J., et al. 1982. *Nature* 298:529–34
75. Michaelson, J. 1981. *Immunogenetics* 13:167–71
76. Moore, K. W., Sher, B. T., Sun, Y. H., Eakle, K., Hood, L. 1982. *Science* 215:679–82
77. Morello, D., Daniel, F., Baldacci, P., Cayre, Y., Gachelin, G., Kourilsky, P. 1982. *Nature* 296:260–62
78. Murphy, D. B. 1980. *The Role of the Major Histocompatibility Complex in Immunology,* ed. M. E. Dorf, pp. 1–32. New York: Garland STPM
79. Nagy, Z., Baxevanis, C. N., Ishii, N., Klein, J. 1981. *Immunol. Rev.* 60:59–83
80. Nairn, R., Yamaga, K., Nathenson, S. G. 1980. *Ann. Rev. Genet.* 14:241–77
81. Nathenson, S. G., Uehara, H., Ewenstein, B. M., Kindt, T. J., Coligan, J. E. 1981. *Ann. Rev. Biochem.* 50:1025–52
81a. Ollo, R., Rougeon, F. 1983. *Cell.* In press
82. Örn, A., Goodenow, R. S., Hood, L., Brayton, P., Woodward, J. G., Frelinger, J. A. 1982. *Nature* 297:415–17
83. Orr, H. T., Bach, F. H., Ploegh, H. L., Strominger, J. L., Kavathas, P., DeMars, R. 1982. *Nature London* 296:454–56
84. Parnes, J. R., Seidman, J. G. 1982. *Cell* 29:661–69
85. Parnes, J. R., Velan, B., Felsenfeld, A., Ramanathan, L., Ferrini, U., et al. 1981. *Proc. Natl. Acad. Sci. USA* 78: 2253–57
86. Pease, L. R., Nathenson, S. G., Leinwand, L. A. 1982. *Nature* 298:382–85
87. Pease, L. R., Schulze, D. H., Pfaffenbach, G. M., Nathenson, S. G. 1983. *Proc. Natl. Acad. Sci. USA* In Press
88. Peterson, P. A., Cunningham, B. A., Berggard, I., Edelman, G. M. 1972. *Proc. Natl. Acad. Sci. USA* 69:1697–1701
89. Ploegh, H. L., Orr, H. T., Strominger, J. L. 1980. *Proc. Natl. Acad. Sci. USA* 77:6081–85
90. Ploegh, H. L., Orr, H. T., Strominger, J. L. 1981. *Cell* 24:287–99
91. Reyes, A. A., Johnson, M. J., Schöld, M., Ito, H., Ike, Y., et al. 1981. *Immunogenetics* 14:383–92
92. Reyes, A. A., Schöld, M., Itakura, K., Wallace, R. B. 1982. *Proc. Natl. Acad. Sci. USA* 79:3270–74
93. Reyes, A. A., Schöld, M., Wallace, R. B. 1982. *Immunogenetics* 16:1–9
94. Robinson, P. J., Lundin, L., Sege, K., Graf, L., Wigzell, H., Peterson, P. A. 1981. *Immunogenetics* 14:449–52

95. Rose, S. M., Cullen, S. E. 1981. *J. Immunol.* 127:1472–77
96. Schreier, P. H., Bothwell, A. L. M., Mueller-Hill, B., Baltimore, D. 1981. *Proc. Natl. Acad. Sci. USA* 78:4495–99
97. Schulze, D. H., Pease, L. R., Wallace, R. B., Nathenson, S. G., Seier, S. S., Reyes, A. A. Submitted for publication
98. Shackelford, D. A., Kaufman, J. F., Korman, A. J., Strominger, J. L. 1982. *Immunol. Rev.* 6:133–87
99. Sharp, P. A. 1981. *Cell* 23:643–46
100. Shin, H. S., Stavnezer, J., Artzt, K., Bennett, D. 1982. *Cell* 29:969–76
101. Silver, J., Hood, L. E. 1976. *Proc. Natl. Acad. Sci. USA* 73:599–603
102. Silver, J., Hood, L. 1976. *Contemporary Topics in Molecular Immunology,* ed. H. N. Eisen, R. A. Reisfeld, 5:35–68. New York: Plenum
103. Silver, L. 1981. *Cell* 27:239–40
104. Silver, L. M. 1982. *Cell* 29:961–68
105. Singer, D. S., Camerini-Otero, R. D., Satz, M. L., Osborne, B., Sachs, D., Rudikoff, S. 1982. *Proc. Natl. Acad. Sci. USA* 79:1403–7
106. Snell, G. D., Dausset, J., Nathenson, S. 1976. *Histocompatibility.* New York: Academic
107. Soloski, M. J., Uhr, J. W., Flaherty, L., Vitetta, E. S. 1981. *J. Exp. Med.* 153:1080–93
108. Sood, A. K., Pereira, D., Weissman, S. M. 1981. *Proc. Natl. Acad. Sci. USA* 78:616–20
109. Steinmetz, M., Frelinger, J. G., Fisher, D., Hunkapiller, T., Pereira, D., et al. 1981. *Cell* 24:125–34
110. Steinmetz, M., Minard, K., Horvath, S., McNicholas, J., Frelinger, J., et al. 1982. *Nature* 300:35–42
111. Steinmetz, M., Moore, K. W., Frelinger, J. G., Sher, B. T., Shen, F.-W., et al. 1981. *Cell* 25:683–92
112. Steinmetz, M., Winoto, A., Minard, K., Hood, L. 1982. *Cell* 28:489–98
113. Stetler, D., Das, H., Nunberg, J. H., Saiki, R., Sheng-Dong, R., et al. 1982.

114. Strominger, J. L., Orr, H. T., Parham, P., Ploegh, H. L., Mann, D. L., et al. 1980. *Scand. J. Immunol.* 11:573–92
115. Suggs, S. V., Wallace, R. B., Hirose, T., Kawashima, E. H., Itakura, K. 1981. *Proc. Natl. Acad. Sci. USA* 78:6613–17
116. Sung, E., Hunkapiller, M. W., Hood, L. E., Jones, P. P. Manuscript in preparation
117. Taniguchi, M., Tokuhisa, T., Kanno, M., Yaoita, Y., Shimizu, A., Honjo, T. 1982. *Nature* 298:172–74
118. Wake, C. T., Long, E. O., Strubin, M., Gross, N., Accolla, R., Carrel, S., Mach, B. 1982. *Proc. Natl. Acad. Sci. USA.* 79:6979–83
119. Weiss, E. H., Mellor, A., Golden, L., Fahrner, K., Simpson, E., et al. Submitted for publication
120. Wieder, K. J., Araneo, B. A., Kapp, J. A., Webb, D. R. 1982. *Proc. Natl. Acad. Sci. USA* 79:3599–3603
121. Wilde, C. D., Crowther, C. E., Cripe, T. P., Gwo-Shu Lee, M., Cowan, N. J. 1982. *Nature* 297:83–84
122. Williams, A. F., Gagnon, J. 1982. *Science* 216:696–703
123. Wilson, P. H., Nairn, R., Nathenson, S. G., Sears, D. W. 1982. *Immunogenetics* 15:225–37
124. Wiman, K., Larhammar, D., Claesson, L., Gustafsson, K., Schenning L., et al. 1982. *Proc. Natl. Acad. Sci. USA* 79:1703–7
125. Winoto, A., Steinmetz, M., Hood, L. 1983. *Proc. Natl. Acad. Sci. USA.* In press
126. Xin, J. H., Kvist, S., Dobberstein, B. 1982. *EMBO J.* 1:467–71
127. Yokoyama, K., Stockert, E., Old, L. J., Nathenson, S. G. 1981. *Proc. Natl. Acad. Sci. USA* 78:7078–82
128. Yokoyama, K., Nathenson, S. G. 1983. *J. Immunol.* In press
129. Zinkernagel, R. M., Doherty, P. C. 1980. *Adv. Immunol.* 27:51–177

Proc. Natl. Acad. Sci. USA 79:5966–70

Ann. Rev. Immunol. 1983 1:569–607

GENETICS, EXPRESSION, AND FUNCTION OF IDIOTYPES

Klaus Rajewsky and Toshitada Takemori

Institute for Genetics, University of Cologne, Cologne, Federal Republic of Germany

1. INTRODUCTION

In this article we speak about the network of idiotypes in the immune system. We start by defining the vocabulary we are going to use. We then discuss the structural basis of idiotypic determinants (i.e. antigenic determinants on the antibody variable region; also called idiotopes) and of their complementary anti-idiotypic antibody binding sites (anti-idiotopes). Idiotopes and anti-idiotopes are encoded by antibody V, D, and J genes. The idiotypic repertoire thus reflects the repertoire of these genes as they are expressed, selected, and somatically modified in the immune system. The idiotypic repertoire, its genetic basis and its selection in ontogeny, is a main theme of the present view. This relates directly to the functional idiotypic network: Which rules do idiotypic interactions in the immune system follow and what is their physiological role? We will see that idiotypic interactions seem to be involved in the growth control of lymphocyte clones with defined receptor specificity and may thus be essential for the generation of a diverse system of interacting cells. They also stabilize the expression of a given receptor repertoire in active immune responses. Idiotypic interactions occur within the compartment of B lymphocytes, but they also involve T cells. We discuss the crucial issue of whether or not recognition of B-cell idiotypes is one of the principles by which T cells are selected in the immune system.

This article cannot deal with all facets of idiotypic research. We restrict ourselves largely to the discussion of experimental work in the mouse and, in particular, we do not discuss in any detail the role of idiotypic interactions in regulatory T-cell circuits controlling immune responses. Detailed models of immunologic control through cell interactions have emerged from recent work in this field. We refer the disappointed reader to a few

0732-0582/83/0410-0569$02.00

reviews on this and other matters that are not discussed here (9, 22, 57, 73a, 78, 95, 184, 198, 204).

2. VOCABULARY AND STRUCTURAL PRINCIPLES OF IDIOTYPIC INTERACTIONS BETWEEN ANTIBODIES

An idiotype (129, 155) is a set of idiotypic determinants or idiotopes expressed on the V region of a particular antibody or the V regions of a set of related antibodies. Idiotypes and idiotopes are defined serologically, i.e. by anti-idiotypic and anti-idiotope antibodies, respectively. Anti-idiotypic antibodies recognize, in general, a set of idiotopes and are therefore polyclonal. Anti-idiotope antibodies recognize individual idiotopes and are therefore, in general, monoclonal reagents. The idiotype of a given antibody distinguishes the latter from most other antibodies, but, as is discussed in more detail below, idiotypic crossreactions are frequently observed.

In Figure 1 we depict schematically the interaction of two complementary antibodies, antibody 1 (ab1) and antibody 2 (ab2). We may consider ab2 as an anti-idiotope, raised against ab1 and binding with its hapten binding site (paratope) to an idiotope of ab1. However, this distinction into anti-idiotope and idiotope-bearing antibody is merely operational: we might as well have immunized an animal with ab2 and obtained, as an anti-idiotope, ab1. In this situation the anti-idiotope would bind with a structure outside of the hapten binding cleft (paratope) to the idiotope-bearing ab2, and the idiotope recognized would be the paratope of ab2 itself. The scheme in Figure 1 indicates that idiotypic interactions between antibodies occur through complementary structures in antibody V regions and must involve amino acid residues on protrusions and clefts of the interacting V regions. Because of the selectivity of idiotypic recognition, the interacting complementary structures must be largely determined by the hypervariable regions

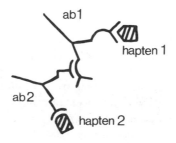

Figure 1   Schematic drawing of two interacting antibodies. Complementary structures are meant to be in the V regions.

of the V domains, an expectation verified by a large body of experimental work (see Section 3). It is satisfying to see that in the available three-dimensional models of antibody V domains, obtained by X-ray diffraction techniques and indirect methods, the hypervariable regions are indeed partly arranged as pockets and partly exposed on the surface of the V domain (reviewed in 75, 159). If this were not the case, idiotypic interactions of the exquisite specificity observed would be impossible.[1]

The structural requirements of idiotypic interactions as depicted in Figure 1 are no more than a restatement of the "sticky end" model of antibody V domains developed by Hoffmann (88). The matter has recently been discussed in detail by Jerne et al (101), and we would like to point out, along with the latter discussion, a few specific features of idiotypic interactions as they can be deduced from the scheme in Figure 1. If we had raised ab2 as anti-idiotope against ab1,[2] we would find that the anti-idiotope would not compete with hapten (hapten 1) for binding to its target (ab1). In classical nomenclature we would call the idiotope recognized by the anti-idiotope non-binding-site related. If, on the other hand, ab1 were the anti-idiotope it would compete with the hapten (hapten 2) for binding to its target (ab2); here the idiotope recognized would be binding-site-related and, in fact, could be the binding site itself. In the original formulation of the network hypothesis (97), the structure on ab1 recognizing ab2 is the internal image of hapten 2. In line with these considerations, binding-site-related (28) and nonbinding-site-related idiotopes (114) have been described in a variety of experimental systems. It is clear, however, that a hapten-inhibitable anti-idiotope must not necessarily be a true internal image of the corresponding hapten, but may recognize an idiotope that is associated with the hapten binding site for structural and/or genetic reasons. Genetic associations even of nonbinding-site-related idiotopes with certain binding specificities have indeed been observed and are discussed below (Section 4.3). On the other hand, there is also good evidence for the existence of true internal images of ligands from the world outside the immune system within the immune system itself, at the level of antibody idiotopes. Striking examples of this are anti-idiotypic antibodies against various hormone-specific antibodies. Sub-

[1]The arrangement of the hypervariable loops in antibody V domains has presumably been selected in evolution not only to permit efficient idiotypic interactions of antibodies. The interaction of antibodies with external antigens such as proteins may often not be restricted to a cleft in the V region, as suggested for small haptens by the available three-dimensional models, but may also involve residues at the V region surface as in the interaction of the V region of protein Kol with its own F_c piece (90, 138).

[2]In the nomenclature of Jerne et al (101), ab2 in Figure 1 would be an anti-idiotope (against ab1) of the α type (ab2α), whereas ab1 would be an anti-idiotope (against ab2) of the β type (ab2β).

sets of such anti-idiotypic antibodies have been shown to be able to mediate hormone-specific effects (183, 193). Likewise, a selected anti-idiotypic antiserum against antibodies with specificity for tobacco mosaic virus (TMV) coat protein was able to induce anti-TMV antibodies in animals of several species (132). There is also evidence for internal images of constant region allotypes at the level of antibody idiotypes in the rabbit (101, 178). As is shown below, the idiotypic repertoire in any given animal is genetically restricted so that not any possible three-dimensional determinant is expressed at the idiotope level (see Section 4.3). Internal images may therefore exist for certain antigenic determinants but not for others.

Let us reiterate that idiotope-bearing and anti-idiotope antibody (idiotype and anti-idiotype) are operational terms by which we label the partners in an idiotypic interaction (Figure 1) on the basis of an ad hoc criterion (mostly considering as "anti-idiotope" the antibody raised by immunization with the "idiotope-bearing" antibody). When one considers the functional role of idiotypic interactions in the immune system, one should bear in mind that the anti-idiotope may control the expression of the idiotope-bearing antibody as well as vice versa. As is shown below, this fundamental point has not been sufficiently taken into consideration in the interpretation of regulatory experiments.

3. STRUCTURAL LOCALIZATION OF IDIOTOPES

Can we structurally localize idiotopes within the V region? This question is of importance not only for the understanding of idiotypic interactions (see above), but also for the geneticist who wants to use idiotopes as markers of the genes that encode antibody V regions (Section 4.3). In the absence of direct information based upon X-ray analysis, one is left with attempts to localize idiotopes on subunits of the antibody molecule and to correlate amino acid sequences with idiotope expression. As one might expect, a complex picture emerges from this kind of analysis. Chain reassociation experiments have shown that idiotypic determinants usually require both heavy (H) and light (L) chains for full expression, although in certain cases some reactivity of isolated chains with the anti-idiotypic antibody is found (e.g. 19a, 30, 33a, 34, 74, 79, 90a, 134, 180a, 190a, 213; see also 57, 89). Anti-idiotypic reagents specific for determinants on either L or H chain can be selectively induced by immunizing with hybrid antibody molecules in which either the H or the L chains are derived from pooled "normal" immunoglobulin (190, 220). Such reagents are particularly useful for studying the genetic basis of V gene expression. Inspection of amino acid sequences in idiotypically related myeloma and hybridoma antibodies support the notion that idiotopes are phenotypic markers of complementarity-determining regions (CDR) (33, 34). In the case of α(1-3)-dextran and (4-hydroxy-3-nitro phenyl)acetyl (NP)-binding λ-chain-bearing murine antibodies, attempts have been made to correlate amino acid sequences of the V, D, and J segments of the H chain with the expression of certain idiotypic determinants. [λ-Chain bearing murine antibodies are particularly useful for this purpose since λ chains show little variability in the mouse (207).] In the case of α(1-3)-dextran antibodies, where the amino acid sequences of a large number of closely related antibodies have been determined (181), it appeared that the expression of a certain idiotypic determinant was dependent on two positions in the D segment, and that of another was dependent on two residues in the second CDR (CDR2 in V_H) (38). However, as the authors noted, the situation is much more complex than would appear at first glance. Table 1 shows the reaction of three monoclonal anti-idiotope antibodies with a series of monoclonal anti-dextran antibodies of defined primary structure. Inspection of the data reveals that the interaction of idiotype and anti-idiotope (which in all cases also depends on the presence of the λ light chain) can in no case be attributed to a single stretch of amino acids, and sequences in V, D, and J seem to be involved in various combinations. Table 1 also contains sequences and idiotypic characteristics of a series of closely related anti-NP antibodies together with a non-NP-binding somatic variant (the sources of these antibodies are given in Table 1). Although the D region is again clearly of importance for idiotope expression, the V_H region is also involved. (Compare the expression of idiotope Ac146 in antibodies B1-8, B1-8.V3, B1-48, and B1-8.V1; and of idiotope Ac38 in antibodies B1-8, B1-48, G100.24, and B1-8.V3. For the contribution of V_H to idiotope Ac38, see also Section 4.3.) Thus, the situation is again more complex than originally anticipated (172). That amino acid residues in both D and V_H can contribute to the construction of a single idiotope is also strongly suggested by the work of Potter and his collaborators on the idiotypic properties of a series of interrelated galactan-binding antibodies (160). One of the idiotopes they analyzed appeared to be influenced by amino acid residues in D and CDR2, and three-dimensional models generated by a computer program indicate that the residues in question are exposed on the surface of the V domain, in reasonable distance from each other. Other idiotypic determinants in the same system also appear to involve the participation of more than a single CDR.

Despite the complex structure of idiotopes, very small changes in primary sequence can cause drastic changes in idiotypic specificity. An example of this is antibody B1-8.V3 in Table 1, where a single amino acid exchange in D leads to the loss of one of the idiotopes, the modification of the second, and, in fact, to the loss of six other idiotopes of the parent molecule. The variant antibody fully retains NP-binding specificity (A. Radbruch, S. Zaiss, K. Rajewsky, K. Beyreuther, to be published).

Table 1   Idiotopes and primary structure of $\alpha(1-3)$-dextran and NP-binding antibody V regions

Antibody[a]	V region segments[b]			Idiotopes[c]		
	V_H	D	J_H	EB3–72	EB3–16	AB3
J558	1	RY	1	+++	+++	+++
Hdex9	3	RY	1	++	+	++
Hdex36	1	RY	2	++	–	–
Hdex2	1	NY	1	++	–	+++
Hdex25	1	SY	1	++	+	++
Hdex12	1	GN	4	++	–	–
Hdex24	1	SS	2	++	–	–
MIOC104E	1	YD	1	–	–	–
				Ac38	Ac146	NP-bdg
B1–8	186–2	YD . YYGS S	2	+	+	+
B1–48	186–2	– L L G	1	+	–	+
G100.16	186–2	Y . . – – – – –	4	+	–	+
G100.24	186–2	– T H – – – – –	2	–	ND[d]	+
B1–8.V3	186–2	– – . – – R – –	2	–	CR[e]	+
B1–8.V1	186–2/102	– – . – – – – –	2	+	–	+

[a] $\alpha(1-3)$-dextran (upper panel) and NP-binding (lower panel) myeloma and hybridoma proteins.

[b] Amino acid sequences of heavy-chain variable regions in abbreviated form (the L chains are germline-encoded $\lambda 1$ chains in all cases). The V_H segment is identical in most of the dextran-binding ($V_H 1$) and the NP-binding molecules (V186–2). $V_H 3$ differs from $V_H 1$ in 4 positions in CDR2. $V_H 186–2/102$ differs from $V_H 186–2$ in 10 positions, most of them in CDR2. The sequences of the D segment are given in the one-letter code. The J_H segment is also indicated. The sequence data on dextran binding molecules are taken from Schilling et al and Clevinger et al (38, 181). The sequences of the NP-binding molecules are from Bothwell et al (23) (B1–8); A. Bothwell, M. Paskind, D. Baltimore, personal communication (B1–48); H. Sakano, K. Karjalainen, personal communication (G100.16 and G100.24); S. Zaiss, K. Beyreuther, personal communication (B1–8.V3); and Dildrop et al (49) (B1–8.V1).

[c] EB3–72, EB3–16, and AB3 are idiotopes of J558. The typing results are from Clevinger (37). Ac38 and Ac146 are idiotopes of B1–8. NP-bdg, NP-binding. The typing of B1–8.V3 was done by A. Radbruch.

[d] Not determined.

[e] Affinity of anti-idiotope 10× lower than on B1–8.

The data in Table 1 also show that the absence of a binding-site-related idiotope [Ac146 (174)] is not necessarily accompanied by a change in hapten binding specificity. Clearly, therefore, the corresponding anti-idiotope and the hapten interact with the target antibody in different ways. The anti-idiotype may bind to a larger surface area than the hapten or to an idiotope that is, for structural and/or genetic reasons, in its close vicinity. An idiotope of the latter kind could, at the level of the total antibody repertoire, be mostly associated with antibodies of the same hapten specificity—as indeed has been observed in the case of idiotope Ac146—although the corresponding anti-idiotope does not carry the "internal image" of the hapten.

To conclude this section, idiotypic determinants appear to be complex structures to which often both H and L chains and, within one chain, more than a single CDR may contribute. They may be distant from, or closely associated with, a hapten-binding site of the antibody, but even an idiotope invariably associated with antibodies of a given hapten specificity and recognized by a hapten-inhibitable anti-idiotope does not necessarily coincide with the hapten binding site.

4. THE IDIOTYPIC REPERTOIRE

4.1 What are the Questions?

Idiotypes are serological markers of antibody V regions, and thus in principle allow the analysis of the expression and selection of antibody V regions in the immune system. In view of the diversity of V regions, any analysis of the repertoire remains a priori fragmentary. However, a number of questions can be asked that allow straightforward answers. (a) Is the idiotypic repertoire generated at the B-cell level restricted and can we correlate it with the genetic information available in the antibody structural genes? (b) Are B cells selected according to their receptor idiotype when they leave the bone marrow and enter the peripheral immune system? (c) How are B cells selected in antigen-driven immune responses and, specifically, are idiotypic interactions involved in this selection? (d) If idiotypic interactions are involved in B-cell selection at any stage, what is its mechanism?

These questions lead to the field of regulatory T cells, which are known to be able to specifically recognize and express idiotypic markers. We discuss later the problems of T-cell idiotypes and their regulatory function.

4.2 Analyzing the Idiotypic Repertoire of B-Cell Receptors

The classical way of studying idiotype expression in the immune system is the idiotypic analysis of antibodies produced in antigen-driven immune responses. This approach is limited in several ways. First, it is restricted to antibodies of a given antigen-binding specificity, and thus does not allow one to study the representation of a given idiotypic marker in the total population of antibodies. This limitation can be overcome by inducing idiotypic responses by immunization with anti-idiotypic antibodies, an approach pioneered by the groups of Urbain (205) and Cazenave (35). The analysis of such responses is complicated by the fact that an anti-idiotypic antibody will induce the production of two types of complementary antibodies, namely (a) antibodies carrying the idiotype defined by the anti-idiotype used for immunization, and (b) anti-anti-idiotypes. The usual serological assays used for idiotype titration do not distinguish between these two types of molecules. This has led to confusion in the literature for the following reason: If any complementary antibody produced in response to immunization with anti-idiotype is considered to express the corresponding idiotype, then the idiotype will be found expressed in any immune system, since each system is sufficiently diverse to allow the induction of complementary antibodies. Several ways are open to distinguish between idiotype and anti-anti-idiotype produced in response to anti-idiotype. One is to identify idiotype by a variety of anti-idiotypic reagents in addition to the one used for immunization (18, 217). Another one is to identify the idiotype by an unusual structural property such as, for example, expression of λ chains in the case of the mouse. Since in mice λ chains are expressed in less than 5% of total immunoglobulin, anti-anti-idiotypes would be expected to express κ light chains. The latter approach has been successfully used (170, 171, 199), and its validity has been recently verified in an independent (third) way that is generally applicable (19), namely, by showing that the idiotypic molecules were indeed encoded by V genes similar to those encoding the original idiotypic antibody (M. Siekevitz, R. Dildrop, K. Beyreuther, K. Rajewsky, to be published). Methods of this kind are essential for the identification of idiotypic molecules produced in response to anti-idiotope. One should bear in mind, however, that they all give a minimum estimate of the idiotypic response and might classify an unknown fraction of idiotypic molecules as anti-anti-idiotopes.

The main problem of analyzing idiotypic repertoires at the level of induced serum antibodies is that the cells producing these antibodies have already been subject to complex regulation, be it before or after immunization. One would therefore like to see how the idiotypic patterns determined at the level of serum antibodies compare to the frequencies of B cells expressing the corresponding idiotypic determinants at the various stages of B-cell ontogeny. Several methods are available for this purpose. One is the limiting dilution assay for mitogen-reactive B cells (4, 5), in which the frequency of mitogen-reactive B cells expressing a given antibody V region can be determined. The assay is a priori limited to mitogen-reactive cells —in the case of bacterial lipopolysaccharide (LPS), roughly every third B cell in the spleen of a young mouse (4). Since LPS-reactive cells can be generated from bone marrow or fetal liver pre-B cells in vitro (66, 141), frequency determinations of, for example, idiotype-expressing cells in the population of newly generated B lymphocytes are possible (147). During ontogeny, certain fractions of the newly generated (LPS-reactive) B cells become responsive to T-cell help (77, 120, 189) and (or?) enter the pool of long-lived, recirculating B cells (64, 191, 192). As T-cell lines delivering help to B cells in vitro in a polyclonal or antigen-specific fashion have recently become available (6, 105, 117, 162), frequency determinations of B cells responsive to T-cell help in an in vitro system appear feasible, but they have not yet been carried out. However, the more complicated technique developed by Klinman and his colleagues (119), in which limiting numbers of B cells are injected into irradiated hosts and the T-cell dependent responses of single cell clones in splenic fragments cultured in vitro are recorded, is thought to allow frequency determinations in just that B-cell population. There is thus a variety of methods available for the determinations of V-region repertoires in B cells, each method determining the repertoire expressed in cells at a particular stage of differentiation.

4.3 The Genetic Basis of the Idiotypic Repertoire

Classical work on the expression of idiotypes at the level of induced antibodies (reviewed in 57) has revealed two types of idiotypic markers, private (minor) idiotypes, expressed irregularly in some individuals of a given species or strain, and recurrent (major, cross-reactive) idiotypes, which regularly appear in certain immune responses in the members of a certain species or strain or even several species (102b, 200a). In the case of strain-specific recurrent idiotypes, genetic analysis has shown that idiotype expression is almost invariably controlled by antibody structural genes only, in most cases by the V_H (reviewed in 208), in some instances by both the V_H and the V_L locus (51, 91, 107, 149). The interpretation of these results was that idiotypes are markers of antibody V genes, that strain-specific idiotypes are due to V-gene polymorphism, and that the predominant control by the V_H locus reflects more extensive polymorphism of V_H than V_L genes. (We now know that D and J region polymorphism must also be taken into account.) By using recombinant mouse strains it was possible to construct rough genetic maps in which regions controlling various idiotypic markers were separated from each other by recombination breakpoints. These subjects have been extensively reviewed (56, 137, 208). The modern recombinant DNA techniques have made it possible to study the genetic basis of idiotype expression at the molecular level. Such studies have largely confirmed the classical work and, in addition, dramatically improved our understanding of the problem. The molecular analysis of several immune responses dominated by recurrent idiotypes showed that each of these responses is based on the expression of a single or very few V_H and a single or very few V_L genes in the germ line. This is the case in the response of mice to arsonate [dominated by the Ars idiotype (148)], $\alpha(1–3)$-dextran [the J558 and MOPC104E idiotype (17, 182)], phosphorylcholine (PC) [dominated by the T15 idiotype (41, 133)], and NP [dominated by the NP^b idiotype (137, 93)]. In these four systems, the antibody response is known to consist of one or a few antibody families, each of which is characterized by the expression of a single V_H gene in combination with several D and J_H segments and of one or a few V_L and J_L genes (23, 24, 33a, 38, 48, 72,

78, 173, 174, 181, 187) [for the molecular biology of antibody structural genes, see Honjo (88a) in this volume]. The members of each family are thus closely related to each other, but the families are very large, as in addition to combinatorial diversity, somatic mutations in the rearranged variable region genes occur, sometimes at high frequency (11, 23, 24, 48, 72, 161, 181, 187, 207; for further families of idiotypically related antibodies see 78).

Thus, at least in responses to certain epitopes, the V-region repertoire is controlled and limited by a few germline genes, despite extensive somatic mutation—a modern version of the "germline theory" of antibody diversity. A simple interpretation of the phenomenon of recurrent and private idiotypes in antigen-induced immune responses emerges: recurrent idiotypes are markers of those germline V genes from which the antibodies in the corresponding immune response are derived and of part of their somatically mutated progeny. Private idiotypes are characteristic for somatic mutants. We have tested this proposition which has been widely entertained [78] with a pair of monoclonal anti-NP antibodies derived from C57BL/6 mice. Antibody B1–8 expresses a germline-encoded V region; antibody S43 uses the same V-region genes but carries somatic point mutations both in V_H and V_L (23, 24, 173). Non-cross-reacting idiotopes on antibody B1–8 and S43, respectively, were defined by monoclonal anti-idiotope antibodies. As expected, the B1–8 idiotopes are regularly expressed in anti-NP responses of C57BL/6 mice (111, 174, 175). In contrast, the S43 idiotopes are only occasionally detectable, and even then at very low levels (G. Wildner, T. Takemori, K. Rajewsky, to be published). The result is consistent with the view that recurrent idiotopes are markers of germline V (and possibly D) genes and that idiotopes specific for somatic variants are irregularly expressed and thus private. This does not, of course, necessarily imply that all private idiotypic markers must be the result of somatic mutation, nor that all germline V genes are regularly expressed in the immune system; we return to these points later.

The finding of extensive somatic mutation in B cells engaged in a specific immune response also provides a potential explanation for a finding that has puzzled immunologists since its original description (156), namely, that of idiotypic cross-reactivity of antibodies directed at different determinants of the immunizing antigen (64a, 87b, 102a, 108, 125, 143, 170, 199, 214a, 217). One might speculate that in a given B-cell clone, activated in response to one of the antigenic sites of the immunizing antigen, somatic variants arise, some of which express specificity to another determinant on the same antigen. Such variants would then be further stimulated by the immunogen and may be selected for higher affinity by subsequent somatic mutation. The mutants may still share idiotypic specificity with the antibody expressed by the original ("wild type") clone, and the idiotypic network might thus also contribute to their selection (97). This picture has recently emerged from the elegant experiments by W. Gerhard and his colleagues in which·mono-clonal antibodies with specificity for influenza hemagglutinin were isolated from individual immunized mice and were compared in terms of primary structure and epitope and idiotypic specificity. The evidence suggests step-wise mutation in single B-cell clones, resulting in the production of antibod-ies with different, fine specificities (194; W. Gerhard, M. Weigert, personal communication).

These considerations add an important aspect to the problem of the generation of the antibody repertoire. If somatic B-cell mutants with altered binding specificity constitute a substantial part of the repertoire of func-tional B cells they might play an important role in responses for which suitable V, D, and J genes have not been selected in the germline, e.g. responses against variants of common pathogens. We can only say at this point that somatic mutation is not extensive enough to obscure the clear pattern of germline determination in responses like those to NP, Ars, and PC.

Genetic determination and restriction of the idiotope repertoire can also be demonstrated when idiotope expression is studied in the total antibody population rather than in populations defined by a given antigen-binding specificity. In the initial experiments of this type, rabbits were immunized with anti-idiotypic antibodies (ab2 in the designation of the authors) to produce antibodies complementary to ab2 (ab3). The population of ab3 contained little or no antibodies with specificity for the antigen, which was recognized by the idiotypic antibody (ab1) against which ab2 has been raised; however, upon immunization with that antigen, the ab2-sensitized animals produced antigen-specific antibodies (ab1') that were idiotypically indistinguishable from ab1 (35, 205). These experiments were later repro-duced in other systems (18, 20, 131, 132, 170, 171, 199, 217). It was also shown that in certain experimental systems the response to ab2 was domi-nated by ab1, as defined by its antigen-binding specificity (178). Thus, a way appeared open to determine the idiotypic repertoire in the functionally active B-cell population. Such experiments required, however, the differen-tiation in the ab3 population of idiotype-bearing antibodies from anti-anti-idiotypes. When this was done (as outlined in Section 4.2), two clear results emerged. In general, the population contains a large component of idiotype-bearing antibodies that do not share antigen binding specificity with the idiotype-bearing antibodies used for the induction of ab2 (18, 170, 171, 199, 217). Together with the work on idiotype sharing between antibodies of different fine specificity, the structural data discussed in Section 3, and other work (52a, 60), this clearly demonstrates that the distributions of idiotopes and antigen-binding specificities in antibody-V regions are overlapping but not identical. This important point is exemplified by the data in Figure 2, where the expression of two idiotopes of the germline-encoded anti-NP antibody B1–8 in the total antibody population is analyzed, upon immunization with the corresponding cross-linked anti-idiotopes. Clearly, one of the two idiotopes (Ac38) is mainly expressed in non-NP binding antibodies, whereas another (Ac146) is largely restricted to antibodies carrying NP binding sites. The appearance of idiotope Ac38 in non-NP binding antibodies is not due only to somatic mutation or to different ways of V-D-J joining. The structural analysis of a somatic variant of antibody B1–8, in which part of the V_H region is encoded by a different germ line V_H gene (30, 49) and of anti-idiotypically induced Ac38+ monoclonal antibodies (R. Dildrop, M. Siekevitz, K. Rajewsky, K. Beyreuther, to be published), demonstrates that a series of closely related V_H genes, only one of which is involved in encoding NP-specific antibodies, encodes V_H regions of Ac38+ molecules. This is direct evidence for idiotype sharing between products of V genes in the germline expressed in antibodies differing in antigen-binding specificity.

A second result can also be read from the data in Figure 2, namely, that also in this approach idiotope expression turns out to be strictly genetically

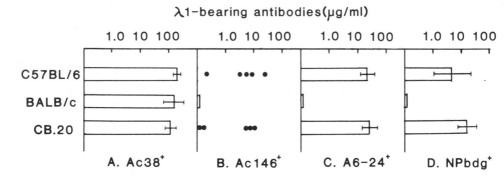

Figure 2 Analysis of idiotope expression by immunization with a cross-linked monoclonal anti-idiotope antibody. This antibody (Ac38) detects an idiotope (Ac38) on the germline-encoded NP-binding antibody B1–8. The latter antibody also carries idiotopes Ac146 and A6–24, defined by monoclonal anti-idiotopes. All these idiotopes are recurrently expressed in anti-NP responses of mice carrying the Igh^b haplotype. Idiotope-bearing and NP-binding molecules in the sera of mice immunized with cross-linked antibody Ac38 were detected on plates coated with either anti-idiotope (left three panels) or NP-bovine serum albumin (right panel), with a radioactive monoclonal anti-λ1 antibody [the anti-idiotopes carry κ chains, the idiotope-bearing molecules carry λ chains (174)]. Absorption on insolubilized antibody Ac38 removed > 98% of all idiotypic and NP-binding molecules, absorption on NP-Sepharose removed 83% Ac146+ and 17% Ac38+ molecules in strain C57BL/6 (Igh$^{b/b}$). Strain CB.20 carries the Igh^b locus on the BALB/c background. BALB/c is Igh$^{a/a}$. The data are taken from Takemori et al (199), where further details can be found.

controlled, again by genes in the Igh linkage group. Thus, $Igh^{b/b}$ but not $Igh^{a/a}$ mice express the Ac146 and A6–24 idiotopes, and although both strains of mice can produce $Ac38^+$ antibodies, it is only in $Igh^{b/b}$ mice that this idiotope is associated in part with NP binding specificity. Since in $(Igh^b \times Igh^a)F_1$ mice, $Ac38^+$ NP-binding antibodies of the IgG1 class all ($> 95\%$) carry the b allotype (199), it would appear that indeed the V_H, D, and J_H genes in the Igh^a haplotype are unable to encode an anti-NP antibody with the Ac38 idiotope. The F_1 experiment, also carried out in another experimental system (53), argues against the possibility that lack of idiotope expression is due to anti-idiotypic (network) control. A strong argument against this possibility has also been made by Siekevitz et al (186) for the Ars idiotype, whose strain-specific expression could be correlated with the presence or absence of the corresponding structural V_H gene. Strains lacking this gene do not express the Ars idiotype, although they are under certain conditions able to produce idiotypically cross-reactive antibodies (136)—presumably products of related, but non-identical, V genes. A similar situation appears to emerge in the case of the M460 idiotype (131). As is discussed in Section 6.1, lack of idiotype expression has been ascribed to anti-idiotypic control in certain other systems.

To summarize this section, recurrent idiotypes have turned out to be encoded by single, or small groups of, germline V_H and V_L genes and some of their somatically mutated progeny. The idiotypic repertoire is thus in part controlled and restricted by the V genes in the germline, and different strains in a species may differ in their repertoire because of V gene polymorphism. The system generates somatic mutants, and since this process appears to be based on random mutation, idiotypes characteristic for such mutants are probably usually nonrecurrent, i.e. private. The total repertoire of idiotopes in the immune system is the sum of the repertoire of recurrent and of private idiotypes. We want to stress again specifically that although certain immune responses like those to PC, NP, Ars, etc, are dominated by recurrent idiotypes (and somatic mutation may in these cases mainly serve to improve the specificity of the response), other responses and cellular interactions may depend entirely on somatic mutants whose generation might therefore be of vital importance for the system. Finally, the data discussed in this section show that idiotopes can be shared between antibodies of different antigen-binding specificity and vice versa, and it provides a clue of how idiotypically related antibodies with different epitope specificity might arise in immune responses.

4.4 Idiotypic Repertoire at the Level of B-Cell Precursors

An important approach to studying the selection of V regions in the immune system is to analyze the V-region repertoire expressed in newly generated B cells, which appear constantly in large numbers in the bone marrow (154), i.e. the "available" repertoire, and to compare it to the V-region repertoire expressed at later stages of B-cell ontogeny and in immune responses—the "functional" repertoire. Unfortunately, studies of this type are still in their infancy. The earliest analyses of frequencies of idiotypically defined B cells showed that T-dependent (188) and T-independent (TI-2) (42) T15 positive precursor cells occur in spleen cells at a frequency of 1–2 $\times 10^{-5}$. These results were later confirmed by Fung & Köhler (71a). Clearly, B cells in the spleen may already be selected from the "available" repertoire. This notion is compatible with subsequent results of Klinman and his colleagues (189), which show that T15-positive, T-dependent precursor cells could be detected in the spleen only when the mice were at least 7 days old. In fact, at the level of B-cell precursors defined by their antigen binding specificity, the late appearance of certain specificities has been observed repeatedly (reviewed in 120a). In line with this result, Fung & Köhler (69) recently found only T-independent (TI-1) $T15^+$ responses in B cells from very young animals; T-dependent responses were observed only later. Frequencies of $T15^+$ precursors in newly arising B cells have not yet been determined. In other systems, frequencies of LPS-reactive, idiotypically defined B precursor cells were measured. Eichmann et al (60) determined the frequency of $A5A^+$ LPS-reactive splenic B cells to be 4×10^{-4}. [A5A is a recurrent idiotype in the response of A/J mice to group A streptococcal carbohydrate (54).] In a recent study, Nishikawa et al (147) measured the frequency of LPS-reactive splenic B cells expressing idiotopes of the germ line-encoded antibody B1–8 (23, 24, 173; see also Table 1 and Figure 2). One of the idiotopes (Ac38, see Section 4.3) known to be present on the products of several V_H genes in combination with the $V_{\lambda 1}$ domain occurred at a frequency of $\sim 10^{-3}$; another one, restricted to antibodies with an NP binding site (Ac146, Section 4.3), was found at a frequency of 10^{-5}. Identical frequencies were observed in LPS-reactive cells from spleens of newborn and adult animals and in cells generated from bone marrow pre-B cells in vitro, in agreement with similar data of Jui et al (106a).

Overall, the frequencies measured for recurrent idiotypes or idiotopes in the various assay systems range from $10^{-3} - 10^{-5}$ (see also 10). There is the indication that in newly arising B cells, the frequency depends on the number of genes involved in the expression of a given idiotypic marker (147). However, to date, there is not a single analysis available in which for a given idiotypic marker frequency determinations have been carried out at the various stages of B-cell ontogeny. LPS-reactive cells may represent immature B cells (70, 139), and it is thus not surprising to find similar precursor frequencies in such cells, be they isolated from the spleen or generated in vitro from pre-B cells. It will be essential for an understanding of B-cell selection in general and the idiotypic network in particular to see how these frequencies compare to those in B-cell populations at later stages of ontogeny, e.g. B cells responsive to specific T-cell help and/or long-lived, recirculating B cells (64, 69, 71, 191, 192). Below (Section 5.3) we give an example that strikingly shows that idiotypic frequencies in LPS-reactive spleen cells may, in certain instances, not at all reflect what is expressed in the immune response generated by the same animal's B-cell population.

Other information that is lacking are the frequencies of precursor cells expressing, at the various stages of ontogeny, idiotypic markers characteristic for somatic mutants and, in the same context, of cells expressing anti-idiotypes against recurrent idiotypic markers. The latter point has to be seen in connection with the so far unresolved question of whether or not the germline directly encodes a network of idiotypes (see Section 4.3). In addition, the frequency of anti-idiotypes may be of importance for the activation of idiotype-bearing cells in the system.

4.5 Immunoglobulin Idiotypes on T Cells

After the initial findings that anti-idiotypic antisera could specifically induce idiotype-bearing T helper (T_H) cells (62) and that alloantibodies shared idiotypes with alloreactive T cells (14), a bulk of experimental evidence has further demonstrated the expression of recurrent antibody idiotypes in functional T cells and antigen-binding material (factors) of T-cell origin. Early literature on this subject has been reviewed (15, 135, 169), and we quote here only more recent work on idiotype or idiotope expression in T_H (59, 104, 144, 194a) and T suppressor (T_S) (81, 132a, 185, 193a, 212) cells, in T cells mediating delayed-type hypersensitivity (DTH) (7, 96, 196, 201, 218), and cytotoxic reactions (68) in antigen-binding material of T-cell origin (46, 47, 127) including suppressor (73, 86, 150) and helper factors (145).

However, with respect to the molecular nature of the T-cell receptor for antigen, idiotypic cross-reactivity of this kind does not have much significance, in particular if one remembers the discussion of "internal images" and the idiotypic cross-reactions between anti-hormone antibodies and hormone receptors (Section 2). It was only after the demonstration of the control of both T- and B-cell idiotypes by the Igh locus that the hypothesis could be advanced that antibody V genes [and specifically V_H genes, since neither V_L determinants nor control of idiotype expression by V_L genes could be identified in T cells (46, 57, 196a)] participate in encoding T-cell receptors for antigen (15, 169). Despite the availability of specific anti-idiotypic reagents and of cloned T-cell lines, little progress has been made over the last years with respect to the elucidation of the structure of T-cell receptors. In addition, although recent immunogenetic data indicate that genes also linked to the Igh complex may encode constant regions of T-cell receptors (158, 202), the available evidence from molecular biology argues against the notion that the same V_H genes that encode antibody V regions are expressed in T cells: regular V_H gene rearrangement has so far not been observed in such cells (128, 130).

We do not wish to go into this complicated matter further in this context, in particular since the molecular basis of T-cell recognition of antigen has not, to date, been resolved. However, we want to make two points, which lead directly into the topic of the next sections, namely, that of idiotypic regulation. First, since T cells express idiotypes [and, consequently, also anti-idiotypic specificities (3, 27, 36, 61, 76, 86, 116, 157, 163, 197, 276)], they are by definition partners in the idiotypic network. Second, if T-cell idiotypes are not encoded by antibody V genes, we have to explain how idiotypic polymorphism develops in parallel in the B- and T-cell compartment. If we do not want to invoke a complicated mechanism of adaptation and rectification of separate sets of V genes in evolution, we have to consider the possibility that T cells are selected in the immune system in ontogeny such that their receptor idiotypically resembles that of B-cell receptors. Such a mechanism would imply a functional idiotypic network in the most fundamental sense. It would be expected to lead to a situation where the idiotypic (and anti-idiotypic) repertoires expressed in the two sets of cells would be similar, but not quite identical, in keeping with substantial experimental evidence (13, 15, 76, 102, 126, 146, 163). We come back to this matter in Section 6.

4.6 Co-Existence of Lymphocytes with Complementary Receptors: The Idiotypic Network

Lymphocytes expressing complementary receptors co-exist in the immune system. At the level of B cells, anti-idiotypic antibodies can be induced in animals known to possess B-cell precursors expressing the corresponding idiotype (reviewed in 57). In addition, animals often produce, in the course of an immune response, anti-idiotypic antibodies against idiotypic determinants of their own antibodies, although mostly in low concentration (12, 39, 67, 77a, 94, 110, 121, 140, 177, 197a, 200). At the level of T cells, idiotype-bearing and anti-idiotypic T helper and suppressor cells can be induced in animals that express similar idiotypes and anti-idiotypes at the B-cell level (see Sections 5 and 6). There is thus no question that self-recognition at the level of receptor idiotopes is possible within the immune system, quite as Ramseier and Lindenmann (171a) pointed out 10 years ago.

What are the physiological consequences of this situation? We can look at the problem from the standpoint of classical immunology and ask how the system is able to establish self-tolerance since it allows the expression of idiotypes in immune responses, undisturbed by anti-idiotypic cells. As we shall see, regulatory mechanisms have been identified that potentially resolve this difficulty. From the standpoint of Jerne's hypothesis (97, 98), the co-existence of idiotypes and anti-idiotypes is an essential feature of the immune system. Cells with complementary binding sites interact with each other and an equilibrium of idiotypes and anti-idiotypes is established. Immune responses are seen as disturbances of this equilibrium and attempts of the system to re-equilibrate. In addition, idiotypic interactions might be essential for the generation of the functional receptor repertoire, in the sense of both promoting somatic diversification and restricting diversity. Quantitative models of idiotypic networks have been developed (2, 87, 88, 176), and the essential points of three of them have been summarized by Jerne (99). In the sections below we discuss some of the experimental evidence that supports the notion of a functional idiotypic network.

5. IDIOTYPIC REGULATION OF THE RECEPTOR REPERTOIRE BY ANTIBODIES

5.1 Effect of Anti-Idiotype on Idiotype Expression

In this and the next section we consider only experiments that approach the physiological situation in that antibodies are applied in native form and in low doses. Under these conditions, anti-idiotypic antibodies are extraordinarily potent regulators of idiotype expression, in the sense of enhancement or suppression. In general, though not without exception (18, 61, 164a, 165, 203), anti-idiotype, when given in native form, did not directly induce synthesis of idiotypic molecules. Rather, its effect appeared to be a long-lasting modification of the functional idiotypic repertoire. The classical work in this area, largely based on xenogeneic and allogeneic conventional anti-idiotypic antibodies, has been extensively reviewed (22, 57). We concentrate the present discussion on more recent work and on the mechanisms through which anti-idiotypes exert their regulatory function.

The monoclonal antibody technique (122) has made it possible to study antibody-mediated idiotypic regulation at the level of individual idiotopes and with isogeneic monoclonal idiotype-bearing and anti-idiotope antibodies. Experiments of this type, carried out in our laboratory, showed that anti-idiotopes of mouse origin injected into the mouse were as efficient in terms of idiotypic regulation as were xenogeneic (guinea pig) anti-idiotypic antibodies injected into the mouse (54, 55, 62): 10–100 ng of anti-idiotope enhanced, whereas 10 μg suppressed the expression of the target idiotope in immune responses induced 4–8 weeks later (111, 112, 175).

The interesting observation that the class of a xenogeneic anti-idiotype determines its regulatory function in the mouse [guinea pig IgG1 enhanced whereas IgG2 suppressed idiotype expression (54, 55, 62)] could not be reproduced with the monoclonal murine anti-idiotopes. In recent experiments (C. Müller, K. Rajewsky, to be published), two families of such antibodies were constructed and employed. The members of each family carry the same V region on different heavy-chain constant regions [spontaneous class switch variants (167)]. It was shown that anti-idiotopes of the IgG1, IgG2a, and IgG2b class all enhanced idiotope expression at nanogram doses and were suppressive at microgram doses. In the isologous situation, the regulatory function of anti-idiotope antibodies of these three classes is thus determined only by the antibody *dose*. One wonders, however, whether the nature of the target idiotope may also play a role, since in earlier experiments we had failed to induce enhancement with a particular anti-idiotope and suspected class-specific regulation (175).

The experiments described above were restricted to the analysis of idiotype expression at the antibody level. Anti-idiotypes have also been successfully used to induce idiotypic effector cells in cellular reactions, such as T cells mediating DTH (7, 96, 196, 201). The class of the anti-idiotope may play a role in these cases.

A striking effect of anti-idiotypic antibodies is also seen when such antibodies are either injected into newborn animals (1, 8, 71, 87a, 109, 115, 116, 124, 182, 209) or enter into the newborn's circulation from the mother, either through the placenta or the milk (210). As a consequence, long-lasting suppression of idiotype expression is generally observed, although a case of enhancement has also been reported (87a, see also Section 6.1). Very small doses of anti-idiotype are required for suppression (1, 8, 71, 109, 115, 116, 182, 209). In our own experiments we have found that 1 μg of monoclonal anti-idiotope injected into the newborn was sufficient for essentially complete idiotope suppression over several months (T. Takemori, K. Rajewsky, submitted for publication).

5.2 Effect of Idiotype on Idiotype Expression

Experiments have also been done in which the effect of injection of an antibody with a given idiotype on the expression of that same idiotype (instead of complementary binding sites, as in the experiments described above) was assessed. All experiments in which soluble "idiotypic" antibody was applied had in principle the same result. Urbain and co-workers immunized a pregnant female rabbit with anti-idiotype to produce complementary antibodies. In the offspring of the immunized animal, the corresponding idiotype was recurrently expressed in the immune response, in contrast to the controls (214). Similarly, Rubinstein et al (179) injected a small dose of an anti-levan antibody whose idiotype is only rarely expressed in the anti-levan response into newborn mice. When the animals had grown up, the same idiotype was dominant in their anti-levan antibodies (179). Enhancement of idiotype expression was also seen when monoclonal antibodies carrying a recurrent idiotype were injected into adult mice. In one study, the induction of idiotype-expressing plaque-forming cells was reported, without deliberate addition of antigen to the system (92; however see also 85a). In an experiment done in our laboratory it was shown that injection of 10 μg of antibody B1–8 led to enhanced expression of B1–8 idiotopes in an anti-NP response induced 6–8 weeks later. An analogous result was obtained by Ortiz-Ortiz et al (153) in the Ars system, but much higher doses of idiotype were used in this case. Formally, these results can be described as *idiotypic memory:* the immune system sees the structure of an antibody variable region and reproduces it later on in an immune response.

Idiotypic antibody has also been injected into animals after it had been chemically coupled to syngeneic thymocytes or spleen cells (50, 195). In contrast to the above experiments, this resulted in suppression of idiotype production either at the level of antibodies or of cells mediating DTH.

5.3 Mechanisms of Idiotypic Regulation by Antibodies

Since enhancement of idiotype expression by idiotype occurs under a variety of experimental conditions, more than a single mechanism might mediate the effect, such as induction of idiotype-specific T-cell help (179) and/or enhancing anti-idiotypic antibody (112) or elimination of suppression by anti-idiotype (153). The suppression induced by cell-bound idiotype is due to the induction of anti-idiotypic suppressor T-cell circuits (50, 195).

How does anti-idiotype exert its effect on idiotype expression? Enhancing doses sensitize idiotype-expressing T_H cells (16, 62, 103), T cells mediating DTH (7, 196, 201), and B cells whose frequency is found enhanced in the population of LPS-reactive spleen cells (60; T. Takemori, K. Rajewsky, unpublished results); it is not clear, however, whether or not the anti-idiotypic antibody sensitizes B cells directly. Alternatively, anti-idiotypic T_H cells (see section 6.1), which might also be induced by anti-idiotypic antibody, possibly via idiotypic T_H cells, could be responsible for this effect (84). Amplification of idiotype-bearing B cells could also be due to interaction with idiotypic T_H cells via bridging anti-idiotypic antibody (61). It is difficult to distinguish experimentally between the latter two possibilities. However, the role of T cells in anti-idiotype-induced B-cell sensitization is indirectly supported by the finding that T-cell populations sensitized by anti-idiotype preferentially collaborate with idiotype-bearing B cells (84).

In the case of idiotype suppression by anti-idiotypic antibody, we find again that both B and T cells are involved in this phenomenon. Suppressor T cells that specifically prevented idiotype expression in the B-cell response were first demonstrated in the A5A idiotypic system (55) and subsequently identified in a variety of cases of adult and neonatal suppression of both B-cell and DTH responses (71, 113, 116, 157, 218). In some instances, the suppressor cells were shown to express anti-idiotypic specificity by virtue of their ability to bind idiotype (116, 157). However, the situation is more complicated in that the T-suppressor pathway includes interacting T-cell subsets, such that, for example, an idiotypic suppressor inducer cell could activate an anti-idiotypic suppressor effector, thus mimicking the effect of the latter (150, 197, 211, 219). An idiotype-expressing cell would of course appear as the natural target of anti-idiotypic antibodies. We do not go further into these complicated matters (which include the problem of the target of suppressor effector cells) in the context of the present discussion. This particular subject has recently been reviewed, and we refer the reader to this literature (9, 73a, 78, 95, 184, 198).

As in the case of enhancement, idiotype suppression can also be demonstrated within the B-cell population itself. In adult mice injected with microgram amounts of anti-idiotype, evidence for transient nonresponsiveness of idiotypic B cells was obtained (85) despite the fact that the frequencies of idiotypic LPS-reactive B-cell precursors in such animals are usually only slightly reduced (60; T. Takemori, K. Rajewsky, submitted for publication). This phenomenon may represent a classical case of B-cell tolerance or "anergy" [see Nossal (149a) in this volume]. Along the lines of the work of Nossal and his colleagues, Nishikawa et al (147) have recently shown that anti-idiotypic antibody in microgram concentrations prevented the maturation of pre-B cells into idiotype-expressing LPS-reactive B cells in vitro. This result suggests that if a given antibody is produced in the immune system in such concentrations, the generation in the bone marrow of functional B cells expressing complementary (anti-idiotypic) receptors may be shut off. The work of Iverson (92a) and of Rowley and his colleagues (179a) together with the inability of Sakato & Eisen (180) to generate anti-T15 antibody in mice in which microgram amounts of $T15^+$ molecules were circulating (133) support this notion (see also section 5.4).

Long-lasting unresponsiveness in the B-cell compartment was observed in systems of neonatal suppression. When newborn mice were suppressed for the T15 idiotype by anti-T15 antibody, $T15^+$ precursor B cells could not be detected in the mice for several months (1, 40, 70). The possibility that this effect was mediated by T cells has been shown to be unlikely in other cases where suppression of the MOPC104E and T15 idiotypes was achieved by injecting anti-idiotype into newborn (T-cell deficient) BALB/ c^{nu}/nu mice (115, 209). On the other hand, there is also some evidence for the existence of idiotype specific suppressor T cells in animals injected with anti-idiotypic antibody at birth (71, 116). The effect of neonatally applied anti-idiotype on both T and B cells is exemplified by an experiment we depict in Figure 3. Mice were injected at birth with 100 μg of anti-idiotope antibody Ac38, recognizing an idiotope recurrently expressed in the primary anti-NP response (see Section 4.3). At 13 weeks of age, splenic T and B cells were isolated from the animals and were injected, together with T and B cells from normal controls, in various combinations into irradiated syngeneic hosts. The hosts were then immunized with NP coupled to chicken gamma globulin, and the expression of the target idiotope in the humoral anti-NP response was recorded (T. Takemori, K. Rajewsky, submitted for publication). As a result, the B cells from the suppressed animals mounted a normal anti-NP response but were unable to express idiotope Ac38 even though they were stimulated in the presence of normal T cells. This is particularly remarkable, since in limiting dilution analysis of LPS-reactive precursor cells, the frequency of Ac38+ precursors was identical to

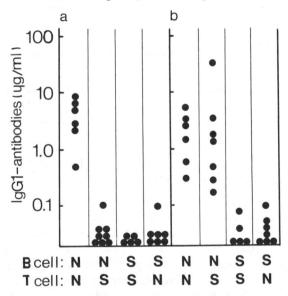

Figure 3   Effect of neonatal injection of a monoclonal anti-idiotope antibody on T and B cells in the adult mouse. Antibody Ac38 (100 μg), which detects a recurrent idiotope (Ac38) in the anti-NP response (174; legend to Figure 2), was injected into newborn C57BL/6 mice. At week 13, T and B cells from the spleens of the injected animals (S) and from control spleens (N) were transfered in various combinations (bottom of Figure) into irradiated hosts (10⁷ cells of each type per host). The hosts were immunized with NP chicken gamma globulin and the sera were titrated 12 days later for total IgG1 anti-NP antibody (which was the same in all groups; data not shown) and for anti-NP antibodies expressing the Ac38 idiotope either together with a second idiotope (Ac146, left panel) or the Ac38 idiotope alone (right panel). Antibody titers in individual sera are indicated. For details of titration methods see Kelsoe et al (112). Data are from T. Takemori, K. Rajewsky, submitted for publication.

that of control animals (data not shown). Thus, as discussed in Section 4.4, the repertoire of LPS-reactive precursors might not be the same as that expressed in the mature, functionally active B-cell population. Alternatively, the idiotypic B-cell response could be suppressed by anti-idiotypic B cells whose frequency in the suppressed animals remains to be determined. Most interesting in this context are recent data of Okumura et al (151), who ascribe a critical role in idiotypic control to a B-cell population expressing the Lyt-1 surface antigen. Coming back to the experiment in Figure 3, the T cells of the suppressed animals were also affected in that they did not allow the expression of the target idiotope Ac38 in a normal B-cell population, although they helped the B cells to mount a normal anti-NP response. Mixing experiments have so far failed to establish whether this deficiency is due to lack of idiotype-specific help, active suppression, or both. Interestingly, although nonresponsiveness in the B-cell compartment comprises all antibodies expressing idiotype Ac38, the T cells seem to see only a subset of these antibodies, namely those co-expressing a second idiotope (Ac146). This not only controls for the purity of the B-cell population in the experiment, it also suggests that the idiotypic determinants seen by T and B cells may not be the same. We come back to this point in Section 6.2.

To conclude this section, small doses of antibodies, applied to the animal in native form, profoundly affect the equilibrium of the corresponding idiotypes and anti-idiotypes in both the T- and the B-cell compartments of the immune system.

5.4 Interpretation of the Regulatory Effects of Idiotypic and Anti-Idiotypic Antibodies

The interpretation of the results described above has been made difficult because of a problem of nomenclature, namely, the terms idiotype and anti-idiotype. Certainly, any idiotype arising in the immune system has to deal with the problem of anti-idiotypes; however, when we inject an "anti-idiotype" into the mouse, it is incorrect to interpret this experiment solely in the sense of control of the corresponding "idiotype," as has usually been done. As discussed in detail in section 2, idiotype and anti-idiotype are operational terms for functionally equivalent complementary binding sites, and any idiotype is equally well an anti-idiotype and vice versa.

When we interpret the data on this basis, a surprisingly simple pattern emerges. At low doses (nanogram range), antibodies enhance the expression of complementary binding sites in the system. This enhancement is not at the level of effector functions, such as high-rate antibody production, but presumably it involves mainly the recruitment of T and B cells expressing complementary binding sites. It is attractive to think that the enhancement phenomenon reflects a fundamental driving force in the immune system in the sense of the network hypothesis. Low level antibody production (natural antibody) could accompany and promote interactions between diverse sets of cells with complementary receptors (61).

Once the concentration of certain antibody binding sites is above a certain threshold (microgram range), for example when the system engages in an active immune response, a different set of mechanisms is set in motion. Now the system is designed to lock itself into the expression of a given antibody repertoire. The "suppression" experiments discussed above reveal some of these mechanisms: microgram quantities of antibodies ("anti-idiotypes" in this case) shut off, in the bone marrow, the production of cells with complementary ("idiotypic") receptors and suppressor T cells suppressing the activation of such cells in the periphery are induced. The system stabilizes the expression of those binding sites (in this case the "anti-idiotype") to which it has committed itself. This is strikingly demonstrated by the finding that idiotype can facilitate the expression of the same idiotype ["idiotypic memory," Section 5.2; the problem of stabilizing immune responses has been discussed from many different angles (e.g. 82, 83, 88)].

The entire set of data discussed above can thus be interpreted in a coherent way that makes sense in terms of the physiology of the system and ascribes the idiotypic network a physiological regulatory role.

A final remark about idiotypic control of antigen-specific antibody responses: As discussed in Section 4.3, such responses consist of antibody families in which sets of idiotopes are expressed on the members in various combinations. An anti-idiotypic response against the more common of these idiotopes would have as its common denominator, because of its heterogeneity, the binding sites for these idiotopes. Because of this asymmetry it could unidirectionally regulate the expression of the idiotypic antibody family, with the implication, however, that regulation could extend to members of the family that do not possess the antigen-binding specificity of the initial immune response. The ab1-ab2-ab3 experiments of Urbain, Cazenave and others (204; section 4.3) can be considered to exemplify this and can be interpreted in this way without complicated genetic assumptions (168). Idiotypic control would be strictly specific for the antigen-induced response if the activated cells were particularly susceptible to the regulatory signals (36), and a priori so in cases where the anti-idiotypic response would see the antigen-binding site as the common denominator in the antigen-induced antibody population, and thus consist largely of "internal images" (100, 101).

5.5 Antibody Idiotypes and Growth Receptors on Pre-B and B Cells

A potentially important issue has not been mentioned so far, namely, that of idiotypic relationship of antibodies and growth receptors expressed in the B-cell lineage. Coutinho and his collaborators (44) have found that certain anti-idiotypic antibodies against recurrent idiotypes can bind to a large fraction of B and pre-B cells and can drive these cells into proliferation. A similar phenomenon was subsequently described for antibody M460 (164), whose idiotype is recurrently expressed in BALB/c anti-dinitrophenyl responses (21, 51, 52, 131). The antibodies are thought to interact with a set of growth receptors expressed in different combinations on different subsets of pre-B and B cells, and a theory was developed according to which antibody idiotypes and B-cell growth receptors have been selected in evolution for efficient interaction, providing a mechanism for self-stimulation within the B-cell system (43). This mechanism would be operating already at the pre-B cell level, that is before the cells express surface Ig [it is interesting in this connection, however, that pre-B cells may carry small amounts of free μ chains on the surface (158a, 206)], and maternal antibodies might thus influence B-cell development in the organism right from the beginning (10). The studies in this field are still in their infancy in terms of molecular analysis. However, together with other evidence (163a, 165a, 183, 193), they point in a direction that may turn out to be of physiological importance in a general way, namely, the possible involvement of antibody idiotypes in interactions with surface receptors other than the antigen receptors of the immune system.

6. T CELLS AND IDIOTYPE REPERTOIRE

6.1 Silent Clones and Idiotype Dominance

The discussion in the previous sections has shown that antibody idiotypes, be they expressed on circulating or cell-bound antibodies, can specifically induce T-cell activities. It is thus not surprising that in antigen-driven immune responses [in which in addition to antigen-specific antibodies spontaneous auto-anti-idiotypes often appear and have been ascribed a regulatory role (39, 67, 77a, 110, 121, 153, 197a, 200; see also Section 4.6)] idiotype-bearing and anti-idiotypic T_H and T_s cells have been observed in a variety of experimental systems, including antibody and DTH responses. The induction of such cells is thus not restricted to manipulations with anti-idiotypic antibodies. Based on sophisticated work with techniques of cellular immunology, idiotypic interactions of regulatory T cells have been incorporated into schemes of T cell circuits regulating immune responses. The discussion of these matters is beyond the scope of the present review and we refer the reader to the literature (9, 73a, 78, 95, 184, 198).

However, since the immune system continuously lives with antibody idiotypes, starting in ontogeny with immunoglobulins of maternal origin, idiotype-dependent modulation of T-cell activities must be expected to occur throughout life. This leads us directly to the issue of silent clones and of idiotype dominance in the B-cell compartment. There is evidence, mainly from the work of Urbain, Cazenave and Bona and their colleagues, that the bone marrow produces clones of cells not expressed as recurrent or dominant idiotypes in immune responses, either because of suppression by T cells or because of the lack of T-cell help (21, 106, 179, 204). Along these lines, several groups have presented data suggesting the dependence of dominant expression of recurrent idiotypes on idiotype-specific T_H cells (3, 25, 36, 85, 216). It seems clear that classical, carrier (i.e. antigen)-specific helper cells are sufficient to drive B cells expressing antibodies of the IgM and IgG class (and also a "dominant" idiotype) into proliferation and antibody production in vitro (31), but this does not exclude the possibility that anti-idiotypic T_H cells often participate in the preselection and/or expansion of certain B-cell clones that dominate certain immune responses. Preferential expansion by anti-idiotypic helpers could be responsible for the dominance of an idiotype in the anti-lysozyme response, being established only late in the response (142). Clearly, such evidence has to be weighed against the classical concept of selection in the immune response of cells expressing antibodies with high affinity for antigen (63). Such selection (in which idiotype-specific T cells could also be involved) may finally explain many cases of idiotype dominance, in particular since the corresponding antibody V regions (often specific for bacterial antigens) might have been already selected in evolution (29, 65, 215). The essential role of antigen in driving immune responses is evident from the fact that the expression of idiotypes in antigen-induced responses is usually, although not always (156), restricted to antigen-binding molecules in the serum (e.g. 54, 148, 170, 217).

6.2 Antibody Idiotypes and the T-Cell Repertoire

The concept that the existence of silent clones and idiotype dominance reflects idiotype-specific T-cell control represents a complication for the discussion of the genetic determination of the idiotypic repertoire in B cells (Section 4.3). Thus, if a given idiotype is not expressed in the B-cell compartment, we might consider this to be unrelated to the absence or presence of certain antibody structural genes and due to suppression by T cells. Let us try to clarify the situation. Almost without exception, idiotype expression is exclusively under the control of the Igh locus (Section 4.3). Therefore, if a given idiotype is expressed in a given strain, but not in another, then the idiotype-negative strain either lacks the structural information for that idiotype (as exemplified in several cases, see Section 4.3) or a regulatory element required for its expression. That element must also be located in the Igh locus. Clear candidates for such hypothetical regulatory elements are genes encoding anti-idiotypes. If such genes were expressed in T cells, these cells would, in principle, be able to interact with idiotype-bearing B cells, either in the sense of suppression (generating silent clones) or enhancement (generating idiotypic dominance). However, this type of strain-specific *indirect* control of idiotype expression via regulatory T cells *would again be due to polymorphism of genes in the Igh locus.*

If, indeed, the repertoire of antibody V-region structural genes were shared between T- and B-cell receptors, the extensive idiotypic interactions between the two types of lymphocytes would not be surprising. In particular, idiotypic polymorphism would be expected to be expressed equally at the level of T and B cells, as is indeed found (Section 4.5). However, as we discussed above, evidence from molecular biology argues against the notion that the V, D, and J genes expressed in antibodies are also expressed in T cells. Furthermore, there is evidence that idiotypic control of B-cell activities by regulatory T cells and within regulatory T-cell circuits is restricted by the Igh locus, suggesting a selection of T cells for the recognition of B-cell idiotypes expressed in the same system (see 9, 26, 106, 198, 219). These considerations lead to the far-reaching hypothesis that the T-cell repertoire might be selected in ontogeny not only for the recognition of self-histocompatibility (MHC) antigens (118, 222), but also for (self) B-cell idiotypes. In its extreme form, this hypothesis explains idiotype sharing between T and B cells by idiotypic interactions between these cells leading to the selection of a T-cell receptor repertoire complementary to that of B-cell receptors. Since the latter contains complementary receptors in itself, and since, in addition, anti-idiotypic T cells selected for B-cell idiotypes may in turn expand idiotypic T cells, a situation would result in which not only efficient idiotypic T-B cell interactions would be guaranteed, but also the receptor repertoire of T cells would be idiotypically (and anti-idiotypically) similar, although not identical, to that of the B cells. The experimental results in Figure 3 can easily be interpreted in this sense.

There is suggestive evidence also in other systems that T and B cells see similar, but not identical, idiotopes on antibodies (76, 80, 102, 163). One might speculate that "regulatory" idiotopes in the sense of Paul & Bona (159a) might be those that T cells are able to see. A priori, a B cell appears to be an ideal target to be seen by T cells: It expresses class I and class II MHC antigens on its surface, as well as a unique structure, the immunoglobulin idiotype. Furthermore, a close structural similarity of immunoglobulin domains and the domains of class I and class II antigens has recently emerged (130a, 152, 221). This has led to the speculation that Ig domains as well as MHC antigens might represent restriction elements for T cells (25, 26, 95, 106), and that V regions could have been selected in evolution for structural similarity with MHC antigens (45). The latter could relate to the finding that helper and suppressor T cells see different idiotypic determinants in certain systems (80, 184), and to the issue of silent clones and idiotype dominance (Section 6.1).

An answer to the question as to what extent the T-cell repertoire is determined by idiotypic interactions with B cells will require the analysis of T-cells that have differentiated in the absence of B-cell idiotypes. Experiments of Bottomly and her co-workers suggest that T cells from Ig-suppressed mice still contain T_H cells specific for conventional antigens (25); but of course the suppressed mice are not free of the antibody idiotypes present on the suppressing antibody. Bottomly and Mosier (27) have also suggested that anti-idiotypic T_H cells are absent from mice lacking the corresponding idiotype in the B cell compartment. This result is controversial at present (32, 166). Further experiments along these lines appear urgently needed. The issue is whether or not idiotypic interactions are the natural way for T cells to communicate with each other and with B cells.

7. CONCLUSIONS

We started our discussion with the principles of idiotypic interactions and the structural and genetic basis of idiotypic determinants. There is clear evidence that the germline controls and limits the available idiotypic repertoire in newly arising B cells.

The cells in the immune system are continuously exposed to idiotypic interactions. A pre-B cell arising in the bone marrow may already be positively selected by antibody idiotypes through interaction with growth receptors on its surface. When the cell expresses its Ig receptor, it becomes subject to idiotypic control of several types. There is evidence that cells arising in the bone marrow and expressing receptors complementary to idiotypes present in the circulation in microgram amounts are turned off. On the other hand, nanogram doses of idiotypes in circulation lead to the expansion of cells with complementary receptors. Idiotypes may thus play a crucial role in establishing the functional repertoire in the system, by recruiting B cells that are otherwise short-lived into the pool of recirculating, long-lived cells and/or that of B cells responsive to specific T-cell help (these two compartments could be overlapping or identical).

In the idiotypic selection of B cells, regulatory T cells can be specifically involved, they themselves being targets of regulation by antibody idiotypes. The interplay of T and B cells via idiotypic interactions will lead to the selection of sets of complementary cells whose receptor repertoire might well differ from that of the newly generated cells. Idiotype sharing between T and B cells and the control of idiotypes by the Igh locus could thus either indicate the involvement of V_H genes in the receptors of both cell types or the adaptation of the T-cell repertoire to B-cell idiotypes in ontogeny and/

or evolution, similar to its adaptation to the (structurally related) MHC antigens. There is experimental evidence that the receptor repertoire of T cells is idiotypically related but not identical to that of antibodies, and hints that T cells might adapt to the recognition of antibody idiotypes in ontogeny.

If it will turn out that idiotypically selected cells represent an important part of the functional repertoire, then immune responses might indeed, quite in the sense of the original network hypothesis (97), depend on the interference of antigen with interactions of complementary cells and reflect release from suppression. Experimental data pointing in this direction have recently been presented (58, 123). Once the system expresses a given repertoire of binding sites in an immune response, the expression of these binding sites is stabilized via idiotypic control ("idiotypic memory").

A functional network of idiotypes may not only exist at the level of T-B (or T-T) cell interactions but also in the B-cell compartment itself. Here it could either have been selected in evolution or reflect an interplay of germline encoded with somatically mutated V regions (168). The question of whether the germline encodes an idiotypic network of antibody V regions can now be directly approached by the tools of molecular biology.

ACKNOWLEDGMENTS

We are grateful to K. Beyreuther, M. Cramer, A. Radbruch, M. Siekevitz, F. Smith, and H. Tesch for valuable criticism and advice, and to E. Siegmund, C. Tomas, and W. Muntjewerf for typing the manuscript. We also thank K. Karjalainen for samples of monoclonal antibodies, and him and H. Sakano, A. Bothwell, M. Paskind, D. Baltimore, S. Zaiss, and K. Beyreuther for letting us have unpublished sequence information. We wish to acknowledge specifically the contributions of M. Reth to our work and thinking. The work from our laboratory was supported by the Deutsche Forschungsgemeinschaft through SFB 74.

Literature Cited

1. Accolla, R. S., Gearhart, P. J., Sigal, N. H., Cancro, M. P., Klinman, N. R. 1977. Idiotype-specific neonatal suppression of phosphorylcholine-responsive B cells. *Eur. J. Immunol.* 7:876–81
2. Adam, G., Weiler, E. 1976. Lymphocyte population dynamics during ontogenetic generation of diversity. In *The Generation of Antibody Diversity: A New Look*, ed. A. J. Cunningham, pp. 1–20. New York: Academic
3. Adorini, L., Harvey, M., Sercarz, E. E. 1979. The fine specificity of regulatory T cells. IV. Idiotypic complementarity

and antigen-bridging interactions in the anti-lysozyme response. *Eur. J. Immunol.* 9:906–9
4. Anderson, J., Coutinho, A., Melchers, F. 1977. Frequencies of mitogen-reactive B cells in the mouse. I. Distribution in different lymphoid organs from different inbred strains of mice at different ages. *J. Exp. Med.* 145:1511–19
5. Anderson, J., Coutinho, A., Melchers, F. 1977. Frequencies of mitogen-reactive B cells in the mouse. II. Frequencies of B cells producing antibodies which lyse sheep or horse erythrocytes,

and trinitrophenylated or nitroiodophenylated sheep erythrocytes. *J. Exp. Med.* 145:1520–30

6. Anderson, J., Schreier, M. H., Melchers, F. 1980. T-cell-dependent B-cell stimulation is H-2 restricted and antigen dependent only at the resting B-cell level. *Proc. Natl. Acad. Sci. USA* 77:1612–16

7. Arnold, B., Wallich, R., Hämmerling, G. J. 1982. Elicitation of delayed-type hypersensitivity to phosporylcholine by monoclonal anti-idiotypic antibodies in an allogenic environment. *J. Exp. Med.* 156:670–74

8. Augustin, A., Cosenza, H. 1976. Expression of new idiotypes following neonatal idiotypic suppression of a dominant clone. *Eur. J. Immunol.* 6:497–501

9. Benacerraf, B., 1980. Genetic control of the specificity of T lymphocytes and their regulatory products. In *Progress in immunology: Immunology 80,* ed. M. Fougereau, J. Dausset, 4:419–31. New York: Academic

10. Bernabé, R. A., Coutinho, A., Cazenave, P.-A., Forni, L. 1981. Suppression of a "recurrent" idiotype results in profound alterations of the whole B-cell compartment. *Proc. Natl. Acad. Sci. USA* 78:6416–20

11. Bernard, O., Hozumi, N., Tonegawa, S. 1978. Sequences of mouse immunoglobulin light chain genes before and after somatic changes. *Cell* 15:1133–44

12. Binion, S. B., Rodkey, L. S. 1982. Naturally induced auto-anti-idiotypic antibodies. Induction by identical idiotopes in some members of an outbred rabbit family. *J. Exp. Med.* 156:860–72

13. Binz, H., Frischknecht, H., Shen, F. W., Wigzell, H. 1979. Idiotypic determinants on T-cell subpopulations. *J. Exp. Med.* 149:910–22

14. Binz, H., Wigzell, H. 1975. Shared idiotypic determinants on B and T lymphocytes reactive against the same antigenic determinants. I. Demonstration of similar or identical idiotypes on IgG molecules and T-cell receptors with specificity for the same alloantigens. *J. Exp. Med.* 142:197–211

15. Binz, H., Wigzell, H. 1977. Antigenbinding idiotypic T-lymphocyte receptors. *Contemp. Top. Immunobiol.* 7:113–77

16. Black, S. J., Hämmerling, G. J., Berek, C., Rajewsky, K., Eichmann, K. 1976. Idiotypic analysis of lymphocytes in vitro. I. Specificity and heterogeneity of B and T lymphocytes reactive with anti-idiotypic antibody. *J. Exp. Med.* 143:846–60

17. Blomberg, B., Geckeler, W. R., Weigert, M. 1972. Genetics of the antibody response to dextran in mice. *Science* 177:178–80

18. Bluestone, J. A., Epstein, S. L., Ozato, K., Sharrow, S. O., Sachs, D. H. 1981. Anti-idiotypes to monoclonal anti-H-2 antibodies. II. Expression of anti-H-2^k idiotypes on antibodies induced by anti-idiotype or H-2^k antigen. *J. Exp. Med.* 154:1305–18

19. Bluestone, J. A., Krutzsch, H. C., Auchincloss, Jr., H., Cazenave, P.-A., Kindt, T. J., Sachs, D. H. 1982. Anti-idiotypes against anti-H-2 monoclonal antibodies: Structural analysis of the molecules induced by in vivo anti-idiotype treatment. *Proc. Natl. Acad. Sci. USA* 24:7847–51

19a. Bluestone, J. A., Metzger, J.-J., Knode, M. C., Ozato, K., Sachs, D. H. 1982. Anti-idiotypes to monoclonal anti-H-2 antibodies. I. Contribution of isolated heavy and light chains to idiotype expression. *Mol. Immunol.* 19:515–24

20. Bona, A., Heber-Katz, E., Paul, W. E. 1981. Idiotype-anti-idiotype regulation. I. Immunization with a levan-binding myeloma protein leads to the appearance of auto-anti-(anti-idiotype) antibodies and to the activation of silent clones. *J. Exp. Med.* 153:951–67

21. Bona, C., Paul, W. E. 1979. Cellular basis of regulation of expression of idiotype. I. T-suppressor cells specific for MOPC 460 idiotype regulate the expression of cells secreting anti-TNP antibodies bearing 460 idiotype. *J. Exp. Med.* 149:592–600

22. Bona, C., Casenave, P.-A. 1980. *Lymphocytic Regulation by Antibodies.* New York: Wiley

23. Bothwell, A. L. M., Paskind, M., Reth, M., Imanishi-Kari, T., Rajewsky, K., Baltimore, D. 1981. Heavy chain variable region contribution to the NP^b family of antibodies: Somatic mutation evident in a γ2a variable region. *Cell* 24:625–37

24. Bothwell, A. L. M., Paskind, M., Reth, M., Imanishi-Kari, T., Rajewsky, K., Baltimore, D. 1982. Somatic variants of murine immunoglobulin λ light chains. *Nature* 298:380–82

25. Bottomly, K., Janeway, C. A. Jr., Mathieson, B. J., Mosier, D. E. 1980. Absence of an antigen-specific helper T cell required for the expression of the T 15 idiotype in mice treated with anti-μ antibody. *Eur. J. Immunol.* 10:159–63

26. Bottomly, K., Maurer, P. H. 1980. Antigen-specific helper T cells required for dominant production of an idiotype (ThId) are not under immune response (Ir) gene control. *J. Exp. Med.* 152:1571–82

27. Bottomly, K., Mosier, D. E. 1979. Mice whose B cells cannot produce the T 15 idiotype also lack an antigen-specific helper T cell required for T 15 expression. *J. Exp. Med.* 150:1399–409

28. Brient, B. W., Nisonoff, A. 1970. Quantitative investigations of idiotypic antibodies. IV. Inhibition by specific haptens of the reaction of anti-hapten antibody with its anti-idiotypic antibody. *J. Exp. Med.* 132:951–62

29. Briles, D. E., Forman, C., Hudak, S., Claflin, J. L. 1982. Anti-phosphorylcholine antibodies of the T 15 idiotype are optimally protective against *Streptococcus pneumoniae. J. Exp. Med.* 156:1177–85

30. Brüggemann, M., Radbruch, A., Rajewsky, K. 1982. Immunoglobulin V region variants in hybridoma cells. I. Isolation of a variant with altered idiotypic and antigen-binding specificity. *EMBO J.* 1:629–34

31. Cammisuli, S., Schreier, M. H. 1981. Individual clones of carrier-specific T cells help idiotypically and isotypically heterogeneous anti-hapten B-cell responses. *Immunology* 43:581–89

32. Cancro, M. P., Klinman, N. R. 1980. B cell repertoire diversity in athymic mice. *J. Exp. Med.* 151:761–66

33. Capra, J. D., Kehoe, J. M. 1975. Hypervariable regions, idiotypy, and the antibody-combining site. *Adv. Immunol.* 20:1–40

33a. Capra, J. D., Slaughter, C., Milner, E. C. B., Estess, P., Tucker, P. W. 1982. The cross-reactive idiotype of A-strain mice. *Immunology Today* 3:332–39

34. Carson, D., Weigert, M. 1973. Immunochemical analysis of cross-reacting idiotypes of mouse myeloma proteins with anti-dextran activity and normal anti-dextran antibody. *Proc. Natl. Acad. Sci. USA* 70:235–39

35. Cazenave, P.-A. 1977. Idiotypic-anti-idiotypic regulation of antibody synthesis in rabbits. *Proc. Natl. Acad. Sci. USA* 74:5122–25

36. Cerny, J., Caulfield, M. J. 1981. Stimulation of specific antibody-forming cells in antigen-primed nude mice by the adoptive transfer of syngeneic anti-idiotypic T cells. *J. Immunol.* 126:2262–66

37. Clevinger, B. 1982. In *Idiotypes: Antigens on the Inside,* ed. I. Westen-

Schnurr, pp. 77–83. Basel: Editiones Roche

38. Clevinger, B., Schilling, J., Hood, L., Davie, J. M. 1980. Structural correlates of cross-reactive and individual idiotypic determinants on murine antibodies to α-$(1 \rightarrow 3)$ dextran. *J. Exp. Med.* 151:1059–70

39. Cosenza, H. 1976. Detection of anti-idiotype reactive cells in the response to phosphorylcholine. *Eur. J. Immunol.* 6:114–16

40. Cosenza, H., Julius, M. H., Augustin, A. A. 1977. Idiotypes as variable region markers: Analogies between receptors on phosphorylcholine-specific T and B lymphocytes. *Immunol. Rev.* 34:3–33

41. Cosenza, H., Köhler, H. 1972. Specific inhibition of plaque formation to phosphorylcholine by antibody against antibody. *Science* 176:1027–29

42. Cosenza, H., Quintáns, J., Lefkovitz, I. 1975. Antibody response to phosphorylcholine in vitro. I. Studies on the frequency of precursor cells, average clone size and cellular cooperation. *Eur. J. Immunol.* 5:343–49

43. Coutinho, A. 1980. The self-nonself discrimination and the nature and acquisition of the antibody repertoire. *Ann. Immunol. (Paris)* 131D:235–53

44. Coutinho, A., Forni, L., Bernabé, R. R. 1980. The polyclonal expression of immunoglobulin variable region determinants on the membrane of B cells and their precursors. *Springer Semin. Immunopathol.* 3:171–211

45. Coutinho, A., Forni, L., Holmberg, D., Ivars, F. 1983. *Is the Network Theory Tautologic?* In *Nobel Symposium 55, Genetics of the immune response,* ed. G. and E. Möller, New York: Plenum. In press

46. Cramer, M., Krawinkel, U., Melchers, I., Imanishi-Kari, T., Ben-Neriah, Y., Givol, D., Rajewsky, K. 1979. Isolated hapten-binding receptors of sensitized lymphocytes. IV. Expression of immunoglobulin variable regions in (4-hydroxy-3-nitrophenyl) acetyl(NP)-specific receptors isolated from murine B and T lymphocytes. *Eur. J. Immunol.* 9:332–38

47. Cramer, M., Reth, M., Grützmann, R. 1981. T Cell V_H versus B cell V_H. In *Immunoglobulin Idiotypes and their Expression,* ed. Janeway, C., Sercarz, E. E., Wigzell, H. 20:429–39. New York: Academic

48. Crews, S., Griffin, J., Huang, H., Calame, K., Hood, L. 1981. A single V_H gene segment encodes the immune re-

sponse to phosphorylcholine: Somatic mutation is correlated with the class of the antibody. *Cell* 25:59–66

49. Dildrop, R., Brüggemann, M., Radbruch, A., Rajewsky, K., Beyreuther, K. 1982. Immunoglobulin V region variants in hybridoma cells. II. Recombination between V genes. *EMBO J.* 1:635–40

50. Dohi, Y., Nisonoff, A. 1979. Suppression of idiotype and generation of suppressor T cells with idiotype-conjugated thymocytes. *J. Exp. Med.* 150:909–18

51. Dzierzak, E. A., Janeway, C. A. Jr., Rosenstein, R. W., Gottlieb, P. D. 1980. Expression of an idiotype (Id-460) during in vivo anti-dinitrophenyl antibody responses. I. Mapping of genes for Id-460 expression to the variable region of immunoglobulin heavy-chain locus and to the variable region of immunoglobulin k-light-chain locus. *J. Exp. Med.* 152:720–29

52. Dzierzak, E. A., Rosenstein, R. W., Janeway, C. A. Jr., 1981. Expression of an idiotype (Id-460) during in vivo anti-dinitrophenyl antibody responses. II. Transient idiotypic dominance. *J. Exp. Med.* 154:1432–41

52a. Dzierzak, E. A., Janeway, C. A. Jr., 1981. Expression of an idiotype (Id-460) during in vivo anti-dinitrophenyl antibody responses. III. Detection of Id-460 in normal serum that does not bind dinitrophenyl. *J. Exp. Med.* 154:1442–54

53. Eichmann, K. 1973. Idiotype expression and the inheritance of mouse antibody clones. *J. Exp. Med.* 137:603–21

54. Eichmann, K. 1974. Idiotype suppression. I. Influence of the dose and of the effector functions of anti-idiotypic antibody on the production of an idiotype. *Eur. J. Immunol.* 4:296–302

55. Eichmann, K. 1975. Idiotype suppression. II. Amplification of a suppressor T cell with anti-idiotypic activity. *Eur. J. Immunol.* 5:511–17

56. Eichmann, K. 1975. Genetic control of antibody specificity in the mouse. *Immunogentics* 2:491–506

57. Eichmann, K. 1978. Expression and function of idiotypes on lymphocytes. *Adv. Immunol.* 26:195–254

58. Eichmann, K. 1982. *Idiotypes-Antigens on the Inside*, ed. I. Westen-Schnurr, pp. 209–21. Basel: Editiones Roche

59. Eichmann, K., Ben-Neriah, Y., Hetzelberger, D., Polke, C., Givol, D., Lonai, P. 1980. Correlated expression of V_H framework and V_H idiotypic determinants on T helper cells and on functionally undefined T cells binding group A streptococcal carbohydrate. *Eur. J. Immunol.* 10:105–12

60. Eichmann, K., Coutinho, A., Melchers, F. 1977. Absolute frequencies of Lipopolysaccharide-reactive B cells producing A5A idiotype in unprimed, streptococcal A carbohydrate-primed, anti-A5A idiotype-sensitized and anti-A5A idiotype-suppressed A/J mice. *J. Exp. Med.* 146:1436–49

61. Eichmann, K., Falk, I., Rajewsky, K. 1978. Recognition of idiotypes in lymphocyte interactions. II. Antigen-independent cooperation between T and B lymphocytes that possess similar and complementary idiotypes. *Eur. J. Immunol.* 8:853–57

62. Eichmann, K., Rajewsky, K. 1975. Induction of T and B cell immunity by anti-idiotypic antibody. *Eur. J. Immunol.* 5:661–66

63. Eisen, H. N., Siskind, G. W. 1964. Variations in affinities of antibodies during the immune response. *Biochemistry* 3:996–1008

64. Elson, C. J., Jablonska, K. F., Taylor, R. B. 1976. Functional half-life of virgin and primed B lymphocytes. *Eur. J. Immunol.* 6:634–38

64a. Enghofer, E. M., Glaudemans, C. P. J., Bosma, M. J. 1979. Immunoglobulins with different specificities have similar idiotypes. *Mol. Immunol.* 16:1103–10

65. Etlinger, H. M., Julius, M. H., Heusser, Chr. H. 1982. Mechanism of clonal dominance in the murine anti-phosphorylcholine response. I. Relation between antibody avidity and clonal dominance. *J. Immunol.* 128:1685–91

66. Fairchild, S. S., Cohen, J. J. 1978. B lymphocyte precursors. I. Induction of Lipopolysaccharide responsiveness and surface immunoglobulin expression in vitro. *J. Immunol.* 121:1227–31

67. Fernandez, C., Möller, G. 1979. Antigen-induced strain-specific autoantidiotypic antibodies modulate the immune response to dextran B512. *Proc. Natl. Acad. Sci. USA* 76:5944–47

68. Frischknecht, H., Binz, H., Wigzell, H. 1978. Induction of specific transplantation immune reactions using anti-idiotypic antibodies. *J. Exp. Med.* 147:500–14

69. Fung, J., Köhler, H. 1980. Late clonal selection and expansion of the TEPC-15 germ-line specificity. *J. Exp. Med.* 152:1262–73

70. Fung, J., Köhler, H. 1980. Mechanism of neonatal idiotype suppression. I.

State of the suppressed B cells. *J. Immunol.* 125:1998–2003

71. Fung, J., Köhler, H. 1980. Mechanism of neonatal idiotype suppression. II. Alterations in the T cell compartment suppress the maturation of B cell precursors. *J. Immunol.* 125:2489–95

71a. Fung, J., Köhler, H. 1980. Immune response to phosphorylcholine. VII. Functional evidence for three separate B cell subpopulations responding to TI and TD PC-antigens. *J. Immunol.* 125:640–46

72. Gearhart, P. J., Johnson, N. D., Douglas, R., Hood, L. 1981. IgG antibodies to phosphorylcholine exhibit more diversity than their IgM counterparts. *Nature* 291:29–34

73. Germain, R. N., Ju, S.-T., Kipps, T. J., Benacerraf, B., Dorf, M. E. 1979. Shared idiotypic determinants of antibodies and T-cell-derived suppressor factor specific for the random terpolymer L-glutamic acid^{60} -L-alanine^{30}-L-tyrosine^{10}. *J. Exp. Med.* 149:613–22

73a. Germain, R. N., Sy, S.-M., Rock, K., Dietz, M. H., Greene, M. I., Nisonoff, A., Weinberger, J. Z., Ju, S.-T., Dorf, M. E., Benacerraf, B. 1981. The role of idiotype and the MHC in suppressor T cell pathways. In *Immunoglobulin idiotypes: ICN-UCLA Symposia on Molecular and Cellular Biology,* ed. C. Janeway, E. E. Sercarz, H. Wigzell, 20:709–23. New York: Academic

74. Ghose, A. C., Karush, F. 1974. Chain interactions and idiotypic specificities of homogeneous rabbit anti-lactose antibodies. *J. Immunol.* 113:162–72

75. Givol, D. 1979. The antibody combining site. In *International Review of Biochemistry, Defense and Recognition: II B,* ed. E. S. Lennox, 23:71–125 Baltimore: University Park

76. Gleason, K., Köhler, H. 1982. Regulatory idiotypes. T helper cells recognize a shared V_H idiotope on phosphorylcholine-specific antibodies. *J. Exp. Med.* 156:539–49

77. Goidl, E. A., Siskind, G. W. 1974. Ontogeny of B-lymphocyte function. I. Restricted heterogeneity of the antibody response of B lymphocytes from neonatal and fetal mice. *J. Exp. Med.* 140: 1285–302

77a. Goidl, E. A., Schrater, A. F., Siskind, G. W., Thorbecke, G. J. 1979. Production of auto-anti-idiotypic antibody during the normal immune response to TNP-Ficoll. II. Hapten-reversible inhibition of anti-TNP plaque-forming cells by immune serum as an assay for auto-anti-idiotypic antibody. *J. Exp. Med.* 150:154–65

78. Greene, M. I., Nelles, M. J., Sy, M.-S., Nisonoff, A. 1982. Regulation of immunity to the Azobenzenearsonate hapten. *Adv. Immunol.* 32:253–300

79. Grey, H. M., Mannik, M., Kunkel, H. G. 1965. Individual antigenic specificity of myeloma proteins. Characteristics and localization to subunits. *J. Exp. Med.* 121:561–75

80. Hannestad, K., Jørgensen, T. 1979. Help and suppression of antibody responses to the idiotype of myeloma protein 315 are mediated by separate V domains. *Scand. J. Immunol.* 10:367

81. Harvey, M. A., Adorini, L., Miller, A., Sercarz, E. E. 1979. Lysozyme-induced T-suppressor cells and antibodies have a predominant idiotype. *Nature* 281: 594–96

82. Herzenberg, L. A., Black, S. J., Herzenberg, L. A. 1980. Regulatory circuits and antibody responses. *Eur. J. Immunol.* 10:1–11

83. Herzenberg, L. A., Tokuhisa, T., Parks, D. R., Herzenberg, L. A. 1982. Epitope-specific regulation. II. A bistable, Igh-restricted regulatory mechanism central to immunologic memory. *J. Exp. Med.* 155:1741–53

84. Hetzelberger, D., Eichmann, K. 1978. Recognition of idiotypes in lymphocyte interactions. I. Idiotypic selectivity in the cooperation between T and B lymphocyte interactions. *Eur. J. Immunol.* 8:846–52

85. Hetzelberger, D., Eichmann, K. 1978. Idiotype suppression. III. Induction of unresponsiveness to sensitization with anti-idiotypic antibody: Identification of the cell types tolerized in high zone and in low zone, suppressor cell-mediated, idiotype suppression. *Eur. J. Immunol.* 8:839–46

85a. Heymann, B., Andrighetto, G., Wigzell, H. 1982. Antigen-dependent IgM-mediated enhancement of the sheep erythrocyte response in mice. *J. Exp. Med.* 155:994–1009

86. Hirai, Y., Nisonoff, A. 1980. Selective suppression of the major idiotypic component of an antihapten response by soluble T cell-derived factors with idiotypic or anti-idiotypic receptors. *J. Exp. Med.* 151:1213–31

87. Hiernaux, J. 1977. Some remarks on the stability of the idiotypic network. *Immunochemistry* 14:733–39

87a. Hiernaux, J., Bona, C., Baker, P. J. 1981. Neonatal treatment with low doses of anti-idiotypic antibody leads to the expression of a silent clone. *J. Exp. Med.* 153:1004–8

87b. Hiernaux, J., Bona, C. A. 1982. Shared idiotypes among monoclonal antibodies specific for different immunodominant sugars of lipopolysaccharide of different Gram-negative bacteria. *Proc. Natl. Acad. Sci. USA* 79:1616–20

88. Hoffmann, G. W. 1975. A theory of regulation and self-nonself discrimination in an immune network. *Eur. J. Immunol.* 5:638–47

88a. Honjo, T. 1983. Immunoglobulin Genes. *Ann. Rev. Immunol.* 1:499–528

89. Hopper, J. E., Nisonoff, A. 1971. Individual antigenic specificity of immunoglobulins. *Adv. Immunol.* 13:57–99

90. Huber, R. 1982. *Idiotypes-Antigens on the Inside,* ed. I. Westen-Schnurr, pp. 96–111. Basel: Editiones Roche 1982

90a. Huser, H., Haimovich, J., Jaton, J.-C. 1975. Antigen binding and idiotypic properties of reconstituted immunoglobulins G derived from homogeneous rabbit anti-pneumococcal antibodies. *Eur. J. Immunol.* 5:206–10

91. Imanishi-Kari, T., Rajnavölgyi, E., Takemori, T., Jack, R. S., Rajewsky, K. 1979. The effect of light chain gene expression on the inheritance of an idiotype associated with primary anti-(4-hydroxy-3-nitrophenyl)acetyl(NP)antibodies. *Eur. J. Immunol.* 9:324–31

92. Ivars, F., Holmberg, D., Forni, L., Cazenave, P.-A., Coutinho, A. 1982. Antigen-independent, IgM-induced antibody responses: Requirement for "recurrent" idiotypes. *Eur. J. Immunol.* 12:146–51

92a. Iverson, G. M. 1970. Ability of CBA mice to produce anti-idiotypic sera to 5563 myeloma protein. *Nature* 227: 273–74

93. Jack, R. S., Imanishi-Kari, T., Rajewsky, K. 1977. Idiotypic analysis of the response of C57BL/6 mice to the (4-hydroxy-3-nitrophenyl)acetyl group. *Eur. J. Immunol.* 7:559–65

94. Jackson, S., Mestecky, J. 1979. Presence of plasma cells binding autologous antibody during an immune response. *J. Exp. Med.* 150:1265–70

95. Janeway, C. A., Jr., Broughton, B., Dzierzak, E., Jones, B., Eardley, D. D., et al. 1981. Studies of T lymphocyte function in B cell deprived mice. In *Immunoglobulin idiotypes: ICN-UCLA Symposia on Molecular and Cellular Biology.* ed. C. Janeway Jr., E. E. Sercarz, H. Wigzell, 20:661–71. New York: Academic

96. Jayaraman, S., Bellone, C. J. 1982. Hapten-specific responses to the phenyltrimethylamino hapten. I. Evidence for idiotype-anti-idiotype interactions in delayed-type hypersensitivity in mice. *Eur. J. Immunol.* 12:272–77

97. Jerne, N. K. 1974. Towards a network theory of the immune system. *Ann. Immunol.* 125C:373–89

98. Jerne, N. K. 1975. The immune system: A web of V-domains. *Harvey lectures* 70:93–110

99. Jerne, N. K. 1976. The immune system: A network of lymphocyte interactions. In *The immune system,* ed. F. Melchers, K. Rajewsky, pp. 259–66. New York: Springer

100. Jerne, N. K. 1982. In *Idiotypes-Antigens on the Inside,* ed. I. Westen-Schnurr, pp. 12–15. Basel: Editiones Roche

101. Jerne, N. K., Roland, J., Cazenave, P.-A. 1982. Recurrent idiotopes and internal images. *EMBO J.* 1:243–47

102. Jørgensen, T., Hannestad, K. 1982. Helper T cell recognition of the variable domains of a mouse myeloma protein (315). Effect of the major histocompatibility complex and domain conformation. *J. Exp. Med.* 155:1587–96

102a. Ju, S.-T., Benacerraf, B., Dorf, M. E. 1980. Genetic control of shared idiotype among antibodies directed to distinct specificities. *J. Exp. Med.* 152:170–82

102b. Ju, S.-T., Benacerraf, B., Dorf, M. E. 1978. Idiotypic analysis of antibodies to poly(Glu^{60} Ala^{30} Tyr^{10}): Interstrain and interspecies idiotypic crossreactions. *Proc. Natl. Acad. Sci. USA* 75:6192–96

103. Julius, M. H., Cosenza, H., Augustin, A. A. 1978. Evidence for the endogenous production of T cell receptors bearing idiotypic determinants. *Eur. J. Immunol.* 8:484–91

104. Julius, M. H., Cosenza, H., Augustin, A. A. 1980. Enrichment of hapten-specific helper T cells using anti-immunoglobulin combining site antibodies. *Eur. J. Immunol.* 10:112–16

105. Julius, M. H., von Boehmer, H., Sidman, C. L. 1982. Dissociation of two signals required for activation of resting B cells. *Proc. Natl. Acad. Sci. USA* 79:1989–93

106. Juy, D., Primi, D., Sanchez, P., Cazenave, P.-A. 1982. Idiotype regulation: evidence for the involvement of Igh-C-restricted T cells in the M-460 idiotype suppressive pathway. *Eur. J. Immunol* 12:24–30

106a. Juy, D., Primi, D., Sanchez, P., Cazenave, P.-A. 1983. The selection and the maintenance of the V region determi-

nant repertoire is germ-line encoded and T cell independent. *Eur. J. Immunol.* In press
107. Karjalainen, K., Hurme, M., Mäkelä, O. 1979. Inherited k chain polymorphism in mouse antibodies to phenyloxazolone. *Eur. J. Immunol.* 9:910–12
108. Karol, R., Reichlin, M., Noble, R. W. 1978. Idiotypic cross-reactivity between antibodies of different specificities. *J. Exp. Med.* 148:1488–97
109. Kearney, J. F., Barletta, R., Quan, Z. S., Quintáns, J. 1981. Monoclonal vs heterogeneous anti-H-8 antibodies in the analysis of the anti-phosphorylcholine response in BALB/c mice. *Eur. J. Immunol.* 11:877–83
110. Kelsoe, G., Cerny, J. 1979. Reciprocal expansions of idiotypic and anti-idiotypic clones following antigen stimulation. *Nature* 279:333–34
111. Kelsoe, G., Reth, M., Rajewsky, K. 1980. Control of idiotope expression by monoclonal anti-idiotope antibodies. *Immunol. Rev.* 52:75–88
112. Kelsoe, G., Reth, M., Rajewsky, K. 1981. Control of idiotope expression by monoclonal anti-idiotope and idiotope-bearing antibody. *Eur. J. Immunol.* 11:418–23
113. Kelsoe, G., Takemori, T., Rajewsky, K. 1981. Generation of specific regulatory T cells with monoclonal anti-idiotope antibody: Induction of suppressor T cells. In *B Lymphocytes in the Immune Response: Functional, Developmental and Interactive Properties,* ed. N. R. Klinman, D. E. Mosier, I. Scher, E. S. Vitetta, pp. 423–30. New York: Elsevier
114. Kelus, A. S., Gell, P. G. H. 1968. Immunological analysis of rabbit anti-antibody systems. *J. Exp. Med.* 127:215–34
115. Kim, B. S., Hopkins, W. J. 1978. Tolerance rendered by neonatal treatment with anti-idiotypic antibodies: Induction and maintenance in athymic mice. *Cell. Immunnol.* 35:460–65
116. Kim, B. S., Greenberg, J. A. 1981. Mechanisms of idiotype suppression. IV. Functional neutralization in mixtures of idiotype-specific suppressor and hapten-specific suppressor T cells. *J. Exp. Med.* 154:809–20
117. Kimoto, M., Fathman, C. G. 1980. Antigen-reactive T cell clones. I. Transcomplementing hybrid I-A-region gene products function effectively in antigen presentation. *J. Exp. Med.* 152:759–70
118. Kindred, B., Shreffler, D. C. 1972. H-2 dependence of co-operation between T and B cells in vivo. *J. Immunol.* 109:940–43
119. Klinman, N. R., Press, J. L. 1975. The B cell specificity repertoire: Its relationship to definable subpopulations. *Immunol. Rev.* 24:41–83
120. Klinman, N. R., Press, J. L. 1975. The characterization of the B-cell repertoire specific for the 2,4-dinitrophenyl and 2,4,6-trinitrophenyl determinants in neonatal BALB/c mice. *J. Exp. Med.* 141:1133–46
120a. Klinman, N. R., Wylie, D. E., Cancro, M. P. 1980. Mechanisms that govern repertoire expression. In *Progress in immunology: Immunology 80,* ed. M. Fougereau, J. Dausset, 4:122–135. New York: Academic
121. Kluskens, L., Köhler, H. 1974. Regulation of immune response by autogenous antibody against receptor. *Proc. Natl. Acad. Sci. USA* 71:5083–87
122. Köhler, G., Milstein, C. 1975. Continuous cultures of fused cells secreting antibody of predefined specificity. *Nature* 256:495–97
123. Köhler, H. 1981. In *Idiotypes-Antigens on the Inside,* ed. I. Westen-Schnurr, pp. 177–82. Basel: Editiones Roche
124. Köhler, H., Kaplan, D. R., Strayer, D. S. 1974. Clonal depletion in neonatal tolerance. *Science* 186:643
125. Kohno, Y., Berkower, I., Buckenmeyer, G., Minna, J. A., Berzofsky, J. A. 1981. Shared idiotopes among monoclonal antibodies to distinct determinants of sperm whale myoglobin. *J. Supramol. Struct. Cell. Biochem.* 5:361
126. Krawinkel, U., Cramer, M., Melchers, I., Imanishi-Kari, T., Rajewsky, K. 1978. Isolated hapten-binding receptors of sensitized lymphocytes. III. Evidence for idiotypic restriction of T-cell receptors. *J. Exp. Med.* 147:1341–47
127. Krawinkel, U., Cramer, M., Kindred, B., Rajewsky, K. 1979. Isolated hapten-binding receptors of sensitized lymphocytes. V. Cellular origin of receptor molecules. *Eur. J. Immunol.* 9:815–20
128. Kronenberg, M., Davis, M. M., Early, P. W., Hood, L. E., Watson, J. D. 1980. Helper and killer T cells do not express B cell immunoglobulin joining and constant region gene segments. *J. Exp. Med.* 152:1745–61
129. Kunkel, H. G., Mannik, M., Williams, R. C. 1963. Individual antigenic specificity of isolated antibodies. *Science* 140:1218–19
130. Kurosawa, Y., von Boehmer, H., Haas, W., Sakano, H., Traunecker, A., Tonegawa, S. 1981. Identification of D

segments of immunoglobulin heavy-chain genes and their rearrangement in T lymphocytes. *Nature* 290:565–70

130a. Larhammar, D., Schenning, L., Gustafsson, K., Wiman, K., Claesson, L., et al. 1982. Complete amino acid sequence of an HLA-DR antigen-like β chain as predicted from the nucleotide sequence: Similarities with immunoglobulins and HLA-A, -B, and -C antigens. *Proc. Natl. Acad. Sci. USA* 79:3687–91

131. Le Guern, C., Ben Aïssa, F., Juy, D., Mariamé, B., Buttin, G., Cazenave, P.-A. 1979. Expression and induction of MOPC-460 idiotopes in different strains of mice. *Ann. Immunol. (Paris)* 130C: 293–302

132. Leo, O., Slaoui, M., Mariamé, B., Urbain, J. 1981. Internal images in the immune network. *J. Supramol. Struct. Cell. Biochem. Suppl.* 5:84

132a. Lewis, G. K., Goodman, J. W. 1978. Purification of functional, determinant-specific, idiotype-bearing murine T cells. *J. Exp. Med.* 148:915–24

133. Lieberman, R., Potter, M., Mushinski, E. B., Humphrey, W., Jr., Rudikoff, S. 1974. Genetics of a new IgV$_H$ (T15 idiotype) marker in the mouse regulating natural antibody to phosphorylcholine. *J. Exp. Med.* 139:983–1001

134. Lieberman, R., Vrana, M., Humphrey, W., Jr., Chien, C. C., Potter, M. 1977. Idiotypes of inulin-binding myeloma proteins localized to variable region light and heavy chains: Genetic significance. *J. Exp. Med.* 146:1294–304

135. Lindahl, K. F., Rajewsky, K. 1979. T cell recognition: genes, molecules and functions. In *International Review of biochemistry, Defence and Regulation: IIA*, ed. E. S. Lennox, 22:97–150. Baltimore: University Park

136. Lucas, A., Henry, C. 1982. Expression of the major cross-reactive idiotype in a primary anti-azobenzenearsonate response. *J. Immunol.* 128:802–6

137. Mäkelä, O., Karjalainen, K. 1977. Inherited immunoglobulin idiotypes of the mouse. *Immunol. Rev.* 34:119–38

138. Marquart, M., Deisenhofer, J., Huber, R., Palm, W. 1980. Crystallographic refinement and atomic models of the intact immunoglobulin molecule Kol and its antigen-binding fragment at 3.0Å and 1.9 Å resolution. *J. Mol. Biol.* 141:369–91

139. McKearn, J. P., Quintáns, J. 1980. Delineation of tolerance-sensitive and tolerance-insensitive B cells in normal and immune defective mice. *J. Immunol.* 124:77–80

140. McKearn, T. J., Hamada, Y., Stuart, F. P., Fitch, F. W. 1974. Anti-receptor antibody and resistance to graft-versus-host disease. *Nature* 251:648–50

141. Melchers, F. 1977. B lymphocyte development in fetal liver. I. Development of reactivities to B cell mitogens "in vivo" and "in vitro." *Eur. J. Immunol.* 7: 476–81

142. Metzger, D. W., Furman, A., Miller, A., Sercarz, E. E. 1981. Idiotypic repertoire of anti-Hen eggwhite lysozyme antibodies probed with hybridomas. *J. Exp. Med.* 154:701–12

143. Metzger, D. W., Miller, A., Sercarz, E. E. 1980. Sharing of an idiotypic marker by monoclonal antibodies specific for distinct regions of hen lysozyme. *Nature* 287:540–42

144. Miller, G. G. P., Nadler, P. I., Hodes, R. J., Sachs, D. H. 1982. Modification of T cell antinuclease idiotype expression by in vivo administration of anti-idiotype. *J. Exp. Med.* 155:190–200

145. Mozes, E., Haimovich, J. 1979. Antigen specific T-cell helper factor cross reacts idiotypically with antibodies of the same specificity. *Nature* 278:56–57

146. Nadler, P. I., Miller, G. G., Sachs, D. H., Hodes, R. J. 1982. The expression and functional involvement of nuclease-specific idiotype on nuclease-primed helper T cells. *Eur. J. Immunol.* 12: 113–20

147. Nishikawa, S., Takemori, T., Rajewsky, K. 1983. The expression of a set of antibody variable regions in LPS reactive B cells at various stages of ontogeny and its control by anti-idiotypic antibody. *Eur. J. Immunol.* In press

148. Nisonoff, A., Ju, S.-T., Owen, F. L. 1977. Studies of structure and immunosuppression of a cross-reactive idiotype in strain A mice. *Immunol. Rev.* 34:89–118

149. Nisonoff, A., Laskin, J. A., Grey, A., Klinman, N., Gottlieb, P. D. 1977. The effect of VL on the inheritance of an idiotype associated with antibodies to p-asophenylarsonate. *Immunogenetics* 5:513–14

149a. Nossal, G. J. V. 1983. Cellular Mechanisms of Immunologic Tolerance. *Ann. Rev. Immunol.* 1:33–62

150. Okuda, K., Minami, M., Sherr, D. H., Dorf, M. E. 1981. Hapten-specific T cell response to 4-hydroxy-3-nitrophenyl acetyl. XI. Pseudogenetic restrictions of hybridoma suppressor factors. *J. Exp. Med.* 154:468–79

151. Okumura, K., Hayakawa, K., Tada, T. 1982. Cell-to-cell interaction controlled

by immunoglobulin genes. Role of Thy-1⁻, Lyt-1⁺, Ig⁺ (B') Cell in allotype-restricted antibody production. *J. Exp. Med.* 156:443–53

152. Orr, H. T., Lancet, D., Robb, R. J., Lopez de Castro, J. A., Strominger, J. E. 1979. The heavy chain of human histocompatibility antigen HLA-B7 contains an immunoglobulin-like region. *Nature* 282:266–70

153. Ortiz-Ortiz, L., Weigle, W. O., Parks, D. L. 1982. Deregulation of idiotype expression. Induction of tolerance in an anti-idiotypic response. *J. Exp. Med.* 156:898–911

154. Osmond, D. G., Nossal, G. J. V. 1974. Differentiation of lymphocytes in mouse bone marrow. II. Kinetics of maturation and renewal of antiglobulin-binding cells studied by double labeling. *Cell. Immunol.* 13:132–45

155. Oudin, J. Michel, M. 1963. Une nouvelle forme d'allotypie des globulines du sérum de lapin, apparemment liée à la fonction et la spécificité anticorps. *Compt. Rend. Acad. Sci.* 257:805–8

156. Oudin, J. Cazenave, P.-A. 1971. Similar idiotypic specificities in immunoglobulin fractions with different antibody functions or even without detectable antibody function. *Proc. Natl. Acad. Sci. USA.* 68:2616–20

157. Owen, F. L., Ju, S.-T., Nisonoff, A. 1977. Presence on idiotype-specific suppressor T cells of receptors that interact with molecules bearing the idiotype. *J. Exp. Med.* 145:1559–66

158. Owen, F. L., Riblet, R., Taylor, B. A. 1981. The T suppressor cell alloantigen Tsuᵈ maps near immunoglobulin allotype genes and may be a heavy chain constant-region marker on a T cell receptor. *J. Exp. Med.* 153:801–10

158a. Paige, C. J., Kincade, P. W., Ralph, P. 1981. Independent control of immunoglobulin heavy and light chain expression in a murine pre-B-cell line. *Nature* 292:631–33

159. Padlan, E. A. 1977. Structural basis for the specificity of antibody-antigen reactions and structural mechanisms for the diversification of antigen-binding specificities. *Q. Rev. Biophys.* 10(1):35–65

159a. Paul, W. E., Bona, C. 1982. Regulatory idiotopes and immune networks: a hypothesis. *Immunol. Today* 3:230–34

160. Pawlita, M., Mushinski, E. B., Feldmann, R. J., Potter, M. 1981. A monoclonal antibody that defines an idiotope with two subsites in galactan-binding myeloma proteins. *J. Exp. Med.* 154:1946–56

161. Pech, M., Höchtl, J., Schnell, H., Zachau, H. G. 1981. Differences between germ-line and rearranged immunoglobulin Vk coding sequences suggest a localized mutation mechanism. *Nature* 291:668–70

162. Pettersson, S., Pobor, G., Coutinho, A. 1982. Ontogenic development of B cell reactivities to cooperative cell signals: dissociation between proliferation and antibody secretion. *Eur. J. Immunol.* 12:653–58

163. Pierce, S. K., Speck, N. A., Gleason, K., Gearhart, P. J., Köhler, H. 1981. BALB/c T cells have the potential to recognize the TEPC 15 prototype antibody and its somatic variants. *J. Exp. Med.* 154:1178–87

163a. Pillemer, E., Weissman, I. L. 1981. A monoclonal antibody that detects a VK-TEPC15 idiotypic determinant. cross-reactive with a Thy-1 determinant. *J. Exp. Med.* 153:1068–79

164. Primi, D., Juy, D., Le Guern, C., Sanchez, P., Cazenave, P.-A. 1981. Recognition of MOPC-460 variable region determinants by polyclonally distributed triggering receptors on B lymphocytes. *J. Immunol.* 127:1714–18

164a. Primi, D., Goodman, J. W., Lewis, G. K. 1981. Idiotype-specific T cell help in the anti-arsonate response of A/J mice. In *Immunoglobulin Idiotypes: ICN-UCLA Symposia on Molecular and Cellular Biology,* ed. C. Janeway Jr., E. E. Sercarz, H. Wigzell, 20:547–62. New York: Academic

165. Primi, D., Lewis, G. K., Goodman, J. W. 1980. The role of immunoglobulin receptors and T cell mediators in B lymphocyte activation. I. B cell activation by anti-immunoglobulin and anti-idiotype reagents. *J. Immunol.* 125:1286–92

165a. Primi, D., Mami, F., Le Guern, C., Cazenave, P.-A. 1982. Mitogen-reactive B cell subpopulations selectively express different sets of V regions. *J. Exp. Med.* 156:181–90

166. Quintáns, J., Quan, Z. S., Arias, M. A. 1981. Mice with the xid defect have helper cells for T 15 idiotype-dominant anti-phosphorylcholine primary and secondary plaque-forming cell responses. *J. Exp. Med.* 155:1245–50

167. Radbruch, A., Liesegang, B., Rajewsky, K. 1980. Isolation of variants of mouse myeloma X 63 that express changed immunoglobulin class. *Proc. Natl. Acad. Sci. USA* 77:2909–13

168. Rajewsky, K. 1983. Symmetry and

asymmetry in idiotypic interactions. *Ann. Immunol.* In press

169. Rajewsky, K., Eichmann, K. 1977. Antigen receptors of T helper cells. *Contemp. Top. Immunobiol.* 7:69–112

170. Rajewsky, K., Reth, M., Takemori, T., Kelsoe, G. 1981. A glimpse into the inner life of the immune system. In *The Immune System,* ed. C. M. Steinberg, I. Lefkovitz, 2:1–11. Basel: Karger

171. Rajewsky, K., Takemori, T., Reth, M. 1981. Analysis and regulation of V gene expression by monoclonal antibodies. In *Monoclonal Antibody and T Cell Hybridoma: Perspective and Technical Advances,* ed. G. J. Hämmerling, U. Hämmerling, J. F. Kearney, pp. 399–409. New York: Elsevier

171a. Ramseier, H., Lindenmann, J. 1972. Aliotypic antibodies. *Transplant. Rev.* 10:57–96

172. Reth, M., Bothwell, A. L. M., Rajewsky, K. 1981. Structural properties of the hapten binding site and of idiotopes in the NP^b antibody family. In *Immunoglobulin Idiotypes: ICN-UCLA Symposia on Molecular and Cellular Biology,* ed. C. Janeway Jr., E. E. Sercarz, H. Wigzell, 20:169–78. New York: Academic

173. Reth, M., Hämmerling, G. J., Rajewsky, K. 1978. Analysis of the repertoire of anti-NP antibodies in C57BL/6 mice by cell fusion. I. Characterization of antibody families in the primary and hyperimmune response. *Eur. J. Immunol.* 8:393–400

174. Reth, M., Imanishi-Kari, T., Rajewsky, K. 1979. Analysis of the repertoire of anti-(4-hydroxy-3-nitrophenyl)acetyl (NP) antibodies in C57BL/6 mice by cell fusion. II. Characterization of idiotopes by monoclonal anti-idiotope antibodies. *Eur. J. Immunol.* 9:1004–13

175. Reth, M., Kelsoe, G., Rajewsky, K. 1981. Idiotypic regulation by isologous monoclonal anti-idiotope antibodies. *Nature* 290:257–59

176. Richter, R. H. 1975. A network theory of the immune system. *Eur. J. Immunol.* 5:350–54

177. Rodkey, L. S. 1974. Studies of idiotypic antibodies. Production and characterization of autoanti-idiotypic antisera. *J. Exp. Med.* 139:712–20

178. Roland, J., Cazenave, P.-A. 1981. Rabbits immunized against b6 allotype express similar anti-b6 idiotopes. *Eur. J. Immunol.* 11:469–74

179. Rubinstein, L. J., Yeh, M., Bona, C. A. 1982. Idiotype-anti-idiotype network. II. Activation of silent clones by treatment at birth with idiotypes is associated with the expansion of idiotype-specific helper T cells. *J. Exp. Med.* 156:506–21

179a. Rowley, D. A., Griffith, P., Lorbach, I. 1981. Regulation by complementary idiotypes. Ig protects the clone producing it. *J. Exp. Med.* 153:1377–90

180. Sakato, N., Eisen, H. N. 1975. Antibodies to idiotypes of isologous immunoglobulins. *J. Exp. Med.* 141:1411–26

180a. Schiff, C., Boyer, C., Milili, M., Fougereau, M. 1981. Structural basis for MOPC 173 idiotypic determinants distinctively recognized in syngeneic and allogeneic immunization: contribution of DH, JH, and JK regions to an idiotype recognized by allogeneic antisera. *Ann. Immunol. Inst. Pasteur* 132C:113–29

181. Schilling, J., Clevinger, B., Davie, J. M., Hood, L. 1980. Amino acid sequence of homogeneous antibodies to dextran and DNA rearrangements in heavy chain V-region gene segments. *Nature* 283:35–40

182. Schuler, W., Weiler, E., Weiler, I. J. 1981. Biological and serological comparison of syngeneic and allogeneic anti-idiotypic antibodies. *Molecular Immunology* 18:1095–105

183. Sege, K., Peterson, P. A. 1978. Use of anti-idiotypic antibodies as cell-surface receptor probes. *Proc. Natl. Acad. Sci. USA* 75:2443–47

184. Sercarz, E. E., Metzger, D. W. 1980. Epitope-specific and idiotype specific cellular interactions in a model protein antigen system. *Springer Semin. Immunopathol.* 3:145–70

185. Sherr, D. H., Dorf, M. E. 1982. Hapten-specific T cell responses to 4-hydroxy-3-nitrophenyl acetyl. XIII. Characterization of a third T cell population involved in suppression of in vitro PFC responses. *J. Immunol.* 128:1261–66

186. Siekevitz, M., Gefter, M. L. Brodeur, P., Riblet, R., Marshak-Rothstein, A. 1982. The genetic basis of antibody production: The dominant anti-arsonate idiotype response of the strain mice. *Eur. J. Immunol.* 12:1023–32

187. Siekevitz, M., Huang, S. Y., Gefter, M. L. 1983. The genetic basis of antibody production: One variable region heavy chain gene encodes all molecules bearing the dominant anti-arsonate idiotype in the strain A mouse. *Eur. J. Immunol.* In press

188. Sigal, N. H., Garbart, P. J., Klinman, N. R. 1975. The frequency of phosphorylcholine-specific B cells in conven-

189. Sigal, N. H., Picard, A. R., Metcalf, E. S., Gearhart, P. J., Klinman, N. R. 1977. Expression of phosphorylcholine-specific B cells during murine development. *J. Exp. Med.* 146:933–48

190. Sogn, J. A., Yarmush, M. L., Kindt, T. J. 1976. An idiotypic marker for the VL region of an homogeneous antibody. *Ann. Immunol. (Paris)* 127C:397–408

190a. Sommé, G., Serra, J. R., Leclercq, L., Moreau, J.-L., Mazié, J.-C., Moinier, D., Fougereau, M., Thèze, J. 1982. Contribution of the H- and L-chains and of the binding site to the idiotypic specificities of mouse anti-GAT antibodies. *Mol. Immunol.* 19:1011–19

191. Sprent, J. 1973. Circulating T and B lymphocytes of the mouse. II. Life span. *Cell. Immunol.* 7:40–59

192. Strober, S. 1972. Initiation of antibody responses by different classes of lymphocytes. V. Fundamental changes in the physiological characteristics of virgin thymus-independent ("B") lymphocytes and "B" memory cells. *J. Exp. Med.* 136:851–71

193. Strosberg, D. A., Couraud, P.-O., Shreiber, A. 1981. Immunological studies of hormone receptors: A two-way approach. *Immunol. Today* 2:75–9

193a. Sugimura, K., Kishimoto, T., Maeda, K., Yamamura, Y. 1981. Demonstration of T15 idiotype-positive effector and suppressor T cells for phosphorylcholine-specific delayed-type hypersensitivity response in CBA/N or (CBA/N × BALB/c) F₁ male mice. *Eur. J. Immunol.* 11:455–61

194. Staudt, L. 1982. In *Idiotypes-Antigens on the Inside*, ed. I Westen-Schnurr, pp. 96–111. Basel: Editiones Roche

194a. Suzan, M., Valstedt, F., Boned, A., Rubin, B. 1982. Genetic chasing of T helper cell idiotype and allotype genes. *Immunogenetics* 16:229–41

195. Sy, M.-S., Bach, B. A., Brown, A., Nisonoff, A., Benacerraf, B., Greene, M. I. 1979. Antigen- and receptor-driven regulatory mechanisms. II. Induction of suppressor T cells with idiotype-coupled syngeneic spleen cells. *J. Exp. Med.* 150:1229–40

196. Sy, M.-S., Brown, A. R., Benacerraf, B., Greene, M. I. 1980. Antigen- and receptor-driven regulatory mechanisms. III. Induction of delayed-type hypersensitivity to azobenzenearsonate with anti-cross-reactive idiotypic antibodies. *J. Exp. Med.* 151:896–909

196a. Sy, M.-S., Brown, A., Bach, B. A., Benacerraf, B., Gottlieb, P. D., et al. 1981. Genetic and serological analysis of the expression of crossreactive idiotypic determinants on anti-p-azobenzenarsonate antibodies and p-azobenzenarsonate-specific suppressor T cell factors. *Proc. Natl. Acad. Sci. USA* 78: 1143–47

197. Sy, M.-S., Dietz, M. H., Germain, R. N., Benacerraf, B., Greene, M. I. 1980. Antigen- and receptor-driven regulatory mechanisms. IV. Idiotype-bearing I-J⁺ suppressor T cell factors induce second-order suppressor T cells which express anti-idiotypic receptors. *J. Exp. Med.* 151:1183–95

197a. Szewczuk, M. R., Campbell, R. J. 1981. Lack of age-associated auto-anti-idiotypic antibody regulation in mucosal-associated lymph nodes. *Eur. J. Immunol.* 11:650–56

198. Tada, T., Okumura, K., Hayakawa, K., Suzuki, G., Abe, R., Kumagai, Y. 1981. Immunological circuitry governed by MHC and V_H gene products. In *Immunoglobulin Idiotypes: ICN-UCLA Symposia on Molecular and Cellular Biology*, ed. C. Janeway Jr., E. E. Sercarz, H. Wigzell, 20:563–72. New York: Academic

199. Takemori, T., Tesch, H., Reth, M., Rajewsky, K. 1982. The immune response against anti-idiotype antibodies. I. Induction of idiotope-bearing antibodies and analysis of the idiotope repertoire. *Eur. J. Immunol.* 12:1040–46

200. Tasiaux, N., Leuwenkroon, R., Bruyns, C., Urbain, J. 1978. Possible occurence and meaning of lymphocytes bearing auto anti-idiotypic receptors during the immune response. *Eur. J. Immunol.* 8:464–68

200a. Thèze, J., Moreau, J.-L. 1978. Genetic control of the immune response to the GAT terpolymer. *Ann. Immunol. (Paris)* 129C:721–26

201. Thomas, W. R., Morahan, G., Walker, I. D., Miller, J. F. A. P. 1981. Induction of delayed-type hypersensitivity to azobenzenearsonate by a monoclonal anti-idiotype antibody. *J. Exp. Med.* 153: 743–47

202. Tokuhisa, T., Komatsu, Y., Uchida, Y., Taniguchi, M. 1982. Monoclonal alloantibodies specific for the constant region of T cell antigen receptors. *J. Exp. Med.* 156:888–97

203. Trenkner, E., Riblet, R. 1975. Induction of antiphosphorylcholine antibody formation by anti-idiotypic antibodies. *J. Exp. Med.* 142:1121–32

204. Urbain, J., Wuilmart, C., Cazenave, P.-A. 1981. Idiotypic regulation in immune networks. In *Contemporary Top-*

ics in Molecular Immunology, ed. F. P. Inwan, W. J. Mandy, 8:113–48. New York: Plenum

205. Urbain, J., Wikler, M., Franssen, J. D., Collignon, C. 1977. Idiotypic regulation of the immune system by the induction of antibodies against anti-idiotypic antibodies. Proc. Natl. Acad. Sci. USA 74:5126–30

206. Vogler, L. B., Preud'homme, J. L., Seligmann, M., Gathings, W. E., Grist, W. M., et al. 1981. Diversity of immunoglobulin expression in leukaemic cells resembling B-lymphocyte precursors. Nature 290:339–41

207. Weigert, M., Riblet, R. 1976. Genetic control of antibody variable regions. Cold Spring Harbor Symp. Quant. Biol. 41:837–46

208. Weigert, M., Riblet, R. 1978. The genetic control of antibody variable regions in the mouse. Springer Semin. Immunopathol. 1:133–69

209. Weiler, E. 1981. In Idiotypes-Antigens on the Inside, ed. I. Westen-Schnurr, pp. 142–49. Basel: Editiones Roche

210. Weiler, I. J., Weiler, E., Sprenger, R., Cosenza, H. 1977. Idiotype suppression by maternal influence. Eur. J. Immunol. 7:591–97

211. Weinberger, J. Z., Germain, R. N., Benacerraf, B., Dorf, M. E. 1980. Hapten-specific T cell responses to 4-hydroxy-3-nitrophenyl acetyl. V. Role of idiotypes in the suppressor pathway. J. Exp. Med. 152:161–69

212. Weinberger, J. Z., Germain, R. N., Ju, S.-T., Greene, M. I., Benacerraf, B., Dorf, M. E. 1979. Hapten-specific T cell responses to 4-hydroxy-3-nitrophenyl acetyl. II. Demonstration of idiotypic determinants on suppressor T cells. J. Exp. Med. 150:761–76

213. Wells, J. V., Fudenberg, H. H., Givol, D. 1973. Localization of idiotypic antigenic determinants in the F_v region of murine myeloma protein MOPC 315. Proc. Natl. Acad. Sci. USA 70:1585–87

214. Wikler, M., Demeur, C., Dewasme, G., Urbain, J. 1980. Immunoregulatory role of maternal idiotypes. Ontogeny of immune networks. J. Exp. Med. 152:1024–35

214a. Wikler, M., Franssen, J.-D., Collignon, C., Leo, O., Mariamé, B., et al. 1979. Idiotypic regulation of the im-

mune system. Common idiotypic specificites between idiotypes and antibodies raised against anti-idiotypic antibodies in rabbits. J. Exp. Med. 150:184

215. Williams, K. R., Claflin, J. L. 1982. Clonotypes of anti-phosphocholine antibodies induced with Proteus morganii (Potter). II. Heterogeneity, class and idiotypic analyses of the repertoires in BALB/c and A/HeJ mice. J. Immunol. 128:600–7

216. Woodland, R., Cantor, H. 1978. Idiotype-specific T helper cells are required to induce idiotype-positive B memory cells to secrete antibody. Eur. J. Immunol. 8:600–6

217. Wysocki, L. J., Sato, V. L. 1981. The strain A anti-p-azophenylarsonate major crossreactive idiotypic family includes members with no reactivity toward p-azophenylarsonate. Eur. J. Immunol. 11:832–39

218. Yamamoto, H., Nonaka, M., Katz, D. H. 1979. Suppression of hapten-specific delayed type hypersensitivity responses in mice by idiotype-specific suppressor T cells after administration of anti-idiotypic antibodies. J. Exp. Med. 150:818–29

219. Yamauchi, K., Chao, N., Murphy, D. B., Gershon, R. K. 1982. Molecular composition of an antigen-specific Ly-1 T suppressor inducer factor. One molecule binds antigen and is I-J⁻, another is I-J⁺, does not bind antigen, and imparts an Igh-variable region-linked restriction. J. Exp. Med. 155:655–65

220. Yarmush, M., Sogn, J. A., Mudgett, M., Kindt, T. J. 1977. The inheritance of antibody V regions in the rabbit: Linkage of an H-chain-specific idiotype to immunoglobulin allotypes. J. Exp. Med. 145:916–30

221. Yang, C., Kratzin, H., Götz, H., Thinnes, F. P., Kruse, T., et al. 1982. Primärstruktur menschlicher Histokompatibilitätsantigene der Klasse II 2. Mitteilung: Aminosäuresequenz der N-terminalen 179 Reste der α-Kette des HLA-Dw2/DR2-Alloantigens. Hoppe-Seyler's Z. Physiol. Chem. 363:671–76

222. Zinkernagel, R. M., Doherty, P. C. 1974. Restriction of in vitro T-cell mediated cytoxicity in lymphocytic choriomeningitis within a syngeneic or semiallogeneic system. Nature 248:701–2

Ann. Rev. Immunol. 1983. 1:609–32

EPITOPE-SPECIFIC REGULATION

Leonore A. Herzenberg, Takeshi Tokuhisa, and Kyoko Hayakawa

Genetics Department, Stanford University School of Medicine, Stanford, California 94305

INTRODUCTION

Long before lymphocytes had been identified with certainty as the precursors of antibody forming cells, immunologists and immunogeneticists were well aware that animals immunized with the complex antigens like bovine serum albumin individually produce antibodies to different subsets of the epitopes (determinants) on the antigen. Over the years, many of the cells and cell interactions that regulate antibody production have been defined in great detail; however, the processes that control the characteristic individuality of antibody responses still remain shrouded in mystery. The epitope-specific regulatory system described here offers a workable explanation of how this variation is generated and maintained. Furthermore, as we shall show, the joint operation of the carrier-specific induction mechanism and the epitope-specific effector mechanism that defines the characteristics of the regulated response provides an explanation for many of the odd observations encountered in studies of immunologic memory and carrier-specific regulation.

Current concepts of how carrier-specific regulation influences antibody responses to the epitopes (immunogenic structures) on carrier molecules date from the early 1970s, when Mitchison (1) and Rajewsky et al (2) first demonstrated that carrier-primed T cells help hapten-primed B cells give rise to adoptive secondary antibody responses. These studies, which showed that ". . . the antigen [hapten-carrier conjugate] is recognized by two receptors, one directed to the hapten and the other to a determinant on the carrier protein" (1), in essence formulated the contemporary definition of a T-dependent antigen, i.e. a macromolecule with at least one "carrier determi-

609

nant" recognized and used by T cells to regulate antibody responses to the various epitopes on the antigen.

The hapten-carrier bridge mechanism suggested for carrier-specific helper activity in these early studies is as viable today as it was when first stated (and still remains unproven); however, the simplistic two-cell model (helper T and memory B) introduced initially has undergone considerable expansion. Two functionally distinct carrier-specific regulatory T cells have been added, one which suppresses antibody responses (3–8) and another which contrasuppresses such responses (9). Furthermore, several carrier-specific cells have been identified as part of developmental cascades leading to the emergence of the functional cell types. Finally, on a theoretical level, a series of carrier-specific regulatory circuits have been proposed locating the various functional cells in a self-limiting ("feed back") type system that controls antibody production by controlling the supply of carrier-specific help (10).

This construct is consistent with most of the available data; however, it fails to explain a number of relevant observations (several of which predate its inception). In particular, because it is predicated on the idea that carrier-specific regulatory interactions do not selectively influence antibody production to individual epitopes on an antigen, it tends to trivialize findings that suggest links between carrier-specific and epitope-specific or other regulatory mechanisms. In essence, it has led to consistent disregard of evidence suggesting that immunizing carrier-primed animals with a "new" hapten coupled to the priming carrier results in the induction of specific suppression for antibody responses to the hapten (see below).

Mitchison (1) and Rajewsky et al (2) overtly chose to ignore this peculiar response failure to simplify consideration of the mechanisms involved in carrier-specific help. Thus, although they carefully cite evidence from several studies (including their own) showing that carrier-primed animals often fail to produce detectable levels of antibodies to haptens introduced subsequently on the priming carrier, they put more trust in opposing evidence showing that significant anti-hapten antibody production can be stimulated by this "carrier/hapten-carrier" immunization protocol. Rajewsky et al (2) make this point quite directly by stating: "We suggest that, in general, an animal pretreated with free carrier and receiving a secondary injection of this carrier complexed with a hapten will be found to produce a better anti-hapten response than without the pre-treatment, if the experimental design is aimed at detecting this effect."

This "leap in faith" (although largely incorrect) was probably crucial to the orderly progress of early studies exploring the mechanisms regulating antibody production. The idea that carrier-priming would augment (rather than suppress) subsequent antibody responses to a hapten presented on the priming carrier cleared away confusion and allowed rational planning of adoptive and in vitro experiments that followed from the newly published carrier-specific help studies. However, although this idea was advanced originally as a prediction and had been shown to be invalid under certain circumstances, it rather rapidly assumed the mantle of truth. Consequently, several years later, we (and most of our colleagues) were quite surprised by the suppressed rather than augmented anti-hapten antibody responses that we obtained following carrier/hapten-carrier immunization (11).

These unexpected findings (e.g. see Figure 1) engendered a rapid series of experiments exploring the mechanism of the suppression and a somewhat more leisurely search through the literature looking for precedents for our observations. Thus, by the time we discovered that previous investigators had ascribed this kind of response failure to impaired anti-hapten memory development (14) or to the presence of anti-carrier antibodies (12, 13), we had already ruled out these possibilities and realized that we were dealing with a previously unrecognized "epitope-specific" regulatory system that selectively controls antibody production to individual epitopes according to the dictates of carrier-specific (and other) regulatory T cells present in the immunologic environment when such epitopes are first introduced (11, 15–22).

In pursuing our studies of this system, we have been concerned primarily with finding out how it works; however, we have also put considerable thought and some experimental effort into determining how it could have escaped notice for so many years. Not surprisingly, the answers to these questions often merge. For example, in attempting to establish adoptive assays for the inducer and effector cells responsible for the suppression, we found that quite strong in situ suppression is difficult to transfer to irradiated recipients, particularly when measured as the ability to suppress a response mounted by a co-transferred cell population (16). This characteristic tendency for help to predominate over suppression in irradiated recipients put several investigators off the track, including Sarvas et al (14) who in 1974 reasoned correctly that carrier-specific suppressor T cells induce suppression for anti-hapten responses in carrier/hapten-carrier immunized

```
              (Ag)                    (Ag)                    (Ag)
    virgin B -------> Early Memory -------> Mature Memory ----------> IgG AFC
    :                     IgD+                   IgD-       : :              :
    :                                                      : :              :
    :                 DEVELOPMENT                          : : EXPRESSION   :
    :................................................... : :..............:
```

Figure 1   Regulation of IgG (memory) responses. Mechanisms that regulate memory B-cell development regulate the potential for IgG antibody production (22). Mechanisms that regulate memory B-cell expression control which and how many of the memory B cells present in a given animal actually give rise to AFC (17).

animals (see below), but discarded the idea because it "requires postulating that the effect of these suppressor cells is abolished in cell transfer experiments."

The general tendency to measure anti-hapten antibody production exclusively while studying the mechanisms involved in carrier-specific regulation also appears to have played a major role in "hiding" the epitope-specific system. Ishizaka & Okudaira (23), for example, demonstrated that anti-hapten responses were suppressed while anti-carrier antibody responses proceeded to secondary levels in carrier-primed animals stimulated with a hapten on the priming carrier. Discussing this work (in 1973), these investigators referred to work by Tada & Okumura (24–26) demonstrating the existence of suppressor T cells and suggested ". . . that immunization with 1 to 10 μg of ovalbumin may result in the formation of so called [carrier-specific] suppressor T cells that might suppress preferentially the primary anti-DNP response" (23). Immuno-history would probably be quite different had Tada (or others interested in carrier-specific suppressor T cells) taken this suggestion seriously enough to test anti-carrier antibody responses in suppression assays routinely.

All in all, the attractive simplicity of the idea that carrier-specific suppressor T cells regulate antibody production by depleting carrier-specific help appears to be sufficient to explain the willingness of many laboratories (including our own) to ignore subtle inconsistencies and leave a few moss-covered stones unturned. Perhaps this has been mainly for the good, since the mechanisms involved in epitope-specific regulation are considerably more complex and would, in fact, have been difficult to explore without the level of technology and theoretical understanding reached with the last few years. In any event, we now appear to have reached the time when re-evaluation of past evidence is essential to present progress.

The sections that follow describe various aspects of our studies of the epitope-specific regulatory system. We begin with an overview section (immediately below) that broadly outlines the system as a whole, documenting statements with references rather than with evidence. The remaining sections, in contrast, discuss aspects of the system in more detail and include much of the evidence upon which our conclusions are based.

The work we discuss here has mainly been conducted in our laboratory at Stanford; however, studies with carrier-specific suppressor T cells and soluble factors were conducted in Dr. Masaru Taniguchi's laboratory in Chiba, Japan. Almost without exception, the evidence we report derives from experiments that repeat earlier work but include key controls aimed at distinguishing carrier-specific from epitope-specific regulatory effects. Thus, we independently measure both the anti-hapten and anti-carrier antibody responses following various antigenic stimulations, and we confirm the induction of epitope-specific suppression for a given anti-hapten antibody response by a final immunization with the hapten conjugated to an unrelated carrier molecule. As we shall show, the application of these rather straightforward techniques provides surprising insights into the older data and reveals the outlines of a highly flexible central regulatory system responsible for controlling all aspects of antibody responses.

EPITOPE-SPECIFIC REGULATION: AN OVERVIEW

The epitope-specific system apparently provides a common channel through which carrier-specific, isotype-specific, allotype-specific, and I-region-defined mechanisms exert control over antibody responses (11, 15–21). It plays a key role in regulating IgG antibody responses to haptens and native epitopes on commonly used carrier molecules such as KLH (keyhole limpet hemocyanin) and CGG (chicken gamma globulin) (11, 15–18). Furthermore, it is active in regulating IgG responses to epitopes on the synthetic ter-polymer TGAL, even in genetic "non-responders" to this antigen (20). Carrier-specific interactions induce it to suppress antibody production (to individual epitopes on the carrier); however, once induced to suppress a given anti-epitope response, it will suppress that response even when the epitope is presented in immunogenic form on a different carrier molecule (11, 15–17).

The effector mechanism in this system controls memory B-cell expression (as opposed to development; see Figure 1) and appears to be the ultimate arbitor of which and how many such B cells will be permitted to differentiate to IgG antibody-forming cells (AFC) in response to a given antigenic stimulus. It is epitope specific in that it independently regulates the amount and affinity of the antibody response produced to each of the epitopes on a complex antigen (11, 15–17); and it is Igh restricted in that it selectively controls the IgCh isotype and allotype expression in such responses (17–19). These properties suggest that the overall system is composed of individually specific elements, each charged with the regulation of a subset of B cells committed to produce unique or closely related IgG molecules (17).

The flexibility of this (compound) regulatory system and the nature of its effects on in situ antibody responses are extraordinary. It is capable of suppressing virtually the entire primary and secondary antibody response to a given epitope. Therefore, it can completely conceal the presence of normal anti-epitope memory B-cell populations clearly demonstrable in adoptive transfer assays (11, 15, 16). Alternatively, it can support the expression of a subset of memory B cells and suppress the expression of others, thereby defining the unique spectrum of an antibody response produced by a given animal (17–19).

Carrier-specific and other regulatory mechanisms operative in the immunological environment when an epitope is first introduced determine which components of the antibody response will be suppressed and which will be supported (11, 15–17). Prior immunization history, genetic predisposition toward responsiveness (20) and perinatal immunologic "conditioning" (18) all contribute to this determination. Thus, the epitope-specific system constitutes an ideal candidate for a central integrative mechanism capable of resolving conflicting signals from various peripheral regulatory systems and translating a coherent decision to the antibody-producing apparatus.

In keeping with this role, this system offers a unique regulatory capability that provides both the stability required to maintain a response pattern once induced and the flexibility to modify that response pattern when stimulatory conditions change dramatically. In essence, the individual epitope-specific, Igh-restricted elements in the system appear to behave similarly to bistable electronic binary ("flip-flop") circuits that can be switched initially into an "on" or "off" position by a small electrical force and then require a substantially larger force to switch them to the opposite position. That is, initial immunization conditions that induce epitope-specific elements to suppress a given IgG antibody response will usually fail to do so once the system has been induced to support that response; and, similarly, conditions that induce the system to support a response will be far less effective once the system has been induced to suppress the response. Nevertheless, either suppression or support can be reversed by sufficient stimulation in the opposite direction (17).

The evidence upon which these conclusions (and hypotheses) are based is presented in detail in several publications that have appeared within the last few months (16–19). Therefore, in the discussion which follows (and in Tables 1 and 2), we summarize these findings and concentrate more intensively on several recent studies defining the T cells that mediate epitope-specific suppression and demonstrating some of the more complex aspects that exist within the system.

CARRIER/HAPTEN-CARRIER IMMUNIZATION INDUCES EPITOPE-SPECIFIC SUPPRESSION FOR IgG ANTI-HAPTEN RESPONSES

The epitope-specific regulatory system can be specifically induced to suppress primary and secondary IgG antibody responses to the dinitrophenyl hapten (DNP) without interfering with antibody responses to epitopes on the carrier molecule on which the DNP is presented (see Table 3). Furthermore, once so induced, it specifically suppresses antibody responses to DNP presented on unrelated carrier molecules. The magnitude of a suppressed primary anti-DNP response is usually about 30% of the normal primary response; however, the affinity of a suppressed response is about 10-fold lower than normal. Suppressed secondary anti-DNP responses are typically less than 10% of normal and have average affinities that are at least 100-fold below normal (11, 16).

For most of our studies, we induce this suppression by immunizing animals sequentially with a carrier molecule such as KLH (keyhole limpet hemocyanin) and then DNP conjugated to the carrier (i.e. with the

Table 1 Induction of epitope-specific suppression occurs under a wide variety of conditions

	Variable tested	Result[a]
Epitope	DNP, TNP, NIP	Suppression induced for both epitopes by carrier/hapten–carrier; suppression inducible for KLH epitopes by other protocols
Carrier	KLH, CGG, OVA, TGAL	All prime for suppression induction; $100\,\mu g$ on alum sufficient; some genetic restrictions (see table 2)
Age	KLH at 8 weeks to >6 months	Suppression equally strong at all ages
Timing	1 to 13 weeks between KLH and DNP–KLH	Suppression equally strong for all intervals between carrier and hapten–carrier
	KLH/DNP–KLH then DNP–KLH or DNP–CGG up to 1 yr later	Suppression equally strong for all intervals between first and second hapten–carrier immunizations
Strains	BALB/C, BAB/14, SJL, SJA, C3H, C3H.SW, A/J, (SJL × BALB/C), C57BL/10, C57BL/6	Suppression inducible in all strains

[a] Summarized from References 1 through 7.

Table 2 Epitope-specific suppression in genetically controlled regulatory conditions

Condition	Immunization	Result
I-region controlled responses	TGAL/TNP–TGAL in C3H (H–2k) and C3H.SW (H–2b)	Suppression induction occurs in responder and non-responder strains (20)
Impaired response to KLH carrier (non-MHC)	KLH priming in A/J or C57BL/10	Poor suppression induction by KLH/DNP–KLH; normal induction by CGG/DNP–CGG (21)
Allotype suppression for Igh–1b (IgG2a) allotype production in (SJL × BALB/C)F1	DNP–KLH at 8 weeks of age (prior to mid-life remission from allotype suppression)	Igh–1b responses to DNP and KLH specifically suppressed during remission (18)

Table 3 Anti-hapten antibody production is specifically suppressed in carrier/hapten-carrier immunized mice

Immunizations[a]			In situ IgG2a antibody responses[b]		
Carrier	First DNP	Second DNP	Anti-DNP μg/ml (Ka)[c]	Anti-KLH units	Anti-CGG units
K	—	—	<3 (<0.3)	20	—
K	D–K	—	5 (<0.3)	170	—
—	D–K	—	35 (5)	15	—
K	D–C	—	20 (2)	—	21
—	D–C	—	13 (1)	—	11
K	D–K	D–K	9 (0.5)	370	—
—	D–K	D–K	120 (300)	130	—
K	D–K	D–C	6 (<0.3)	—	8
—	D–K	D–C	60 (100)	—	9
—	D–C	D–C	85 (400)	—	100

[a] K = KLH (keyhole limpet hemocyanin); C = CGG (chicken gamma globulin); DNP = 2,4-dinitrophenyl hapten; D–K = DNP–KLH; D–C = DNP–CGG. (BALB/c X SJL)F1 mice injected i.p. with 100 μg of the indicated antigen on alum at approximately six week intervals.

[b] Serum antibody levels measured by RIA two weeks after last indicated immunization. Anti-carrier antibody expressed as percentage of antibody in a "standard" secondary response serum pool.

[c] $K_aM^{-1} \times 10^6$ measured by RIA (22).

KLH/DNP-KLH immunization sequence). Several weeks later we again immunize with DNP, this time either on KLH or on an unrelated carrier molecule such as CGG (chicken gamma globulin). We inject 100 μg of each antigen on alum, usually at 4-week intervals, and compare the IgG anti-hapten and anti-carrier responses obtained 2 weeks after each immunization to the responses raised in (age, sex, and strain matched) control groups that are not primed initially with the carrier protein and thus are immunized only with the appropriate hapten-carrier conjugates.

This carrier/hapten-carrier immunization protocol induces marked suppression for IgG anti-hapten antibody production (as indicated above) but does not interfere with anti-carrier antibody responses or with the development of normal anti-hapten memory B-cell populations. That is, splenic B cells from either KLH/DNP-KLH immunized animals or DNP-KLH primed animals give rise to equivalent anti-hapten memory responses in adoptive recipients supplemented with an appropriate source of carrier-specific help. Furthermore, IgG anti-DNP responses in KLH/DNP-KLH immunized animals are suppressed to below primary level while anti-KLH responses are equivalent to the (secondary) anti-KLH responses obtained from control animals immunized a similar number of times with the carrier protein, i.e. twice with KLH or DNP-KLH (11–16).

Comparison of responses in KLH/DNP-KLH/DNP-CGG immunized animals and DNP-KLH/DNP-CGG immunized controls demonstrates the specificity of the suppression for anti-DNP responses even more dramatically. Control animals produce typical high magnitude, high affinity in situ secondary IgG anti-DNP responses. The experimental group produces IgG anti-DNP responses that are still below primary levels. Nevertheless, all animals (experimental and control) produce equivalent primary IgG antibody responses to the CGG epitopes on the second hapten-carrier conjugate. Thus, carrier-primed animals fail to produce antibodies to a "new" hapten presented subsequently on the priming carrier and animals immunized in this way develop a persistent suppression specific for responses to the hapten, even when presented next on a different carrier molecule (11, 15, 16).

Although the failure of anti-hapten responses in carrier/hapten-carrier immunized animals appears novel from a contemporary perspective, this phenomenon was well known in an earlier era (ca 1970), having been described in the landmark papers demonstrating adoptive carrier-specific help interactions (1, 2). It was later attributed to interference with memory B-cell development (12) and then largely forgotten as attention shifted to using adoptive and in vitro assays for characterizing the carrier-specific (and other) mechanisms regulating antibody responses. We view the loss of this key "immunologic fact" as understandable (17) but nonetheless regrettable, since some serious misconceptions (discussed below) could have been avoided if it had not fallen from sight.

CARRIER-SPECIFIC SUPPRESSOR T CELLS INDUCE EPITOPE-SPECIFIC SUPPRESSION

Carrier-specific suppressor T cells (CTs) that arise shortly after priming with a carrier molecule (4–6) are responsible for inducing the epitope-specific system to suppress IgG anti-hapten responses to "new" epitopes presented on the carrier molecule (15, 16). These well-known regulatory T cells were commonly believed to regulate antibody production by interfering with carrier-specific help; however, by repeating the original CTs transfer experiments (4, 5) with additional controls that define the specificity of the mechanism mediating the suppression in CTs recipients, we have shown that KLH-specific CTs regulate responses by inducing typical epitope-specific suppression for anti-DNP responses when the recipients are immunized with DNP-KLH (15, 16). Thus, whether KLH-primed animals are immunized directly with DNP-KLH (KLH/DNP-KLH immunization sequence) or whether T cells from KLH-primed animals are challenged with

DNP-KLH in (non-irradiated) recipients, anti-DNP responses are persistently suppressed whereas anti-carrier responses proceed normally.

The demonstration of CTs in KLH-primed animals generated some confusion initially because splenic T cells from such animals provide an excellent source of carrier-specific help (CTh) rather than suppression in (irradiated) adoptive transfer recipients. This difference appeared to be due to the use of aqueous KLH for generating CTs and alum-precipitated KLH for generating CTh; however, our results indicate that priming either with aqueous or alum-precipitated antigen induces both CTh and CTs in the immunized animals, that both kinds of regulatory T cells are active in such animals, and that each can be demonstrated independently in the in vitro or adoptive assays developed to reveal its activity.

The aqueous KLH priming protocols usually used to generate CTs did prove to be somewhat more effective in priming for in situ suppression induction than the alum-KLH priming protocols commonly used to generate KLH-specific helper T cells; however, as a rule, we have used alum-KLH priming in the carrier/hapten-carrier sequence and in adoptive studies with KLH-primed T cells, and have generated strong epitope-specific suppression. In fact, as we shall show below, roughly equivalent suppression is induced (by DNP-KLH) in animals primed with KLH on alum, in Freund's adjuvant or in aqueous form.

In essence, these studies collectively demonstrate that the epitope-specific system constitutes the major, if not the only, effector mechanism through which CTs control antibody production. Thus, they define a new role for CTs (as inducers of epitope-specific suppression) and cast these cells as "conditioners" of the immunologic environment that, when present, alter how the epitope-specific system responds to (carrier-borne) epitopes that it has not "seen" before. Coupled with the bistable properties of the epitope-specific system, this construct explains how priming with an antigenic (carrier) molecule can simultaneously prepare the animal to produce typical secondary antibody responses to epitopes encountered initially on the priming antigen and yet to specifically suppress antibody production to "new" epitopes encountered subsequently on the same antigenic molecule. We return to this point in a later section exploring the consequences of bistable regulation.

The revised view of how CTs regulate antibody production discussed above is consistent with recent evidence from studies with T-cell lines and hybrids demonstrating two carrier-specific inductive pathways that terminate in epitope-specific cells (7, 8), one that suppresses antibody production and the other that augments antibody production. Relationships (if any) between these cells and the epitope-specific effector mechanism described here have yet to be established; however, the separate carrier-specific pathways they define support the inductive role our evidence assigns to carrier-specific regulatory cells in in situ antibody responses.

EPITOPE-SPECIFIC REGULATION IS IGH-RESTRICTED

The induction and maintenance of suppression (in carrier/hapten-carrier immunized animals) varies in efficiency for individual isotype anti-hapten responses (17–19, 21). IgM responses show no evidence of suppression. IgG2a, IgG2b, and IgG3 responses are easily suppressed whereas IgG1 responses are more refractory to suppression in that they tend to be suppressed in fewer animals under suboptimal suppression-induction conditions and to escape from suppression more frequently than the other IgG isotypes after a given number of restimulations with the hapten. This characteristic isotype hierarchy prevails in animals in which the induction of epitope suppression is either genetically impaired or experimentally minimized by immunizing initially with low doses of the carrier protein. In fact, whenever suppression is weak initially or begins to wane after repeated antigenic stimulation, IgG1 antibody responses are always the first to appear (17, 21).

IgG2a, IgG2b, and IgG3 responses show about the same susceptibility to suppression; however, when suppression is weak or waning, these isotypes often "escape" individually or in random pairs (very occasionally in animals that remain suppressed for IgG responses). This selective expression demonstrates the independent control exerted by the eptitope-specific system with respect to isotype representation in antibody responses. Studies with allotype-suppressed mice similarly show that the epitope-specific system can specifically suppress Igh-1b (IgG2a allotype) responses to individual epitopes in an allotype heterozygote without interfering with production of the (allelically determined) Igh-1a responses to the same epitopes (18). Thus, the individual elements that mediate epitope-specific regulation appear to be restricted to controlling the production of antibodies with the same or closely related combining-site structures and a single heavy chain constant-region structure (allotype/isotype).

Theoretical considerations suggest that this Igh constant-region restriction may be based exclusively on the recognition of allotypic (rather than isotypic) structures. That is, since isotypic structures are shared between allotypically different heavy chains, isotype-restricted regulation cannot explain the selective regulation of Igh-1b allotype antibodies. Allotype-restricted regulation, in contrast, can clearly account for selective isotype regulation since nearly all Igh allotypic structures are unique to (and thus can identify) the heavy chain isotype on which they are found (27). Thus, it is likely that the selective regulation of both isotype and allotype representation in individual anti-epitope responses derives from a requirement for recognition of polymorphic (allotypic) regions of Igh heavy chain constant regions.

EPITOPE-SPECIFIC REGULATION IS BISTABLE

Bistable systems, by definition, have two alternative steady states with mutually exclusive functions. When confronted initially with a stimulus favoring one state or the other, these systems move rapidly to the favored state. Stabilization mechanisms then maintain the initially induced state, so that a substantially stronger signal is required to move to the other steady state than would have been required to establish that state initially. Thus, bistable systems tend to remain as initially induced but nonetheless remain capable of shifting to the alternate state if stimulatory conditions so dictate (17, 28).

The characteristics of the regulation provided by the epitope-specific system meet these criteria (17). As we have indicated, carrier/hapten-carrier immunization induces suppression for IgG responses to the hapten. Once induced, this suppression tends to be maintained (especially for IgG2a, IgG2b, and IgG3 responses). Repeated stimulation with the hapten (on any carrier), however, eventually induces IgG anti-hapten antibody production (more quickly for IgG2 than for the "more suppressible" isotypes).

Antibody production, once initiated, also tends to be maintained. Carrier/hapten-carrier protocols that induce strong suppression for anti-hapten antibody responses in virgin animals are substantially less effective in animals producing an ongoing primary IgG anti-hapten response (e.g. due to stimulation with the hapten on an unrelated carrier prior to completion of the carrier/hapten-carrier sequence). Under these conditions, detectable suppression is induced in about half the animal and, when induced, mainly affects the more suppressible isotypes. Thus, the initiation of antibody production impairs the subsequent induction of suppression, and the initial induction of suppression tends to prevent subsequent initiation of antibody production (17).

This reciprocal relationship defines a bistable regulatory mechanism that fixes long-term antibody response patterns according to the conditions under which it first "sees" individual epitopes. Thus, it provides a vehicle through which even quite transient conditions in the initial regulatory environment can strongly influence the characteristics of subsequent anti-epitope responses.

CONSEQUENCES OF BISTABLE REGULATION

Although memory B-cell induction and development are required for anamnestic (memory) responses, the epitope-specific system and the mechanisms that induce it to suppress or support memory B-cell expression play a key role in determining which and how many of these B cells will be expressed when an animal reencounters a given epitope. In fact, as a general rule, the characteristics of in situ memory responses provide a much better measure of the status of the epitope-specific system than of the memory B-cell populations that have been generated by a particular priming protocol.

Our evidence on this point suggests that much of the information from studies using in situ secondary responses to evaluate the effects of priming conditions on the development of memory B cells requires re-evaluation to distinguish conditions that truly influence B-cell development from those that influence the in situ expression (epitope-specific regulation) of memory B-cell responses. Carrier/hapten-carrier immunization, for example, induces normal anti-hapten memory populations that can be revealed in adoptive recipients (11, 15, 16); however, these memory cells remain entirely cryptic in situ under the conditions usually used to test for the presence of immunologic memory, e.g. repeated boosting with $1 \mu g$ aqueous antigen (unpublished observations). Stimulatory conditions that overcome the initially induced epitope-specific suppression reveal the presence of the cryptic memory populations, but this generally requires several immunizations with priming doses of the antigen (e.g. 100 μg of alum-precipitated hapten-carrier conjugate were used for each immunization in the suppression-reversal experiments discussed above).

Adoptive studies in which transferred spleen cells (co-resident B plus T) are used to evaluate memory development suffer from much the same problem, since suppression is maintained when spleen cells from suppressed animals are transferred to adoptive recipients (16, 21). In essence, we have found that the only reliable way to reveal anti-hapten memory that is not expressed in situ is to transfer T-cell depleted splenic (B cell) populations into recipients supplemented with carrier-primed T cells (preferably from animals primed with alum-precipitated carrier protein plus *B. pertussis*).

These considerations lead us to question the idea that variation in the memory B-cell populations generated in individual animals immunized with a typical protein antigen accounts for the well-known tendency for such animals to produce antibodies to different subsets of the epitopes on the immunizing antigen (17). B-cell "clonal dominance" mechanisms may contribute to this individualization of antibody responses; however, considerable variability will also be introduced by stochastic processes inherent in the operation of a bistable regulatory system in which the major force inducing the system to suppress antibody production (CTs) matures to full function several days after the initial immunization with a carrier protein and its associated epitopes (17, 19).

That is, since a bistable mechanism tends to maintain itself in its initially induced configuration, anti-epitope responses established prior to the emergence of a functional CTs population will tend to continue despite the presence of these cells; however, anti-epitope responses that could not (or did not) become stabilized rapidly enough will tend to be suppressed once CTs become active. Therefore, the probability that a given epitope on a priming antigen will induce stable antibody production (or stable suppression) in a particular animal will be a function of the rate at which the epitope induces support for antibody production in comparison with the rate at which CTs mature (17).

In practice, this "horserace" can be expected to result in the induction of suppression for responses to essentially random subsets of priming antigen epitopes in individual animals. Some epitopes, however, will tend to be more like the DNP hapten, which induces support so rapidly that stable antibody production is established well before CTs mature in virtually all immunized animals (perhaps because such epitopes bind to a very wide variety of antibody combining sites and consequently can stimulate a large number of B cells). Other epitopes will be notably less successful in establishing antibody production. Thus (in accord with common serologic experience), the frequency of responses to individual epitopes obtained in a group of immunized animals will be relatively reproducible while the combination of epitopes detected by the antibodies produced by individual animals will vary from animal to animal.

TESTING THE HYPOTHESIS

If CTs induce suppression for responses to carrier-borne epitopes that have not as yet induced stable support for antibody production, then antibody responses to the all of the epitopes on a priming antigen (even DNP) should fail if the level of antigen-specific CTs activity is sufficient to initiate the induction of epitope-specific suppression immediately after priming. This condition is rarely (if ever) met naturally since the developmental cascade that results in the appearance of functional CTs only begins after the first (priming) encounter with an antigen; however, it can be approximated experimentally by priming animals with a hapten-carrier conjugate shortly after they have been injected with appropriately specific CTs-secreted factors (CTsF) that mediate the induction of epitope-specific suppression. Under these conditions (if our concept of bistable immunoregulation is correct), antibody responses to the hapten and to the native epitopes on the carrier should all be suppressed.

Data in Table 4 show that, as predicted, anti-DNP and anti-KLH antibody responses are suppressed in animals that received KLH-specific CTsF shortly before being primed with DNP-KLH. Furthermore, anti-DNP responses predictably remain suppressed in these animals when the hapten is presented subsequently on an unrelated carrier molecule (DNP-CGG), whereas antibody responses to the (CGG) epitopes on the second carrier molecule proceed normally. Thus, this immunization protocol results in the induction of typical epitope-specific suppression for antibody responses to DNP and (by inference) to all other epitopes presented on the priming carrier.

These findings (T. Tokuhisa, M. Tagawa, M. Taniguchi, manuscript in preparation) directly demonstrate that all epitopes on hapten-carrier conjugate are treated equivalently to the hapten in the carrier/hapten-carrier immunization sequence when CTs activity is artificially introduced prior to priming with a hapten-carrier conjugate. Nevertheless, under normal circumstances, the emergence of active CTs shortly after priming does not interfere with antibody production to (at least some of) the epitopes on the priming antigen. This apparent paradox confirms the existence of a bistable mechanism that allows priming antigen epitopes to induce specific protec-

Table 4 Carrier-specific suppressor T-cell factor (CTsF) induces epitope-specific suppression[a]

BALB/c KLH–TsF[b]	Immunizations[c]		Strain	IgG2a antibody responses[d]		
				DNP (µg/ml)	KLH (units)	CGG (units)
—	D–K		BALB/c	73	24	—
+	D–K		BALB/c	18	3	—
—	D–K	D–C	BALB/c	102	—	7
+	D–K	D–C	BALB/c	19	—	6
—	D–C		BALB/c	22	—	—
+	D–C		BALB/c	30	—	—
—	D–K		C57BL/6	71	—	—
+	D–K		C57BL/6	102	—	—

[a] T. Tokuhisa, M. Tagawa, M. Taniguchi, manuscript in preparation.

[b] KLH-specific suppressor factor (CTsF) prepared from BALB/c thymocytes as previously described (5); yield from 10^8 thymocytes injected per animal 24 hours prior to first immunization. CTsF prepared from BALB/c (H–2d) is not active in C57BL/6 (H–2b) (5).

[c] 100 µg each antigen on alum at two week intervals.

[d] Measured by RIA 2 weeks after last indicated immunization; anti-carrier responses expressed as percentages of "standard" adoptive secondary responses to the indicated antigen.

tion (support) for antibody production before CTs gain sufficient strength to induce specific suppression for unprotected responses.

EXCEPTIONS TO THE RULES

During the course of studies characterizing the epitope-specific system and the conditions that induce it to suppress or support antibody production, we tested the effects of a wide variety of protocol modifications (different antigens, doses, timing, adjuvants, mouse strains, etc). In nearly all cases, the results we obtained (summarized in Tables 1 and 2) were consistent with the properties of the bistable regulatory system defined by our original carrier/hapten-carrier immunization studies; however, we noted two striking exceptions, one concerning adjuvant effects on the carrier-specific suppression induction mechanism and the other (somewhat more surprising) concerning the maintenance of suppression for anti-DNP responses in sequential immunizations with DNP on different kinds of carriers.

The adjuvant studies show that although KLH/DNP-KLH immunization results in the induction of typical epitope-specific suppression when animals are primed with aqueous KLH, KLH on alum or KLH on alum plus complete Freunds adjuvant (CFA), priming with KLH on alum plus *Bordatella pertussis* completely prevents the subsequent induction of suppression for anti-DNP responses (see Table 5). Coupling the *B. pertussis*

Table 5 Carrier immunization with *Bordatella pertussis* prevents subsequent suppression induction

Immunizations with KLH (K) or DNP-KLH (D–K)[a]			Anti-DNP in serum (μg/ml)	
First	Second	Third	IgG2a	IgG1
—	—	D–K alum	64	145
K alum + PV	—	D–K alum	60	125
K alum	—	D–K alum	<13	125
K alum	K alum	D–K alum	<6	15
K CFA	—	D–K alum	<15	63
K aqueous	—	D–K alum	<6	55
K aqueous	K aqueous	D–K alum	<7	<28
—	—	D–K CFA	70	200
K alum	—	D–K CFA	<9	128
—	—	D–K alum + PV	90	225
K alum	—	D–K alum + PV	33	225
K alum	K alum	D–K alum + PV	<13	45

[a] 100 μg each antigen at 4 week intervals; antibody responses measured by radioimmune assay 2 weeks after last immunization.

with the DNP-KLH immunization in the sequence, in contrast, does not appear to interfere substantially with suppression induction since anti-DNP responses are clearly lower than in non-preimmunized controls that received DNP-KLH plus *B. pertussis.*

The interference with suppression induction does not appear to be due to the increased amounts of anti-KLH antibody produced by the KLH/*pertussis* immunized animals since KLH/CFA immunization stimulates roughly the same amount of antibody but does not interfere with suppression induction (see Table 6). These findings suggest that KLH/*pertussis* induces a carrier-specific cell population that prevents epitope-specific suppression induction when animals are subsequently immunized with DNP-KLH. Since this population appears to be functionally similar to the carrier-specific "contra-suppressor" population described by Gershon and colleagues (9), we wonder whether carrier immunization with *B. pertussis* might not be an excellent way to stimulate contra-suppressor cells and whether, in fact, such cells might not be the effector mechanism through which *B. pertussis* acts as an adjuvant to augment antibody responses.

Results from current carrier/hapten-carrier immunization studies with DNP coupled to sheep erythrocytes (D-SRBC) introduce another, more disquieting, exception to the otherwise consistent behavior of the epitope-specific system (19). SRBC/D-SRBC immunization induces what appears to be typical epitope-specific suppression in that the IgG anti-DNP response is substantially smaller and has a lower affinity than control responses (in D-SRBC immunized animals), whereas the IgG anti-SRBC response climbs to secondary levels comparable to those in SRBC/SRBC-immunized controls. Similarly, IgG anti-DNP responses remain suppressed after a second immunization with D-SRBC, and anti-SRBC responses proceed normally. However, when SRBC/D-SRBC-immunized animals are immunized with DNP-KLH, their IgG anti-DNP responses are barely suppressed!

Table 6 Priming with KLH in complete Freund's adjuvant or plus *Bordatella pertussis* yields similar responses

Immunizations[a]		IgG anti-KLH response[b]		
First	Second	Status	IgG2a	IgG1
KLH on alum	DNP–KLH on alum	primary	8	6
		secondary	100	106
KLH plus CFA	DNP–KLH on alum	primary	48	36
		secondary	87	140
KLH on alum plus *B. pertussis*	DNP–KLH on alum	primary	15	18
		secondary	99	236

[a] See legend to Table 5.
[b] Percentage of a "standard" adoptive secondary anti-KLH response.

Several other carriers (including ficoll and certain synthetic amino acid copolymers) yield similar results vis a vis their failure to induce suppression that extends to responses to DNP on KLH or CGG (unpublished observations). Furthermore, although we have not extensively cross-checked these carriers, the suppression they induce does not appear to extend to DNP presented on any of the others. Thus, we are faced with a heterogeneous group of carriers (cells, carbohydrates, artificial proteins) that apparently induce a suppression specific for anti-DNP responses (since anti-carrier antibody responses proceed normally) and yet do not induce suppression for such responses per se (since presentation of DNP on other carriers induces IgG anti-DNP antibody production).

This puzzling set of findings would be explained if the suppression induced in each case affected production of a different subset of the combining sites in the anti-DNP repertoire. This would mean that DNP presented on each of these carriers evokes a substantially different antibody response and a correspondingly unique set of regulatory cells specific for that response. This idea is consistent with theoretical considerations concerning the structure of combining sites likely to bind DNP in different structural environments (29–31; A. Edmondsen, personal communication); however, whether it will prove correct remains to be seen.

EFFECTOR CELLS IN THE EPITOPE-SPECIFIC SYSTEM

Early attempts to characterize the cell(s) responsible for epitope-specific suppression were stymied by difficulties in reliably measuring suppressive activity in adoptive recipients. Recent studies (T. Tokuhisa, M. Tagawa, M. Taniguchi, manuscript in preparation) have had more success using an in vitro assay in which cells from CGG/DNP-CGG immunized animals suppress anti-DNP antibody production by spleen cells from DNP-KLH or DNP-OVA (ovalbumin)-primed mice. These rather elegant studies demonstrate that the epitope-specific suppression is mediated by Thy-1 positive cells that have a specific receptor for DNP and (as Table 7 shows) can be removed on DNP-BSA coated plates.

At present, data are consistent with one or more DNP-specific suppressor T cells being required for suppression in this assay and with the suppressor cell(s) serving as direct mediators of suppression or as inducers of cells that mediate suppression (since the responding spleen cells provide a source of T cells that potentially can be induced to suppress antibody production). However, in either event, these studies clearly define a new category of regulatory cells by demonstrating the existence of epitope-specific suppressor T cells (ETs) that control antibody production independently of the carrier on which the epitope is presented.

Table 7   Epitope-specific suppressor T cells (detected in vitro) bind specifically to DNP-coated plates[a]

DNP–KLH primed spleen[b]	CGG/DNP–CGG immunized spleen		IgG1 anti-DNP (ng/ml)
	Cell population[c]	Cells added[b]	
3	—	None	238
3	T cells (anti-MIg depleted)	1	25
3	T cells (anti-MIg depleted)	2	13
3	T cells bound to DNP–BSA	0.4	50
3	T cells bound to BSA	0.4	288
3	Unseparated spleen	1	88
3		2	100
4	—	None	450
5	—	None	625

[a] T. Tokuhisa, M. Tagawa, M. Taniguchi, manuscript in preparation.
[b] Cells cultured ($\times 10^5$).
[c] Spleen cells that failed to bind to anti-MIg coated plates (anti-MIg depleted fraction) were applied to DNP–BSA or BSA coated plates and incubated 30 minutes at $37°C$. Plates were then washed, chilled and the bound cells were eluted by gentle pipetting. About 1 percent of the cells in the original spleen-cell suspension were recovered from the DNP–BSA coated plate.

A MODEL FOR A BISTABLE REGULATORY MECHANISM

Several years ago, we introduced a set of hypothetical immunoregulatory circuits whose operation provides bistable, Igh-restricted, epitope-specific regulation for antibody responses (28). This model proposed a system of central (Core) "circuits," each composed of two helper and two suppressor T cells and each providing the basic "on-off" regulation for production of antibodies carrying a particular antibody-combining site coupled to a particular Ig heavy chain (isotype/allotype). By allowing each suppressor cell in the circuit to attack one of the helper cells and be helped (to differentiate and expand) by the other (see Figure 1), we arrived at a set of cell interactions that approximates the behavior of an electronic binary ("flip-flop") circuit in that the circuit tends to stabilize in a help or suppression mode but nevertheless can shift to the opposite configuration in response to a dramatic shift in stimulatory conditions.

The decision as to whether a given Core circuit in the model permits or suppresses production of the antibody it regulates is partially internal to the circuit but also depends on the relative strengths of positive and negative signals transmitted to the Core circuit from a series of functionally distinct auxiliary regulatory circuits that directly sense the antigenic environment. Thus, this model places Ig-specific regulation to the B cell being regulated and establishes a system whereby a variety of positive and negative auxiliary

CIRCUIT

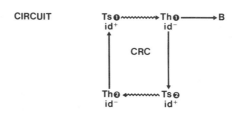

Figure 2 Model for a bistable regulatory circuit. In this theoretical cell-interaction circuit (28), a B cell carrying Id+ VH Ig surface receptors is helped by a T cell, Th1, that has complementary (id−) receptors. Th1 (analogous to an idiotype-specific helper T cell) is depleted by a suppressor T cell, Ts1, (analogous to an idiotype suppressor T cell) which carries id+ receptors similar to the B cell and therefore tends to bind the same haptenic determinant as the B cell. Ts1 is helped by Th2, and Th2 is depleted by Ts2 (which is helped by Th1). Th2 and Ts2 are distinguished from Th1 and Ts1 by non-VH related surface determinants. The configuration of this circuit is such that it will tend to stabilize either with Th1 and Ts2 dominant or Th2 and Ts1 dominant, since either of these pairs will decrease the activity of the other. Thus this circuit will tend to maintain itself either in a help or suppression configuration depending on how it is induced initially by conditions of antigenic stimulation (for further explanation, see 28).

stimulatory signals are integrated (by the Core circuits) to determine whether one or more of the possible antibody populations shall be represented in a response.

In framing the model, we deliberately avoided considering auxiliary circuits containing CTs since we could see no straightforward way of rationalizing the then-current view (that CTs regulate antibody production by depleting carrier-specific help) with the central regulatory system we were proposing. Working now, we would draw a CTs-containing circuit that induces the Core circuits to suppress antibody production (to new epitopes on the carrier). Similarly, concepts of contrasuppression (9) were nascent (or unknown) when this model was drafted but would now be included in auxilary circuits that favor help rather than suppression. Thus, although the model as published "shows its age," the basic principles it embodies are more viable than ever.

We would be pleased to claim that our studies on the epitope-specific system were directly instigated by this "theoretical exercise;" however, serendipity had more to do with the initiation of these studies than rationality. That is, having forgotten the earlier work (1, 2) showing that in situ anti-hapten responses fail following immunization with (what we now call) the "carrier/hapten-carrier" sequence, we set up an allotype suppression experiment in which we hoped to augment anti-hapten antibody production by pre-immunizing with carrier (32). The minimal anti-hapten and normal anti-carrier responses we obtained in control animals intrigued us sufficiently to trigger a further series of experiments. Thus, we inadvertantly reopened an old question and, armed with modern methods for measuring memory B-cell development and expression, wound up some three months later with evidence (11) outlining a previously cryptic (epitope-specific) regulatory system whose properties, we found, were largely predicted by the theoretical regulatory circuits we had proposed earlier.

The coincidence of these properties adds credibility to the proposed model (28) and opens the way for a direct test of some of its more specific predictions (e.g. if the model is correct, epitope-specific regulation as described here should be based on co-ordinated idiotype-specific regulatory interactions that can be dissected by introducing conditions that perturb idiotype representation in anti-epitope antibody responses). However, aside from its value as a guide to future experimentation, this model (correct or incorrect) serves a clear and current purpose. That is, it demonstrates that a workable set of cell interactions can be devised to account for the bistable regulation of antibody responses demonstrated in the epitope-specific system. Thus it de-mystifies this novel regulatory capability and brings it into the realm of possibility for a system that has evolved complex cellular mechanisms to protect the animal against invasion by deleterious agents.

GENERALITY OF EPITOPE-SPECIFIC REGULATION

In essence, the evidence we have presented casts the epitope-specific system as an integrative central mechanism responsible for shaping antibody responses according to the dictates of the regulatory environment when an epitope is first introduced. The status of the carrier-specific regulatory system, as we have shown, plays a key role defining the properties of this environment (11, 15, 16). The status of allotype-specific and I-region defined regulatory mechanisms contribute significantly, apparently by preventing rapid initiation of antibody production and hence allowing CTs activity to predominate (18, 20). Thus, through the agency of this central system, the activities of a variety of independently studied regulatory interactions come together as co-ordinated influences on the antibody responses produced by a given animal (17).

Perusal of the literature suggests that an analogous epitope-specific system centrally regulates cellular immune responses. For example, recent studies demonstrate that the induction of allergic encephalomyelitis (AE) by an encephalitogenic peptide-carrier conjugate can be inhibited (suppressed) by prior immunization with the carrier protein, i.e. by carrier/-hapten-carrier immunization (33). Similarly, the mechanisms regulating delayed-type hypersensitivity (34) show a specificity for epitopes not unlike the mechanisms described here. Therefore, it is reasonable to suspect that each of the major types of immune responses are controlled by "Core" systems that duplicate the properties of the epitope-specific system controlling antibody responses.

Taking this supposition one step further, the existence of such Core systems would constitute an overall mechanism through which cellular and humoral responses could be co-ordinated, perhaps by direct communication or perhaps by differential responsiveness to common "auxiliary" systems. The complexity inherent in such a mechanism is staggering; however, given the extraordinary versatility of the immune system as it operates even in its earliest evolved form, we (as investigators) will indeed be lucky if immune response regulation proves to be based on such a simplistic view.

ACKNOWLEDGMENTS

We thank David Parks, who played a major role in developing the concepts of bistable regulatory circuits discussed here, and Leonard A. Herzenberg, who has continuously provided invaluable scientific criticism, advice, support, and editorial help for these studies. This work was supported in part by NIH grants HD 01287, AI-08917, and CA-04681.

Literature Cited

1. Mitchison, N. A. 1971. The carrier effect in the secondary response to hapten-protein conjugates. I. Measurement of the effect and objections to the local environment hypothesis. *Eur. J. Immunol.* 1:10–17

2. Rajewsky, K., V. Schirrmacher, S. Nase and N. K. Jerne. 1969. The requirement of more than one antigenic determinant for immunogenicity. *J. Exp. Med.* 129:1131–1143

3. Gershon, R. K. 1974. T Cell Control of Antibody Production. In *"Contemporary Topics in Immunobiology"*, Vol. 3., ed. M. D. Cooper and N. L. Warner, pp 1–35. New York: Plenum

4. Tada, T., K. Okumura and M. Taniguchi. 1972. Regulation of Homocytotrophic Antibody Formation in the Rat: VII. Carrier Functions in the Antihapten Homocytotrophic Antibody Response. *J. Immunol.* 108:1535

5. Tada, T. and K. Okumura. 1979. The role of antigen-specific T cell factors in the immune response. *Adv. Immunol.* 28:1–87

6. Sercarz, E. E., R. L. Yowell, D. Turkin, A. Miller, B. A. Araneo and L. Adorini. 1978. Different Functional Specificity Repertoires for Suppressor and Helper T Cells. *Immunological Rev.* 39:109–136

7. Tada, T. and K. Hayakawa. 1980. Antigen-specific Helper and Suppressor Factors. In *"Immunology 80; Progress in Immunology IV"*. ed. M. Fougereau and J. Dausset, New York: Academic. 389 pp.

8. Taniguchi, M., T. Saito and T. Tada. 1979. Antigen-specific Suppressive Factor Produced by a Transplantable I-J bearing T cell Hybridoma. *Nature* 278:555

9. Yamauchi, K., D. R. Green, D. D. Eardley, D. B. Murphy and R. K. Gershon. 1981. Immunoregulatory Circuits that Modulate Responsiveness to Suppressor Cell Signals. Failure of B10 Mice to Respond to Suppressor Factors Can be Overcome by Quenching the Contrasuppressor Circuit. *J. Exp. Med.* 153:1547–1561

10. Green, D. R., R. K. Gershon, and D. D. Eardley. 1981. Functional deletion of different Ly-1 T cell inducer subset activities by Ly-2 suppressor T lymphocytes. *Proc. Nat. Acad. Sci. USA* 78:3819–23

11. Herzenberg, L. A., T. Tokuhisa and L. A. Herzenberg. 1980. Carrier-priming leads to hapten-specific suppression. *Nature* 285:664–666

12. Katz, D. H., W. E. Paul, E. A. Goidl, and B. Benacerraf. 1970. Carrier function in anti-hapten immune responses: I. Enhancement of primary and secondary anti-hapten antibody responses by carrier preimmunization. *J. Exp. Med.* 132:261–282

13. Paul, W. E., D. H. Katz, E. A. Goidl and B. Benacerraf. 1970. Carrier function in anti-hapten immune responses. II. Specific properties of carrier cells capable of enhancing anti-hapten antibody responses. *J. Exp. Med.* 132:283–304

14. Sarvas, H., O. Makela, P. Toivanen and A. Toivanen. 1974. Effect of carrier preimmunization on the anti-hapten response in chicken. *Scand. J. Immunol.* 3:455–60

15. Herzenberg, L. A. and T. Tokuhisa. 1981. Carrier Specific Suppression Operates Through the Hapten-Specific System. In: *Immunoglobulin Idiotypes and Their Expression, ICN-UCLA Symposia on Molecular and Cellular Biology,* Volume XX, ed. Charles Janeway, Eli E. Sercarz, Hans Wigzell and C. Fred Fox. pp. 695–707 New York: Academic

16. Herzenberg, L. A., T. Tokuhisa. 1982. Epitope-specific regulation: I. Carrier-specific induction of suppression for IgG anti-hapten antibody responses. *J. Exp. Med.* 155:1730

17. Herzenberg, L. A., T. Tokuhisa, D. R. Parks and L. A. Herzenberg. 1982. Epitope-specific Regulation: II. Bistable Control of Individual IgG Isotype Responses. *J. Exp. Med.* 155:1741

18. Herzenberg, L. A., T. Tokuhisa, and L. A. Herzenberg. 1982. Epitope-specific Regulation: III. Induction of allotype-restricted suppression for IgG antibody responses to individual epitopes on complex antigens. *Eur. J. Immunol.* 12:814–8

19. Herzenberg, L. A., K. Hayakawa, R. R. Hardy, T. Tokuhisa, V. T. Oi and L. A. Herzenberg. 1982. Molecular, cellular and systemic mechanisms for regulating IgCH expression. *Immunol. Rev.* 67:5–31

20. Herzenberg, L. A., T. Tokuhisa and K. Hayakawa. 1982. Lack of immune response (IR) gene control for the induction of epitope-specific suppression by the TGAL antigen. *Nature.* 295:329–331

21. Herzenberg, L. A. 1981. *New perspectives on the regulation of memory B cell development and expression.* Thesis, Doctorate d'etat et sciences, University de la Sorbonne Paris VI, France. 247 pp.

22. Herzenberg, L. A., S. J. Black, T. Tokuhisa and L. A. Herzenberg. 1980. Memory B cells at successive stages of differentiation: Affinity maturation and the role of IgD receptors. *J. Exp. Med.* 141:1071–1087

23. Ishazaka, K. and H. Okudaira. 1973. Reaginic antibody formation in the mouse. II. Enhancement and suppression of anti-hapten antibody formation in priming with carrier. *J. Immunology* 110:1067–1076

24. Tada, T., M. Taniguchi and K. Okumura. 1971. Regulation of homocytotropic antibody formation in the rat. II. Effect of X-irradiation. *J. Immunology* 106:1012

25. Okumura, K. and T. Tada. 1971. *J. Immunology.* Regulation of homocytotropic antibody formation in the rat. III. Effect of thymectomy and splenectomy. 106:1682

26. Tada, T., K. Okumura and M. Taniguchi. 1972. Regulation of homocytotropic antibody formation in the rat. VII. Carrier funtions in the anti-hapten homocytotropic antibody response. *J. Immunology* 108:1535–41

27. Herzenberg, L. A. and L. A. Herzenberg. 1978. Mouse Immunoglobulin Allotypes: Description and Special Methodology. In *"Handbook of Experimental Immunology"* ed. D. M. Weir, ed. pp. 12.1–12.23. Oxford: Blackwell Sci. 3rd ed.

28. Herzenberg, L. A., S. J. Black and L. A. Herzenberg. 1980. Regulatory circuits and antibody responses. *Eur. J. Immunol.* 10:1–11

29. Singer, S. J. 1964. On the heterogeneity of anti-hapten antibodies. *Immunochem.* 1:15–20

30. Haber, E., F. F. Richards, J. Spragg, K. F. Austen, M. Vallotton and L. B. Page. 1967. Modifications in the heterogeneity of the antibody response. In *Antibodies,* Cold Spring Harbor Symposia on Quantitative Biology, Vol. 32. pp. 233–310

31. Levine, B. B. 1965. Studies on delayed hypersensitivity. I. Inferences on the comparative binding affinities of antibodies mediating delayed and immediate hypersensitivity reactions in the guinea pig. *J. Exp. Med.* 121:873–888

32. Herzenberg, L. A. 1983. Allotype Suppression and Epitope-specific Regulation. *Immunol. Today* 32:In press

33. Rausch, H. C., I. N. Montgomery and R. H. Swanborg. 1981. Inhibition of Experimental Allergic Encephalomyelitis by Carrier Administered Prior to Challenge with Encephalitogenic Peptide-Carrier Conjugate. *Eur. J. Immunol.* 11:335–338

34. Bullock, W. W., D. H. Katz and B. Benacerraf. 1975. Induction of T-Lymphocyte Responses to a Small Molecular Weight Antigen. *J. Exp. Med.* 142:275–287

Ann. Rev. Immunol. 1983. 1:633–55

T-LYMPHOCYTE CLONES

C. Garrison Fathman and John G. Frelinger

Division of Immunology, Department of Medicine, Stanford University, Stanford, California 94305

INTRODUCTION AND HISTORICAL OVERVIEW

Several discoveries and technological developments have revolutionized the field of cellular immunology during the past decade. Of primary importance was the development by Kohler & Milstein of hybridoma cell lines which immortalized production of monoclonal antibodies (1,2). Another important advance was the utilization of molecular genetics to study regulatory and effector products of cellular immunology. A third important development has been the adaptation of culture conditions to allow the growth and maintenance of T-lymphocyte clones in long-term culture (3). This rapid advance is all the more remarkable when one considers that the field of cellular immunology has developed so very recently. In fact, less than 25 years ago the function of lymphocytes was still unknown. A 1958 review by Trowell (4) on the lymphocyte included the following:

> "The small lymphocyte seems a poor sort of cell characterized by mostly negative attributes: small in size with especially little cytoplasm, unable to multiply, dying on the least provocation, surviving *in vitro* for only a few days, living *in vivo* perhaps a few weeks . . . Although lymphocytes have been studied for over a hundred years, not a single function can yet be ascribed to them with any confidence, and their role in the body remains both an enigma and a challenge. Broadly speaking, two rival and mutually exclusive views have been entertained. The one is that lymphocytes are primitive stem cells capable of differentiating into all the other types of blood cells, and also into reticuloendothelial cells and fibroblasts. The other is that they are end cells dying after a few hours without having performed any known function."

Perhaps none of the advances has been as potentially far reaching as the recent adaptation of tissue culture techniques that allow cloning and functional immortalization of T lymphocytes. Because this was primarily a technical advance, the areas in which T-cell clones have played a critical role have been diverse. In part, this was a reflection of the methods used to generate T-lymphocyte clones, each of which has allowed different T-cell functional properties to be retained. In this review, we will present a brief

0732-0582/83/0410-0633$02.00

overview of how T-cell clones were derived, describe the as yet unsuccessful attempts to isolate T-cell receptors, and survey the advances made in quite varied areas of cellular immunology that were made possible by the use of T-cell clones.

METHODS OF IMMORTALIZING T CELLS

There are basically three ways to develop T-lymphocyte clones. One may obtain clonal derivatives of T cells by the fusion of a T-cell lymphoma and a normal T lymphocyte so as to establish an immortalized cell line that maintains functions of interest. Or one may choose one of two other approaches that allow the extended growth of T lymphocytes in the absence of somatic cell fusion. T cells can be selected and grown continuously in media containing large amounts of T-cell growth factor(s) (TCGF), now designated Interleukin 2 (IL-2). The final, and perhaps the most laborious technique involves maintaining T cells in media containing little or no exogenous IL-2, and requires repeated antigenic stimulation and serial reculture. This method requires the presence of antigen-presenting cells for helper T-cell clones, or the presence of appropriate target cells for cytolytic T-cell clones.

Hybridomas

One of the first successful fusions between functional T cells and a thymoma line was reported by Tanaguchi & Miller (5). Taking advantage of the fact that suppressor cells bind antigen and express I-J determinants, these investigators utilized two selective procedures which increased the probability of their isolating a suppressor T-cell hybridoma. Antigen-binding T cells were enriched from the spleens of mice that had been previously rendered tolerant to human gammaglobulin (HGG) by absorption to and elution from HGG-coated plastic dishes. These antigen-binding T cells were fused to a T-cell tumor line. Using the fluorescence-activated cell sorter, Tanaguchi & Miller selected and ultimately propagated those hybrid cells which were stained with antibodies directed against products of the I-J subregion of the murine major histocompatibility complex (MHC). There have been many subsequent reports of the establishment of functional T-cell hybridomas by the fusion of T lymphoma cell lines and primed lymphocytes (6–10).

Despite the relative ease with which suppressor T-cell hybridomas were developed, it was initially difficult to isolate other types of functional T-cell hybridomas following fusion. In retrospect, this may have been due to the lack of an appropriate method of assay. Thus, although helper T cells have antigen specificity, such cells "see" antigen in the context of an "MHC-restricted" recognition. Antigen-specific helper T cells do not bind free antigen, and thus cannot be selected simply by antigen-binding criteria. This problem was circumvented in a very elegant series of experiments by Marrack & Kappler and their coworkers (11, 12). Although helper T cells do not bind free antigen or MHC products, helper T cells do bind to antigen-pulsed presenting cells of the appropriate haplotype. The laboriousness of studying antigen-binding-cell:antigen-presenting-cell interactions following physical association by enumeration precluded this as a conventional assay technique. However, Marrack & Kappler noted that hybridomas formed between a helper T-cell line and a drug-marked thymoma secrete lymphokines in response to challenge with the appropriate combination of antigen and MHC. These investigators were able to isolate helper T-cell hybridomas using the production of TCGF-IL-2 by the hybridomas as an assay method.

Nabholz and his collaborators have utilized somatic cell hybridization to develop cytotoxic T-lymphocyte (CTL) hybrids derived between CTL lines and drug-marked thymomas (13, 14). These workers are developing techniques for analysis of CTL function by somatic cell genetics and are studying the molecular basis of the most distinctive traits of the CTL phenotype: specificity, cytolytic competence, and TCGF dependence.

Thus, utilizing the technique of somatic cell hybridization, it has been possible to generate stable hybridoma lines of suppressor, helper and cytolytic cell lineages. These hybrid cells are relatively easy to grow and maintain, and they can be obtained in large numbers. The major disadvantages of this technique for the generation of T-cell clones are the unknown contributions of the fusion partner and, at least in some cases, the instability of the differentiated function of the T-cell hybrids.

IL-2 Dependent Lines

The second approach which has allowed the isolation and propagation of T-cell clones is the extensive use of TCGF. As early as 1965, it was demonstrated that medium obtained from lymphocytes that had been stimulated in mixed lymphocyte culture caused unrelated third party lymphocytes to proliferate (15, 16). It was not until much later that the significance of these observations was appreciated by cellular immunologists. In fact, credit for the concept of TCGF is often attributed to Gallo and his colleagues, who were studying the effects of growth-promoting substances on human granulopoiesis. Their studies showed that conditioned medium from phytohemagglutinin(PHA)-stimulated human lymphocytes had not only conventional colony-stimulating factor activity, but also caused extended growth of lymphoblasts (17). Initially, it was assumed that these were B lymphoblasts which had been transformed as a result of infection with Epstein-Barr virus. Later, when Gallo and his associates were again attempting to bring about long-term growth of human granulopoietic cells, they examined lymphoblasts which grew out of such cultures and found, to their surprise, that these lymphoblasts were mature T lymphocytes. They subsequently characterized the active agent in such conditioned medium, the lymphokine IL-2, and named it "T-cell growth factor" (17, 18). With the discovery of TCGF, the stage was set for cloning T lymphocytes. Within the following year, long-term lines of mouse cytotoxic T cells were being routinely maintained in the presence of TCGFs (19), and soon a variety of cells were cloned (20–24). The source of the TCGF used in these cultures (IL-2) is usually conditioned media obtained following the stimulation of lymphocytes with mitogen, although the "induced" supernates of certain T-cell lymphoma lines can also be used. There is a variety of products in such conditioned media in addition to TCGF.

One of the remarkable findings which has recently come to light concerns the karyotypic evolution of cytolytic T-cell lines maintained by TCGFs. Johnson et al (35) have karyotyped a variety of murine CTL lines, and have shown that there is a remarkably high frequency of chromosomal rearrangement in many of the lines. In fact, none of the clones that they studied were karyotypically normal. Their studies suggested that rearrangements of chromosomes were more frequent than numerical aberrations. This is in contrast to previous reports on the spontaneous evolution of other murine cell cultures (26). Thus, it must be borne in mind that T cells which are maintained in long-term tissue culture in the presence of TCGFs will undoubtedly develop karyotypic abnormalities. One must be aware of this possibility when using IL-2 dependent T-cell lines in designing experiments which depend upon a normal karyotype.

Serial Antigenic Stimulation

The third methodology which has allowed the growth and maintenance of T-cell clones depends upon repeated antigenic stimulation of cloned populations of antigen-specific T cells. This method was an outgrowth of in vitro studies of mixed lymphocyte reactions in which lymphocytes were repeatedly challenged with stimulator cells (27, 28). T-cell lines (29–31), and subsequently clones which were alloreactive (32, 33), cytolytic (33–36) or nominal antigen-reactive (37, 38), were isolated using antigenic restimulation and TCGFs. This method has the advantage that the selection and growth conditions continuously require the retention of antigenic specificity of the T cells. The major disadvantages of this technique are that: (a) maintaining these lines is relatively laborious; (b) it is difficult to obtain large numbers of cells; and (c) there is obligate contamination of T-cell clones with the antigen-presenting cells. However, utilizing serial antigenic challenge as the method of selection and propagation, T-cell clones have been developed with functions which are perhaps more characteristic of normal cells.

THE HELPER T-CELL RECEPTOR FOR ANTIGEN: A CONUNDRUM

The helper T-cell receptor is virtually uncharacterized biochemically, as can be seen by the list of its properties in Figure 1. Included in this Figure for comparison are the properties of immunoglobulins, which are the antigen-specific receptors of B cells, and the properties of soluble factors which are secreted from suppressor T cells (these entities are discussed in other reviews in this volume). As can be seen from this list, very little is currently known about the properties of antigen-specific receptors on helper T-cells. It has not been possible to demonstrate that helper T cells bind free soluble antigen. Rather, helper T cells seem to recognize antigen only as it is displayed on the surface of antigen-presenting cells in association with an MHC-encoded cell surface protein, the Ia molecule (39). Such associative recognition has been termed "MHC-restricted recognition," and it is a feature not only of helper T cells which exhibit Ia restriction, but is also a hallmark of effector T cells. Effector T cells, such as CTLs, recognize target antigens only in association with the K/D-like (class I) molecules of the MHC (40–43).

One basic tenet of the clonal selection theory is that on each antigen-specific cell there are receptors with specificity for antigen. Although receptors have not yet been isolated from helper T cells, there is evidence that an individual helper T lymphocyte and its clonal progeny recognize antigen specificity. The most thoroughly studied antigen-specific recognition system

Properties of T-Cell Receptors for Antigen

Property	T_H	T_S	B cell (Ig)
Soluble Ag binding	N	Y?	Y
Clonally Restricted (Idiotype)	Y	Y	Y
Allelic exclusion	?	?	Y
Ig-like Idiotype	Y?	Y?	Y
MHC Restriction	Y	Y?	N?
Associated with Soluble Factor	Y?	Y?	Y
Functional Domains	?	?	Y

Figure 1   A comparison of the properties of the receptors on helper T cells, suppressor T cells, and B cells. Y = Yes, N = No, ? = Unknown, (? following Y or N indicates uncertainty or conflicting reports). (From 2.)

is the immunoglobulin receptor on B cells. One of the major questions concerning T-cell receptors for antigen concerns the homology that such receptors might share with immunoglobulin molecules. A hallmark of antigen-specific receptors on B cells is the genetic mechanism of allelic exclusion, which allows only one immunoglobulin genotype to be expressed as the cell surface receptor. Allelic exclusion is a sine qua non for surface immunoglobulin, and may be a necessity for clonally-restricted receptors for antigen on the surface of T cells. Another question still not completely answered is whether there are immunoglobulin-like idiotypes present on T-cell receptors for antigen. Finally, are such T-cell receptors expressed as soluble factors which are secreted or shed from helper T cells, as they seem to be from suppressor T lymphocytes? Are there functional domains within each molecule analogous to the functional domains of immunoglobulin molecules? These questions have by no means been addressed adequately, yet should now be approachable through the utilization of helper T-cell clones.

The ease with which B-cell receptors for antigen were identified once cloned lines were available raised false expectations concerning the relative ease with which T-cell receptors for antigen might be characterized once cloned T-cell lines became available. As mentioned previously, the major problem with understanding the nature of helper T-cell receptors for antigen is that helper T cells do not recognize free soluble antigen, but yet recognize antigen in association with Ia molecules. This associative recognition is shown in Figure 2, where the three methods by which such a ternary complex can be formed have been diagrammatically illustrated. There are three potential methods in which a ternary complex composed of the "T-cell receptor," the antigen, and the MHC-encoded Ia molecule can combine. Our ignorance concerning the manner in which these three elements interrelate is practically complete. The Ia molecule may in some way specifically combine with antigen to form a binary complex which is then recognized by the T-cell receptor. Alternatively, it is possible that the T-cell receptor has some affinity for either the free Ia molecules or for free antigen, and that one of these binary complexes is sufficiently stable to allow the third member of the ternary complex to bind, resulting in T-cell activation. Until we understand either the basis for the association of Ia molecules and antigen or the nature of the antigen-specific receptor on helper T cells, we will not be able to understand what underlies the recognition and subsequent activation of antigen-specific helper T cells.

Immunoglobulin Homology

One of the most logical approaches to studying T-cell receptors was to study immunoglobulin expression in cloned T cells. Theoretically, the use of

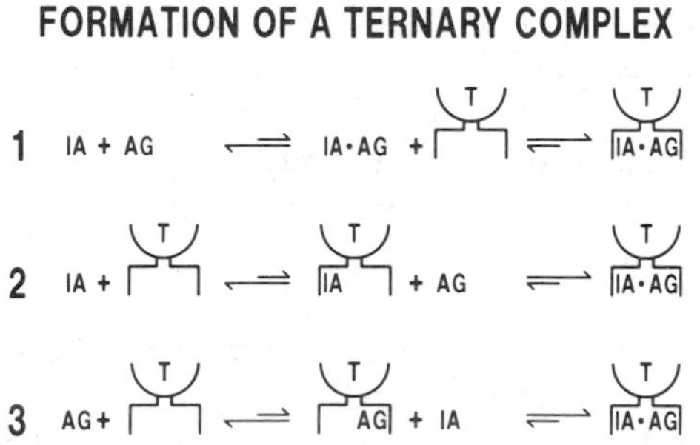

FORMATION OF A TERNARY COMPLEX

Figure 2 Possible ways that the recognition of Ia and antigen by the T-cell receptor(s) may take place. IA = Ia or class II molecules, AG = antigen, T = T cell. (suggested by J. Kappler, personal communication.)

B-cell V-region genes in T cells would eliminate the need for two entirely separate antigen-binding systems. Additionally, it has been shown that T cells which respond to a variety of antigens, including carbohydrates (44), haptens (45), MHC alloantigens (46, 47), proteins (48), and synthetic polypeptides (49, 50), express idiotypic determinants in common with antibodies to the same antigens. Perhaps the most elegant studies have been those by Eichman and his collaborators, who have shown that anti-idiotypic sera in some manner affect T-cell functions (51).

The advent of cloning strategies and technologies has allowed the isolation of relatively large numbers of homogeneous cell populations for studies on the elusive T-cell receptor for antigen. These cells should provide the starting material for the isolation of T-cell receptors, much as myelomas provided material for the isolation of immunoglobulins. Unfortunately, studies by molecular geneticists have demonstrated that the rearrangements of the constant regions of light-chain (52–56) or of heavy-chain (52–55, 57) immunoglobulin genes are not obligatory in either functional cloned helper or cloned killer T-cell lines. Additionally, there was no transcription of light-chain constant region sequences in any homogeneous T-cell line tested (52). Further, Kronenberg (52) found no evidence for transcription of J_H, $C\mu$ or $C\alpha$ sequences in cloned T-cell lines, although $C\mu$ transcripts have been found in some T-cell lymphomas (58), and in thymocytes (59) and hybridomas (54). These data taken as a body are rather convincing, and suggest that constant regions of immunoglobulins are not utilized by T cells. If the explanation for shared idiotype is a shared structural gene, then the expression of immunoglobulin idiotypes in T lymphocytes must be due to the expression of either all or part of a V_H gene. Studies are currently underway in several laboratories in an attempt to demonstrate V_H gene expression in T cells utilizing homogeneous T-cell lines which have an idiotype (idiotope) in common with immunoglobulins. The general approach is to utilize a V_H gene probe derived from a hybridoma cell line which expresses a characterized idiotype. This V_H gene probe would then be used to assay transcription and DNA rearrangements in cloned T-cell lines which express the same idiotype. Thus, it should be possible in the near future to answer the question of whether or not T-cell lines utilize any immunoglobulin gene segments as a portion of the T-cell receptor for antigen.

Additional questions to be addressed concern the potential for allelic exclusion of T-cell receptors on homogeneous T-cell lines derived from heterozygote mice. Studies to address this question are currently in progress in our laboratory, as well as elsewhere. Basically, these studies make use of monoclonal antibodies reactive with murine T-cell surface products which are controlled by genes residing on the murine twelfth chromosome (Ig1 linked) (60–62). Such antibodies are produced by purposeful immunization between immunoglobulin congenic strains of mice using an inoculum containing T cells, for instance BALB/c anti-CB.20 con A blasts. Subsequently, monoclonal antibodies which have specific reactivity for T cells are selected. Evidence to date would suggest that these antibodies recognize a product which, for simplicity, has been named C_t (62). It should be possible, utilizing antigen-specific T-cell clones from mice heterozygous for the Ig1 locus (carrying both the Ig^a and Ig^b allotype), to address the question of whether or not F_1 T-cell clones can be divided into clones which express only C_t^a, versus T-cell clones which express only C_t^b products. Evidence for allelic exclusion of a gene product expressed on cloned T cells would indeed be intriguing. Despite the fact that T-cell clones have been used in attempts to characterize the T-cell receptor for antigen, these attempts have not yet been successful.

Dual versus Single Receptor

T-cell cloning offers a method of studying whether there are single or dual receptors on T cells to account for MHC restriction. Two groups have addressed this topic, using T-cell hybridomas, and have come to conflicting results, though the general approach utilized by each was very similar (11, 12, 63). In both cases, a T-cell hybridoma specific for a given antigen and a Ia molecule (restriction site) were fused to another T cell with a different restriction specificity. The resultant hybrids were screened to see if any would respond to the first antigen when presented by cells expressing the second restriction element (Ia molecule). Such recognition would imply that the receptors involved in binding of antigen, and the restriction element, were distinct, and thus favor a dual receptor model for T cells. The opposite result would be most compatible with a single receptor model. Kappler and Marrack constructed a series of T-cell hybridomas by the fusion of two already existing hybridomas each restricted to a different Ia molecule and having a different antigenic specificity. Of 27 resultant hybridomas tested, 16 inherited the antigen and MHC specificities of both T-cell parents. None of the 16 hybrids tested could recognize a mixture that had parental-antigen-recognition and MHC-restriction specificities (11, 12). Thus, the prediction that antigen and MHC could be recognized independently by two totally different T-cell receptors did not appear to be correct. However, data recently reported by Lonai et al (63), who used the production of H-2 restricted helper T-cell factors as an assay system, are contradictory. In fact, their results support the dual receptor model. Hybridomas were derived by the fusion of T cells which recognized chicken gamma globulin and were H-2^b restricted to BW-5147, an H-2^k thymoma line. Lonai et al reasoned that although BW-5147 has no known antigen specificity, since it is an H-2^k thymoma, its restriction specificity should be for self MHC—that is, H-2^k. Lonai and his collaborators isolated hybridomas from a fusion of BW-5147 and chicken gamma-globulin-specific H-2^b-restricted T cells which produced helper factors specific for chicken gamma globulin, but were H-2^k restricted. This scrambling of antigen specificity and MHC restriction suggested that there were two independent receptors (63). These contradictory data have not yet been resolved, although the solution may lie with the different assay methods utilized. If there *are* two independent receptors, they must be closely associated and recognize a complex of antigen and MHC products, because even Lonai et al (64) did not find the binding of soluble antigen in the absence of MHC products.

Therefore, we must leave questions on the nature of, and the number of, T-cell receptor(s) for antigen unanswered at the present time. It is to be hoped that the methodologies outlined above will soon allow resolution to this perplexing problem.

SUCCESSFUL AREAS OF T-CELL CLONE RESEARCH

Rather than dwell on other questions which have been inadequately addressed utilizing T-cell clones, it is pertinent to discuss research topics which have been successfully studied by such techniques. There has been substantial progress in understanding several critical areas of cellular immunology using T-cell clones. For example, the dual specificity of T cells for nominal antigen as well as alloantigen has been conclusively demonstrated using T-cell clones. T-cell clones have been used to examine the repertoire of T cells. Antibodies to T-cell clones have allowed the biochemical and functional definition of cell surface markers on T cells. Studies with T-cell clones have led to the observation that a single helper T-lymphocyte clone can provide help for at least two independent pathways of B-cell differentiation. Further, T-cell and antigen-presenting cell interactions have been examined using T-cell clones to show clearly that B cells can present antigen. It has been shown that T-cell clones can recognize, in a functional manner, hybrid or combinatorial Ia molecules. Finally, the in vivo activity of T-cell clones has been examined with the aim of studying both the biology of T cells and the potential uses of T-cell clones as therapeutic agents. These areas will be discussed below.

Alloreactivity

The use of cloned T cells has allowed investigators to address interesting questions about the alloreactivity of antigen-specific T-cell clones. One of the most perplexing questions concerned the extraordinarily high incidence of T cells which recognize and respond to antigens of the MHC of other members of the same species. Due to this high incidence, it had been suggested that the cell population which responded to antigen must overlap to a large degree with the T-cell population responding to alloantigens (65, 66). T-cell cloning has allowed the direct and unequivocal demonstration that a single T cell has specificity for both an alloantigen and another foreign antigen in association with a self MHC gene product. The first such demonstration was by von Boehmer and his colleagues (22). They isolated a cytotoxic T-cell clone with specificity for the male antigen (H-Y) in association with self H-2D^b MHC products. At least one clone which was H-Y specific and restricted by H-2D^b additionally killed allogeneic targets which were H-2D^d, irrespective of the sex of the mouse from which the target cells were derived. Kanagawa et al (67) have confirmed and extended this observation by finding that approximately 20% of the clones specific for H-Y and H-2D^b exhibit this same crossreactivity with H-2D^d. In a similar vein, Braciale and coworkers have isolated influenza-specific H-2-restricted cytolytic T-cell clones (35), one of which also was alloreactive. This clone recognized the immunizing virus associated with H-2K^d, yet could also lyse normal (uninfected) H-2K^k-expressing targets (68). Additional and convincing data from the laboratory of Schwartz and his colleagues have shown that antigen-specific proliferating T-cell clones also possess alloreactivity (69, 70). They isolated a clone derived from B10.A mice which recognized DNP-OVA in association with B10.A antigen-presenting cells, but was equally well able to recognize allogeneic B10.S spleen cells in the absence of DNP-OVA. Further, these antigen-reactive clones could be passaged for 68 days on alloreactive stimulators and still retain their original antigen specificity (70). The high frequency of finding T-cell clones with such dual specificity demonstrated not only by Schwartz and his collaborators but also by others (71; M. Kimoto, unpublished data) has proven that a large percentage of T cells perform both functions. Although we can state with certainty that a single T cell can be both antigen reactive and alloreactive, it has not yet been possible to determine unequivocally whether such alloreactive and antigen-reactive T-cell clones use two independent receptors, or use one receptor unit with crossreactivity.

T-Cell Repertoire

The repertoire of T-cell receptors has been most extensively analyzed at the clonal level using CTL clones derived by limited dilution cloning. Obtaining a representative sample of all the possible reactivities of T cells is critical to an analysis of repertoire. Hence, the use of short-term cloned lines may be preferable to using clones derived from long-term bulk cultures. The latter cell lines may have undergone more selection, so that potentially reactive clones have been eliminated. Sherman has analyzed alloantigen-specific CTL clones using a panel of target cells from K^b mutant mice (72). The mice had previously been shown to possess distinct and defined amino acid substitutes in the K^b molecule (73). The reactivity patterns on this panel distinguish between CTL clones that use different receptors for antigen. Using clones derived from a B10.D2 anti H-2K^b response, Sherman found 47 distinct reactivity patterns (72). Thus, there must be at least 47 different types of clonally distributed cytolytic T-cell receptors used by the B10.D2 mouse to recognize the B6 K^b molecule. Using these data in analyzing the frequency of K^b-restricted CTL precursors, the minimum size of the CTL repertoire for a given antigenic determinant can be calculated to be about 20,000 specificities. Because this is a minimum estimate, the repertoire of T-cell receptors may be extensive.

The nature of the determinants seen by T-cell receptors is uncertain. It has been reported that helper T cells respond to both native and denatured antigen equally well (74), whereas suppressor cells and antibodies react predominantly with the native form of the molecule originally used as the immunogen (74, 75). Helper T-cell clones can respond equally well, on a molar basis, to sperm whale myoglobin or to cyanogen bromide peptides of this molecule (76). This finding is consistent with the generalization raised above. However, other studies suggest that many murine alloreactive helper T-cell clones directed against I-A molecules recognize the combinatorial association of the α- and β-chain Ia polypeptides (32, 77, 78). Although not made in the identical strain combination, monoclonal antibodies raised following alloimmunization against these Ia molecules seem to recognize the α or the β chain independently, in the sense that most monoclonal anti-Ia antibodies do not see combinatorial determinants which require the specific pairing of a given α and β chain. Using a panel of monoclonal antibodies reactive with the Ia^k molecule, derived by Pierres et al (79), we were able to precipitate hybrid Ia molecules from cells of a (b X k)F_1 heterozygous mouse (J. Frelinger, unpublished data). This confirms the lack of preferential recognition by anti Ia antibodies of the "parental" Ia conformation (Figure 3). These results suggest that recognition of Ia molecules by T cells is more dependent upon tertiary conformation than is the recognition of Ia molecules by antibodies. However, this is not true in every case, since monoclonal antibodies can react with combinatorial Ia determinants (80).

In a series of experiments utilizing B6-derived CTL clones induced in response to a point mutation in the H-2K^b molecule, it was found that a single clone could frequently recognize two or more mutants. These mutants differed from each other with respect to the position of their amino acid substitution (73). These data suggested that CTL clones respond to conformational changes in H-2 molecules, and that the same determinant recognized by a CTL clone can be generated by amino acid substitutions at separate locations within the molecule. An analogous panel of monoclonal antibodies in a parent anti-mutant combination has not yet been successfully generated. Interestingly, even MHC restriction, once thought to be the exclusive province of T cells, has recently been shown to be a property of certain antibodies. Selected monoclonal antibodies raised against influenza-infected cells would only bind to infected cells and not to free virus (82). It is clear from these data that T cells can recognize subtle conformational changes of MHC molecules, perhaps better than antibodies do. Helper T cells may be less sensitive to the denaturation of nominal antigen than are antibodies. However, these apparent preferences might also be related to the exact experimental design and previous selection in the system (e.g., mutants are selected by skin graft rejection, which requires a vigorous T-cell response).

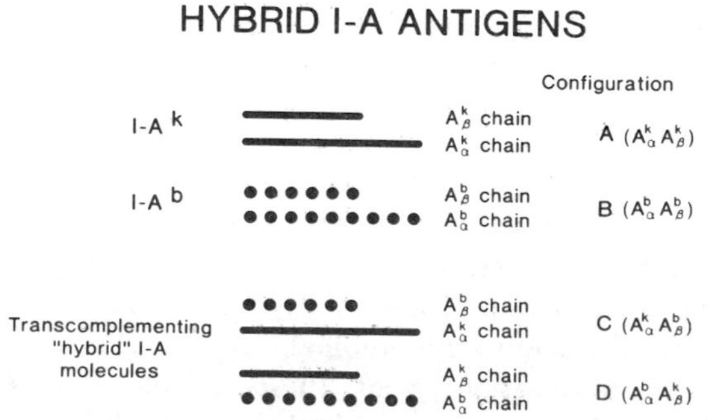

HYBRID I-A ANTIGENS

Configuration

I-A^k A_β^k chain   A ($A_\alpha^k A_\beta^k$)
A_α^k chain

I-A^b A_β^b chain   B ($A_\alpha^b A_\beta^b$)
A_α^b chain

Transcomplementing "hybrid" I-A molecules

A_β^b chain   C ($A_\alpha^k A_\beta^b$)
A_α^k chain

A_β^k chain   D ($A_\alpha^b A_\beta^k$)
A_α^b chain

Figure 3 Schematic representation of the hybrid I-A antigens of an (H-2^b X H-2^k) F_1 mouse. Configurations A and B indicate the parental I-A molecules. Configurations C and D indicate the unique "hybrid" I-A molecules formed by the combinatorial association of the α- and β-chain polypeptides (reprinted by permission of the editors of *Immunological Reviews*).

Antibodies to Murine T-Cell Clones

The combination of monoclonal antibodies and T-cell clones affords a powerful tool for studying cell interaction in the immune system at a biochemical level. The two general approaches for examining this interaction have been either to look for monoclonal antibodies which bind to the immunizing clone, or alternatively to screen for an effect of the monoclonal antibodies on a functional activity of the T-cell clone. As a general method, this will give a battery of reagents which can be used to study the structure and function of the cell surface molecules on T cells. However, in a more directed sense, it was hoped that some monoclonal antibodies would be clonally specific, and thus might be reactive with the T-cell receptor.

Several investigators have made antibodies directed at CTL clones. In screening monoclonal antibodies by the inhibition of cytotoxicity, a repeated finding has been that anti-Lyt-2 antibodies inhibit the cytotoxic function of many clones (83–86). However, the clones were heterogenous in their ability to be blocked by anti-Lyt-2 antibodies (84–86). Variant clones lacking Lyt-2 had markedly reduced levels of cytotoxic activity, although this diminished functional activity could be overcome by the addition of lectins (87). However, Lyt-2 may not be a universal requirement for cytolytic cells, since an H-2-unrestricted CTL clone does not require Lyt-2 for its lytic function (88). It was thus proposed that Lyt-2, while perhaps not obligatory in the cytolytic process, might be involved in the stabilization of effector-cell:target-cell interaction. Very recently, in a similar screen for inhibition of cytotoxicity, a monoclonal antibody—GK 1.5 —was found which may be directed to the murine analog of the human OKT4 antigen (D. Dialynas, personal communication). It is likely that this approach will generate a series of monoclonal antibodies which may ultimately define all the components involved in the process of T-cell-mediated reactivity. This approach can be generalized to include the definition of molecules involved in any cell:cell interaction in which homogenous cells take part and for which functional assay techniques are available.

Antibodies to T-cell clones which react in a clonally specific manner are rare. We have, however, succeeded in making antibodies to alloreactive T-cell clones which stimulate the immunizing clones specifically (89). Since the antibodies can distinguish between or among clones derived from the same strain of mouse, we have called these antisera "anti-idiotypic." Our initial antibodies were made by injecting a strain A alloreactive clone which responds to B6 into a (B6 X A) F_1 mouse. This F_1 anti-parent response should generate antibodies which are highly restricted. By a variety of immunization schemes, including a totally syngeneic immunization protocol, we have been able to make similar antibodies to antigen-reactive T-cell clones (J. Frelinger, unpublished data). Thus, the ability to make such antisera is not an aberrant finding resulting from the alloreactive immunization scheme. Although these antisera are functionally clonally specific, they bind to inappropriate clones, probably because there are other antibodies in the antisera. To date, we have not generated monoclonal antibodies with exactly the same functional activities. A monoclonal antibody—384.5—which both binds and specifically blocks a cytolytic clone has recently been described by others (86). However, these investigators were unable to immunoprecipitate proteins with this antibody. They suggested that the difficulty resulted from the low level of expression of the molecule on the T-cell clone. By flow cytofluorometry, it was estimated that monoclonal antibody 384.5 recognized only 10^4 molecules per cell. Thus, extremely sensitive assay techniques will need to be developed to analyze molecules of such limited expression. Although the approaches utilizing clonally-specific antibodies have not yet produced any solid data addressing the biochemical nature of the T-cell receptor, the general method remains a promising one.

Helper Activity of T-Cell Clones

A single helper T-cell clone can provide help for two independent pathways of B-cell maturation into antibody secreting cells (90). Asano et al had originally shown that a KLH-specific T-cell clone could provide help to hapten (TNP)-primed B cells, as measured in a plaque assay. Further, the source of the TNP-primed B cells could be varied, using either CBA/CaHN (Lyb-5$^+$ and Lyb-5$^-$ B cells) or CBA/N (Lyb-5$^-$ B cells) mice as donors. The results of these experiments can be summarized as follows. At low concentrations of antigen, the T-cell clone cooperated with hapten-primed B cells from either CBA/CaHN or CBA/N mice to generate predominantly an IgG response. Such help was hapten-carrier linked, and the T-cell:B-cell interaction was MHC restricted. At high concentrations of antigen, the T-cell clone could provide help to CBA/CaHN B cells, but not to the CBA/N B cells. The T-cell help generated under these conditions is "nonspecific." Although the interaction of the helper T-cell clone with the antigen-presenting cell was still MHC restricted, the T-cell:B-cell interaction was not genetically restricted. This "high-dose antigen" interaction resulted in a predominantly IgM response, and in the presence of other antigens the help was not antigen specific. Thus, the same T-cell clone, under conditions of different concentrations of antigen, could help two distinct B-cell subsets produce antibody.

Since the "nonspecific" help seen with high antigen concentrations seemed to be mediated by soluble factors, we were interested in developing a system in which we could assess the helper activity of the T-cell clone while simultaneously assaying lymphokine production and T-cell proliferation. Using an in vitro antibody formation system, we found that the production of antibodies could be monitored from supernates by a sensitive enzyme-linked immunoassay technique. Because the amounts required for assay were so small, the same culture supernates could be tested for a variety of lymphokines. Our preliminary analysis, in collaboration with E. Vitteta and coworkers, indicated that the production of lymphokines, like the proliferation of T-cell clones, was related to the antigen concentration. Further, in the same system we could clearly separate proliferation of the T-cell clone from its helper activity. At lower concentrations of antigen, the clone did not proliferate, yet it could provide sufficient help to allow B-cell differentiation and production of IgG (M. Shigeta, unpublished data). Using a cloned helper line specific for influenza virus, Lamb and coworkers (91, 92) showed that the most efficient helper activity of the clone was at an antigen concentration approximately 10-fold lower than that required for maximal proliferation. These data also support the notion that T-cell proliferation and T-cell help are independent.

The Nature of T-Cell: Antigen-Presenting-Cell Interaction

Work with T-cell clones has provided definitive insight into antigen-presenting cell function. The general question of what circumstances are necessary and sufficient to present antigen to T cells has been surrounded by controversy. For example, can B cells present antigen to T cells? In part, the question arose because of the difficulty of isolating homogeneous cell populations. In particular, the possibility of contaminant-antigen-presenting cells being mixed in with either the T- or B-cell populations always existed. With the advent of T-cell clones and T-cell hybridomas, and the availability of B-cell tumor lines, investigators were able to address this question more exactly. McKean et al showed that Ia$^+$ B-cell tumor lines could present antigen and induce proliferation in an antigen-specific T-cell clone (93). Similarly, Glimcher et al showed that Ia$^+$ tumors could present antigen to long-term T-cell lines (94). Kappler et al showed that B-cell hybridomas could present antigen to T-cell hybridomas (95). Finally, human T-cell clones specific for tetanus toxoid were induced to proliferate using Epstein-Barr virus B lymphoblastoid cells as a source of antigen-presenting cells for the tetanus toxoid (96). These data conclusively demonstrate that some Ia$^+$ cells of the B-cell lineage can effectively present antigen to T cells. This type of analysis, using cloned populations of interacting cells, is still in its infancy, but it will eventually provide clear answers to many puzzles of cell interaction in the immune system.

Definition and Functional Expression of Hybrid Ia Antigens

The description of hybrid Ia antigens as functional entities was made possible using T-cell clones. Initially, using alloreactive clones, investigators were able to demonstrate the existence of unique neoantigens which were present only on F_1 cells (hybrid antigens) and were not present on the cells of either parent (32, 77, 78, 97). These hybrid Ia antigens were formed by the combinatorial association of α and β chains (Figure 2). Later, these hybrid Ia molecules were shown to allow effectively the recognition of nominal antigen by antigen-reactive T-cell clones (37, 69, 76, 77, 98). These results were complemented by the biochemical demonstration of hybrid Ia molecules (99, 100). By using anti-IA monoclonal antibodies, both in immunoprecipitation assays and in T-cell proliferation-blocking assays, it was possible to determine which of the F_1I-A complexes a T-cell clone recognized (101). Further, on a given molecular complex the existence of multiple functional sites for "restriction" of antigen presentation could be demonstrated. On a panel of T-cell clones restricted to the $A_\alpha^k A_\beta^b$ complex, three different blocking patterns with a panel of anti-I-A monoclonal antibodies were seen (102). These results indicated that there are multiple functional sites on a given Ia molecule. Additionally, using a I-A mutant mouse—bm12—it was possible to demonstrate multiple functional sites on a given IA complex (103). The crucial observation was the identification of clones which would respond to B6 and the mutant bm12, whereas other clones would respond only to cells from B6 and not bm12. Since the bm12 mutation is a β-chain mutation, this suggested that there are two distinct sites on the $A_\alpha A_\beta$ complex determined by the β chain. These observations demonstrate that there are two epigenetic mechanisms for increasing antigen presentation. First, the number of restriction sites is expanded by the creation of hybrid antigens formed by the combinatorial association of α and β chains. Second, on a given Ia molecule, there are multiple functional sites or "domains." These mechanisms greatly expand the ways through which a limited number of Ia polypeptides can present antigen to T cells.

T-cell clones can respond to quantitative as well as qualitative changes in Ia expression. The level of expression of Ia molecules in a variety of strains expressing the $A_e E_\alpha$ complex was measured by quantitative fluorescence-activated cytofluorometry and by immunoprecipitation and two-dimensional gel analysis. There was shown to be a direct correlation between the level of Ia expression and the ability of cells to present antigen to T-cell clones (104, 105). A similar conclusion was reached from studies using T-cell clones in a proliferation assay which made use of F_1 or homozygous antigen-presenting cells to vary genetically the amount of Ia on the surface of the antigen-presenting cells. The amount of Ia on the surface again correlated with the ability to present antigen (106).

Cloned T Cells Used for Reconstitution In Vivo

The examination of the ways in which T-cell clones function in vivo is important in understanding how T cells work, as analysis of cellular interactions in vitro may or may not allow extrapolation regarding their performance in vivo. In particular, studies on T-cell clones in vivo are an important step in the potential development of T-cell clones as a therapeutic tool. CTL clones are effective in reducing the size of seeded tumors when both are injected in the same site—for example, into the peritoneal cavity —but are relatively ineffective when clones and tumors are administered separately by different routes (107, 108). This has been attributed to two major problems. First, the homing or trafficking patterns of these clones are abnormal; second, the clones are dependent on IL-2 for growth (112, 113). Thus, their relative inefficiency in vivo might be due to an inability to circulate to the site of the tumor, or to their relative short lifetime in vivo. One apparent exception is a clone specific for influenza virus, which protected mice from a respiratory infection with virus (111), perhaps by the abnormal homing of the cloned T cells to the lungs (109, 110). The problem of IL-2 dependence of T-cell lines studied in vivo may be partially overcome by the addition of exogenous IL-2 (112). For example, the intravenous injection of certain Ly2$^+$ T-cell clones which do not require exogenous IL-2 to grow in vitro induced the destruction of allogenic tumor cells within the peritoneal cavity of immunosuppressed histocompatible mice (113). The activity of helper T cells in vivo seems somewhat more efficient, although this may reflect the assay systems used. Tees & Shreier have shown that nude mice reconstituted with T-cell clones will make antibodies specific for the antigen the clones recognize (114); no "bystander" effect is seen. Studies in our laboratory (P. Nelson, unpublished data) on the reconstitution of lethally irradiated syngeneic recipients using cloned antigen-specific T-cell clones and hapten-primed B cells demonstrate essentially the same results. The anti-hapten response requires hapten-carrier linkage, and no bystander effects are observed. The homing patterns, as in the case of CTL clones, are abnormal (110). However, B cells can cooperate with the T-cell clones to produce antibody. This may be due to the fact that although migration of these cells is abnormal, significant numbers of both T and B cells home to the spleen, where they can interact effectively.

SUMMARY

To date, the most successful uses of T-cell clones have been in the demonstration that a single type of cell can perform multiple functions. However, their potential usefulness is enormous, and the study of cell interactions using clonal populations has just begun. The development and study of more cloned populations will surely lead to a clearer analysis of cellular interactions in the immune system. The use of T-cell clones and hybridomas to analyze T-cell receptors and/or factors is well under way, and will continue to be an area of intense investigation. Molecular biologists will undoubtedly make more extensive use of T-cell clones in the future, both as a source of cloning material and as transfection recipients. The most exciting area for development, from a medical point of view, is the potential for use of these cell lines or their products in immunotherapy and in providing a mechanism for *specifically* modulating the immune response.

ACKNOWLEDGMENTS

We thank Frank Fitch for directing us to Trowell's delightful review. We thank John Kappler for providing us with the outline for Figure 2, and Patsi Nelson, J. A. Frelinger, and Jane Parnes for helpful criticisms of this manuscript. Special thanks also go to Karen Carpenter for careful and patient secretarial assistance.

C. G. Fathman is supported by grant RCDA AI-00485 and J. G. Frelinger is a Fellow of the Jane Coffin Childs Foundation. We are also supported by grants AI-18716 and AI-18705.

Literature Cited

1. Kohler, G., Milstein, C. 1975. Continuous cultures of fused cells secreting antibody of predefined specificity. *Nature* 256:495–497
2. Kohler, G., Milstein, C. 1976. Derivation of specific antibody-producing tissue culture and tumor line by fusion. *Eur. J. Immunol.* 6:511–519
3. Fathman, C. G., Fitch, F. W., eds. 1982. *Isolation, Characterization and Utilization of T Lymphocyte Clones.* New York:Academic
4. Trowell, O. A. 1958. The lymphocyte. *Int. Rev. Cytol.* 7:235–293
5. Taniguchi, M., Miller, J.F.A.P. 1978. Specific suppressive factors produced by hybridomas derived from the fusion of enriched suppressor T cells and a T lymphomas cell line. *J. Exp. Med.* 148:373–382
6. Kapp, J. A., Araneo, B. A., Clevinger, B. L. 1980. Suppression of antibody and T cell proliferative responses to L-glutamic acid^{60}-L-alanine^{30}-L-tyrosine^{10} by a specific monoclonal T cell factor. *J. Exp. Med.* 152:235–240
7. Kontiainen, S., Simpson, E., Bohrer, E. M., Beverley, P. C. L., Herzenberg, L. A., et al. 1978. T-cell lines producing antigen-specific suppressor factor. *Nature* 274:477–480
8. Hewitt, J., Liew, F. Y. 1979. Antigen-specific suppressor factors produced by T cell hybridomas for delayed-type hypersensitivity. *Eur. J. Immunol.* 9: 572–575
9. Taussig, M. J., Holliman, A. 1979. Structure of an antigen-specific suppressor factor produced by a hybrid T-cell line. *Nature* 277:308–310
10. Taniguchi, M., Saito, T., Tada, T. 1978. Antigen-specific suppressive factors produced by transplantable I-J bearing T cell hybridoma. *Nature* 278:555–558
11. Kappler, J. W., Skidmore, B., White, J., Marrack, P. 1980. Antigen-inducible, H-2 restricted, interleukin-2-producing T cell hybridomas: Lack of independent antigen and H-2 recognition. *J. Exp. Med.* 153:1198–1214
12. Marrack, P., Graham, S., Leibson, H. J., Roehm, N., Wegmann, D., Kappler, J. 1982. In *Isolation, Characterization and Utilization of T Lymphocyte Clones,* ed. C. G. Fathman, F. W. Fitch, pp. 119–126. New York: Academic
13. Nabholz, M., Conzelmann, A., Acuto, O., North, M., Hass, W., et al. 1980. Established murine cytolytic T-cell lines as tools for a somatic cell genetic analysis of T-cell functions. *Immunol. Rev.* 51:125–156
14. Nabholz, M., Cianfriglia, M., Acuto, O., Conzelmann, A., Hass, W., et al. 1980. Cytolytically active murine T-cell hybrids. *Nature* 287:437–440
15. Gordon, J., MacLean, L. D. 1965. A lymphocyte-stimulating factor produced *in vitro. Nature* 208:795–796
16. Kasakura, S., Lowenstein, L. 1965. A factor stimulating DNA synthesis

derived from the medium of leucocyte cultures. *Nature* 208:794–795

17. Morgan, D. A., Ruscetti, F. W., Gallo, R. C. 1976. Selective *in vitro* growth of T lymphocytes from normal human bone marrows. *Science* 193:1007–1008

18. Ruscetti, F. W., Gallo, R. C. 1981. Human T lymphocyte growth factor: Regulation of growth and function of T lymphocytes. *Blood* 57:379–394

19. Gillis, S., Smith, K. A. 1977. Long-term culture of tumour-specific cytotoxic T cells. *Nature* 268:154–156

20. Nabholz, M., Engers, H. S., Collavo, D., North, M. 1978. Cloned T-cell lines with specific cytolytic activity. In *Current Topics in Microbiology and Immunology*, Vol 81, ed. F. Melcher, M. Potter, N. Warner, pp. 176–187. New York: Springer

21. Watson, J. 1979. Continuous proliferation of murine antigen-specificity helper T lymphocytes in culture. *J. Exp. Med.* 150:1510–1519

22. von Boehmer, H., Hengarner, H., Nabholz, W., Lenhardt, W., Schreier, M. H., Hass, W. 1979. Fine specificty of a continuous growing killer cell clone specific for H-Y antigen. *Eur. J. Immunol.* 9:592–597

23. Baker, P. E., Gillis, S., Smith, K. A. 1979. Monoclonal cytolytic T-cell lines. *J. Exp. Med.* 149:273–278

24. Fresno, M., Nabel, G., McVay-Boudreau, L., Furthmayer, H., Cantor, H. 1981. Antigen-specific T lymphocyte clones. I. Characterization of T lymphocyte clones expressing antigen-specific suppressive activity. *J. Exp. Med.* 153:1246–1259

25. Johnson, J. P., Cianfriglia, M., Nabholz, M. 1982. Karyotype evolution of cytolytic T-cell lines. In *Characterization and Utilization of T Lymphocyte Clones*, ed. C. G. Fathman, F. W. Fitch, pp. 183–191. New York: Academic

26. Terzi, M. 1974. *Genetics and the Animal Cell.* London: Wiley

27. MacDonald, H. R., Engers, H. D., Cerottini, H. C., Brunner, K. T. 1974. Generation of cytotoxic T lymphocytes *in vitro*. II. Effect of repeated exposure to alloantigens on the cytotoxic activity of long-term mixed leukocyte cultures. *J. Exp. Med.* 140:718–730

28. Hayry, P., Anderson, L. C. 1974. Generation of T memory cells in one-way mixed lymphocyte culture. *Scand. J. Immunol.* 3:823–832

29. Dennert, G., Rose, M. 1976. Continuously proliferating T killer cells specific for H-2^b. Targets, selection, and characterization. *J. Immunol.* 116:1601–1612

30. Glasebrook, A. L., Fitch, F. W. 1979. T-cell lines which cooperate in generation of specific cytolytic activity. *Nature* 278:171–173

31. Watanabe, T., Fathman, C. G., Coutinho, A. 1977. Clonal growth of T cells *in vitro*: preliminary attempts to a quantitative approach. *Immunol. Rev.* 35:3–37

32. Fathman, C. G., Hengartner, H. 1978. Clones of alloreactive T cells. *Nature* 272:617–618

33. Glasebrook, A. L., Fitch, F. W. 1980. Alloreactive cloned T cell lines. I. Interactions between cloned amplifier and cytolytic T cell lines. *J. Exp. Med.* 151:876–895

34. Glasebrook, A. L., Sarmiento, M., Loken, M. R., Dialynas, D. P., Quintans, J., et al. 1981. Murine T lymphocyte clones with distinct immunological functions. *Immunol. Rev.* 54:225–266

35. Braciale, T. J., Andrew, M. E., Braciale, V. L. 1981. Heterogeneity and specificity of cloned lines of influenza virus-specific cytotoxic T lymphocytes. *J. Exp. Med.* 153:910–923

36. Weiss, A., Brunner, K. T., MacDonald, H. R., Cerottini, J.-C. 1980. Antigenic specificity of the cytolytic T lymphocyte response to murine sarcoma virus-induced tumors. III. Characterization of cytolytic T lymphocyte clones specific for Maloney leukemia virus associated cell surface antigens. *J. Exp. Med.* 152:1210–1225

37. Kimoto, M., Fathman, C. G. 1980. Antigen reactive T cell clones. I. Transcomplementing hybrid I-A region gene products function effectively in antigen presentation. *J. Exp. Med.* 152:759–770

38. Sredni, B., Tse, H. Y., Schwartz, R. H. 1980. Direct cloning and extended culture of antigen-specific MHC-restricted, proliferating T lymphocytes. *Nature* 283:581–583

39. Benaceraff, B., Germain, R. N. 1978. The immune response genes of the major histocompatibility complex. *Immunol. Rev.* 38:70–119

40. Zinkernagal, R. M., Doherty, T. C. 1974. Restriction of an *in vitro* T-cell mediated cytotoxicity in lymphocytic chorio-meningitis within a syngeneic or semi-allogeneic system. *Nature* 248:701–702

41. Shearer, G. M. 1974. Cell mediated cytotoxicity to trinitrophenyl modified

syngeneic lymphocytes. *Eur. J. Immunol.* 4:527–533

42. Bevan, M. J. 1975. The major histocompatibility complex determines suseptibility to cytotoxic T-cells directed against minor histocompatibility antigens. *J. Exp. Med.* 142:1349–1364

43. Zinkernagel, R. M., Dougherty, P. C. 1979. MHC-restricted cytotoxic T cells studies on the biological role of polymorphic major transplantation antigens determining T-cell restriction specificity, function and responsiveness. *Adv. Immunol.* 27:51–177

44. Eichmann, K., Rajewsky, K. 1975. Induction of T and B cell immunity by anti-idiotypic antibody. *Eur. J. Immunol.* 5:661–671

45. Cosenza, H., Julius, M. H., Augustin, A. A. 1977. Idiotypes as variable region markers: analogies between receptors on phosphorylcholine-specific T and B lymphocytes. *Immun. Rev.* 34:3–33

46. Binz, H., Wigzell, H. 1975. Shared idiotypic determinants on B and T lymphocytes reactive against the same antigenic determinants. I. Demonstration of similar or identical idiotypes on IgG molecules and T cell receptors with specificity for the same alloantigens. *J. Exp. Med.* 142:197–211

47. Binz, H., Wigzell, H. 1975. Shared idiotypic determinants on B and T lymphocytes reactive against the same antigenic determinants. II. Determination of frequency and characteristics of idiotypic T and B lymphocytes in normal rats using direct visualization. *J. Exp. Med.* 142:1218–1240

48. Harvey, M. A., Adorini, L., Miller, A., Sercarz, E. 1979. Lysozyme induced T suppressor cells and antibodies have a predominant idiotype. *Nature* 281:594–596

49. Germain, R. N., Ju, S.-T., Kipps, T. J., Benacerraf, B., Dorf, M. E. 1979. Shared idiotypic determinants on antibodies and T-cell-derived suppressor factor specific for the random terpolymer L-glutamic acid^{60}-L-alanine^{30}-L-tyrosine^{10}. *J. Exp. Med.* 149:613–622

50. Kapp, J. A., Araneo, B. A., Ju, S. T., Dorf, M. E. 1983. Immunogenetics of monoclonal suppressor T cell products. In press

51. Eichmann, K. 1978. Expression and function of idiotypes on lymphocytes. *Adv. Imm.* 26:195–254

52. Kronenberg, M., Davis, M. M., Early, P. W., Hood, L. E., Watson, J. D. 1980. Helper and killer T cells do not express B cell immunoglobulin joining and constant region gene segments. *J. Exp. Med.* 152:1745–1761

53. Kronenberg, M., Kraig, E., Horvath, S. J., Hood, L. E. 1982. Cloned T cells as a tool for molecular geneticists: approaches to cloning genes which encode T cell antigen receptors. In *Isolation, Characterization and Utilization of T Lymphocyte Clones*, ed. C. G. Fathman, F. W. Fitch, pp. 467–491. New York: Academic

54. Zuniga, M. C., D'Eustachio, P., Ruddle, N. H. 1983. Immunoglobulin heavy chain gene rearrangement and transcription, In *Murine T Cell Hybrids and T Lymphomas.* In press

55. Cayre, Y., Palladino, M. A., Marcu, K. B., Stavnezer, J. 1981. Expression of an antigen receptor on T cells does not require recombination at the immunoglobulin J_H-C_μ locus. *Proc. Natl. Acad. Sci. USA* 78:3814–3818

56. Forster, A., Hobart, M., Hengartner, H. M., Rabbitts, T. H. 1980. An immunoglobulin heavy-chain gene is altered in two T cell clones. *Nature* 286:897–899

57. Kurosawa, Y., von Boehmer, H., Haas, W., Sakano, H., Traurieker, A., Tonegawa, S. 1981. Identification of D segments of immunoglobulin heavy chain genes and their rearrangements in T lymphocytes. *Nature* 290:566–570

58. Kemp, D. J., Harris, A. W., Cory, S., Adams, J. M. 1980. Expression of the immunoglobulin C_μ gene in mouse T and B lymphoid and myeloid cell lines. *Proc. Natl. Acad. Sci. USA* 77:2876–2880

59. Kemp, D. J., Wilson, A., Harris, A. W., Shortman, K. 1980. The immunoglobulin μ constant region gene is expressed in mouse thymocytes. *Nature* 286:168–170

60. Owen, F., Spurll, G. W., Panageas, E. 1982. Tthy^d, A new thymocyte alloantigen linked to Igh-1. *J. Exp. Med.* 155:52–56

61. Spurll, G. M., Owen, F. L. 1981. A family of T cell alloantigens linked to Igh-1. *Nature* 293:742–744

62. Tokuhisa, T., Komatsu, Y., Uchida, Y., Taniguchi, M. 1982. Monoclonal alloantibodies specific for the constant region of T cell antigen receptors. *J. Exp. Med.* 153:888–897

63. Lonai, P., Bitton, S., Savelkoul, H. F. J., Puri, J., Hammerling, G. J. 1981. Two separate genes regulate self-Ia and carrier recognition in H-2 restricted helper factors secreted by hybridoma cells. *J. Exp. Med.* 154:1910–1921

64. Lonai, P., Steinman, L., Freidman, V., Drizlikh, G., Puri, J. 1981. Specificity of antigen binding T cells: competition between soluble and Ia associated antigen. *Eur. J. Immunol.* 11:382–387

65. Simonsen, M. 1967. The clonal selection hypothesis evaluated by grafted cells reactive against their host. *Cold Spring Harbor Symp. Quant. Biol.* 32:517–523

66. Wilson, D. B. 1974. Immunologic reactivity to major histocompatibility alloantigens: HARC, effector cells and the problem of memory. *Prog. Immunol.* 2:145–156

67. Kanagawa, O., Louis, J., Cerottini, J. 1982. Frequency and cross reactivity of cytolytic T lymphocyte precursors reacting against male antigen. *J. Immunol.* 128:2362–2366

68. Braciale, T. J., Andrew, M. E., Braciale, V. L. 1981. Simultaneous expression of H-2-restricted and alloreactive recognition by a cloned line of influenza virus-specific cytotoxic T lymphocytes. *J. Exp. Med.* 153:137–13761

69. Sredni, B., Schwartz, R. 1981. Antigen-specific, proliferating T lymphocyte clones. Methodology, specificity, MHC restriction and alloreactivity. *Immuno. Rev.* 54:187–224

70. Schwartz, R. H., Sredni, B. 1982. Alloreactivity of antigen specific T cell clones. In *Isolation, Characterization and Utilization of T Lymphocyte Clones,* ed. C. G. Fathman, F. W. Fitch, pp. 375–384. New York: Academic

71. Janeway, C. A. Jr., Conrad, P. J. 1983. Degeneracy in the response of individual T cells to non-self major histocompatibility complex antigens. *Behrin. Inst. Res. Comm.* In press

72. Sherman, L. A. 1980. Dissection of the B10.D2 anti H-2K^b cytolytic T lymphocyte receptor repertoire. *J. Exp. Med.* 151:1386–1397

73. Nairn, R., Yamaga, K., Nathenson, S. G. 1980. Biochemistry of the gene products from murine MHC mutants. *Ann. Rev. Genet.* 14:241–277

74. Endres, R. O., Grey, H. M. 1980. Antigen recognition by T cells. I. suppressor T cells fail to recognize cross-reactivity between native and denatured ovalbumin. *J. Immunol.* 125:1515–1520

75. Maizels, R. M., Clarke, J., Harvey, M. A., Miller, A., Sercarz, E. E. 1980. Epitope specificity of the T Cell proliferative response to lysozyme: proliferative T cells react to predominantly different determinants from those recognized by B cells. *Eur. J. Immunol.* 10:509–515

76. Infante, A. J., Atassi, M. Z., Fathman, C. G. 1981. T cell clones reactive with sperm whale myoglobin:isolation of clones with specificity for individual determinants on myoglobin. *J. Exp. Med.* 154:1342–1346

77. Fathman, C. G., Kimoto, M. 1981. Studies utilizing murine T cell clones: Ir genes, Ia antigens and MLR determinants. *Immun. Rev.* 54:57–80

78. Fathman, C. G., Infante, P. D. 1979. Hybrid I region antigen and I region restriction of recognition in MLR. *Immunogenetics* 8:577–581

79. Pierres, M., Devaux, C., Dosseto, M., Marctetto, S. 1981. Clonal analysis of B- and T-cell responses to Ia Antigens. I. Topology of epitope regions in I-A^k and I-E^k molecules analyzed with 35 monoclonal alloantibodies. *Immunogenetics* 14:481–495

80. Lerner, E. A., Matis, L. A., Janeway, C. A. Jr., Jones, P. P., Schwartz, R. H., Murphy, D. B. 1980. A monoclonal antibody against an immune response (Ir) gene product? *J. Exp. Med.* 152:1085–1101

81. Sherman, L. A. 1982. Recognition of conformational determinants on H-2 by cytolytic T lymphocytes. *Nature* 297:511–513

82. Wylie, D. E., Sherman, L. A., Klinman, N. R. 1982. Participation of the major histocompatability complex in antibody recognition of viral antigens expressed on infected cells. *J. Exp. Med.* 155:403–414

83. Sarmiento, M., Glasebrook, A. L., Fitch, F. W. 1980. IgG or IgM monoclonal antibodies reactive with different determinants on the molecular complex bearing Lyt-2 antigen block T cell mediated cytolysis in the absence of component. *J. Immunol.* 125:2665–2672

84. MacDonald, H. R., Thiernesse, N., Cerottini, J.-C. 1981. Inhibition of T cell-mediated cytolysis by monoclonal antibodies directed against Lyt-2: heterogeneity of inhibition at the clonal level. *J. Immunol.* 126:1671

85. Golstein, P., Goridis, C., Schmit-Verhulst, A., Hayot, B., Pierres, A., et al. 1982. Lymphoid cell surface I interaction structures detected using cytolysis inhibition monoclonal antibodies. *Immuno. Rev.* 68:5–42

86. Loken, M. R., Fitch, F. W. 1982. Cloned T lymphocytes and monoclonal antibodies as probes for cell surface molecules active in T cell mediated cytolysis. *Immuno. Rev.* 68:135–170

87. Dialynas, D. P., Loken, M. R., Glasebrook, A. L., Fitch, F. W. 1981. Lyt-2/Lyt-3 variants of a cloned cytolytic T cell line lack an antigen receptor functional in cytolysis. *J. Exp. Med.* 153: 595–604

88. Giorgi, J. V., Zawadzki, J. A., Warrer, N. L. 1982. Cytotoxic T lymphocyte lines reactive against murine plasma cytoma antigens: dissociation of cytotoxicity and Lyt-2 expression. *Eur. J. Immuno.* 12:831–837

89. Infante, A. J., Infante, P. D., Gillis, S., Fathman, C. G. 1982. Definition of T cell idiotypes using anti-idiotypic antisera produced by immunization with T cell clones. *J. Exp. Med.* 155:1100–1107

90. Asano, Y., Shigeta, M., Fathman, C. G., Singer, A., Hodes, R. J. 1982. Role of the major histocompatibility complex in T cell activation of B cell subpopulations. A single monoclonal T helper cell population activates different B cell subpopulations by distinct pathways. *J. Exp. Med.* 156:350–360

91. Lamb, J. R., Eckels, D., Lake, P., Johnson, A. H., Hartzman, R. J., Woody, J. N. 1982. Antigen-specific human T lymphocyte clones: induction, antigen specificity, and MHC restriction of influenza virus-immune clones. *J. Immunol.* 128:233–238

92. Lamb, J. R., Woody, J. N., Hartzman, J., Eckels, D. 1982. *In vitro* influenza virus specific antibody production in man: antigen specific and HLA-restricted induction of helper activity mediated by cloned human T lymphocytes. *J. Immunol.* 129:1456–1470

93. McKean, D. J., Infante, A. J., Nelson, A., Fathman, C. G., Walker, E., Warner, N. 1981. MHC restricted antigen presentation to antigen reactive T cells by lymphocyte tumor cells. *J. Exp. Med.* 154:1419–1431

94. Glimcher, Z. H., Kim, K. J., Green, I., Paul, W. E. 1982. Ia antigen bearing B cell tumor lines can present protein antigen and alloantigen in a major histocompatibility complex restricted fashion to antigen reactive T cells. *J. Exp. Med.* 155:445–459

95. Kappler, J., White, J., Wegmann, P., Mustain, E., Marrack, P. 1982. Antigen presentation by Ia$^+$ B cell hybridomas to H-2 restricted T cell hybridomas. *Proc. Natl. Acad. Sci. USA* 79:3604–3607

96. Issekutz, T., Chu, E., Geha, R. 1982. Antigen presentation by human B cells: T cell proliferation induced by Epstein-Barr virus B lymphoblastoid cells. *J. Immunol.* 129:1446–1450

97. Fathman, C. G., Hengartner, H. 1979. Clones of alloreactive T cells. II. Cross-reactive MLR determinants recognized by cloned alloreactive T cells. *Proc. Natl. Acad. Sci. USA.* 76:5863–9

98. Shigeta, M., Fathman, C. G. 1981. I region genetic restriction imposed upon the recognition of KLH by murine T cell clones. *Immunogenetics* 14:415–422

99. Silver, J., Swain, S. L., Hubert, J. J. 1980. Small subunit of I-A subregion antigens determines the allospecificity recognized by a monoclonal antibody. *Nature* 286:272–274

100. Jones, P. P., Murphy, P. B., McDevitt, H. O. 1978. Two gene control of the expression of a murine Ia antigen. *J. Exp. Med.* 148:925–939

101. Beck, B. N., Frelinger, J. G., Shigeta, M., Infante, A. J., Cummings, D., et al. 1982. T cell clones specific for hybrid I-A molecules. Discrimination with monoclonal anti I-A^k antibodies. *J. Exp. Med.* 156:1186–1194

102. Frelinger, J. G., Shigeta, M., Infante, A. J., Nelson, P., Pierres, M., Fathman, C. G. 1983. Multiple restriction sites per Ia molecule recognized by T cell clones. In *Ir genes: Past, Present and Future,* ed. C. W. Pierce, S. E. Cullen, J. A. Kapp, B. D. Schwartz, and D. C. Shreffler. Clifton, NJ: Humana. In press

103. Nelson, P. A., Beck, B. N., Fathman, C. G. 1983. Functional evidence for two antigen presentation sites for a single I-A molecule. In *Ir genes: Past, Present and Future,* ed. C. W. Pierce, S. E. Cullen, J. A. Kapp, B. D. Schwartz, and D. C. Shreffler. Clifton, NJ: Humana. In press

104. McNicholas, J. M., Murphy, D. B., Matis, L. A., Schwartz, R. H., Lerner, E., et al. 1982. Immune response gene function correlates with the expression of an Ia antigen. I. Preferential association of certain A$_e$ and E$_α$ chains results in a quantitative deficiency in expression of an A$_e$E$_α$ complex. *J. Exp. Med.* 155: 490–507

105. Matis, L. A., Jones, P. P., Murphy, D. B., Hedrick, S. M., Lerner, E. A., et al. 1982. Immune response gene function correlates with cell surface expression of an Ia antigen. II. A quantitative deficiency of an A$_e$E$_α$ complex expression causes a corresponding defect in antigen presenting cell function. *J. Exp. Med.* 155:508–523

106. Infante, A. J., Fathman, C. G., Atassi, M. Z. 1982. Myoglobin reactive T cell clones. *Adv. Exp. Med. Biol.* 150:159–168

107. Giorgi, J. V., Warner, N. L. 1981. Continuous cytotoxic T cells reactive against murine plasmacytoma-associated antigens. *J. Immunol.* 126:322–330

108. Dailey, M. O., Pillemer, E., Weissman, I. 1982. Protection against syngeneic lymphoma by a long-term cytotoxic T cell. *Proc. Natl. Acad. Sci. USA* 79:5384–5387

109. Rosenberg, S. A. 1982. Potential use of expanded T lymphoid cells and T cell clones for the immunotherapy of Cancer. In *Isolation, Characterization and Utilization of T lymphocyte Clones,*. ed. C. G. Fathman, F. W. Fitch, pp. 451–466. New York: Academic

110. Dailey, M. O., Fathman, C. G., Butcher, E., Pillemer, E., Weissman, I.

1982. Abnormal migration of T lymphocyte clones. *J. Immunol.* 128:2134–2136

111. Lin, L. Y., Askonas, B. A. 1980. Cross reactivity for different type A influenza viruses of a cloned T-killer cell line. *Nature* 288:164–164

112. Cheever, M. A., Greenburg, P. D., Tefer, A., Gillis, S. 1982. Augmentation of the anti-tumor therapeutic efficiency of long-term culture T lymphocytes by *in vitro* administration of purified interleukin-2. *J. Exp. Med.* 155:968–980

113. Engers, H. D., Glasebrook, A. L., Sorenson, G. D. 1982. Allogeneic tumor rejection induced by the intravenous injection of Ly-2+ cytolytic T lymphocyte clones. *J. Exp. Med.* 156:1280–1285

114. Tees, R., Schreier, M. H. 1980. Selective reconstitution of nude mice with long-term cultured and cloned specific helper T cells. *Nature* 283:780–781

SUBJECT INDEX

A

a-Acid glucosidase
mouse *Neu-1* locus and, 136
Acid phosphatase liver
mouse *Neu-1* locus and, 136
Acquired immunity
to malaria, 364–80
Adenoviruses
complex immunoglobulin
transcription units in,
407
Agretopes
limitations and hierarchies
of, 477
Aleutian mink virus
systemic lupus erythematosus
and, 201
Alkaline phosphatase
distribution in tissue, 12
Allelic exclusion
immunoglobulin gene
expression and, 509–10
Allergic encephalomyelitis
induction of
inhibition of, 627
Alloantigens
B-cell tolerance and, 49–50
Allogeneic effect factor
B-cell activation and, 228–29
Allograft reaction
tissue antigen and, 143–44
see also Graft rejection
Alloreactivity, 164–66
T-cell clones and, 642–43
Anaphylatoxin
histamine release from mast
cells and, 346
vasopermeability and, 348
Androgens
autoimmune disease and,
203
murine lupus and, 193
Antibodies
antigen binding to
consequences of, 102–3
anti-idiotopic, 570
anti-idiotypic, 570
effects of, 586
regulatory effects of,
590–91
in autoimmune thyroiditis,
183–84
cells secreting, 211–13
combining sites of, 94–95
cross-reactivity for protein
antigens and, 470–71
graft rejection and, 144
high/low affinity
cross reactivity of, 458

idiotypic, 486
pre-B and B cells and,
591–92
regulatory effects of,
590–91
T-cell repertoire and,
594–95
idiotypic interactions
between, 570–72
idiotypic regulation by,
587–90
of receptors, 585–92
malaria and, 364
murine T-cell clones and,
645–46
phosphocholine-binding
sequences of, 92–93
Antibody-forming cells
effector cell blockade and,
46
Antibody function, 87–113
Antibody genes
encoding idiotopes and
anti-idiotopes, 569
Antibody idiotypes
pre-B and B cells and,
591–92
regulatory effects of, 590–91
T-cell repertoire and, 594–95
Antibody responses
antigen-specific
idiotypic control of, 591
epitope-specific regulation of,
613–14
Antibody-secreting cells
production of
T-cell replacing factors
and, 309–12
Antigen-antibody complexes
complement pathway
activation and, 110–13
immunologic tolerance
induction and, 45
mechanism of, 103
mediators of inflammation
and, 336
Antigen E
helper T-cell induction and,
471
Antigenic stimulation
T-cell cloning and, 636–37
Antigen recognition
allograft reaction and,
143–44
Antigens
B-cell proliferation and
maturation and, 394
cytolytic T-cell induction
and, 292
differentiation of, 126

hybrid Ia
expression of, 647–48
multideterminant
determinant selection on,
475–77
protein
cross-reactivity for by T
cells and antibodies,
470–71
recognition as fragments
by T cells, 471–73
recognition of
helper/suppressor T cell,
481–82
suppressor T cell, 477–82
T-cell recognition of
MHC restriction and,
467–68
T cell-T cell interactions and,
479–80
see also specific type
Anti-idiotopes
encoding of, 569
Anti-idiotypes
idiotype expression and,
585–86
Arachidonic acid
mediators of inflammation
and, 341
Arthritis
autoimmune disease and, 180
see also specific type
Autoantibodies
in murine lupus, 190–92
in systemic lupus
erythematosus, 199–200
Autoimmune disease
antibody production and, 47
autoimmune thyroiditis,
183–85
autoimmunity and, 202–4
classification of, 177
infectious agents and, 176
mechanisms of, 180–82
self-antigens in, 179
spontaneous thyroiditis,
182–83
systemic lupus
erythematosus, 185–202
human, 194–202
murine, 186–94
Autoimmune thyroiditis
antibodies in, 183–84
cells in, 184–85
classification of, 177–78
Autoimmunity, 175–204
autoimmune disease and,
202–4
contrasuppression and, 455
high immunity antibody
responses and, 459–60

657

juvenile-onset diabetes and,
157
polyclonal B-cell activators
and, 458
see also Autoimmune disease

B

B-cell activation, 315–16
antigen-nonspecific T-cell
factors and, 228–32
antigen-specific T-cell factors
and, 225–28
helper T cells and, 220–23
B-cell activators
polyclonal
autoimmunity and, 458
B-cell growth factor
characteristics of, 312–15
B-cell hyperactivity
in murine lupus, 186–88
B cells
antibody formation and,
36–37
antibody idiotypes and
growth receptors on,
591–92
antibody production and
T-cell Ly-1 marker and,
443
antigen binding to, 103–5
antigen presentation to T
cells by, 469–70
antigen-specific
activation of, 211–12
self-MHC determinants
and, 232–35
anti-idiotypic antibodies and,
587
C3b receptors and, 261
clonal selection and, 36
development of
differential
immunoglobulin
expression during,
395–97
differentiation of
MHC-unrestricted, 319–24
receptor cross-linking and,
40–42
T-cell replacing factors
and, 319–23
idiotypic interactions in, 569
immunity induction and,
38–39
immunoglobulin gene
expression and, 413–17
interleukins and, 307–27
long-term lines of, 318–19
maturation of
helper T-cell clones and,
646–47

membrane immunoglobulins
on, 100–2
MHC expression, 126
MHC-restricted recognition
of virally infected cells
and, 77–80
monoclonal antibody
derivation and, 65–66
negative signaling of, 43–46
precursors of
idiotypic repertoire of,
581–83
proliferation cofactors for,
309–12
receptors of
idiotypic repertoire of,
575–77
recognition of viral
determinants and, 68–69
responses of
IL2 and, 323–24
MHC-restricted, 324–26
to viral antigens at clonal
level, 63–82
self-antigens and, 48–51
systemic lupus erythematosus
and, 195
triggering of
MHC-unrestricted, 309–19
see also specific type
Basophils
C5 fragments chemotactic
for, 350
lymphokines chemoattractant
for, 350–51
Blastogenic factor, 286
Bone marrow
antibody production and,
211–12
Bradykinin
inflammation mediation and,
342
vasopermeability and, 347
t-Butylhydroperoxide
oxidant effects on
erythrocytes, 384

C

Cathepsin G
inflammatory process and,
352
cDNA clones
class I murine MHC gene,
534–35
class II murine MHC
polypeptide structure
and, 555
Cell-mediated immunity
resistance to malaria in
children and, 362
suppressor factors blocking,
427

Cell-mediated
lymphocytotoxicity
MHC C locus and, 129
Chemotactic factor inactivator
immune complex lung
disease and, 352–53
Chemotaxis
leukocyte emigration and,
349–51
Chloroquine
heme chelation and, 383
Chromatin
structure in immunoglobulin
gene expression, 404–5
Clonal abortion
self-MHC elimination and,
38
T cells and, 55–56
Clonal anergy
immunologic tolerance and,
42–43
self-tolerance and, 46–48
Clonal selection
helper T cells and, 637–38
immune responses and, 309
immunologic tolerance and,
36
T cell-B cell interaction and,
216
Colchicine
effector cell blockade and, 46
Collagenases
inflammatory process and,
352
Complement
skin graft susceptibility to,
144
Complement activation
C3 and, 244–47
factor H and, 247
ligand receptors and, 249–58
ligands and, 244–47
diffusible, 247–48
Complement components
MHC-controlled
humoral immunity and,
120–21
systemic lupus erythematosus
and, 198
Complement genes
MHC genes and, 121
Complement receptors, 243–66
C1q, 249–50
biological function of,
258–59
C3, 250–55
biological function of,
259–62
factor H, 255–56
biological function of, 262
Complementarity determining
regions
idiotopes and, 572–75

ORDER FORM

A NONPROFIT SCIENTIFIC PUBLISHER

Annual Reviews Inc.

4139 EL CAMINO WAY • PALO ALTO, CA 94306 USA • (415) 493-4400

...ase list the volumes you wish to order by volume number. If you wish a standing order (the latest volume
...t to you automatically each year), indicate volume number to begin order. Volumes not yet published will
... shipped in month and year indicated. All prices subject to change without notice. Prepayment required
...m individuals. Telephone orders charged to VISA, MasterCard, American Express, welcomed.

ANNUAL REVIEW SERIES

		Prices Postpaid per volume USA/elsewhere	Regular Order Please send: Vol. number	Standing Order Begin with: Vol. number
...nual Review of ANTHROPOLOGY				
Vols. 1-10	(1972-1981)	$20.00/$21.00		
Vol. 11	(1982)	$22.00/$25.00		
Vol. 12	(1983)	$27.00/$30.00		
Vol. 13	(avail. Oct. 1984)	$27.00/$30.00	Vol(s). _____	Vol. _____
...nual Review of ASTRONOMY AND ASTROPHYSICS				
Vols. 1-19	(1963-1981)	$20.00/$21.00		
Vol. 20	(1982)	$22.00/$25.00		
Vol. 21	(1983)	$44.00/$47.00		
Vol. 22	(avail. Sept. 1984)	$44.00/$47.00	Vol(s). _____	Vol. _____
...nual Review of BIOCHEMISTRY				
Vols. 29-50	(1960-1981)	$21.00/$22.00		
Vol. 51	(1982)	$23.00/$26.00		
Vol. 52	(1983)	$29.00/$32.00		
Vol. 53	(avail. July 1984)	$29.00/$32.00	Vol(s). _____	Vol. _____
...nual Review of BIOPHYSICS AND BIOENGINEERING				
Vols. 1-10	(1972-1981)	$20.00/$21.00		
Vol. 11	(1982)	$22.00/$25.00		
Vol. 12	(1983)	$47.00/$50.00		
Vol. 13	(avail. June 1984)	$47.00/$50.00	Vol(s). _____	Vol. _____
...nual Review of EARTH AND PLANETARY SCIENCES				
Vols. 1-9	(1973-1981)	$20.00/$21.00		
Vol. 10	(1982)	$22.00/$25.00		
Vol. 11	(1983)	$44.00/$47.00		
Vol. 12	(avail. May 1984)	$44.00/$47.00	Vol(s). _____	Vol. _____
...nual Review of ECOLOGY AND SYSTEMATICS				
Vols. 1-12	(1970-1981)	$20.00/$21.00		
Vol. 13	(1982)	$22.00/$25.00		
Vol. 14	(1983)	$27.00/$30.00		
Vol. 15	(avail. Nov. 1984)	$27.00/$30.00	Vol(s). _____	Vol. _____

...E ORDERING INFORMATION ON PAGE 4.

		Prices Postpaid per volume USA/elsewhere	Regular Order Please send:	Standing Order Begin with:
			Vol. number	Vol. number

Annual Review of ENERGY

Vols. 1-6	(1976-1981)	$20.00/$21.00		
Vol. 7	(1982)	$22.00/$25.00		
Vol. 8	(1983)	$56.00/$59.00		
Vol. 9	(avail. Oct. 1984)	$56.00/$59.00	Vol(s). _____	Vol. _____

Annual Review of ENTOMOLOGY

Vols. 7-16, 18-26	(1962-1971; 1973-1981)	$20.00/$21.00		
Vol. 27	(1982)	$22.00/$25.00		
Vol. 28	(1983)	$27.00/$30.00		
Vol. 29	(avail. Jan. 1984)	$27.00/$30.00	Vol(s). _____	Vol. _____

Annual Review of FLUID MECHANICS

Vols. 1-13	(1969-1981)	$20.00/$21.00		
Vol. 14	(1982)	$22.00/$25.00		
Vol. 15	(1983)	$28.00/$31.00		
Vol. 16	(avail. Jan. 1984)	$28.00/$31.00	Vol(s). _____	Vol. _____

Annual Review of GENETICS

Vols. 1-15	(1967-1981)	$20.00/$21.00		
Vol. 16	(1982)	$22.00/$25.00		
Vol. 17	(1983)	$27.00/$30.00		
Vol. 18	(avail. Dec. 1984)	$27.00/$30.00	Vol(s). _____	Vol. _____

Annual Review of IMMUNOLOGY

Vol. 1	(1983)	$27.00/$30.00		
Vol. 2	(avail. April 1984)	$27.00/$30.00	Vol(s). _____	Vol. _____

Annual Review of MATERIALS SCIENCE

Vols. 1-11	(1971-1981)	$20.00/$21.00		
Vol. 12	(1982)	$22.00/$25.00		
Vol. 13	(1983)	$64.00/$67.00		
Vol. 14	(avail. Aug. 1984)	$64.00/$67.00	Vol(s). _____	Vol. _____

Annual Review of MEDICINE: Selected Topics in the Clinical Sciences

Vols. 1-3, 5-15	(1950-1952; 1954-1964)	$20.00/$21.00		
Vols. 17-32	(1966-1981)	$20.00/$21.00		
Vol. 33	(1982)	$22.00/$25.00		
Vol. 34	(1983)	$27.00/$30.00		
Vol. 35	(avail. April 1984)	$27.00/$30.00	Vol(s). _____	Vol. _____

Annual Review of MICROBIOLOGY

Vols. 17-35	(1963-1981)	$20.00/$21.00		
Vol. 36	(1982)	$22.00/$25.00		
Vol. 37	(1983)	$27.00/$30.00		
Vol. 38	(avail. Oct. 1984)	$27.00/$30.00	Vol(s). _____	Vol. _____

Annual Review of NEUROSCIENCE

Vols. 1-4	(1978-1981)	$20.00/$21.00		
Vol. 5	(1982)	$22.00/$25.00		
Vol. 6	(1983)	$27.00/$30.00		
Vol. 7	(avail. March 1984)	$27.00/$30.00	Vol(s). _____	Vol. _____

Annual Review of NUCLEAR AND PARTICLE SCIENCE

Vols. 12-31	(1962-1981)	$22.50/$23.50		
Vol. 32	(1982)	$25.00/$28.00		
Vol. 33	(1983)	$30.00/$33.00		
Vol. 34	(avail. Dec. 1984)	$30.00/$33.00	Vol(s). _____	Vol. _____

SEE ORDERING INFORMATION ON PAGE 4.